BOTANICAL PEN-PORTRAITS

BOTANICAL
PEN-PORTRAITS

BY

Dr. J. W. MOLL
Professor of Botany in the University of Groningen

AND

Pharm. Dr. H. H. JANSSONIUS
Botanist in charge of the description of Indian woods

WITH 111 ILLUSTRATIONS IN THE TEXT

Springer-Science+Business Media, B.V
1923

ISBN 978-94-017-5833-8   ISBN 978-94-017-6287-8 (eBook)
DOI 10.1007/978-94-017-6287-8
Softcover reprint of the hardcover 1st edition 1923

TABLE OF CONTENTS

    †) With macroscopic characters.
    *) With macroscopic characters and micrography of the powder.

†) With macroscopic characters,
*) With macroscopic characters and micrography of the powder.

PREFACE

The primary object of this book is to give somewhat elaborate descriptions of the microscopic characters of a certain number of vegetable drugs, representing as far as possible all parts of plants used in pharmacy. Nevertheless, for reasons to be mentioned below, several descriptions containing the macroscopic characters have been added.

The macroscopic characters have been described almost from the beginning of pharmacognostical studies and may be found in various stages of development in several of the older pharmacopoeae. But it was not until the middle of the 19th century that the development of microscopy allowed the use of this method in pharmacognosy. The introduction of microscopical research is principally due to two investigators. C. A. J. A. Oudemans published in 1854—1856 his „Aanteekeningen op het systematisch- en pharmacognostisch-botanische gedeelte der Pharmacopoea Neerlandica'' and it may be said that this work marks the introduction of microscopy into pharmacognosy. Some years afterwards, in 1865, O. Berg published his well known „Anatomischer Atlas zur Pharmaceutischen Waarenkunde''. This book with its pleasing, though somewhat conventional figures, had also a very decided influence in the same direction.

Since then a large number of papers and books appeared containing the results of similar investigations. Of what has been done in this direction the reader may form some idea by consulting the bibliography added to this work. It mentions not only a large number of papers containing special information, but gives moreover a fairly complete list of the general works on the subject. Since the foundation of this branch of Pharmacognosy by the works of Oudemans and Berg, important results were gained as well from a theoretical as from a practical point of view. Our knowledge of facts concerning many drugs was enlarged, their identification has risen to a higher degree of certainty and many of the existing books may be considered as valuable and faithful guides for students occupied with the microscopical examination of drugs. Such text-books are the works of Koch, Arthur Mayer, Tschirch, Tschirch und Oesterle. It may be asked why it was necessary to write yet another book on a subject, elaborately treated of by many eminent authors. The answer is that of late a modified method of describing objects of natural history has been developed and that it was thought interesting to try its application in pharmacognosy, especially as it had already been tried with much success in another branch of science, viz. the description of the anatomical structure of wood, of which something more will be said below. I will call it the method of pen-portraits or of portraying descriptions. It tends to improve the results in the three directions mentioned above, materially augmenting our morphological knowledge, leading to greater certainty of identification and furnishing an aid to the worker in practical microscopy, far superior to that supplied by existing works. The present book may be considered as an example of the introduction of pen-portraits into Pharmacognosy, a large number of drugs remaining for future investigations by the same method.

I will now in the first place try to give the reader some notion of what is meant by speaking of the method of pen-portraits or portraying descriptions, and for this purpose it is necessary first to recall some of the general principles on which the scientific description of plants should be based.

It is true that botanical gardens, herbaria, museums, collections of slides for microscopical examination, collections of pictures, photographs and models render invaluable services to the study of botany. Yet here are only found the original phenomena of nature, but slightly affected by scientific conceptions. But these are not sufficient for many purposes of investigation, education and practice. For all these purposes there are moreover wanted records of a more theoretical character, containing the results of scientific investigations concerning the natural phenomena or in other words descriptions are wanted.

In making descriptions of objects of natural history the chief difficulty is their complex structure. Chemists often say, that for identifying some substance a single well chosen reaction ought to be sufficient. But there is no biologist I think who would defend the opinion that a plant or animal or even a part of them might be identified by mentioning a single character. To be sure it is true that among living organisms many examples are to be found of perfect identity of all the characters with the exception of a single one, and such cases may be of the utmost scientific or practical importance. But even in such cases there is wanted for identification more than the single differential character. In organisms there exists by no means the same absolute and perfect correlation between their several characters as in chemical substances. Therefore in the case of the above organisms the characters common to both are as important for their identification as the single one in which they differ. In other words, although in chemistry a large number of characters are of course necessary for a thorough knowledge, yet for identification a single well chosen character may suffice; in biology on the other hand even for this purpose a complex of as many characters as possible should be used.

The view that descriptions wanted in biology ought to contain the largest possible number and variety of characters, is strongly corroborated by the frequent use made of the term „habit" (habitus) and the important place occupied by it in biological science. J. D. Hooker says on this head: „habit" of a plant, of a species, a genus, etc., consists of such general characters as strike the eye at first sight, such as size, colour, ramification, arrangement of the leaves, inflorescence, etc., and are chiefly derived from the organs of vegetation" [1]).

Auguste Pyrame de Candolle possessed the gift of distributing the seedlings to be planted out in the botanical garden of Geneva and giving them their proper places according to their order as they were shown to him only making use of their „habit" and his son Alphonse, who records the fact, was afterwards able to do the same [2]).

It is often thought that „habit" is the essential and real factor for thoroughly knowing an organism and that trying to use descriptions instead of the more intuitive knowledge of habit is a hopeless undertaking.

Thus cases might be cited in which investigators have not even tried what might be done with the help of a description, but have referred the reader once for all to „habit" of which the knowledge is only to be got in the field or in the laboratory.

It is doubtless true that in the present state of science the knowledge of „habit" is in many cases of invaluable service, but I have at the same time the firm conviction that we ought by no means to remain contented with this state of things.

Two arguments may be advanced in support of this proposition. Habit has two very important drawbacks compared with description. In the first place, as has been worked out in the example cited above, the knowledge of habit being of an intuitive kind and considered as unfit for being expressed in words, it cannot be used in a scientific discussion between botanists, which discussion is however essential to scientific clarity. In the second place if a man has a thorough and wide knowledge of habit, like the two de Candolle's, this may be indeed of infinite service to himself in his studies, and thus undoubtedly promote science, but yet he takes that knowledge with him into the grave, whilst every description which a man has made can be used by posterity.

Therefore it is at least worth while to make a serious attempt at replacing habit in science by descriptions. I am moreover convinced that it is very well possible to go a long way in this direction and in corroboration I offer the following considerations.

In my youth coming back from excursions in the field and bringing home plants quite unknown to me, I have often tried to ascertain their names by giving merely a very superficial description to an accomplished botanist who

[1]) Flora of British India. Vol. I. 1875, p. XXIV, 183.
[2]) Alphonse de Candolle. La Phytographie. p. 182.

knew those plants very well — but without the slighest success. This experience led me to the conclusion that in such cases „habit" is the essential thing. But it is evident that here „habit" seemed so all important a factor because practically no record worthy of the name was supplied by me. If a minute description could have been offered and if it could have been compared with living objects, not the slightest difficulty would have arisen.

Another example may be cited. Every student of botany on entering his career finds that in general plants cannot be identified if their flowers are unknown, that in the order *Umbelliferae* no determination can take place if the fruits are wanting and that there are many such handicaps in field botany. But at the same time he knows very well that all plants may be known at first sight and recognized ever afterwards without flowers, fruits, etc. These facts often lead to the conclusion that the knowledge of „habit" is not only far more important than that which books can give, but also that it is impossible that books ever should take the place now held in science by „habit". The latter proposition however will not be confirmed by a closer inspection of the facts. Every student who has occupied himself with the determination of plants knows very well that in all difficult or impossible cases his plants show a great variety of the most important characters which might lead at once and with absolute certainty to a determination if only the author had condescended to mention them. It is perfectly true that the flowers of the higher plants are the parts best adapted for rapidly arriving at a determination, and that the same holds good in the case of *Umbelliferae* for fruits. But it is also perfectly true that the smallest piece of a leaf blade or a stem and even a single cell might be sufficient for the same purpose if only its characters had been duly analyzed and registered. Nevertheless for a long time to come „habit" will remain a very important factor in botany, and a gift in this direction will be a very decided advantage for a botanist. But it is the task of science to break the ban of „habit" by substituting for it descriptions of well defined characters. What we call „habit" is in fact but the summation of characters adapted and necessary for identification and even for a thorough knowledge, but without these characters having been duly analysed, compared with others and classified. It is clear that science ought at least to make this attempt, and I am sure that much more may be done than is commonly thought. At the same time it is evident that in the present imperfect state of our knowledge of all organisms a certain number of characters will defy all attempts at analysis and record, so that a certain scope for „habit" will remain.

But it will diminish as science proceeds.

The foregoing remarks lead to the conclusion that our principal care should be to strive after the highest possible completeness. It is moreover of the utmost importance that our descriptions should be adapted for easy and rapid consultation and can be readily compared with one another. Lastly it is necessary that our descriptions though complete should take a minimum amount of labour and time in their construction.

Picture and description. By way of introduction to what I have to say concerning pen-portraits I will now first make some remarks on the difference between descriptions and pictures, remarks equally applicable to plastic models. Much of what I will say is to be found in de Candolle's „Phytographie" [1]). Pictures were mentioned above among the records but slightly affected by science, such as herbaria etc., but they are used so often in close relation to descriptions that it is worth while to say a few words about them.

In the first place drawings are much less closely related to allied domains of human knowledge than verbal descriptions and this is all the more so, when technical scientific terms replace detailed explanations. Whilst a description can only give what has been understood at least to a certain extent, a drawing can give many characters not understood by any living man. Faithful drawings may be made even by persons ignorant of the first principles of science. Thus a drawing will lead with the utmost ease to some knowledge of habit which a description may only attain with the greatest difficulty or even not at all. Therefore drawings are to be preferred for purposes of superficial examination, comparison and identification. From a scientific point of view we can say that drawings require simpler means and this explains why pictures of great exactness and beauty were made some three hundred years ago accompanied, if at all, by most primitive descriptions. Pictures bear a concrete character, descriptions are abstract and where correctness is wanted pictures are in some measure to be preferred to descriptions.

This last remark leads to an other point of difference of somewhat wider scope, viz. that generalization is possible in a description, but impossible in a drawing from nature and it is only with this that we are concerned here, schematic figures being out of the question.

A drawing can only represent a single individual, but a description can give the characters of a species as well as those of a genus or of any other group. There are yet other things tending in the same direction. That a drawing can give colour only by means of an additional process is of small moment. But it can give in no way more than one aspect of an object, all other sides of it remaining unknown. Still less can it give what is hidden in the interior of the plant and comes to light if several sections are made. In a description on the other hand all these various characters are combined with the greatest ease and in this respect also it may be said that a wider generalization is possible.

These considerations lead to the conclusion that drawings can indeed give very interesting and even essential characters which cannot appear in a description. Descriptions on the other hand may attain the highest possible degree of completeness compatible with the present state of science.

[1]) l. c. p. 312.

The description can give in many respects the same characters as the picture
and even a greater number of them, but the latter can give in addition many
characters constituting habit and in this respect only a drawing is superior to
a description. I will show that the object of the method of pen-portraits is to
bring into the description some of the essential features of drawings, and thus
somewhat more of habit and of individuality then is generally the case now. In
the latter respect description has the enormous advantage over drawing that it
can sketch the individuality of groups as well as that of single specimens.
It is a decided advantage of drawings as compared with descriptions that it
is both easier to consult them as to compare them with one another. On the
other hand drawings for publication are more costly than descriptions. Much
time and labour is necessary to obtain both and it would be difficult to say
whether in this respect there is a decided difference.
It is impossible to make real improvements in the art of drawing for botanical
purposes.
The figures in the works of Dodonaeus and Fuchs are vastly superior to any
product of the present age, and it is probable that ancient Egyptian artists
would have even surpassed them. The progress of science may furnish fresh subjects for drawings but it cannot in the least alter their essential features as
indicated above, closely allied as these are with artistic conceptions. Therefore
nothing more remains to be said on this head.
With descriptions the case is quite the reverse. They have steadily improved
with the progress of science during a long lapse of time and they are susceptible
of further improvement. As descriptions have been improved in systematical
botany the number of drawings has diminished very sensibly and if at this date
publications on subjects of plant anatomy abound in figures, this is caused by
the fact that the art of micrography is still in its infancy. This possibility of
improvement is the most important advantage of descriptions as compared
with drawings and botanists must try to make drawings more and more superfluous, although at present they are often indispensable and will remain so
for a very long time to come.
The principal purpose of this preface is to show which improvements the
art of describing has undergone in course of time, and how further improvements may be made. Of these the text of this book will give a series of examples.
In treating of this subject it will be well to distinguish between Phytography
and Micrography as I understand these terms. Phytography is the art of describing those characters of plants constituting the subject of Organography. In
a general way they may be said to be characters of an outward organisation.
Micrography, on the other hand, is the art of describing those structures which
are the subjects of Anatomy; they are in general internal structures.
The Linnean method in general and its application to Phytography. If Linnaeus has been able to lay the foundations of modern Classifying Systematic

Botany, this is due in a large measure to his having introduced into science not only the binomial nomenclature but also an organographical terminology and a phytographical method, largely based on his terminology and in several respects much superior to what existed before him. In de Candolle's admirable work, „la Phytographie", cited above, the reader will find a detailed exposition of the Linnean method and I will give here only an abstract of it. The principal feature of the Linnean method is perhaps its conformity to a sequence of characters fixed beforehand. This sequence is by no means an arbitrary one; it is founded on the very nature of things. If description is to be made of a structure consisting of a certain number of parts of the same kind, for instance a stem with leaves, it is most natural to begin by ascertaining the number of those parts and then to consider their arrangement with respect to one another before coming to a description of the leaf itself. In most cases however the parts to be described have shapes of so complex a character that it is impossible to find adequate geometrical terms for the description. Then it becomes necessary to make an analysis of such a part seeking out the various characters constituting it and perhaps to make a fresh analysis of some of the characters thus gained, until at last characters are found allowing of a simple geometrical explanation. In many cases where further analysis is impossible, such an explanation will however remain imperfect and it is then that our knowledge of characters passes into that of „habit". The order of the several characters coming to light one after another in the analysis just mentioned is at the same time the order to be followed in the description. This is in fact the general principle of Linnean Organography. The principal advantages secured by the rigorous order of the characters in a Linnean description are in the first place a certain completeness, because nothing is omitted by chance, and in the second place repetitions will be avoided. In the description of so complex structures as plants generally offer, those advantages are indeed of the utmost importance.

As mentioned above the introduction of organographical terminology was an important factor in the Linnean method. It was developed in his „Philosophia Botonica" and has remained the foundation of our present terminology.

The advantages and disadvantages of terminology are obvious. On the one hand terminology allows us to express many things by a single word, connecting together many facts and ideas playing a part in science. But on the other hand terminology necessarily introduces a certain abstraction and generalisation leading to the inverse of the knowledge of „habit" which we have recognised as so very important. Nevertheless the use of terminology is of the utmost advantage and we will see below how the drawback mentioned here may be avoided. Linnean descriptions are further characterized by their peculiar style. Linnaeus allowed a sentence for every part of the plant, for instance for the calyx, corolla, stamens, etc. Moreover he rigorously avoided the use of verbs and thus a sort of telegram style was originated in which no superfluous word appears.

To write well in this style is indeed difficult, even more difficult than ordinary writing. But it is a fact that with care and good taste it is quite possible to make a Linnean description a thing of beauty in its own way. A man having a bad style in ordinary writing will never succeed. The advantage of this telegram style is principally its brevity and so to say its automatic order. Thus Linnean descriptions may be consulted with much ease without being read entirely, and this generally speaking is their proper use. It is often thought that by this telegram-style the description becomes unreadable. Such however is not the case. A long description of a complex structure will always be difficult to read, in whatever style it may be written and with the exception of a few systematic botanists the present generation has not been educated in this direction. Experience has shown me that this defect of education cannot be remedied by the use of a seemingly more readable style. The shortness and concreteness of the telegram style on the contrary make the description more digestible, and every beginner, after a few trials will find that this is true. Some training however is necessary.

In order to ensure a rigorous and in all cases constant order of the characters to be described it is absolutely necessary when making a Linnean description to make use of what I call „guiding schemes". Experience has amply shown me that most men, even if daily occupied with this sort of work, will be apt to make omissions and breaks in the regular order if they are not constantly reminded of it by such a guide. In phytography they are necessary, in micrography they are indispensable. Therefore I will return to this subject below. Moreover these schemes allow the student to use all his faculties for the description itself. Lastly I mention a very important advantage of the Linnean method, which in my opinion has been overlooked. Linnean descriptions offer the possibility of gradual enlargement and improvement. We have seen that the order in which the characters are mentioned and described is not arbitrary, but fixed by logical considerations. Therefore if an author has rigorously followed the rules on which the method is founded, his successor can make additions to his description rendered necessary or possible by the progress of science. But this successor will not try to make a fresh description, because he knows that he would come to results identical in all important respects with those of his predecessor. Thus much unnecessary work may be avoided and the fund of facts given in the descriptions of vegetable structures may be gradually enlarged whilst at the same time continuity remains assured. In Phytography this gradual progress often occurs.

Thus we come to the result that the Linnean method of description leads to completeness, avoids unnecessary repetitions, gives a maximum of information by a minimum of words and in a form essentially capable of being consulted in whole or in part, while being as readable as such compositions can be. Moreover he who has made a good Linnean description may hope that an intelligent successor following the same path will make use of the results of his work. *The Linnean method as applied to Micrography.* In the year 1880 de Candolle [1]) drew the attention of botanists to the fact that Micrography had in many respects reached no higher level than Phytography did one or more centuries ago. He pointed out that the Linnean method of describing had not been applied in Micrography and that this was the principal cause of so unsatisfactory a state of things.

This is not the place to trace the causes which have led to this astonishing result, and a few words concerning the fact itself may suffice. It might indeed have been supposed that the method of Linnaeus who is considered to be in so many ways the father of modern botany would have been applied at once to micrography, as soon as attempts were made to employ anatomical characters in systematic botany. But nothing of the sort occurred although anatomical characters were used in many systematic investigations, even as early as the beginning of the 19th century in France and afterwards also in Germany. Valeton [2]) published in 1888 a most interesting historical paper on this subject. Radlkofer [3]) in 1883 delivered an oration in the Royal Academy of Munich in which the anatomical method was more or less officially introduced into systematic botany. His pupil Solereder even wrote extensive works on systematic anatomy [4]). But neither he nor any of his predecessors tried to apply the Linnean method of description to their researches. Had they done so, the results would no doubt have been still even more interesting.

In papers dealing with anatomy, not from a systematic but from a general morphological or from a physiological point of view, there are generally no traces of the Linnean method and at the same time illustrations are used to an enormous extent. But there is no cause for wonder here, in as far as the majority of these authors are not systematic botanists and are practically unacquainted with phytographical work.

Nevertheless in two fields of applied research some attempts have been made to employ the Linnean method. First this occurred in the micrography of woods. The names of Gamble (1902), Herbert Stone (1904) and Perrot (1907) might be cited for our present purpose; it would be superfluous to give details. More interesting for us is the work of Janssonius on the micrography of Javanese woods. Of this work however it will be better to speak below after having given a survey of some conditions, which ought to be fulfilled if the Linnean method shall really show its usefulness in micrography.

---

[1]) l. c. p. 221.

[2]) Th. Valeton. De anatomische methode in de classificatie der planten. Programma van het Gymnasium te Groningen over het jaar 1888—1889. It is a pity that this paper has only appeared in Dutch and then only in so obscure a periodical.

[3]) L. Radlkofer. Ueber die Methoden der Botanischen Systematik, insbesondere die ˚anatomische Methode. Festrede.

[4]) H. Solereder. Ueber den systematischen Wert der Holzstructur bei den Dicotyledonen. Diss. München, 1885 and Systematische Anatomie der Dicotyledonen. 1899, with an Ergänzungsband, 1908.

In Pharmacognosy, the field of practical botany with which we are more particularly concerned, the Linnean method was used with some success and on a large scale by L. Koch. In the first place this author published a great work on the micrography of powdered drugs [1]. Two of Linnaeus' principles were applied, viz. a rigorously an ordered arrangement of the facts described and the use of a telegram style.

Both were applied with considerable success. The book is profusely illustrated and a convenient work of reference.

The same author published another valuable work, which has an even greater bearing on our present purpose. It is entitled: „Pharmakognostischer Atlas'', is richly illustrated, as the title indicates, and two volumes have appeared in 1911 and 1914. It contains however moreover very detailed descriptions of the anatomical structure and here again the Linnean principles of strict order and simplicity of language are applied. This work of Koch doubtless marks a decided progress in the art of micrography, but at the same time it is deficient in other respects, for the mere introduction of regular order and a telegram style into micrography are not sufficient. On the contrary there is needed at the same time the introduction of two other principles as natural consequences of two important differences between phytography and micrography. Therefore a really efficient application of the Linnean method cannot take place without their help. The objects of phytography as well as of micrography having three dimensions, what the reader wants is a description combining at once the several aspects of the object from various sides, in short what I will call here a „perspective description''. Objects large enough to be examined with the naked eye present no difficulties in this respect. The aspects from all sides are combined without the slightest difficulty into a description of the object as a body with three dimensions, and it will mostly be easy to add some observations about the interior e. g. internal cavities, etc. Phytography generally being concerned only with larger objects seldom meets with difficulties in this respect. But in Micrography the case is quite the reverse. For anatomical investigations it is generally necessary to use the microscope and even relatively high powers of magnification, necessitating in their turn the use of very thin sections, thus practically reducing the observation to that of a single plane, not only of the larger organs but even of the smallest cells. Thus it is out of the question to obtain at once a truly perspective description, and the only way to get it is to combine the observations gathered from several sections, at least one for each dimension of space, but in some cases many more. In other words, if a somewhat complex anatomical structure is to be described, there is no help but to make descriptions of the several sections of the same object, and afterwards to combine these observations into a single perspective description. These descriptions of sections I will call „empirical descriptions''.

[1] L. Koch. Die mikroskopische Analyse der Drogenpulver. 4 Bände, 1901—1908.

It is perfectly clear that if the introduction of the Linnean method into micrography is intended it must be applied to perspective descriptions, for these only can give the reader an adequate idea of anatomical structures. The author who gives only an empirical description of a transverse section of some organ, only furnishes the material for a perspective description. And if an author publishes three empirical descriptions of sections in three directions at right angles to one another, without combining them into a perspective description, he leaves to the reader a part of the work necessary for bringing his description into the form which it wants if it is to be of real service. Nevertheless many examples might be cited of both cases in the litterature, either without or with the application of the Linnean method.

A second fundamental condition in micrography and again one which ought to be fulfilled if the introduction of the Linnean method shall really be useful is the preliminary introduction of the principles expressed by the terms Cytology, Histology and Microscopical Anatomy.

The term Cytology has a well defined meaning; it means the knowledge of the different parts constituting the cell: cytoplasm, nucleus, plastids, vacuoles, cell wall, etc. The terms Histology and Microscopical Anatomy are also generally used, but they are not generally understood in the same way. I therefore mention that I take them in the sense Pekelharing has claimed for them in his work on Histology [1]) and I have found that his interpretation is one of the utmost importance for a clear understanding of anatomical facts and ought therefore to be considered as one of the fundamental principles in micrography. By Histology we will understand the knowledge of the several kinds of cells as they are characterised by their shapes, thickness and chemical constitution of walls, prominent contents, etc., and their arrangement into tissues. The latter may be either simple, i. e. consisting of a single kind of cells, but they may also be compound, i. e. consisting of two or more kinds of cells thus forming relatively simple organs such as stomata, glands, vascular bundles, secondary xylem, etc. In one word how cells form tissues is the subject of histology as it is the subject of cytology how cells are built up out of their parts and organs.

Microscopical Anatomy lastly teaches how by tissues simple and compound the parts or organs of the plant are formed, designated as roots, leaves, flowers, etc. This chapter of anatomy treats of the internal architecture of the vegetable body. Epidermis, ground tissue and stele are the principal component tissue systems distinguished at present. We further distinguish General Microscopical Ana omy which treats of the rules of architecture prevailing in all organs of the plant; Special Microscopical Anatomy on the contrary treats of the differentiation proper to special kinds of organs, such as stems, leaves, stamens, seeds, etc.

---

[1]) C. A. Pekelharing. Voordrachten over weefselleer. Haarlem, 1905.

If the reader realises what has been explained here perhaps in a too summary way, he will see at once that this is a principle of the utmost importance for micrography. It ought to be kept in mind when the introduction of the Linnean method is intended.

The foregoing considerations have their influence not only on the character of the descriptions but also on their outward form. Experience has shown that it is necessary to give separate cell descriptions containing only cytological data. The rest, containing the observations on microscopical anatomy and histology, may be united together and called Topography. It is improbable that all readers consulting such a description want to know all cytological particulars contained in it and for the majority of students the topographical part is by far the most interesting. On the mutual arrangement of topography and cell descriptions a few remarks will be made below.

Summarising we may say that in micrography all descriptions ought to be perspective ones and based on the fact that the internal structure of plants cannot be understood if its most characteristic articulations represented by cell, tissue and tissue system are not allowed to exert their proper influence. These considerations are more important than the introduction of the Linnean method itself; if micrography is to be raised to the level of modern phytography, full attention ought to be given to them.

Now returning to the consideration of the works cited above, which tried to introduce the Linnean method into micrography, we find that neither the principle of perspective descriptions, nor that of the separate topography and cell descriptions have been introduced along with the Linnean method. With a single exception the empirical descriptions of three sections are given separately.

The only work in which the Linnean description was introduced at the same time as the principle of perspective descriptions and that of separate topography and cell description, the former taking into consideration the difference between histology and microscopical anatomy, is that of Janssonius on Javanese woods, already mentioned above [1]).

In this book we find at the same time introduced the method of pen-portraits of which I will speak below.

Before coming to this subject however, it will be well to remark that the necessity for the use of guiding schemes, already noticed when treating of the Linnean method in phytography, is even more indicated in micrography. Long experience has amply shown us the impossibility of making a micrography according to the Linnean method without the constant use of such guiding schemes.

---

[1]) H. H. Janssonius. Mikrographie des Holzes der auf Java vorkommenden Baumarten. Im Anschluss an „Additamenta ad cognitionem florae arboreae Javanicae, auct. S. H. Koorders et Th. Valeton. Up to this date there have appeared three volumes, respectively in the years 1906, 1908 and 1918 at Leiden.

The method of portraying description or Pen-portraits. This method is intended for use both in phytography and in micrography. In both cases it is combined with, or rather added to, the Linnean method. In the case of micrography the principles of perspective descriptions, of histology and microscopical anatomy and of separate topography and cell descriptions are moreover applied. Thus it is in some sort an extension of the existing methods on which it is based. The principle on which it is founded and its value for science will be discussed here. Preliminary accounts of this method as applied to micrography were given in my „Handboek der Botanische Micrographie" in 1907, a work intended as a vade-mecum in practical work for students of medicine, pharmacy and biology; further in a paper published in the year 1916, and entitled: „De beschrijvingen der simplicia in de Nederlandsche Pharmacopee (4th edition)" [1].

The term pen-portraits by no means implies that only portraits of individual plants are intended; on the contrary portraying descriptions of varieties, species, genera, etc. may be made with the help of this method and will in most cases be its object.

The leading principle of the method is most simple indeed. Its purpose is to bring into phytography and micrography as great a deal of „habit" as possible, and to do this not in a casual and sporadical manner but according to a preconcerted and well defined plan. It tends to bring into the description as much as possible of what is now generally considered as belonging to the domain of drawings and plastic models, in other words it tends to make pictures more and more superfluous.

The method of pen-portraits starts from the Linnean description, using its regular and strict order, its telegram style and its terminology. In this last respect however there is a somewhat essential difference. In phytography, the only part of Botany in which the Linnean description is hitherto generally used, there is a tendency to bring terminology into the foreground. Many descriptions consist almost wholly of organographical terms and in a general way the development of this method of description has been accompanied and characterized by the addition of new organographical terms. The method of pen-portraits on the other hand tends to the restriction of terminology within more moderate bounds and new terms are introduced only in cases of absolute necessity. The following considerations may show the plausibility of this assertion.

In the first place there is a consideration of more general import, equally applicable to descriptions of the old school and to pen-portraits. Limitation of the terminology is always advisable, because the use of many less generally known terms increases the difficulty of understanding the descriptions; in such

---

[1] Pharmaceutisch Weekblad. 1916, n°. 42 and 43.

cases it is preferable to give an explanation instead of a term. In the second place there is a certain contrast between terminology and the method of pen-portraits. As I remarked before terminology leads to a generalization, whilst the knowledge of „habit" is the aim of pen-portraits and this has a near relation to individuality. Therefore in portraying descriptions we ought to start from a basis formed with the help of a simple and restricted terminology and then to strike out as soon as possible into explanations in which no technical terms are used and which are available for the description of characters as yet unfit for generalization. In doing so the student leaves as soon as possible the well trodden path of orthodox science, in order to give full scope to his own original powers of observation. This must of course be combined with the use of a telegram style. As science proceeds and the knowledge of general rules increases the number of technical terms may however be increased from time to time, and the starting point of portraiture may thus be advanced, enabling it to go a longer way.

Thus expressly and consciously going beyond terminology the method of pen-portraits may give not only the utmost completeness allowed by the present state of science, but will moreover tend to enlarge our knowledge beyond the existing bounds.

The method of pen-portraiture, in common with the Linnean method, has at present not been applied in many branches of phytography and vegetable anatomy, but nevertheless it has been amply proved that the extra labour spent in its application leads to better results than can be attained without its help. Problems may be solved which till now were held to be insoluble with the help only of simple descriptions. In many cases it is quite sufficient to make a pen-portrait in order to come at once to the solution of some problem.

The method of pen-portraits has been used by me in the fourth edition of the Pharmacopoea Neerlandica, in describing the macroscopic characters of the drugs and the microscopic characters of the powders and although it is a fact that by these simple means in the Dutch Pharmacopaea a power of identification has been attained, which is not to be found in any other Pharmacopaea, this fact has as yet not been generally recognised.

An experiment on a somewhat larger scale was conducted by Dr. Janssonius in his „Mikrographie des Holzes der auf Java vorkommenden Baumarten", already cited before. In this work it was shown that with the help of pen-portraits the anatomical structure of secondary xylem affords ample material for drawing systematical conclusions concerning families, genera and species and moreover for identification and determination even of species. In the latter respect the book of Dr. Janssonius has already been found to possess a considerable practical value. Moreover it proved to be an easy task to show that the oldest fossil wood of *Angiosperms*, found in the Cretaceous formation and described by Mrs. Stopes under the name of *Aptiana radiata*, belongs in fact to the recent genus

Eurya and is nearly related or even identical with the recent species *Eurya japonica* or *glabra* [1]. .

In another paper it was shown that the wood of the well known graft-hybrid *Cytisus Adami* is the same as that of *Cytisus Laburnum* with only a few insignificant deviations certainly due to the combination with the epidermis of *Cytisus purpurea* [2].

The present volume has as object the introduction of the method of pen-portraits into theoretical Pharmacognosy as it was introduced into practical Pharmacognosy by the Pharmacopoea Neerlandica in its fourth edition. The micrography of some basts described in the present book was already published in the Dutch language by Dr. Janssonius in his dissertation [3].

There is no doubt that much time and much labour are required for making good pen-portraits and the psychical energy of the student is taxed rather heavily because in his mind he ought to unite all the characters described into a plastic image of the organs and structures, which are the subjects of his investigations. But there is no help for this, as there exists no other way of attaining the same results, which are nevertheless necessary for gaining a thorough knowledge of vegetable structures, leading to their ultimate explanation. It may indeed be a subject for consideration in how far it will be possible in a not too distant future to extend the use of the method of pen-portraits over the whole vegetable kingdom. As for the present work, it may suffice to say that it remains within the bounds of what is now possible, so that there is a fair prospect that it will be in future completed by similar pen-portraits of drugs now omitted. Anyhow a sufficient number of examples chosen from all principal organs of plants has been given here. I feel assured that they will contribute somewhat to the better knowledge of the anatomical characters of drugs and will materially facilitate identification.

II. PRACTICAL REMARKS

About the year 1901 the plan of writing this book was first conceived by me. For executing it a great deal of labour was necessary and it was impossible for me at that period of my life to spare the time for working out the plan by own hand. Neither was this unavoidable, for it is evident that if the method of pen-portraits is to be of any use in the future a large number of scientific workers ought to take it in hand and I deemed it not improbable that a student might be found who could under my direction commence by making pen-por-

---

[1] H. H. Janssonius and J. W. Moll. The Linnean method of describing anatomical structures. Proc. Kon. Akad. v. Wetensch. Amsterdam, Oct. 26th. 1912, p. 620 and Recueil des Trav. Botaniques Néerlandais. Vol. 9, 1912, p. 452.

[2] H. H. Janssonius und J. W. Moll. Der anatomische Bau des Holzes der Pfropfhybride Cytisus Adami und ihrer Komponente. Rec. d. Trav. Bot. Néerl. Vol. VIII, 1911.

[3] H. H. Janssonius. De tangentiale groei van eenige pharmaceutische basten. Groningen, 1918.

traits of drugs, or at least collect the observations necessary to make them, thus
gradually becoming independent in his work. Thus it was in the first place
necessary to obtain some finantial aid, and I was happy enough to find this
in a sufficient measure. In the first place I received during three years a not
inconsiderable subsidy from the „Groninger Universiteitsfonds", an institution
which has for many years promoted the scientific interests at our University.
In the second place a certain yearly sum was added during three years by the
well known institution, called „Teyler's Genootschap" at Haarlem. Lastly the
Royal Academy of Sciences at Amsterdam assigned the income for the year
1902 of the „Korthals fonds" to this work. In this manner an annual sum,
which would not seem large at the present day, was secured for a period of three
consecutive years.

Dr. H. H. Janssonius, who at that time had just concluded his pharmaceutical studies, was found willing to undertake the work, and from the beginning
of the year 1901 till the end of the year 1903 he amassed a large amount of
material, in fact the major part of that which is now offered to the public.
Several causes however concurred in making the material thus collected, unfit
for immediate publication. Among these one certainly was the handicap of a
foreign language, which from the beginning was intended to be english. Moreover at the end of the year 1903 Dr. Janssonius was entrusted by the Dutch
Colonial Department with the important but at the same time very difficult
task of writing his book on the anatomical structure of Javanese woods, mentioned above. In the course of the following years Dr. Janssonius and I gave as
much of our time as was possible not only to the elaboration of the raw material
collected during the three first years, but also to making new observations and
putting these at once in the form of pen-portraits. These new observations
especially contributed to the knowledge of secondary phloem. Thus we were
able to put a number of 13 pen-portraits into a form ready for publication, but
then we were obliged to let the work rest for some years.

A new impulse was given to the work by my colleague and successor Dr. J. C.
Schoute, and my best thanks are due to him for the proposal he made to me,
at the same time procuring the means for executing it. He proposed that a
person should be sought, well versed in the english language and at the same
time able to bring the data collected by Dr. Janssonius into the proper form
for pen-portraits. It seemed to me that botanical knowledge was not so much
required, because I could give myself the necessary guidance in this respect. We
were happy enough to find in the late Miss A. B. Ledeboer a most intelligent
and thoroughly competent help. Unhappily her health failed; she was obliged
to lay down her work when it was only partly finished. Nevertheless we may
thank her for a large amount of labour faithfully performed and excellent in
every respect.

She was succeeded by Mrs C. van Eck-de Wiljes, who was soon able to bring her work up to the same high level as that attained by Miss Ledeboer and she was happy enough to finish it. To both these ladies many thanks are due. Thus after many vicissitudes our work can at last be brought before the scientific public, and though of course in many respects it would have been preferable to have published it sooner, yet this late publication has also some advantages. During the twenty years elapsed since the beginning of the work the method of pen-portraits has been much developed, and important anatomical investigations have materially strengthened the terminological basis on which these descriptions rest. Thus the work is now in several respects decidedly better than it could have been fifteen years ago. Though of course an additional experience of many years would have enabled Dr. Janssonius to have occasionally attained greater completeness and perfection, yet the material brought together by him has in all cases proved to be of sterling quality, and has always enabled us to come to an interesting pen-portrait.

It was originally my intention to make a complete set of pen-portraits of all drugs, occurring in the British, United States and Dutch Pharmacopoea. This plan would however have meant the expenditure of a considerable amount of labour without giving for the present adequate advantages. The choice of drugs to be mentioned in a Pharmacopoea is in some respects an arbitrary one, and very different according to the several countries and editions.

What we now need in the first place is rather a set of pen-portraits of drugs, complete in this respect that all the several organs and products of plants used as drugs fairly represent the principal forms in which they occur. And we may confidently say that this object has been reached in the present work. There are to be found here not only a fair number of pen-portraits of the principal and most commonly used drugs, but moreover the student, who wishes to make a pen-portrait for himself, will readily find an example of the same kind which may be of much use to him and will tend to facilitate his work. Such a help may often be acceptable and in some cases even necessary. Though the chief object of this work is to give as far as possible complete descriptions of the anatomy of drugs, yet there have been added a certain number of two other kinds of descriptions in which the method of pen-portraits could be used with advantage. I have mentioned before that in the 4th edition of the Pharmacopoea Neerlandica pen-portraits were introduced for the first time in describing the macroscopic characters of the drugs and in the micrography of some powders. Translations of these articles have been added in the present work in the cases presenting occasion for it and I trust that this addition though incomplete may be welcome. In both these kinds of descriptions the chief object was a purely practical one, viz. a reliable and easy identification. In this respect there is a decided contrast between them and the bulk of the work.

In order to reach this purpose the order in which the elements are mentioned in the micrography of the powder is that in which they attract the students

2

attention under the microscope. Thus the elements, which are most numerous or especially attract the attention of the microscopist by their size or shape, are described in the first place, and so on till at the end are noticed those elements occurring only in small numbers and often only to be found after some search. This empirical arrangement of the elements has proved of considerable value for identification.

The description of the macroscopic characters is not pure organography or phytography, but anatomical characters are freely mentioned, provided that they are to be observed with the naked eye or with the help of a very low power. The macroscopic characters of a few basts were described expressly for this book.

But in most cases the descriptions of macroscopic characters belong to drugs mentioned in the fourth edition of the Pharmacopoea Neerlandica.

The micrography of the powder was added in the Pharmacopoea Neerlandica to a restricted number of drugs and only when these drugs have been included in the present work the reader will find it here [1]).

Here and there some differences of small consequence occur between the microscopic characters of the drug described by Dr. Janssonius and those of the powder described by me. The principal causes are the following:

1. the different purposes of the two descriptions. Starch for instance is much more important in the powder than in the drug itself and may therefore be described more completely in the former case.

2. the material used for the powder and for making sections was of different origin.

3. the investigation of sections and of more or less isolated fragments may lead to slightly different results. Stratification of cell walls may be better observed in sections; for the observation of occasional pit-canals isolated fibres are to be preferred.

I hope that the addition of macroscopic characters of the drug and the micrography of the powder, though only occurring in a restricted number of articles, may be of some interest for many who will use this work.

Nomenclature. Many drugs have a rather large number of names, but in this book only those names are mentioned at the heads of the articles, which are to be found in the Pharmacopoeae of Great Brittain and the United States of America.

As for plant names, those considered by the Index Kewensis as the proper ones were used in all cases.

---

[1]) Under the title: „Materia medica Monographs" there appeared in „The Chemist and Druggist, March 10th, 1917, Vol. 89, p. 98 a translation of the article Capsici Fructus in the Pharmacopoea Neerlandica, Ed. IV. This article contains a description of the macroscopic characters of the fruit, as well as a micrography of the powder. It is a very good example of what I intended in the 4th Edition of the Pharmacopoea Neerlandica and has met accordingly with some criticism. It has not been reproduced in this work, which does not contain Fructus Capsici.

The arrangement of the articles is alphabetical, but at the same time so that drugs showing the same organographical character have been placed together as much as possible. For this purpose however it was necessary not to write the names of the drugs beginning with that of the plant followed by the morphological character, as for instance Coriandri Fructus, Hydrastis Rhizoma, etc. This system was adopted in the British and United States Pharmacopoea in an evil hour, scattering the drugs over these works in the most hopeless way. In the present work the names are placed in the inverse order, as Fructus Coriandri, Rhizoma Hydrastis, etc. as is the case in the majority of Pharmacopoeae. Thus amyla, folia, cortices, flores, etc. are for the greater part placed together and further arranged alphabetically according to the plant names. It is often necessary to compare descriptions of homologous organs with one another and for this purpose the system followed here is greatly to be preferred.

Literature. In each article the description of the microscopic characters is preceded by some preliminary notices, and among these in the first place by a list of literature containing data concerning that particular drug. Only papers offering anatomical facts are cited here, and in the first place those used by Dr. Janssonius and consulted at the time when he was making his observation. Others were added afterwards but completeness was not aimed at. Nevertheless the student will find here much which can be of service. For the rest he will do well to consult Tschirch's large „Handbuch" [1]) and Solereder's „Systematische Anatomie" cited above; both these works often contain additional references. Evidently a pen-portrait will in most cases greatly surpass a micrography of the old school, both in correctness and completeness. Therefore the reader will not often find in the literature cited anything of importance wanting in our pen-portraits. This however does not mean that the existing literature has not been of service to Dr. Janssonius or may not be so to any one making a pen-portrait. On the contrary the investigator's attention may be, and was in our case, often drawn towards characters of particulars, which would have escaped it, notwithstanding the use of his guiding schemes, had he not consulted previous descriptions.

A complete alphabetical list of the works and papers, cited in the several articles, will be found after this preface.

Material and reagents. In the second and third place there are mentioned in each article the character of the material used for the anatomical investigation and as far as possible its origin, together with the several media and reagents used. Both are necessary for understanding the descriptions or at least certain parts of them. For the material this seems obvious. A knowledge of media and reagents is perhaps of still more importance, as it gives the reader the best means of controlling the value of the results reached in the descriptions. If for example

---

[1]) Tschirch. Handbuch der Pharmakognosie. 1919.

no phloroglucine and hydrochloric acid or similar reagent has been used and lignification is not mentioned it does not follow that it does not exist. If on the other hand a reagent for lignin is mentioned, the reader can be sure that all cell walls for which lignification is not mentioned are without it, whilst in the positive cases he will know on what grounds the observation was recorded. I am convinced that such a control is most valuable, and fully repays the trouble of mentioning these particulars in every case.

Following this preface the reader will find a complete alphabetical list of the media and reagents used, with a precise account of their character or composition.

It was not thought necessary to mention in each article all the several sections and preparations used for building up the perspective description. The reader will understand that in all cases where tissues are described, and these are the great majority, at least three sections according to the three directions of space were used, and that several of these sections were treated with different reagents and media. But I cannot see that an enumeration of these preparations can give the student any hint for the better understanding of the descriptions as is the case with media and reagents. It would be otherwise if there occurred in the work also less complete descriptions made with the help of a restricted number of sections. This being the case in „Die Mikrographie des Holzes", cited above, Dr. Janssonius has in this work added to each pen-portrait a list of preparations used for it.

Micrography. The student who uses this book is supposed to possess some elementary knowledge of vegetable anatomy and organography, which will enable him to understand the terminology used in the descriptions. The system of organographical terms has since Linné been worked out by many eminent botanists and brought to a state of completeness and consolidation. In many well known works the exact meaning of these terms is explained and therefore I have assumed that in the present book, moreover containing but a limited number of these terms, the student will easily help himself, if his knowledge should prove insufficient in some particular instance.

With respect to anatomical terms the case is somewhat different. It is true that a sufficient number of these has come into general use. Therefore the reader who possesses some elementary knowledge will find only exceptionally in our descriptions terms quite unknown to him. But concerning the exact meaning of many terms there may often exist some want of unanimity. No complete system of anatomical terms, taking into account the principles of histology and microscopical anatomy, has yet been published. I have worked out such a system and we have made use of it as a foundation for terminology in the present work, but it is impossible to publish this system here. Therefore it is necessary in another way to enable the reader to find in all cases the exact sense in which a term is used.

Moreover, as I have remarked before, this book is intended to serve as an example for making pen-portraits of all kinds of organs of the most different plants. Those who wish to follow our example will of course study in the first place the articles offered in this book, and this no doubt will be of much service to them.

But this will by no means suffice, and obviously these students will often need to know the exact meaning of the terms they are to use. I have tried to supply these wants by placing an alphabetical glossary at the end of the work. In this glossary a fairly complete set of anatomical terms has been explained according to the views prevailing in this work. Some of them may perhaps not occur in any of the articles published here. After each term in the glossary it has been mentioned whether it belongs to cytology (Cyt.), histology (Hist.), or microscopical anatomy (Micr. Anat.). This was done on account of the vast importance of these subdivisions for a clear understanding of anatomy.

In the first part of this preface I have mentioned the use of guiding schemes in making pen-portraits. Ample experience has taught us their absolute necessity; indeed neither of us would be able to make a good description without them. Therefore a student wishing to make a trial with the pen-portrait method must be provided with the guiding schemes we have used. They are the fruits of long years of experience, and have been added after this preface.

The anatomical guiding schemes reproduced here will give the student at the same time some insight into the system of anatomical terms mentioned above. There have been also added a few schemes for the description of the macroscopic characters of some drugs, for which experience has shown that this is useful or even necessary.

In all descriptions the dimensions of the several parts have been given as often as possible. Especially in micrography this is of great importance. If in describing a cell for instance the dimensions in the three directions of space are mentioned at the outset, the reader will at once obtain an idea of its shape, which will be completed by adding some descriptive term, such as prism, ellipsoid, etc. Therefore measures of this kind, expressed in micro-millimeters (μ), are abundant in all articles. Since three dimensions are always noticed, it was of course necessary to mention the direction of each. For this purpose however abbreviations were introduced. In the large majority of cases the directions mentioned must be understood as recorded with respect to the whole organ of which the structure is described, such as stem, root, leaf, etc. Only for these cases the following abbreviations were used.

For organs of all shapes:

L. = in a longitudinal direction, i. e. in the direction of an axis uniting the organic base and apex.

For organs showing radial symmetry:

R. = in a radial direction, i.e. transverse or at a right angle to the longitudinal direction, and at the same time extending from the centre to the periphery;

T. = in a tangential direction, i. e. transverse, and at the same time at a right angle to the radial direction.

For organs showing bilateral symmetry:

H. = height, i. e. in a direction perpendicular to the surface of the organ, therefore being different in different parts of the organ.

Lev. = in the direction of the surface of the organ. Therefore differing in different parts of the organ.

Lev. L. = parallel to the surface and at the same time in a longitudinal direction.

Lev. B. = parallel to the surface of the organ and at the same time in a transverse direction.

In a certain number of cases it is impossible to make use of the abbreviations mentioned in the above table, for instance if the measurements of the epithelium cells of schizogenous glands are to be recorded or those of wood parenchyma cells extending over the surface of xylem vessels. In such cases no abbreviations have been used, but instead of them the words length, depth, width, etc. These terms however are not meant in reference to the whole organ, but in reference to the special part or elements being under consideration.

The measures were taken either by chosing what seemed to be a mean value or a minimum and a maximum value. Many examples of both kinds are to be found in all articles. These measures are to be considered as rough statistical evaluations, which in many cases would have to be modified if more exact methods were to be employed. It is probable that in future measures resulting from the application of simple statistical methods will be more generally introduced into our descriptions.

At present however it is only in the fourth edition of the Pharmacopcea Neerlandica, that a few statistical data concerning several drugs are to be met with. Some of these have been reproduced in this work, though in a somewhat modified and more elaborate form. It is therefore necessary to say some words concerning the abbreviations used in these cases: n = number of variates, used for the construction of the frequency curve; M (indicated in the Pharmacopoea Neerlandica by x) = arithmetic mean; Med = median, dividing the observed variates into two classes of equal frequency; q_1 = lower quartile, dividing the observed variates beneath the median into two classes of equal frequency; Q_1 = lower quartile deviation, obtained by substraction of the lower quartile from the median value; q_3 = upper quartile, dividing the observed variates above the median into two classes of equal frequency; Q_3 = upper quartile deviation, obtained by substraction of the median value from the upper quartile; min. = minimum of the observed values of the variates; max. = maximum of the observed values of the variates.

Thus in the article on Amylum Solani the reader will find for the length of the grains the following data: $n = 320$; $M = 23 \mu$; $Med = 17.75 \mu$; $Q_1 = 7.90 \mu$; $Q_3 = 14.15 \mu$; min. in the class 0—5 μ.; max. in the class 80—85 μ. It is obvious that from these data a fairly complete representation of the frequency curve may be given and this both in the case of the normal and in that of the skew curve.

The difference between the two quartile deviations indicates the existence of skewness, but is not an exact measure for the degree of skewness.

Compound curves (bimodal—multimodal) may of course be represented in the same manner, but in these cases only the minimum of the lowest component of the compound curve and the maximum of its upper component is known (see Amylum Tritici).

In the description of starch grains showing an excentric hilum, the position of the hilum is indicated by means of a fraction of which the numerator and denominator correspond to the relation of the distances between the excentric hilum and the margin of the grain on both sides. Thus the fraction $^1/_3$ means that the hilum is to be found at a distance from the margin of one fourth part of the diameter in the direction of the excentricity, etc.

The cellulose reaction with iodine and sulphuric acid was applied in all cases without a single exception. The result in the large majority of cases was of course positive, but it would have been tedious to mention this in almost all cell descriptions. Therefore the chemical composition of the cell wall has been mentioned only in those cases in which the presence of other substances has been observed. But at the same time the student may be sure that cellulose has been found in all cell walls of which the chemical constitution is not mentioned.

A similar case is that of crystals consisting of calcium oxalate. As the large majority of crystals occurring in vegetable tissues consists of this substance, it was thought superfluous to mention it in so vast a number of cases. Therefore, if crystals are mentioned without further addition, this means that they consist of calcium oxalate and this has been almost always expressly controled by means of a reaction with acetic and hydrochloric acid.

Concerning the colour of the objects mentioned in the descriptions a similar remark may be made. In most cases cell walls, protoplasm, etc. are colourless, but it would have been tedious to have mentioned this in each particular case. Therefore if no colour is mentioned this means that the object is colourless. If the denomination „cortex" is used without further addition, it may be taken for granted that the bast of stems and not of roots is meant.

If detailed descriptions of complex structures are made in rather large numbers in the course of some years, it is clear that by and by some modifications and improvements will be introduced. Yet, even from a merely practical point of view, it will be often impossible afterwards to modify descriptions already finished. Therefore in a work like the present one there will always be found a certain lack of uniformity, even though it might be only detected by experienced eyes. In this respect it will conduce to a fairer criticism of our work if the chronological order of the several descriptions can be stated. Therefore a date is added at the end of each article, showing at what time the original description was made by Dr. Janssonius and in the case of some cortices by both of us. Furthermore it has been mentioned above that a certain number of articles are due to the collaboration of Dr. Janssonius and myself, but that in the editing of a rather large number we have had the help of Miss Ledeboer and Mrs. van Eck-de Wiljes. Each article has been signed at the end, as the reader will see, with the initials of those who have contributed to it by their labour, viz. J for Dr. Janssonius; L for Miss Ledeboer; v. E. d. W for Mrs. van Eck-de Wiljes and M for myself.

One or more figures have been added to the majority of the articles and I am sure that the reader who has followed my exposition will not be surprised at this fact. He will know, that, although a pen-portrait strives to eliminate habit, this purpose at the present state of our knowledge can never be quite achieved and that illustrations can give much help in this respect. In most cases the figures represent transverse sections because in these the largest number of important and at the same time ununderstood characters are to be found. All figures were of course made with the help of the camera lucida, and they generally only represent characters belonging to the domain of microscopical anatomy. To each of them a scale has been added, so that their magnification may at once be known. The drawings, which have been reproduced in the figures, were mostly made in two stages. The first sketches were made by Mr. L. van Wolde and after his death they were worked up by Mr. R. Hoeksema, both in their capacity of assistant in the Botanical Laboratory at Groningen.

On page 21 I mentioned the alphabetical glossary of anatomical terms to be formed at the end of this work and in this respect it may suffice to refer the reader to what has been said here.

Moreover a general alphabetical index has been added, referring the reader to the pages on which the words are to be found.

Alphonse de Candolle in his „Phytographie" page 96 has warned against the use of more than one alphabetical index at the end of the book and his argumentation is very strong.

Nevertheless I have not united the glossary with the general index because the former often contains rather long definitions of anatomical terms and might prove to be an obstacle to the practical use of a general index combined with it, especially as the latter in this case contains only a relatively small number of words.

Groningen, 1920—'21.

GUIDING SCHEMES

These guiding schemes are bad masters but good servants. They will show you a safe and easy way through the trodden paths of science, leaving your mind free to make new and original observations. Do not object to them on account of their prolixity, you may object to the use of a railway guide on the same account. Yet a man going to the railway station without having consulted his guide and hoping to find there his train just starting has a better chance of success than a botanist has of making a good pen-portrait without using a guiding scheme.

The following guiding schemes are sufficient for the description of all parts of the higher plants. Scale-leaves, tepals, sepals and petals ought to be described with the help of the guiding scheme for the leaf; the fruit with the help of that for the pistil.

No. 1. Guiding scheme for the description of cells in a general sense : unicellular organisms provided with a cell wall, parenchyma cells, fibres, hyphae (pseudoparenchyma), etc., naked cells.

1a. Cells provided with cell walls.
    2a. Dimensions in the three directions of space.
    2b. Shape.
    2c. Wall.
        3a. Thickness.
        3b. Coarser structure.
            4a. Middle lamella, containing the primary layers.
            4b. Secondary thickening layers.
            4c. Tertiary thickening layers.
        3c. Finer structure.
            4a. Stratification.
            4b. Striation.
        3d. Colour.
        3e. Chemical character.
        3f. Partial thickenings of the wall.
            4a. Pits.
                5a. Simple pits.
                    6a. Number and arrangement.
                    6b. Shape.
                        7a. Circular.
                        7b. Slit-like.
                    6c. Pit canal.
                    6d. Closing membrane.
                5b. Bordered pits.

              6*a*. Number and arrangement.
                  7*a*. Combined bordered pits.
              6*b*. Shape.
                  7*a*. Circular.
                  7*b*. Slit-like.
                  7*c*. Scalariform.
              6*c*. Border.
              6*d*. Not widened part of the pit canal.
                  7*a*. Slit.
                  7*b*. Canal of the bordered pit.
                      8*a*. Inner aperture.
                      8*b*. Outer aperture.
              6*e*. Closing membrane with torus.
           5*c*. Unilateral or one-sided bordered pits, semi-bordered pits. See 5*a* and 5*b*.
        4*b*. Annular bands.
        4*c*. Spiral bands.
        4*d*. Reticulate thickenings.
        4*e*. Cystoliths, rod-shaped thickenings and transverse bars.
        4*f*. Collenchymatous thickenings.
    3*g*. Crystals within the cell wall.
        4*a*. Simple crystals.
        4*b*. Raphides.
        4*c*. Cluster crystals.
        4*d*. Crystal sand.
        4*e*. Sphaerites.
    3*h*. Perforations of the cell wall.
        4*a*. Ordinary perforations.
           5*a*. Number, arrangement and shape.
        4*b*. Perforations for connecting threads of protoplasm.
           5*a*. Number, arrangement and shape.
        4*c*. Sieve structure.
           5*a*. Number, arrangement and shape.
    3*i*. Intercellular spaces.
        4*a*. Dimensions and shape.
        4*b*. Connecting frames. (câdres d'union of Mangin).
        4*c*. Rodlets. (bâtonnets of Mangin).
  2*d*. Contents.
    3*a*. Protoplast.
        4*a*. Parietal cytoplasm and strings of protoplasm.
           5*a*. Thickness.
           5*b*. Microsomes.
           5*c*. Internal streaming movements.
               6*a*. Rapid.
                  7*a*. Rotation.
                  7*b*. Circulation.
                      8*a*. Parietal.
                      8*b*. Central.
                  7*c*. Streaming.
               6*b*. Slow movements of orientation.

         7*a*. Epistrophe.
         7*b*. Apostrophe.
         7*c*. Systrophe.
      5*d*. Connecting threads. (Plasmodesms).
4*b*. Nucleus.
    5*a*. Number and arrangement.
    5*b*. Shape and dimensions.
    5*c*. Nucleoli; number, shape and dimensions.
    5*d*. Nuclear membrane.
    5*e*. Nuclear network (chromosomes).
4*c*. Plastids.
    5*a*. Number and arrangement.
    5*b*. Shape and dimensions.
    5*c*. Colour.
        6*a*. Chloroplasts.
           7*a*. Chloroplasts proper.
           7*b*. Spurious chloroplasts.
        6*b*. Leuco- or amyloplasts.
        6*c*. Chromoplasts.
    5*d*. Starch grains.
        6*a*. Simple.
           7*a*. Shape, dimensions and position within the plastid.
           7*b*. Stratification.
           7*c*. Hilum.
               8*a*. Central.
               8*b*. Excentric.
        6*b*. Compound.
           7*a*. Number of component grains.
           For the rest see 6*a*.
        6*c*. Half-compound. See 6*b*.
    5*e*. Albumen crystals.
4*d*. Vacuoles.
    5*a*. Ordinary vacuoles.
        6*a*. Sap cavities.
           7*a*. Number and arrangement.
           7*b*. Shape and dimensions.
           7*c*. Tonoplast.
           7*d*. Contents.
               8*a*. Chemical character of fluid contents.
               8*b*. Aleurone grains.
                  9*a*. Albumen crystals.
                  9*b*. Globoids.
                  9*c*. Crystals of calcium oxalate.
               8*c*. Crystals.
                  9*a*. Simple crystals.
                  9*b*. Raphides.
                  9*c*. Cluster crystals.
                  9*d*. Crystal sand.
                  9*e*. Sphaerites.
        6*b*. Adventitious vacuoles. See 6*a*.

             5*b.* Contractile vacuoles.
                 6*a.* Rhytm of contraction.
      3*b.* Contents of dead cells.
      3*c.* Thyloses.
        To be described like cells.
1*b.* Naked cells (Locomotion).
    2*a.* Pseudopodia and amoeboid creeping movements.
    2*b.* Cilia, flagella and swimming movements.
      3*a.* Monotrichous cells.
      3*b.* Ditrichous cells.
      3*c.* Lophotrichous cells.
      3*d.* Peritrichous cells.
    2*c.* Eye spot.
    See for the rest 1*a.*

No. 2. Guiding scheme for the description of vessels.

1*a.* Xylem vessels.
    2*a.* With respect to the principal sculptures on the lateral walls.
      3*a.* Annular vessels.
      3*b.* Spiral vessels.
      3*c.* Reticulate vessels.
      3*d.* Pitted vessels.
      3*e.* Scalariform vessels.
    2*b.* Dimensions.
      3*a.* Thickness.
      3*b.* Length of articulations and total length of vessel.
    2*c.* Shape.
    2*d.* Partition walls.
      3*a.* Direction.
        4*a.* Horizontal.
        4*b.* Oblique.
      3*b.* Perforations.
        4*a.* Shape.
          5*a.* Circular or oval.
          5*b.* Sclarariform.
        4*b.* Number, place, arrangement, and dimensions. Dimensions of the remaining parts of the transverse walls.
        4*c.* Margin.
          5*a.* Smooth.
          5*b.* Bordered pit like.
    2*e.* Lateral walls. See guiding scheme No. 1: 1*a,* 2*c.*
    2*f.* Contents. See guiding scheme No. 1: 1*a,* 2*d.*
1*b.* Sieve-tubes.
    2*a.* Dimensions, also length of articulations.
    2*b.* Shape.
    2*c.* Partition walls.
      3*a.* Direction.
        4*a.* Horizontal.
        4*b.* Oblique.

      3*b*. Sieve-plates.
          4*a*. Number, place, arrangement and dimensions.
          4*b*. Shape.
    2*d*. Lateral walls.
      3*a*. Sieve-plates. See for the rest of lateral walls guiding scheme No. 1: 1*a*, 2*c*.
    2*e*. Contents. See guiding scheme No. 1: 1*a*, 2*d*, 3*a*.
1*c*. Latex vessels.
    2*a*. Dimensions and shape.
    2*b*. Transverse walls and perforations.
    2*c*. Lateral walls. See guiding scheme No. 1: 1*a*, 2*c*.
    2*d*. Contents. See guiding scheme No. 1: 1*a*, 2*d*, 3*a*.

No. 3. Guiding scheme for the description of meristems in their relation to the adult tissues.

1*a*. Only meristem present (primitive meristem).
1*b*. Meristem and adult tissue both present.
    2*a*. Primary meristem.
      3*a*. Apical (vegetative cone).
      3*b*. Basal.
      3*c*. Intercalary.
          4*a*. Intercalary proper.
          4*b*. Cambium.
              5*a*. One-sided, monopleuric.
              5*b*. Two-sided, dipleuric.
    2*b*. Secondary, tertiary etc. meristem.
      Subdivision like 2*a*.

No. 4. Guiding scheme for the description of meristems themselves.

1*a*. Storied or tier-like meristem.
    2*a*. Number and construction of the stories.
1*b*. Initial celled meristem.
    2*a*. Initial cells.
      3*a*. Apical cell.
      3*b*. Vegetative point.
      3*c*. Large number of initial cells.
    2*b*. Segments of the apical cell and ordinary meristem cells.
    2*c*. Procambium.
    2*d*. Adult elements.

No. 5. Guiding scheme for the description of stomatic apparatus.

1*a*. Ordinary stomata.
    2*a*. Cryptoporous.
      3*a*. Vestibule.
      3*b*. Guard-cells.
      3*c*. Porus.
          4*a*. Front chamber.
          4*b*. Slit.

                 4c.   Back chamber.
            3b.   Respiratory cavity.
            3c.   Subsidiary cells.
       2b.   Phaneroporous.
            See *2a*, with the exception of the vestibule.
1b.   Water-pores.
     See *1a, 2b*.

No. 6. Guiding scheme for the description of the appendages of the outer surface and the surfaces of internal cavities.

1a.   Trichomes.
       2a.   Papillae.
            3a.   Ordinary papillae.
            3b.   Stigmatic papillae.
            3c.   File papillae.
       2b.   Ordinary hairs.
            3a.   Ramification.
            3b.   Number of cells.
            3c.   Uni- to multiseriate.
       2c.   File hairs.
            3a.   Ramification.
            3b.   Number of cells.
            3c.   Uni- to multiseriate.
       2d.   Root hairs.
       2e.   Rhizoids.
       2f. Clinging hairs.
       2g.   Stellate hairs.
            3a.   Number of cells.
       2h.   Stellate file hairs.
            3a.   Number of cells.
       2i.   Balance hairs.
       2k.   Epidermal emissaries.
       2l.   Paraphyses.
       2m. Capitate hairs.
            3a.   Ordinary capitate hairs.
                 4a.   Unicellular.
                 4b.   Pluricellular.
                     5a.   Basal part.
                     5b.   Stalk.
                     5c.   Head.
            3b.   Glandular hairs.
                 4a.   Ordinary glandular hairs.
                     5a.   Unicellular.
                         6a.   Cuticular bladder absent.
                         6b.   Cuticular bladder present.
                     5b.   Pluricellular.
                         6a.   Basal part.
                         6b.   Stalk.
                         6c.   Head.

             6*d*. Cuticular bladder with contents.
        4*b*. Type of Labiatae. See 5*b*.
        4*c*. Type of Compositae. See 5*b*.
        4*d*. Colleters. See 5*b*.
    2*n*. Stings.
    2*o*. Bristle trichomes.
    2*p*. Scurves (scales).
        3*a*. Ordinary scurves.
        3*b*. Ramenta.
    2*q*. Prickle trichomes.
1*b*. Emergences.
    2*a*. Bristle emergences.
    2*b*. Prickle emergences.
    2*c*. Fruit pulp emergences.

No. 7. Guiding scheme for the description of emissaries (hydathodes).

1*a*. Epidermal emissaries.
    2*a*. Ordinary emissaries.
        3*a*. Unicellular.
        3*b*. Pluricellular.
    2*b*. Trichome emissaries.
1*b*. Emissaries provided with the ends of xylem bundles.
    2*a*. Water pores wanting.
        3*a*. Excreting cells.
        3*b*. Ends of xylem bundles.
    2*b*. Water pores present.
        3*a*. Water pores.
            4*a*. Number and arrangement.
            4*b*. Shape. See guiding scheme No. 5.
        3*b*. Epithema.
        3*c*. Ends of xylem bundles.
    2*c*. Apical openings.
        3*a*. Cuticle.
        3*b*. Cavity.
        3*c*. Ends of xylem bundles.

No. 8. Guiding scheme for the description of internal glands.

1*a*. Schizogenous glands, including vittae.
    2*a*. Dimensions and shape.
    2*b*. Epithelium.
    2*c*. Sheath.
        3*a*. Parenchyma sheath.
        3*b*. Sclerenchyma sheath.
            4*a*. Sclerenchymatic elements.
            4*b*. Passage cells.
    2*d*. Cavity and contents.
1*b*. Oblito-schizogenous and schizo-lysigenous glands.
    2*a*. Dimensions and shape.

2b. Epithelium.
2c. Sheath.
    3a. Parenchyma sheath.
    3b. Sclerenchyma sheath.
        4a. Sclerenchymatic elements.
        4b. Passage cells.
2d. Cavity and contents.
2e. Cover and evacuation slit.

No. 9. Guiding scheme for the description of a primary vascular bundle.

1a. Simple vascular bundle.
  2a. Closed.
    3a. Collateral.
      4a. Phloem.
        5a. Exarch.
          6a. Protophloem.
            7a. Primitive or primordial sieve tubes.
            7b. Flattened elements.
          6b. Metaphloem.
            7a. Cribral elements.
              8a. Sieve-tubes. See guiding scheme No. 2.
              8b. Companion cells.
            7b. Cambiform cells.
        5b. Endarch. See 5a.
        5c. Mesarch. See 5a.
      4b. Connecting parenchyma.
      4c. Xylem.
        5a. Exarch.
          6a. Protoxylem.
            7a. Primitive or primordial vessels.
            7b. Intercellular space.
            7c. Flattened elements.
          6b. Metaxylem.
            7a. Tracheal elements.
              8a. Xylem vessels. See guiding scheme No. 2.
              8b. Tracheids.
            7b. Xylem parenchyma.
    3b. Bicollateral.
      4a. Outer phloem. See 3a, 4a.
      4b. Connecting parenchyma.
      4c. Xylem. See 3a, 4c.
      4d. Connecting parenchyma.
      4e. Inner phloem. See 3a, 4a.
    3c. Concentric.
      4a. Amphivasal. See 3a.
      4b. Amphicribral. See 3a.

        3*d*. Radial [1]). See 3*a*.
   2*b*. Open.
        3*a*. Collateral. See 2*a*, 3*a*, but cambium replacing the connecting parenchyma.
        3*b*. Bicollateral. See 2*a*, 3*d*, but cambium at least partly replacing the connecting parenchyma.
        3*c*. Radial. See 2*a*, 3*d*, but cambium replacing the connecting parenchyma.
1*b*. Compound vascular bundle.
   2*a*. Simple bundles composing the compound vascular bundles.
        3*a*. Number and arrangement within the compound bundles. See for the rest 1*a*.
   2*b*. Medullary commissures within the compound bundles.
        3*a*. Parenchyma.
        3*b*. Idioblasts.

No. 10. Guiding scheme for the description of a fibrovasal bundle.

1*a*. Sclerenchyma sheath.
   2*a*. Elements of the above.
1*b*. Vascular bundle. See guiding scheme No. 9.

No. 11. Guiding scheme for the description of secondary cork tissue and lenticels.

1*a*. Ordinary secondary cork tissue.
   2*a*. Initial celled cork.
        3*a*. Phellem.
            4*a*. Growth rings.
            4*b*. Part taken by the several elements in the construction of the phellem.
                5*a*. Periderm.
                5*b*. Phelloid (common parenchyma).
                5*c*. Sclerenchyma.
                5*d*. Aerenchyma.
        3*b*. Phellogen.
        3*c*. Phelloderm.
            4*a*. Ordinary phelloderm cells.
            4*b*. Idioblasts.
   2*b*. Storied cork.
        3*a*. Number and construction of the stories.
            4*a*. Periderm.
1*b*. Lenticels.
   2*a*. Phellem.
        3*a*. Complementary tissue.
        3*b*. Closing membranes or intermediate bands.
        3*c*. Aerenchyma.
   2*b*. Phellogen.
   2*c*. Phelloderm.

---

[1]) This term is used by de Bary, Vergl. Anatomie, 1877, p. 361, Comparative Anatomy, 1884, p. 348 for the whole stele, only with the exception of the pericycle, in roots, etc. Though this term agrees with the letter of the definition of vascular bundle, its use is inconsistent with the stelar theory and it ought therefore to be considered as obsolete. It is only mentioned here, because the authority of de Bary has procured it a general use.

   3a. Ordinary phelloderm cells.
   3b. Idioblasts.

No. 12. Guiding scheme for the description of secondary phloem.

1a. Rhytidoma or bark.
 2a. Scaly rhytidoma.
  3a. Ordinary secondary phloem. See *1b.*
  3b. Secondary cork layers and lenticels. See guiding scheme No. 11.
 2b. Annular or ringed rhytidoma. See *2a.*
1b. Secondary phloem proper.
 2a. Growth rings.
 2b. Part taken by the several elements in the construction of the phloem.
  3a. Phloem without storied arrangement.
   4a. Elements of the cribral system.
    5a. Sieve-tubes.
    5b. Companion cells.
    5c. Keratenchyma.
   4b. Elements of the system of bast fibres.
    5a. Non septate bast fibres.
     6a. Sclerenchymatic.
     6b. Thin-walled.
    5b. Septate bast fibres. See *5a.*
   4c. Elements of the parenchymatic system.
    5a. Bast parenchyma.
     6a. Arrangement.
      7a. Metacribral.
     6b. Part taken by the several elements in the construction of the bast parenchyma.
      7a. Septate bast parenchyma fibres.
       8a. Regular bastparenchyma.
        9a. Common parenchyma.
        9b. Conjugated parenchyma.
        9c. Sclerenchyma.
        9d. Idioblasts.
        9e. Internal glands. See guiding scheme No. 8.
       8b. Irregular bast parenchyma. See *8a.*
      7b. Substitute fibres. See *9a—e* under *7a.*
    5b. Medullary rays.
     6a. Simple medullary rays.
      7a. Procumbent cells.
       8a. Common procumbent cells.
       8b. Sclerenchyma.
       8c. Idioblasts.
      7b. Upright cells. See *7a.*
      7c. Internal glands. See guiding scheme No. 8.
     6b. Compound medullary rays. See *6a.*
   4d. Latex vessels. See guiding scheme No. 2.
  3b. Phloem with storied arrangement. See *3a.*

No. 13. Guiding scheme for the description of secondary xylem.

1a. Sap-wood (alburnum).
    2a. Growth rings.
        3a. Early wood (spring wood).
        3b. Middle layer.
        3c. Late wood (autumn wood).
    2b. Part taken by the several elements in the construction of the xylem.
        3a. Xylem without storied arrangement.
            4a. Elements of the tracheal system.
                5a. Vessels.
                    6a. Number and arrangement. See for the rest guiding scheme No. 2.
                5b. Tracheids.
                    6a. Vascular tracheids.
                        7a. Ordinary.
                        7b. Conjugated.
                    6b. Fibre tracheids.
            4b. Elements of the libriform system.
                5a. Non septate libriform fibres.
                5b. Septate libriform fibres.
            4c. Elements of the parenchymatic system.
                5a. Wood parenchyma.
                    6a. Arrangement.
                        7a. Paratracheal (circumvasal).
                        7b. Metatracheal (laminar).
                        7c. Scattered among the other elements.
                    6b. Part taken by the several elements in the construction of the wood parenchyma.
                        7a. Septate wood parenchyma fibres.
                            8a. Regular wood parenchyma.
                                9a. Ordinary.
                                9b. Conjugated.
                                9c. Idioblasts.
                                9d. Internal glands. See guiding scheme No. 8.
                          8b. Irregular wood parenchyma. See 8a.
                      7b. Substitute fibres. See 9a—d under 8a.
            5b. Medullary rays.
                6a. Simple medullary rays.
                    7a. Procumbent cells.
                        8a. Ordinary.
                        8b. Idioblasts.
                    7b. Upright cells.
                        8a. Ordinary.
                        8b. Conjugated.
                        8c. Idioblasts.
                    7c. Tile-shaped cells.
                    7d. Tracheidal medullary ray cells.
                    7e. Internal glands. See guiding scheme No. 8.
                6b. Compound medullary rays. See 6a.

4*d*. Phloem bundles. See guiding scheme No. 12.
     3*b*. Xylem with storied arrangement. See 3*a*.
1*b*. Heart wood (duramen). See 1*a*.

No. 14. Guiding scheme for the description of the root.

1*a*. Root-cap (calyptra).
     2*a*. Columella.
     2*b*. Outer calyptra cells.
     2*c*. Calyptra cambium. See guiding scheme No. 4.
1*b*. Epidermis.
     2*a*. Epidermal cells.
     2*b*. Root hairs.
     2*c*. Water tissue.
1*c*. Cortex.
     2*a*. Exodermis.
          3*a*. Primary cork layer.
          3*b*. Sclerenchyma.
     2*b*. Primary cortex layers proper.
          3*a*. Common parenchyma.
          3*b*. Aerenchyma.
          3*c*. Sclerenchyma.
          3*d*. Internal glands. See guiding scheme No. 8.
          3*e*. Latex vessels. See guiding scheme No. 2.
          3*f*. Vascular bundles. See guiding scheme No. 9.
          3*g*. Idioblasts.
     2*c*. Secondary tissues.
          3*a*. Secondary cork and lenticels. See guiding scheme No. 11.
          3*b*. Secondary cortical parenchyma.
          3*c*. Secondary vascular bundle tissue.
               4*a*. Secondary parenchyma formed towards the periphery.
               4*b*. Cambium. See guiding schemes No. 3 and 4.
               4*c*. Secondary parenchyma formed towards the centre.
               4*d*. Secondary vascular bundles. See guiding scheme No. 9.
     2*d*. Endodermis.
          3*a*. Endodermis cells proper.
          3*b*. Passage cells.
          3*c*. Idioblasts.
          3*d*. Internal glands. See guiding scheme No. 8.
          3*e*. Secondary vascular bundle tissue. See 2*c*, 3*c*.
1*d*. Stele.
     2*a*. Pericycle.
          3*a*. Primary pericycle.
               4*a*. Common parenchyma.
               4*b*. Idioblasts.
               4*c*. Internal glands. See guiding scheme No. 8.
               4*d*. Sclerenchyma.
               4*e*. Latex vessels. See guiding scheme No. 2.
          3*b*. Secondary pericycle.
               4*a*. Cork tissue and lenticels. See guiding scheme No. 11.

        4b. Phloem and xylem. To be described with the secondary phloem and xylem (*2b, 3a* or *3b*).

        4c. Lateral roots.

        4d. Secondary vascular bundle tissue. See *1c, 2c, 3c.*

2b. Phloem and xylem bundles.

    3a. Secondary phloem and xylem either wanting or, if present, considered in its relation to the original phloem and xylem bundles.

        4a. Number and arrangement of phloem and xylem bundles.

        4b. Secondary tissues wanting.

            5a. Phloem.

                6a. Sclerenchyma.

                6b. Phloem proper. See guiding scheme No. 9.

            5b. Xylem. See guiding scheme No. 9.

        4c. Secondary tissues present.

            5a. Primary phloem. See *4b, 5a.*

            5b. Secondary phloem. Parts formed by pericycle or medullary commissures also to be described here.

                6a. Produced by the phloem arches of the cambium. See guiding scheme No. 12.

                6b. Produced by the xylem arches of the cambium. See guiding scheme No. 12. If consisting only of parenchymatous elements forming a medullary ray, and then not to be mistaken for a medullary commissure. See also *5d, 6b.*

            5c. Cambium.

                6a. Phloem arches. See guiding schemes No. 3 and 4.

                6b. Xylem arches. See guiding schemes No. 3 and 4.

            5d. Secondary xylem. Parts formed by pericycle or medullary commissures also to be described here.

                6a. Produced by the phloem arches. See guiding scheme No. 13.

                6b. Produced by the xylem arches. See guiding scheme No. 13. If consisting only of parenchymatous elements forming a medullary ray, and then not to be mistaken for a medullary commissure. See also *5b, 6b.*

            5e. Primary xylem. See guiding scheme No. 9.

    3b. Secondary xylem and phloem present, but not considered in its relation to the original phloem and xylem bundles.

        4a. Primary phloem. See guiding scheme No. 9.

        4b. Secondary phloem. See guiding scheme No. 12.

        4c. Cambium. See guiding scheme No. 3 and 4.

        4d. Secondary xylem. See guiding scheme No. 13.

        4e. Primary xylem. See guiding scheme No. 9.

2c. Medullary commissures.

    3a. Primary tissue.

        4a. Common parenchyma.

        4b. Idioblasts.

    3b. Secondary tissue. If present already described with secondary phloem and xylem.

2d. Medulla.

    3a. Common parenchyma.

    3b. Idioblasts.

No. 15. Guiding scheme for the description of the stem.

1*a.* Epidermis.
    2*a.* Primary epidermal tissue.
        3*a.* Epidermis proper.
            4*a.* Epidermal cells.
            4*b.* Stomata. See guiding scheme No. 5.
            4*c.* Idioblasts.
            4*d.* Appendages of the surface. See guiding scheme No. 6.
        3*b.* Hypoderma.
            4*a.* Hypodermal cells.
    2*b.* Secondary epidermal tissue.
        3*a.* Cork tissue and lenticels. See guiding scheme No. 11.
1*b.* Cortex.
    2*a.* Primary cortex layers proper.
        3*a.* Collenchyma.
        3*b.* Common parenchyma.
        3*c.* Aerenchyma.
        3*d.* Periderm.
        3*e.* Sclerenchyma.
        3*f.* Idioblasts.
        3*g.* Latex vessels. See guiding scheme No. 2.
        3*h.* Internal glands. See guiding scheme No. 8.
        3*i.* Appendages of internal surfaces. See guiding scheme No. 6.
        3*k.* Meristeles. See for the vascular or fibrovasal bundles of these: guiding
                schemes No. 9 and 10.
    2*b.* Secondary tissues.
        3*a.* Secondary cork and lenticels. See guiding scheme No. 11.
        3*b.* Aerenchyma.
    2*c.* Endodermis.
        3*a.* Primary endodermis.
            4*a.* Protective sheath.
                5*a.* Ordinary cells.
                5*b.* Passage cells.
                5*c.* Idioblasts.
            4*b.* Starch sheath. See 4*a.*
            4*c.* Crystal sheath. See 4*a.*
            4*d.* Enzyme sheath, etc. See 4*a.*
        3*b.* Secondary endodermal tissue.
            4*a.* Cork tissue. See guiding scheme No. 11.
            4*b.* Secondary vascular bundle tissue.
                5*a.* Secondary parenchyma formed towards the periphery.
                5*b.* Cambium. See guiding scheme No. 3 and 4.
                5*c.* Secondary parenchyma formed towards the centre.
                5*d.* Secondary vascular bundles. See guiding scheme No. 9.
1*c.* Stele.
    2*a.* Pericycle.
        3*a.* Primary pericycle.
            4*a.* Common parenchyma.
            4*b.* Sclerenchyma.

           4*c*. Latex vessels. See guiding scheme No. 2.

           4*d*. Idioblasts.

           4*e*. Internal glands. See guiding scheme No. 8.

        3*b*. Secondary pericycle.

           4*a*. Cork tissue. See guiding scheme No. 11.

           4*b*. Phloem and xylem. To be described with the interfascicular secondary phloem and xylem. See 2*c*, 3*b*, 4*a*, 5*b*, 6*b* and 6*d*.

           4*c*. Adventitious roots. See guiding scheme No. 14.

           4*d*. Secondary vascular bundle tissue. See 1*b*, 2*c*, 3*b*, 4*b*.

    2*b*. Vascular or fibrovasal bundles.

        3*a*. Vascular bundles.

           4*a*. Secondary phloem and xylem either wanting or, if present, considered in its relation to the vascular bundles.

               5*a*. Simple vascular bundles.

                  6*a*. Number and arrangement.

                  6*b*. Secondary tissues wanting. See guiding scheme No. 9.

                  6*c*. Secondary tissues present.

                      7*a*. Primary phloem. See guiding scheme No. 9.

                      7*b*. Secondary phloem (fascicular phloem). See guiding scheme No. 12.

                      7*c*. Cambium (fascicular cambium). See guiding schemes No. 3 and 4.

                      7*d*. Secondary xylem (fascicular xylem). See guiding scheme No. 13.

                      7*e*. Primary xylem. See guiding scheme No. 9.

               5*b*. Compound vascular bundles.

                  6*a*. Number and arrangement.

                  6*b*. Simple vascular bundles composing the compound bundles.

                      7*a*. Number and arrangement within the compound bundles. See for the rest 5*a*.

                  6*c*. Medullary commissures within the compound bundles. See 2*c*.

            4*b*. Secondary phloem and xylem present, but not considered in its relation to the vascular bundles.

               5*a*. Primary phloem (see guiding scheme No. 9) and corresponding primary part of the medullary commissures. See 2*c*, 3*b*, 4*a*, 5*b*, 6*a*.

               5*b*. Secondary phloem. See guiding scheme No. 12.

               5*c*. Cambium. See guiding schemes No. 3 and 4.

               5*d*. Secondary xylem. See guiding scheme No. 13.

               5*e*. Primary xylem (see guiding scheme No. 9) and corresponding primary part of the medullary commissures. See 2*c*, 3*b*, 4*a*, 5*b*, 6*e*.

        3*b*. Fibrovasal bundle.

            4*a*. Sclerenchyma sheath.

               5*a*. Elements of the above.

            4*b*. Vascular bundle. See 3*a*.

  2*c*. Medullary commissures.

        3*a*. Monocotyledonous stem; to be described with the medulla.

        3*b*. Dicotyledonous stem.

            4*a*. Between the compound vascular bundles.

               5*a*. Secondary tissues wanting.

                  6*a*. Common parenchyma.

        6*b*. Idioblasts.
     5*b*. Secondary phloem and xylem present.
        6*a*. Primary part corresponding with primary phloem of vascular bundles. See 5*a*.
        6*b*. Secondary part corresponding with secondary phloem of vascular bundles (interfascicular secondary phloem). See guiding scheme No. 12, and if a part of the cambium is formed by the pericycle 1*c*, 3*b*, 4*b*.
        6*c*. Cambinal part (interfascicular cambium). See guiding schemes No. 3 and 4.
        6*d*. Secondary part, corresponding with secondary xylem of vascular bundles (interfascicular secondary xylem). See guiding scheme No. 13, and if a part of the cambium is formed by the pericycle 1*c*, 3*b*, 4*b*.
        6*e*. Primary part corresponding with primary xylem of vascular bundles. See 5*a*.
    4*b*. Between the simple vascular bundles. See 4*a*.
2*d*. Medulla.
    3*a*. Common parenchyma.
    3*b*. Sclerenchyma.
    3*c*. Idioblasts.
    3*d*. Latex vessels. See guiding scheme No. 2.
    3*e*. Internal glands. See guiding scheme No. 8.
    3*f*. Phloem bundles. See guiding scheme No. 9.
    3*g*. Vascular bundles, sometimes with secondary growth. See guiding schemes No. 9, 12 and 13.
    3*h*. Secondary parenchyma.
    3*i*. Lysigenous cavity.

No. 16. Guiding scheme for the description of the leaf.

1*a*. Blade.
    2*a*. Intervenium.
        3*a*. Epidermis.
            4*a*. Epidermis proper.
                5*a*. Epidermis of upper side.
                    6*a*. Epidermal cells.
                    6*b*. Stomata. See guiding scheme No. 5.
                    6*c*. Water pores. See guiding scheme No. 5.
                    6*d*. Idioblasts.
                    6*e*. Appendages of the outer surface. See guiding scheme No. 6.
                    6*f*. Periderm.
                5*b*. Epidermis of under side. See 5*a*.
            4*b*. Hypoderma or water tissue.
                5*a*. Common parenchyma.
                5*b*. Idioblasts.
         3*b*. Mesophyll.
            4*a*. Water tissue.
            4*b*. Palisade chlorenchyma.
            4*c*. Spongy chlorenchyma.

        4*d*. Appendages of inner surfaces.
        4*e*. Sclerenchyma.
        4*f*. Internal glands. See guiding scheme No. 8.
        4*g*. Tissues containing reserve material.
        4*h*. Idioblasts.
    2*b*. Margin.
        3*a*. Epidermis. See 2*a*, 3*a*.
        3*b*. Collenchyma.
        3*c*. Sclerenchyma. See for the rest 2*a*, 3*b*.
    2*c*. Veins.
        If the stele is fully developed the guiding scheme for the stem is used for the main part. In all other cases the following scheme is used.
        3*a*. Epidermis. See 1*a*, 2*a*, 3*a*.
        3*b*. Mesophyll.
            4*a*. Water tissue.
            4*b*. Collenchyma.
            4*c*. Colourless parenchyma.
            4*d*. Palisade chlorenchyma.
            4*e*. Spongy chlorenchyma.
            4*f*. Sclerenchyma.
            4*g*. Idioblasts.
            4*h*. Internal glands. See guiding scheme No. 8.
            4*i*. Appendages of internal surfaces. See guiding scheme No. 6.
            4*k*. Endodermis.
                5*a*. Protective sheath.
                    6*a*. Ordinary cells.
                    6*b*. Passage cells.
                    6*c*. Idioblasts.
                5*b*. Starch sheath. See 5*a*.
                5*c*. Crystal sheath. See 5*a*.
                5*d*. Enzyme sheath, etc. See 5*a*.
        3*c*. Meristeles.
            4*a*. Pericycle.
                5*a*. Common parenchyma.
                5*b*. Sclerenchyma.
                5*c*. Idioblasts.
            4*b*. Vascular or fibrovasal bundles.
                5*a*. Number and arrangement. See for the rest guiding schemes No. 9 or 10.
            4*c*. Medullary commissures between compound bundles.
                5*a*. Common parenchyma.
                5*b*. Idioblasts.
            4*d*. Medulla.
                5*a*. Common parenchyma.
                5*b*. Idioblasts.
1*b*. Petiole. See 1*a*, 2*c*.

No. 17. Guiding scheme for the description of the stamen.

1*a*. Filament.

    2a. Epidermis.
        3a. Epidermal cells.
        3b. Stomata. See guiding scheme No. 5.
        3c. Idioblasts.
        3d. Appendages of the surface. See guiding scheme No. 6.
    2b. Mesophyll.
        3a. Common parenchyma.
        3b. Idioblasts.
        3c. Internal glands. See guiding scheme No. 8.
        3d. Endodermis.
    2c. Meristele.
        3a. Pericycle.
            4a. Elements of the above.
        3b. Vascular or fibrovasal bundle. See guiding schemes No. 9 or 10.
1b. Anther.
    2a. Connective.
        3a. Epidermis. See 1*a*, 2*a*. ·
        3b. Mesophyll. See 1*a*, 2*b*.
        3c. Meristele. See 1*a*, 2*c*.
    2b. Thecae.
        3a. Epidermis. See 1*a*, 2*a*.
        3b. Layer of fibrous cells.
        3c. Intermediate parietal layer.
        3d. Tapetal layer.
        3e. Partition wall between the loculi.
        3f. Pollen.
            4a. Exine.
            4b. Intine.
            4c. Pores.
            4d. Contents.

No. 18. Guiding scheme for the description of the pistil.

1a. Ovary.
    2a. Wall.
        3a. Epidermis.
            4a. Epidermis of outer side.
                5a. Epidermal cells.
                5b. Stomata. See guiding scheme No. 5.
                5c. Idioblasts.
                5d. Appendages of the surface. See guiding scheme No. 6.
            4b. Epidermis of inner side. See **4a**.
        3b. Ground tissue.
            4a. Common parenchyma.
            4b. Collenchyma.
            4c. Sclerenchyma.
            4d. Idioblasts.
            4e. Internal glands. See guiding scheme No. 8.
            4f. Endodermis.
        3c. Meristeles.

        4*a*. Pericycle.
            5*a*. Elements of the above.
        4*b*. Vascular or fibrovasal bundles. See guiding schemes No. 9 or 10.
    2*b*. Septae. See 2*a*.
    2*c*. Placentae. See 2*a*; moreover conductive tissue.
1*b*. Style.
    2*a*. Epidermis. See 1*a*, 2*a*, 3*a*, 4*a*.
    2*b*. Ground tissue. See 1*a*, 2*a*, 3*b*.
    2*c*. Meristeles. See 1*a*, 2*a*, 3*c*.
    2*d*. Conductive tissue and stylar canal.
1*c*. Stigma.
    2*a*. Epidermis.
        3*a*. Epidermal cells.
        3*b*. Appendages of the surface. See guiding scheme No. 6.
    2*b*. Conductive tissue.

No. 19. Guiding scheme for the description of the ovule.

1*a*. Funicle.
    2*a*. Epidermis.
    2*b*. Ground tissue.
    2*c*. Meristeles.
1*b*. Ovule proper.
    2*a*. Atropous (orthotropous) ovule.
        3*a*. Nucellus.
            4*a*. Epidermis.
            4*b*. Ground tissue with chalaza.
            4*c*. Embryo sack.
            4*d*. Egg cell.
            4*e*. Synergidae.
            4*f*. Antipodal cells.
        3*b*. Integuments and micropyle.
            4*a*. Outer integument with exostomium.
                5*a*. Outer epidermis.
                5*b*. Ground tissue.
                5*c*. Meristeles.
                5*d*. Inner epidermis.
            4*b*. Inner integument with endostomium. See 4*a*.
    2*b*. Anatropous ovule. See 2*a*. Moreover the raphe is to be described here.
    2*c*. Hemianatropous ovule. See 2*b*.
    2*d*. Campylotropous ovule. See 2*a*.
    2*e*. Camptotropous ovule. See 2*a*.

No. 20. Guiding scheme for the description of the seed.

In the first place it ought to be determined which kind of ovule has given rise to the seed.
Micrography.
1*a*. Funiculus.
    2*a*. Aril.

1*b*. Seed coat.
 2*a*. Seed coat proper.
  3*a*. Layers produced by the outer integument.
   4*a*. Outer epidermis.
    5*a*. Epidermal cells.
    5*b*. Stomata. See guiding scheme No. 5.
    5*c*. Idioblasts.
    5*d*. Appendages of the surface. See guiding scheme No. 6.
   4*b*. Ground tissue.
    5*a*. Common parenchyma.
    5*b*. Idioblasts.
    5*c*. Endodermis.
   4*c*. Meristeles.
   4*d*. Inner epidermis. See 4*a*.
  3*b*. Layers produced by the inner integument. See 3*a*.
  3*c*. Layers produced by the nucellus.
  3*d*. Layers produced by the endosperm.
 2*b*. Hilum.
 2*c*. Raphe.
 2*d*. Chalaza.   See for the main part 2*a*.
 2*e*. Micropyle, arillode, caruncle.
 2*f*. Operculum.
1*c*. Nucleus of seed.
 2*a*. Albumen.
  3*a*. Perisperm.
   4*a*. Epidermis.
   4*b*. Ground tissue.
    5*a*. Parenchyma.
    5*b*. Idioblasts.
  3*b*. Endosperm.
   4*a*. Epidermis.
   4*b*. Ground tissue.
    5*a*. Parenchyma.
    5*b*. Idioblasts.
 2*b*. Embryo.
  3*a*. Radicle. See guiding schemes No. 3 and 4.
  3*b*. Plumule. See guiding schemes No. 3 and 4.
  3*c*. Cotyledons. See guiding schemes No. 16, 3 and 4.

No. 21. Guiding scheme for the description of the macroscopic characters of drugs. To be used in combination with the necessary organographical and anatomical schemes.

1*a*. Shape. For instance: in bundles, cylindrical rolls, cone-like masses, slices, sheets, strips, shreds, flakes, cakes; curved, flattened, arched, tortuous, twisted, quilled (quills), channelled, gutter-shaped, barrel-shaped, laminated; etc.
1*b*. Dimensions.
1*c*. General elementary physical characters. For instance: hard, soft; tough, flexible, elastic, brittle, pulverisable; light, heavy, dense, compact; translucent; etc.
1*d*. Outer surface. For instance: rough, smooth, dull, shining, fibrous, silky; striated, wrinkled, furrowed, pitted, reticulated, fissured; colour; etc.

1*e*. Inner surface. See outer surface.

1*f*. Fracture. For instance: clean, uneven, short, splintery, fibrous; radiate; resinous, mealy, laminated; colour; etc.

1*g*. Section. See fracture.

1*h*. Odour. For instance: aromatic, fragrant, warm, irritating, nauseous; etc.

1*i*. Taste. For instance: bitter, sweet, acid, astringent, aromatic, warm, hot, burning, pungent, acrid, nauseous; oily, mucilaginous; gritty between the teeth; etc.

COMPLETE ALPHABETICAL LIST OF CHEMICALS
AND REAGENTS

Alcohol. 96 per cent. and absolute.
Alcoholic soda solution.
Ammonia. 10 per cent.
Aniline sulphate. Watery solution 1 per cent. to which a few drops of sulphuric acid are added.
Chloral hydrate. Chloral hydrate 5 parts and water 2 parts.
Chloroform.
Chromic acid. In watery solution of $^1/_2$, 1 and 50 per cent.
Concentrated sulphuric acid.
Crystalized phenol and oil of cloves. Küster. Bot. Centralbl. Bd. 69. 1897. 50; for discovering siliceous bodies the sections are heated on the slide with a few small crystals of phenol and after removal of the phenol oil of cloves is added.
Cupric acetate. Saturated watery solution.
Cupric sulphate. Saturated watery solution.
Eosine. Concentrated solution in absolute alcohol.
Fehling's solution. a. Cupric sulphate in watery solution 3.5 per cent. b. Sodium potassium tartrate in watery solution 17.3 per cent. c. Sodium hydroxide in watery solution 12 per cent. Dilute immediately before use equal volumes of a and b and c each with 2 volumes of water and mix them together.
Ferric acetate. Liquor Ferri acetici. Immediately before use a few drops are added to a whatch glass filled with water.
Ferric chloride.
Glacial acetic acid.
Glycerine.
Glycerine jelly. Dissolve 50 grammes glycerine, 7 grammes gelatine and 1 gramme phenol in 42 grammes water.
Hydrochloric acid. 25 per cent.
Iodine in chloral hydrate. Chloral hydrate 5 parts, water 2 parts, iodine in excess.
Millon's reagent. Dissolve 1 part of metallic mercury in 2 parts of nitric acid specific gravity 1.42 and dilute afterwards with 2 parts of water.
Nitric acid. 50 per cent.
Oil of cloves.
Origanum oil.
Osmic acid. In watery solution of 1 per cent.
Phloroglucin and hydrochloric acid. Phloroglucin and hydrochloric acid in separated solutions: I. 1 gramme phloroglucin in 125 grammes alcohol 96 per cent. II. Hydrochloric acid 25 per cent.
Potash. $12^1/_2$ and 50 per cent.
Potassium dichromate. Saturated watery solution.

Potassium iodide and cadmium iodide. Dissolve 10 parts cadmium iodide and 20 parts potassium iodide in water 70 parts.

Potassium iodide iodine. 1 gramme iodine and 0.5 gramme potassium iodide in 2 grammes water. When solution has been effected dilute with water to 100 grammes.

Safranin. Saturated solution in alcohol 96 per cent.

Schulze's macerating mixture. Small parts of the tissue in a test tube with the same volume of potassium chlorate and a few drops of nitric acid. Heat slightly for a few moments in order to induce the process and wait for some hours or even in the case of woody and sclerotic tissues for some days till the desired maceration has taken place. Then wash out in abundance of water.

Sugar. Watery solution 3 per cent.

Sulphuric ether.

Tartaric acid. Alcoholic solution 5 per cent.

Tincture of iodine. Iodine 1 per cent. in alcohol 96 per cent. Iodine in water may be prepared immediately before use by adding some drops of tincture of iodine to a whatch glass filled with water.

Vanillin. Alcoholic solution.

Water.

ALPHABETICAL LIST OF LITERATURE

ABRAHAM. Bau und Entwicklung der Wandverdickungen in den Samenoberhautzellen einiger Cruciferen. Jahrbücher für wissenschaftliche Botanik. Bd. XVI. 1885.

Mc. ALPINE. The Fibro-vascular system of the Quince fruit, compared with that of the Apple and Peer. Proc. Linn. Soc. N. S. Wales. Vol. XXXVII. 1912.

ANEMA. De zetel der alcaloiden bij enkele narcotische planten. Diss. Utrecht, 1892.

D'ARBAUMONT. Nouvelles observations sur les cellules à mucilage des graines de Crucifères. Ann. d. Sc. nat. Bot. Série 7. T. XI. 1890.

ATTEMA, J. J. De zaadhuid der Angiospermae en Gymnospermae en hare ontwikkeling. Diss. Groningen, 1901.

BAILLON. Sur l'origine du macis de la Muscade et des arilles en général. Compt. Rend. T. 98.

BARTH, H. Studien über den Nachweis von Alcaloiden in pharmaceutisch verwendeten Drogen. Bot. Centr. Bd. 75. 1898. 338.

BARY, A. DE. Vergleichende Anatomie der Vegetationsorgane der Phanerogamae und Farne. Leipzig. 1877.

BEAUVERIE, J. Le Bois. Paris. 1905.

BENECKE, W. Mikroscopisches Drogenpraktikum. In Anlehung an die 5e Ausgabe des Deutschen Arzneibuches. Jena. 1912.

BERG, O. Anatomischer Atlas zur Pharmaceutischen Waarenkunde. Berlin. 1865.

BEYERINCK, M. W. Beobachtungen über die ersten Entwicklungsphasen einiger Cynipidengallen. Amsterdam. 1883. 153.

BIECHELE, M. Mikroskopische Prüfung der offizinellen Drogen. Regensburg. 1904.

BIERMANN, M. Beiträge zur Kenntnis der Entwickelungsgeschichte der Früchte von Citrus vulgaris, Risso. Diss. Bern. 1896.

BIERMANN, R. Ueber Bau und Entwicklungsgeschichte der Oelzellen und die Oelbildung in ihnen. Diss. Bern. 1897.

BIGELOW. Glands in the Hop tree. Proceed. Jowa Acad. of Sc. Vol. II. 1895.

BLIESENICK, H. Ueber die Obliteration der Siebröhren. Diss. Erlangen. 1891.

BLOIS, LE. Canaux sécréteurs et poches sécrétrices. Ann. des Sciences. Série 7. Botanique. T. VI. 1887.

BOCKMANN, F. Beiträge zur Entwicklungsgeschichte officineller Samen und Früchte. Diss. Bern. 1901.

BÖHM, R. Ueber homologe Phloroglucine aus Filixsäure und Aspidin. Liebigs Annalen der Chemie. Bd. 302.

BÖHM, R. Ueber Filicinsäure. Liebigs Annalen der Chemie. Bd. 307.

BOHNY, P. Beiträge zur Kenntnis des Digitalisblattes und seiner Verfalschungen mit Berücksichtiging des Pulvers. Diss. Zürich. 1906.

BORTSCH. Beiträge zur Anatomie und Entwicklung der Umbelliferenfrüchte. Diss. Breslau. 1882.

BOULGER, G. S. Wood. London. 1908.

BRISSI, G. Intorno all'anatomia delle foglie dell' Eucalyptus globulus, Lab. Milano. 1891.

BURCHARD. Ueber den Bau der Samenschalen einiger Brassica- u. Sinapis-Arten. Journal für Landwirte. Jahrg. XLII. 1894.

CABANNES, EUG. Etude de quelques espèces du genre Rhamnus. Diss. Montpellier. 1896.

CHALON, JR. La graine des Légumineuses. Mons. 1875.

CHATIN. Développement de l'Ovule et de la Graine dans les Scrophularinées, Solanées, Boraginées et Labiées. Ann. d. Sc. nat. Série V. T. XIX. 1894.

CLAUTRIAU. Localisation et signification des alcaloides dans quelques graines. Extrait des Annales de la Société belge de Microscopie. (Mémoires). T. XVIII. 1894.

COCX. Valeriana officinalis. Pharm. Weekblad. Jrg. 56. 1919.

COHN, F. Kryptogamen-Flora von Schlesien. Bd. II. Hälfte 2. Flechten. Breslau. 1879.

COL. Appareil intérieur sécréteur des Composées. Thèse. Paris. 1903. Copy of Journal de Bot. 1903.

COLIGNON. Canaux sécréteurs dans les Ombellifères. 1874.

COLLIN, EUG. Sur la Digitale. Journ. Pharm. et Chim. Série 6. T. XXII. 1905. 56.

COMPTON. An investigation of the seedling structure in the Leguminosae. The Journ. of Linn. Soc. T. 411. 1912.

COPPER, A. C. Beiträge zur Entwicklungsgeschichte der Samen und Früchte offizineller Pflanzen. Diss. Bern. 1908.

DAFERT und MIHLAUZ. Untersuchungen über die Kohle-ähnliche Masse der Compositen. (Chem. Teil). Wien. Denkschrifte. Bd. 87. 1912.

DANIEL. Recherches anatomiques sur les Bractées de l'Involucre des Composées. Ann. des Sc. nat. Série 7. T. XI. 1890.

DEIVEURE. Recherches sur le cubèbe et sur les Piperaceae. 1894.

DEUTSCHES ARZNEIBUCH. 5e Ausgabe. Berlin. 1910.

DOULIOT, H. Recherches sur le Periderme. Ann. des Sc. nat. Série 7. Tome 10. 1889. 325.

DUMONT, A. Recherches sur l'anatomie comparée des Malvacées, Bombacées, Tiliacées, Sterculiacées. Ann. d. Sc. Nat. Série 7. Tome 6. 1887. 129.

DYE. Unterirdische Organen von Valeriana, Rheum und Inula. Diss. Bern. 1901.

ELLRODT, G. Ueber die Verteilung des Gerbstoffes in offizinellen Blättern, Kräutern und Blüten. Diss. Würzburg. 1913.

ENGLER u. PRANTL. Die natürlichen Pflanzenfamilien. Leipzig. 1897.

ERDMANN-KÖNIG's Grundriss der allgemeinen Warenkunde unter Berücksichtigung der Technologie. Für Handels- und Gewerbeschulen sowie zum Selbstunterrichte entworfen und fortgesetzt von Prof. Drs. Otto Linné Erdmann und Chr. Rud. König. Zwölfte volständig neubearbeitete und umgeänderte Auflage van Prof. Eduard Hanausek. Leipzig. 1895.

ERRERA. Sur la distinction microchimique des alcaloides et des matières proteiques. Ann. d. l. Soc. Belge de Microsc. (Mémoires). T. XIII. 1889.

ERRÉRA, MAISTRIAU et CLAUTRIAU. Premières Recherches sur la Localisation et la Signification des Alcaloides dans les Plantes. Journal de médicine, de chirurgie et de pharmacie publié par la Société royale des Sciences médic. et natur. de Bruxelles. 1887.

FANFANI. Morfologia ad istologia del frutto e del seme della Apiacee. N. G. B. G. Vol. XXIII. 1891. 451. Abstract Bot. Jahresber. 1891.

FEDDE. Vergleichende Anatomie der Solanaceen. Diss. Breslau. 1896.

FELDHAUS. Quant. Untersuchungen der Verteilung des Alkaloides in den Organen von Datura Stramonium. Diss. Marburg. 1903.

FEUILLOUX, J. Contribution à l'étude de l'appareil tecteur et glandulair des Composées. Thèse. Paris. 1901/02.

FLORENCE A. MC. CORMICK. Notes on the Anatomy of the Young Tuber of Ipomoea

Batatas, Lam. Bot. Gazette. Vol. 61. 1916.

FLÜCKIGER, F. A. Pharmakognosie des Pflanzenreiches. 3e Auflage. Berlin. 1891.

FLÜCKIGER, F. A. and D. HANBURY. Pharmacographia. A History of the principal Drugs of vegetable origin, met with in Great Britain and British India. Second edition. London. 1879.

FOXWORTHY, F. W. Indo-Malayan Woods. The Philippine Journ. of Science. C. Botany. Vol. IV. 1909.

FRANK. Anat. Bedeutung und Entstehung der vegetabilischen Schleime. Jahrbücher für wissenschaftliche Botanik. Bd. 4. 1865. 161.

GAMBLE, J. S. Ind. Timbers. 1902.

GAMPER. Beiträge zur Kenntnis der Angosturarinden. Diss. Zürich. 1900.

GARCIN. Apocynacées. Diss. Lyon. 1889.

GEIGER, H. Beiträge zur pharmakognostischen und botanischen Kenntnis der Jaborandi-blätter. Diss. Zürich. 1897.

GÉRARD, G. Recherches sur les Bois de diff. espèces de Légumineuses africaines. Diss. Paris. 1907.

GERDTS, C. L. Bau und Entwickelung der Compositenfrucht, mit besonderer Berück-sichtigung der off. Arten. Diss. Bern. 1904.

GEROCK, J. E. und SKIPPARI, F. J. Ueber den Sitz der Alkaloide im Strychnossamen. Archiv d. Pharm. Bd. CCXXX. 1892.

GEYGER. Ueber den Gefässbundelverlauf in den Laubblattregionen der Coniferen. Jahr-bücher für Wissenschaftliche Botanik. Bd. 6.

GILG, E. Ber. Pharm. Ges. 1901. 166.

GILG, E. Lehrbuch der Pharmakognosie. 2e Aufl. Berlin. 1910.

GLASER, L. Mikroskopische Analyse der Blattpulver von Arzneipflanzen. Verhand-lungen der Phys-Med. Gesellschaft zu Würzburg. N. F. Bd. XXXIV. Würzburg. 1901.

GOFFART. Recherches sur l'anatomie des feuilles dans les Renonculacées. Archives de l'Institut botanique de l'Université de Liège. T. III. 1901.

GORIS, A. Die anatomische Unterscheiding der wichtigsten Aconitumarten. Bulletin des Sciences pharmacologiques. No. 4. 1901.

GOSSE, L. A. Monographie de l'Erythroxylon Coca. 1861.

GREENISH. The structure of Coca Leaves. Pharmac. Journal. 1904. 493.

GREENISH, H. G. E. and COLLIN. An Anatomical Atlas of Vegetable Powders. Designed as an Aid to the Microscopic Analysis of Powdered Foods and Drugs. London. 1904.

GRÈS, L. Contribution à l'étude anatomique et microchimique des Rhamnées. Diss. Paris. 1901.

GRIMM. Beiträge zur vergleichenden Anatomie der Compositen-Blätter. Diss. Kiel. 1904.

GUÉRIN, P. Cellules à mucilage chez les Urticées. Bull. Soc. bot. de France. T. LVII.

GUIGNARD, L. Recherches sur la localisation de Principes actifs de Crucifères. Journal de Botanie. T. IV. 1890.

GUIGNARD, L. Sur la localisation dans les Amandes et le Laurier-Cerise des principes qui fournissent l'acide cyanhydrique. Journal de Botanique. Tome IV. 1890. 3.

GUIGNARD, L. Recherches sur le développement de la graine. Journal de Botanie. T. VII. 1893.

GÜNTHER, W. Beiträge zur Anatomie der Myrtifloren mit besonderer Berücksichtigung der Lythraceae. Diss. Breslau. 1905.

GURNIK. Kernholzbildung. Diss. Bern. 1915.

HAASE, P. Pharmakognostisch-chemische Untersuchung der Ipomoea fistulosa, Mart. Diss. Straszburg. 1908.

HABERLANDT, G. Vergleichende Anatomie des assimilatorischen Gewebesystems der

Pflanzen. Jahrbücher für wissenschaftliche Botanik. 1882.

HABERLANDT, G. Physiologische Pflanzenanatomie. 3e Auflage. Leipzig. 1904.

HAGERS. Handbuch der Pharmaceutischen Praxis für Apotheker, Aerzte, Drogisten und Medicinalbeamte. Unter Mitwirkung von Max Arnold-Chemnitz, G. Christ-Berlin, K. Dietrich-Helfenberg, Ed. Gildemeister-Leipzig, P. Janzen-Perleberg, C. Schriba-Darmstadt, vollständig neu bearbeitet und herausgegeben von B. Fischer, Breslau und C. Hartwich, Zürich. Bd. I 1900, Bd. II 1902. Berlin.

HALLSTRÖM, K. TH. Vergleichend-anatomische Studien über die Samen der Myristicaceen und ihre Arillen. Archiv der Pharmacie. 1895.

HANAUSEK. Untersuchungen über die Kohle-ähnliche Masse der Compositen. Anz. k. Akad. Wiss. Bd. XLVII. 1910.

HAROLD MATHEWS, E. The Vittae of Caraway fruits. Pharmaceutical Journal. Series IV. 1898. No. 1446.

HARTIG, TH. Beiträge zur vergleichende Anatomie der Holzpflanzen. Bot. Zeitung. Jahrg. 17. 1859.

HARTIG, R. Unterscheidungsmerkmale der wichtigeren in Deutschland wachsenden Hölzer. 1898.

HARTWICH, C. Ueber Gerbstoffkugeln und Ligninkörper in der Nahrungsschicht der Infectoriagalle. Ber. d. d. bot. Ges. Bd. III. 1885. 146.

—— Ueber die Meerzwiebel. Archiv d. Pharmacie. Bd. 227. 1889. 577.

—— Ueber die Schleimzellen von Althaea officinalis, L. Pharm. Centralhalle. 1891.

—— Strychnos-Drogen. Festschrift 50 jähr. Stift. des Schweizer Apotheker Vereins in Zürich. 1893.

—— Ueber die Samenschale der Solaneen. Vierteljahrschrift der Naturf. Gesellsch. Zürich. XLI. Jubelb. 1896.

—— Ueber einige bei Aconitumknollen beobachtete Abnormitäten. Bot. Centralblatt. 1897. 70.

—— Beiträge zur Kenntnis des Zimmt. Archiv d. Pharmacie. Bd. 239. 1901. 181.

—— Beiträge zur Kenntnis der Cocablätter. Archiv d. Pharmacie. 1903. 617.

HARTWICH und GAMPER. Beiträge zur Kenntnis der Angosturarinden. Archiv. d. Pharmacie. Bd. 238. 1900.

HEINECK. Beitrag zur Kenntnis des feineren Baues der Fruchtschale der Compositen. Diss. Giessen. 1899.

HÉRAIL, J. Ann. d. Sc. nat. Série 7. 1885.

—— Traité de Matière Médicale. Pharmacographie. 2 edition. Paris. 1912.

HEUT, G. Beiträge zur Kenntnis des Emulsins. Archiv der Pharmacie. 1901.

HEYL und TUNMANN. Santoninfreie Flores Cinae. Apotheker Zeitung. XXVIII.

HOAZ, C. S. A comparison of the stem anatomy of the Cohort Umbelliflorae. Annals of Botany. T. 29. 1915.

HÖHLKE, F. Ueber die Harzbehälter und die Harzbildung bei den Polypodiaceen und einigen Phanerogamen. Beihefte zum Botanischen Centralblatt. Bd. XI 1901/02. 8.

HÖHNEL, F. R. v. Anatomische Untersuchungen über einige Secretionsorgane der Pflanzen. Sitzungsberichte der Mathematisch-Naturwissenschaftliche Classe der Kaiserlichen Akademie der Wissenschaften zu Wien. Bd. LXXXIV. Abtheilung 1. 1881. 565.

—— Ueber stockwerkartig aufgebaute Holzkörper. Ber. Wiener Akademie. Bd. 89. Abteilung 1. 1884. 30.

HOLZNER und LERMER. Hopfen. Zeitschr. ges. Brauwesen. Jhrg. 1892—95.

INGERMANN, D. Mikroscopie der voornaamste handelswaren. Bewerkt naar K. Hassack, Physikalische und mikroskopische Warenprüfungen. Amsterdam 1910.

Iroside. The anatomical structure of the New-Zealand Piperaceae. Trans. N. Zealand Inst. XLIV.

Jacquemin, Alb. Sur la localisation des alcaloides chez les Légumineuses. Rec. de l'Inst. Erréra. Univ. Bruxelles. VI. 1906.

Jaensch, Th. Anatomie einiger Leg. Hölzer. Ber. d. d. bot. Ges. Bd. II. 1884. 268.

Janssonius, H. H. Mikrographie des Holzes der auf Java vorkommenden Baumarten. 1906.

Jensen. Ueber den Bau der Rinde von Hamamelis Virginica. Pharmaceutical Archives. 1901. No. 7.

Jentsch. Der Urwald Kameruns. Beih. z. Tropenpflanzer. Bd. XII. 1911. 1.

Jenzer. Pharmakologische Untersuchungen über Pilocarpus pennatifolius Lemaire und Erythroxylon Coca Lamarck mit besonderer Berücksichtigung ihrer Alcaloiden. Diss. Zürich. 1910.

Johannsen, W. Sur la localisation d'émulsine dans les amandes. Ann. d. Sc. nat. Bot. Sér. VI. 1887.

Johnson. Studies in the development of the Piperaceae. Bot. Centralblatt. Bd. 117.

Karsten, G. und Oltmanns, Fr. Lehrbuch der Pharmakognosie. Zweite vollständig umgearbeitete Auflage von G. Karstens Lehrbuch der Pharmakognosie. Jena. 1909.

Kayser. Beiträge zur Kenntniss der Entwickelungsgeschichte der Samen mit besonderer Berücksichtigung des histogenetischen Aufbaues der Samenschale. Jahrbücher für wissenschaftliche Botanik. 1893.

Klein. Ueber den mikrochemischen Nachweis von Strychnin und Brucin im Samen von Strychnos Nux-vomica. Anz. ksl. Akad. Wiss. Wien. math.-nat. Kl. III. 1914.

Koch, L. Die mikroskopische Analyse der Drogenpulver. Ein Atlas für Apotheker, Drogisten und Studierende der Pharmacie. Leipzig. Bd. 1. 1901; Bd. 2. 1903; Bd. 3. 1906; Bd. 4. 1908.

—— Pharmakognostischer Atlas. Ein Atlas für Apotheker, Grossdrogisten, Sanitätsbeamte, Studierende der Pharmacie u. s. w. Leipzig. Bd. 1. 1911; Bd 2. 1914.

—— und Gilg, E. Pharmakognostisches Praktikum. Eine Anleitung zur mikroskopischen Untersuchung von Drogen und Drogenpulvern zum Gebrauche in praktischen Kursen der Hochschulen. Berlin. 1907.

Kraemer, H. The crystals in Datura Stramonium. Bull. of the Torrey Bot. Club. 1900.

—— A Text-book of Botany and Pharmacognosy. Intended for the use of students of Pharmacy, as a reference book for Pharmacists, and as a handbook for food and Drug analysts. Fourth revised and enlarged edition. Philadelphia and London. 1910.

Kramer, H. Mikroskopisch-pharmakognostische Beiträge zur Kenntnis von Blättern und Blüten. Diss. Würzburg. 1907.

Krah, F. W. Ueber die Verteilung der par. Elemente im Xylem etc. Diss. Berlin. 1883.

Kuntz, H. A Hyoscyamus niger alkaloidatartalmának szövetrendszerbeli eloszlása. Bot. körlem. XVII. 1918. 1/3. 1—16. With a German summary.

Kuntze, G. Beiträge zur vergleichenden Anatomie der Malvaceen. Bot. Centralblatt. Bd. 45. 1891. 161.

Lampe. Zur Kenntnis des Baues und der Entwickelung saftiger Früchte. Diss. Halle-Wittenberg. 1884.

Lange, J. Ueber die Entwickelung der Oelbehälter in den Früchten der Umbelliferen. Schriften der physikalisch-oekonomischen Gesellschaft zu Königsberg in Preussen. Jhrg. 25. 1884.

Lavialle, P. Recherches sur le développement de l'ovaire et du fruit des Composées. Ann. d. Sc. nat. Bot. Série 9. T. XV. 1912.

Leonhard, M. Beiträge zur Anatomie der Apocynaceen. Bot. Centralbl. Bd. XLV. 1891. 1.

LETH. A comparative morphological and anatomical examination of Ramuli Sabinae and substitutes. Archiv d. Pharmacie. Bd. 13 (63). 1906.

LIGNIER, O. Recherches sur l'anatomie comparée des Calycanthées, des Melastomacées et des Myrtacées. Diss. Paris. 1887.

LINDEN, E. V. D. Recherches microchimiques sur la présence des alcaloides et des glycosides dans la famille des Renonculacées. Recueil de l'Institut botanique de Bruxelles. T. V. 1902.

LINDT, O. Ueber den mikrochemischen Nachweis von Brucin und Strychnin. Zeitschrift für Wissenschaftliche Mikroskopie. Bd. I. 1884.

LLOYD, F. E. Abscission in Mirabilis Jalapa. Bot. Gazette. Vol. 61. 1916.

LONAY. Structure anatomique du pericarpe et du spermoderme chez les Renonculacées. Liège, Arch. Inst. botan. T. III. 1907. 4.

LOTSY, J. P. Ueber die Auffindung eines neuen Alkaloids in Strychnos-Arten auf mikrochemischem Wege. Rec. d. Travaux Bot. Neerlandais. Vol. II. 1905.

LUERSSEN, CHR. Handbuch der Systematischen Botanik mit besonderer Berücksichtigung der Arzneipflanzen. Leipzig. Bd. I. 1879; Bd. II. 1882.

LUTZ, G. Ueber die oblito-schizogenen Secretbehälter der Myrtaceen. Diss. Bern. 1895. The same in Bot. Centralbl. Bd. 64. 1895. 145.

MAHLERT. Anatomie der Laubblätter der Coniferen. Bot. Centralbl. Bd. 24. 1885.

MAISEL. Recherches anatomiques et taxinomiques sur le tégument de la graine des Légumineuses. Bull. soc. hist. nat. T. 22. 1909.

MARIÉ, P. Semen-contra. Thèse. Paris. 1884.

—— Recherches sur la structure des Renonculacées. Ann. d. Sc. nat. Bot. Série 6. T. 20. 1885. 5.

MARMÉ, W. Lehrbuch der Pharmacognosie des Pflanzen- und Thierreichs. Im Anschlusz an die zweite Ausgabe der Pharmacopoea germanica für Studierende der Pharmacie, Apotheker und Medicinalbeamte. Leipzig 1886.

MARTEL. Note sur l'anatomie de la fleur des Ombellifères. Bot. Centr. Bd. 101. 1906.

MATTIROLI E BASCALIONI. Ricerche anatomico-fisiologiche sui tegumenti seminali delle Papilionacee. Torino. 1892.

MAZURKIEWICZ, W. Die anatomischen Typen der Zimtrinden. Eine vergleichend anatomische Studie. Bulletin international de l'Académie des Sciences de Cracovie. Classe des Sciences mathématiques et naturelles. Serie B. Sciences naturelles. 1910. 140.

MEUNICHET. Ueber eine Verfälschung des Pfeffers. Journal de Pharmacie durch Pharm. Ztg. 1901. No. 15.

MEYER, A. Aconitum Napellus L. und seine wichtigsten nächsten Verwandten. Archiv d. Pharmacie. 1881.

—— Ueber die Entstehung der Scheidewände in dem secretführenden plasmafreien Intercellularraume der Vittae der Umbelliferen. Bot. Zeitung. Jhrg. 47. 1889.

—— Wissenschaftliche Drogenkunde. Berlin. Bd. 1. 1891; Bd. 2. 1892.

—— Die Grundlagen und die Methoden für die mikroskopische Untersuchung von Pflanzenpulvern. Jena. 1901.

MIKA. Beiträge zur Morphologie und mikroskopischen Nachweisung des Hesperidins. Majyar Növénytany hupok. I. 93. Summary in Just's Bot. Jahresbericht. Bd. 1. 1878. 20.

MIKOSCH, K. Untersuchungen über die Entstehung des Kirschgummi. Ber. Wiener Akad. Bd. 115. Abth. 1. 1906. 911.

MITLACHER, W. Toxikologisch oder forensisch wichtige Pflanzen und vegetabilische Drogen mit besonderer Berücksichtigung ihrer mikroskopischen Verhältnisse.

MOELLER, J. Vergleichende Anatomie des Holzes. Denkschrifte Wiener Akademie. Bd. 36. 1876.

MOELLER, J. Anatomie der Baumrinden. Vergleichende Studien. Berlin. 1882.

—— Mikroskopie der Nahrungs- und Genussmittel aus dem Pflanzenreiche. Berlin. 1886.

—— Lehrbuch der Pharmakognosie. Wien. 1889.

—— Leitfaden zu mikroskopisch-pharmakognotischen Uebungen für Studierende und zum Selbstunterricht. Wien. 1901.

—— Digitalis und Verbascum. Pharmazeutische Post. Jahrg. 37. 1904. 677. The same in Apothekerzeitung. 1904. 953.

MOLISCH, H. Mikrochemie der Pflanze. 1913.

—— Ueber das Verhalten der Zystolithen gegen Silber und andere Metallsalze. Ber. d. d. bot. Ges. Bd. XXXVI. 1918.

MOLLE. Recherches de microchimie compar. sur la localisation des Alcaloides dans les Solanées. Mémoires de l'Academie d. Sc. de Belgique. T. LIII. 1896.

MOREAU. Etude sur la Hachich. Paris. 1904.

MÜLLER, A. Beiträge zur Kenntnis des Baues und der Inhaltstoffe der Compositenblätter. Diss. Göttingen. 1912.

NADELMANN. Ueber Schleimendosperme der Leguminosen. Jahrbücher für wissenschaftliche Botanik. Bd. XXI. 1890.

NESTEL. Beiträge zur Kenntnis der Stengel- und Blattanatomie der Umbelliferen. Diss. Zürich. 1905.

NETOLITZKY, F. Ein Kennzeichen der Cannabis-Frucht. Arch. Chem. Mikrosk. Bd. 5. 1912.

NÖRDLINGER. Querschnitte. Bd. VI. 1874.

NUSSBAUM, KARSTEN und WEBER. Lehrbuch der Biologie. 1914.

OBERLIN et SCHLAGDENHAUFFEN. Etude hist. et chim. d. diff. écorces de la fam. d. Diosmées. Journ. d. Pharm. et d. Chimie. Série 4. T. 28. 1878.

OUDEMANS, C. A. J. A. Aanteekeningen op het Systematisch- en Pharmacognostisch-Botanische gedeelte der Pharmacopoea Neerlandica. Met een atlas van 2 morphologische en 35 anatomische platen. Rotterdam. 1854—1856.

—— Handleiding tot de Pharmacognosie van het Planten- en Dierenrijk. 2e druk. Amsterdam. 1880.

PARMENTIER. Recherches sur l'influence d'un mouvement continu regulier imprimé à une plante en végétation normale. Revue gen. Bot. T. 22. 1910.

PÄTZOLDT. Harz und Holz von Guaiacum. Diss. Straszburg. 1902.

PECKE, K. Mikroch. Nachweis der Cyanwasserstoffsäure in Prunus Laurocerasus. Ber. Wiener Akad. Bd. CXXI. 1912.

—— Mikrochemischer Nachweis des Myrosins. Ber. d. d. bot. Ges. Bd. XXXI. 1913.

PERROT, E. Anatomie comparée des Gentianacées. Annales d. Sciences naturelles. Série 8. Botanique. T. VII. 1898. 105.

—— Sur l'anatomie du fruit de Coriandre. (C. sativum). Bull. Sc. pharmacol. Ann. 3. 1901.

—— et MOREL. Quelques remarques sur l'anatomie des Ombellifères. Bull. Société bot. de France. LX.

—— et GÉRARD, G. Recherches sur les bois de diff. espèces de Légumineuses africaines. 1907. The same as Diss. Paris. 1907 of Gérard, see Gérard.

PFAEFFLIN. Untersuchungen über Entwickelungsgeschichte, Bau etc. Papilionaceen Samen. Diss. Bern. 1897.

PFEFFER. Hesperidin, ein Bestandteil einiger Hesperideen. Bot. Zeitung. 1874. 529.

PFEIFFER, H. Zur Methode der mikroskopische Anatomie ruhender Umbelliferenfrüchte. Mikrokosmos. 1918/19.

PICCIOLI. I caraterri anatomici per conoscere i principali legnami adoperati in Italia. 1906.

PLANCHON, G. et COLLIN, E. Les drogues Simples d'origine végétale. Paris. T. I. 1895; T. II. 1896.

PRAËL. Vergleichende Untersuchungen über Schutz- und Kernholz der Laubbäume. Jahrbücher für wissenschaftliche Botanik. Bd. 19. 1888.

PRINS, J. J. De fluctueerende variabiliteit van microscopische structuren bij planten. Diss. Groningen. 1904.

PRODINGER, MARIE. Das Periderm der Rosaceen. Denkschr. d. Wiener Akad. Bd. 84. 1909. 329.

RANT, A. De Gummosis der Amygdalaceae. Diss. Amsterdam. 1906.

RECORD, S. J. Economic Woods of the United States. 1912.

REENS, EMMA. Sur la cire des feuilles de Coca de Java. Trav. Lab. Mat. méd. Paris. XI. 1917—1919.

—— La Coca de Java. Diss. Paris. 1919.

REINSCH, A. Ueber die anatomischen Verhältnisse der Hamamelidaceae mit Rücksicht auf ihre systematische Gruppierung. Englers bot. Jahrbücher. Bd. 11. 1890. 347.

ROMPEL, J. Krystalle von Kalkoxalat in der Fruchtwand der Umbelliferen und ihre Verwertung für die Syst. Ber. Wiener Akademie. Bd. 104. Abt. I. 1895.

ROSENTHALER und STADLER. Ein Beitrag zur Anatomie von Cnicus benedictus L. Arch. d. Pharm. Bd. 246. 1908.

ROSOLL, A. Beiträge zur Histochemie der Pflanzen. Ber. d. Wiener Akad. der Wissenschaften. 1883.

SANIO, C. Vergleichende Untersuchungen über den Bau und die Entwicklung des Korkes. Jahrbücher für wissenschaftliche Botanik. Bd. 2. 1859.

SARGENT, C. S. The Woods of the United States. New York. 1885.

SAUPE, A. Der anatomische Bau des Holzes der Leguminosen und sein syst. Werth. Flora. Bd. 70. 1887. 259.

SAUVAN, M. L. Untersuchungen über die Lokalisation von Brucin und Strychnin in den Samen von Strychnos Nux-vomica u. a. Journal de Botanie. X. 1896.

SCHACHT, H. Ueber ein neues Secretions-Organ im Wurzelstock von Nephrodium Filix mas. Jahrbücher für wissenschaftliche Botanik. Bd. III. 1863. 352.

SCHAD. Entwicklungsgeschichtliche Untersuchungen über den Malabar Cardamomen und vergleichende anatomische Studien über die Samen einiger anderen Amomum- und Elettaria-Arten. Diss. Bern. 1897.

SCHIMPER, A. F. W. Anleitung zur mikroskopischen Untersuchung der vegetabilischen Nahrungs- und Genussmittel. Zweite umgearbeitete Auflage. Jena. 1900.

SCHLOTTERBECK. Beiträge zur Entwickelungsgeschichte pharmacogn. wichtiger Samen. Diss. Bern. 1896.

SCHMITZ. Sitzungsberichte der Naturforschende Gesellschaft zu Halle. Botanische Zeitung. 1875.

SCHNEIDER, A. Powdered Vegetable Drugs. Pittsburgh Pa. 1902.

SCHORN. Ueber Schleimzellen bei Urticaceen und über Schleimcystolithen von Girardinia palmata. Ber. Wiener Akad. Bd. 116. Abth. 1. 1906.

SCHULZE. Beiträge zur Blattanatomie der Rutaceen. Diss. Heidelberg. 1902.

SCOTT, D. H. On some points in the Anatomy of Ipomoea versicolor, Meissn. Ann. of Botany. Vol. 5. 1890.

SENFT, E. Ueber die sogenannten Inklusen in Glyzirrhiza glabra L. und über ihre Function. Berichte d. d. botanischen Gesellschaft. Bd. XXXIV. Heft 9. 1916.

SHIMOYANA. Beiträge zur Kenntnis der Buchublätter. Archiv der Pharmacie. 1888.

SIECK. Die schizolysigenen Secretbehälter. Diss. Bern. 1895. See also Jahrbücher für wissenschaftliche Botanik. 1895. 197.

SIEVERS, A. F. Verdeeling der Alcaloiden in de verschillende deelen der Atropa Bella-donna, L. Am. J. Pharm. 1914.

SMALL, J. On the floral anatomy of some Compositae. Linn. Soc. Journ. Bot. XLIII. 1917.

SOLEREDER, H. Holzstructur. Diss. München. 1885.

—— Bemerkenswerte anatomische Vorkommnisse u. s. w. Archiv der Pharmacie. Bd. 245. 1907.

—— Systematische Anatomie der Dicotyledonen. Stuttgart. 1899. Ergänzungsband. 1908.

—— Zur mikroskopischen Pulveranalyse der Folia Salviae. Archiv der Pharmacie. Bd. 249. 1911. 123.

SPATZIER. Ueber das Auftreten und die physiologische Bedeutung des Myrosins in der Pflanze. Jahrbücher für wissenschaftliche Botanik. Bd. 25. 1893.

SPEYER, J. Beiträge zur Entwicklungsgeschichte der Rinde pharmakognostisch interes-santer Pflanzen. Diss. Bern. 1907.

SPIRE, C. Contributions à l'étude des Apocynées et en particulier des lianes indo-chinoises. Trav. Labor. mat. méd. Ecole sup. Pharm. Paris. T. II. The same as Diss. Paris. 1905.

STONE, H. Timbers of Commerce. 1904.

—— Les Bois utiles de la Guyane Française. Ann. du Musée Colonial de Marseille. Année 25. 3e Série, 4e Vol. 1916.

—— and FREEMAN. The timbers of British Guiana. 1914.

STRASSBURGER, E. Zellbildung und Zellteilung. 1880.

STSCHERBATSCHEFF. Beiträge zur Entwickelungsgeschichte einiger offizinellen Pflanzen. Archiv d. Pharmacie. Bd. 245. 1907.

STYGER, J. Beiträge zur Anatomie der Umbelliferenfrüchte. Schweiz. Apoth. Ztg. LVII. 1919.

TANFANI. Morfologia ed Istologia del frutto e del seme delle Apiacee. N. G. B. J. Vol. XXIII. 1891. Report Bot. Jahresberichte. 1891.

TIEGHEM, v. Canaux Sécréteurs des Plantes. Ann. des Sciences naturelles. Série 5. Bota-nique. T. XVI. 1872.

—— Seconde Mémoire sur les Canaux sécréteurs des Plantes. Ann. des Sciences naturelles. Série 7. Botanique. T. I. 1885.

—— Sur les canaux à mucilage des Piperacées. Ann. d. Sc. nat. Série 9. Botanique. T. 7. 1908.

TJADEN, M. E. H. Microscopisch onderzoek van Hout. Amsterdam. 1919.

TREIBER, K. Ueber den anatomischen Bau des Stammes der Asclepiadeen. Bot. Cen-tralblatt. Bd. 48. 1891. 209.

TRIEBEL. Oelbehälter in Wurzeln von Compositen. Nova Acta. Leop. Carol. Ak. Naturf. Bd. L. No. 7. 1885. Report in Botanische Jahresberichte. 1885.

TSCHIRCH, A. Angewandte Pflanzenanatomie. Ein Handbuch zum Studium des Anato-mischen Baues der in der Pharmacie, den Gewerben, der Landwirtschaft und dem Haus-halte benutzten pflanzlichen Rohstoffe. Wien und Leipzig. 1889.

—— Indische Fragmente. Archiv der Pharmacie. 1890.

—— Die Harze und Harzbehälter. Historisch-kritische und experimentelle in gemein-schaft mit zahlreichen Mitarbeitern ausgeführte Untersuchungen. Leipzig. 1900.

—— Schweiz. Wochenschr. 1901. No. 15.

—— Sind die Antheren der Compositen verwachsen oder verklebt? Flora. Bd. 93. 1904.

—— Handbuch der Pharmacognosie. Bd. I—II, 2. Leipzig. 1917.

—— und HOLFERT. Ueber den Süssholz. Archiv d. Pharmacie. Bd. 26. Heft 2. 1888.

—— und OESTERLE, O. Anatomischer Atlas der Pharmakognosie und Nahrungsmittel-kunde. Leipzig. 1900.

Tschirch, A. und Schad. Schweizer Wochenschrift für Chemie und Pharmacie. 1897.

Tunmann, O. Untersuchungen über die Secretbehälter (Drüsen) einiger Myrtaceen, speziell über ihren Entleerungsapparaten. Archiv der Pharmacie. Bd. 248. 1900.

—— Ueber die Alkaloide in Strychnos Nux-vomica L. während der Keimung. Archiv d. Pharm. Bd. CCXLVIII. 1900.

—— Ueber die Secretdrüsen. Diss. Bern. 1900.

—— Ueber die Kristalle in Herba Conii. Pharm. Zeitung. L. 1905. 100.

—— Ueber Folia Uvae Ursi und den mikrochemischen Nachweis des Arbutin. Pharm. Centralhalle. Bd. 47. 1906. 945—947.

—— Ueber die resinogene Schicht der Secretbehälter der Umbelliferen. Ber. d. pharmaceutischen Gesellschaft. Bd. 17. 1907.

—— Pharmac. Untersuchungen von Pilocarpus pennatifolius und Erythroxylon Coca mit besonderer Berücksichtigung ihrer Alkaloiden. Pharmaz. Post. Bd. 42. 1909. 768.

—— Anatomie und Inhaltsstoffe von Chondrus crispus Stachhouse. Apotheker Zeitung. Bd. 24. 1909. 151—154.

—— Pflanzenmicrochemie. 1913.

—— Ueber Jalapa Knollen. Apotheker Zeitung. Bd. XXXI. 1916.

Unger, W. Zum Kapittel „Folia Belladonnae". Apotheker Zeitung. Bd. XXVII. 1912.763.

Ven, A. J. van de. Over het Cyaanwaterstofzuur bij de Prunaceae. Diss. Amsterdam. 1898.

Vesque, J. L'Anatomie comparée de l'écorce. Annales d. Sciences Naturelles. Série 6. Bot. T. II. 1875. 82.

—— L'Anatomie des tissus appliquée à la classification des Plantes. Nouv. arch. du Musée d'Histoire naturelle. Série 2. Tome IV. 1881.

Verschaffelt, E. De imbibitie van Strychnoszaad. Pharm. Weekblad. 1913. No. 24.

Virchow, H. Ueber Bau und Nervatur der Blattzähne und Blattspitzen. Diss. Bern. 1895.

Vogl, A. Ueber die Intercellularsubstanz und die Milchsaftgefäsze in der Wurzel des gemeinen Löwenzahns. Ber. Wiener Akad. Bd. 48. Abth. 1. 1863. 668.

Vogl, A. E. Anatomischer Atlas zur Pharmakognosie. Wien und Leipzig. 1887.

—— Die wichtigsten vegetabilischen Nahrungs- und Genussmittel mit besonderer Berücksichtigung der mikroscopischen Untersuchung auf ihre Echtheit, ihre Verunreinigungen und Verfälschungen. Berlin und Wien. 1899.

Vogl, K. Anatomische Studien über Blatt und Achse der einheimischen Daphne-Arten mit besonderer Berücksichtigung der Bastfasern. 40. Jahresber. k. k. Staatsgymnasiums in Oberhollabrunn am Schlusse des Schuljahres 1909/10. Oberhollabrunn. Verl. der Anstalt. 1910. 3—29.

Voigt. Ueber den Bau und die Entwicklung des Samens und des Samenmantels von Myristica fragrans. Diss. Göttingen. 1885.

Vuillemin. De la valeur des Caractères anatomiques au point de vue de la classification des végétaux (Tige des Composées). Paris. 1884.

Walliczek, H. Studien über die Membranschleime vegetativer Organe. Jahrbücher für wissenschaftliche Botanik. Bd. 25. 1893. 209. The same as Diss. Bern. 1893.

Wasicky, R. Der mikroskopische Nachweis von Strychnin und Brucin im Samen von Strychnos Nux-vomica L. Zeitschrift allgemein oesterreichischer Apotheker Verein. LII. 1914.

Wèvre, A. de. Localisation de l'Atropine. Bulletin des séances de la Société belge de Microscopie. Octobre 1887.

Wiesner, J. Die Rohstoffe des Pflanzenreiches. Versuch einer technischen Rohstofflehre des Pflanzenreiches. Zweite gänzlich umgearbeitete und erweiterte Auflage. Leipzig. Bd. I. 1900; Bd. II. 1903.

WIGAND, A. Lehrbuch der Pharmakognosie. Mit besonderer Rücksicht auf die Pharmacopoea germanica sowie als Anleitung zur naturhistorischen Untersuchung vegetabilischer Rohstoffe. Dritte vermehrte Auflage. Berlin. 1879.

WILKE, K. Ueber die anatomischen Beziehungen des Gerbstoffes zu den Secretbehälter der Pflanze. Diss. Halle. 1883.

WINTON. Anatomie des Hanfsamens. Zeitschr. Unters. Nahr- und Genussmittel. 1904.

WISSELINGH, C. v. Vittae d. Umbellif. Verh. der Koninklijke Academie van Wetenschappen. Amsterdam. (2de sect.). T. IV. No. 1. 1894.

—— Mikrochemische Untersuchungen über die Zellwände der Fungi. Jahrbücher für wissenschaftliche Botanik. Bd. 31. 1898. 619.

—— Bijdragen tot de Kennis van de Zaadhuid. 1ste Bijdrage. Compositae. Pharm. Weekblad. Jaargang 55. 1918. 871.

—— Bijdragen tot de Kennis van de Zaadhuid. Bijdrage II. Umbelliferae. Pharm. Weekblad. Jaargang 55. 1918.

—— Bijdragen tot de Kennis van de Zaadhuid. Bijdrage IV. Cruciferae. Pharm. Weekblad. Jaargang 56. 1919.

—— Bijdragen tot de Kennis van de Zaadhuid. Bijdrage V. Linaceae. Pharm. Weekbl. Jaargang 56. 1919. 1437.

WITTLIN, J. Ueber die Bildung der Kalkoxalat-Taschen. Bot. Centralblatt. Bd. 67. 1896. 33. The same as Diss. Bern. 1896.

WITTMACK und BUCHWALD. Die Unterscheidung der Mandeln von ähnlichen Samen. Ber. d. d. Bot. Gesellschaft. 1901.

ZACCHARIAS. Ueber Secretbehälter mit verkorkten Membranen. Bot. Zeitung. 1879.

ZIEGENSPECK, H. Die chem. Zusammensetzung der Raphiden von Scilla maritima. Ber. d. d. Bot. Ges. Jhrg. 32. 1914. 630.

ZIJLSTRA, K. Ueber Carum Carvi L. Rec. d. Travaux Bot. néerl. Vol. XIII. 1916. 159.

———

PEN-PORTRAITS

AMYLUM MANIHOT.

Cassava Starch. Tapioca Starch.

The starch obtained from the tubers of Manihot utilissima, Pohl. Pl. Bras. Ic. I. 32. t. 24 and some allied species of the same genus.

Macroscopic characters.

White powder, gathering into small lumps, odourless and tasteless.

Anatomical characters.

LITERATURE. Berg. Anat. Atl. 1865. 100. Taf. 50. Erdmann-König's Grundr. d. allg. Warenkunde. 1895. 219. Greenish & Collin. Anat. Atl. o. veget. powders. 1904. 18. Hager. Handb. d. pharm. Praxis. Bd. 1. 1900. 296. Hérail. Mat. Méd. 1912. 65. Ingerman. Mikrosk. d. voorn. handelsw. 1910. 72. Karsten u. Oltmanns. Pharmakogn. 1909. 323. Kraemer. Botany a. Pharmacogn. 1910. 789. Marmé. Pharmacogn. 1886. 493. Moeller. Mikroskopie d. Nahr.- u. Genussmittel. 1886. 200. Moeller. Pharmacogn. 1889. 349. Moeller. Leitf. z. mikrosk.-pharmacogn. Übungen. 1901. 50. Oudemans. Pharmacogn. 1880. 450. Planchon et Collin. Drogues simples. T. 1. 1895. 333. Schimper. Anl. z. mikr. Unters. d. veget. Nahr.- u. Genussmittel. 1900. 34. Schneider. Powdered veget. drugs. 1902. 115. Tschirch. Angew. Pf.-anat. 1889. 86. Tschirch u. Oesterle. Anat. Atl. 1900. 226. Taf. 51. Tschirch. Handb. d. Pharmakogn. T. 2. 1910. 174. Vogl. Anat. Atl. 1887. Taf. 56. Vogl. Die wicht. veget. Nahr.- u. Genussmittel. 1899. 185. Wiesner. Rohstoffe. Bd. 1. 1900. 620. Wigand. Pharmakogn. 1897. 326. MATERIAL. The drug; dried tubers; fresh tubers from the passar at Buitenzorg (Java) in alcohol. REAGENT. Water.

MICROGRAPHY. In the tubers by far the largest part of the grains compound, 2- to 8-adelphous, moreover a small number of smaller simple grains; in the drug almost exclusively separate component grains, mixed with only a very few compound, mostly 2-adelphous grains. The separate component grains tolerably uniform in size, the largest very numerous and often 20 μ in diameter; also a pretty large number of smaller grains. Component grains joined together by even planes but the originally free part of their surfaces rounded; for the greater part having belonged to 2-adelphous grains, therefore having a flat inner surface and being truncately ovate, also showing a smaller flat surface and being sometimes even nearly spherical; sometimes having belonged to poly-adelphous grains, in these cases the inner surface angular showing 2 or more facets. Hilum central, often indicated by a simple, also 3- or 4- or even poly-rayed cleft. Stratification nearly wanting.

January 1909, September 1911.   J; M.

AMYLUM MARANTAE.
Maranta Starch. Maranta Arrowroot.
The starch obtained from the rhizomes of some species of the genus Maranta, especially M. arundinacea, Linn. Sp. Pl. 2, cultivated in the East- and West-Indies.

Macroscopic characters.
Moderately fine, white, odourless and tasteless powder, gathering into small masses.

Anatomical characters.
LITERATURE. Berg. Anat. Atl. 1865. 100. Taf. 50. Erdmann-König's Grundr. d. allg. Warenkunde. 1895. 217. Flückiger. Pharmakogn. 1891. 244. Flückiger & Hanbury. Pharmacographia. 1879. 630. Gilg. Pharmakogn. 1910. 70. Greenish & Collin. Anat. Atl. o. veget. powders. 1904. 12. Hager. Handb. d. pharm. Praxis. Bd. 1. 1900. 296. Hérail. Mat. Méd. 1912. 60. Ingerman. Mikrosk. d. voorn. handelsw. 1910. 73. Karsten u. Oltmanns. Pharmakogn. 1909. 323. Kraemer. Botany a. Pharmacogn. 1910. 786. Marmé. Pharmacogn. 1886. 492. Moeller. Mikroskopie d. Nahr.- u. Genussmittel. 1886. 197. Moeller. Pharmacogn. 1889. 348. Moeller. Leitf. z. mikrosk.-pharmacogn. Übungen. 1901. 49. Oudemans. Pharmacogn. 1880. 449. Planchon et Collin. Drogues simples. T. 1. 1895. 237. Schimper. Anl. z. mikr. Unters. d. veget. Nahr.- u. Genussmittel. 1900. 33. Schneider. Powdered veget. drugs. 1902. 115. Tschirch. Angew. Pfl.-anat. 1889. 79. Tschirch u. Oesterle. Anat. Atl. 1900. 225. Taf. 51. Tschirch. Handb. d. Pharmakogn. II. 1. 1910. 170. Vogl. Die wicht. veget. Nahr.- u. Genussmittel. 1899. 179. Wiesner. Rohstoffe. Bd. 1. 1900. 612. Wigand. Pharmakogn. 1897. 325. MATERIAL. The drug. REAGENT. Water.

MICROGRAPHY. Grains simple; for the greater part large; length: $n = 300$; $M = 25.5$ μ; $Med = 25{,}68$ μ; $Q_1 = 4.25$ μ; $Q_3 = 3.50$ μ; min. within the class 4.17—8.34 μ; max. within the class 37.53—41.70 μ [1]). Shape oval, often irregularly oblong, somewhat flattened. Hilum excentric ($^1/_1$ to $^1/_3$), commonly at the broadest end and indicated by a cleft filled with air and often having the shape of the wings of a poised bird. Stratification faint. Not more than a fourth part of the grains without cleft, and no grains showing a distinct stratification allowed (potato starch).

October 1901, September 1911. J; M.

AMYLUM MAIDIS.
Maize Starch. Corn Starch.
The starch obtained from the seed of Zea Mays, Linn. Sp. Pl. 971.

Macroscopic characters.
Fine, white, odourless and almost tasteless powder, crackling between the fingers.

Anatomical characters.
LITERATURE. Berg. Anat. Atl. 1865. 99. Taf. 50. Erdmann-König's Grundr. d. allg. Warenkunde. 1895. 193. Greenish & Collin. Anat. Atl. o. veget. powders. 1904. 8. Hager. Handb. d. pharm. Praxis. Bd. 1. 1900. 295. Hérail. Mat. Méd. 1912. 55. Ingerman. Mikrosk. d. voorn. handelsw. 1910. 67. Karsten u. Oltmanns. Pharmakogn. 1909. 319. Kraemer. Botany a. Pharmacogn. 1910. 787. Marmé. Pharmacogn. 1886. 488. Moeller. Mikros-

---

[1]) Calculated from observations of Dr. J. J. Prins. De fluctueerende variabiliteit van microscopische structuren bij planten. Diss. Groningen. 1904. p. 32, no. 7.

kopie d. Nahr.- u. Genussmittel. 1886. 118. Moeller. Pharmacogn. 1889. 348. Moeller. Leitf. z. mikrosk.-pharmakogn. Übungen. 1901. 47. Oudemans. Pharmacogn. 1880. 448. Planchon et Collin. Drogues simples. T. 1. 1895. 116. Schimper. Anl. z. mikr. Unters. d. veget. Nahr.- u. Genussmittel. 1900. 26. Schneider. Powdered veget. drugs. 1902. 113. Tschirch. Angew. Pfl.-anat. 1889. 82. Tschirch u. Oesterle. Anat. Atl. 1900. 222. Taf. 43 & 50. Tschirch. Handb. d. Pharmakogn. T. 2. 1910. 196. Vogl. Die wicht. veget. Nahr.- u. Genussmittel. 1899. 124. Wiesner. Rohstoffe. Bd. 1. 1900. 601. Wigand. Pharmakogn. 1897. 325. MATERIAL. The drug; starch from a fresh seed. REAGENTS. Water, iodine in chloral hydrate.

MICROGRAPHY. Grains simple, having an average diameter of 10—15 μ. Shape from polygonal to spherical, the smaller grains being generally more rounded than the larger ones. Hilum central, indicated by a generally stellate cleft, having 3 or more rays. Stratification wanting, many grains showing a radial striation or even fine radial fissures.

September 1911.   J; M.

AMYLUM ORYZAE.
Rize Starch.
The starch obtained from the seeds of Oryza sativa, Linn. Sp. Pl. 333.

Macroscopic characters.
Very fine, white, odourless and tasteless powder.

Anatomical characters.
LITERATURE. Deutsch. Arznb. 5. Ausg. 1910. 52. Erdmann-König's Grundr. d. allg. Warenkunde. 1895. 193. Gilg. Pharmakogn. 1910. 28. Greenish & Collin. Anat. Atl. o. veget. powders. 1904. 10. Hager. Handb. d. pharm. Praxis. Bd. 1. 1900. 295. Hérail. Mat. Méd. 1912. 53. Ingerman. Mikrosk. d. voorn. handelsw. 1910. 69. Karsten u. Oltmanns. Pharmakogn. 1909. 319. Kraemer. Botany and Pharmacogn. 1910. 788. Marmé. Pharmacogn. 1886. 487. Moeller. Mikroskopie d. Nahr.- u. Genussmittel. 1886. 114. Moeller. Pharmacogn. 1889. 348. Moeller. Leitf. z. mikrosk.-pharmakogn. Übungen. 1901. 48. Oudemans. Pharmacogn. 1880. 448. Planchon et Collin. Drogues simples. T. 1. 1895. 115. Schimper. Anl. z. mikr. Unters. d. veget. Nahr.- u. Genussmittel. 1900. 28. Schneider. Powdered veget. drugs. 1902. 114. Tschirch. Angew. Pfl.-anat. 1889. 85. Tschirch u. Oesterle. Anat. Atl. 1900. 223. Taf. 50. Tschirch. Handb. d. Pharmakogn. T. 2. 1910. 192. Vogl. Die wicht. veget. Nahr.- u. Genussmittel. 1899. 134. Wiesner. Rohstoffe. Bd. 1. 1900. 599. MATERIAL. The drug. REAGENTS. Water, potassium iodide iodine, iodine in chloral hydrate.

MICROGRAPHY. Grains compound, poly-adelphous, ellipsoids, having an average length of 25 μ. Component grains polyhedric, having an average diameter of 5 μ. Hilum central, sometimes indicated by a cleft filled with air and often invisible. Besides many loose component grains and compound grains also many larger masses consisting of the latter, and sometimes entire cells quite filled with starch. The occurrence of unicellular hairs not permitted (oat starch).

November 1901, September 1911.   J; M.

AMYLUM SOLANI.

Potato Starch. Farina. Potato Flour.

The starch obtained from the tubers of Solanum tuberosum, Linn. Sp. Pl. 185.

Macroscopic characters.

Glossy, white, odourless and almost tasteless powder, crackling between the fingers.

Anatomical characters.

LITERATURE. Berg. Anat. Atl. 1865. 99. Taf. 50. Erdmann-König's Grundr. d. allg. Warenkunde. 1895. 213. Flückiger & Hanbury. Pharmacographia. 1879. 633. Gilg. Pharmakogn. 1910. 311. Greenish & Collin. Anat. Atl. o. veg. powders. 1904. 12. Hager. Handb. d. pharm. Praxis. Bd. 1. 1900. 296. Hérail. Mat. Méd. 1912. 67. Ingerman. Mikrosk. d. voorn. handelsw. 1910. 21, 69. Kraemer. Botany a. Pharmacogn. 1910. 787. Karsten u. Oltmanns. Pharmakogn. 1909. 323. Marmé. Pharmacogn. 1886. 486. Moeller. Mikroskopie d. Nahr.- u. Genussmittel. 1886. 193. Moeller. Pharmacogn. 1889. 347. Moeller. Leitf. z. mikrosk.-pharmakogn. Übungen. 1901. 49. Oudemans. Pharmacogn. 1880. 451. Planchon et Collin. Drogues simples. T. 1. 1895. 568. Schimper. Anl. z. mikr. Unters. d. veget. Nahr.- u. Genussmittel. 1900. 32. Schneider. Powdered veget. drugs. 1902. 115. Tschirch. Angew. Pfl.-anat. 1889. 80. Tschirch u. Oesterle. Anat. Atl. 1900. 233. Taf. 35. Tschirch. Handb. d. Pharmakogn. T. 2. 1910. 163. Vogl. Die wicht. veget. Nahr.- u. Genussmittel. 1899. 175. Wiesner. Rohstoffe. Bd. 1. 1900. 625. Wigand. Pharmakogn. 1897. 327. MATERIAL. The drug. REAGENTS. Water, iodine in chloral hydrate.

MICROGRAPHY. Most grains simple; compound and semi-compound grains to be found in every sample, though in relatively small numbers. Length of the grains: $n = 320$; $M = 23\ \mu$; $Med = 17.75\ \mu$; $Q_1 = 7.90\ \mu$; $Q_3 = 14.15\ \mu$; min. within the class 0—5 μ; max. within the class 80—85 μ [1]). Grains ovate, but often of a more irregular shape, not flattened. Hilum excentric ($^1/_4$), situated towards the narrow end of the grain. Stratification generally very distinct.

May 1901, September 1911. J; M.

AMYLUM TRITICI.

Wheat Starch.

The starch obtained from the seed of Triticum vulgare, Vill. Hist. Pl. Dauph. II. 153.

Macroscopic characters.

Very fine, white, odourless and almost tasteless powder.

Anatomical characters.

LITERATURE. Berg. Anat. Atl. 1865. 99. Taf. 50. Deutsch. Arznb. 5. Ausg. 1910. 52. Erdmann-König's Grundr. d. allg. Warenkunde. 1895. 189. Gilg. Ber. Pharm. Ges. 1901. 166. Gilg. Pharmakogn. 1910. 29. Greenish & Collin. Anat. Atl. o. veget. powders. 1904. 6. Hager. Handb. d. pharm. Praxis. Bd. 1. 1900. 294. Hérail. Mat. Méd. 1912. 44. Ingerman. Mikrosk. d. voorn. handelsw. 1910. 54. Karsten u. Oltmanns. Pharmakogn. 1909. 317. Koch u. Gilg. Pharmak. Praktik. 1907. 19. Koch. Die mikr. Analyse d. Drogenpulver. Bd. 4. 1908. 156. Kraemer. Botany a. Pharmacogn. 1910. 788. Luerssen. Med.-Pharm. Bota-

---

[1]) Calculated from observations of Dr. J. J. Prins. De fluctueerende variabiliteit van microscopische structuren bij planten. Diss. Groningen. 1904. p. 25, no. 12.

nik. Bd. 2. 1882. 369. Marmé. Pharmacogn. 1886. 485. Moeller. Mikroskopie d. Nahr.- u. Genussmittel. 1886. 94. Moeller. Pharmacogn. 1889. 346. Moeller. Leitf. z. mikrosk.-pharmakogn. Übungen. 1901. 46. Oudemans. Pharmacogn. 1880. 447. Planchon & Collin. Drogues simples. T. 1. 1895. 110. Schimper. Anl. z. mikr. Unters. v. veget. Nahr.- u. Genussmittel. 1900. 14. Schneider. Powd. veget. drugs. 1902. 116. Tschirch. Angew. Pfl.-anat. 1889. 79. Tschirch u. Oesterle. Anat. Atl. 1900. 221. Taf. 50. Tschirch. Schweiz. Wochenschr. 1901. no. 15. Tschirch. Handb. d. Pharmakogn. T. 2. 1910. 187. Vogl. Die wicht. veg. Nahr.- u. Genussm. 1899. 68. Wiesner. Rohstoffe. Bd. 1. 1900. 595. Wigand. Pharmakogn. 1897. 324. MATERIAL. The drug. REAGENT. Water.

MICROGRAPHY. Grains in 2 kinds: 1. large grains 15 per cent. of the whole number; simple; having the shape of a double-convex, circular, oval or reniform lens; with a central hilum seldom to be distinguished, as well as stratification and clefts; 2. small grains 85 per cent. of the whole number; generally spherical or angular, the latter being components of compound grains; intact compound grains very scarce. Variation curve of diameter: compound, min. $= 0.45\,\mu$, max. $= 51.25\,\mu$. Large grains 15 per cent. of the whole number: n $=$ 237; Med $= 20{,}57\,\mu$; $Q_1 = 4.17\,\mu$; $Q_3 = 4.40\,\mu$. Small grains 85 per cent. of the whole number: n $= 272$; Med $= 4.6\,\mu$; $Q_1 = 0.38\,\mu$; $Q_3 = 1.48\,\mu$ [1]).

November 1901, September 1911.   J; M.

BULBUS SCILLAE.
Squill.

Slices of the middle scales of the bulb of Urginea Scilla, Steinh. in Ann. Sc. Nat. Sér. II. 1. (1834) 321, obtained by cutting thin transverse slices from the bulb, after removal of the exterior membranaceous scales; the inner parts of the bulb to be rejected.

Macroscopic characters.

Up to 5 c.M. in length and 5 m.M. thick, cubes or flat pieces; curved or more or less twisted, often crescent shaped; in a transverse section scattered vascular bundles. Yellowish white, semitranslucent. In a dry state horny, brittle, but very hygroscopic, and after exposure to air soon flexible and tough. Inodorous; taste nauseously bitter.

Anatomical characters.

LITERATURE. de Bary. Vergl. Anat. 1877. 145. Deutsch. Arzneib. 5. Ausg. 1910. 87. Flückiger. Pharmakogn. 1891. 624. Flückiger & Hanbury. Pharmacographia. 1879. 691. Gilg. Pharmakogn. 1910. 48. Greenish & Collin. Anat. Atl. o. veget. powders. 1904. 274. Hager. Pharm. Praxis. Bd. 2. 1902. 857. Hartwich. Üb. d. Meerzwiebel. Arch. d. Pharmac. Bd. 227. 1889. 577. Hérail. Mat. Méd. 1912. 521. Karsten u. Oltmanns. Pharmakogn. 1909. 148. Koch. Mikr. Anal. d. Drogenpulver. Bd. 3. 1906. 229. Taf. 21. Koch u. Gilg. Pharmakogn. Praktik. 1907. 156. Marmé. Pharmacogn. 1886. 103. Meyer. Wissensch. Drogenk. Bd. 2. 1892. 238. Moeller. Mikr.-pharm. Üb. 1901. 283. Oudemans. Pharmacogn. 1880. 124. Planchon et Collin. Drogues simples. T. 1. 1895. 154. Prins. De fluctueerende variabiliteit van microscopische structuren bij planten. Diss. Groningen. 1904. 29. Schneider. Powdered veget. Drugs. 1902. 288. Tschirch. Angew. Pfl.-anat. 37, 105, 122. Vogl. Anat. Atl. 1887.

---

[1]) Calculated from observations of Dr. J. J. Prins. De fluctueerende variabiliteit van microscopische structuren bij planten. Diss. Groningen. 1904. p. 25, no. 13.

5

Taf. 46. Ziegenspeck. Die chem. Zusammensetzung der Raphiden von Scilla maritima. Ber. Bot. Ges. Jhrg. 32. 1914. 630. MATERIAL. The drug; bulbs of the white variety, gathered March 1902, from the Botanic Garden at Groningen, fresh and in alcohol. REAGENTS. Water, glycerine, safranin, potassium iodide iodine, phloroglucin and hydrochloric acid, iodine and sulphuric acid 66 per cent., concentrated sulphuric acid, potash, Schulze's macerating mixture, Fehling's solution.

MICROGRAPHY.

Epidermis. Upper (inner) side. Stomata wanting. Cells in longitudinal rows. Epidermal cells proper. H. 30 μ, Lev. B. 35 μ, Lev. L. 80—180 μ, the cells becoming smaller towards the margin of the scale; tetra- to heptagonal tables. Outer walls very thick; showing stratification; sometimes showing a transverse, parallel cuticular striation, the cuticle becoming red in phloroglucin and hydrochloric acid; lateral walls very thin. Cell contents: in material preserved in alcohol often some sphaerocrystals, 25 μ in diameter, having a more transparent centre, not showing double refraction, persisting in sulphuric acid 66 per cent., fusing together in potash; needle-shaped crystals, sometimes twins, adjoining the outer walls, showing double refraction, insoluble in sulphuric acid and potash.

Under (outer) side. Stomata very rare; lying in the longitudinal rows of the epidermal cells; Lev. B. 30 μ, Lev. L. 40 μ. Epidermal cells proper. H. 50 μ, Lev. B. 50 μ, Lev. L. 150—300 μ, tetra- or pentagonal tables.

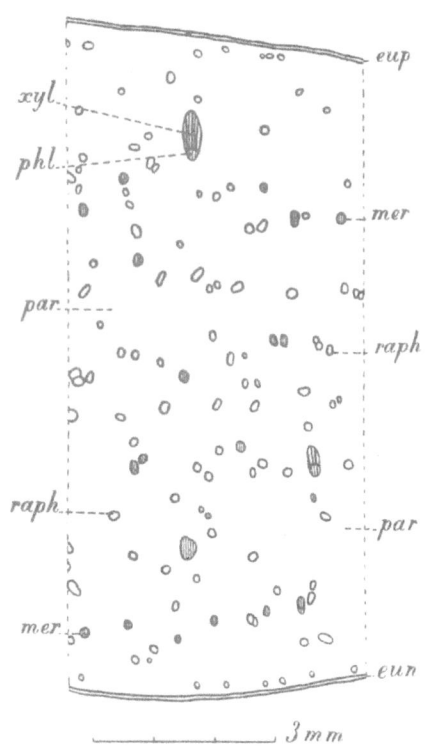

Fig. 1. *Urginea Scilla*. Scale of bulb, transverse section. eun Epidermis under side; eup Epidermis upper side; mer Thin meristeles; par Colourless parenchyma of mesophyll; phl Phloem; raph Raphide cells; xyl Xylem.

Mesophyll. Entirely consisting of a colourless common parenchyma, containing a large number of idioblasts. Colourless parenchyma. Cells arranged in longitudinal rows.

Cells of the above. H. 100—150 μ, Lev. 100—150 μ, smaller near the epidermis on both sides and especially towards the margins of the scales; poly-, generally hexagonal prisms with a longitudinally directed axis and rounded edges. Walls pitted; with intercellular spaces and showing connecting frames, especially in the outermost cell layers. Cell contents: glucose; sometimes starch grains; the vacuoles of many cells, especially those of the outermost cell layer, containing anthocyan; sometimes a granular mass; in material preserved for some time in alcohol the same sphaerocrystals as in the epidermis; in fresh material recently put into alcohol irregular masses, soluble in glycerine.

Idioblasts. Raphide cells, many times more numerous in the inner than in the outer scales [1]), varying much in length, having for the rest the same shape as

[1]) P r i n s. l. c. 29.

those of the colourless parenchyma. Walls the same as those of the colourless parenchyma. Cell contents: in each cell mucilage, homogeneous in fresh material, granular and somewhat contracted in alcoholic material, surrounding a bundle consisting of many raphides, their number increasing with their decreasing length and thickness. Length of the raphides of the outer scales: n = 400; Med = 351.7 μ; Q_1 = 163.8 μ; Q_3 = 163.1 μ; min. within the class 0 — 65 μ; max. within the class 845 — 910 μ; length of the raphides of the inner narrow scales: n = 375; Med = 200.2 μ; Q_1 = 87.7 μ; Q_3 = 117.0 μ; min. within the class 0 — 65 μ; max. within the class 715—780 μ [1]).

Meristeles. Scattered through the mesophyll; almost wholly consisting of a cylindrical vascular bundle.

Pericycle. Not indicated in the figure, only represented by a string of parenchyma cells accompanying the xylem on its outer side; these cells thicker than those of the xylem parenchyma.

Vascular bundle. Collateral, closed.

Phloem.

S i e v e-t u b e s. Diameter 12—18 μ, L of articulations 120—190 μ; polygonal prisms with a longitudinally directed axis. Walls sometimes a little thickened; transverse walls horizontal and somewhat curved, entirely occupied by a very distinct sieve-plate; intercellular spaces wanting. Contents: often a granular mass on the hollow side of the sieve-plate. C a m b i f o r m c e l l s. Diameter 12—18 μ, L. 70—100 μ; polygonal prisms with a longitudinally directed axis. Walls sometimes a little thickened; intercellular spaces wanting. Cell contents: nucleus very distinct; sometimes a granular mass.

Xylem. Consisting of annular and spiral vessels scattered through a parenchyma and of 2 layers of parenchyma cells without vessels on the phloem side. V e s s e l s. R. 30 μ, T. 20 μ; annular and spiral thickenings lignified. P a r e n c h y m a c e l l s. Diameter 10—15 μ, L. 75—100 μ; polygonal prisms with a longitudinally directed axis. Intercellular spaces wanting.

February 1902, October 1911.   J; M.

CARRAGEEN.

Chondrus. Irish Moss. Carragheen.

The entire plant of Chondrus crispus, Lyngb. Hydr. dan. p. 15, Tab. 5, A. B. and Gigartina mamillosa. J. Ag. Alg. med. p. 104, bleached by the action of the sun.

Macroscopic characters.

Thallus up to 17 c.M. in height; cartilaginous; semi-translucent; becoming in water swollen, slippery and soft; very variable in shape; stalk longer or shorter, cylindrical at its base; the upper parts flat, repeatedly bifurcate, fan-shaped, but in the dry state curled and entangled. Lobes 3—10 m.M. broad, linear or more or less cuneate; apex acute, obtuse or emarginate, incisions obtuse or almost acute; margin sometimes showing more or less parted prolifications. Surface incrustated here and there with chalky matter of animal origin, for the rest

[1]) Calculated from observations of Dr. J. J. Prins. De fluctueerende variabiliteit van microscopische structuren bij planten. Diss. Groningen. 1904. p. 29, no. 21; p. 30, no. 22.

smooth, dull, yellowish white; the stalk brownish-yellow at its base. The two species mentioned not to be distinguished from each other by the thallus, but only by the cystocarps; these however often wanting in commercial material. Cystocarps of Chondrus scattered, about 2 m.M. in length, embedded in the thallus, flat, oblong, forming on one surface a very insignificant elevation, and on the other a corresponding shallow pit; after the discharge of the spores a scar, often a hole of the same shape in the thallus. Cystocarps of Gigartina forming spherical or oblong protuberances. Odour peculiar, faint; taste mucilaginous.

Anatomical characters.

LITERATURE. Berg. Anat. Atl. 1865. 3. Taf. 2. Flückiger & Hanbury. Pharmacographia. 1879. 748. Karsten u. Oltmanns. Pharmakogn. 1909. 8. Marmé. Pharmacogn. 1886. 4. Oudemans. Aanteek. o. d. Pharmac. Neerl. 1854—56. 4. Oudemans. Pharmacogn. 1880. 2. Tschirch. Angew. Pflanzenanatomie. 1889. 100. Tschirch. Handb. d. Pharmakogn. T. 2. 1910. 288, where some papers are cited which were inaccessible to us. Tunmann. Anatomie und Inhaltsstoffe von Chondrus crispus Stackhouse. Apoth. Ztg. Berlin. Bd. 24. 1909. 151—154. MATERIAL. The drug, soaked in alcohol 50 per cent. during 24 hours; fresh material — only of Chondrus crispus — and after fixation in chromic acid. REAGENTS. Water, glycerine, iodine in chloral hydrate, potassium iodide iodine, phloroglucin and hydrochloric acid, iodine and sulphuric acid 66 per cent., tincture of iodine and sulphuric acid 66 per cent., concentrated sulphuric acid.

MICROGRAPHY.

Thallus.

Cortical part. Pseudoparenchyma consisting of rows of some 6 oblong cells, standing perpendicular to the surface.

Cells of the above. H. 5—6 µ, Lev. 4 and 4 µ. Walls very thick, middle lamella not to be distinguished, the outer walls of the outermost cells covered by a thin cuticle not disappearing in concentrated sulphuric acid, the walls parallel to the surface pitted. Cell contents: some brown plastids.

Medullary part. Consisting of longitudinally directed hyphae; the cells increasing in length towards the inner parts.

Cells of the above. Walls very thick; showing in the swollen state a very thick inner part and a distinct much thinner outer part, resembling a middle lamella; the inner part showing stratification and cellulose reaction; the middle lamella remaining yellow. Cell contents: in fresh material a peripheral protoplasm, containing some light brown plastids; in the drug the contents contracted into a worm-like very granular mass, becoming brown to violet by the action of potassium iodide iodine, black or deep-blue by iodine and sulphuric acid 66 per cent.

Cystocarps. The cortical part in all respects resembling that of the thallus; the medullary part consisting of globular or ellipsoidal bodies, showing granular contents and embedded in a tissue of thread-like hyphae.

June 1911.   J; M.

CARYOPHYLLI.
Caryophyllum. Caryophyllus. Cloves.
The dried flower-buds of Eugenia caryophyllata, Thunb. Diss. I.

Macroscopic characters.

Long about 16 m.M. and up to 21 m.M. Flower complete, actinomorphous, epigynous; hypanthium much prolonged under the ovary, long 2 c.M., wide 4 m.M., thick 2 m.M., flatly cylindrical. Calyx gamosepalous, 4-sect, wheel-shaped; tube long about 0.5 m.M.; teeth long 2 m.M., deltoid. Corolla 4-phyllous; petals alternating with the sepals and decussate; petals without a claw, sticking together and forming an almost spherical closed cap; limb long and wide 5 m.M., bowl-shaped, with an obtuse apex, a somewhat darker midrib and an entire margin. Stamina indefinite in number, tetradelphous; the numerous stamina of each bundle only at the base slightly grown together; the bundles inserted in superposition to the petals; filaments long 2.5 to 3.5 m.M.; anthers long about 0.5 m.M., innate, introrse. Pistil compound; consisting of 2 carpels; with a single cylindrical style, long about 3 m.M. and a single slightly developed stigma. Ovary with cavities long 2 to 3 m.M., inferior, 2-locular, with a complete true dissepiment and thick axile placentae quite covered with about 20 small ovules in each cavity. External surface of all parts smooth and dark-brown, that of the petals somewhat paler brown. Odour and taste strongly aromatic.

Anatomical characters.

LITERATURE. Berg. Anat. Atl. 1865. Tafel 41. Benecke. Mikr. Drogenpract. 1912. 64. Biechele. Mikr. Prüf. d. off. Drogen. 1904. 20. Erdmann-König's allg. Warenk. 1895. 307. Flückiger. Pharmakogn. 1891. 796. Flückiger & Hanbury. Pharmacographia. 1879. 280. Gilg. Pharmakogn. 1910. 238. Hager. Pharm. Praxis. Bd. 1. 1900. 663. Hérail. Pharmacol. 1912. 316. Karsten u. Oltmanns. Pharmakogn. 1909. 212. Koch. Mikr. Anal. d. Drogenpulver. Bd. 3. 1906. 235. Koch u. Gilg. Pharmakogn. Praktik. 1907. 171 Kraemer. Botany a. Pharmacogn. 1910. 549, 772. Luerssen. Syst. Bot. Bd. 2. 1882. 818. Lutz. Die oblitoschizogenen Secretbeh. d. Myrtaceen. Diss. Bern. 1895; also Bot. Centralblatt. Bd. 64. 1895. 292. Marmé. Pharmacogn. 1886. 254. Meyer. Wiss. Drogenk. Bd. 2. 1892. 331. Meyer. Mikr. Unters. v. Pflanzenpulv. 1901. 64. Moeller. Mikr. d. Nahr.- u. Genussm. 1886. 68. Moeller. Pharmakogn. 1889. 102. Moeller. Mikr. pharm. Ueb. 1901. 118. Molisch. Mikrochemie. 1913. 133. Oudemans. Aanteek. o. d. Pharmac. neerl. 554. Oudemans. Pharmacogn. 1880. 338. Planchon et Collin. Drogues simples. T. 2. 1896. 335. Schimper. Mikr. Unters. d. veget. Nahr.- u. Genussm. 1900. 101. Solereder. Bemerkenswerte anat. Vorkomnisse u. s. w. Archiv d. Pharmacie. Bd. 245. 1907. 410. Tschirch. Angew. Pfl. Anat. 1889. 254. Tschirch. Die Harze u. Harzbeh. 1900. 367. Tschirch. Pharmakogn. Bd. 2. Abt. 2. 1917. 1223. Tschirch u. Oesterle. Anat. Atl. 1900. Tafel 13. Tunmann. Unters. ü. d. Secretbeh. (Drüsen) einiger Myrtaceen, speziell ü. ihren Entleerungsapparaten. Arch. d. Pharmacie. Bd. 248. 1900. 23—42. Tunmann. Pflanzenmikrochemie. 1913. 243. Vogl. Veget. Nahr.- u. Genuszm. 1899. 364. Wigand. Pharmakogn. 1879. 246. MATERIAL. The drug. REAGENTS. Water, glycerine, potash 50 per cent., potash 12½ per cent., chloral hydrate, phloroglucin and hydrochloric acid, iodine and sulphuric acid 66 per cent., concentrated sulphuric acid, iron acetate, chromic acid 50 per cent.

MICROGRAPHY.

Hypanthium below the ovary.

Epidermis. Stomata phaneroporous; lying somewhat above the level of the surrounding epidermal cells; showing a large air-cavity.

Epidermal cells proper. R. 18 μ, T. and L. 12 μ; polygonal prisms. Outer walls strongly thickened, 12 μ; somewhat yellow; in potash 50 per cent. forming large yellow drops, the inner part remaining and also filled with large drops; after 24 hours the first mentioned yellow drops coagulated into yellow solid masses, diminished in volume and still covered with a membrane; near the yellow masses often probably some colourless crystals; in potash 12½ per cent. drops formed only after a longer time e. g. 12 hours; these globules not distinctly yellow. These outer walls moreover colouring brown in iodine and sulphuric acid 66 per cent., remaining unaltered in chromic acid 50 per cent. and concentrated sulphuric acid, though somewhat swollen in the latter reagent; in each outer wall a single pit-canal. Stomata. R. 18 μ, long 25 μ, wide 20 μ; outer and inner walls strongly thickened.

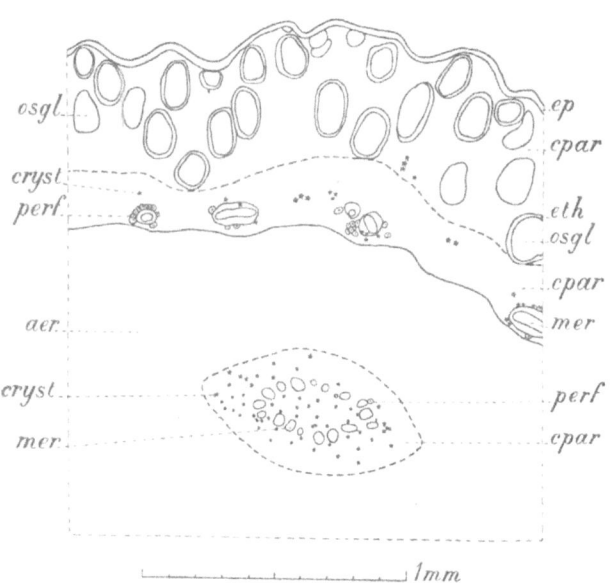

Fig. 2. *Eugenia caryophyllata*. Alabastrum, transverse section of hypanthium below the ovary. aer Aerenchyma; cpar Common parenchyma; cryst Cluster crystal idioblasts; ep Epidermis; eth Epithelium; mer Meristeles; osgl Oblito-schizogenous glands; perf Pericyclic sclerenchyma fibres.

Ground tissue. Consisting of 4 parts:

1. An outer part consisting of common parenchyma cells arranged more or less in radial rows and diminishing in size towards the periphery, the outermost cells again being larger. This outer part containing: a. hemispherical groups of cells lying here and there close to the epidermis and turning their flat sides towards it, containing a homogeneous mass; this mass insoluble in hot water, showing double refraction and colouring yellow in potash 50 per cent.; b. oblito-schizogenous glands [1]), 70—80 in number in a cross section in the basal half of the hypanthium and forming 2 or 3 more or less intermixing layers; these glands ellipsoidal and radially elongated, with an epithelium consisting of 2 or 3 cell layers [2]); these cells not strongly obliterated, often torn asunder. Glands of the outermost layer R. 120 μ, T. and L. 80 μ; of the innermost

[1]) See Lutz and also Tschirch. Harze u. Harzbeh. l. c.
[2]) According to Lutz 1 layer of secreting cells and 2 of mechanical cells.

layer R. 150 μ, T. and L. 100 μ. Contents of the cavity: partly consisting of a homogeneous colourless mass, soluble in alcohol 96 per cent.; moreover sometimes 1 or more drops; these contents colouring red in concentrated sulphuric acid but not in phloroglucin and hydrochloric acid (caryophylline?), colouring black in iron acetate or osmic acid 1 per cent. in consequence of the presence of eugenol.

Common parenchyma cells. R. 30 μ, T. and L. 20 μ; polyhedra, mostly radially elongated, round the glands flat. Walls somewhat brown in consequence of infiltration; intercellular spaces wanting. Walls and cell contents: colouring black in iron acetate, probably only in consequence of an infiltration with eugenol; the latter substance colouring black in iron acetate. Epithelium cells. Thick 20 μ, wide 25 μ, long 40 μ; polygonal tables with rounded edges. Walls very thin; somewhat brown in consequence of infiltration; colouring red in phloroglucin and hydrochloric acid, brown in iodine and sulphuric acid 66 per cent., remaining in concentrated sulphuric acid or chromic acid 50 per cent. Intercellular spaces wanting.

2. A second part consisting of common parenchyma cells lying more or less in longitudinal rows; cluster crystal idioblasts here and there. The innermost 1 or 2 layers containing meristeles; these meristeles each surrounded by a very incomplete endodermis, developed as a crystal-sheath and consisting of some crystal fibres divided into cells, each containing a cluster crystal.

Common parenchyma cells. R. 20 μ, T. 25 μ, L. 30 μ; in the vicinity of the meristeles smaller; polygonal prisms with a longitudinally directed axis and rounded edges, mostly a little flattened, especially in the radial direction. Walls collenchymatously thickened, somewhat brown; showing intercellular spaces. Walls and cell contents: colouring blue-black in iron acetate; see first part of ground tissue. Fibres of crystal-sheath. R. and T. 8 μ, L. of partitions 10 μ; polygonal prisms with a longitudinally directed axis. Walls coloured brown. Walls and cell contents: in iron acetate colouring blue-black, see first part of ground tissue.

3. A third part consisting of aerenchyma; the cavities, smaller in the innermost part, being separated from one another by lamellae of 1 cell thick; cluster crystal idioblasts numerous in the innermost part of this tissue.

Aerenchyma cells. R. and T. 25 by 14 μ, L. 30 μ; rectangular prisms with a longitudinally directed axis. Walls, especially those adjoining the cavities, somewhat thickened; all walls brown in consequence of infiltration. Walls and cell contents: colouring blue-black in iron acetate, see first part of ground tissue.

4. A central columella ending in the placentae and consisting in the peripheral parts of common parenchyma intermixed with cluster crystal idioblasts, in the central part almost exclusively of flattened cluster crystal cells; small intercellular spaces in the peripheral parts. Meristeles, about 17 in number, arranged in a ring in the peripheral part of the columella, these meristeles not surrounded by an endodermis.

Cells of central part. Walls coloured brown. Walls and cell contents: colouring blue-black in iron acetate, see first part of ground tissue.

Meristeles. In the second part of ground tissue. In the basal half of the hypanthium in a cross section about 20 in number, lying in 1 or 2 layers.

Pericycle. Developed especially at the inner and at one of the lateral sides; consisting of sclerenchyma fibres.

F i b r e s . R. 20 μ, T. 25 μ, L. 250—450 μ; polygonal. Walls thickened; lignified, especially the clearly discernible middle lamella.

Vascular bundle. A single compound one; the component simple vascular bundles bicollateral.

Outer phloem. Exarch; in iodine and sulphuric acid 66 per cent. colouring blue.

Xylem. Endarch; consisting of a single radial row of 3—5 spiral vessels.

S p i r a l   v e s s e l s . R. 6 μ, T. 7 μ. Walls showing lignification; coloured brown; in iron acetate colouring blue-black, see first part of ground tissue.

Inner phloem. Endarch; often separated from the xylem by parenchyma cells; for the rest see outer phloem.

S e p a r a t i n g   p a r e n c h y m a   c e l l s . R. 3 μ, T. 7 μ; polygonal in a cross section. Walls coloured brown; in iron acetate colouring blue-black, see first part of ground tissue.

Medullary commissures. Between the simple vascular bundles mostly uniseriate; consisting of common parenchyma cells, here and there mixed with cluster crystal idioblasts.

P a r e n c h y m a   c e l l s . R. 6—8 μ, T. 3—5 μ; polyhedra. Walls coloured brown; in iron acetate colouring blue-black, see first part of ground tissue. Cell contents: sometimes in the xylem part a homogeneous mass, showing double refraction and colouring yellow in potash 50 per cent.

In the columella. Pericycle represented by a very few sclerenchyma fibres with lignified walls. Vascular bundle small; in the xylem spiral vessels showing lignification.

U p p e r   p a r t   o f   H y p a n t h i u m containing the ovary.

Outer epidermis. See hypanthium below the ovary.

Ground tissue. Consisting of 3 parts: 1 and 2. These outer parts closely resembling the same in the hypanthium below the ovary. 3. A third part, consisting of sclereids intermixed with a few cluster crystal idioblasts. Middle lamellae not discernible, hence form and dimensions of the cells not easily distinguished.

S c l e r e i d s . Walls strongly thickened; the layer adjoining the cell cavity showing strong double refraction and colouring blue in iodine and sulphuric acid 66 per cent.; the rest of the wall remaining colourless, but strongly swelling up in water; cell cavity slit-like.

Inner epidermis. Consisting of cells, rectangular in a cross section and walls adjoining the cavities of the ovary showing a cuticle; cluster crystal idioblasts numerous.

Ovary. Dissepiments. Ground tissue resembling the outer part of the ground tissue in the upper part of the hypanthium. Inner epidermis see that of upper part of hypanthium. Placentae containing many cluster crystal idioblasts and numerous meristeles.

C a l y x. Epidermis: see that of hypanthium; stomata wanting at the upper side. Ground tissue forming a continuous layer with the first and second part of ground tissue of the hypanthium; at the under side containing 1 or 2 layers of oblito-schizogenous glands, at the upper side those glands not numerous;

cluster crystal idioblasts numerous at the upper side. Meristeles about 10 in number in a cross section through the middle part of the tooth.

Corolla. Epidermis. Upper side consisting of cells of very irregular shapes in a surface view; stomata wanting.

Epidermal cells proper. Polygonal tables either longitudinally elongated or with sinuous lateral walls etc. Outer walls much less thickened than those of the hypanthium.

Under side consisting of somewhat longitudinally elongated cells; see for the rest epidermal cells of hypanthium; stomata wanting. Ground tissue consisting of longitudinally elongated common parenchyma cells; in the thicker parts of the petal containing oblito-schizogenous glands along the upper and under side; those along the under side mostly somewhat larger, in the thinner parts of the petal glands smaller and reaching from epidermis to epidermis; see for the rest those of hypanthium; cluster crystal idioblasts numerous. Meristeles very small, only present in the midrib portion of the petal.

Stamina. Filament. Epidermis.

Cells mostly longitudinally elongated tetragonal tables, at the base of the filament with sinuous walls.

Ground tissue consisting of common parenchyma cells intermixed with numerous glands, thick 70 µ, L. 80 µ, see for the rest glands of hypanthium; cluster crystal idioblasts numerous, lying in longitudinal rows. Meristele, a single one, containing spiral vessels. Anther, on the line of dehiscence a band of cluster crystal cells. Connective containing a single meristele; at the top a large gland, thick 200 µ, L. 200 µ, with a brown epithelium. Thecae. Fibrous layer very distinct. Pollen grains 15 µ in diameter; tetrahedra with flattened angles, each showing a pore; cell contents: showing at each angle a yellow globule.

Pistil. Ovary, see upper part of hypantium containing the ovary.

Disc. Epidermis consisting of cells with thickened outer walls. Ground tissue consisting of common parenchyma cells, containing in the outermost part along the epidermis numerous glands like those of the hypanthium.

Style. Epidermis consisting of cells with thickened outer walls. Ground tissue consisting of common parenchyma cells; containing in the outermost part numerous glands, like those of the hypanthium; cluster crystal idioblasts especially in the innermost part numerous, lying in longitudinal rows. Meristeles in the innermost part of the ground tissue forming a ring. Conductive tissue in the centre of the style consisting of thick-walled cells; in the uppermost part of the pistil more or less torn asunder, showing a central cavity.

Stigma discernible at the apex of the style by the presence of cells without thickened walls.

Micrography of the powder.

Common parenchyma containing elliptical glands with some layers of flat epithelium cells. Sclerenchyma fibres, long 350 µ. Spiral vessels. Many cluster crystal idioblasts, often in longitudinal rows. Anthers with a clearly discernible fibrous layer and tetrahedral pollen grains. Epidermal cells of hypanthium and ovary penta- or hexagonal with much thickened outer walls; stomata. Starch wanting. The presence of short sclerenchyma cells and of reticulate vessels indicating a mixture with petioles; the presence of starch a mixture with anthophylli.

Februari 1903. J; M; v. E. d. W.

CORTEX ALYXIAE.
Alyxia Bark.
The dried bark of stem and branches of Alyxia stellata, Roem & Schult. Syst. IV. 439, after removal of the dark outer layers.

Macroscopic characters.

Quills or gutter-shaped pieces of varying length and breadth; the bark being thick up to 4 m.M., rather heavy, firm, hard and brittle. Outer surface almost quite deprived of a thin outer dark brown layer; rather smooth, dull, of a light ochre colour, here and there with irregular fissures. Inner surface finely wrinkled in a longitudinal direction and fissured, of a still lighter colour. Transverse fracture uneven, somewhat fibrous and splintery, of a very light ochre colour. Odour of cumarin; taste the same, bitterish.

Alyxia stellata is called in Dutch East India, among other names, by that of poelasari.

The use of thin quills, taken from very young branches — poelasari dedès —, is permitted, but thick pieces taken from old stems and branches are of small worth.

Anatomical characters.

LITERATURE. We have found no papers treating of the anatomy of Alyxia stellata. The following deal with the anatomy of the family of the Apocynaceae. Garcin. Apocynacées. Diss. Lyon. 1889. Leonhard. Beitr. z. Anat. d. Apocynaceen. Bot. Centralbl. Bd. 45. 1891. Solereder. Syst. Anat. d. Dicot. 1899. 597 and Ergänzungsb. 1908. 211. Spire. Contrib. à l'ét. d. Apocynées &c. Trav. Labor. mat. méd. Ec. sup. pharm. Paris. T. 2; also Diss. Paris. 1905. MATERIAL. The drug, received Jan. 23th. 1904 from the Botanic Garden at Buitenzorg; young leaf-bearing shoots thick some 2 m.M., collected in the Botanic Garden at Buitenzorg and received Nov. 2d. 1911. REAGENTS. Water, glycerine, potash, potassium iodide iodine, phloroglucin and hydrochloric acid, iodine and sulphiric acid 66 per cent.

MICROGRAPHY.

Epidermis.

Epidermis proper. Present only in young shoots, soon thrown off. Cells arranged in longitudinal rows, doubtless often derived from 1 cell divided by transverse walls into many cells; stomata wanting. Somewhat numerous smaller epidermis cells grown out, without separating wall, into unicellular, conical, somewhat curved, very thick-walled hairs — long 60 to 250 μ — showing short longitudinal cuticular striae.

E p i d e r m a l c e l l s p r o p e r. R. 20 μ, T. 16—30 μ, L. 6—17 μ; tetra- to octo-, mostly hexagonal prisms with a radially directed axis. Outer walls very thick, side walls somewhat cuneiform and pitted; intercellular spaces wanting, also on the side of the cortex.

Secondary cork tissue. In shoots 1.25 m.M. thick, tangential partition walls appearing here and there in the epidermical cells and hairs. In the drug the secondary cork being removed, only here and there in the neighbourhood of a lenticel a small portion remaining, thick some 300 μ.

S e c o n d a r y c o r k t i s s u e p r o p e r. Consisting of many layers of tangentially elongated quite flattened cells, having thin yellow walls; some 5 inner layers of thick-walled yellow cells constituting perhaps a phelloderm; some of these cells containing a simple crystal, the innermost layer consisting wholly of such cells often containing some crystals, or partitioned by radial walls; all crystals in crystal skins.

L e n t i c e l s. Having almost quite lost their phellem, the remaining parts being shriveled up; some layers of phelloderm cells, in part only thick-walled.

C o r t e x.

Primary cortex proper. Consisting of 10—12 layers of parenchyma cells, in young shoots this number being smaller, i. e. 5 to 6 layers. The outer 5—8 layers containing a very large number of tangentially elongated sclereids, often arranged in longitudinal rows, and here and there a crystal idioblast, with 1 or more simple crystals in crystal skins; in young shoots only the 1 or 2 outer layers of the cortex showing sclerotized cells. The inner 5 or 6 layers of the cortex consisting principally of tangentially elongated common parenchyma cells, often arranged in longitudinal rows and often divided by a thin radial wall; this tissue containing: 1. less tangentially elongated sclereids, isolated or in groups, the larger ones of these being longitudinally elongated; 2. laticiferous tubes only in the innermost part of the cortex, running in a longitudinal direction, sinuous in a tangential plane, conspicuous by their granular contents and often consisting of very long parts joining sideways more or less together; 3. a few crystal idioblasts as described above.

S c l e r e i d s. R. 15—40 μ, T. 50—100 μ, L. 15—40 μ; polyhedra but rounded when adjoining thin-walled parenchyma, especially those of the inner cortex showing irregular protuberances wedged between adjoining parenchyma cells. Walls strongly thickened, leaving only a small cavity; showing stratification; lignified, especially in the outer parts; with numerous, often branched pit canals. C o m m o n p a r e n c h y m a c e l l s. R. 25—40 μ, T. 40—90 μ, L. 30—70 μ; polyhedra with rounded edges. Walls a little thickened; intercellular spaces present and often large. Cell contents: starch, especially in the cells lying in the vicinity of the laticiferous tubes; often a granular mass. L a t i c i f e r o u s t u b e s. R. 60—70 μ, T. 100—110 μ, but often in several places much thinner; elliptical cylinders. Walls thickened. Contents: a granular, dirty yellow, contracted mass, often broken in pieces.

Endodermis. Developed in the youngest shoots examined as a distinct starchsheath, commonly thick 1 layer of cells. Some of these cells much larger, containing no starch and probably in the way of being developed into laticiferous

tubes, afterwards joining the layer of cortical laticiferous tubes, described a-
bove. In the drug not to be distinguished.

S t e l e.

Pericycle. In shoots, thick 1.5 m.M., chiefly consisting of longitudinal, often
tangentially broader bundles of sclerenchyma fibres, separated from each other
by radial bands of pa-
renchyma, broad 1 layer of
cells and containing in
their outermost parts here
and there an incipient
laticiferous tube; the
bundles of fibres thick in
a radial 4—5, in a tan-
gential direction 2—8 elements, having a smooth
outer and a more or less
uneven inner surface. In
the course of growth an
abundant development of
parenchymatic tissue tak-
ing place; in the drug
all sclerenchyma fibres
separated from each other,
becoming rounded at the
edges and running iso-
lated in sinuous lines—
though more or less uni-
ted into groups corre-
sponding with the bundles
formerly present or parts
of these — through a

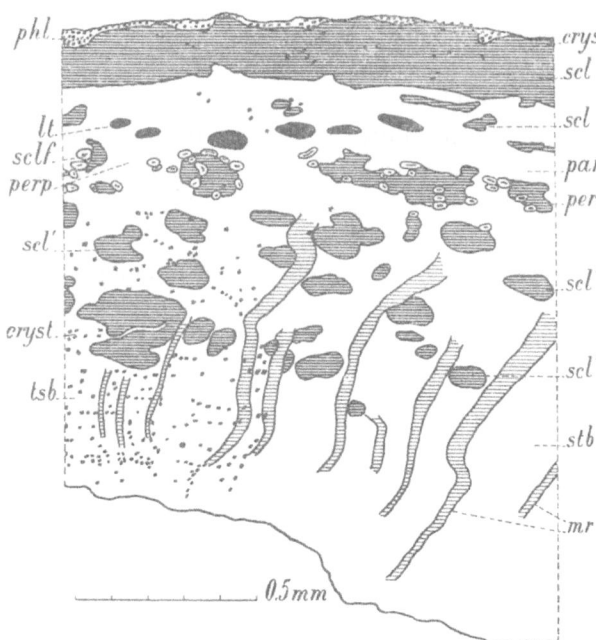

Fig. 3. *Alyxia stellata*. Bark, transverse section. cryst Crystal idioblasts; lt Laticiferous tubes; mr Medullary rays; par Colourless
cortical parenchyma; perp Parenchyma of pericycle; pers Sclereids
of pericycle; phl Phelloderm, inner layer containing crystals; scl
Sclereids of primary cortex; scl' Clusters of sclereids in secondary
phloem; sclf Sclerenchyma fibres of pericycle; stb Sieve-tubes and
bast parenchyma; Crystal idioblasts drawn only in the left part of
the figure in the inner part of the cortex, the pericycle and secondary
phloem.

parenchymatic ground mass, not to be distinguished from the surrounding
tissues, belonging to cortex and outer secondary phloem and containing as well
as these, groups of sclereids and crystal idioblasts.

S c l e r e n c h y m a f i b r e s. R. 15—18 μ, T. 18—25 μ; circular or elliptical in a cross
section; walls very thick, almost quite closing up the cavity; showing stratification; showing no lignification but a distinct cellulose reaction; pit canals wanting.

Phloem.

P r i m a r y p h l o e m. In the drug not to be distinguished.

S e c o n d a r y p h l o e m. In the drug very much shriveled, especially the
inner layers and thereby a detailed description becoming impossible.

Cribral system. Represented by tangential bands, thick in a radial direction

2—4 elements, and alternating with metacribral bast parenchyma bands. Sieve-tubes very thin-walled. **System of bast fibres** wanting. **Parenchymatic system.** M e t a c r i b r a l   b a n d s thick 3—4 elements in a radial direction, consisting of common parenchyma containing starch, and very many crystal fibres arranged more or less in tangential layers, occupying the middle portions of the metacribral bands and containing mostly simple crystals in a crystal skin, sometimes cluster crystals. Sclereids becoming less numerous towards the inner parts of the bark, isolated or in generally longitudinally extended groups; the elements themselves often being very elongated, e.g. R. and T. 80 μ, L. up to 500 μ. See for the rest the sclereids of primary cortex, only the walls being now and then thinner and the cavity always larger. M e d u l l a r y   r a y s separated from each other in a tangential direction by 2—8 layers of elements; uni- to biseriate; the cells outwards becoming somewhat broader in a tangential and shorter in a radial direction, mostly containing starch, often crystals, and these crystal cells lying in the same tangential rows with those of the metacribral bands.

September 1912. J; M.

CORTEX CINNAMOMI.
Cinnamom Bark.
The dried inner layer of the bark of shoots of cultivated plants of Cinnamomum zeylanicum, Nees, in Wall. Pl. As. Rar. II; 74; III. 32.

Macroscopic characters.
Somewhat flattened cylinders, up to 1 M. in length and 1 c.M. thick, composed of 8—10 layers of quills, closely rolled up especially on both margins. The bark about 0.25 m.M. thick, brittle. Outer surface smooth; dull; pale-yellow brown; with lighter coloured, longitudinal, parallel, straight or more or less wavy, sometimes anastomosing lines. Here and there small scars or holes, indicating the places formerly occupied by leaves or twigs. Inner surfaces darker brown, less smooth, finely striated. Transverse fracture short-splintery. Odour and taste very aromatic.

Anatomical characters.
LITERATURE. De Bary. Vergl. Anat. 1877. 150, 545. Berg. Anat. Atl. 1865. 71. Taf. 36. Biermann. Üb. Bau u. Entw.-gesch. d. Ölzellen u. d. Ölbild. in ihnen. Diss. Bern. 1897. 24. Deutsch. Arzneib. 5. Ausg. 1910. 136. Erdmann-König. Allg. Warenk. 1895. 301. Flückiger. Pharmakogn. 1891. 603. Flückiger & Hanbury. Pharmacographia. 1879. 525. Gilg. Pharmakogn. 1910. 120. Greenish & Collin. Anat. Atl. o. veget. powders. 1904. 194. Hager. Pharm. Praxis. Bd. 1. 1900. 841. Hartwich. Beitr. z. Kenntn. d. Zimmts. Arch. d. Pharmac. 1901. 182. Hérail. Mat. Méd. 1912. 288. Ingerman. Mikroskopie d. voorn. handelswaren. 1910. 160. Karsten u. Oltmanns. Pharmakogn. 1909. 122. Koch. Pharmakogn. Atl. Bd. 1. 1909. 22. Luerssen. Syst. Bot. Bd. 2. 1882. 564. Marmé. Pharmacogn. 1886. 127. Mazur-Kiewicz. Die anat. Typen d. Zimmtrinden. Bull. d. l'Acad. de Cracovie. Sér. B. 1910. 146. Meyer. Wissensch. Drogenk. Bd. 2. 1892. 145. Meyer. Mikr. Unters. v. Pflanzenpulver. 1901. 190. Moeller. Mikr. d. Nahr.- u. Genussm. 1886. 346.

Moeller. Mikr.-pharm. Üb. 1901. 250. Oudemans. Aanteek. o. d. Pharmac. neerl. 1854—56. 129. Oudemans. Pharmacogn. 1880. 211. Planchon & Collin. Drogues simples. T. 1. 1895. 363. Prins. De fluct. variabilit. v. microsc. structuren bij planten. Diss. Groningen. 1904. 33. Schimper. Mikr. Unters. d. veget. Nahr.- u. Genussm. 1900. 126. Schneider. Powdered Veget. Drugs. 1902. 163. Tschirch. Angew. Pfl.-anat. 1889. 199, 200. Tschirch u. Oesterle. Anat. Atl. 1900. 132. Taf. 32. Vogl. Veget. Nahr.- u. Genussm. 1899. 511. Wiesner. Rohstoffe. 1900. Bd. 1. 774. Wigand. Pharmakogn. 1879. 178. MATERIAL. The drug; fresh shoots, thick 0.18, 2.5, and 5 c.M., gathered March 1901 in the Botanic Garden at Groningen. REAGENTS. Glycerine, chloral hydrate, potash, potassium iodide iodine, phloroglucin and hydrochloric acid, iodine and sulphuric acid 66 per cent., concentrated sulphuric acid, ferric acetate, Schulze's macerating mixture.

MICROGRAPHY.

E p i d e r m i s. Present even in the oldest shoots used, a considerable increase in number and size of the epidermal cells taking place during secondary growth. Stomata wanting. Epidermal cells in longitudinal rows.

E p i d e r m a l   c e l l s   p r o p e r. R. 18 μ, T. 13 μ, L. 10—20 μ; tetragonal prisms, with transverse walls directed obliquely in respect to the radial plane. Outer walls much thickened, radial walls wedge-shaped. Cell contents: very small starch grains.

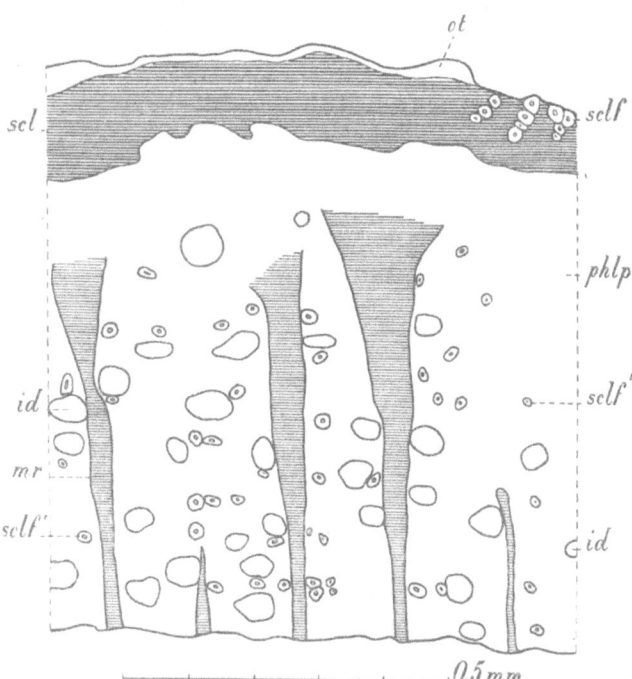

Fig. 4. *Cinnamomum zeylanicum.* The drug, transverse section. id Idioblasts, mucilage and oil cells; mr Medullary rays; ot Shriveled outer tissues; phlp Bast parenchyma, mixed with bundles of sieve-tubes; scl Sclereids of pericycle; sclf Sclerenchyma fibres of pericycle; sclf′ Sclerenchyma fibres of secondary phloem.

C o r t e x.

Secundary cork tissue. Appearing in the shape of irregular scales, having a surface of 1—2 sq. m.M., arranged more or less in longitudinal rows, much increasing in number but not much in thickness during the secondary growth of the shoots. Developed from the primary cortex, at the margin of the scales immediately below the epidermis, at their centre some layers of cells below the surface. Lenticels wanting.

P h e l l e m. 1—6 layers of radially arranged periderm cells. Often only the innermost layer, consisting of dome-shaped cells, remaining and seeming to form a new epidermis.

Cells of the above. R. 9 μ, T. 17 μ, L. 20 μ; tetra- to hexagonal tables, with a radially directed axis. Walls of the outer cell layers generally thickened, the innermost cell layer showing only much thickened dome-shaped outer walls. Cell contents: brown; in many cells tannin and always in those of the inner layer but one; the innermost layer moreover containing starch.

Phellogen. One layer of cells, containing starch and sometimes tannin.

Phelloderm. 2—3 layers of radially arranged cells, resembling in shape and contents the periderm cells.

Primary cortex proper. Consisting of 8—16 layers of colourless parenchyma cells, this number of layers not increasing during secondary growth of the shoots; some of these cells sclerenchymatous, especially in the oldest shoots, the outer cell walls mostly however remaining thin. Idioblasts in 2 kinds: 1. oil cells, numerous, often arranged in longitudinal rows consisting of 2—3 cells; 2. mucilage cells few in number.

Cells of colourless parenchyma. At first R. 15 μ, T. 15 μ, L. 40—50 μ; penta- or hexagonal prisms with a longitudinally directed axis. Afterwards transverse divisions of the cells and tangential growth having taken place R. 15 μ, T. 15—30 μ, L. 20 μ. Walls somewhat collenchymatous; pitted; intercellular spaces wanting. Cell contents: starch and oblong crystals. Oil cells having the same size and shape as the surrounding parenchyma cells. Cell contents: granular; sometimes tannin. See for the rest the secondary phloem. Mucilage cells larger than the surrounding cells; ellipsoids. See for the rest the secondary phloem.

All tissues mentioned above have been removed from the drug.

Endodermis. Developed as a starch-sheath, distinct in shoots of the first year; in the drug some remnants being present now and then in the shape of a brown outer layer.

Cells of the above. Size and shape the same as those of the adjacent cells of the colourless cortical parenchyma, but keeping in older shoots their original length. Cell contents: more starch than in the parenchyma, all starch grains thrown to the lower part of the cells.

Stele.

Pericycle. 3—4 layers of sclereids, mixed with bundles of sclerenchyma fibres, containing 3—4 layers in a radial, 1—5 in a tangential direction, producing the lighter coloured, longitudinal lines mentioned among the macroscopic characters. In young shoots the pericycle consisting almost wholly of these bundles with small strips of thin-walled parenchyma between.

Sclereids. R. 32 μ, T. 40—70 μ, L. 35 μ; polyhedra, generally elongated in the tangential direction. Walls very thick, outer walls sometimes thin; showing stratification; lignified; having pit canals; distinctly visible cavity. Cell contents: sometimes starch. In young shoots the sclereids R. 13 μ, T. 16 μ, L. 54—60 μ, evidently afterwards transverse divisions taking place; penta- or hexagonal prisms with a longitudinally directed axis. Cell contents: starch and oblong crystals, the latter perhaps [1]) disappearing after-

---

[1]) Because it should be kept in mind, that the young shoots examined belonged to a tree cultivated in the Botanic Garden at Groningen.

wards; sometimes tannin. S c l e r e n c h y m a   f i b r e s. R. and T. 9 μ, L. 1000—2500 μ; tetra- or pentagonal. Walls very thick, almost no cavity remaining; showing stratification; lignified, especially the middle lamellae; showing pit canals; intercellular spaces wanting.

Phloem.

P r i m a r y   p h l o e m and m e d u l l a r y   c o m m i s s u r e s not to be distinguished in the drug.

S e c o n d a r y   p h l o e m. Consisting in a radial direction of 35—45 layers of elements. **Cribral system** represented only by bundles of soon obliterated sieve-tubes. **System of bast fibres** represented only by non-septate bast fibres, generally isolated but arranged in tangential and often also in longitudinal rows; increasing in number towards the inner part of the bark. **Parenchymatic system** represented by b a s t   p a r e n c h y m a, containing 2 kinds of idioblasts and by m e d u l l a r y   r a y s. Bast parenchyma being the chief constituent of the bark; the idioblasts often adjoining the medullary rays; very numerous mucilage cells and a few oil cells. Medullary rays separated from each other in a tangential direction by 9—14 layers of parenchyma cells, uni- or biseriate, 7—13 cells in height, cuneiform, becoming broader outwards in consequence of tangential growth of the cells.

S i e v e - t u b e s. R. 14 μ, T. 16 μ; tetragonal prisms. Sieve-plates on the transverse walls sometimes to be distinguished. Cell walls yellowish brown. N o n - s e p t a t e   b a s t f i b r e s. R. 15 μ, T. 25 μ, L. 200—600 μ; rectangular, seldom with forked ends. Walls very thick, almost no cavity; showing stratification; lignified; pit canals wanting. B a s t p a r e n c h y m a   c e l l s. R. 22 μ, T. 30 μ, L. 40—50 μ; tetragonal prisms with a longitudinally directed axis. Walls yellowish brown; intercellular spaces wanting. Cell contents: simple starch grains and sometimes oblong crystals or tannin. M u c i l a g e c e l l s. R. 25 μ, T. 60 μ, L. 170—280 μ; ellipsoids. Walls very thick, almost no cavity remaining; no cellulose reaction; consisting of a thin outer layer inward lignified and a very thick inner layer showing stratification. O i l   c e l l s only differing from the mucilage cells by thin walls and brown resinous contents. C e l l s   o f   m e d u l l a r y   r a y s radially directed cylinders; intercellular spaces in all directions. Cell contents: either a brownish yellow mass and many oblong crystals or starch grains and crystals, sometimes tannin. A few cells somewhat larger than the rest and without contents.

Micrography of the powder. Bast fibres, length: n = 300; Med = 541,80 μ; Q_1 = 56,49 μ; Q_3 = 56,09 μ; min. within the class 288,68—278,46 μ; max. within the class 676,26—716,04 μ; transverse diameter: n = 300; Med = 24,56 μ; Q_1 = 2.86 μ; Q_3 = 2.86 μ; min. within the class 12.07—15.52 μ; max. within the class 36.22—39.67 μ. [1] Generally quite intact, some united into bundles; walls thick, showing only a very few pit canals and no stratification; cell cavity small. Sclerenchyma cells separate or in groups; walls relatively less thickened than in the fibres, with many often branched pit canals, without distinct stratification. Parenchyma cells having thin brown walls. Dark red-brown

---

[1] Calculated from observations of Dr. J. J. Prins. De fluctueerende variabiliteit van microscopische structuren bij planten. Diss. Groningen. 1904. p. 34, no. 27 and 28.

resinous masses or parts of these, derived from the oil cells, sometimes still contained in them. Small, prismatic crystals of oxalate of lime not very conspicuous (examined in water). Starch grains separate and contained in many sclerenchyma and parenchyma cells; grains simple or compound, 2- to 5-adelphous; the simple and component grains isodiametric, rounded, about 6 μ in diameter, with central hilum, without distinct stratification.

February 1901, October 1911. J; M.

CORTEX CONDURANGO.
Condurango Bark.
The dried bark of a South-American species of the genus Marsdenia.

Macroscopic characters.
Quills or gutter-shaped pieces; long up to 13 c.M., broad up to 3 c.M.. The bark thick 1—7 m.M.; sometimes slightly curved; hard; brittle. Outer surface somewhat wrinkled in a longitudinal direction; showing small warts, partly lenticels, partly small scales of rhytidoma; light greyish brown; often covered with white or grey crustaceous lichens. Inner surface longitudinally striated, very light grey-brown. Transverse fracture of the outer, somewhat 1 m.M. thick layer grey-brown and mostly fibrous; that of the inner part uneven, granular, somewhat mealy, light brown-yellow, with numerous roundish darker brown-yellow patches or corresponding small pits. On the transverse section a thin, grey-brown cork layer; the darker patches somewhat distinctly in tangential rows and, especially in the outer part, also somewhat radially arranged. Odour slightly aromatic; taste bitterish, somewhat acrid.

Anatomical characters.
LITERATURE. Deutsch. Arzneib. 1910. 138. Flückiger. Pharmakogn. 1891. 591. Gilg. Pharmakogn. 1910. 283. Greenish & Collin. Anat. Atl. o. veget. powders. 1904. 196. Hager. Pharm. Praxis. Bd. 1. 1900. 941. Karsten u. Oltmans. Pharmakogn. 1909. 139. Koch. Mikr. Anal. d. Drogenpulver. Bd. 1. 1901. 91. Koch. Einf. i. d. mikr. Anal. d. Drogenpulver. 1906. 24. Koch. Pharmakogn. Atlas. Bd. 1. Lief. 2. 1910. 27. Koch u. Gilg. Pharmakogn. Praktik. 1907. 39. Marmé. Pharmakogn. 1886. 143. Meyer. Wiss. Drogenk. Bd. 2. 1892. 132. Moeller. Anat. d. Baumrinden. 1882. 173. Moeller. Pharmacogn. 1889. 241. Moeller. Mikr.-pharm. Üb. 1901. 259. Planchon & Collin. Drogues simples. T. 1. 1895. 690. Speyer. Beitr. z. Entw.-gesch. d. Rinde pharm. interess. Pfl. Diss. Bern. 1907. 69. Treiber. Ueb. d. anat. Bau d. Stammes d. Asclepiadeen. Bot. Centralbl. Bd. 48. 1891. 214. Tschirch. Angew. Pfl.-anat. 1889. 347 u. 529. Tschirch u. Oesterle. Anat. Atl. 1900. 267. Vesque. L'anat. comp. de l'écorce. Ann. sc. Sér. 6. T. 2. 1875. 192. Vogl. Anat. Atl. 1887. Taf. 38 & 39. MATERIAL. The drug, thick 4 m.M. REAGENTS. Water, glycerine, potassium iodide iodine, phloroglucin and hydrochloric acid, iodine and sulphuric acid 66 per cent., concentrated sulphuric acid.

MICROGRAPHY. According to the researches of Treiber [1]) the secondary cork, covering the drug on the outside, originates in Gonolobus Condurango in the epidermis, in all Marsdenieae examined, in the outermost cortical layer.

---

[1]) l. c. p. 214.

Moeller [1]) mentions that in Asclepiadeae the secondary cork tissue originates in the epidermis or in the adjoining cortical layer. The structure of the drug, examined here, quite coincides with the results mentioned above. Under the phelloderm there occur some 8 or 10 layers of cortex-like parenchyma cells and within these a zone containing many bundles of fibres and isolated fibres, resembling those commonly to be met with in the pericycle.

Therefore we will describe the drug, starting from the supposition that the above mentioned results of Treiber and Moeller are correct.

E p i d e r m i s. Thrown off in the drug.
C o r t e x.

Secondary cork tissue proper. Initial celled cork.

P h e l l e m. Consisting of up to 100 layers of flattened periderm cells, more and less flattened portions alternating in a radial direction, especially in the vicinity of a lenticel. Cells showing a strict radial arrangement, but not arranged in tangential layers. In many places in transverse sections the phellem showing lenticular portions of tissue — R. 175—200 μ, T. about 1 m.M. — consisting of cells arranged in the same radial rows with the periderm cells, but resembling in some respects — e.g. in the occurrence of simple crystals — phelloderm cells, though sometimes much larger in a radial direction. This structure perhaps the result of a rhytidoma-like development of cambium in the phelloderm, the latter being thinner in places corresponding with the lenticular portions.

P e r i d e r m c e l l s. T. 10—30 μ, L. 18—30 μ; penta- or hexagonal prisms with a radially directed axis. Walls dingy brown and generally lignified. Cell contents: in the most flattened cells a dark black-brown mass. **C e l l s o f l e n t i c u l a r p o r t i o n s.** R. 20—40 μ, T. 10—40 μ; rectangular in a transverse section. Walls somewhat thicker and often somewhat collen chymatous, more feebly coloured than those of the periderm

Fig. 5. *Marsdenia spec.* Bark, transverse section. coll Collenchyma; lt Non-articulated laticiferous tubes; par Common parenchyma; perf Sclerenchyma fibres of pericycle; perp Pericyclic parenchyma; phg Phellogen; phm Phellem; scl Clusters of sclereids in secondary phloem.

---

[1]) l. c. 170.

cells; intercellular spaces wanting. Contents of the phelloderm-like cells sometimes a simple crystal; for the rest wanting.

Phellogen. Some layers of very flattened cells.

Phelloderm. Consisting of 8—10 layers of radially arranged common parenchyma cells. The portions of the phelloderm, corresponding with the lenticular portions of the phellem, mentioned above, thick only 3—4 layers of cells. Cells of the above. R. 10—25 μ, T. 15—40 μ, L. 12—25 μ; penta- or hexagonal prisms with a radially directed axis. Walls somewhat thickened, often somewhat collenchymatous; often lignified and in these cells the simple crystals in a lignified crystal skin; intercellular spaces wanting. Cell contents: many starch grains; in many cells a simple, often perforated crystal in a crystal skin.

Lenticels. The periderm surrounding the lenticels much thicker; its flattened cell layers corresponding to the closing bands of the lenticels. The phellem of the lenticel — corresponding inward to an area of the primary cortex, showing common parenchyma in stead of collenchyma — thicker than that of the cork tissue proper and consisting of alternating layers of complementary cells and closing bands, each up to 10 in number but often less by the disappearance of the outer layers. Layers of complementary cells thick up to 12 cells and having two or more times the thickness of closing bands; the latter consisting of 1—4 layers of more or less flattened cells. Phelloderm thick about 12 cell layers. Complementary cells. R. 15—35 μ, T. 25—35 μ, L. often 20 μ; rounded in shape. Walls very thin, somewhat dingy brown; intercellular spaces very large. Cell contents: mostly wanting, sometimes a yellow mass. Cells of closing bands. T. 10—35 μ; flattened in a radial direction. Walls dingy brown. Cell contents: in most cells a dark brown mass. Phelloderm cells. R. 15—35 μ, T. 20—30 μ, L. 25—30 μ; hexagonal prisms with a radially directed axis and rounded edges. Walls not thickened; colourless; with large intercellular spaces. Cell contents: in most cells a cluster crystal.

Primary cortex proper.

Collenchyma. Thick some 3 cell layers; wanting under the lenticels; with scattered idioblasts containing each a cluster crystal in a crystal skin and in its inner part some non-articulated laticiferous tubes. Collenchyma cells. R. 20—25 μ, T. 30—50 μ, L. about 50 μ; penta- to octogonal prisms with a longitudinally directed axis. Walls thick about 3 μ; sometimes somewhat yellow; pitted. Cell contents: starch grains. Laticiferous tubes: see under secondary phloem.

Common parenchyma. Thick 8—10 layers of cells, some divided by 1 or 2 thinner radial walls; many crystal idioblasts, often divided by 1 or 2 thinner radial walls, each cell containing 1 large cluster crystal in a crystal skin, filling almost the whole cell cavity; non-articulated laticiferous tubes fairly numerous. Common parenchyma cells. R. 30—40 μ, T. 50—110 μ, L. 30—50 μ; polyhedra with often strongly rounded edges. Walls thick about 1.5 μ, sometimes with slight collenchymatous thickening; with many intercellular spaces. Cell contents: many starch grains, simple and compound, up to 5-adelphous. Laticiferous tubes: see secondary phloem.

Endodermis. Developed as a starch-sheath; in many places very distinct, at least in radial section; consisting of 1 cell layer; in other places not to be distinguished from the surrounding parenchyma.

C e l l s o f t h e a b o v e. R. 30—50 μ, L. 35—50 μ; penta- to octogonal prisms with a longitudinally directed axis and rounded edges. Walls thick 2 μ; transverse walls pitted; with intercellular spaces when adjoining other cells. Cell contents: starch grains as in the surrounding parenchyma, but larger and up to 6-adelphous.

S t e l e.

Pericycle. Chiefly consisting of common parenchyma, thick about 7 cells, more or less regularly arranged in transverse layers; the cells often divided by 1 or 2 thinner radial walls. Crystal idioblasts and non-articulated laticiferous tubes like in the cortical parenchyma. Numerous sclerenchyma fibres, isolated and in bundles, and constituting together a tangential reticulum; the bundles consisting of up to 15 elements in a transverse section.

P a r e n c h y m a c e l l s. The same as in the cortex, only of somewhat smaller dimensions. L a t i c i f e r o u s t u b e s. See secondary phloem. S c l e r e n c h y m a f i b r e s. R. 18—25 μ, T. 25—30 μ; the isolated ones circular or elliptical, those of the bundles angular when adjoining each other, rounded when adjoining parenchyma cells. Walls very much thickened, the cavity having quite disappeared or being extremely small; not lignified; in potassium iodide iodine red, in iodine and sulphuric acid 66 per cent. lighter coloured than the parenchyma cell walls; showing stratification; pit canals wanting.

Phloem.

P r i m a r y p h l o e m not to be distinguished.

S e c o n d a r y p h l o e m. Constituting the largest portion of the drug, showing a radial arrangement of elements only in its innermost part.

Cribral system. Constituting in the innermost parts of the bark the ground mass of the tissue between the medullary rays; outward becoming very soon flattened and constituting irregular radial bands of keratenchyma, the cells of the surrounding bast parenchyma becoming larger and taking its place. Companion cells wanting. **Parenchymatic system.** Represented by bast parenchyma and medullary rays. B a s t p a r e n c h y m a regular, with clusters of sclereids and latex tubes. Bast parenchyma fibres. The dimensions of the cells much increasing towards the outer part of the secondary phloem, especially in a tangential direction; some cells in the outermost part divided by 1—4 thinner radial walls. Clusters of sclereids occurring in the largest number in the middle part of the secondary phloem; more or less arranged in abrupt tangential layers; often some arranged in longitudinal rows and separated from each other by some bast parenchyma mixed up with laticiferous tubes. Clusters much varying in size, the longitudinal dimension being generally the largest; consisting of a few up to a very large number of sclereids — the largest R. and T. 300 μ, L. up to 1700 μ. Non-articulated laticiferous tubes [1]) very numerous, often branching in a tan-

---

[1]) d e B a r y. Vergl. Anat. 1877. 195; English edition 187.

gential plane, often adjoining clusters of sclereids; in tangential sections here and there a blind end. The surrounding parenchyma cells having a somewhat modified shape. M e d u l l a r y r a y s only to be distinguished in the inner part of the secondary phloem; outward the cells growing out irregularly in a tangential direction. Separated from each other by 2 or 3 layers of elements; uni- sometimes biseriate, 5—25 cells in height. Many idioblasts each containing a single cluster crystal in a crystal skin or divided by a transverse wall and containing a crystal in each division. Sometimes clusters of sclereids interrupting the course of the medullary rays.

S i e v e-t u b e s. R. 25 μ, T. 20 μ, L. of articulations 170—270 μ; tetra- to octogonal. Transverse walls placed obliquely on the radial side walls; sieve-plates distinct, callus plates now and then. Walls thick 1—2 μ. B a s t p a r e n c h y m a f i b r e s. In the vicinity of the cambium R. 20 μ, T. 18 μ, in the middle of the secondary phloem R. 30—40 μ, T. 40—55 μ, near the pericycle R. 30—40 μ, T. up to 150 μ; of all cells L. about 90 μ; tetra- to octogonal. Cell contents: starch grains. S c l e r e i d s. R. 30—100 μ, T. 65—100 μ, L. 25—100 μ; polyhedra, the side walls adjoining soft tissue curved, convex or concave. Walls thick up to 20 μ; yellow; lignified; showing distinct stratification and numerous pit canals. L a t i c i f e r o u s t u b e s non articulated; R. 20—40 μ, T. 30—60 μ; hexa- to decagonal with rounded edges. Walls often somewhat swollen, thick 3 μ; somewhat blue in potassium iodide iodine. Contents: a dingy brown, finely granular mass. C e l l s o f m e d u l l a r y r a y s. R. 20—35 μ, T. 15—25 μ, L. 10—45 μ; tetragonal prisms with a radially directed axis and rounded edges. Cell contents: starch grains and often a cluster crystal.

October 1912.   J; M.

CORTEX CUSPARIAE.

Cusparia Bark. Cortex Angosturae. C. A. verus. Angostura Bark. Carony Bark.
The dried bark of Cusparia febrifuga, Humb. Tabl. Géogr. ex DC. in Mém. Mus. Par. IX. 144.

Macroscopic characters.

Flat or transversely curved pieces, sometimes longitudinally curved outwards; long up to 13 c.M., wide up to 5.5 often 3 c.M., thick up to 3, mostly 2 m.M. Hard and brittle. Outer surface dull, uneven owing to the presence of very numerous generally flat warts of greatly varying dimensions; thicker pieces of bark showing a velutinous, uneven surface; greyish, the velutinous parts somewhat lighter coloured; sometimes covered with lichens. Inner surface smooth; somewhat dull; showing fine, short, longitudinal, glittering striae; sometimes slightly exfoliating; light brown; with lighter coloured, undulating, transverse bands, up to 1 m.M. in height and at a distance from each other of up to 2 m.M.; strips of yellowish wood occasionally attached to it. Transverse fracture smooth, showing especially in the outer part numerous white glittering points; the thin outer layer whitish, followed by a thin brown layer; the much thicker inner part showing wedge-like figures, alternately turning their edges in- and outward, the former being of a lighter colour. Transverse section showing the same characters as the transverse fracture, only somewhat less distinct. Odour musty and somewhat nauseous; taste very bitter.

Anatomical characters.

LITERATURE. Berg. Anat. Atl. 1865. 73. Douliot. Recherches sur le Périderme. Ann. d. Sc. nat. Série 7. T. 10. 1889. 347. Flückiger & Hanbury. Pharmacographia. 1879. 107. Gamper. Beitr. z. Kenntnis der Angosturarinden. Diss. Zürich 1900. 27—31. Greenish & Collin. Anat. Atl. o. veget. powders. 1904. 198. Hartwich u. Gamper. Beitr. z. Kenntnis d. Angosturarinden. Archiv d. Pharmacie. Bd. 238. 1900. 573. Hager. Pharm. Praxis. 1. 1900. 309. Marmé. Pharmacogn. 1886. 129. Mitlacher. Toxik. od. Forens. wicht. Pfl. u. Drogen. 1904. 136. Moeller. Anat. d. Baumrinden. 1882. 330. Oberlin et Schlagdenhauffen. Étude hist. et chim. d. diff. écorces de la fam. d. Diosmées. Journ. d. Pharm. et d. Chimie. Sér. 4. T. 28. 1878. 228. Oudemans. Aanteek. 1854—56. 543. Oudemans. Pharmacogn. 1880. 203. Planchon et Collin. Drogues simples. T. II. 1896. 614. Solereder. Syst. Anat. 1899. 204. Solereder. Syst. Anat. Ergänzungsband. 1908. 65. Wigand. Pharmakogn. 1879. 151. MATERIAL. The drug, pieces thick 2 and 3 m.M. REAGENTS. Water, glycerine, potassium iodide iodine, phloroglucin and hydrochloric acid, iodine and sulphuric acid 66 per cent., potash.

MICROGRAPHY. Young branches of this plant not being available, we had no means to determine in what part of the stem the first secondary cork tissue is developed. But probably this takes place in the outermost cell layer of the cortex or in the epidermis, this being the case in all other Rutaceae hitherto examined, according to Douliot [1]), Moeller [2]) and Solereder [3]). The bark described here however consists only of secondary phloem. Thus there must have taken place formation of rhytidoma, to which the outer cork layer here present belongs. In a few cases some instances of the formation of scaly rhytidoma, extending into the secondary phloem proper, could be observed.

Secondary phloem.

Rhytidoma. Probably annular, only the innermost cork layer remaining.

Secondary cork tissue. Lenticels wanting.

Phellem. Thick in the parts, showing a velutinous outer surface, about 2 m.M., for the rest about 0.5 m.M. The anatomical structure in both cases somewhat different and to be described separately; but here and there, in the thinner phellem, parts occurring with the same structure as that of the thicker phellem. The thicker phellem consisting of radially arranged periderm cells, all thin-walled, but of different radial dimensions; thus alternating layers being formed, thick mostly 2 or 3 cells shorter in a radial direction and 3 to 5 cells of about 5 times larger radial dimensions. These layers showing longitudinal undulating folds; the layers, consisting of cells shorter in a radial direction, anastomosing or sometimes 2 fusing together. The thinner phellem consisting of radially arranged periderm cells, partly thin-walled and partly sclerotic and somewhat shorter radial dimensions; thus alternating layers being formed, thick 1—5 thin-walled and 1 or 2 thick-walled cells. The transition from the thinner phellem into the thicker parts, mentioned above, somewhat abrupt, often the thicker parts

---

[1]) l. c. 347.
[2]) l. c. 392.
[3]) l. c. 204.

being detached from the surrounding thinner phellem; no correspondence occurring between the sclerotic layers of the thinner and thicker phellem.

Phellogen. One or 2 layers of cells; showing longitudinal undulating folds.

Phelloderm. Consisting of 3 to 12 layers of common parenchyma cells; the innermost one or two layers in some cases here and there showing groups of sclereids. Moreover here and there idioblasts, containing either raphides or sometimes simple crystals, both directed tangentially; a few other idioblasts being mucilage cells, the latter to be fully described under the head secondary phloem proper. The phelloderm corresponding in place with the thicker phellem, described above, showing the following differences with that under the thinner phellem: generally a larger number of cells in a radial direction, sometimes larger radial dimensions of the same, a longitudinally undulatingly folded outer surface and in the principal folds the arrangement of the cells often resembling that of storied cork, moreover the intercellular spaces being somewhat more numerous there.

P e r i d e r m c e l l s. Thicker phellem, cells with shorter radial dimensions: R. 10—15 μ, T. 25—30 μ, L. 15—30 μ; mostly hexagonal prisms with a radially directed axis. Walls thick 1 μ; yellow; lignified; intercellular spaces wanting. Cell contents: sometimes a yellow mass. Cells with larger radial dimensions: R. 30—60 μ. Walls lighter yellow. Cell contents: wanting. For the rest the same as the radially shorter cells. Thinner phellem, thin-walled cells: R. 18—30 μ, T. 18—40 μ; mostly hexagonal prisms with a radially directed axis. Walls thick 1 μ; yellow; lignified; intercellular spaces wanting. Cell contents: a granular yellow mass. S c l e r o t i c c e l l s. R. 10—25 μ. Walls thickened, sometimes on all sides, sometimes especially the inner and radial ones; often pitted. For the rest the same as the thin-walled cells. P h e l l o d e r m c e l l s. C o m m o n p a r e n c h y m a c e l l s. R. 15—30 μ, T. 18—35 μ, L. 15—30 μ; mostly hexagonal prisms with a radially directed axis and often rounded edges. Walls thick 1 μ; intercellular spaces often present. S c l e r e i d s. Walls thick 4 μ; lignified; pitted. See for the rest the common parenchyma cells. R a p h i d e c e l l s. R. 25—32 μ, T. 30—40 μ; prisms to ellipsoids. Walls somewhat swollen. Cell contents: many thin raphides. I d i o b l a s t s w i t h s i m p l e c r y s t a l s. Excepting their contents, in all respects the same as the common parenchyma cells.

In a few cases here and there some scaly rhytidoma, resembling lenticels. The scales having the following dimensions: R. about 300 μ, T. about 1.5 m.M., and consisting of a small portion of secondary phloem proper, somewhat protruding from the surface and cut off from the rest by a layer of secondary cork tissue, resembling in every respect that described above. The portions of secondary phloem showing yellow cell walls and contents; see for the rest the secondary phloem proper.

Secondary phloem proper.

T a n g e n t i a l e x p a n s i o n, necessary for keeping up with the increase in thickness of the branch, principally due to division and growth of cells belonging to the medullary rays; these cell divisions restricted to one of the two radial layers, constituting the mostly 2-seriate medullary, rays and amounting towards the outer part of the secondary phloem to the number of 7 or more.

The tangential extension by growth amounting for each partition to 10 times the tangential dimension of the original undivided cells; thus by the combined action of division and growth a total extension of 70 times and even more resulting. Abundant tangential growth and cell division only occurring in those rays, extending in a peripheral direction beyond the plane occupied by the outermost bundles of bast fibres; thus these rays much widened in their outer half and becoming distinctly cuneiform; moreover these same rays being much higher towards the periphery. The bast parenchyma cells between the medullary rays also showing some slight tangential growth without cell division. Finally in the outer half of the secondary phloem in a transverse section here and there cuneiform groups of parenchyma cells, turning their edges inward and resembling in all respects the cuneiform parts of the medullary rays, but showing no correspondence with these. These parts in all probability however being only transverse sections of the upper and lower portions of the higher outer parts of the medullary rays mentioned above.

R a d i a l   a r r a n g e m e n t of the elements very distinct in the inner part of the secondary phloem, becoming less distinct outward, being lost in the outermost part in consequence of the subsequent growth of the elements.

T a n g e n t i a l   a r r a n g e m e n t showing layers of bast parenchyma thick about 3 elements and alternating with layers of cribral elements, flattened already at a small distance from the cambium and surrounded by parenchyma cells, often containing a long prismatic simple crystal; the latter layers thick about 3—5 elements. This tangential arrangement distinct through the whole of the secondary phloem, but most striking where the medullary rays show an important tangential growth. Bundles of bast fibres, lying at a distance of about 500—700 µ from the secondary cork, isolated but lying in a tangential plane; the same phenomenon in some cases being repeated a few times at the same distance inward. C r i b r a l   s y s t e m consisting of sieve-tubes somewhat thick-walled and a few companion cells; see for the rest the tangential arrangement.

S y s t e m   o f   b a s t   f i b r e s represented by bundles thick 17—50 fibres, showing about the same number in a radial and in a tangential direction and being connected together by single fibres, thus in a transverse section an isolated fibre here and there appearing; see for the rest the tangential arrangement.

B a s t   f i b r e s. R. 10—15 µ, T. 10—12 µ; penta- to octogonal. Walls very thick, leaving almost no cavity; yellow; with a distinct middle lamella more lignified than the rest of the wall; generally the walls of the fibres lying at the circumference of a bundle being somewhat more lignified.

P a r e n c h y m a t i c   s y s t e m consisting of bast parenchyma and medullary rays. **Bast parenchyma** regular, with idioblasts in 3 kinds: 1. Mucilage cells[1]) isolated; numerous; lying in the tangential bands of not flattened paren-

---

[1]) By most authors these cells are described as oil cells. As however we have found, neither in literature, nor in the facts observed by us, a single argument supporting this opinion and as moreover the drug has by no means an aromatic odour or taste, whilst we observed a mucilage layer in some of these cells we have thought it better to call them mucilage cells.

chyma, mentioned above; much larger than the common parenchyma cells, moreover their dimensions becoming larger outward; the greater part of these cells having no contents, but some, especially in the vicinity of the cambium, showing a mucilaginous layer adjoining the wall. 2. Crystal cells containing each 1 simple elongated prismatic crystal; always 2 or more of these cells in a longitudinal row and accompanying the elements of the cribral system as described above. These cells being very rare in the outer part of the bark. 3. Raphide cells numerous, isolated, lying in the tangential bands consisting of bast parenchyma only and somewhat larger than the surrounding common bast parenchyma cells; the raphides longitudinally directed. **Medullary rays** separated from each other in a tangential direction by 2—15 elements. In the intervals between the much widened rays reaching to the periphery as described above, a varying number of not much widened medullary rays, often more or less curved towards each other in their outer parts. In the inner half of the secondary phloem 1- to 3-, mostly 2-seriate; the 1-seriate generally only a few—up to 5—cells in height; the 2- and 3-seriate up to 20, generally 10 to 15 cells in height. Consisting of procumbent cells, only the upper- and undermost rows of upright cells; moreover idioblasts, viz. raphide cells with radially directed raphides in the inner half of the secondary phloem and large mucilage cells, especially in the upper- and undermost cell rows, these being much larger towards the periphery, especially in the much widened parts of the rays and here in a tangential direction.

Cells of regular parenchyma. Common parenchyma cells. R. 15—30 μ, near the cambium 10—15 μ; T. 15—70 μ, near the cambium 10—20 μ; L. 50—115 μ; mostly tetragonal prisms with a longitudinally directed axis and edges rounded a little or not at all. Walls thick 1 μ; with small intercellular spaces. Contents: often a yellowish green granular mass, generally containing starch grains. Mucilage cells. R. 45—60 μ, T. 50—85 μ, L. 45—50 μ; ellipsoids. Walls somewhat swollen. Contents: in some cases a mucilage layer covering the wall and not or only slightly coloured in potassium iodide iodine. Cells containing a simple crystal. Crystals long up to 100 μ, thick up to 8 μ, with cuneiform ends. Raphide cells. R. 40—45 μ, T. 22—90 μ, L. 105—200 μ; somewhat ellipsoidical. Quite filled up with many raphides, thick up to 1 or 2 μ. Cells of medullary rays. Procumbent cells. R. 20—35 μ, T. near the cambium 12 μ, not far from the cuneiform parts of the medullary rays 30 μ, in the cuneiform parts up to 120 μ, L. 12—30 μ; tetra- to polygonal prisms with a radially directed axis and rounded radial edges; sometimes the outer walls of the cells, covering the lateral sides of the medullary rays, curved outward. Walls thick 1 μ; sometimes slightly pitted; intercellular spaces small, occurring not only between the cells of the rays, but also between these and the surrounding cells. Contents: often starch grains and a somewhat greenish yellow mass. Upright cells. R. 10—30 μ, L. 28—30 μ. Walls, especially the tangential, somewhat thickened; tangential walls distinctly pitted. See for the rest the procumbent cells. Raphide cells often somewhat larger than the surrounding cells; raphide bundles generally radially arranged. Mucilage cells. R. 40—70 μ, T. 50—120 μ, L. 40—90 μ; ellipsoidical or spherical. Walls somewhat swollen. Cell contents: sometimes a mucilaginous layer covering the wall.

January 1914.   J; M.

CORTEX FRUCTUS AURANTII.

Cortex Aurantii Amari. Bitter Orange Peel.

The dried two outermost layers of the ripe fruit of Citrus vulgaris, Risso, in Ann. Mus. Par. XX.
190, divided into 4 nearly elliptical pieces.

Macroscopic characters.

Pieces long 5—8 c.M., wide 3—5 c.M., thick 2—5 m.M.; on the outer surface yellowish brown, wrinkled to finely furrowed, on the inside spongy white. Odour aromatic; taste aromatic and bitter.

Anatomical characters.

LITERATURE. Berg. Anat. Atl. 1865. 89. Pl. XXXXV. Benecke. Mikr. Drogenprakt. 1912. 39. Biermann. Entwickl. gesch. d. Früchte v. Citrus vulgaris Risso. Diss. Bern. 1896. Flückiger. Pharmakogn. 1891. 837. Flückiger & Hanbury. Pharmacogr. 1879. 124. Gilg. Pharmacogn. 1910. 186. Hérail. Pharmacogr. 1912. 542. Koch. Mikr. Anal. d. Drogenpulver. Bd. I. 1901. 56. Pl. I. Kraemer. Botany a. Pharmacogn. 1910. 175. Marmé. Pharmacogn. 1886. 301. Meyer. Wiss. Drogenk. Bd. II. 1891. 408. Mika. Beitr. z. Morphol. u. mikr. Nachw. d. Hesperidins. Maj. Növenytany hup. I. 93. Summary in Just. 1878. Bd. I. 20. Molisch. Mikrochemie d. Pfl. 1913. 164. Oudemans. Pharmacogn. 1880. 374. Planchon et Collin. Drogues simples. T. II. 1896. 653. Pfeffer. Hesperidin, ein Bestandt. ein. Hesperideen. Bot. Ztg. 1874. 529. Sieck. Schizolysig. Sekretbeh. Diss. Bern. 1895. 26. See also Pringsheims Jahrb. 1895. 222. Tschirch u. Oesterle. Anat. Atl. 1900. 70. Pl. 69. Tschirch. Harze u. Harzbeh. 1900. 370, 371, 372, 381. Tschirch. Pharmakogn. Bd. II. 1911. 858. Tunmann. Pfl. microchemie. 1913. 150,369. MATERIAL. The drug. Ripe and unripe fruits, collected in the Botanic Garden at Groningen in the year 1891, in alcohol; fresh fruits of Citrus vulgaris Risso. REAGENTS. Water, glycerine, potash, phloroglucin and hydrochloric acid, iodine and sulphuric acid 66 per cent., concentrated sulphuric acid, Schulze's macerating mixture, potassium dichromate, copper acetate and iron acetate, nitric acid.

MICROGRAPHY.

Wall of the pericarp without the pulp.

Epidermis outer side. Cells showing sometimes a tangential partition wall; growth of the epidermis by cell division apparent; in the full grown fruit cells often forming groups of 2 belonging together, in the not full grown

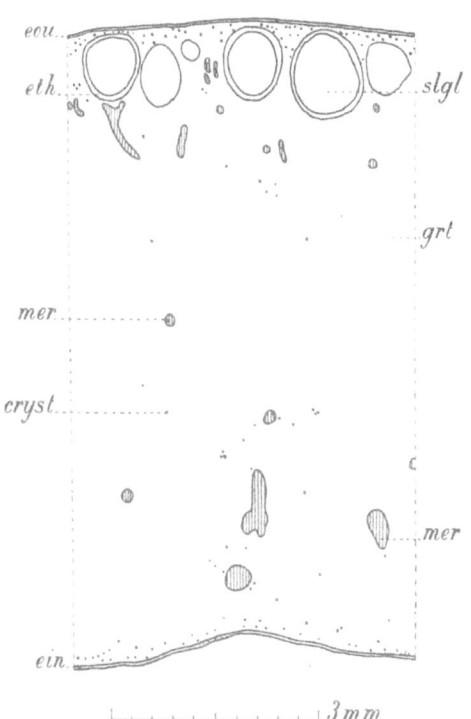

Fig. 6. *Citrus vulgaris.* Alcoholic material of the fruit, transverse section. cryst Crystal idioblasts; ein Epidermis inner side; eou Epidermis outer side; eth Epithelium of glands; grt Ground tissue; mer Meristeles; slgl Schizo-lysigenous glands.

fruit groups of 4. Above the glands cells somewhat larger in tangential directions. Stomata 10 to the square m.M.; phaneroporous; lying in the same level as the epidermal cells; surrounded by mostly 7 subsidiary cells, those cells often showing a tangential partition wall.

E p i d e r m a l   c e l l s   p r o p e r. R., T. and L. 10 μ; polygonal tables. Outer walls thickened, here and there lateral walls strongly thickened; outer walls cuticularized, lateral walls showing a cuneiform cuticularized part. Cell contents: chromoplasts, orange-red, but not coloured above the glands; small oil drops [1]). S t o m a t a. 22 μ by 20 μ. Outer and inner walls of the guard-cells very thick.

Ground tissue. Chiefly consisting of spongy parenchyma; intercellular spaces extending into all directions; the innermost part of this tissue in the drug torn by the peeling of the fruit. Towards the outer epidermis this tissue loosing its spongy character, the cells becoming smaller and polyhedral with rounded edges, nearest to the epidermis edges least rounded; intercellular spaces. Outermost cell layers containing very numerous orange-red chromoplasts; especially in these layers crystal idioblasts. In the outermost layer or layers here and there scales of secondary periderm, the epidermis bursting open on the lateral sides of the scales. Glands: schizo-lysigenous [2]); often impinging on the epidermis; ovate; R. up to 1 m.M., T. and L. 800 μ; showing an epithelium of some layers of cells. Cavity containing oil as a homogeneous mass or in larger or smaller drops; disorganized cells 30—40 μ in diameter, very thin-walled, provided with a few small pale yellow or colourless granules.

C e l l s   o f   t h e   l a y e r s   n e a r e s t   t h e   o u t e r   e p i d e r m i s. Walls pitted. Cell contents: chromoplasts mentioned above. C r y s t a l   i d i o b l a s t s. Containing simple crystals, sometimes with curved sides and surrounded by a crystal skin. In alcoholic material hesperidine throughout the whole ground tissue, nearly wanting in the drug; in sulphuric acid its sphero-crystals fusing into a yellow liquid; in nitric acid their structure showing up more clearly. E p i t h e l i u m   c e l l s. Thick 20 μ, wide 80 by 100 μ. Walls thickened; pitted.

Meristeles. Consisting of a pericyclic sheath of parenchyma and a collateral closed vascular bundle. Phloem containing crystal fibres, sometimes composed of 20 cells, each containing a simple crystal. Xylem consisting of annular, spiral and pitted vessels.

P a r e n c h y m a   c e l l s   o f   s h e a t h. R. and T. 20—25 μ, L. 60—80 μ; polygonal prisms with a longitudinally directed axis. Walls somewhat thickened; pitted. C o m-p o n e n t   c e l l s   o f   c r y s t a l   f i b r e s. Long 5 μ, wide 10 μ. X y l e m   v e s s e l s. Transverse walls very oblique and showing bordered pits; walls showing lignification.

Epidermis inner side. Only to be seen in the alcoholic material. Cells arranged in tangential rows; tetragonal tables or fibre-shaped. Sometimes a fibre with thickened, yellow, lignified walls, showing slit-like pits. |

C e l l s   o f   t h e   a b o v e. R. 20 μ, T. 10 μ, L. 150—300 μ. Outer walls somewhat thickened; showing a cuticle.

Micrography of the powder. Colourless parts of the spongy parenchyma

---

[1]) See for the character of the oil T s c h i r c h, Harze und Harzbeh. 381.
[2]) See T s c h i r c h   u n d   O e s t e r l e, l. c. 300 and S i e c k, l. c. 28 or 224. In both cases the leaves only were investigated.

of the innermost layer, often showing not even cell structure. Common parenchyma of the outermost yellowish brown layer, often flattened and containing tetra- to hexagonal oxalate crystals. Epidermis consisting of tetra- to hexagonal cells with a sharp outline, often divided into 2 by a thin wall, brownish with thickened outer walls. Annular and spiral vessels. Schizo-lysigenous glands not to be seen. Very fine starch grains to very small amount to be found in all the parenchyma cells; in the parenchyma of the flavedo sometimes cells with many starch grains.

March 1902.   J; M; L.

CORTEX GRANATI.
Pome granate Bark.
The dried bark of root and stem of Punica Granatum, Linn. Sp. Pl. 472.

Macroscopic characters.

Root bark: channelled, sometimes only somewhat curved pieces of very varying and irregular shape, or quills; up to 7 c.M. in length, up to 2.5 c.M. in width; bark up to 3 m.M. thick, hard and brittle. Outer surface yellow-brown, often with local scaling of the cork layer, thence the larger pieces having an uneven surface; after removal of the cork layer light brown, not green. Inner surface more yellowish, also often with dark longitudinal stripes, rough by likewise longitudinal prominent stripes. Transverse section pale yellow; with indistinct, fine, radial striations. Transverse fracture pale brownish yellow, even, dull, mealy. Bark of stems and branches: more quills than channelled pieces; up to 15 c.M. in length. Outer surface grey-brown; longitudinally wrinkled; with many longitudinally elongated, yellow-brown lenticels, often irregularly arranged in longitudinal rows; generally bearing lichens and black dots, the fructifications of lichens or fungi; after careful removal of the cork layer a yellowish green layer appearing. Inner surface less rough.

Anatomical characters.

LITERATURE. De Bary. Vergl. Anat. 1877. 546, 556 u. 575. Berg. Anat. Atl. 1865. 79. Taf. 40. Deutsches Arzneibuch. 1910. 140. Flückiger. Pharmakogn. 1891. 515. Flückiger & Hanbury. Pharmacographia. 1879. 291. Gilg. Pharmakogn. 1910. 234. Greenish & Collin. Anat. Atl. o. veget. powders. 1904. 202. Günther. Beitr. z. Anat. d. Myrtifloren mit besonderer Berücksichtigung d. Lythraceae. Diss. Breslau. 1905. 24. Hager. Pharm. Praxis. Bd. 1. 1900. 1248. Hérail.Pharmacol. 1900. 753. Karsten u. Oltmans. Pharmakogn. 1909. 135. Koch. Mikr. Anal. d. Drogenpulver. Bd. 1. 1901. 103. Koch. Einf. i. d. mikr. Anal. d. Drogenpulver. 1906. 21. Koch. Pharmakogn. Atlas. Bd. 1. Lief. 2. 1910. 39. Koch u. Gilg. Pharmakogn, Praktik. 1907. 33. Kraemer. Botany and Pharmacogn. 1910. 535. Luerssen. Syst. Bot. Bd. 2. 1882. 823. Marmé. Pharmacogn. 1886. 135. Meyer. Wiss. Drogenk. Bd. 2. 1892. 134. Meyer. Mikr. Unters. v. Pflanzenpulv. 1901. 181. Moeller. Anat. d. Baumrinden 1882. 353. Moeller. Pharmakogn. 1889. 236. Moeller. Mikr.-Pharm. Ueb. 1901. 256. Oudemans. Pharmacogn. 1880. 230. Planchon et Collin. Drogues simples. T. 2. 1896. 325. Schneider. Powdered veget. drugs. 1902. 205. Tschirch u. Oesterle. Anat. Atl. 1900. 83. Taf. 21. Wigand. Pharmakogn. 1879. 156. Wittlin. Ueb. d. Bildung d. Kalk-oxalattaschen. Bot. Centralbl. Bd. 67. 1896. 100; the same in Diss. Bern. 1896. 20.

MATERIAL. The drug; the root bark thick 1—2 m.M., that of stem and branches thick 1 m.M. Young roots, thick about 0,8 m.M., collected in the Botanic Garden at Groningen Nov. 5th 1912 and preserved in alcohol. Young branches, thick about 1 m.M., collected in the Botanic Garden at Groningen June 21th 1911 and preserved in alcohol. RE-AGENTS. Water, glycerine, potassium iodide iodine, phloroglucin and hydrochloric acid, iodine and sulphuric acid 66 per cent., ferric acetate.

MICROGRAPHY.

OF THE ROOT BARK.

E p i d e r m i s. Present in the young root thick 0.8 m.M., thrown off in the drug.

C o r t e x. Consisting of 10—12 layers of cells; a primary cork layer, adjoining the epidermis, to be distinguished by its colourless cell walls and want of intercellular spaces, the latter and a brown colour of the walls being present in the adjoining layers of the cortex. The inner cell layers radially arranged. Endodermis distinct by the regular arrangement of its cells, consisting of smaller cells and having less coloured walls. Cortex quite thrown off in the drug.

S t e l e.

Pericycle. In the young root consisting of at most 3 layers of common parenchyma cells, the secondary cork tissue developing in the outermost of these. Primary phloem of tetra- to hexarch root thrown off in the drug.

The parts now to be described are present in the drug.

Secondary phloem.

R h y t i d o m a. Scaly rhytidoma, the scales being thrown off very soon, to be distinguished only here and there, thick about 250 μ and enclosing about 15 cell layers of the phloem. Secondary cork tissue running parallel to the layers of secondary phloem and only curving outward at the margin of the scales or traversing these layers obliquely. Layers of secondary phloem having brown cell walls and contents; see for the rest the secondary phloem proper. Layers of secondary cork tissue. Initial celled cork.

Secondary cork tissue proper. Phellem. Thick about 5 layers of periderm cells, showing a strict radial arrangement and more or less distinct tangential layers. The outer cells often ruptured. Phellogen. Generally 1, sometimes 2 layers of cells. Phelloderm. 1, sometimes 2 layers of common parenchyma cells, containing some starch grains; here and there an idioblast, containing a single cluster crystal.

P e r i d e r m c e l l s. R. 8—20 μ, T. and L. 18—30 μ; tetra- to octogonal prisms with a radially directed axis. Walls: inner walls much thickened and pitted, radial and transverse walls somewhat thickened towards the inner walls, outer walls thin; the outer 1 to 3 layers of periderm cells without thickening of walls and thence more liable to be ruptured; dingy yellow; all walls lignified, especially the thickening layers.

Lenticels [1]). T. 1.5 m.M., L. about 0.8 m.M.; not protruding from the surface of

---

[1]) Not numerous in the root bark.

the bark, the phellogen lying about 200 μ deeper than that of the secondary cork tissue proper. Phellem consisting of alternating layers of complementary cells and closing bands, each up to 5 in number. Closing bands thick generally 1, sometimes 2 layers of cells resembling in all respects the periderm cells of the cork tissue proper. Layers of complementary tissue thick up to 8 fairly radially arranged cells.

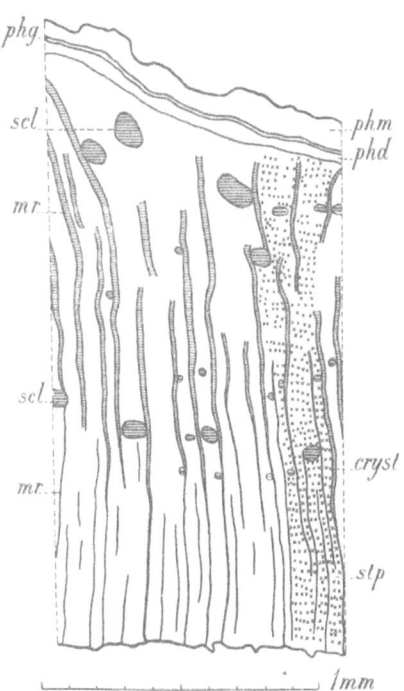

Fig. 7. *Punica Granatum*. Root bark, transverse section. cryst Crystal fibres, drawn only in a narrow radial strip of the section; mr Medullary rays; phd Phelloderm; phg Phellogen; phm Phellem; scl Sclereids; stp Sieve-tubes, companion cells and bast parenchyma.

Cells of complementary tissue. R. 15 μ, T. 25 μ, L. 20—25 μ; ellipsoids. Walls thickened, but in a much less marked manner, in the same way as those of the periderm cells proper; yellow; with very large intercellular spaces. Cell contents: brown-yellow.

Secondary phloem proper. Showing, except in the outmost portion of the bark, a radial and a very distinct tangential arrangement. The tangential growth of the outer part of the bark, in order to keep up with the increase in thickness of the root, resulting from the growth of all elements of the parenchymatic system, accompanied only by comparatively rare thin radial walls, especially in the cells of the medullary rays. Under the rare lenticels moreover cuneiform strips of parenchyma being developed in the same way as those occurring in the stem bark and fully described there. The secondary phloem consisting for the greater part of tissues belonging to the parenchymatic system; the system of bast fibres wanting and the cribral system constituting in the vicinity of the cambium only ¹/₅th of the bark and moreover being soon flattened. **Cribal system** more or less distinctly arranged in tangential layers, thick 1 element, soon flattened into tangential layers of keratenchyma. Consisting of sieve-tubes, companion cells appearing only here and there. **Parenchymatic system.** Consisting of bast parenchyma and medullary rays. Bast parenchyma consisting of: 1. Bast parenchyma fibres, more or less distinctly arranged in tangential layers, thick 1 or 2 cells; 2. Crystal fibres, most numerous and arranged in tangential layers, nearly always thick 1 element; divided into about 15 short cells, each containing a single cluster crystal, in a very few cases a simple crystal. The crystals only in the nearest vicinity of the cambium being somewhat smaller than the rest; 3. Sclereids, present only in the outer half of the bark, single or in longitudinally elongated clusters, consisting of up to 6 cells, often constituting together more or less distinctly a fibre. Medullary rays separated from each other in a tangential direction by 1—4, mostly 3 radial layers of elements; uni- to bi-, mostly uniseriate; the uniseriate ones being 1—8 cells in height, the biseriate ones somewhat higher, e. g. 13 cells; consisting of cells having generally the same radial and longitudinal dimensions.

Sieve-tubes. R. 10 μ, T. 10—15 μ; tetra- to hexagonal. Transverse walls placed obliquely on the radial side walls; sieve-plates fairly distinct. Companion cells. R. and T. 5 μ; tri- or tetragonal. Bast parenchyma fibres. In the vicinity of the cambium R. and T. 10 μ, in the middle of the secondary phloem R. 10 μ, T. 20 μ, near the cork tissue R. 10—15 μ, T. 20—28 μ, L. of the cells 60—80 μ, near the cork tissue about 20 μ; tetra- to hexagonal. Intercellular spaces only in the vicinity of the cork tissue. Cell contents: simple and compound, often 2-adelphous starch grains; tannin. Crystal fibres. Near the cambium R. and T. 10—12 μ, in the middle part of the phloem R. 10—12 μ, T. 15 μ, near the cork tissue R. 10—12 μ, T. 18—25 μ, L. of the cells 10—20 μ; tetra- to hexagonal. Walls: transverse and radial walls thin, tangential ones somewhat thicker; intercellular spaces wanting. Cell contents: mostly a single cluster crystal in a very thin crystal skin. Sclereids. R. 20—70 μ, T. 10—60 μ, L. 15 —160 μ; polyhedra, often with convex, sometimes also concave planes. Walls very variable in thickness, often the cavity very small; showing stratification, especially the inner part; lignified; with pit canals. Cells of medullary rays. Of all cells: R. 20—30 μ, L. 18—30 μ, of the cells near the cambium T. 8—10 μ, of those in the middle part T. 15—18 μ, of those near the cork tissue T. 25 μ; tetra- to octogonal prisms with a radially directed axis. Walls: the tangential ones pitted; intercellular spaces in all directions, but only present in the outmost layers of the secondary phloem. Cell contents: simple and compound, 2-adelphous starch grains; tannin.

OF THE BARK OF STEM AND BRANCHES.

Epidermis. Present in the young branches thick about 1 m.M., thrown off in the drug. Stomata scarce, longitudinally directed. Cells arranged in longitudinal rows; very long and often 2, sometimes 3 together of a somewhat fibrous shape. Cuticle distinct and showing longitudinal striation.

Cortex. Consisting of about 5 layers of slightly elongated cells. The outer cell layer with somewhat collenchymatously thickened walls and cells of the same description constituting the body of the 4 wings present on the young branches. About 3 layers of common parenchyma cells with thin walls and large intercellular spaces; some idioblasts containing mostly a cluster crystal, sometimes a simple crystal. Endodermis developed as a starch-sheath, thick 1 cell layer of somewhat short cells. Cortex quite thrown off in the drug.

Stele.

Pericycle. In a branch thick about 1 m.M. — just developing its secondary cork tissue — consisting of bundles of sclerenchyma fibres, separated from each other by some common parenchyma; moreover 1 layer of common parenchyma cells, somewhat shorter than those of the cortex, constituting the innermost part of the pericycle and in this layer the secondary cork tissue being developed. In transverse sections some isolated sclerenchyma fibres, the bundles often thick 6 fibres, 1—3 in a radial, up to 6 in a tangential direction. During the growth in thickness of the branch, parenchyma cells growing out between the fibres; thence in more advanced stages the bundles becoming thinner and more isolated from each other. Some parenchyma cells adjoining the starch-sheath, containing the same large starch grains as the latter.

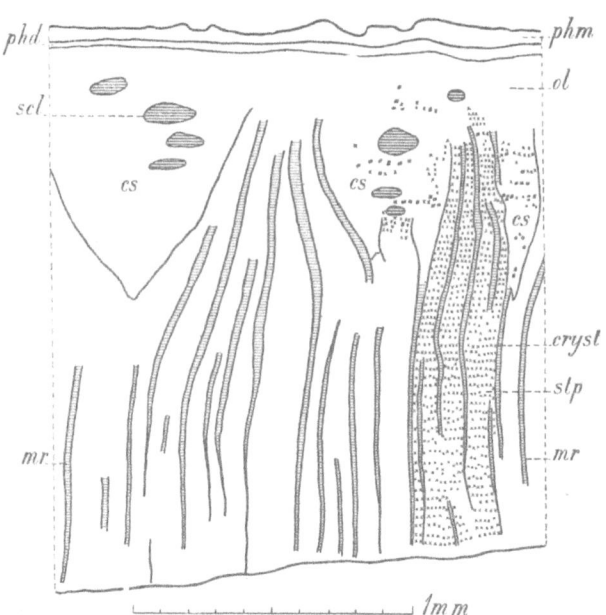

Fig. 8. *Punica Granatum.* Stem bark, tranverse section. cryst Crystal cells, drawn only in a narrow radial strip of the section; cs Cuneiform strips of parenchyma; mr Medullary rays; ol Outer layers without radial arrangement; phd Phelloderm; phm Phellem; scl Sclereids; stp Sieve-tubes, companion cells and bast parenchyma.

The parts now to be described are present in the drug.

Primary phloem. Not to be distinguished.

Secondary phloem. In very many respects resembling that of the root; the differences only mentioned here.

R h y t i d o m a. Only present in a few of the thickest specimens of the drug. Thence the secondary cork tissue of the drug generally being of pericyclic origin, for the rest resembling in all respects that of the root bark.

S e c o n d a r y   p h l o e m   p r o p e r. The tangential growth of the outer part of the secondary phloem, for keeping up with the initial strong increase in thickness of the branch, resulting in the outermost 10—15 cell layers from an increase in tangential dimension of all elements of the parenchymatic system accompanied by the formation of many radial walls, all traces of radial arrangement having disappeared here. For the rest of the secondary phloem this tangential growth resulting in part from an inconsiderable tangential dilatation of parenchyma cells, not disturbing the radial arrangement, the medullary rays becoming somewhat broader outward. This growth, in the middle portion of the secondary phloem however, resulting principally from the formation of cuneiform longitudinal strips of parenchyma, adjoining the 10—15 outer layers of cells, mentioned above and resembling them in all respects; these strips occurring at fairly regular tangential intervals, being broadest under lenticels and originating from a local tangential growth of the same kind as that of the outer 10—15 layers, covering the whole circumference, but in most cases only taking place in the bast parenchyma, in a few in the medullary rays. In the also cuneiform portions of the secondary phloem between the cuneiform strips described above, the medullary rays curved and approaching each other outward; the number of separating layers between the medullary rays more numerous inward, being 1 or 2 in the outermost part and up to 7 in the innermost part. In the outer 10—15 cell layers and in the cuneiform strips of parenchyma crystal clusters rare, and simple crystals not numerous. The sclereids in these tissues and especially in the cuneiform strips, tangentially much elongated. Chlorophyl occurring in the parenchyma tissue of the stem bark.

Micrography of the powder. Crystal clusters having a diameter of about 10 μ; isolated or in short, thin-walled, colourless parenchyma cells, each containing 1 cluster crystal, arranged in longitudinal rows, and alternating with rows of cells without crystal clusters. Starch in isolated grains or in the parenchyma cells without crystals; grains mostly simple, about globular or oblong, up to 10 μ in diameter; hilum and stratification not very apparent; sometimes compound, 2- to 4-adelphous grains. Sclereids long up to 400, broad up to 200 μ, most variable and often irregular in shape, but many somewhat spindle-shaped with acute ends; walls very thick, showing stratification and pit canals; cell cavity often insignificant. Except these large sclereids, especially root bark being present, also somewhat smaller more bar-shaped, with truncate ends. Almost colourless cork cells, penta- or hexagonal in a tangential view, with pitted tangential wall; in a side view showing a strongly thickened inner wall; cork cells and loose inner thickened layers always to be found in an isolated state. Some few simple crystals.

November 1912. J; M.

CORTEX HAMAMELIDIS.
Hamamelis Bark. Witch Hazel Bark.
The dried bark of Hamamelis virginiana, Linn. Sp. Pl. 124.

Macroscopic characters.
Flat, usually curved pieces, exhibiting now and then some wood on their inner side, long up to 20 c.M., broad up to 2 c.M., thick up to 1.5 m.M. but generally much smaller in all directions; hard and brittle. Outer surface here and there covered with lichens, dull, smooth, often with small longitudinal clefts, showing fairly numerous transversely somewhat elongated protruding lenticels, here and there united into groups; colour silvery gray or dark gray. Outer surface of pieces freed from the cork layer smooth, light red-brown. Inner surface smooth, finely striated longitudinally; colour light red-brown. Transverse fracture smooth or somewhat granular, sometimes the inner part somewhat fibrous. Transverse section in a moistened state exhibiting from the dark grey cork layer inward, a very light brown layer of the same thickness, a thinner dark brown layer and a much thicker red-brown inner layer. Odour slightly sweetish; taste astringent.

Anatomical characters.

LITERATURE. De Bary. Vergl. Anat. 1877. 149. Hager. Pharm. Praxis. Bd. 2. 1902. 14. Jensen. Üb. d. Bau d. Rinde v. Hamamelis virginica. Pharmac. Arch. 1901. n°. 7, from Just. Bot. Jahresber. Jg. 29. Abt. 2. 1901. 58. Kraemer. Botany a. Pharmacogn. 1910. 526. Planchon et Collin. Drogues simples. T. 2. 1896. 363. Reinsch. Üb. d. anat. Verhältn. d. Hamamelidaceae &c. Englers Jahrb. Bd. 11. 1890. 366. Speyer. Beitr. z. Entw.-gesch. d. Rinde pharm. Pfl. Diss. Bern. 1907. 67. MATERIAL. The drug; shoots collected in the Botanic Garden at Groningen, May 29th 1911, some weeks old, thick 2.75 m.M. REAGENTS. Water, glycerine, potassium iodide iodine, phloroglucin and hydrochloric acid, iodine and sulphuric acid 66 per cent., ferric acetate.

MICROGRAPHY.

Epidermis. Intact in the shoots some weeks old, afterwards thrown off, the cells arranged irregularly, only here and there a longitudinal group consisting of 3—5 cells. Stomata wanting. Stellate hairs 1 or 2 to the sq. m.M.; consisting of 6—8 thick-walled, conical cells, grown out from epidermal cells without the formation of a partition wall; the top sometimes somewhat curved; 100—200 μ in length, and about 20 μ thick at the base; the surrounding epidermal cells being somewhat smaller than the rest.

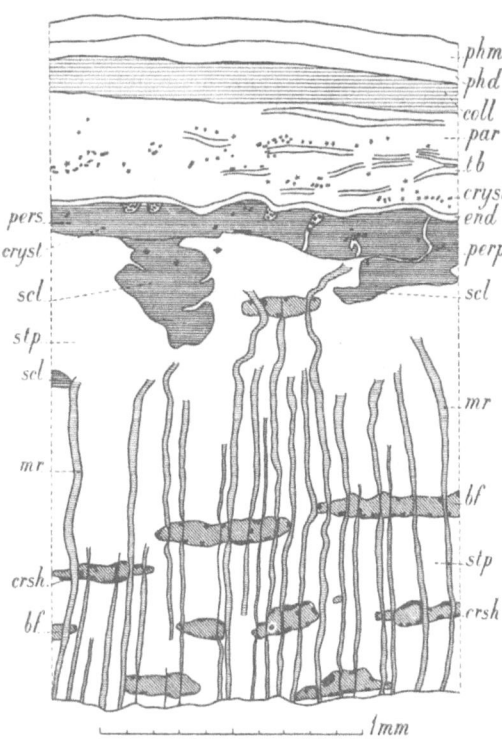

Fig. 9. *Hamamelis virginiana.* Bark, transverse section. bf Bundles of bast fibres; coll Collenchyma; crsh Crystalsheaths; cryst Crystal cells; end Endodermal crystalsheath; mr Medullary rays; par Colourless cortical parenchyma; perp Pericyclic parenchyma; pers Pericyclic sclereids; phd Phelloderm; phm Phellem; scl Clusters of sclereids in secondary phloem; stb Sieve-tubes and bast parenchyma; tb Tangential bands of thicker walled parenchyma cells.

Epidermal cells proper. R. 10—12 μ, T. 10—20 μ, L. 15—70 μ, often 50 μ; tetra- to hexagonal prisms with a radially directed axis. Outer walls much thickened, showing a thick cuticle and cuticular layers, inner walls collenchymatously thickened; side walls pitted.

Cortex.

Secondary cork tissue proper. Initial celled cork, developed from the outer layer of cortical cells directly after the completion of longitudinal growth. **Phellem.** Consisting of up to 20 layers of radially arranged periderm cells, all more or less flattened.

Cells of the above. R. 5—8 μ, T. and L. 20—30 μ; penta- or hexagonal prisms with a radially directed axis; the tangential walls curved outward. Walls generally thin, here and there a cell having thicker inner tangential and side walls; colourless; of the inner cells sometimes a little lignified. Cell contents: a red-brown homogeneous mass, filling the whole cavity; tannin.

P h e l l o g e n wanting. **P h e l l o d e r m.** Some 5 or 6 layers of radially arranged cells.

C e l l s o f t h e a b o v e. R. 6—12 μ; tangential walls not curved. Walls: see the periderm cells; the inner walls of the innermost layer of cells somewhat collenchymatously thickened. Cell contents: a few small plastids containing starch grains; tannin.

Lenticels. The anatomical structure of the lenticel extending on all sides over an area much larger than the externally visible lenticel; this area having at least the same breadth as the lenticel. The central portion of the phellem, corresponding with the lenticel proper and inward with an area of the primary cortex showing ordinary parenchyma in stead of collenchyma, much thicker, consisting of a larger number of cell layers, and sometimes more or less detached from the surrounding phellem. The corresponding central portion of the phelloderm showing a somewhat more irregular arrangement of its cells; the peripheral portions mostly consisting of a larger number of cell layers, the inner ones being collenchymatous.

P h e l l e m. Consisting of alternating layers of complementary cells and closing bands, each up to 12 in number in the central detached portion mentioned above. Bands of complementary cells having twice or more times the thickness of the closing bands; consisting of cells with thin and brown walls, being more or less flattened, especially the inner bands. Closing bands consisting of 1—4, generally 2 or 3 tangential layers of thick-walled radially arranged cells.

C e l l s o f c l o s i n g b a n d s. R. 9—12 μ, T. and L. 10—30 μ; tetra- to octogonal prisms with a radially directed axis. Walls thick 2—3 μ; colourless; distinctly pitted; intercellular spaces wanting. Cell contents: a dark red-brown mass quite filling the cavity.

P h e l l o g e n very indistinct. **P h e l l o d e r m.** Some 4—6 layers of cells. See for the rest the phelloderm of cork tissue proper; only the walls generally somewhat thicker.

Primary cortex proper.

C o l l e n c h y m a. Thick 5—6 layers of cells, wanting under the central portions of the lenticels mentioned above; more or less distinctly arranged in tangential layers; some cells divided by a thin radial wall.

C e l l s o f t h e a b o v e. R. 10—15 μ, T. 50—80 μ, L. 20—25 μ; prisms with a tangentially directed axis. Walls: the radial ones less thickened than the rest; intercellular spaces wanting. Cell contents: some plastids containing starch grains; tannin; sometimes a red-brown mass, quite filling the cavity.

C o l o u r l e s s p a r e n c h y m a. Thick some 18—20 layers of irregularly arranged cells, showing many large tangentially and longitudinally elongated intercellular spaces, the cells of the inner layers being somewhat elongated in a tangential direction; in this tissue fairly numerous, not very extensive tangential bands, generally thick 1 layer of tangentially elongated, somewhat thick-walled cells; moreover especially in the middle region of the cortex many crystal idioblasts, often arranged in tangentially elongated groups, often very thin-walled and containing mostly simple, sometimes cluster crystals, but in the youngest shoots only cluster crystals.

Cells of colourless parenchyma proper. R. 15—22 μ, T. 15—45 μ, L. 15—22 μ; spherical to ellipsoidical, mutually flattening each other when adjoining. Walls sometimes a little thickened; especially the radial walls pitted; besides the large intercellular spaces mentioned above many small ones. Cell contents: often some small plastids containing starch grains; tannin; sometimes the cells quite or partly filled with a yellow-brown mass. Cells of tangential bands mentioned above. R. 10—20 μ, T. 15—50 μ, L. 10—20 μ; tetra- to octogonal prisms with a tangentially directed axis, the transverse and tangential walls sometimes curved inward. Walls thickened; sometimes pitted; intercellular spaces almost wanting. Cell contents: tannin; generally the cavity quite filled up with a yellow-brown mass.

Endodermis. Developed in the youngest shoots examined as a distinct starch-sheath; thick almost always 1 layer of cells, some of these containing 1 or 2 simple crystals, a very few differentiated as sclereids. In the drug remaining conspicuous, almost every cell containing 1 or more simple crystals.

Cells of the above. R. 10—15 μ, T. 15—40 μ, L. 12—20 μ; polyhedra. Walls thin. Each crystal surrounded by a somewhat lignified crystal skin.

S t e l e.

Pericycle. In shoots some weeks old chiefly consisting of longitudinal, tangentially elongated bundles of sclerenchyma fibres, thick in a radial direction 5—7, in a tangential 3—20 elements; these bundles having a smooth outer and a more or less uneven inner surface and containing inside here and there a parenchyma cell, a few times a sclereid. Moreover consisting of parenchyma, separating in radial bands of 1 or 2 cells broad and mixed with some sclereids the bundles of fibres and coating the uneven inside of these bundles. Here and there a sclereid on the outside of the bundles, more or less mingling with the endodermis.

In the drug thick 5—7 elements, those of the middle region being the largest. Chiefly consisting of rounded groups of sclereids, the groups measuring R. 400—700 μ, T. 275—300 μ, L. 140—1900 μ; the largest sclereids to be found in the central parts of these groups and sometimes more or less radially arranged; the inner layer of each group consisting of tangentially elongated, somewhat more thin-walled sclereids. Bundles of sclerenchyma fibres very scarce and far between. Parenchyma separating the groups of sclereids by layers broad up to 4 cells, often containing crystals similar to those of the endodermis, for the rest with brown contents.

Sclereids. R., T. and L. 20—80 μ; polyhedra. Walls much thickened; lignified; showing many pit canals; intercellular spaces wanting. Cell contents: sometimes the cavity quite filled with a red-brown mass. Sclerenchyma fibres. R. 10—15 μ, T. 25—30 μ; penta- to octogonal. Walls much thickened; middle lamella indistinct; lignified; pit canals wanting; intercellular spaces very scarce. Contents: often a red-brown mass filling the cavity.

Phloem.

Primary phloem. In the drug not to be distinguished.

Secondary phloem. Thick up to 1 m.M., showing a fairly distinct radial arrangement of cells. **Cribral system.** Represented by more or less distinct tangential bands, alternating with metacribral bast parenchyma bands and constituting with these the ground mass of the secondary phloem; companion cells not very conspicuous, perhaps wanting; elements in the outer parts of the secondary phloem more or less flattened. **System of bast fibres.** Longitudinal bundles of bast fibres, more or less arranged in tangential bands; the bundles thick in a radial direction 2—3 elements, in a tangential 10—36; the medullary rays passing through them. **Parenchymatic system.** Metacribral bands of bast parenchyma fibres, these fibres consisting of 3—5 cells, containing here and there an isolated crystal fibre. Crystal-sheaths nearly completely surrounding the bundles of bast fibres and often filling out irregularities of the surface of these bundles. Clusters of sclereids, much varying in size and occurring only in the outermost parts of the secondary phloem; sometimes fusing with groups of sclereids, belonging to the pericycle and described there; consisting of cells equal to the larger pericambial sclereids mentioned

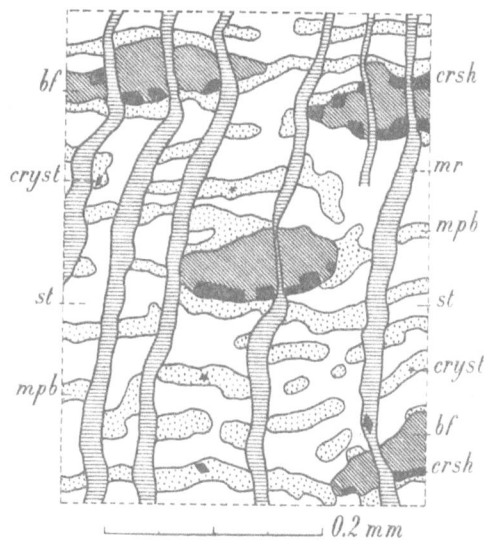

Fig. 10. *Hamamelis virginiana.* Secondary phloem, transverse section showing the alternating bands. bf Bundles of bast fibres; crsh Crystal-sheath; cryst Crystal fibres; mpb Metacribral parenchyma bands; mr Medullary rays; st Sieve-tubes.

above. Medullary rays separated from each other in a tangential direction by 2—10 layers of elements; generally uniseriate, sometimes the middle portion biseriate, 5—25 cells in height; the cells becoming somewhat broader outward and being mostly procumbent, those of the under- and uppermost 1—3 radial rows and some others here and there being upright; some scattered crystal idioblasts, especially upright cells, sometimes partitioned by a transverse wall.

Sieve-tubes. R. 20 μ, T. 25 μ; tetra- to hexagonal. Transverse walls placed obliquely on the radial side walls; sieve-plates not very conspicuous, seen only a few times. Walls somewhat thickened. Bast fibres. R. 20—30 μ, T. 15—25 μ; penta- to octogonal. Walls much thickened, only a small cavity remaining; lignified, especially the middle lamella; sometimes showing stratification; pit canals scarce and only in the radial walls. Intercellular spaces and contents wanting. Bast parenchyma fibres. R. 10—18 μ, T. 10—20 μ, L. e. g. 90 + 70 + 110 + 90 + 75 μ; tetra- to hexagonal. Walls pitted; intercellular spaces wanting. Cell contents: often some starch grains; tannin; here and there an isolated cell filled with a brown mass. Cells of crystal idiob l a s t s and c r y s t a l - s h e a t h. The fibres of the same dimensions as the bast parenchyma fibres, but the cells constituting them much more numerous and shorter, L. 10—20 μ. Walls often very thin. Contents of the sheath-cells: always a simple crystal, often perforated; of the idioblasts: sometimes a cluster crystal; all simple crystals enclosed in a sometimes lignified crystal skin, being distinctly connected with the wall. C e l l s o f m e d u l l a r y r a y s. P r o c u m b e n t c e l l s. R. and T. 15—20 μ, L. 8—20 μ; mostly tetragonal prisms with a radially directed axis and rounded radial edges. Tangential walls pitted; intercellular spaces between these cells and between them and the surrounding tissue. Cell contents: some starch grains; tannin; sometimes a simple crystal, perhaps a few times a cluster crystal. U p r i g h t c e l l s. R. 12—20 μ, T. 8—20 μ, L. 15—50 μ; see for the rest the procumbent cells.

January 1912.　　　　　　　　　　　　　　　　　　　　　　　　　　　J; M.

CORTEX MEZEREI.
Mezereon Bark. Mezereum.
The dried bark of Daphne Mezereum, Linn. Sp. Pl. 356.

Macroscopic characters.
Strips long from 0.50 up to 1 M., broad 2—4 c.M., thick up to 1.5 m.M.; more or less flattened; tapering towards the top, the top-half being moreover split into several narrow bands; showing here and there a hole corresponding with the place of a bough. Two or 3 strips together rolled up into a bundle, the insides being turned outward, and the broad bases forming the outer part; the bundles of various dimensions, often long 8—13 c.M. and thick some 5 c.M.; each tied up with a narrow strip of bark. A ware also occurring in commerce, consisting of loose narrower shorter and thinner strips, curled into quills. Bark readily separated into 2 layers; the outer of these very thin; the inner one much thicker, flexible, too tough to be broken; fibrous outer surface with a dull gloss, often somewhat blistery, with transverse wrinkles and clefts and transversely directed striae, resembling lenticels, but occasioned by locally reiterated cork formation, varying in colour from alive-brown or reddish-brown to deep purplish brown. Inner surface satiny; showing longitudinal clefts with fibrous margins; whitish with yellow longitudinal stripes. Transverse section, in a moistened state, showing a thin dark-brown outer layer; the much thicker inner layer of a lighter colour, with somewhat darker more or less triangular radially arranged figures proceeding from the inner side and with a fine tangential striation. Odour, when fresh, disagreeable; in a dry state not marked. Taste persistently acrid and burning.

Anatomical characters.
LITERATURE. De Bary. Vergl. Anat. 1877. 138, 544. Berg. Anat. Atl. 1865. 77. Taf. 39. Flückiger & Hanbury. Pharmacographia. 1879. 541. Hager. Pharm. Praxis. Bd. 2. 387. Luerssen. Syst. Bot. Bd. 2. 1882. 827. Mitlacher. Toxic. od. Forens. wicht. Pfl. u. Veget. Drogen. 1904. 114. Moeller. Baumrinden. 1882. 115. Oudemans. Aanteek. o. d. Pharmac. neerl. 1854—56. 143. Oudemans. Pharmacogn. 1880. 228. Planchon et Collin. Drogues simples. T. 1. 1895. 354. Schneider. Powd. Veget. Drugs. 1902. 244. Speyer. Beitr. z. Entw.-gesch. d. Rinde pharmakogn. interess. Pfl. Diss. Bern. 1907. 72. Vogl. Anat. Stud.

ü. Blatt u. Achse d. einheim. Daphne-Arten. 40 Jahrber. Staatsgymn. Oberhollabrunn. 1909/10. 3. Wigand. Pharmakogn. 1879. 177. MATERIAL. The drug, 0.5 m.M. thick; shoots, collected May 1902, thick 1.8 and 2.4 m.M., collected Oct. 1893, thick 3.5, 9, 11, 14 m.M. and 3 c.M., all from the Botanic Garden at Groningen, fresh and in alcohol. REAGENTS. Water, glycerine, chloral hydrate, potash, iodine in chloral hydrate, phloroglucin and hydrochloric acid, iodine and sulphuric acid 66 per cent., concentrated sulphuric acid, Schulze's macerating mixture, potassium dichromate.

MICROGRAPHY.

Epidermis.

Epidermis proper. Present only in the thinnest shoots; here and there small epidermis cells grown out, without separating wall, into unicellular somewhat conical hairs, curved towards the top of the shoot; L. up to 150 μ.

Epidermal cells proper. R. 30 μ, T. 25 μ, L. 30 μ; polygonal tables with a radially directed axis. Outer walls with a cuticle, showing a longitudinal parallel cuticular striation; the other walls showing by the cellulose reaction a net-work of cellulose bands.

Secondary cork tissue. In shoots thick 2.4 m.M. tangential partition walls appearing in the epidermis cells; some hairs showing the same phenomenon, others remaining without partition walls.

Phellem. In shoots thick 14 m.M. 15—20 layers of radially arranged periderm cells; the outermost layers often flattened; each cell of the innermost layers generally showing a radial partition wall.

Cells of the above. R. 8 μ, T. 30—70 μ, L. 20 μ; generally hexagonal tables with a radially directed axis. Walls yellow; cork lamella and inner cellulose lamella often distinct, the latter sometimes lignified.

Phellogen. One layer of cells; inner walls collenchymatous. Phelloderm and lenticels wanting.

Cortex.

Primary cortex proper. Consisting of collenchyma; in the thinnest shoots 10—15 layers of cells, in the shoot thick 1.4 c.M. only about 5 layers; the cells having the largest tangential dimensions generally with radial partition walls. In the shoots thick 3 c.M., here and there secondary cork layers adjoining with their margins in a rhytidoma-like manner the secondary cork tissue originated in the epidermis and occasioning the transversely directed striae, mentioned among the macroscopic characters.

Collenchyma cells. R. 25 μ, T. 80 μ and much larger if partitioned, L. 30 μ; ellipsoids or polyhedra mostly with rounded edges. Walls thickened; pitted; with intercellular spaces. Cells contents: in fresh shoots, collected in October, chloroplasts containing large globular starch grains; in the drug starch grains wanting or only small.

Endodermis. Developed as a starch-sheath; distinct only in the thinnest shoots; generally 1 layer, sometimes 2 layers of cells; here and there a cell showing a tangential partition wall.

Stele.

Pericycle. About 6 layers of sometimes more or less flattened parenchyma cells; this parenchyma mixed with longitudinal bundles of often somewhat flattened sclerenchyma fibres, undulating in a tangential plane, consisting of 1, up to 20 fibres in a transverse direction. In older shoots the sclerenchyma bundles remaining the same but the parenchyma much increased.

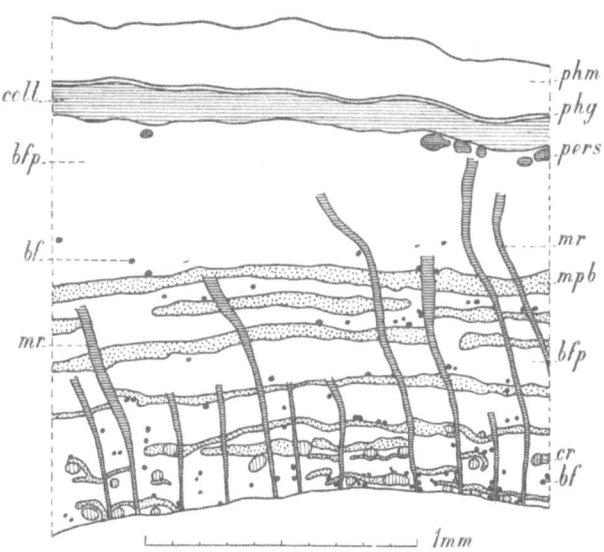

Parenchyma cells. R. 35—40 μ, T. 40—45 μ, L. 35—40 μ; globes, ellipsoids or polyhedra with strongly rounded edges. Walls pitted; with intercellular spaces. Cell contents: the same as those of the collenchyma cells of cortex. Sclerenchyma fibres. R. 9 μ, T. 18 μ, the transverse dimension varying in the same fibre, L. up to 3000 μ; edges strongly rounded. Walls much thickened, sometimes almost no cavity remaining; showing stratification; not lignified; with intercellular spaces.

Fig. 11. *Daphne Mezereum*. Bark, transverse section. bf Bast fibres with much thicker walls; bfp Mixture of bast fibres and bast parenchyma fibres; cr Bundles of elements of the cribral system; coll Collenchyma of primary cortex proper; mpb Metacribral parenchyma bands; mr Medullary rays; pers Pericylic sclerenchyma fibres; phg Phellogen; phm Phellem.

Phloem.

Primary phloem not to be recognized in the drug and the thicker shoots.

Secondary phloem.

Cribral system. To be distinguished as bundles consisting of sieve-tubes and companion cells, only in the vicinity of the cambium and there always as a constituent of the afterwards to be mentioned parenchyma bands or in contact with these. **System of bast fibres.** A mixture of more or less flattened bast fibres and bast parenchyma fibres being the chief constituent of the bast; some fibres with much thicker walls. **Parenchymatic system.** Besides the bast parenchyma fibres mixed with bast fibres mentioned above, represented by metacribral bands, consisting of some radially arranged layers of cells, and medullary rays. Medullary rays almost always uni- a few times biseriate, high up to 12 cells, clearly to be distinguished only in the inner parts of the bark.

Sieve-tubes. R. and T. 10—15 μ, L. of articulations 110 μ; polygonal prisms with a longitudinally directed axis. Walls somewhat thickened; transverse walls often somewhat oblique, sieve-plates distinct. Companion cells. R. and T. 4—8 μ, L. corresponding with that of the articulations of the sieve-tubes; mostly tetragonal prisms with a longitudinally directed axis. Cell contents: a granular mass. Sclerenchyma fibres. R. and T. 10—12 μ, L. up to 3000 μ; polygonal with rounded edges. Walls of nearly all fibres very slightly thickened; here and there a fibre with strongly thickened walls, stratified and sometimes somewhat lignified; intercellular spaces. Bast parenc h y m a   c e l l s. Cells mixed with bast fibres: R. 20—30 μ, L. 50—100 μ, in the outer-most part of the secondary phloem often much larger; polygonal prisms with a longi-tudinally directed axis. Cells of the metacribral bands: R. 15 μ, T. and L. 25 μ; generally rectangular prisms with a longitudinally directed axis and rounded edges. Walls somewhat thickened; with intercellular spaces. Contents of both kinds of cells: a few small chloro-plasts; in the material collected in October the cells filled with globular starch grains. C e l l s   o f   m e d u l l a r y   r a y s. R. 15—35 μ, T. 15 μ, L. 20 μ; rectangular prisms with a radially directed axis. Walls pitted. Cell contents: see the collenchyma cells of primary cortex proper.

May 1902, October 1911.   J; M.

CORTEX PRUNI VIRGINIANAE.
Virginian Prune Bark. Wild Cherry.
The dried bark of Prunus serotina, Ehrh. Beitr. III. 20, collected in autumn.

Macroscopic characters.

Irregular, flat or somewhat transversely curved pieces; up to 13 c.M. in length, up to 5 c.M. in width, up to 4 m.M. thick. Hard and brittle. Outer surface very uneven in consequence of the throwing off of thick rhytidoma scales, cinnamom coloured; remaining rhytidoma scales dark grey-brown. Inner surface reticu-lately striated, light brown, covered here and there with wood. Transverse fracture splintery, light brown; radial fracture showing the medullary rays as square, red patches. Transverse section showing a somewhat coarse radial striation. Odour somewhat sweetish acid, very faintly bitter almond-like; when macerated in water strongly bitter almond-like. Taste after some time agreea-ble bitter. For the above description material was used furnished by Caesar & Loretz, Halle a/S. and consisting of bark derived from somewhat thick stems and branches. In the English and American Pharmacopoea younger bark is also described.

Anatomical characters.

LITERATURE. De Bary. Vergl. Anat. 1877. 554, 559, 564, 566, 577 (some other species of Prunus). Bliesenick. Ueb. die Obliteration der Siebröhren. Diss. Erlangen. 1891. 48 (P. Padus). Flückiger & Hanbury. Pharmacographia. 254. Kraemer. Botany and Pharma-cogn. 1910. 538, 759. Mikosch. Unters. üb. d. Entsteh. d. Kirschgummi. Ber. Wien. Akad. Abt. 1. Bd. 115. 1906. 941—951. (some other species of Prunus). Moeller. Anat. d. Baum-rinden. 1882. 369 (4 other species of Prunus). Planchon et Collin. Drogues simples. T. 2. 1896. 414. Rant. De gummosis der Amygdalaceae. Diss. Amsterdam. 1906. 75 (some other species of Prunus) Schneider. Powdered veget. drugs. 1902. 264. v. d. Ven. Over het cyaan-waterstofzuur bij de Prunaceae. Diss. Amsterdam. 1898. 24, 25, 38, 39 (some other species of Prunus). MATERIAL. The drug, thick 2—4 m.M.; shoots 1 year old of Prunus virgini-ana Linn., collected in the Botanic Garden at Groningen, Dec. 6th 1912 and preserved in alcohol. REAGENTS. Water, glycerine, potassium iodide iodine, phloroglucin and hy-drochloric acid, iodine and sulphuric acid 66 per cent., Millon's reagent.

MICROGRAPHY. No young shoots of Prunus serotina being available, we examined shoots 1 year old of P. virginiana. The secondary cork tissue was found to be developed in the outmost layer of the cortex. Moeller [1]) obtained the

---
[1]) M o e l l e r, l. c. 369.

same result by the investigation of 4 other species of the same genus. Moreover Kraemer [1]) gives a figure of the young bark of P. serotina showing the bundles of pericyclic fibres, and between these and the secondary cork tissue from 4—6 layers of cortical parenchyma. From these several data we conclude, that in

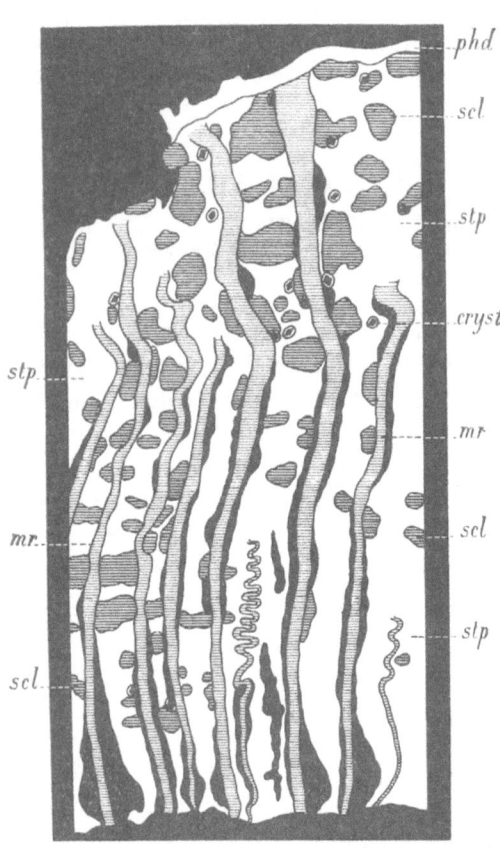

Fig. 12. *Prunus serotina*. Stem bark, transverse section. cryst Crystal cells; mr Medullary rays; phd Phelloderm; scl Clusters of sclereids; stp Cribral system and bast parenchyma.

P. serotina also the secondary cork tissue is developed in the outermost layer of the cortex. In the material examined by us, this cork layer however was thrown off long ago in consequence of an abundant formation of rhytidoma.

Secondary phloem.

Rhytidoma. Scaly, the scales thrown off in the drug, only here and there a very few remaining. The outside of the drug always consisting of a phelloderm layer, in consequence of the removal of the phellem and phellogen together with the thrown off scales.

Layer of secondary phloem. If present showing colourless walls and red-brown contents of the cells. See for the rest the secondary phloem proper.

Layer of secondary cork tissue. Initial celled cork.

Phellem. Thick about 250 μ: 60—70 layers of periderm cells. The parts developed in the medullary rays identical to those between the rays. The outer $^2/_3$ parts of the phellem showing dark brown cell contents.

Phellogen indistinct. **Phelloderm** thick 3—6 layers of common parenchyma cells, mixed with a few sclereid idioblasts and somewhat more numerous crystal cluster idioblasts. The outer cell layers showing citrine or brown walls.

Periderm cells. R. 4—10, mostly 4—5 μ, T. 10—25 μ. Walls somewhat thickened, colourless. Phelloderm cells. R. 10—25 μ, T. and L. 15—30 μ; tetra- to octogonal prisms with a radially directed axis. Walls thickened, especially the tangential ones; pitted.

Secondary phloem proper. Tangential growth of the outer part of the se-

---

1) K r a e m e r, l. c. 538, Fig. 235B.

condary phloem, owing to the dilatation of the bark, caused principally by the bast parenchyma, the cells of the medullary rays having in the outermost part of the bark at most only double the tangential dimensions of those near the cambium. Moreover the medullary rays always almost quite separated from the surrounding tissue by radial clefts of some breadth on both sides; into these clefts bast parenchyma cells, sometimes also ray parenchyma cells here and there growing [1]). Consisting for the greater part of tissues belonging to the parenchymatic system; the system of bast fibres wanting.

Cribral system. Flattened even in the immediate neighbourhood of the cambium; occurring in the shape of thin, often anastomosing and irregular bands. These bands more numerous in the inner than in the outer part of the bark. **Parenchymatic system.** B a s t p a r e n c h y m a consisting principally of common bast parenchyma fibres; in the inner part of the bark more or less flattened, in the outer part the cells showing always transverse and often radial walls, thus the longitudinal and radial dimensions of the cells becoming the same. Large intercellular spaces with here and there cells protruding into them; these intercellular spaces perhaps in part gum cavities. Crystal fibres very numerous and mixed up with the bast parenchyma fibres; the fibrous arrangement of the cells distinct only in the innermost part of the bark, the fibres consisting of up to 7 cells. Many crystal cells adjoining the clusters of sclereids, to be mentioned beneath and wedged in between the cells protruding from the surface of these clusters. Crystal clusters to be found principally in the innermost part of the bark, for the rest nearly exclusively simple crystals. Clusters of sclereids often adjoining the medullary rays and having about the same height as these; wanting in the innermost part of the bark; thick about 0.5 m.M., in the outer part of the bark larger and often more or less tangentially elongated, inward somewhat smaller, though usually consisting of the same number of elements, sometimes here radially elongated. Shape of the clusters ellipsoids with a longitudinally directed longest axis; outer surface often irregular, many superficial cells being branched and the branches protruding between the surrounding parenchyma cells. The largest clusters of sclereids consisting of up to 250 elements, the smallest only of 3 or 4 and these often showing a distinct fibrous shape. L. of clusters up to 500 μ. M e d u l l a r y r a y s in 2 kinds: uni- or bi-, usually uniseriate and the other more numerous kind tri- to penta-, mostly tetraseriate. Lateral distance between the rays about 150—250 μ; the rays being often placed more or less distinctly in longtidunal rows. The uniseriate rays up to 10 and often 5 cells in height; the broader rays up to 40, often 25—30 cells. The cells of all rays procumbent; nevertheless those of the

---

[1]) These clefts have been described in full by M i k o s c h, l. c. 942 and contain gum in the species of *Prunus*, examined by him. The drug, examined by us, did not contain gum, perhaps because it was some years old. d e B a r y, l. c. 554 is of opinion, that there is a causal connection between the formation of these radial clefts and the tangential dilatation of the bark. We cannot think, that this is the right explanation.

uniseriate and sometimes those on the radial surfaces of the broader rays somewhat shorter in a radial direction. The uniseriate rays showing very numerous sharp longutidunal folds, causing the interchange of radial and tangential aspects on the tangential and radial sections. Sometimes radially running small gum pockets in the broader rays.

C e l l s o f c r y s t a l f i b r e s. Walls much varying in thickness, from thin to sclerotic; often lignified, the sclerotic ones always. Cell contents: simple crystals, augmenting in size towards the outer part of the bark, R., T. and L. 10—40 μ, in cross sections of the bark often of a rhombic shape, generally surrounded by a crystal skin, this being lignified in cells with lignified walls; the simple crystals sometimes perforated, the cluster crystals always. S c l e r e i d s. R. 25—40 μ, T. 25—50 μ, L. 25—40 μ; polyhedra, more or less rounded on the outside of the clusters. Walls varying in thickness, up to the total disappearance of the cavity; showing stratification; lignified; with pit canals. Cell contents: often a brown mass. C e l l s o f m e d u l l a r y r a y s. R. 50—70 μ, T. near the cambium about 12 μ, near the secondary cork layer 15 μ, in the immediate vicinity of the same 20 μ, L. 10—20 μ; tetra- to octogonal prisms with a radially directed axis and rounded radial edges. Walls thick 1.5 μ, the tangential ones somewhat thicker; showing pits, especially on the tangential walls; radial intercellular spaces. Cell contents: some granules, probably chloroplasts.

December 1912.   J; M.

CORTEX QUILLAIAE.
Quillaia Bark. Panama Bark. Soap Bark.
The dried bark of Quillaia Saponaria, Molina, Sagg. Chile, 175, 354, deprived of its brown outer parts.

Macroscopic characters.
Large, flat or slightly and generally inwardly curved pieces, up to 1 M. long, up to 12 c.M. wide, generally about 4 m.M. thick. Hard and brittle. Outer surface smooth; dull; showing especially in direct sunlight many glistering particles; colour white to more or less brown-red; here and there showing longitudinal dark brown patches, thick up to 1 m.M.; sometimes the white surface showing longitudinal red-brown striae, wide up to 0.5 m.M. and often about 2 m.M. distant from each other. Inner surface very smooth, dull, showing the same glistering particles as on the outer surface, yellowish white, sometimes passing into red-brown. Transverse fracture coarsely splintery, laminated, white, showing many glistering particles as mentioned above. Transverse section marked with fine radial and sometimes slightly coarser tangential striae. Odour slightly sweetish; taste at first somewhat sweet, afterwards acrid and slightly sour.

Anatomical characters.
LITERATURE. Benecke. Mikr. Drogenprakt. 1912. 43. Deutsches Arzneib. 5 Ausg. 144. Douliot. Rech. s. l. Periderme. Ann. Sc. Nat. 7. Sér. T. 10. 1889. 354. Flückiger. Pharmakogn. 1891. 615. Gilg. Pharmakogn. 1910. 138. Greenish & Collin. Anat. Atl. o. veget. powders. 1904. 204. Hager. Pharm. Praxis. Bd. 2. 1902. 717. Karsten u. Oltmanns. Pharmakogn. 1909. 124. Koch. Mikr. Anal. d. Drogenpulver. Bd. 1. 1901. 119. Koch. Pharmakogn. Atl. Bd. 1. Lief. 2. 1910. 51. Koch u. Gilg. Pharmakogn. Praktik. 1907. 46. Krämer.

Botany and Pharmacogn. 1910. 782. Moeller. Anat. d. Baumrinden. 1882. 368. Moeller. Pharmacogn. 1889. 243. Moeller. Mikr.-Pharm. Üb. 1901. 243. Planchon et Collin. Drogues simples. T. 2. 1896. 430. Prodinger. Das Periderm d. Rosaceen. Denkschr. Wien. Ak. Bd. 84. 1909. 341. Schneider. Powdered veget. drugs. 1902. 270. Speyer. Beitr. z. Entw.-gesch. d. Rinde pharm. Pfl. Diss. Bern. 1907. 76. Tschirch. Angew. Pfl. Anat. 1889. 104. Wiesner. Rohstoffe. Bd. 1. 1900. 765. MATERIAL. The drug, pieces thick 4—5 m.M. REAGENTS. Water, glycerine, potassium iodide iodine, phloroglucin and hydrochloric acid, iodine and sulphuric acid 66 per cent., hydrochloric acid [1]), Schulze's macerating mixture.

MICROGRAPHY.

No young shoots of Quillaia Saponaria were available. According to Douliot [2]), Prodinger [3]) and Speyer [4]) the secondary cork tissue is developed in the outermost layer of the cortex, sometimes locally in the next layer. The original secondary cork has however been removed from the drug long ago by formation of rhytidoma, this rhytidoma being equally removed by the scaling of the drug; only here and there some parts of it remaining in the shape of the dark brown patches, mentioned among the macroscopic characters.

S e c o n d a r y p h l o e m.

Rhytidoma. The remaining patches consisting of many cell layers belonging to the secondary phloem and a secondary cork layer on the inside.

L a y e r o f s e c o n d a r y p h l o e m. Showing red-brown cell walls and often red-brown contents in the cells of the parenchymatic system. See for the rest the secondary phloem proper.

L a y e r o f s e c o n d a r y c o r k t i s s u e. Storied cork consisting of about 15 layers of periderm cells, generally arranged in 2 stories of about equal thickness and enclosing here and there small clusters of bast sclereids, small bundles of bast fibres and crystal idioblasts. Parts of the periderm corresponding with the medullary rays having the same breadth as these. The outer 7—10 periderm layers usually filled with a yellow to red-brown fine granular mass.

P e r i d e r m c e l l s. R. 6—20 μ, T. 15—30 μ, L. 20—40 μ; mostly hexagonal discs with a radially directed axis. Walls thick 1—2 μ; colourless.

Secondary phloem proper. In this bast, gathered from very thick stems, the tangential growth of the outer part of the secondary phloem, resulting from the dilatation of the bark, slight and caused by the medullary rays, bast parenchyma and bast fibres. Consisting of more or less distinct tangential layers, belonging to the several tissues constituting the secondary phloem, and arranged in the following order: bast fibres, bast parenchyma, cribral system, bast parenchyma, bast fibres, etc.

C r i b r a l s y s t e m. Consisting chiefly of sieve-tubes and a few companion cells; the articulations of the sieve-tubes and the companion cells showing a

---

[1]) We were unable to confirm the observation of some authors that the walls of the sclerotic elements take a red colour by hydrochloric acid only.

[2]) L. c. 354.

[3]) L. c. 341.

[4]) L. c. 76.

somewhat storied arrangement; the whole tissue becoming gradually more and more flattened towards the outer part of the bark.

S i e v e - t u b e s. R. 10—30 μ, T. 20—30 μ, L. of articulations 70—220 μ; penta- to octogonal prisms with a longitudinally directed axis. Transverse walls placed obliquely in a radial or a tangential direction; sieve-plates numerous on each transverse wall and often to be also found on the radial walls. C o m p a n i o n c e l l s. R. and T. 8—10 μ, L. 70—220 μ; tri- or tetragonal prisms with a longitudinally directed axis.

S y s t e m o f b a s t f i b r e s wanting in the innermost, 0.5 m.M. thick part of the bark; very abundant in the outer and by far the largest part, constituting here more than half of the tissue between the medullary rays and having the same appearence through all parts of the bark. The tangential layers of bast fibres thick 8—15 elements, showing perforations for the medullary rays and moreover containing here and there some parenchyma fibres, often also broken up into smaller bundles or even isolated fibres.

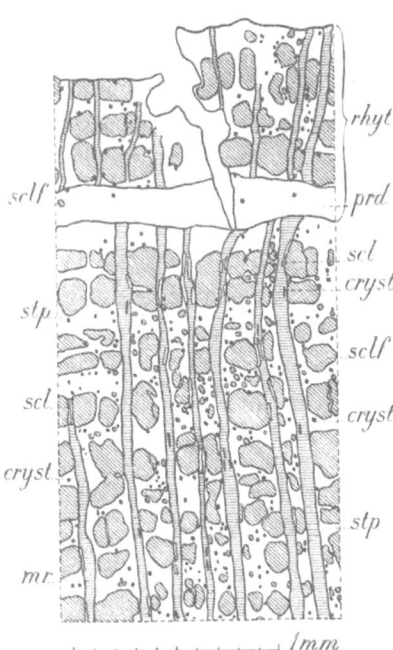

Fig. 13. *Quillaia Saponaria*. Bark, transverse section of the outer part. cryst Crystal cells; mr Medullary rays; prd Periderm; rhyt Rhytidoma; scl Sclereids; sclf Bundles of sclerenchyma fibres with adjoining sclereids; stp Sieve-tubes, companion cells and bast parenchyma fibres.

B a s t f i b r e s. R. and T. 20—25 μ, L. 500—1000 μ; penta- to octogonal, but adjoining the parenchyma with rounded edges and sides; the ends not acute, but generally somewhat obtuse or even swollen. Walls thick 9 μ, cavity very small; showing indistinct concentric layers; lignified; with rare pit canals.

P a r e n c h y m a t i c s y s t e m. B a s t p a r e n c h y m a. Consisting of bast parenchyma fibres, often divided into 3 cells; these cells in the outermost part of the bark often showing an increasing number of transverse walls, the cells thus becoming of the same height and breadth. Crystal fibres numerous; mostly isolated; scattered through the bast parenchyma; especially numerous in the immediate neighbourhood of the bast fibres; divided e.g. into 3 or 4 cells, each containing a very large simple crystal, a few cells up to 5 or 6 smaller crystals; here and there a cell with a cluster crystal and in the outermost part of. the bark a few filled with crystal sand. Sclereids not numerous and occurring only in the immediate vicinity of the bast fibres and between them in the bundles. M e d u l l a r y r a y s. Lateral distance between the rays 100—300 μ; occupied by 4—16, generally 12—15 cell layers. Rays uni- to tetra-, mostly tri- or tetraseriate; height of the uni- and biseriate rays 4 or 5 cells, height of the tri- and tetraseriate 9—30 cells. Nearly all cells procumbent; those of the 1 or 2 upper- and undermost radial rows somewhat shorter in a radial and longer in a longitudinal direction, often upright. Many cells on the radial sides of the rays, adjoining bast fibres, developed into sclereids. Here and there a crystal idioblast, containing a single long radially directed simple crystal; these cells very scarce in the neighbourhood of the cambium and here a very few cells containing a cluster crystal or crystal sand.

C e l l s   o f   b a s t   p a r e n c h y m a   f i b r e s. R. 10—30 μ, T. 20—40 μ, L. 20— 150 μ; penta- to decagonal prisms with a longitudinally directed axis and sometimes rounded edges. Walls thick 1 μ; pitted on the radial walls; intercellular spaces sometimes distinct. Cell contents: globular starch grains, up to 6 μ in diameter. C e l l s   o f   c r y s- t a l   f i b r e s. Transverse dimensions somewhat inferior to those of the cells of bast

parenchyma fibres. Cell contents: the simple crystals, mentioned above, thick 10 μ, L. 100 μ, tetragonal prismatic, with pointed ends; these crystals being surrounded by a lignified crystal skin. See for the rest the cells of bast parenchyma fibres. S c l e r e i d s. R. 20—50 μ, T. 20—30 μ, L. 50—80 μ. Walls thick 2—5 μ, sometimes much thicker; lignified; often with crossing slit-like pits, especially numerous on the transverse and radial walls, and on the latter sometimes arranged in a radial row. Cell contents: in a very few cases a simple crystal, resembling those of the crystal fibres; starch wanting. See for the rest the cells of bast parenchyma fibres. P r o c u m b e n t c e l l s. R. 50—100 μ, T. near the periderm 18—20 μ, near the cambium 12—15 μ, L. 10—20 μ; penta- to octo-, often hexagonal prisms, with a radially directed axis and rounded radial edges. Walls at most thick 1 μ; intercellular spaces

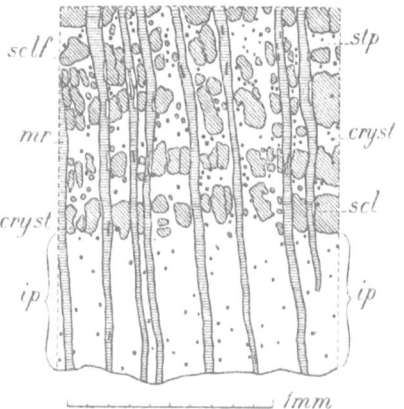

Fig. 14. *Quillaia Saponaria*. Bark, transverse section of the inner part. ip Inner part of bark, wanting sclerenchyma; see for the rest Fig. 13.

distinct in a radial direction. Cell contents: simple globular starch grains, up to 8 μ in diameter; crystals the same as those in cells of bast parenchyma fibres, but radially directed. U p r i g h t   c e l l s. R. 20—50 μ, L. about 30 μ. See for the rest the procumbent cells. S c l e r e i d s   o f   m e d u l l a r y   r a y s. Walls thick 1—3 μ; lignified; with many elliptical pits, e. g. 4 by 6 or 5 by 6 μ. See for the rest parenchyma cells of medullary rays.

April 1913.   J; M.

CORTEX RHAMNI FRANGULAE.
Alder Buckthorn Bark.
The dried bark of the stem and branches of Rhamnus Frangula, Linn. Sp. Pl. 193.

Macroscopic characters.
Quills of various lengths, broad up to 2 c.M. The bark thick up to 2 m.M.; brittle; rolled up from one or both sides; here and there holes, indicating the places formerly occupied by branches. Outer surface longitudinally somewhat wrinkled, sometimes here and there covered with lichens, greyish or blackish brown, dark red after removal of the outer cork layers; many lenticels, in general

transversely oblong, with some elevated light coloured small ridges in a direction perpendicular on their longest axis. Inner surface finely striated, dull red-brown or orange, if moistened with alkalies cherry-coloured. Transverse fracture irregular; the inner part somewhat coarsely fibrous. Inodourous; taste somewhat sweetish and bitterish; by chewing the bark the saliva is coloured yellow.

Anatomical characters.

LITERATURE. De Bary. Vergl. Anat. 1877. 51, 544, 545, 563, 579. Berg. Anat. Atl. 1865. 79. Taf. 40. Cabannes. Ét. d. q. espèces d. Genre Rhamnus. Diss. Montpellier. 1896. 10. Deutsch. Arzneib. 5. Ausg. 1910. 139. Flückiger. Pharmakogn. 1891. 520. Gilg. Pharmakogn. 1910. 213. Greenish & Collin. Anat. Atl. o. veget. powders. 1904. 184. Grès. Contr. à l'ét. anat. et microchim. d. Rhamnées. Diss. Paris. 1901. Hager. Pharm. Praxis. Bd. 1. 1900. 1179. Hérail. Mat. Méd. 1912. 596. von Höhnel. Anat. Unters. ü. einige Sekretionsorg. d. Pfl. Ber. d. Ak. d. Wiss. Wien. Bd. 84. Abth. 1. 1881. 591. Karsten u. Oltmanns. Pharmakogn. 1909. 130. Koch. Mikr. Anal. d. Drogenpulver. Bd. 1. 1901. 97.

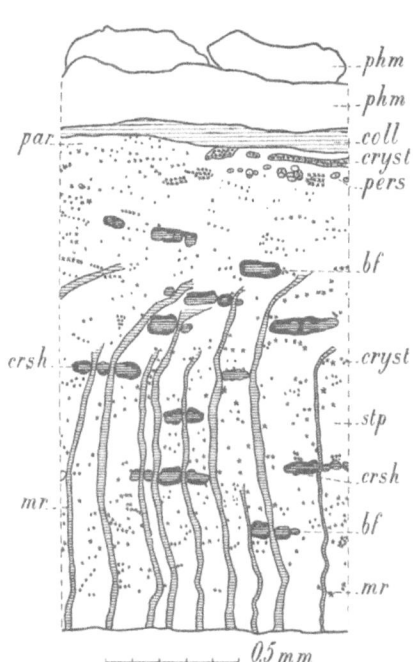

Fig. 15. *Rhamnus Frangula.* Bark, transverse section. bf Bast fibes; coll Collenchyma of primary cortex proper; crsh Crystal-sheaths; cryst Crystal cells; mr Medullary rays; par Colourless parenchyma of primary cortex proper; pers Pericyclic sclerenchyma fibres; phm Phellem; stp More or less distinct alternating bands of sieve-tubes and bast parenchyma.

Koch. Pharmakogn. Atl. Bd. 1. 1910. 33. Taf. 7. Koch u. Gilg. Pharmakogn. Praktik. 1907. 37. Kraemer. Botany a. Pharmacogn. 1910. 522, 735. Luerssen. Syst. Bot. Bd. 2. 1882. 731. Marmé. Pharmacogn. 1886. 130. Meyer. Wiss. Drogenk. Bd. 2. 1892. 128. Moeller. Mikr.-pharm. Üb. 1901. 239. Oudemans. Pharmacogn. 1880. 226. Planchon et Collin. Drogues simples. T. 2. 1896. 590. Schneider. Powdered Veget. Drugs. 1902. 195. Speyer. Beitr. z. Entw.-gesch. d. Rinde pharm. Pfl. Diss. Bern. 1907. 77. Walliczek. Stud. ü. d. Membranschleime veget. Org. Pringsheim's Jahrb. Bd. 25. 1893. 260; the same in Diss. Bern. 1893. 52. Wigand. Pharmakogn. 1879. 153. MATERIAL. The drug; shoots 1 year old and shoots up to 12 m.M. thick, collected in the month of May, preserved in alcohol; shoots up to 15 m.M. thick, collected Febr. 1901 from the Botanic Garden at Groningen, fresh and in alcohol. REAGENTS. Water, glycerine, chloral hydrate, potash, potassium iodide iodine, phloroglucin and hydrochloric acid, iodine and sulphuric acid 66 per cent., concentrated sulphuric acid, Schulze's macerating mixture.

MICROGRAPHY.

Epidermis. Still partly remaining in shoots of 1 year old, afterwards thrown off. Stomata wanting. Hairs numerous, occasioning a pubescence of the young shoots; arranged in pairs, these directed transversely or obliquely; long cylindrical with an acute apex; curved; consisting of 1 row of from 2 to 5 cells.

Epidermal cells proper. R. 12—15 μ, T. 15 μ, L. 10 μ; tetra- or polygonal prisms with a radially directed axis. Outer walls lignified. C e l l s o f h a i r s with thickened side walls and granular yellow contents.

C o r t e x.

Secondary cork tissue proper. Developed from the outer layer of cortical cells.

P h e l l e m. Consisting of 15—20 layers of radially arranged and in this direction somewhat flattened periderm cells.

C e l l s o f t h e a b o v e. R. 10 μ, T. 16 μ, T. 14 μ; pentagonal prisms with a radially directed axis. Walls thin and colourless. Cell contents: brown or brick-red.

P h e l l o g e n. 1 layer of cells. Phelloderm wanting.

Lenticels.

P h e l l e m. Not much protruding from the surface, but extending cushion-shaped into the cortex; consisting of somewhat 50 layers of cells, showing a regular alternation of thick-walled colourless complementary tissue thick 4—5 globular cells and closing membranes thick about 7 periderm cells with intercellular spaces, but for the rest quite the same as the periderm cells of the common cork tissue. [1]) Phellogen: 1 layer of cells. Phelloderm wanting.

Primary cortex proper.

C o l l e n c h y m a. Thick 3—5 layers of cells, wanting under the lenticels; the cells often divided by 1 or 2 thin radial partition walls.

C e l l s o f t h e a b o v e. R. 16 μ, T. 50 μ, L. 80 μ; polygonal prisms with a longitudinally directed axis. Walls rather thickened; colourless in the drug, red in alcoholic material; pitted; intercellular spaces wanting. Cell contents: starch and some granular yellow mass; see for the rest the colourless parenchyma.

C o l o u r l e s s p a r e n c h y m a. Thick 10—12 layers of cells, many cells divided by 1 or 2 thin radial partition walls; in tangential sections some cells more elongated and united into longitudinal, somewhat sinuous bands, broad 2—5 cells. In this tissue large clefts, extending in a tangential and longitudinal direction over many cells — R. 20—75 μ — and often containing crystal clusters and remains of cells; these clefts, though less in number, to be found already in shoots some months old. Two kinds of idioblasts: crystal cells and mucilage cells. Crystal cells, containing 1 or 2 crystal clusters, arranged in tangential rows, especially very numerous in the inner part of the colourless parenchyma and here constituting in shoots 1 year old almost the whole tissue; for the rest the same as the ordinary cells of the colourless parenchyma. Mucilage cells to be found only in young shoots, afterwards disappearing, quite obliterated in the drug, generally lying in the inner part of the colourless parenchyma, 20—30 in every transverse section of a shoot, often some united into groups; R. 85 μ, T. 90 μ, L. 180 μ.

C e l l s o f c o l o u r l e s s p a r e n c h y m a. R. 15 μ, T. 35 μ, L. 19 μ; tangentially elongated ellipsoids. Walls somewhat thickened; colourless in the drug, red in alcoholic

---

[1]) d e B a r y, l. c. 579 describes the anatomical structure of lenticels, divided into some parts. Probably this description corresponds with the macroscopic character of the lenticels mentioned above, as showing elevated light-coloured ridges.

material; pitted; small intercellular spaces in all directions. Cell contents: starch; a yellow granular mass, coloured red by concentrated sulphuric acid and alkali (frangulin).

Endodermis. Developed as a starch-sheath in shoots one year old; 1 sometimes 2—3 layers of cells, arranged in longitudinal rows; the cells of adjoining rows alternating with one another like brick-work. In the drug not to be distinguished. Cells of the above. R. 20 µ, T. 30 µ, L. 25 µ; prisms. Intercellular spaces wanting.

S t e l e.

Pericycle. In young shoots 2—3 layers of sclerenchyma fibres, and between these here and there some longitudinal parenchyma bands, much increasing in tangential dimension with the growth in thickness of the shoot. Thus in the drug only here and there tangentially elongated longitudinal bundles, consisting in a cross section of 5—14 fibres. Sclerenchyma fibres. R. 5 µ, T. 20 µ, L. 500—600 µ; somewhat flattened, often with forked ends. Walls not very much thickened; showing distinct stratification; in alcoholic material the middle lamella coloured violet; not lignified; showing pit canals. Parenchyma cells not to be distinguished from those of the inner layers of colourless cortical parenchyma.

Phloem.

P r i m a r y   p h l o e m. Originally in the shape of a continuous layer; in the drug not to be distinguished from the surrounding tissues.

S e c o n d a r y   p h l o e m. Consisting of about 50 layers of cells, showing in the inner part of the bark a distinct radial arrangement.

Cribral system. Sieve-tubes in tangential bands, thick 1—3 elements in a radial direction; often much flattened in the same direction; sieve-plates distinct, often also on the lateral walls. **System of bast fibres.** Bast fibres in longitudinal bundles, flat in a tangential direction, broad in this direction 1—10 fibres, thick in a radial direction 1—3 fibres. **Parenchymatic system.** Consisting of bast parenchyma und medullary rays. B a s t   p a r e n c h y m a represented by metacribral bands of bast parenchyma mixed with many crystal fibres; crystal-sheaths completely surrounding the bundles of bast fibres. Metacribral bands, thick in a radial direction 1—3 cells, showing radial arrangement only in transverse sections; the crystal fibres often adjoining the medullary rays, also occurring more or less in tangential bands and having in each cell 1 or 2 cluster crystals. Crystal-sheaths thick 1 layer of cells and consisting of parenchyma fibres, divided into about 30 cells, each containing a simple crystal. M e d u l l a r y   r a y s separated from each other by 3—11 elements, uni- to triseriate, 8—20 cells in height; all cells procumbent.

Sclerenchyma fibres. R. 17 µ, T. 15 µ, L. 930 µ; tetra- to hexagonal. Walls strongly thickened, almost no lumen remaining; showing stratification; yellow; middle lamella lignified, secondary layers without cellulose reaction; showing pit canals. B a s t   p a r e n c h y m a   c e l l s. R. 15 µ, T. 25 µ, L. 40 µ; tetragonal prisms with a longitudinally directed axis. Walls the same as those of the colourless cortex parenchyma, but without pits and intercellular spaces. Cell contents: starch and a yellow, granular mass (frangulin). C e l l s   o f   c r y s t a l   f i b r e s. Dimensions in a transverse direction

somewhat larger than those of the cells of the crystal-sheaths. C e l l s o f c r y s t a l-
s h e a t h s. R. 11 μ, T. 10 μ, L. of parenchyma fibre 370 μ; tetra- or pentagonal prisms
with a longitudinally directed axis. Walls thin. C e l l s o f m e d u l l a r y r a y s. R.
25 μ, T. and L. 14 μ; tetragonal prisms with a radially directed axis. Walls pitted; small
intercellular spaces running in all directions. Cell contents: the same as those of the bast
parenchyma cells.

February 1901, November 1911.   J; M.

CORTEX RHAMNI PURSHIANAE.
Cascara Sagrada. Sacred Bark. Bearberry Bark.
The dried bark of stem and branches of Rhamnus Purshiana, DC. Prod. 2. 25.

Macroscopic characters.
Pieces flat or somewhat chanelled, derived from stems and branches of very
varying age, seldom quills; of varying length, broad up to 5 c.M.; the bark
thick 1—7 m.M., tolerably brittle, spreading dust when fractured. Outer surface with longitudinal, somewhat sinuous fissures and often distinct, transversely elongated, lighter coloured lenticels; in many places covered with mosses
and especially lichens, the latter often covering large areas and sometimes pos-
sessing distinct fructifications; colour red-brown to grey-brown, but partly
white by the lichens. Inner surface smooth; sometimes even somewhat polished;
with fine, longitudinal, lighter coloured striae; colour in general red-brown, sometimes very dark, sometimes much lighter up to yellow-brown, when moistened with ammonia dark brown-red. Transverse fracture tolerably smooth, the
outer part light yellow-brown, the inner part somewhat darker and coarsely
fibrous. In a transverse section the outer part showing small, somewhat darker
spots; the inner part sometimes showing some radial striation. Odour feebly
acid; taste acrid and bitter.

Anatomical characters.
LITERATURE. Cabannes. Étude d. q. espèces du Genre Rhamnus. Diss. Montpellier.
1896. 32. Deutsch. Arzneib. 5. Ausg. 145. Flückiger. Pharmakogn. 1891. 524. Gilg. Pharmakogn. 1910. 215. Greenish & Collin. Anat. Atl. o. veget. powders. 1904. 186. Hager.
Pharm. Praxis. Bd. 2. 1902. 727. Hérail. Traité d. Mat. Med. 1912. 599. Karsten u. Oltmanns. Pharmakogn. 1909. 133. Koch. Pharmakogn. Atl. Bd. 1. 1910. 57. Taf. 11. Kraemer.
Botany a. Pharmacogn. 1910. 525, 759. Moeller. Pharmakogn. 1889. 235. Moeller. Mikr.-
pharm. Üb. 1901. 240. Schneider. Powdered Veget. Drugs. 1902. 272. Speyer. Beitr. z.
Entw.-gesch. d. Rinde pharm. Pfl. Diss. Bern. 1907. 81. MATERIAL. The drug; shoots
collected in the Botanic Garden at Leyden November 1911 1 year old, thick 2.5 m.M.
and thicker, up to 8.5 m.M., fresh and in alcohol. REAGENTS. Water, glycerine, potassium iodide iodine, phloroglucin and hydrochloric acid, iodine and sulphuric acid 66 per
cent., concentrated suphuric acid, potash.

MICROGRAPHY.
E p i d e r m i s. Intact in shoots of 1 year old, afterwards thrown off, showing
no distinct regular arrangement of cells, only here and there some cells in small
longitudinal groups. Stomata wanting. Hairs tolerably numerous, only here

and there arranged in pairs, for the rest solitary, the surrounding epidermal cells in some cases somewhat radiating from the place where the hair is inserted, up to 200 μ in length, and about 10 μ thick at the base, slightly conical with an acute apex, curved, consisting of 1 row of 2—6 cells varying with the length of the hair.

E p i d e r m a l c e l l s p r o p e r. R. 15 μ, T. 8—10 μ, L. 8—20 μ; tetra- to polygonal, mostly penta- and hexagonal prisms with a radially directed axis. Outer walls much thickened, here and there cells obviously divided by a radial wall after the thickening of the outer wall; radial walls pitted. Cell contents: a homogeneous brown-red mass. C e l l s o f h a i r s. L. 20—30 μ, the longest toward the base; side walls somewhat thickened. Cell contents: partly filled with a yellow to brown mass.

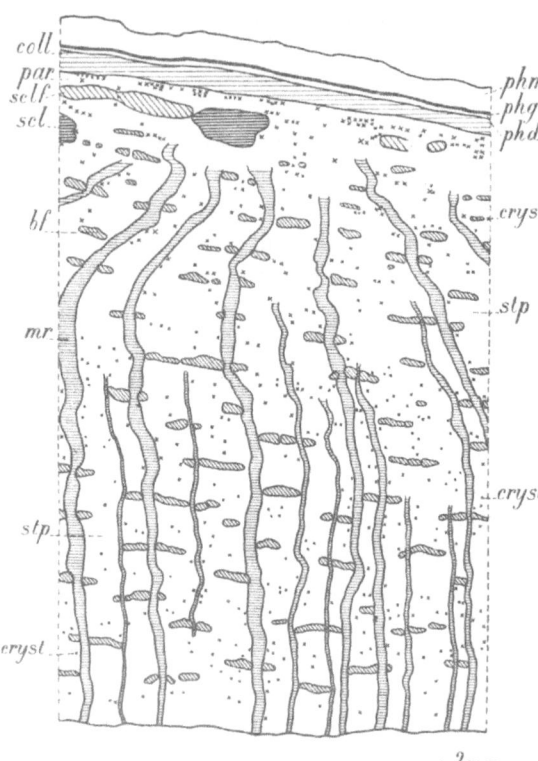

Fig. 16. *Rhamnus Purshiana*. Bark, transverse section. bf Bundles of bast fibres; coll Collenchyma; cryst Crystal cells; mr Medullary rays; par Colourless parenchyma; phd Phelloderm; phg Phellogen; phm Phellem; scl Clusters of sclereids; sclf Sclerenchyma fibres of pericycle; stp Sieve-tubes and bast parenchyma.

C o r t e x.

Secondary cork tissue proper. Initial celled cork developed from the outer layer of cortical cells, already present in the youngest shoots examined.

P h e l l e m consisting of up to 15 layers of radially arranged cells, particularly the outer layers more or less flattened. C e l l s o f t h e a b o v e. R. 4—8 μ, T. 10—20 μ, L. 8—20 μ; tetra- to hexagonal prisms with a radially directed axis. Walls thin and colourless. Cell contents: particularly that of the outer cells a yellow-brown homogeneous mass.

P h e l l o g e n 1 layer of flat cells.

P h e l l o d e r m 1 layer of cells.

C e l l s o f t h e a b o v e. R. 10—14 μ; inner walls collenchymatously thickened, side walls inwards thickened wedge-like; no intercellular spaces. Cell contents: a very few small chloroplasts.

Lenticels. Already present in the youngest shoots examined, growing out very much in all directions during the thickening of the shoots, losing by excoriation the central portions of the older phellem, thus appearing in the drug in the shape of shallow excavations, coated in their outer part with remnants of alternating closing bands and bands of complementary cells, and showing only at the bottom some intact bands. The surrounding cortical tissue meanwhile having acquired a very distinct tangential arrangement of the tangentially elongated cells, the phellogen and phelloderm in the drug adjoining these cortical layers in an oblique direction.

P h e l l e m. Bands of complementary cells twice as thick as the closing bands, and consisting of thin-walled colourless cells; closing bands consisting of thick-walled cells with red-brown contents. Phellogen 1 layer of cells. Phelloderm consisting of 1 or 2 layers of cells.

Primary cortex proper.

C o l l e n c h y m a. Thick some 6 layers of cells, wanting under the lenticels; often showing a somewhat regular arrangement of the cells in tangential rows; many cells divided by 1 or 2 thin, radial partition walls; the inner layers of cells having a less collenchymatous character.

Cells of the above. R. 10—15 μ, T. 25—80 μ, the partitioned cells 140 μ, L. 10—30, generally 15 μ; tetragonal prisms with a longitudinally directed axis. Walls thick; colourless; transverse and radial walls pitted; intercellular spaces wanting. Cell contents: some chloroplasts, filled with starch; becoming red in potash.

C o l o u r l e s s   p a r e n c h y m a. Thick some 5 or 6 layers of cells, in shoots 1 year old some 8 layers; the cells arranged more or less regularly in transverse layers and in these layers in tangential rows, the layers and rows having between them many distinct and large tangentially running intercelllular spaces; in shoots of 1 year old arranged in longitudinal rows, with large tangential intercellular spaces in the outer part of the tissue; some cells divided by 1 or 2 thin radial partition walls. In this tissue clefts R. 20 μ, T. 130 μ, L. 20 μ, often containing crystal clusters and remains of cells; these clefts, though less in absolute numbers, to be found already in shoots of some months old. Clusters of sclereids consisting of some 2000 cells and less, scattered through the whole of the parenchyma in the drug, here and there in groups of 2 or 3 together, wanting in all shoots up to 8.5 m.M. thick; shape of the clusters irregularly globular, sometimes a little elongated in a tangential direction; clusters more or less completely surrounded by a sheath of crystal cells, consisting of 1 or 2 layers. Idioblasts in 2 kinds: cluster crystal cells and very large mucilage cells. Crystal cells divided by 1—3 thin radial walls, each division containing 1 cluster crystal; in shoots of 1 year old these cells very numerous in the inner part of the cortical parenchyma, and those in the outer part often very large. Mucilage cells to be found in young shoots; being partly obliterated already in a shoot 8.5 m.M. thick, and quite so in the drug; arranged in single longitudinal rows containing some 15 cells, and here and there in the middle of the rows 2 cells in a transverse direction; generally more numerous in the inner part of the colourless parenchyma, scattered here and there, some 20 in every transverse section of a shoot.

Cells of colourless parenchyma. R. 12—18 μ, T. 30—35 μ, of the radially partitioned cells 80—100 μ, L. 10—20 μ; cylinders with a tangentially directed axis. In shoots thick 2.5 m.M.: R. and T. 18 μ, L. 25—50 μ; those cells adjoining the mucilage cells transversely elongated and shorter; polygonal prisms with a longitudinally directed axis and strongly rounded edges. Walls somewhat thickened; radial and transverse walls pitted. Cell contents: some chloroplasts adjoining the walls, more numerous than in the collenchyma cells. S c l e r e i d s. R. 15 μ, T. 30—50 μ, L. 15 μ; polyhedra. Walls thicken-ed, almost to the disappearance of the lumina; middle lamella distinct; sometimes show-ing stratification; lignified, especially the middle lamella; intercellular spaces wanting. C e l l s o f c r y s t a l-s h e a t h s. R. 10—18 μ, T. and L. 10—20 μ; polyhedra, some-times with rounded edges. Walls thin. Cell contents: 1 simple crystal, corresponding in its dimensions with those of the cells, surrounded by a crystal skin. M u c i l a g e c e l l s. R. and T. up to 120 μ, L. up to 90 μ; cylinders with a longitudinally directed axis. Walls very thick, almost to the disappearance of the lumina; stratification somewhat indistinct during the process of solution in glycerine, the middle lamellae quite disappearing during this process; some radial striation apparent in alcohol.

Endodermis. Developed as a not very distinct starch-sheath, in shoots 1 year old the cells being conspicuous by dark red-brown contents; 1—3 layers of cells, arranged in longitudinal rows, the cells of adjoining rows alternating with one another like brick-work. In the drug not to be distinguished.

C e l l s o f t h e a b o v e. R. and T. 10—15 μ, L. 10—25 μ; polygonal prisms with a longitudinally directed axis. Longitudinal intercellular spaces distinct.

S t e l e.

Pericycle. In young shoots somewhat gutter-shaped longitudinal bundles, thick some 3—6, broad up to 10 or 15 layers of sclerenchyma fibres, mixed here and there with some parenchyma cells; these bundles corresponding with the original vascular bundles and separated from each other by longitudinal parenchyma bands, consisting of 2—4 cells in a tangential direction, differ-entiated much in the same manner as the endodermis cells. In older shoots the parenchymatic tissue of the pericycle much increasing. In the drug the pericycle consisting almost wholly of parenchyma with a bundle of sclerenchyma fibres, scattered here and there. Parenchyma cells in the drug the same as those of the surrounding tissue. Sclerenchyma fibres in the drug R. 30 μ, T. 12 μ; flattened in a tangential direction. Walls slightly thickened, the secondary layers often detached from the middle lamella; not lignified.

Phloem.

P r i m a r y p h l o e m. In the drug not to be distinguished.

S e c o n d a r y p h l o e m. Thick some 2 m.M.; composed of tangential bands alternately belonging chiefly to the cribral and wholly to the parenchymatic system; some of the latter, perhaps one in five especially in the inner parts of the phloem, containing a tangential band of bast fibres.

Cribral system. Tangential bands of sieve-tubes thick 1—2 elements, very soon becoming more or less flattened and being quite so in the outer parts of the phloem. Companion cells wanting. **System of bast fibres.** The outer bands consisting of radially flattened and anastomosing bundles, thick 1—3 elements in a radial direction; the inner bands consisting of more or less continuous layers of fibres, — here and there a few fibres adjoining the clusters of sclereids to be mentioned beneath. **Parenchymatic system.** Consisting of bast parenchyma and medullary rays. B a s t   p a r e n c h y m a represented by metacribral bands of bast parenchyma fibres, mixed more or less with crystal fibres; the outermost bands moreover containing clusters of sclereids, much resembling those of the colourless cortical parenchyma, — bast parenchyma fibres between the sieve-tubes–crystal-sheaths generally completely surrounding the bands and bundles of bast fibres — and medullary rays. Metacribral bands thick 2—4 cells; the cells not being much differentiated near the cambium, outward becoming larger and showing more contents and intercellular spaces, corresponding to the increasing flattening of the sieve-tubes. The outermost bands, in the parts of the phloem not showing distinct medullary rays, consisting of a parenchyma much resembling that of the primary cortex and exhibiting the same clefts mentioned there. Bast parenchyma fibres consisting of about 8—10 cells, but in the outer parts of the phloem each cell divided again by a transverse wall and the fibres less distinct. Crystal

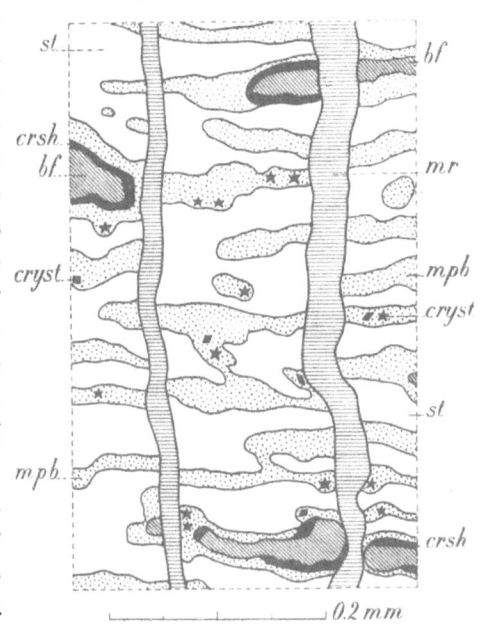

Fig. 17. *Rhamnus Purshiana*. Secondary phloem, transverse section showing the alternating bands. bf Bundles of bast fibres; crsh Crystal-sheath; cryst Crystal fibres; mpb Metacribral parenchyma bands; mr Medullary rays; st Sieve-tubes.

fibres most conspicuous in the outer metacribral bands; the fibres often containing in each cell some cluster crystals or simple crystals, but in many cases the cells being again divided by thin transverse walls into 3—4 short cells, each containing 1 cluster crystal, sometimes 1 simple crystal. Crystal-sheaths consisting of crystal fibres surrounding in 1 layer the bands and bundles of bast fibres; the whole fibres divided into short cells, each containing 1 simple crystal. M e d u l l a r y   r a y s separated from each other in a tangential direction by 2—14 layers of elements; uni- to tetraseriate, 5—50 cells in height, the broader ones being also the highest; outward often becoming somewhat undulated, with broader and radially shorter cells and a less regular radial arrangement; all cells procumbent, but in the outermost parts becoming even larger in a tangential than in a radial direction. Here and there 2 medullary rays placed in the same radial plane and separated from each other by 1 or 2 layers of obliquely arranged elements.

S i e v e - t u b e s. R. and T. 20—30 μ, L. of articulations 350 μ; polygonal prisms with rounded edges. Transverse walls placed very obliquely on the radial side walls, and each showing a large number of sieve-plates. Walls somewhat thickened; that of the partly flattened sieve-tubes sometimes becoming red in phloroglucin and hydrochloric acid; the tangential walls showing here and there a small sieve-plate; intercellular spaces wanting. Cell contents: here and there, adjoining the walls, some dirty-yellow masses. B a s t f i b r e s. R. and T. 8—12 μ, but the tangential dimension generally the largest; mostly hexagonal. Walls very thick, almost no cavity remaining; middle lamella distinct; lignified; pit canals and intercellular spaces wanting. B a s t p a r e n c h y m a f i b r e s a n d c r y s t a l f i b r e s mixed with these. R. 8—12 μ, T. 12—18 μ, L. of cells 20—60 μ, of crystal cells 10—15 μ; usually hexagonal with rounded edges. Radial walls sometimes pitted; intercellular spaces. Cell contents: many yellow-green plastids filled with starch; becoming red in potash; for the crystal cells see above. C r y s t a l - s h e a t h f i b r e s. R. 5—8 μ, T. 10—15 μ; L. of cells somewhat 10 μ; tetra- to hexagonal. Walls very thin; showing intercellular spaces on the side of the bast parenchyma. Crystals corresponding in size with that of the cells, surrounded by a somewhat lignified crystal skin. S c l e r e i d s somewhat smaller than those of the cortical parenchyma, but for the rest being wholly the same. C e l l s o f m e d u l l a r y r a y s. R. 20—40 μ, T. 10—25 μ, L. 10—20 μ; tetra- to octogonal prisms with a radially directed axis and rounded radial edges. Intercellular spaces in a radial direction between the cells and on the side of the bast parenchyma. Cells contents: the same as that of the parenchyma cells but plastids more numerous and smaller.

Micrography of the powder. In many fragments of tissue a yellow colouring matter, soluble in chloral hydrate. Parenchyma cells having a more or less oblong, somewhat variable shape; walls mostly thin, sometimes with collenchymatous thickenings, always colourless; cluster crystals in a tolerable number of these cells. Sometimes medullary rays running across the other tissue. Bundles of very long fibres or pieces of these; walls of fibres very thick, colourless, with distinct pit canals; cell cavity very narrow. These bundles of fibres generally surrounded by a sheath of crystal fibres with thin walls and partitioned into cells, each containing a square crystal; the walls of the crystal fibres generally being torn by the powdering process and the crystals remaining, the bundles of thick-walled fibres generally seeming to be covered on all sides with crystals. Isolated simple crystals and cluster crystals. Bright brown, more or less oblong corpuscules, the contents of periderm cells. Periderm cells penta- or hexagonal flat prisms; walls thin and colourless; contents a homogeneous, bright brown mass, like that mentioned above. Clusters of sclereids catching the eye as large dark masses, but only on careful inspection (chloral hydrate) appearing to consist of sclereids; also some isolated sclereids. The sclereids irregularly oblong and varying in shape, with rounded ends; walls very thick, almost colourless, showing pit canals; cell cavity very small. Starch in tolerable quantity; small globular grains, generally somewhat numerous in some parenchyma cells.

September 1912.　　　　　　　　　　　　　　　　　　　　　　　　　　J; M.

CORTEX SYZYGII.

Syzygium Bark.

The dried bark of Eugenia Jambolana, Lam. Encyc. III. 189 (Syzygium Jambolanum, DC. Prod. III. 259).

Macroscopic characters.

Pieces flat or somewhat gutter-shaped, long up to 15 c.M., broad up to 7 c.M., thick up to 2 c.M., light, tough, fibrous at the sides; the rhytidoma often being removed. Outer surface, the rhytidoma being present, uneven; with larger and smaller longitudinal, also transverse fissures; light brown, sometimes with irregular lighter yellow-brown spots; the rhytidoma being removed surface more even, red-brown, somewhat longitudinally striated. Inner surface fairly even, with fine longitudinal striae; red-brown to black-brown. Transverse fracture coarsely fibrous, splintery to granular. Transverse section, the rhytidoma being present, divided by an irregularly broken line in an outer darker brown half — the rhytidoma — and an inner lighter coloured, red-brown part; in both parts scattered, irregularly roundish, lighter coloured spots; moreover, especially in the inner part, very many exceedingly fine darker spots, causing together a very fine, somewhat irregular tangential striation. The rhytidoma being removed, the transverse section only showing the inner, lighter coloured part mentioned above. Eugenia Jambolana is called in Dutch East-India, among other names, with that of djamblang. Odourless; taste somewhat adstringent.

Anatomical characters.

LITERATURE. Hager. Pharm. Praxis. Bd. II. 1902. 1010. Lignier. Rech. s. l'anat. compar. d. Calycanthées, d. Melastomacées et d. Myrtacées. Diss. Paris. 1887. 358—374. Planchon et Collin. Drogues simples. T. 2. 1896. 340. MATERIAL. The drug, received from Gehe & Co. June 30th 1902, and M. P. Spillenaar Bilgen, Apothecary at den Haag; young leaf bearing shoots thick 2.5 m.M., and older shoots thick 1.5 c.M., collected in the Botanic Garden at Buitenzorg, and received Nov. 2nd 1911. REAGENTS. Water, glycerine, potassium iodide iodine, phloroglucin and hydrochloric acid, iodine and sulphuric acid 66 per cent.

MICROGRAPHY.

E p i d e r m i s. Thrown off very soon, always wanting in the drug. Cells often arranged in longitudinal rows of a somewhat fibrous shape and composed of some 12 cells, without doubt derived from a single cell by repeated transverse divisions. Stomata wanting.

E p i d e r m a l   c e l l s. R. 10 μ, T. 12—18 μ, L. 12—25 μ; tetragonal prisms with a radially directed axis. Outer walls thickened, showing a cuticle; lateral walls pitted.

C o r t e x. Thrown off in the drug.

Secondary cork tissue. Initial celled cork, derived generally from the first layer of cortical cells, sometimes from the second. Lenticels not very numerous.

Primary cortex proper. Thick some 8 layers of cylindrical cells. The outer part, thick some 3 layers, consisting of common parenchyma cells, showing here and there a longitudinally directed ellipsoidical oblito-schizogenous [1])

---

[1]) L u t z. Über die oblito-schizogenen Secretbehälter der *Myrtaceen*. Bot. Centralbl. Bd. 64. 1895. 145. Also Diss. Bern. 1895.

gland between them; moreover here and there some large more thin-walled parenchyma idioblasts. The inner part of the cortex, thick some 5 cell layers, consisting of sclereids as a chief constituent, mixed up with a few common parenchyma cells and showing, especially inward, more numerous large thin-walled parenchyma idioblasts, arranged in long longitudinal rows.

S c l e r e i d s with thin outer walls, sometimes also the outer parts of radial and transverse walls being thin; for the rest the walls mostly much thickened and showing distinct pits.

Endodermis. Developed as an interrupted starch-sheath; the starch grains heaped up against one of the transverse walls, in all cells in the same direction.

S t e l e.

Pericycle. Thrown off in the drug. In shoots thick 2.5 m.M. consisting of longitudinal bundles of sclerenchyma fibres and common parenchyma. The bundles of sclerenchyma fibres thick in a radial direction 1—5 mostly 2, in a tangential 6—20 elements, having a smooth outer and a more or less uneven inner surface. Some isolated sclerenchyma fibres moreover mixed with the parenchyma on the inner side of the bundles. Common parenchyma covering the bundles of fibres on their outside as an often interrupted layer of 1 cell thick; separating the bundles in a tangential direction by radial bands,

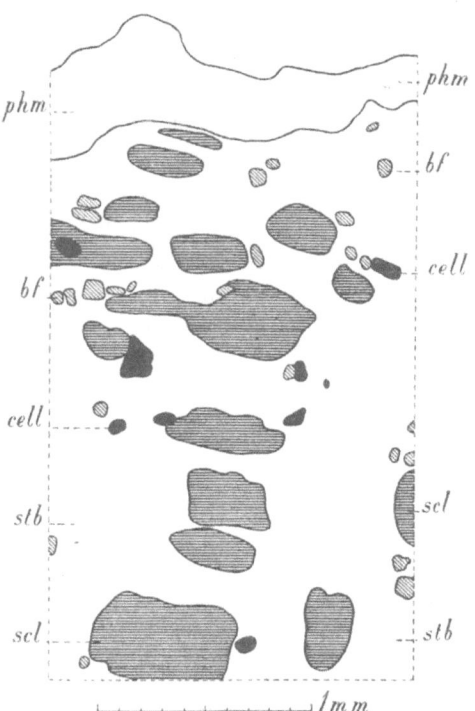

Fig. 18. *Eugenia Jambolana*. Bark, part with outermost layer of secondary cork tissue, transverse section. bf Bundles of bast fibres; cell Cells with more dark brown and granular contents; phm Phellem; scl Clusters of sclereids; stb Sieve-tubes and bast parenchyma.

broad 1 layer of often tangentially elongated and sometimes more or less sclerotized cells; covering the bundles on their inside as a layer thick in a radial direction 1—3 cells. Intercellular spaces between the parenchyma cells; these cells showing red homogeneous contents.

Primary phloem thrown off in the drug.

The parts now to be described are present in the drug.

Secondary phloem.

R h y t i d o m a thick up to 4 m.M.; annular, sometimes scaly; generally consisting of a thick layer of secondary phloem and a thin inner layer of secondary cork tissue, here and there traversing the secondary phloem in a an oblique direction and often not quite full grown. In some cases on the outside of the drug 1 or even 2 preceding secondary cork layers with corresponding layers of secondary phloem between them, being present.

Layer of secondary phloem coloured more or less red-brown by the contents of many and the walls of some cells. See for the rest the secondary phloem proper.

Layers of secondary cork tissue. Initial celled cork.

P h e l l e m thick some 6—15 layers of periderm cells, showing not only a strict radial, but moreover an arrangement in alternating tangential layers of radially longer and radially much shorter cells, each layer being generally 1 cell thick, the layers of larger cells in the outer parts of the phellem sometimes 2 cells.

P h e l l o g e n often indistinct, in other cases consisting of 1 layer of cells.

P h e l l o d e r m thick some 1—4 layers of parenchyma cells, with here and there an idioblast containing a cluster crystal.

P e r i d e r m c e l l s. R. of the radially larger cells 20—40 μ, of the smaller 10 μ; of both cells T. 20—35 μ, L. 15—30 μ; penta- to octogonal prisms with a radially directed axis; the radially larger cells of the outer layers having somewhat rounded edges and radial intercellular spaces. Walls thin; especially those of the radially smaller cells brownish yellow; often lignified; those of the larger cells pitted.

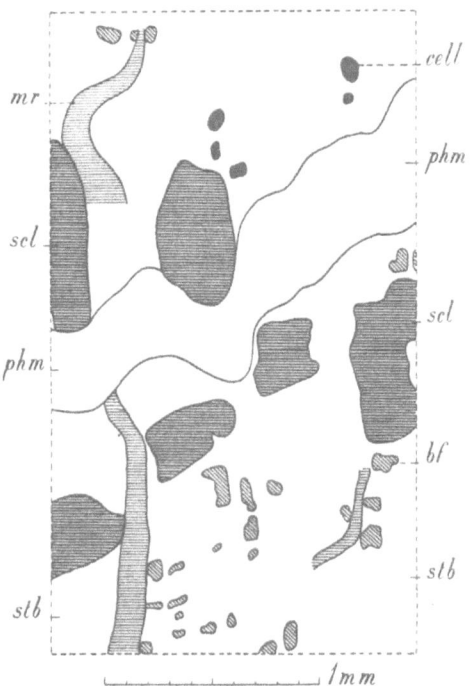

Fig. 19. *Eugenia Jambolana*. Bark, part with inner-most layer of secondary cork tissue, transverse section. bf Bundles of bast fibres; cell Cells with more dark brown and granular contents; mr Medullary rays; phm Phellem; scl Clusters of sclereids; stb Sieve-tubes and bast parenchyma.

S e c o n d a r y p h l o e m p r o p e r showing only in its inner parts a distinct radial arrangement. In the older parts groups of parenchyma cells sclerotizing and much increasing in size and also the common parenchyma cells much grown and hence the radial arrangement much less distinct. The tangential growth of the bast only resulting from this growth of sclereids and parenchyma cells, the medullary rays remaining outwards about of the same breadth. Consisting of a ground mass belonging to the parenchymatic system, enclosing bundles of bast fibres, each adjoining the inside of a bundle of elements belonging to the cribral system. These combined tangentially elongated bundles, especially in the vicinity of the cambium, more or less distinctly arranged in tangential rows.

Cribral system always more or less flattened; only sieve-tubes distinguished. **System of bast fibres.** Bundles often anastomosing in a tangential plane; thick in a radial direction 1—3, mostly 1 or 2 elements; in a tangential 1—5. **Parenchymatic system.** Consisting of bast parenchyma and medullary rays. B a s t p a r e n c h y m a consisting of common parenchyma fibres, mixed up with numerous crystal fibres, divided in many short cells, each containing a single

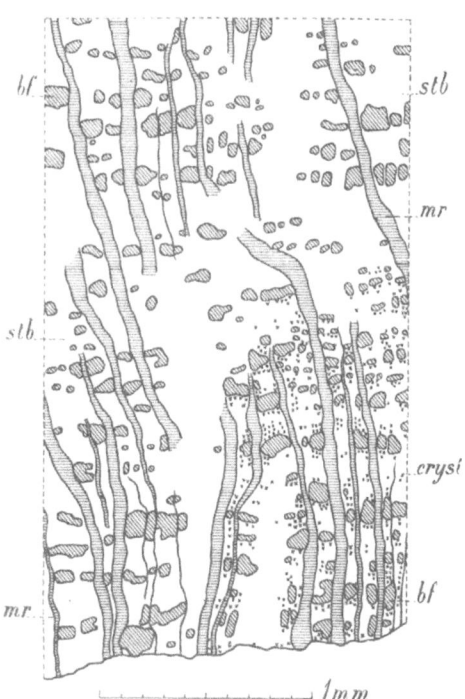

cluster crystal; these crystal fibres towards the cambium constituting the ground mass. Moreover sclereids in short, often radially elongated groups consisting of a smaller or larger number of often tangentially elongated elements. These groups of sclereids wanting in the vicinity of the cambium in a layer thick up to 3 m.M.; outwards increasing in number, also in dimensions by the growth of the constituting elements. Here and there in the older parts of the phloem a cell with more dark brown and granular contents. M e d u l l a r y r a y s separated from each other in a tangential direction by 3—6 layers of elements; uni- to penta-, mostly tri- to pentaseriate, 4—35, mostly 20 cells in height; in many cases compound and showing 3 stories, the upper and under one being some cells high. All constituting cells

Fig. 20. *Eugenia Jambolana*. Bark, inner part, transverse section. bf Bundles of bast fibres; cryst Crystal cells, only drawn in a small part of the figure; mr Medullary rays; stb Sieve-tubes and bast parenchyma.

procumbent, those of the uniseriate rays and parts, sometimes also those on the radial sides of the broader parts, somewhat larger in a tangential and shorter in a radial direction.

S i e v e - t u b e s. Cross walls placed very obliquely; showing very numerous transverse elliptical sieve-plates, arranged in a vertical row. B a s t f i b r e s. R. 40 μ, T. 30 μ; penta- to octogonal, the outmost fibres of the bundles showing rounded edges. Walls very thick, cavity small; lignified, especially the middle lamella; showing distinct pit canals. In the outer parts of the rhydidoma sometimes the secondary layers of the walls more or less dissolved. C e l l s o f c o m m o n p a r e n c h y m a f i b r e s. R. 20—60 μ, T. 40—100 μ, L. about 50 μ; penta- to octogonal prisms with a longitudinally directed axis and rounded edges. Intercellular spaces often distinct. Cell contents: simple starch grains, afterwards a brown mass. C e l l s o f c r y s t a l f i b r e s. R. and T. 10—30 μ, L. 15—35 μ; tetra- to hexagonal prisms with a longitudinally directed axis. Walls thin; in the younger parts of the phloem intercellular spaces wanting. Cell contents: 1 cluster crystal, quite filling the cavity. S c l e r e i d s. R. 25—120, generally 70 μ, T. 200—350, generally 260 μ, L. generally 70 μ; shape variable, irregular, often more or less curved. Walls thick 1.5—3.5 μ; often showing stratification; lignified; with cross slit-like pits. In the outer part of the rhytidoma often spirally wound slits on the inside of the walls, probably resulting from partial solution. C e l l s o f m e d u l l a r y r a y s. R. 40—80 μ, T. 20—35 μ, L. 15—35 μ; penta- to octogonal prisms, with a radially directed axis and strongly rounded edges. Walls very thin, at most 1 μ; showing distinct radial intercellular spaces. Cell contents: almost always a red-brown mass.

1913.   J; M.

CORTEX VIBURNI PRUNIFOLII.
Viburnum Prunifolium. Black Haw.
The dried bark of Viburnum prunifolium, Linn. Sp. Pl. 268.

Macroscopic characters.

Strips, more or less transversely curved pieces or small quills of various length, the bark thick up to 2 m.M., but mostly much thinner; the thicker pieces brittle, the thinner ones more tough. Outer surface darker or lighter grey-brown; that of the thinner pieces more smooth, somewhat purplish, with oblong transversely elongated lenticels; that of the thicker pieces more fissured, especially in an irregularly longitudinal direction and sometimes covered with lichens and fungi. Inner surface more or less uneven and longitudinally striated; light red-brown; pretty often showing chips of white wood. Transverse fracture of the thicker pieces uneven; the outer part scaling off in the shape of thin plates, darker; the inner part coloured lighter, with irregularly scattered coarse dots or corresponding small pits, formed by the removal of tissue. Odour faint; taste somewhat adstringent and bitter.

Anatomical characters.

LITERATURE. Douliot. Recherches sur le périderme. Ann. Sc. Nat. Sér. 7. T. 10. 1889. 385. Hérail. Pharmacol. 1900. 745. Karsten u. Oltmanns. Pharmakogn. 1909. 146. Kraemer. Botany and Pharmacogn. 1910. 774. Moeller. Anat. d. Baumrinden. 1882. 147. Planchon et Collin. Drogues simples. T. 2. 1896. 199. Sanio. Vergl. Unters. üb. den Bau u. d. Entwick. d. Korkes. Pringsheims Jahrb. Bd. 2. 1860. 59. Schneider. Powdered veget. drugs. 1902. 319. MATERIAL. The drug, pieces thick 0.5 to 2 m.M. REAGENTS. Water, glycerine, potassium iodide iodine, phloroglucin and hydrochloric acid, iodine and sulphuric acid 66 per cent., potash.

MICROGRAPHY.

The secondary cork tissue is developed in the epidermis, according to Sanio [1]) and Moeller [2]). In the drug all pieces still showing this original cork layer, covering the primary cortex.

E p i d e r m i s.
Secondary cork tissue.

C o r k t i s s u e p r o p e r. Initial celled cork consisting in the case of the thinner pieces, mentioned under macroscopic characters, of only about 10 layers

---

[1]) l. c. 59.
[2]) l. c. 147.

of phellem, in the thicker pieces phellem thick up to 1 m.M., developed as periderm cells; showing a radial arrangement, often somewhat indistinct in consequence of the subsequent growth of the cells. The 1 or 2 inmost layers of cells being much wider in a radial direction.

P e r i d e r m   c e l l s. R. 8—30 μ, T. 50—85 μ, L. 30—40 μ; originally penta- or hexagonal prisms with a radially directed axis, but the cells of the outer layers having become afterwards cushion-shaped, in consequence of subsequent lateral growth causing an outward folding of the transverse and radial walls. Walls colourless; pitted. Cell contents: a yellow-brown homogeneous mass, becoming red-brown in potash.

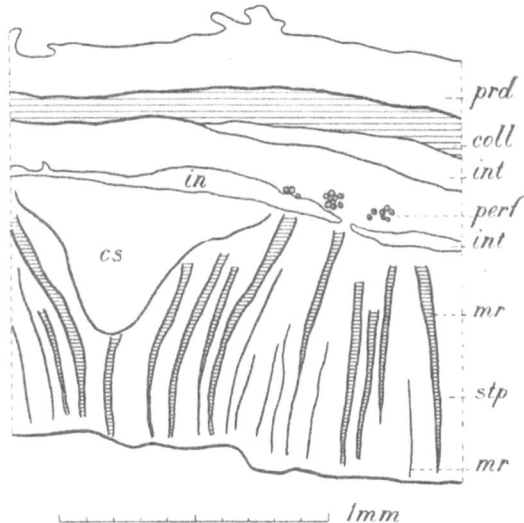

Fig. 21. *Viburnum prunifolium*. Bark, transverse section. coll Collenchyma; cs Cuneiform strips of parenchyma; int Intercellular spaces; mr Medullary rays; perf Pericyclic fibres; prd Periderm; stp Sieve-tubes, companion cells and bast parenchyma.

L e n t i c e l s.

Phellem. The central portion only consisting of about 10 cell layers belonging to the innermost layer of complementary cells; all other layers of complementary tissue and all closing bands having disappeared. The peripheral and persistent portion of the lenticel consisting of about 5 closing bands, each thick about 2 of 3 cells conspicuous by their brown contents, and alternating with layers of large thin-walled parenchyma cells, without intercellular spaces. All cells radially arranged.

C o m p l e m e n t a r y   c e l l s. R. 15—30 μ, T. 20—40 μ, L. 20 —35μ; prisms with rounded edges or cylinders, both with a radially directed axis. Walls thick up to 1 μ; often somewhat yellow; pits wanting. C e l l s   o f   t h e   l a y e r s   b e t w e e n   t h e   c l o s i n g   b a n d s   o f   t h e   p e r i p h e r a l   p o r t i o n. Prisms with a radially directed axis. Walls thin; intercellular spaces wanting. C e l l s   o f   c l o s i n g   b a n d s. R. 6—10 μ, T. and L. 30 μ; prisms with a radially directed axis. Walls thick 1.5 μ; no pitting. Cell contents: a yellow to red-brown mass.

Phellogen. Consisting of 4 or 5 cell layers.

C e l l s   o f   t h e   a b o v e. R. 6 μ, T. and L. 20—30 μ.

Phelloderm. Consisting of 5—10 layers of radially arranged cells.

C e l l s   o f   t h e   a b o v e. R. 6—12 μ, T. and L. 20—30 μ. Walls thick 1—1.5 μ; colourless; with distinct intercellular spaces.

C o r t e x.

Collenchyma. Thick about 5 cells in a radial direction; these cells generally arranged in horizontal layers and a few times in longitudinal rows, the transverse walls in these rows being thinner than elsewhere. Many cells divided by thin radial walls in 2, sometimes in 3 or 4 partitions.

Cells of the above. R. 16—20 µ, T. 70—125 µ, L. 20—25 µ. Walls thick 4 µ, the thinner radial ones 1.5 µ; intercellular spaces wanting. Cell contents: a greenish granular mass containing many very small starch grains and tannin.

Common parenchyma. Thick about 5 cells in a radial direction, the cells being arranged in horizontal layers. Intercellular spaces in all parts of this tissue and often very large and elongated in a tangential direction, the largest ones being found in the neighbourhood of the collenchyma e. g. R. 50 µ, T. 900 µ, L. 750 µ.; the cells adjacent to these spaces often protruding into them. Many cells divided by thin radial walls in 2, sometimes 3 or 4 partitions. Here and there a few idioblasts containing each a cluster crystal and sometimes arranged in short tangential rows.

Cells of common parenchyma. R. 20—30 µ, T. 45—120 µ, L. 20—30 µ; polyhedra with rounded edges. Walls sometimes collenchymatous, thick 4—6 µ, the thin radial ones 1.5 µ; often pitted. Cell contents: a yellowish green granular mass containing many very small starch grains and tannin. I d i o b l a s t s mostly wholly filled up by a cluster crystal, having a diameter of up to 35 µ. Walls thin.

Endodermis not to be distinguished in the drug.

S t e l e.

Pericycle. Consisting of a few layers of parenchyma cells, resembling in all respects those of the inner layers of the primary cortex and the outer layers of the secondary phloem and showing the same idioblasts and intercellular spaces. Imbedded in this parenchyma here and there some longitudinal bundles of sclerenchyma fibres consisting in a cross section of 1—8 elements; the bundles as well as their constituting fibres showing larger tangential than radial dimensions. These bundles having a tendency to aggregate into groups of 2 or 3 bundles and often anastomosing by means of a single fibre. The bundles often running more or less separated from the surrounding parenchyma and often their constituting fibres being also more or less loosened from each other.

Sclerenchyma fibres of the above. R. 12—20 µ, T. 28—40 µ; hexa- to octogonal with more or less rounded edges or even elliptical. Walls thick at least 6 µ, often thicker, up to the disappearance of the cavity; chemical reactions omitted in this case; pits wanting.

Phloem.

Primary phloem not to be distinguished.

Secondary phloem. Radial arrangement of the elements very distinct in the vicinity of the cambium, but somewhat broken outward by the growth of the crystal cells and in the outermost layers by the tangential growth of all elements. The relatively enormous tangential expansion of the outer part of the secondary phloem, necessary for keeping up with the first increase in thickness of the thin branch, chiefly resulting from tangential growth, combined with cell divisions in cells of the parenchymatic system, and leading to the formation of longitudinal cuneiform strips of parenchyma, turning their edges inward, and resembling in all respects the parenchymatic parts of the pericycle; these strips most probably corresponding in place with the original medullary commissures. The same tangential growth in the parts between the above cuneiform strips due to two causes: 1. tangential growth without cell division of the elements of the parenchymatic system, the outermost cells of the medullary rays, near the pericycle, showing in consequence even the same dimensions and shapes as the surrounding bast parenchyma cells; 2. cell divisions in the cambium, increasing the number of radial rows of this tissue. The latter two kinds of tangential growth leading to the formation of longitudinal cuneiform strips of tissue with their edges outward, wedged in between the parenchymatic strips mentioned above. The medullary rays enclosed in the latter wedges and numbering in each 8—12, thus in a transverse section more or less curved and approaching each other outwards; the number of separating cells increasing inward from 2—5 to 6—15. The less intense later tangential growth of the inner part of the bark due only to the two causes mentioned above as operating in the cuneiform strips, turning their edges outwards. Elements of the system of bast fibres wanting. Clusters of sclereids wanting in the thinner pieces of bark, present in the thicker and scattered through a zone, reaching from the vicinity of the cambium outward to a distance from the pericycle smaller than the whole secondary phloem of the thinner pieces; the clusters having everywhere approximately the same dimensions and consisting of the same number of cells, all the sclereids themselves showing the same thickness of the walls.

Cribral system. More or less distinctly arranged in tangential layers, thick 1 element; flattened only in the outer part of the bark; consisting of sieve-tubes and a very few companion cells.

S i e v e - t u b e s. At a distance of 150 μ from the cambium R. 20 μ, T. 35 μ, L. of articulations e. g. 210 μ; with slightly oblique transverse walls; sieve-plates not very distinct and occurring also here and there on the side walls. Sometimes a callus mass near the transverse walls.

Parenchymatic system. Consisting of bast parenchyma and medullary rays. B a s t   p a r e n c h y m a. Consisting of bast parenchyma fibres, showing here and there conjugated cells; crystal fibres divided into 10—25 cells, each containing a single cluster crystal. Sclerenchyma, clusters of sclereids rather equally distributed but not arranged in longitudinal rows, L. up to 2 m.M., generally spindle-shaped and consisting of a few up to very many cells; the adjoining bast parenchyma cells often more or less flattened, but especially on both ends of the spindles the parenchyma cells more or less wedged in between the sclereids. M e d u l l a r y   r a y s. Separated from each other in a tangential direction by 1—5, mostly 3 elements. Rays in 2 kinds: the first exclusively simple, uniseriate, mostly 7—8 cells in height and built up of upright cells; the second kind consisting of simple and compound rays, the simple ones bi-, rarely triseriate, 6—10 cells in height and, with the exception of the upper and lower radial row, built up of procumbent cells; the compound rays showing mostly 3, rarely 5 stories; the broader ones bi-, rarely triseriate, 5—6 cells in height and for the rest resembling in all respects the biseriate simple rays; the uniseriate stories 2—6 cells in height and resembling for the rest in all respects the rays of the first kind. In both kinds of rays here and there clusters of sclereids, resembling in all respects those described for the bast parenchyma. In the triseriate rays and stories rarely a radially directed schizogenous duct.

Cells of bast parenchyma fibres. R. 10—12 μ, T. 20—25 μ, L. 50—90 μ; tetra- to hexagonal prisms with a longitudinally directed axis. Walls thick 2 μ. Cell contents: granular, somewhat yellow, in potash coloured somewhat darker yellow, moreover in this reagent oil drops appearing. Cells of crystal fibres. R. and T. in the vicinity of the cambium 10—15 μ, L. 15—25 μ. Partition walls very thin. Cell contents: 1 cluster crystal, enclosed in a crystal skin. Sclereids. R., T. and L. 50—80 μ; polyhedra, those on the outside of the clusters with outwardly curved outer walls. Walls thick 12—30 μ, thickest on those sides of the cells perpendicular to their largest diameter; lignified, especially the middle lamellae and the walls lying on the outside of the clusters; showing many pit canals. Procumbent cells. T. 12—20 μ, L. 10—25 μ; penta- to octogonal prisms with a radially directed axis and at least the radial edges rounded. Walls thick 1—2 μ; pitted. Intercellular spaces especially in a radial direction. Cell contents: a granular yellow-green mass, becoming brown in potash and then showing the same oil-like globules as mentioned in the parenchyma cells. Upright cells. R. 30 μ, L. 60 μ, T. near the cambium 15 μ, at a distance of 0.5 m.M. from the cambium 25—30 μ, in the outer parts of the secondary phloem 45 μ; tetragonal prisms with a longitudinally directed axis, this shape changing during the subsequent tangential growth to that of a tangentially elongated common parenchyma cell with rounded edges. Walls thick 2 μ. Cell contents: the same as those of the procumbent cells. Sclereids the same as those of the bast parenchyma.

September 1913. J; M.

FLORES ARNICAE.

Arnica.

The dried flowers of Arnica montana, Linn. Sp. Pl. 884.

Macroscopic characters.

Heads up to 6 c.M. in diameter. Receptacle up to 9 m.M. in diameter, flat, afterwards and always in a dry state convex; the margin of the flower-scars showing a ring of short hairs. Involucre long 1.5 c.M., cylindrical, afterwards thrown backwards; bracts 20—24 in number, arranged in 2 ranks; the outer ones lanceolate; the inner ones linear, green with darker coloured midrib; outer surface and margin with many hairs and between these shorter, petiolate, red glandular hairs. Ray-flowers, number: n = 346, Med = 16.69, Q_1 = 1.58, Q_3 = 1.53, min. = 10, max. = 24, arranged in a single rank, long 3 c.M., female, epigynous; pappus white; corolla ligulate, yellow, 3-dentate at the top and 7-to 12-veined, tube showing hairs and glandular hairs, limb with scanty hairs on its outer side; stamens wanting or represented by 3—4 separate filiform rudiments; pistil compound, consisting of 2 carpels, with 1 long style and 2 yellow stigmata curved outward; ovary somewhat cylindrical, unilocular, white, afterwards of a

9

dark colour, with numerous short acute bristles, directed upward and short glandular hairs, 1 ovule. Disk-flowers numerous, up to 90, long 15 m.M., complete, actinomorphous; corolla gamopetalous, 5-lobed, tubularly campanulate, yellow; stamens 5 in number, inserted on the corolla, syngenesious; anthers showing on their top a flat oblong triangular prolongation of the connective, at first yellow, afterwards brown; pollen yellow; for the rest the same as the ray-flowers. Odour slightly aromatic; taste slightly bitter.

Flores Arnicae may not contain receptacles, involucres and fruits.

Anatomical characters.

LITERATURE. Benecke. Mikr. Drogenprakt. 1912. 69. Biechele. Mikr. Prüf. d. off. Drogen. 1904. 33. Daniel. Rech. anat. sur l. bractées de l'Inv. des Comp. Ann. d. Sc. nat. Sér. 7. T. VI. 1890. 17. Ellrodt. Verteil. d. Gerbstoffs in off. Blätt., Kräut. u. Blüten. Diss. Würzburg. 1903. 25. Flückiger. Pharmakogn. 1891. 819. Gerdts. Bau u. Entw. d. Kompos.-frucht m. bes. Berücks. d. off. Arten. Diss. Bern. 1904. 11, 37. Gilg. Pharmakogn. 1910. 356. Karsten u. Oltmanns. Pharmakogn. 1909. 222. Koch. Mikr. Anal. d. Drogenpulver. Bd. 3. 1906. 177. Pl. XVI. Koch u. Gilg. Pharmacogn. Prakt. 1907. 161. Kraemer. Botany a. Pharmacogn. 1910. 746. Marmé. Pharmacogn. 1886. 269. Meyer. Wiss. Drogenk. Bd. 2. 1892. 296. Oudemans. Pharmacogn. 1880. 326. Schneider. Powd. veg. Drugs. 1902. 123. Tschirch. Sind die Antheren der Compositen verwachsen oder verklebt? Flora. Bd. 93. 1904. 51. Pl. 2. van Wisselingh. Bijdr. t. d. Kennis v. d. Zaadh. 1e Bijdr. Comp. Pharm. Weekblad. 1918. 871. MATERIAL. Flower heads gathered June 1902 in the neighbourhood of Groningen (Peizer-maden), fresh and in alcohol. REAGENTS. Water, glycerine, chloral hydrate, phloroglucin and hydrochloric acid, iodine and sulphuric acid 66 per cent., cupric acetate and ferric acetate, potassium dichromate.

MICROGRAPHY.

D i s k-f l o w e r.

Pappus. Nearly always having the same length as the corolla. Axis of the cells longitudinally directed; uppermost ends of cells bending off, the longest cells deviating least.

C e l l s o f t h e a b o v e. R. and T. 15 μ, L. 80—160 μ; cylinders. Walls pitted. Cell contents: mostly one or more sap cavities, on treatment with potassium dichromate filled with a reddish brown precipitate; moreover some adventitious vacuoles remaining colourless in the same reagent.

Corolla.

E p i d e r m i s. Inner side. Cells in longitudinal rows, especially at the base of the corolla. Showing smaller or larger papillae, most distinct above the veins and on the teeth. Above the veins the cells much larger than elsewhere.

C e l l s o f t h e a b o v e. H. 20 μ, Lev. B. 15—20 μ, Lev. L. 70—120 μ, the largest cells in the middle part of the corolla; on the veins H. 20 μ, Lev. B. 40 μ, rectangular; cells of the teeth Lev. 35 μ, polygonal. Outer walls showing radial cuticular striation. Cell contents: in buds and newly expanded flowers yellow chromoplasts; in older flowers a finely granular yellow mass; nucleus distinct. At the base of the corolla and especially above the veins and on the teeth in each cell one or more larger vacuoles, on treatment with potassium dichromate filled with a reddish brown precipitate; moreover one or more smaller ones remaining colourless in the same reagent. In the papillae fine granules, colouring red in phloroglucin and hydrochloric acid; to a certain extent the same in the other cells.

Outer side. Cells in longitudinal rows, especially at the base of the corolla. Above the veins cells often larger than elsewhere; in the middle part of the corolla cells longest. Stomata. Some on each tooth, phaneroporous, lying in the same level as the surrounding epidermal cells. Trichomes in 2 kinds. 1°. Glandular hairs of the Compositae type on the undermost half of the corolla and on the teeth, consisting of 4—6 stories of 2 cells, each with a cuticular bladder covering only the uppermost stories. 2°. Conical hairs, numerous on the undermost half of the corolla, on the uppermost half only here and there on the veins, arranged in longitudinal rows, consisting of 1 row of 5—7 cells, at their base 50 μ in diameter, up to 600 μ long.

E p i d e r m a l c e l l s p r o p e r. H. 18 μ, Lev. B. 15—25 μ, Lev. L. 50—150 μ; rectangular; longitudinal lateral walls sinuous. Outer walls showing longitudinal cuticular striation. Cell contents: see the inner side; but here throughout the epidermis all the cells showing the precipitate with potassium dichromate, clearliest to be seen at the base and the top of the corolla and above the veins. S t o m a t a. Lev. 30 by 40 μ. G u a r d-c e l l s. Contents: see the pappus. C e l l s o f g l a n d u l a r h a i r s. 30 μ in diameter, 60 μ high. Cell contents: see the pappus. C e l l s o f c o n i c a l h a i r s. At the base of the hair cells much broader and shorter than at the top. Outer walls showing a cuticle. Cell contents: see the pappus.

M e s o p h y l l. Consisting at the base of the corolla of about 5 layers of common parenchyma; for the rest wanting with the exception of some layers in the veins. Showing intercellular spaces.

C e l l s o f t h e a b o v e. Lev. B. 10—15 μ; cylinders with a longitudinally directed axis. Cell contents: in the veins often, in the rest of the parenchyma more rarely small vacuoles, on treatment with potassium dichromate filled with a dark brown precipitate.

M e r i s t e l e. At the base of the corolla 2 concentric rings of 5 meristeles each. Meristeles of the inner ring continued into the stamens; those of the outer ring continued to the margin of the corolla between the teeth and there bifurcating; the branches running along the margin of the teeth and anastomosing at the top; sometimes here again showing a continuation. Spiral vessels lignified.

Stamen.

F i l a m e n t.

Epidermis. Cells in longitudinal rows. For the greater part more or less regularly rectangular; thin-walled; sometimes with sinuous longitudinal lateral walls. In the upper part, adjoining the anther, the cells of the outer and lateral sides strictly rectangular, shorter and showing triangular collenchymatous thickenings of their lateral walls (so-called articulation).

T h i n-w a l l e d c e l l s. R. 18 μ, T. 10—15 μ, L. 30—80 μ, L. largest in the undermost half of the filament. Outer walls showing a cuticle. Cell contents: see the pappus, especially for the undermost half of the filament. C e l l s o f t h e a r t i c u l a t i o n. T. 15 μ, L. 15—30 μ. Walls thickened; lignified; pitted.

Mesophyll. Consisting of common parenchyma cells.

C e l l s o f t h e a b o v e. R. and T. 8—10 μ; circular cylinders with a longitudinally directed axis.

Meristele. Containing a vascular bundle and extending to the top of the filament.
A n t h e r .

Connective.

Epidermis. Some glandular hairs, as described above, on the outer side of the top of the connective. For the rest see the thecae.

Mesophyll. Consisting of some layers of cylindrical common parenchyma cells; the outer layer thick-walled, continuous with the fibrous layer and showing intercellular spaces.

Meristele. Wanting or only represented by a few elements showing a more distinct cellulose reaction than the rest.

C e l l s o f t h i c k - w a l l e d o u t e r l a y e r . R. 4 μ, T. 8 μ, L. 50—100 μ; polygonal tables. Walls thickened; lignified; pitted. Cell contents: see the pappus. C o m m o n p a r e n c h y m a c e l l s . R. and T. 8—10 μ. Cell contents: see the pappus.

Thecae.

Epidermis. Cells practically arranged in longitudinal rows; separate thecae connected together by epidermis cells [1]).

C e l l s o f t h e a b o v e . R. 6 μ, T. 15 μ, L. 70 μ; mostly rectangular. Outer walls showing a cuticle. Cell contents: see the pappus.

Fibrous layer. Consisting of 1 layer of cells; wanting at the line of dehiscence of the thecae.

C e l l s o f t h e a b o v e . H. 20 μ, T. 10 μ, L. 10—20 μ; H. largest near the connective; rectangular prisms. Walls with lignified reticulate thickenings. Cell contents: see the pappus.

Pollen. Grains 25 μ in diameter, globular, triporous; yellow. Exine finely dotted.

Pistil.

O v a r y .

Wall.

Epidermis. Outer side. Cells in longitudinal rows. Trichomes in 2 kinds. 1°. Glandular hairs; see the outer epidermis of the corolla of the disk-flower. 2°. Twin-hairs, pointing upwards; consisting of 2 adjacent cells each implanted on a small epidermal cell, or a few times of 2 rows of cells; at their base 10 μ in diameter, length 120 μ; top somewhat acute.

E p i d e r m a l c e l l s p r o p e r . R. 15 μ, T. 18 μ, L. 50 μ; cells bearing hairs R. 20 μ, T. 16 μ, L. 16 μ; rectangular tables. Outer walls showing a cuticle; the rest pitted. Cell contents: when treated with potassium dichromate a yellow mass showing colourless cavities.

Inner side.

C e l l s . 25 by 30 μ; polygonal. Walls very thin; outer walls showing a cuticle.

Mesophyll. Constituted of about 12 layers of elements. Outermost layer consisting of much wider common parenchyma cells. Between this outer layer and the inner layers of narrower elements a very spacious intercellular fissure, in most places bordered on its outside by the outermost layer of wider cells but in a few places by 1 or 2 layers of the narrower elements remaining attached to the

---

[1]) T s c h i r c h . l. c. 51, comes to a different conclusion.

outermost layer. On the inner surface of this fissure and covering the walls of the elements bordering it a black mass, remaining unaltered in water, alcohol 96 per cent., chloral hydrate, phloroglucin and hydrochloric acid, and sulphuric acid 66 per cent., and showing dendritic figuration on a tangential view; this mass increasing during the ripening of the fruit. The elements within the outer layer of wide cells to distinguish into 3 parts; 1°. 2 layers of strongly elongated sclerotic elements with lignified walls, partly lying on the outside of the cleft as mentioned above; walls of the elements bordering the fissure showing protuberances, projecting into it; 2°. about 5 layers of elements in some respects resembling those of the preceding layers, but without thickened walls, this tissue containing the meristeles showing spiral vessels with lignification of the spiral thickenings; 3°. about 5 layers of cylindrical common parenchyma cells with large intercellular spaces.

Elements of the mesophyll. Outermost wide cells. R. 16 μ, T. 14 μ, L. 30 μ; rectangular prisms with rounded edges and a longitudinally directed axis. Sclerotic elements. R. and T. 5 μ, L. very great; hexagonal. Cell contents: some chloroplasts and tannin. Common parenchyma cells of the innermost 5 layers. R. and T. 10—15 μ, L. 100—200 μ.

Conducting tissue. Forming 2 ridges, diametrically opposite to each other on the flat sides of the fruit and immediately adjoining the ovule. Each ridge containing 1 outer layer of common parenchyma enclosing some layers of thick-walled elements.

Common parenchyma cells. R. and T. 5 μ. Thick-walled elements. R. and T. 6 μ.

Ovule. Attached to the base of the ovarial cavity; anatropous. Integument very thick; the epidermal cells of the outer side containing tannin in globules; raphe being continued beyond the chalaza. Endosperm not discernible.

S t y l e. Epidermis. Cells exactly arranged in longitudinal rows. Cells of the uppermost end showing somewhat sinuous lateral walls.

Cells of the above. R. 20 μ, T. 20 μ, L. 40—65 μ. Outer walls showing a cuticle. Cell contents: see the pappus.

Mesophyll. Consisting of some layers of common parenchyma cells. Showing intercellular spaces.

Cells of the above. In the outermost 3 layers R. and T. 20 μ, L. 50 μ; polygonal prisms with rounded edges and a longitudinally directed axis; in the layers surrounding the meristeles R. and T. smaller, L. larger. Cell contents: in some cells in the vicinity of the meristeles tannin.

Meristeles. 2 on each side of the central conducting tissue. Phloem divided into bundles by intermediate parenchyma cells. Xylem containing spiral vessels with lignification of the spiral thickenings. Cell contents: see the pappus.

Conducting tissue. In the centre of a style; 8-shaped on a transverse view; consisting of very thick-walled elements.

Cells of the above. R. and T. 5 μ. Cell contents: see the pappus.

S t i g m a t a. Each stigma showing on the upper side 2 bands of thin, relatively long papillae; the longest ones at the margin. Conducting tissue immediately under the papillae consisting of 2—3 layers of elements. Moreover in each stigma a meristele containing spiral vessels with lignification. See for the rest the style.

R a y-f l o w e r.

Pappus. See the disk-flower.

Corolla.

Epidermis. Inner side. Cells in longitudinal rows; mostly somewhat papillaform; longitudinal lateral walls undulated, 1 wave covering 2 cells. Trichomes. At the base of the corolla and near the veins; in 2 kinds, like those of the disk-flower; for the rest see there.

E p i d e r m a l c e l l s p r o p e r. H. 35 μ, Lev. B. 20 μ, Lev. L. 30 μ; rectangular prisms. Outer walls showing mostly radial, at the base of the corolla longitudinal cuticular striation. Cell contents: in buds and newly expanded flowers yellow plastids; in elder flowers a yellow granular mass; nucleus distinct.

Outer side. Cells in longitudinal rows. Stomata, some on the top of each tooth. Trichomes more numerous than at the inner side; for the rest see there.

E p i d e r m a l c e l l s p r o p e r. H. 20 μ, Lev. B. 20 μ, Lev. L. 35 μ. L. greater on the veins and at the base of the corolla. Outer walls with longitudinal cuticular striation. Cell contents: see epidermal cells of inner side. On the veins and at the top of the corolla in each cell 1 or more large vacuoles on treatment with potassium dichromate filled with a reddish brown precipitate; moreover 1 or more smaller vacuoles remaining colourless in the same reagent. G u a r d-c e l l s o f s t o m a t a. Contents: see the epidermal cells proper.

Mesophyll. Consisting of 5—6 layers of common parenchyma. Cells often somewhat stellate.

C e l l s o f t h e a b o v e. In the veins polygonal prisms with a longitudinally directed axis and rounded edges. Cell contents: in many cells in potassium dichromate the same reaction as in the disk-flower; in some cells tannin in globules.

Meristeles. Mostly 9 in number; parallel to each other. The 2 outermost soon ending blind; the 2 next ones running along the margin to the top of the 2 outermost teeth; 3 of them, each running to the top of a tooth; the 2 last running to the incision of the teeth, bifurcating there, running along the margin of the teeth and there repeatedly anastomosing with each other and with the 2 marginal ones; here and there small side branches. Xylem containing spiral vessels with lignified spirals.

Stamen. Mostly represented by 3—4 rudimentary filaments.

Pistil. See the disk-flower.

June 1902.   J; M; L.

FLORES CHAMOMILLAE ROMANAE.
Chamomile Flowers. Anthemis. Flores Anthemidis.
The double heads of a cultivated variety of Anthemis nobilis, Linn. Sp. Pl. 849.

Macroscopic characters.
About 2 c.M. in diameter. Involucre in the shape of the undermost part of a hemisphere; bracts about 30 in number, more or less arranged in ranks, long 3—5 m.M., wide up to 1 m.M., lanceolate; base of the outermost bracts widened, of the innermost ones narrowed; only the middle part of the bracts pale green and on the outer side hairy; the margin very broad, especially towards the top, colourless, membranous, entire and glabrous. Receptacle high 3—4 m.M., wide 2—3 m.M., conical with a curved surface, internally filled with tissue, externally covered with numerous paleae; those oblong ob-ovate, towards the top finely serrate, at the dorsal side hairy. Ray-flowers very numerous, incomplete, without a pappus, female, zygomorphic, epigynous. Corolla ligulate, the limb bent outwards at about a right angle; tube long 2 m.M., very pale yellow, when magnified 50 times showing distinct short obtuse glandular hairs on the outside, especially near the throat; limb long 8—9 m.M., wide above 2 m.M., at the top with 2 minute incisions, with 8 veins, white. Pistil compound, composed of 2 carpels, with 1 style and 2 stigmata; the latter slightly projecting above the tube, flat, pale yellow, somewhat curved outwards, ligulate; ovary long 1 m.M., inferior, very pale yellowish green, when magnified 50 times showing a larger or smaller number of glandular hairs and a few longitudinal thin shining projecting ridges. Disk-flowers to a larger or smaller number, long 3.5 m.M., incomplete, without a pappus, hermaphrodite, actinomorphous, epigynous. Corolla gamopetalous, pentamerous, tubular; tube somewhat narrowing in the middle, yellowish white, lobes yellow; on the tube, especially above the stricture, glandular hairs; corolla of some flowers showing transitions into that of the ray-flowers. Stamens 5, inserted in the undermost part of the tube of the corolla, syngenesious; filaments rather long; anthers innate, opening introrsely by fissures; connectives continued into 5 projecting parts curved inwards. Pistil essentially the same as that of the ray-flower. Odour strongly aromatic; taste very bitter.

Anatomical characters.
LITERATURE. Daniel. Rech. anat. s. l. Bract. d. l'Involucre d. Comp. Ann. d. Sc. nat. Sér. 7. T. XI. 1890. 17. Flückiger. Pharmakogn. 1891. 833. Gerdts. Bau u. Entw. d. Comp.-frucht, m. bes. Berücks. d. off. Arten. Diss. Bern. 1904. 11, 37. Greenish & Collin. Anat. Atl. o. veget. Powders. 1904. 94. Kraemer. Bot. a. Pharmacogn. 1910. 746. Nussbaum, Karsten u. Weber. Lehrb. d. Biologie. 1914. 203. Oudemans. Pharmacogn. 1880. 328. Planchon et Collin. Drogues simples. T. 2. 1896. 75. Schneider. Powd. veget. Drugs. 1902. 119. Tschirch u. Oesterle. Anat. Atl. 1900. 7. MATERIAL. Flowers, gathered July 1902 in the Botanic Garden at Groningen, fresh and in alcohol. REAGENTS. Water, glycerine, chloral hydrate, phloroglucin and hydrochloric acid, potassium dichromate.

MICROGRAPHY. Only mentioning the differences between Anthemis nobilis and Matricaria Chamomilla. For the rest see Matricaria.

Receptacle.

Epidermis, walls more thickened; lateral walls pitted. Mesophyll, walls of the 2 outermost cell layers of the parenchyma collenchymatous; throughout the parenchyma walls comparatively thick; the inner part consisting of parenchyma cells with somewhat thickened walls and showing large intercellular spaces; glands wanting. Meristeles only in the outer part of the receptacle.

Bracts.

Epidermis. Inner side. Cells in longitudinal rows, also those of the membranous margin; this margin consisting at the top only of 1 layer of cells, the free ends of which mostly acute.

Epidermal cells above the green part of the bract. H. 12 μ, Lev. B. 12 μ, Lev. L. 120 μ; mostly tetragonal. Cell contents, except above the midrib: in potassium dichromate colouring brick-red and showing dark red granules, especially at the base of the bract. Cells of the membranous margin. H. 10 μ, Lev. B. 12 μ, Lev. L. 100—150 μ. Outer walls a little thickened. Cell contents wanting. Under side. Above the uppermost end of the green part of the bract cells small and with sinuous lateral walls; longer than those of the inner side. See moreover the inner side. Stomata. Only above the uppermost end of the green part. Trichomes. Conical hairs mostly 5 in number; only at the uppermost end of the green part; diameter at the base 12 μ, length 500 μ; consisting at the base of 1 row of 3—4 small cells and on top of these 1 very long cell bent towards the top of the bract. Glandular hairs only on and near the midrib.

Stomata. Lev. 20 by 25 μ. In the ends of the guard-cells in potassium dichromate a reddish brown mass. Conical hairs. Small cells long 10—15 μ. Contents: in potassium dichromate a yellow to brown mass.

Mesophyll.

Chlorenchyma cells. Diameter 8—10 μ; rectangular prisms with rounded edges; in the midrib polygonal prisms with rounded edges or cylinders, both with a longitudinally directed axis.

Meristele. Bifurcating at the top.

Pericyclic fibres. Diameter 16 by 8 μ, length 200—250 μ; tetragonal with rounded sides.

Xylem. Showing 1—2 spiral vessels in a transverse section.

Ray-flower.

Corolla.

Epidermis inner and outer side. Cell contents: in potassium dichromate a cluster of some dark brown globules, especially at the base of the limb.

Pistil.

Ovary. Epidermis: the longitudinal rows of tabular cells much less developed. Ovule: the epidermal cells of the integument showing sinuous lateral walls.

July 1902.   J; M; L.

FLORES CHAMOMILLAE VULGARIS.

Matricaria.

The heads of Matricaria Chamomilla, Linn. Sp. Pl. 891.

Macroscopic characters.

About 2 c.M. in diameter. Involucre in the shape of the undermost part of a hemisphere; bracts about 37 in number, arranged in 2 rather distinct ranks, lanceolate and inserted with a broad base, long 2.5 m.M., wide 0.75 m.M., glabrous, only the middle part green. Margin-much broader, that of the inner bracts-membranaceous, colourless, entire and undulated. Receptacle high 4—5 m.M., wide 3—4 m.M., conical with rounded surface, internally hollow, wall thick about 0.14 m.M., internally covered with some white remainders of tissue, glabrous. Ray-flowers, number: n = 125, Med = 13.35, Q_1 = 1.75, Q_3 = 1.83, min. = 10, max. = 21 [1]), in 1 whorl, incomplete, pappus wanting, female, zygomorphic, epigynous. Corolla ligulate, limb turned outwards at a right angle; tube long 2 m.M., very pale yellowish green; when magnified 50 times showing short obtuse glandular hairs on the outside, especially distinct near the throat. Pistil compound, consisting of 2 carpels and with 1 style and 2 stigmata somewhat standing out above the tube, flat pale yellow, somewhat bent outwards, ligulate. Ovary long 1 m.M., inferior, very pale yellowish green; when magnified 50 times showing glandular hairs to a larger or smaller number and a few longitudinal thin, shining, projecting ridges. Disk-flowers numerous, long 3 m.M., incomplete, pappus wanting, hermaphrodite, actinomorphous, epigynous. Corolla gamopetalous, pentamerous, tubular; tube somewhat narrowing in the middle, yellowish white, lobes yellow; on the tube, especially on the stricture glandular hairs. Stamens 5, inserted in the undermost part of the tube of the corolla, syngenesious: filaments rather long; anthers innate, opening introrsely by fissures; connectives continued into 5 projecting parts curved inwards. Pistil essentially the same as that of the ray-flower. Odour peculiar, strongly aromatic; taste aromatically bitter.

Anatomical characters.

LITERATURE. Benecke. Mikrosk. Drogenprakt. 1912. 68. Daniel. Rech. anat. s. l. bract. d. l'Involucre d. Composées. Ann. d. Sc. nat. Sér. 7. T. XI. 1890. 17. Ellrodt. Verteilung d. Gerbstoffs i. Off. Blätt. Kräut. u. Blüten. Diss. Würzburg. 1903. 25. Flückiger. Pharmakogn. 1891. 830. Gilg. Pharmakogn. 1910. 348. Gerdts. Bau u. Entw. d. Komp.-frucht mit bes. Ber. d. off. Arten. Diss. Bern. 1904. 11, 37. Karsten u. Oltmanns. Pharmakogn. 1909. 220. Koch. Mikr. Anal. d. Drogenpulver. Bd. 3. 1906. 187. Kraemer. Botany a. Pharmacogn. 1910. 746. Marmé. Pharmacogn. 1886. 266. Meyer. Wiss. Drogenk. Bd. 2. 1892. 301. Nussbaum, Karsten u. Weber. Lehrb. d. Biologie. 1914. 200. Oudemans. Pharmacogn. 1880. 329. Schneider. Powd. veg. Drugs. 1902. 240. Tschirch. Angew. Pfl.-Anat. 1889. 466. Tschirch. Harze u. Harzbeh. 1900. 384. Tschirch. Pharmakogn. Bd. 2. 1911. Tschirch u. Oesterle. Anat. Atl. 1900. 5. Pl. 2. van Wisselingh. Bijdr. t. d. Kennis v. d. Zaadhuid. 1ste Bijdr. Compositae. Pharm. Weekbl. 1918. 871. MATERIAL. Heads, gathered July

---

[1]) In an other variety of the same plant Med = about 21.

1902 near Groningen, fresh and in alcohol. REAGENTS. Water, glycerine, chloral hydrate, iodine in chloral hydrate, phloroglucin and hydrochloric acid, iodine and sulphuric acid 66 per cent., concentrated sulphuric acid, Schulze's macerating mixture, copper acetate and iron acetate. potassium dichromate.

MICROGRAPHY.

Receptacle.

Epidermis.

C e l l s. R. 25 μ, T. 20 μ, L. 30—40 μ; polygonal prisms. Outer walls a little thickened, inner walls somewhat collenchymatous; outer walls showing a cuticle, strongly swelling in water and then even throwing off the cuticle; showing distinct stratification in iodine and sulphuric acid 66 per cent.

Ground tissue. Consisting of common parenchyma containing schizogenous glands. Cells of the outermost 6 layers the smallest, increasing in size towards the interior, the parenchyma becoming spongy at the same time. G l a n d s. Lying close to the phloem of most of the meristeles; R. and T. 40—60 μ; mostly partly filled with a light green mass, in iodine looking like a yellowish emulsion with colourless drops.

Meristeles. Numerous; much anastomosing; the largest in the innermost part surrounding the cavity of the receptacle.

C o m m o n p a r e n c h y m a c e l l s. In the outermost 6 layers R. 15 μ, T. 25 μ, L. 40—50 μ; in the following layers R. 25 μ, T. 30 μ, L. 50—70 μ; in the innermost layers still larger; polyhedra with rounded edges. Walls of the outermost layers somewhat collenchymatous; of the innermost layers sometimes colouring red in phloroglucin and hydrochloric acid. Cell contents: chloroplasts. E p i t h e l i u m c e l l s. Thick 10 μ, wide 25 μ; in a transverse section rectangular.

P e d i c e l s. Numerous; surrounded by polygonal parenchyma cells of 15 μ in diameter, with rounded edges. Each containing a meristele.

B r a c t.

Epidermis. Inner side. In the vicinity of the margin cells arranged in rows at right angles to it, and here the bract often only 1 cell thick.

C e l l s o f t h e a b o v e. H. 15 μ, Lev. B. 20 μ, Lev. L. 250 μ; polygonal tables. In the vicinity of the margin H. 10 μ, Lev. B. 15 μ, Lev. L. 50—100 μ; often with rounded edges. Outer walls showing a cuticle; lateral walls sometimes pitted. Cell contents: in the vicinity of the margin in potassium dichromate here and there some small yellow to brown granules.

Outer side. Cells in longitudinal rows, especially at the base of the bract; for the margin see the inner side. Stomata here and there on the middle part of the bract containing the mesophyll; longitudinally directed; phaneroporous, lying in the same level as the epidermal cells. Glandular hairs of the Compositae type not numerous; occurring on the middle part of the bract; consisting of 4—5 stories of 2 cells each; the partition wall at right angles to the longitudinal axis of the bract; the cuticle of the uppermost stories raised into a bladder.

E p i d e r m a l c e l l s p r o p e r. H. 15 μ, Lev. B. 25 μ, Lev. L. 30—70 μ; rectangular; lateral walls sometimes sinuous. Outer walls with longitudinal cuticular striation. Cell contents: in the middle part and especially towards the top of the bract in potassium dichromate yellow to brown granules. S t o m a t a. Lev. B. 18, Lev. L. 20 μ. G u a r d-

c e l l s. Contents: in potassium dichromate a brown mass. C e l l s o f g l a n d u l a r
h a i r s. H. 15 μ, Lev. B. 15 μ, Lev. L. 30 μ. Contents: in potassium dichromate a brown
mass.

Mesophyll. Only in the middle part of the bract. Consisting of 5—8 layers of
chlorenchyma cells with intercellular spaces and glands.

C h l o r e n c h y m a c e l l s. H. and Lev. B. 15—30 μ, Lev. L. 100—150 μ; polygonal
prisms with rounded edges and a longitudinally directed axis. Cell contents: chloroplasts;
in potassium dichromate in some cells yellow to brown granules.

G l a n d s. 1 above the meristele of the midrib; diameter 20 μ; containing a
colourless mass; showing a single layer of epithelium cells.

C e l l s o f e p i t h e l i u m. Thick 10 μ, wide 15 μ; rectangular. Cell contents: when
treated with potassium dichromate a yellow to brown mass.

Meristeles. Pericycle. Developed in the shape of 1 or 2 layers of sclerenchyma
fibres only at the tops of the principal meristele and its branches. These fibres
spreading into the mesophyll, especially towards the margin.

P e r i c y c l i c f i b r e s. Diameter 6 μ. Walls thickened; lignified; with pit canals.

D i s k - f l o w e r.

Corolla.

Epidermis. Inner side. Cells arranged in longitudinal rows at the base of the
corolla; becoming larger towards the top. Glandular hairs sometimes 1 or 2 on
the teeth; in the uppermost stories of the hairs cells often much smaller than
in the others. See for the rest the under epidermis of the bracts.

E p i d e r m a l c e l l s p r o p e r. At the base of the corolla H. 20 μ, Lev. B. 25 μ,
Lev. L. 30—40 μ; rectangular; farther towards the top Lev. B. 20 μ, Lev. L. up to 100 μ;
on the teeth Lev. B. 15 μ, Lev. L. 60 μ, lateral walls somewhat sinuous. Outer walls of
all cells somewhat thickened; with transverse cuticular striation. Cell contents: nucleus
distinct; spherical yellow plastids, especially on the teeth; in potassium dichromate the
cells near the teeth containing a yellow mass with colourless cavities, the cells on the
teeth some small yellow globules. A similar reaction with copper acetate and iron acetate:
yellowish brown granules, some of them somewhat darker.

Outer side. Glandular hairs, especially in the widening parts of the tube; see
for the rest the inner side.

E p i d e r m a l c e l l s p r o p e r. On the teeth the lateral walls more sinuous; lateral
walls of all the cells often a little thickened; pitted. For the rest see the inner side.

Mesophyll. Consisting of about 6 layers of common parenchyma at the base of
the corolla, wanting at the top. At the base cells cylindrical, farther towards
the top more or less stellate. Showing intercellular spaces.

C e l l s o f t h e a b o v e. H. 10 μ, Lev. B. 10 μ. Cell contents: small chloroplasts,
containing 1 or 2 needle-shaped starch grains.

Meristeles. 5 in number; ending blind at a short distance from the incisions
of the teeth. At the top of the corolla on both sides of the meristeles a large intercellular space. Xylem containing spiral vessels with lignification of the
spirals.

Stamen. Placed opposite to the meristeles of the corolla. In transverse section
semi-circular, the flat side turned inwards.

Filament.

Epidermis. Cells in longitudinal rows; thin-walled. In the upper part, adjacent to the anther, cells on the outside rectangular prisms, somewhat collenchymatous (so-called articulation).

T h i n - w a l l e d c e l l s. R. 10 μ, T. 8 μ; rectangular. Outer walls with a cuticle. C e l l s o f a r t i c u l a t i o n. R. and T. 10 μ, L. 15 μ, the shortest cells in the middle. Walls thickened, especially at the corners; somewhat yellow; lignified.

Mesophyll. Consisting of common parenchyma.

C e l l s o f t h e a b o v e. R. and T. 5—7 μ, elliptical cylinders or polygonal prisms with rounded edges and a longitudinally directed axis.

Meristeles. Represented by a branch of the corresponding meristele of the corolla. Spiral vessels with lignification.

A n t h e r.

Connective.

Epidermis. After the dehiscence the outermost parts of the anthers joined together by the epidermis. [1]

C e l l s o f t h e a b o v e. R. and T. 10 μ. Walls very thin.

Mesophyll. Consisting of common parenchyma, adjoining the epidermis; also at the top of the anther 1 layer of cells; in a transverse section distinctly joining the fibrous layer.

C e l l s o f t h e a b o v e. R. and T. 10 μ, L. 30—50 μ; tetragonal tables. Walls thickened, except the inner walls and the inner part of the lateral walls; lignified; pitted.

Meristele. Very small. Containing here and there a cluster crystal.

Thecae.

Epidermis.

C e l l s. R. 3 μ, T. 6 μ. Walls very thin.

Fibrous layer. 1 layer of cells.

C e l l s o f t h e a b o v e. R. 20 μ, T. 8 μ, L. 30—60 μ, R. greatest near the connective; rectangular prisms, mostly with rounded inner walls. Inner and lateral walls showing thickened yellow lignified bands.

Pollen. Grains 20 μ in diameter; globes or tetrahedra; showing 3 pores. Exine showing dots and a distinct layer composed of rods.

Pistil.

O v a r y. Showing 2—5 longitudinal ribs.

Wall.

Epidermis. Outer side. Showing 12—15 projecting longitudinal rows of very flat cells, all the rows running on the ribs of the ovary. Glandular hairs wanting on the ribs; see for the rest the under epidermis of the involucre.

C e l l s o f t h e p r o j e c t i n g r o w s. R. 20 μ, T. 30 μ, L. 4 μ; rectangular tables with rounded inner and outer walls and a longitudinally directed axis. Outer walls a little thickened; swelling in water; with a cuticle. Cell contents: sometimes a cluster crystal.

C e l l s b e t w e e n t h e r i b s. R. 5 μ, T. 7 μ. Outer walls showing a cuticle. Cell contents: mostly 1 cluster crystal.

---

[1]) Tschirch. l. c. 51, comes to a different conclusion.

Inner side. Cells much resembling the epidermal cells between the ribs of the outer side; somewhat larger. Cell contents: sometimes starch grains.

Mesophyll. Consisting of common parenchyma; only occurring in and near the ribs. Cells of the above. L. 30—50 μ; cylinders with a longitudinally directed axis.

Meristeles. In each rib 1 meristele, the rest of the meristeles in the vicinity of the ribs. Xylem containing spiral vessels with lignification.

Ovule. Anatropous; attached to the base of the ovarial cavity.

Integument. Epidermis outer side containing rather much starch; in potassium dichromate cell contents colouring reddish brown and showing colourless cavities. Inner part already more or less reduced. Raphe in the integument clearly to see. Endosperm and embryo distinct.

At the base of the ovary a disk consisting of common parenchyma cells containing many crystals.

S t y l e.

Epidermis. Cells arranged in longitudinal rows. Crystal cluster idioblasts at the upper end of the style.

E p i d e r m a l   c e l l s   p r o p e r. R. 18 μ, T. 16 μ, L. 50 μ, L. smaller at the base and greater at the top of the style; rectangular prisms. Outer walls a little thickened; with a cuticle. Cell contents: in potassium dichromate some yellow globules. At the base T. 30 μ, L. 25—40 μ; rectangular, sometimes with rounded sides. Walls thickened; a little yellow; pitted.

Mesophyll. Consisting of some layers of common parenchyma cells. Adjoining the epidermis at the top of the style 1 layer, at the base some layers showing larger cells than the rest.

L a r g e r   o u t e r m o s t   c e l l s. R. and T. 12 μ; rectangular with rounded edges. Walls a little thickened. S m a l l e r   c e l l s. R. and T. 8 μ; polygonal prisms with a longitudinally directed axis and rounded edges.

Meristeles. Xylem containing spiral vessels with lignification; 1 meristele on each side of the central conducting tissue, ending blind in the top of the style; sinuous in not full-grown styles. Conducting tissue in the shape of an elliptical cylinder; cell walls very thick, middle lamella not to distinguish.

S t i g m a t a.

Epidermis. Inner side and margin bearing 2 bands of conical papillae. The epidermal cells between these bands very long. Hairs only on the top of the stigmata; somewhat cudgel-shaped; unicellular; diameter 20 μ, length 80 μ.

P a p i l l a e. Thick 8 μ, long 20 μ. Showing a cuticle. Contents: yellow plastids; in potassium dichromate yellow globules. C e l l s   o f   h a i r s. Showing cuticular striation at the top. Contents: see the papillae.

Conducting tissue. Lying immediately under the papillae and containing a long schizogenous gland.

R a y - f l o w e r.

Corolla.

Epidermis. Inner side. Cells at the base of the corolla in longitudinal rows. More towards the top distinctly papillaform.

C e l l s o f t h e a b o v e. H. 22 μ, Lev. B. 18 μ, Lev. L. 22 μ; hexagonal. P a p i l l a-
f o r m c e l l s. H. 30 μ, Lev. B. 20 μ, Lev. L. 25 μ; polygonal with sinuous lateral
walls. Outer walls of all cells showing radial cuticular striation. Cell contents: some small
drops; nucleus distinct.

Outer side. Cells practically arranged in longitudinal rows; lateral walls strongly
sinuous, not so on the tube. Glandular hairs very rare on the limb, more
numerous on the tube; see for the rest the epidermis of the disk-flower.
E p i d e r m a l c e l l s p r o p e r. H. 25 μ, Lev. B. 20—25 μ, Lev. L. 35—75 μ; rectangular prisms. On the tube Lev. B. 20 μ, Lev. L. 40—70 μ. Outer walls mostly with
wavy longitudinal cuticular striation. Cell contents: some small drops; nucleus.

Mesophyll. Consisting of 4—5 layers of spongy parenchyma; cells stellate, very
long, at the base of the corolla somewhat cylindrical; cell contents: some small
chloroplasts. Meristeles mostly 4 in number; parallel to each other; anastomosing in the top of the corolla; 2 of them ending just below the incisions
between the teeth; sometimes showing 2 or more branches; xylem containing
spiral vessels with lignification.

Pistil. See the disk-flower.

Ovary with 5 ribs and 5 meristeles, more distinct than those of the disk-flower.[1)]

July 1902.   J; M; L.

FLORES CINAE.
Santonica.
The unexpanded heads of Artemisia Cina, Berg, Darstell. IV. t. 29c., imported from Central Asia,
especially from Turkestan.

Macroscopic characters.

Long 2—4 m.M., thick up to 1.5 m.M., often with a short petiole, cylindrical,
somewhat angular, towards the top and the base rounded. Involucre of 12—20
bracts, closely imbricated. Bracts long up to 2.5 m.M., the outer ones much
shorter and oval, the higher placed lanceolate; apex blunt; midrib, especially
that of the outermost bracts, projecting on the outer side; large membranous
margin, also at the top, without veins; surface greyish green, afterwards yellowish brown; on the outer side of the not-membranous middle part many cylindrical, sessile, yellow glandular hairs and moreover especially on the outermost
bracts some slender, sinuous hairs. Receptacle slender, cylindrical, glabrous.
In each head 2—6 flowers, sometimes in a very early stage of development,
long not over 2 m.M., epigynous, hermaphrodite; pappus wanting; corolla tubular, beset with glandular hairs. Odour peculiar, aromatic; taste nauseously
aromatic, bitterish, cooling.

Anatomical characters.

LITERATURE. Benecke. Mikr. Drogenprakt. 1912. 68. Biechele. Mikr. Prüf. d. off. Drogen. 1904. 34. Daniel. Rech. anat. s. l. Bract. d. l'Involucre d. Composées. Ann. d. sc. nat.
Sér. 7. T. 11. 1890. 17. Flückiger. Pharmakogn. 1891. 82. Gerdts. Bau u. Entw. d. Kompos.-

---

[1)] Meyer. Drogenkunde, mentions only 4 ribs and meristeles and the same number for the
corolla.

frucht m. bes. Berücks. d. off. Arten. Diss. Bern. 1904. 11, 37. Gilg. Pharmakogn. 1910. 350. Greenish & Collin. Anat. Atl. of veget. Powders. 1904. 108. Hérail. Pharmacogn. 1912. 227. Heyl u. Tunmann. Santoninfr. Fl. Cinae. Apoth. Zeit. XXVIII. 248. Karsten u. Oltmanns. Pharmakogn. 1909. 226. Koch. Mikr. Anal. d. Drogenpulver. Bd. III. 1906. 197. Pl. XVIII. Koch. Einf. i. d. Anal. d. Drogenpulver. 1906. 89. Koch u. Gilg. Pharmakogn. 1909. 157. Kraemer. Botany a. Pharmacogn. 1910. 550. Marié. Semen contra. Thèse. Paris. 1884. Marmé. Pharmakogn. 1886. 59. Meyer. Wiss. Drogenk. 1891. Bd. II. 308. Mitlacher. Toxik. od. Forens. wicht. Pfl. u. Drogen. 1904. 186. Moeller. Pharmakogn. 1889. 90. Moeller. Mikr.-pharmac. Üb. 1901. 116. Molisch. Mikrochemie der Pflanze. 1913. 144. Oudemans. Pharmacogn. 1880. 330. Planchon et Collin. Drogues simples. T. II. 1896. 82. Schneider. Powder. veget. Drugs. 1902. 284. Tschirch. Angew. Pfl.-Anat. 1889. 154, 293, 442. Tschirch. Pharmakogn. Bd. II. 1911. 1012. Tschirch u. Oesterle. Anat. Atl. 1900. 316. Pl. 73. Tunmann. Pfl. - micr. - Chemie. 1913. 248. Vogl. Anat. Atl. 1887. 44. MATERIAL. The drug. REAGENTS. Water, glycerine, iodine in chloral hydrate, phloroglucin and hydrochloric acid, iodine and sulphuric acid 66 per cent., concentrated sulphuric acid, Schulze's macerating mixture, alcoholic soda solution.

MICROGRAPHY.

Petiole and receptacle.

Epidermis. Stomata relatively numerous close to the head; phaneroporous; lying in the same level as the epidermal cells. Here and there glandular hairs of the Compositae type.

Epidermal cells proper. Rectangular in a transverse section. Outer walls thickened, showing a cuticle.

Cortex. Consisting of some layers of common parenchyma cells. Cells mostly radially elongated.

Cells of the above. Cell contents: chloroplasts, sometimes containing starch grains; here and there a cluster crystal; in alcoholic soda solution colouring yellow, sometimes the reagent producing in each cell a yellow globule.

Endodermis. Mostly distinct.

Stele.

Pericycle. Consisting of 1 or 2 layers of fibres with lignified walls.

Vascular bundles. 2 larger bundles, alternating with 2 smaller ones; in the upper part of the receptacle mostly 3 bundles often widely diverging. Xylem containing radially arranged vessels, showing lignification.

Medullary commissures. Mostly uni- to biseriate; consisting of parenchyma. Cells mostly with thickened and lignified walls.

Medulla. See the medullary commissures.

Bract. Structure, except that of the epidermis and the membranous margin, varying according to the lower or higher insertion of the bract in the involucre.

Thicker middle part.

Epidermis. Inner side. Consisting of fibres. Stomata occurring on the uppermost part of several bracts; phaneroporous; lying in the same level as the epidermal fibres; surrounded by 4—5 subsidiary cells.

Epidermal fibres. H. 10 μ, Lev. B. 8—10 μ, Lev. L. 50—200 μ. Outer walls thickened; yellow; lignified. Cell contents: a few minute chloroplasts, colouring yellow in alcoholic soda solution. Stomata. Lev. 18 by 20 μ.

Outer side. Cells arranged in longitudinal rows. Stomata especially occurring above the outer border of the thicker part; phaneroporous; lying in the same level as the epidermal cells; generally surrounded by 4—5 subsidiary cells. Trichomes above the outer border of the thicker part; in 2 kinds: 1°. Glandular hairs of the Compositae type, especially occurring on the uppermost half of the bract on the re-entering curves; consisting of 2—3 stories of 2 cells each; partition walls of the stories at right angles to the midrib; the cuticle of the outside of the uppermost cells raised into a bladder; thick 40 μ, long 55 μ. 2°. Conical hairs, often 2 hairs inserted immediately next to each other; very sinuous; unicellular; diameter 6—10 μ, lengt 200—500 μ.

Epidermal cells proper. H. 15 μ, Lev. B. 9—12 μ, Lev. L. 20—60 μ; tetra- or pentagonal prisms. Outer walls thickened, especially those of the middle rows; lateral walls mostly thickened and pitted, sometimes like a string of beads; outer walls with a cuticle. Cell contents: sometimes a cluster crystal or a prismatic crystal; for the rest see the inner side. Stomata. H. 9 μ, Lev. 18 by 22 μ. Glandular hairs. Cells thick 8 by 20 μ, height of a story 28 μ.

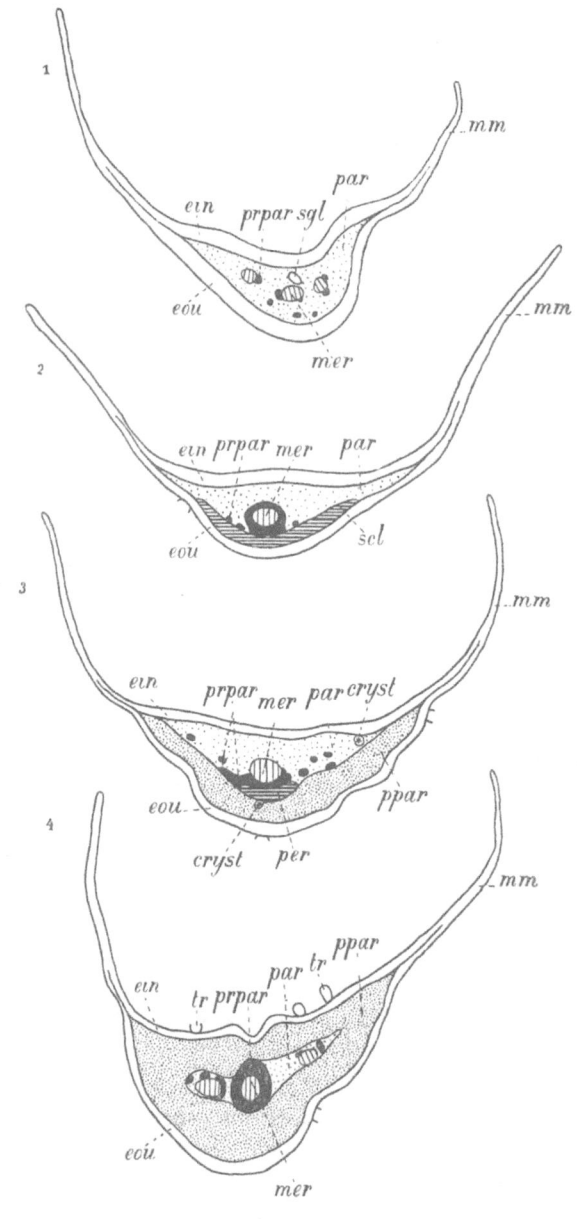

Fig. 22. *Artemisia Cina*. Bracts, transverse sections: 1. uppermost bracts; 2. higher middle bracts; 3. lower middle bracts; 4. undermost bracts. cryst Cells containing cluster crystals; ein Epidermis inner side; eou Epidermis outer side; mer Meristele; mm Membranaceous margin; par Common parenchyma containing few chloroplasts; per Pericycle; ppar Palisade chlorenchyma; prpar Prismatic parenchyma cells; scl Sclerenchyma; sgl Schizogenous glands; tr Trichomes.

Cell contents: colouring red in alcoholic soda solution; those of the bladder sometimes crystalline.

M e s o p h y l l.

Sclerenchyma. 3 layers of sclerotic elements under the outer epidermis, only in the higher middle bracts (fig. 2).

E l e m e n t s   o f   t h e   a b o v e. H. 10 μ, Lev. B. 10 μ, Lev. L. 30—100 μ; polygonal prisms, often showing a small branch. Walls thickened; yellow; lignified, especially the middle lamella; pitted.

Palisade chlorenchyma. In the undermost bracts constituting the bulk of the mesophyll on both sides of the bract (fig. 4); in the lower middle bracts only represented by 1—2 cell layers adjoining the outer epidermis (fig. 3); in the higher bracts wanting (figs. 1 and 2).

C e l l s   o f   t h e   a b o v e. H. 20 μ, Lev. 6 by 8 μ; polygonal prisms with rounded edges or cylinders with a radially directed axis. Cell contents: forming a green to brown mass, containing starch grains; in alcoholic soda solution colouring yellow and in each cell a yellow globule; sometimes a cluster crystal.

Common parenchyma containing less chloroplasts. In the undermost bracts forming with the meristeles a central layer between the 2 layers of palisade chlorenchyma (fig. 4). Gradually increasing in quantity in the higher placed bracts; first appearing also at the inner side of the meristeles; in the lower middle bracts quite taking up the place of the inner layer of palisade chlorenchyma (fig. 3) and in the higher bracts filling up all the remaining parts of the mesophyll (figs. 2 and 1).

C e l l s   o f   t h e   a b o v e. Lev. 10 by 12 μ, 10 by 15 μ, etc.; in a transverse section polygonal with rounded edges. Cell contents: mostly chloroplasts, containing a few starch grains; sometimes a cluster crystal.

Prismatic common parenchyma cells, containing a dark brown mass. Mostly adjoining the meristeles, sometimes 1 or more of these cell layers surrounding them (figs. 4 and 3).

Schizogenous glands. To a small number at the inner side of the meristeles (fig. 1), e. g. H. 7 μ, Lev. B. 10 μ.

M e r i s t e l e. In the undermost bracts 1 large meristele giving off numerous flattened branches. In the higher bracts, especially in the base, these branches decreasing in number.

Pericycle. In the shape of a sclerenchyma bundle present in the lower of middle bracts (fig. 3).

S c l e r o t i c   e l e m e n t s. See those of the mesophyll.

Xylem. Containing annular and spiral vessels with lignification.

Membranous margin. In the undermost bracts narrow and much smaller than the thicker middle part; consisting of 1 layer of fibres in all respects similar to the fibres of the inner epidermis of the thicker middle part, but with a cuticle at the under side. Stomata and hairs wanting.

F l o w e r.

Flowers in various stages of development, except full-grown ones.

Pappus wanting.

Corolla. Epidermis. Cells of the outer and the inner epidermis showing a cuticle. Glandular hairs numerous on the outer side of the teeth and the undermost part of the tube; for the rest see those of the bracts. Mesophyll. Some layers of parenchyma cells, only in the undermost half of the corolla. Meristeles 5.

Stamen.

F i l a m e n t. In the uppermost part adjoining the anther, cells of the outermost half showing thickened lateral walls (so-called articulation).

A n t h e r.

Top of connective and base of both thecae. Showing a rather large continuation of fibre-shaped cells; cells of these appendages often showing a cluster crystal. Thecae. Fibrous layer mostly distinct. Pollen: grains globular, sometimes trigonal, diameter 14 μ; walls somewhat thickened, showing 3 slit-like pores.

Pistil. Compound, consisting of 2 carpels.

Ovary. Wall showing several rows of longitudinal tabular cells. Ovule: anatropous; inserted at the base of the ovarial cavity on a ring of thick-walled cells.

Stigma. Bearing 2 bands of papillae at its inner and lateral sides; on the top long sweep hairs.

Micrography of the powder.

Consisting of fragments of all the parts of the heads, especially of those of the bracts; the following ones most striking the eye. Pollen grains often united into oblong columns, nearly spherical, glabrous, with re-entering curves. Glandular hairs consisting of 3—4 stories of 2 cells and a cuticular bladder, often still connected with the epidermis. Pieces of the middle parts of the bracts, containing annular and spiral vessels, also sclereids, the latter sometimes much elongated, also knotty, with a yellowish pitted wall. Pieces of the membranous margin of the bracts; cells prosenchymatic, thin-walled. Epidermis of the thicker middle part of the bracts; cells oblong, lateral walls thickened, like a string of beads from many pit canals; stomata. Parts of hairs, slender, sinuous, mostly 1-celled, thin-walled. Fragments of the fibrous layers of the anthers. Starch nearly quite wanting; here and there globular parenchyma cells with some small starch grains. In alcohólic soda solution in all the fragments bright yellow masses, the secretion of the glandular hairs even yellowish orange. Leaves and stems should not be mixed with it.

October 1902. J; M; L.

FLORES TILIAE.

Linden Flowers. Lime-tree Flowers.

The dried inflorescences of Tilia platyphyllos, Scop. Fl. Carn. ed. II. 1. 373 and Tilia cordata, Mill. Gard. Dict. ed. VIII. n. 1.

Macroscopic characters.

Description of Tilia platyphyllos: dichotomous cyme, number of flowers:

n = 179, M = 2.652, Med = 2.84, Q_1 = 0.52, Q_3 = 0.37, min. = .1, max. 5; at the base of the peduncle often a small leaf bud pulled off with the inflorescence; peduncle long up to 7 c.M., bearing a bract commencing at a distance of at most 1 c.M. from the base and grown together for about one third of its length with the peduncle. This bract long up to 10 c.M., membranaceous, undulated somewhat gutter-shaped, yellowish green, lanceolate, irregularly feather-veined with reticulate very conspicuous veinlets; apex blunt; base acute and unequal sided; margin entire; surface smooth, only on the veins with fine hairs. Pedicels long up to 1.5 c.M., showing scars of bractlets thrown off at an early period. Flower up to 2 c.M. in diameter, complete, actinomorphous, hypogynous. Calyx 5-phyllous, wheel-shaped, yellowish white; sepals oblong with an acute apex, keel-shaped, pilose on the margin and inside, containing honey in the fresh flower. Corolla 5-phyllous, wheel-shaped, light green; petals 1½ times the length of the sepals, lanceolate-spatulate, somewhat gutter-shaped, at the base thickened and excreting honey; apex acute and curved upwards. Stamina indefinite in number (30—40), united into 5 groups, placed in superposition to the petals; anthers with a triangular or compactly bifurcated connective; thecae with longitudinal lateral dehiscence. Pistil compound, consisting of 5 carpels, with a single cylindrical style, long up to 8 m.M., but very short in the bud and a somewhat 5-lobate stigma; ovary spherical, 5-lobular, with axile placentae, densely tomentous; 2 ovules in each cavity.

Tilia cordata resembles in all main points T. platyphyllos; the principal difference being that the number of flowers in the inflorescence is 5—15.

Odour very weakly aromatic; taste mucilaginous.

Anatomical characters of Tilia cordata.

LITERATURE. Attema. De zaadhuid der Angiosp. en Gymnosp. Diss. Groningen. 1901. 68. Benecke. Mikr. Drogenprakt. 1912. 66. Ellrodt. Ueber die Verteilung des Gerbstoffes in off. Blättern, Kräutern u. Blüten. Diss. Würzburg. 1903. 28. Flückiger. Pharmakogn. 1891. 790. Gilg. Pharmakogn. 1910. 215. Hager. Pharm. Praxis. Bd. II. 1902. 1051. Hérail. Pharmacol. 1912. 100. Karsten u. Oltmans. Pharmakogn. 1909. 207. Kraemer. Botany a. Pharmakogn. 1910. 328. Luerssen. Syst. Bot. Bd. 2. 1882. 655. Marmé. Pharmacogn. 1886. 249. Meyer. Wiss. Drogenk. Bd. 2. 1892. 279. Meyer. Mikr. Unters. v. Pflanzenpulv. 1901. 156. Moeller. Anat. d. Baumrinden. 1882. 246. Moeller. Pharmakogn. 1889. 97. Molisch. Mikrochem. d. Pfl. 1913. 215, 313—315. Oudemans. Aanteek. o. d. Pharmac. neerl. 1854—56. 457—459. Oudemans. Pharmacogn. 1880. 335. Planchon et Collin. Drogues simples. T. II. 1896. 696. Tschirch. Angew. Pfl. Anat. 1889. 203. Tschirch. Pharmakogn. Bd. II. Abt. 1. 1912. 366. Tschirch u. Oesterle. Anat. Atl. 1900. Taf. II. Tunmann. Pflanzenmikrochemie. 1913. 576. Walliczek. Studien über die Membranschleime vegetativer Organe. Diss. Bern. 1893. 39, 44; also in Pringsheims Jahrbücher. Bd. 25. 1893. 228, 231, 247, 252. Wigand. Pharmakogn. 1879. 246. MATERIAL. The drug. Inflorescences gathered July 1903 in the Botanic Garden at Groningen, fresh, fixed with chromic acid 1 per cent. and in alcohol. REAGENTS. Water, glycerine, chloral hydrate, iodine in chloral hydrate, phloroglucin and hydrochloric acid, iodine and sulphuric acid 66 per cent., concentrated sulphuric acid, osmic acid, Schulze's macerating mixture, potassium dichromate, copper acetate and iron acetate, the latter reagent for the detection of tannin only used for the bract.

MICROGRAPHY.
Bract.
Intervenia.

Epidermis. Upper side. Along the margin, especially at the top of the bract 2—3 rows of longitudinally elongated cells. Stomata and trichomes wanting. Epidermal cells proper. H. 18 μ, Lev. B. 18—20 μ, Lev. L. 40—50 μ; polygonal tables with sinuous lateral walls; near the top smaller, Lev. B. 20 μ, Lev. L. 25 μ, on these places nearly always rectangular tables without sinuous lateral walls. Outer walls with a cuticle, sometimes with a cuticular striation. Cell contents: some chloroplasts and tannin; in material treated with potassium dichromate small red-brown globules clustered together.

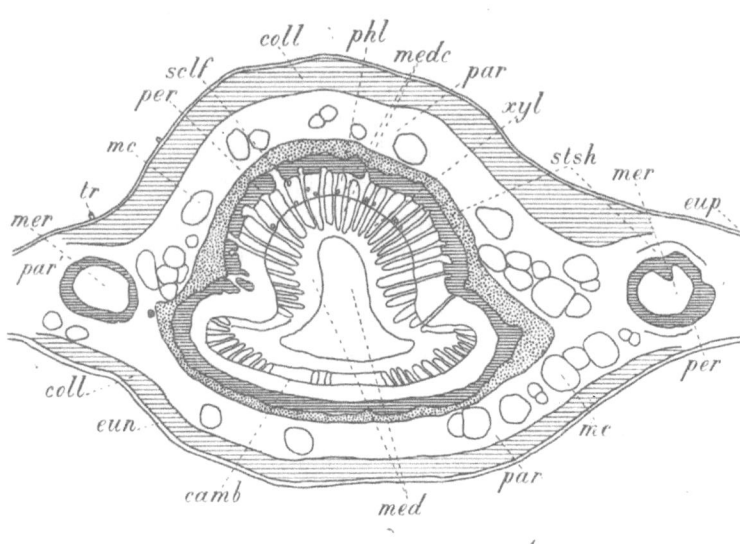

Fig. 23. *Tilia cordata*. Bract, transverse section of the basal part, containing also the peduncle, through the midrib and the neighbouring portions of the lamina. camb Cambium; coll Collenchyma; eun Epidermis under side; eup Epidermis upper side; mc Mucilage cells; med Medulla; medc Medullary commissures; mer Meristele; par Colourless parenchyma; par' Common parenchyma; per Pericycle; phl Phloem; sclf Sclerenchyma fibres of phloem; stsh Starch-sheath; tr Trichomes; xyl Xylem.

Under side. Along the margin here also some rows of longitudinally elongated cells. Stomata. 60—70 to the sq. m.M., phaneroporous, lying in the same level with the epidermal cells and showing 4—5 subsidiary cells. Trichomes wanting. Epidermal cells proper. H. 18 μ, Lev. B. 20 μ, Lev. L. 40 μ; near the top Lev. B. 25 μ, Lev. L. 30 μ; polygonal tables with sinuous lateral walls. Outer walls with an undulating cuticular striation. Cell contents: see those of upper side. Stomata long 30 μ, wide 20 μ.

Mesophyll. Consisting of common parenchyma, mostly 3 cell layers, near the midrib often 4 cell layers; cells of the uppermost layer more or less palisade-shaped and somewhat tapering inwards; cells of the other layers more or less branched; large intercellular spaces, increasing towards the under side.

Cells of the above. Cells of the uppermost layer H. 25 μ, Lev. at the outer side 15 μ, at the inner side 10 μ; those of the undermost layers H. 15 μ, Lev. 10 by 30 μ. Cell contents: chloroplasts, each containing some small needle-shaped starch grains; tannin.

Margin. Consisting of some layers of cells with somewhat thickened walls; chloroplasts wanting.

Veins.

Epidermis. Upper side. Cells arranged in longitudinal rows, also on the veinlets, but not on the middle part of the midrib. Stomata wanting. Trichomes. Only a few pluricellular uniseriate club-shaped hairs, at both sides of the base of the midrib; long 70 μ, thick at their base 15 μ, at their top 25 μ.

Epidermal cells proper. H. 15 μ, Lev. B. 10 μ, Lev. L. 15—50 μ; polygonal prisms, on the veinlets tetragonal prisms. Walls somewhat thickened, with a cuticle; all walls often becoming brown in iodine and sulphuric acid 66 per cent. Cell contents: some chloroplasts and tannin; in material treated with potassium dichromate some red-brown globules, on the veinlets cells then nearly entirely filled with a red-brown mass with the exception of 1 or 2 colourless globular spots.

Under side. See the upper side, but trichomes wanting.

Epidermal cells proper. Lev. B. mostly 8 μ; see for the rest those of upper side.

Mesophyll examined in the basal part of the bract, still united with the peduncle. Collenchyma at the upper side 4, at the under side mostly 2 cell layers. Colourless parenchyma at the upper side 10, at the under side 6 cell layers; these layers in there middle part, but especially in that of the upper side, throughout the whole length of the midrib, interrupted by 1 layer of mucilage cells mixed with common parenchyma cells; the flat parenchyma cells, surrounding the mucilage cells, containing somewhat more chloroplasts. Cluster crystal idioblasts in small number.

Collenchyma cells. R. and T. 10 μ, L. 35—40 μ; polygonal prisms with a longitudinally directed axis. Walls collenchymatous, especially in the second cell layer of the upper side; at the upper side in iodine and sulphuric acid 66 per cent. colouring yellow and not blue; here and there small intercellular spaces. Cell contents: some chloroplasts with small starch grains; tannin; in material treated with a solution of potassium dichromate some red-brown globules. Parenchyma cells. At the upper side R. 25 μ, T. 30 μ, L. 60—70 μ; at the under side R. and T. 20 μ, L. 60—70 μ; flat parenchyma cells surrounding the mucilage cells thick 12 μ, wide 30 μ. All cells polygonal prisms with a longitudinally directed axis and rounded edges. Walls very thin; showing intercellular spaces. Cell contents: some small chloroplasts, each with some starch grains; tannin in some cells. Mucilage cells. R. and T. 60 μ; polygonal prisms with a longitudinally directed axis and mostly strongly rounded edges. Walls very strongly often excentrically thickened, leaving only a small often branched cavity; consisting of mucilage and showing a distinct stratification in material treated with alcohol, chromic acid or copper acetate if examined in alcohol; in dilute glycerine the mucilage, except of the chromic acid material, strongly swelling up and not discernible. [1]) Cell contents: some protoplasm. Cluster crystal idioblasts. R. and T. 20 μ, L. 25 μ.

Endodermis. Developed as a starch-sheath; consisting of 2—3 layers of common parenchyma cells; here and there shorter cluster crystal idioblasts, arranged in short longitudinal rows.

Common parenchyma cells. R. 10 μ, T. 20 μ, L. 30—50 μ; polygonal prisms with a longitudinally directed axis and somewhat rounded edges. Walls somewhat

---

[1]) According to W a l l i c z e k the mucilage is formed by secondary thickenings of the walls and where 2 or more mucilage cells join upon each other the separating cellulose walls often disappear by slow modification into mucilage.

thickened; pitted; showing intercellular spaces. Cell contents: especially of the cells of the upper side numerous chloroplasts, filled with mostly compound 3-adelphous starch grains; often tannin.

S t e l e in the basal part of the bract containing also the peduncle: 1. a complete central stele belonging to the peduncle and surrounded by a complete endodermis; 2. two lateral meristeles belonging to the bract and surrounded only by a partially developed endodermis (starch-sheath); the meristeles in the primary veins of the basal part of the bract arising from these meristeles. In the free apical part of the bract the midrib containing some meristeles running parallel to one another and towards the top successively bending outwards, the last 2 leaving the midrib close to the top. In all parts of the bract anastomoses and free ends occurring.

Central stele.

P e r i c y c l e. Consisting chiefly of sclerenchyma fibres, forming an almost continuous layer, thick at the upper side 5—6 at the under side 4 fibres, interrupted by common parenchyma only in a few places, corresponding with medullary commissures.

S c l e r e n c h y m a   f i b r e s. R. and T. 10 μ, L. very considerable; polygonal. Walls strongly thickened, leaving often only a very small cavity; lignified; pits rare; with a clearly discernible middle lamella; intercellular spaces wanting. P a r e n c h y m a   c e l l s. R. and T. 10 μ, L. 100 μ; polygonal prisms with a longitudinally directed axis. Walls thickened; lignified; with numerous circular pits. Cell contents: often tannin.

V a s c u l a r   b u n d l e s. Numerous; simple; collateral. Phloem exarch; the outer part consisting in a cross section of smaller and larger elements; sieve-tubes being not discernible; in the outermost part here and there a cell containing tannin; in the innermost part the elements fairly well arranged in radial rows and close to the cambium here and there in a cross section at the end of a radial row of elements a single sclerenchyma fibre with strongly thickened, lignified walls. Cambium consisting of some layers of elements. Xylem endarch; consisting of uniseriate and near the cambium of biseriate radial rows of pitted, reticulate and spiral vessels, close to the cambium intermixed with parenchyma cells.

V e s s e l s. R. 15—18 μ, T. 10—15 μ; polygonal prisms with often slightly rounded edges. Walls thickened; showing lignification. P a r e n c h y m a   c e l l s. R. 5 μ, T. 8 μ, L. 40—50 μ; tetragonal prisms with a longitudinally directed axis. Walls thickened; lignified; with unilateral bordered pits when adjoining a vessel.

M e d u l l a r y   c o m m i s s u r e s. Mostly uniseriate; consisting of elements with thin and others with thickened walls. Phloem part consisting of fibres, divided by cross walls into cells, sometimes 1 or more of those cells again divided into smaller cells, each containing a cluster crystal; cluster crystal cells either thick- or thin-walled. Xylem part consisting of cells containing chloroplasts and tannin.

F i b r e s   o f   p h l o e m   p a r t. R. and T. 12 μ, L. 50—90 μ; L. of the crystal cells of the divided fibres 10—15 μ. Walls usually thickened; with numerous pits. C e l l s   o f x y l e m   p a r t. R. 12 μ, T. 8 μ; polygonal prisms with a longitudinally directed axis. Walls thickened; lignified.

M e d u l l a. Consisting of common parenchyma cells, mixed with a few muci-lage cells.

C o m m o n   p a r e n c h y m a   c e l l s. Those of the outermost part. R. and T. 9 μ, L. 40—50 μ; polygonal prisms with a longitudinally directed axis. Walls somewhat thickened; pitted; intercellular spaces wanting. Cell contents: some chloroplasts; here and there tannin. Those of the central part. R. and T. 10—20 μ, L. 30—50 μ; polygonal prisms with a longitudinally directed axis and somewhat rounded edges. Walls somewhat thickened; pitted; showing small intercellular spaces. Cell contents: some chloroplasts; here and there tannin. M u c i l a g e   c e l l s. Larger than the parenchyma cells; see for the rest those of colourless parenchyma.

Lateral meristeles. Much smaller than the central stele, but showing a well developed pericycle.

Meristeles in free apical part of bract. Pericycle containing bundles of small thin-walled elements.

S e p a l.

E p i d e r m i s. Upper side. Stomata wanting. Hairs unicellular or stellate, the latter in consequence of 2—4 unicellular hairs [1]) being joined together; very numerous at the base, towards the margin and towards the apex of the sepal; stellate especially along the margin and at the apex; branches increasing in number towards the apex, decreasing in length towards the apex and the margin but becoming more twisted, especially towards the margin; length of hairs up to 1.5 m.M., especially at the base of the sepal; in the bud the sepals being united together by means of these hairs.

E p i d e r m a l   c e l l s   p r o p e r. H. 15 μ, Lev. B. 15—30 μ, Lev. L. 30—45 μ; poly-gonal tables. Outer walls a little thickened, with a cuticle. Cell contents: some small drops of liquid; tannin in the shape of a small brown-red lumb. C e l l s   o f   h a i r s. Walls not thickened. Cell contents: in material treated with potassium dichromate several red-brown lumbs.

Under side. Stomata 25 to the sq. m.M.; wide 20 μ, long 25 μ; phaneroporous, lying in the same level with the epidermal cells. Hairs at the apex of the sepal sometimes 2- to 4-, mostly 4-branched; those with 2 branches being the longest, e. g. 250 μ. See for the rest hairs of upper side.

E p i d e r m a l   c e l l s   p r o p e r. H. 18 μ, Lev. B. 25 μ, Lev. L. 35 μ; poly- often tetragonal. Outer walls a little thickened, with a cuticle. Cell contents: some chloro-plasts each containing some needle-shaped starch grains; tannin in the shape of small red-brown globules.

M e s o p h y l l. Thick up to the margin of the sepal about 8 layers of elements: 1. Common parenchyma, forming about 3 layers at the upper and 3 at the under side of the sepal; the size of the cells diminishing inwards; in the innermost layers often cluster crystal idioblasts. 2. Mucilage cells forming 1—2 central, not continuous layers.

C o m m o n   p a r e n c h y m a   c e l l s. Cell contents: especially in the outermost layers

---

[1]) T s c h i r c h mentions 2 as the highest number of branches.

some chloroplasts, each containing some needle-shaped starch grains; in the outermost layers sometimes tannin. M u c i l a g e c e l l s. See those of mesophyll of bract.

M e r i s t e l e s. 3 in number; branching upwards, free ends not or nearly not discernible.

P e t a l.

E p i d e r m i s. Upper side. Stomata wanting. Trichomes only present at the middle part of the undermost half of the petal; mostly uniseriate 7-cellular; curved and somewhat club-shaped hairs; resembling those of the veins of the bract upper side; thick at the base 15 μ, in the thickest part 20 μ in diameter, long 70 μ.

E p i d e r m a l c e l l s p r o p e r. H. 8 μ, Lev. B. 10—18 μ, Lev. L. 20—40 μ; polygonal tables with somewhat sinuous lateral walls; near the base and somewhat more upwards still on the veins Lev. B. 10 μ, Lev. L. 30—60 μ; rectangular tables. Outer walls with longitudinal cuticular striation. Cell contents: tannin; in material treated with potassium dichromate and also in fresh material a colourless vacuole discernible; in osmic acid many small drops not colouring brown or black, the rest of the contents colouring brown.

Under side. Stomata a few near the top; phaneroporous, lying in the same level with the epidermal cells; long 30 μ, wide 22 μ. Trichomes wanting.

E p i d e r m a l c e l l s p r o p e r. See those of upper side.

M e s o p h y l l. Chiefly consisting of common parenchyma; thick 4 cell layers in the central part of the petal. Here and there large mucilage cells arranged in a row, reaching from epidermis to epidermis and surpassing in length the thickness of the parenchyma; in these places the petal thick 100 μ, and the epidermis on both sides bulging out. Cluster crystal idioblasts in the middle layers of parenchyma cells, often towards the base, a few towards the margin of the petal.

C o m m o n p a r e n c h y m a c e l l s. Of the outermost layer at the upper side H. 8 μ, Lev. B. 15 μ, Lev. L. 20 μ; polyhedra with somewhat rounded edges and sides. M u c i l a g e c e l l s. See those of mesophyll of bract.

M e r i s t e l e s. 3 in number; branching towards the apex; anastomoses not numerous.

S t a m e n. Filament with a central meristele. Anther: connective with a single meristele; thecae: fibrous layer clearly discernible, middle layer consisting of 1, sometimes 2 layers of cells, filled with a granular mass; partition walls of the thecae with numerous cluster crystal idioblasts and also some mucilage cells. Pollen grains: having the shape of a tetrahedral biconvex lens with very strongly rounded angles, 25 μ in diameter, thick 16 μ; exine rather thick with a crenate inner outline and without any structure, a pore discernible at each of the 3 angles; intine in the vicinity of the pores rather thick, for the rest thin, yellow and granular; contents: contracted beneeth the pores, somewhat yellow, granular, often containing numerous small starch grains.

P i s t i l.

Ovary.

W a l l.

Epidermis. Outer side. Stellate hairs very numerous, with 1—4 branches; mostly very long and often much twisted.

E p i d e r m a l   c e l l s   p r o p e r. Rectangular in a cross section. Cell contents: tannin. C e l l s   o f   h a i r s. Walls not thickened. Cell contents: tannin.

Inner side. Cells in a cross section rectangular.

Ground tissue. Consisting of 3 parts: 1. An outer part, thick 250 μ, consisting of 14 layers of common parenchyma cells; these cells often divided into 2 by a tangential partition wall; cluster crystal idioblasts here and there. 2. A middle part, thick 130—140 μ, also consisting of common parenchyma cells, containing some meristeles; cluster crystal idioblasts here and there. 3. An innermost part, thick about 200 μ, consisting of common parenchyma and mucilage cells; cluster crystal idioblasts numerous, especially in the parts corresponding with the cavities of the ovary. In the middle part and in the outermost layers of the innermost part meristeles running in all directions.

C o m m o n   p a r e n c h y m a   c e l l s. Of the outermost part polyhedra with rounded edges. Cell contents: often tannin and several small granules. Of the middle part, cell contents tannin. Of the innermost part, cell contents often tannin. M u c i l a g e c e l l s. See those of mesophyll of the bract.

D i s s e p i m e n t s. Epidermis. See inner epidermis of wall. Ground tissue resembling the innermost part of the ground tissue of the wall; containing meristeles.

C e n t r a l   a x i s. Formed by common parenchyma cells with large intercellular spaces and containing tannin; moreover many cluster crystal idioblasts and some meristeles.

O v u l e s. Anatropous; with 2 integuments; here and there a cluster crystal idioblast.

Style.

Ground tissue. Under the epidermis some layers of parenchyma cells mixed with mucilage cells; towards the centre meristeles surrounding the conductive tissue. Conductive tissue towards the top surrounding a stylar canal; this canal star-shaped in a transverse section with mostly 5 rays and sometimes a few smaller additional ones.

A l l   c e l l s   o f   g r o u n d   t i s s u e   containing tannin. C e l l s   o f   c o n d u c t i v e t i s s u e. Walls much thickened, showing a strongly refringent inner layer.

Stigmata. Papillae slightly developed.

July 1903.   J; M; v. E. d. W.

FOLIA ALTHAEAE.

Marshmallow Leaves.

The dried leaves of Althaea officinalis, Linn. Sp. Pl. ed. I. 686, collected when the plant is in flower.

Macroscopic characters.

Simple, petiolate. Petiole from 2—6 c.M. in length, somewhat flattened on the upper side. Blade from 5—10 c.M. in length, 3—8 c.M. broad; ovate or cordate, more or less palmately 3- to 5-lobed; apex acute or obtuse; midrib and veins on the under side strongly protruding; margin irregularly serrate. Surface of petiole and blade entirely grayish pubescent; hairs appearing when magnified 50 times stellate, 2- to 5-rayed. Inodorous; taste mucilagenous.

Anatomical characters.

LITERATURE. Bohny. Beitr. z. Kenntn. d. Digitalisblattes &c. Diss. Zürich. 1906. 41. Deutsch. Arzneib. 5. Ausg. 1910. 224. Dumont. Rech. s. l'anat. comp. des Malvacées, etc. Ann. sc. nat. Sér. 7. T. 6. 1887. 145. Ellrodt. Üb. d. Vert. d. Gerbst. in off. Blättern, Kräutern u. Blüten. Diss. Würzburg. 1903. 11. Flückiger. Pharmakogn. 1891. 633. Gilg. Pharmakogn. 1910. 220. Glaser. Mikr. Anal. d. Blattpulver. Verh. d. Phys.-Med. Ges. zu Würzburg. N. F. Bd. 34. 1901. 249. Karsten u. Oltmanns. Pharmakogn. 1909. 158. Koch. Mikr. Anal. d. Drogenpulver. Bd. 3. 1906. 82. Koch u. Gilg. Pharmakogn. Praktik. 1907. 135. Kuntze. Beitr. z. vergl. Anat. d. Malvaceen. Bot. Centrbl. Bd. 45. 1891. 231. Marmé. Pharmacogn. 1886. 165. Meyer. Wiss. Drogenk. Bd. 2. 1892. 208. Moeller. Pharmakogn. 1889. 61. Moeller. Mikr.-pharm. Üb. 1901. 99. Oudemans. Pharmacogn. 1880. 305. Planchon et Collin. Drogues simples. T. 2. 1896. 702. Tschirch. Angew. Pfl.-anat. 1889. 263, 441. Tschirch. Pharmakogn. Bd. 2. 1911. 355. Tschirch u. Oesterle. Anat. Atl. 1900. 310. Taf. 71. Vogl. Anat. Atl. 1887. Taf. 15. Walliczek. Stud. üb. d. Membranschleime veget. Org. Pringsheim's Jahrb. Bd. 25. 1893. 258. The same as Diss. Bern. 1893. 50. Wigand. Pharmakogn. 1879. 202. MATERIAL. Leaves collected June 1902 in the Botanic Garden at Groningen, fresh, in alcohol and after fixation in chromic acid 1 per cent. REAGENTS. Water, glycerine, chloral hydrate, iodine in chloral hydrate, phloroglucin and hydrochloric acid, iodine and sulphuric acid 66 per cent., concentrated sulphuric acid, Schulze's macerating mixture, cupric acetate and ferric acetate, potassium dichromate.

MICROGRAPHY.

Intervenia.

Epidermis. Upper side. Stomata 40 to the sq. m.M., phaneroporous, lying somewhat above the level of the epidermis, nearly always surrounded by only 3 subsidiary cells, viz. 2 larger and 1 much smaller. Many epidermal cells developed as mucilage cells, not differing in size and shape from the normal cells. Trichomes in 2 kinds, especially the first surrounded by 6—8 smaller and stellately arranged epidermal cells: 1. stellate hairs L. 150—600 μ, consisting of 1—6 generally 4 conical cells with a somewhat enlarged base of 10 μ in diameter; the basal part of these hairs protruding somewhat beneath the epidermis into the palisade chlorenchyma. 2. Glandular hairs, implanted in shallow depressions of the epidermis, having a single basal cell — L. 26 μ, diameter 18 μ — somewhat protruding into the palisade chlorenchyma and a somewhat curved head — L. 30 μ, diameter 25 μ — consisting of 3 stories of cells, the uppermost of these having 4 cells; between the cells of the head sometimes intercellular spaces; sometimes vestiges of a cuticular bladder.

Epidermal cells proper. H. 25 μ, Lev. 25—35 and 40—50 μ; tabular, lateral walls sinuous. Outer walls having a cuticle. Cell contents: some chloroplasts; by potassium dichromate in the vacuole a yellow-brown precipitate, mixed with some dark brown granules. Mucilage cells. Inner walls much thickened by a mucilaginous matter, occupying up to one half or more of the height of the cell. These thickenings to be distinguished particularly well after fixation in chromic acid, coloured green by cupric acetate; in material treated with alcohol instead of these thickenings sphero-crystals, soluble in glycerine. Cell contents: by potassium dichromate a yellow-brown precipitate. G u a r d - c e l l s of s t o m a t a. H. 10 μ, Lev. 10 and 25 μ. C e l l s of s t e l l a t e h a i r s. Walls of the bases of the cells thickened; in a few cases lignified; showing pit canals; the rest of the walls having a cuticle. C e l l s of h e a d of g l a n d u-l a r h a i r s. Cell contents: much protoplasm, often containing chloroplasts; by potassium dichromate a dark brown precipitate.

Under side. Stomata 160 to the sq. m.M., lying in the same level with the epidermis. See for the rest the upper side; precipitates by potassium dichromate however wanting.

Mesophyll.

P a l i s a d e c h l o r e n c h y m a. Generally 2 layers of cells, the cells of one layer corresponding in place with that of the other and thus prolonging each other. Intercellular spaces.

C e l l s of t h e a b o v e. H. 30—40 μ, Lev. 10—15 μ; polygonal prisms with strongly rounded edges or cylinders. Cell contents: poly-, generally hexagonal tabular chloroplasts, paving the walls, thick 2 μ, diameter 3—4 μ.

S p o n g y c h l o r e n c h y m a. Four layers of stellate cells, only the under-most layer consisting of cells more or less resembling in shape those of the palisade chlorenchyma.

C e l l s of t h e a b o v e. Stellate cells, diameter 20 μ. Those of undermost layer H. 30 μ, Lev. 10 μ. See for the rest the palisade chlorenchyma.

V e i n s.

Epidermis. Upper side. In the lateral parts of the veins, covering palisade chlorenchyma, in all respects the same as in the intervenia, only the lateral walls less sinuous. A middle band, covering collenchyma, consisting of elongated cells and without stomata. Mucilage cells the same as in intervenia. Both kinds of trichomata somewhat more numerous on both parts of this epidermis than on that of the intervenia; stellate hairs somewhat larger and having often lignified walls of the bases.

E p i d e r m a l c e l l s of m i d d l e b a n d. R. 30 μ, T. 15—20 μ, L. 50—120 μ; penta- or hexagonal prisms with a radially directed axis. Outer walls a little thickened, having a cuticular striation; inner walls collenchymatous. Cell contents: after treatment with potassium dichromate a yellow-brown precipitate containing numerous dark brown granules.

Under side. Epidermal cells in longitudinal rows. Stomata rare, lying in the level of the epidermis. See for the rest the epidermis of the upper side of the intervenia, only the dark brown precipitate after treatment with potassium dichromate wanting here in mucilage cells and trichomes.

E p i d e r m a l c e l l s p r o p e r. R. 20 μ, T. 10—15 μ, L. 25—100 μ; tetra- to hexagonal prisms with a radially directed axis. Outer walls having a cuticle. Cell contents: some chloroplasts; after treatment with potassium dichromate a yellow-brown precipitate, containing some dark brown granules. G u a r d - c e l l s. T. 12 μ, L. 30 μ.

Mesophyll.

C o l l e n c h y m a. In the midrib on the upper side a narrow band thick 4—5 layers of cells; on the under side 1 layer of collenchyma cells, some of these being divided by a tangential wall.

C e l l s   o f   t h e   a b o v e. Those of upper side R. and T. 25 μ, L. 130—200 μ, those of under side R. and T. 15 μ, L. 70 μ. Cells of both sides polygonal prisms with a longitudinally directed axis; the tangential walls of those of the under side being strongly curved. Walls showing stratification. Cell contents: chloroplasts in the cells of the under side.

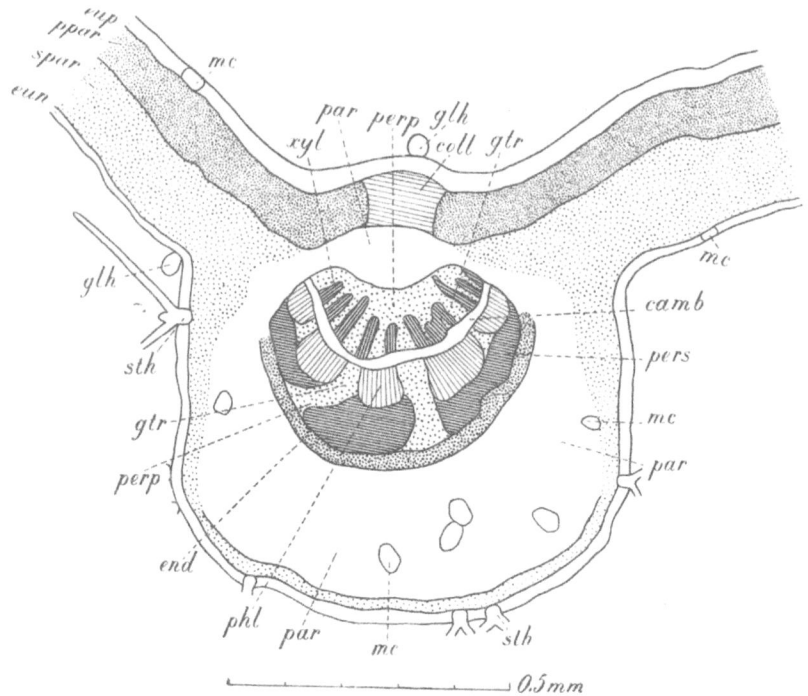

Fig. 24. *Althaea officinalis*. Leaf, transverse section through the midrib and the neighbouring portions of the lamina. camb Cambium; coll Collenchyma; end Starch-sheath; eun Epidermis under side; eup Epidermis upper side; glh Glandular hairs; gtr Medullary commissures; mc Mucilage cells; par Colourless parenchyma; perp Pericyclic parenchyma; pers Pericyclic sclerenchyma; phl Phloem; ppar Palisade chlorenchyma; spar Spongy chlorenchyma; sth Stellate hairs; xyl Xylem.

P a l i s a d e   c h l o r e n c h y m a. In all respects the same as in the adjoining intervenia on both sides of the upper collenchyma band.

C o l o u r l e s s   p a r e n c h y m a. On the upper side 4—5 layers, on the under side 6 layers of cells, having intercellular spaces between them. Many cells, arranged in longitudinal rows, divided by transverse walls into smaller cells, each containing a cluster crystal. Mucilage cells fairly numerous, in longitudinal rows, in size and shape not differing from the surrounding cells.

C o m m o n   p a r e n c h y m a   c e l l s. On the upper side R. and T. 10—15 μ, L. 50—80 μ, on the under side R. and T. 35 μ, L. 70 μ; polygonal prisms with a longitudinally directed axis, the cells of the under side having rounded edges. Walls of the cells of the upper side sometimes a little collenchymatous. Cell contents: at the under side in the undermost layer of cells some chloroplasts containing needle-shaped starch grains.

M u c i l a g e c e l l s. Walls very much thickened, almost to the disappearance of the lumen; consisting of a thin outer cellulose layer, except in the transverse walls, and for the rest of mucilage layers showing particularly in alcohol a stratification. These cells in all respects very similar to the mucilage cells of the root.

E n d o d e r m i s. Developed as a starch-sheath of 1 layer of cells, only on the under side of the meristele.

C e l l s o f t h e a b o v e. Cell contents: chloroplasts often filled with 1 globular starch grain. See for the rest the common parenchyma cells of the colourless parenchyma.

Meristele. In the midrib slightly gutter-shaped, the concave side turned upwards.

P e r i c y c l e. On the upper side some layers of parenchyma cells, without intercellular spaces. On the under side more distinct and consisting of some large sclerenchyma bundles separated from each other and from the phloem bundles by ordinary parenchyma.

P a r e n c h y m a c e l l s o f u p p e r s i d e. R. and T. 10 μ, L. 120—160 μ; polygonal prisms with a longitudinally directed axis. Cell contents: some small chloroplasts. For parenchyma cells of under side see those of colourless parenchyma of mesophyll.

S c l e r e n c h y m a f i b r e s. R. and T. 15 μ, L. 400—600 μ; polygonal. Walls slightly and somewhat collenchymatously thickened; sometimes a little lignified.

V a s c u l a r b u n d l e. A single compound one, containing about 10 simple bundles. Simple bundles collateral, open; phloem bundles broader and less in number than the xylem bundles, thus the correspondence of the phloem and xylem bundles with one another somewhat irregular.

P h l o e m. Bundles consisting of sieve-tubes and cambiform cells.

S i e v e - t u b e s. R. and T. 5—10 μ, L. of articulations 100 μ; polygonal prisms, distinct sieve-plates. C a m b i f o r m c e l l s. R. and T. 5—10 μ; polygonal prisms. Cell contents: some chloroplasts.

C a m b i u m. Some layers of elements.

E l e m e n t s o f t h e a b o v e. R. 5 μ, T. 10 μ, L. 70 μ, in the medullary commissures the radial dimension much larger; tetragonal prisms with a longitudinally directed axis.

X y l e m. Bundles consisting of spiral vessels and parenchyma cells.

S p i r a l v e s s e l s. R. 15 μ, T. 12 μ, L. of articulations 200—300 μ.

Medullary commissures. Uni- or biseriate; in the phloem parts many cells, arranged in longitudinal rows, divided by transverse walls into smaller cells, each containing a cluster crystal.

C e l l s o f t h e a b o v e. R. 10—15 μ, T. 8—15 μ, L. 50—80 μ; poly-, generally tetragonal prisms with a longitudinally directed axis. Cell contents: some small chloroplasts.

June 1902, November 1911.   J; M.

FOLIA BUCHU.
Buchu Leaves. Buchu.
The dried leaves of Barosma betulina, Bartl. a. Wendl. f. Diosm. 102.

LITERATURE. Flückiger & Hanbury. Pharmacographia. 1879. 109. Greenish & Collin. Anat. Atl. o. veget. Powders. 1904. 60. Kraemer. Botany a. Pharmacogn. 1910. 719. Kramer. Mikr.-pharm. Beiträge z. Kenntn. v. Blättern u. Blüten. Diss. Würtemberg. 1907. 9. Oudemans. Aant. o. d. Pharm. neerl. 1854—56. 548. Oudemans. Pharmacogn. 1880.

289. Planchon et Collin. Drogues simples. T. 2. 1896. 618. Schneider. Powder. veget. Drugs. 1902. 137. Shimoyana. Beitr. z. Kenntn. d. Buchublätter. Archiv d. Pharm. 1888. 64. Sieck. Schizo-lys. Secretbeh. Diss. Bern. 1895. 20. See also Pringsheim's Jahrb. 1895. 216. Solereder. Syst. Anat. d. Dicot. 1899. 199. Schulze. Beitr. z. Blattanat. d. Rutaceen. Diss. Heidelberg. 1902. 7, 8. (referring to other Barosma species). Tschirch. Angew. Pfl. Anat. 1889. 510. Tschirch. Pharmacogn. Bd. II. 2. 1917. 1172. Walliczek. Stud. ü. d. Membranschl. veget. Organe. Diss. Bern. 1893. 31,34. See also Pringsheim's Jahrb. 1893. 216. Wigand. Pharmakogn. 1879. 201. MATERIAL. The drug. Leaves of Diosma alba, gathered November 1902 in the Botanic Garden at Groningen, in alcohol. REAGENTS. Water, glycerine, chloral hydrate, potash, iodine in chloral hydrate, iodine in alcohol, phloroglucin and hydrochloric acid, iodine and sulphuric acid 66 per cent., concentrated sulphuric acid, Schulze's macerating mixture.

MICROGRAPHY.

Intervenia.

Epidermis upper side. Stomata wanting. Trichomes wanting.

Epidermal cells proper. H. 60—70 μ, Lev. B. 45 μ, Lev. L. 50—70 μ; poly-, mostly hexagonal prisms with rounded inner walls. Outer walls strongly thickened, consisting of a thicker outermost and a thinner innermost part; outermost part covered with a cuticle, for the rest composed of cuticular layers, containing crystal sand and becoming yellow on the addition of potash; innermost part consisting of pure cellulose; inner walls showing thin secondary layers of pure cellulose, covered by a mucilage layer, this layer covered again by a thin cellulose layer, a very few times followed again by a mucilage and a cellulose layer [1];

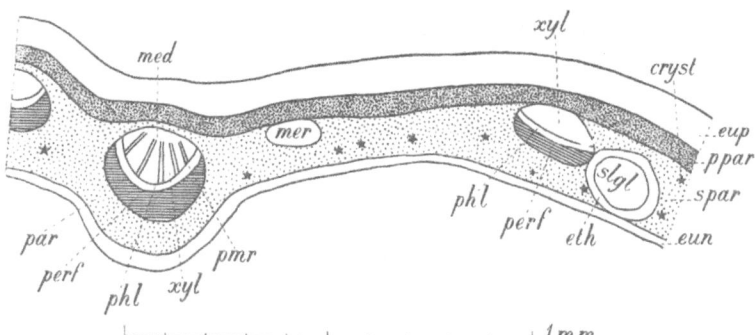

Fig. 25. *Barosma betulina*. Leaf, transverse section through the midrib and the neighbouring portions of the lamina, the mucilage layers of the epidermal cells swollen by the action of glycerine. cryst Cluster crystal; eth Epithelium; eun Epidermis under side; eup Epidermis upper side; med Medulla; mer Meristele; par Colourless parenchyma; perf Pericyclic sclerenchyma fibres; phl Phloem; ppar Palisade chlorenchyma; pmr Medullary commissures; slgl Schizo-lysigenous glands; spar Spongy chlorenchyma; xyl Xylem.

---

[1] According to W a l l i c z e k l. c. the protoplasma first secretes the mucilage and afterwards the cellulose layer. This view agrees with the way in which these layers join the lateral walls, as may be observed in potash and sulphuric acid 66 per cent.; further with the fact, that in young leaves of *Diosma alba* much mucilage is already found along the inner walls of the epidermis, whilst the cellulose layers have not yet been formed, the epidermal cells of older leaves exhibiting a structure similar to that of *Folia Buchu*. W a l l i c z e k founds his opinion on the following facts:

1°. that he never has seen division of epidermal cells;

2°. that the cellulose layers are developed at a later stage than the mucilage;

3°. that he has been able to observe the development of the cellulose layer on the outside of the protoplast.

On the contrary S h i m o y a n a holds the opinion that the cell division takes place first and that the mucilage is formed afterwards from the bottom of the undermost cell. He mentions never to have seen a nucleus or protoplasm in the undermost cell.

the mucilage layer in alcohol showing stratification in the shape of hyaline and dark granular layers alternating with each other; in water and potash the mucilage swelling strongly and becoming invisible; in glycerine the same process, but the swelling less strong; in iodine and in iodine and sulphuric acid 66 per cent. the mucilage nearly always remaining colourless, sometimes becoming slightly yellow and stratified; the part of the lateral walls corresponding with the mucilage very thin and mostly torn by the mucilage swelling in water and the other reagents. Cell contents: in the small vacuoles (15 μ) yellow often spherical sphero-crystals of hesperidine showing distinct radiate structure, especially in sections treated with alcohol; in the mucilage also some sphero-crystals and dendrites of hesperidine; a few sphero-crystals persisting in potash and hot water.

Epidermis under side. Cells covering the glands much larger than the other ones, except those covering the middle part of the smaller glands. Stomata 45 to the square m.M.; in the same level as the epidermal cells and surrounded by 5—7 subsidiary cells, partly lying under them; showing a distinct vestibulum.

Epidermal cells proper. H. 35 μ, Lev. B. 30 μ, Lev. L. 35 μ; polygonal prisms with rounded inner walls. The mucilage not developed in all the cells and always to a much smaller amount than in the epidermal cells of the upper side; the lateral walls not torn by the mucilage swelling in water. For the rest see the upper side. Stomata. H. 15μ, Lev. B. 25 μ, Lev. L. 28 μ.

Mesophyll. Consisting of palisade chlorenchyma and spongy chlorenchyma, the latter containing glands.

Palisade chlorenchyma. 1 layer of cells; here and there 2 layers together of the same height. Showing intercellular spaces.

Cells of the above. H. 30—40 μ, Lev. 10 μ; in the cells half high as the others Lev. often somewhat larger, e. g. 15 μ. Cell contents: chloroplasts containing starch grains; sometimes a cluster crystal, especially in the cells half as high as the others.

Spongy chlorenchyma. Consisting of 5—6 layers of cells; the uppermost layer somewhat palisade-shaped, the rest, except the undermost layer, stellate. Showing large intercellular spaces. **Glands.** Schizo-lysigenous [1]); ellipsoidal; H. 120 μ, Lev. 200 μ; surrounded by an epithelium of 2 layers of cells; the epithelium cells adjoining the epidermal cells of the under side and the palisade tissue.

Cells of spongy chlorenchyma. Uppermost layer H. 15 μ, Lev. 10 μ; undermost layer H. 20 μ, Lev. 10 μ; stellate cells H. 30 μ, Lev. B. 20 μ, Lev. L. 25 μ. Cell contents: chloroplasts containing starch grains; sometimes a cluster crystal. Epithelium cells of glands. Thick 15 μ, length and width 20 μ; polygonal.

Margin.

Epidermis. H. of cells 35 μ. Mucilage and cellulose layers covering it, generally wanting.

Glands. More numerous and larger than in the intervenia; every incision with a corresponding gland, sometimes between 2 glands a smaller one. H. 180 μ, Lev. 250 μ. The epithelium cells adjoining the epidermis on both sides, the glands

[1]) Sieck. p. 20 or 216.

even somewhat bulging out, especially at the under side. Contents: a dark yellow granular mass often containing globules.

Veins.

Epidermis upper side. Consisting of 2—4 rows of cells. Trichomes at the base 12 μ in diameter, 50—80 μ in length; not numerous and only occurring on the undermost half of the midrib; conical; not divided from the epidermal cells by a wall.

Epidermal cells proper. H. 40 μ, Lev. B. 30 μ, Lev. L. 40—80 μ; longitudinally elongated polygonal prisms. Outer walls somewhat more thickened than those of the cells of the intervenia; see for the rest those of the intervenia; the quantity of the mucilage strongly varying, sometimes the mucilage nearly or quite wanting and then the covering cellulose layers wanting too; the lateral walls not torn by the mucilage swelling in water. Trichome cells. Walls very thick; cuticularized.

Epidermis under side. Cells lying in longitudinal rows.

Cells of the above. H. 30 μ, Lev. B. 22 μ, Lev. L. 40—80 μ; rectangular prisms. Outer walls more strongly thickened than those of the cells of the upper side; pitted, the pits reaching only as far as the middle of the wall; showing stratification; lateral walls somewhat thickened and pitted; inner walls thickened and collenchymatous; mucilage and covering cellulose layers often wanting. See for the rest the cells of the upper epidermis of the intervenia.

Mesophyll. Consisting of colourless parenchyma containing glands; an endodermis, moreover at the upper side a layer of palisade chlorenchyma continuing that of the intervenia and identical with it.

Colourless parenchyma. At the upper side 1—2 layers of cells, the innermost layer showing intercellular spaces. At the under side 3 layers of cells.

Parenchyma cells of the upper side. R. 10—15 μ, T. 15—20 μ, L. in the outermost layer up to 50 μ, in the innermost layer up to 80 μ; polygonal prisms with a longitudinally directed axis and more or less rounded edges. Walls of the innermost layer somewhat thickened; pitted. Cell contents: a few chloroplasts with starch grains, in the innermost layer only occurring along the outer cell walls. Parenchyma cells of the under side. In the outermost layer H. 20 μ, Lev. B. 20 μ, Lev. L. 40—100 μ; in the innermost layer H. 20 μ, Lev. B. 20 μ, Lev. L. 40—80 μ; polygonal prisms with a longitudinally directed axis and rounded edges. Walls of the outermost layer collenchymatous. Cell contents: here and there a starch grain.

Glands. See those of the intervenia.

Endodermis. Developed as a starch-sheath, only to be distinguished from the colourless parenchyma by a larger amount of starch grains.

Meristele. About cylindrical.

Pericycle. Only present at the lower side of the veins as a gutter-shaped bundle of sclerenchyma fibres; this bundle thick in its middle part from 5 to 6 fibres, the outermost ones the largest; number of fibres decreasing towards the lateral sides.

Sclerenchyma fibres. Middlemost fibres R. 12 μ, T. 12 μ, L. 100—250 μ; mostly penta- or hexagonal. Walls strongly thickened; middle lamella distinct; not ligni-fied; showing stratification; cross wise slit-like pits.

V a s c u l a r   b u n d l e. A single compound one, containing about 6 simple bundles. Simple bundles collateral; open.

P h l o e m. Adjoining the bundle of pericyclic fibres and forming a thin layer of flattened elements. Here and there containing a cluster crystal.

X y l e m. Consisting of a secondary part containing pitted vessels lying in radial rows and an uppermost primary part containing spiral vessels.

V e s s e l s. R. 11 μ, T. 9 μ, L. of articulations 150—250 μ; uppermost vessels with smaller radial and tangential dimensions; poly-, mostly hexagonal prisms. Walls strongly thickened; lignified, especially the middle lamella.

M e d u l l a r y   c o m m i s s u r e s. Consisting of radial strips of common parenchyma. Intercellular spaces wanting.

C e l l s   o f   t h e   a b o v e. In the uppermost part R. 6 μ, T. 9 μ, L. 50—70 μ; mostly penta- or hexagonal prisms with a longitudinally directed axis. Walls somewhat thickened; pitted. In the undermost part R. 13 μ, T. 10 μ, L. 50—70 μ; mostly hexagonal. Walls thickened; pitted.

M e d u l l a. Represented by some common parenchyma adjoining the primary xylem and the medullary commissures. Intercellular spaces wanting.

C e l l s   o f   t h e   a b o v e. Outermost cells R. 15 μ, T. 22 μ, L. 80—100 μ; innermost cells R. 12 μ, T. 15 μ, L. up to 200 μ; polygonal prisms with a longitudinally directed axis. Walls thickened; hardly ever lignified; showing cross-wise slit-like pits.

October 1902.   J; M; L.

FOLIA COCAE.
Coca Leaves. Coca.
The dried leaves of Erythroxylum Coca, Lam. Encyc. II. 393 and its varieties.

LITERATURE. Benecke. Mikr. Drogenprakt. 1912. 50. Erdmann-König's Allg. Warenkunde. 1895. 262. Flückiger. Pharmakogn. 1891. 635. Gilg. Pharmakogn. 1910. 179. Glaser. Mikr. Anal. d. Blattpulver. Verh. d. Phys.-med. Ges. Würzburg. N. F. Bd. XXXIV. 1901. 253. Gosse. Monogr. d. l'Erythroxylon Coca. 1861. Greenish. Struct. o. Coca leaves. Pharm. Journ. 1904. 493. Greenish & Collin. Anat. Atl. o. veget. Powders. 1904. 62. Hager. Pharm. Prax. Bd. 1. 1900. 868. Hartwich. Beitr. z. Kenntn. d. Cocablätter. Arch. d. Pharm. 1903. 617. Hérail. Pharmacogr. 1912. 746. Jenzer. Pharm. Unters. ü. Pilocarpus pennatifolius Lemaire u. Erythroxylon Coca Lamarck m. bes. Berücks. d. Alkaloiden. Diss. Zürich. 1910. 80. Pl. 3. Karsten u. Oltmanns. Pharmakogn. 1909. 150. Kraemer. Botany a. Pharmacogn. 1910. 606, 724. Moeller. Mikr. d. Nahr. u. Genussm. 1886. 45. Moeller. Pharmakogn. 1889. 74. Moeller. Mikr.-pharm. Üb. 1901. 88. Planchon et Collin. Drogues simples. T. 2. 1896. 688. Reens. La Coca de Java. Diss. Paris. 1919. Reens. Sur la cire des feuilles de Coca de Java. Trav. Lab. Mat. méd. Paris. XI. 1917—1919. 54. Schneider. Powdered veget. Drugs. 1902. 165. Tschirch. Angew. Pfl.-Anat. 1889. 323, 438. Tschirch u. Oesterle. Anat. Atl. 1900. 263. Tunmann. Pharm. Unters. v. Pilocarpus pennatifolius u. Erythroxylon Coca m. bes. Berücks. d. Alkaloiden. Pharm. Post. Bd. 42. 1909. 768. Tunmann. Pfl. mikrochemie. 1913. 305. Vogl. Veget. Nahr. u. Genussm. 1899. 272. MATERIAL. The drug. Leaves, collected November 1902 in the Botanic Garden at Groningen, fresh, in alcohol and fixed with 1 per cent. chromic acid. REAGENTS. Water, glycerine, chloral hydrate, potash, potassium iodide iodine, iodine and choral hydrate, phloroglucin and hydrochloric acid, iodine and sulphuric acid 66 per cent., Schulze's macerating mixture, copper acetate and iron acetate, potassium dichromate.

MICROGRAPHY.

Intervenia.

Epidermis upper side. Some cells containing mucilage, as described for Folia Buchu; the number of these cells varying greatly in different leaves. Near the margin of the leaf cells elongated in the direction of the margin. Stomata wanting. Trichomes wanting.

Epidermal cells proper. H. 18 μ, Lev. B. 20 μ, Lev. L. 20—40 μ; mucilage cells somewhat higher; elongated cells e. g. Lev. 40 μ; polygonal prisms. Outer walls somewhat thickened, showing a granular cuticle; lateral walls pitted; structure of the inner walls of the mucilage cells: thin secondary cellulose layer, covered by a thicker mucilage layer, and this again covered by a very thin layer of pure cellulose; the mucilage also in alcohol showing darker and lighter granular layers [1]); often in a section of the drug treated with iodine and choral hydrate the protoplasm and the nucleus distinct in the cavity of the mucilage cell and adjoining the thin covering cellulose layer, the mucilage being dissolved in the reagent. Cell contents: alcaloid, occurring in

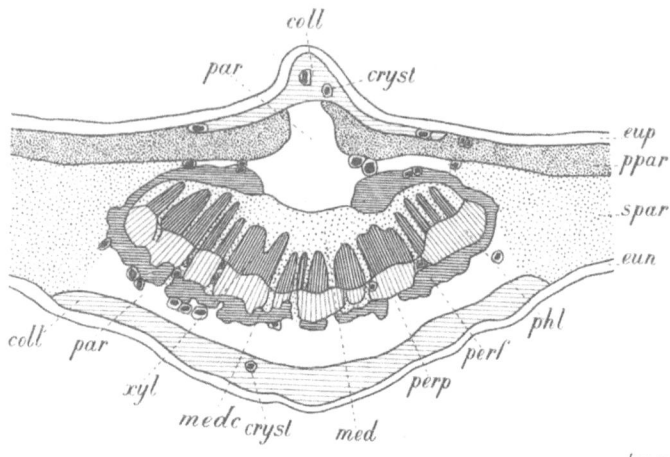

Fig. 26. *Erythroxylum Coca*. Leaf, transverse section through the midrib and the neighbouring portions of the lamina. coll Collenchyma; cryst Crystal fibres; eun Epidermis under side; eup Epidermis upper side; med Medulla; medc Medullary commissures; par Colourless parenchyma; perf Pericyclic sclerenchyma fibres; perp Pericyclic parenchyma; phl Phloem; ppar Palisade chlorenchyma; spar Spongy chlorenchyma; xyl Xylem.

all the cells; in fresh material treated with potassium iodide iodine in all the cells several small reddish brown grains; in sections of leaves, treated previously with alcohol during 3—4 hours, this precipitate wanting on the addition of potassium iodide iodine; some choroplasts; in the drug here and there a cell filled with a yellow to brown mass; in sections treated with potassium dichromate in a few cells a dark brown globe.

Epidermis under side. Consisting of parenchyma and prosenchyma, the latter only on the curved line on the lower surface of the leaf on either side of the midrib and near to it. Stomata somewhat sunken, with 2 subsidiary cells, 1 on either side (see Tschirch u. Oesterle, fig. 5). Trichomes wanting.

Parenchymatic elements. H. 14 μ, Lev. B. 15 μ, Lev. L. 20—30 μ, low polygonal papillae, with slightly sinuous lateral walls. Outer walls somewhat thickened and having a granular cuticle. Cell contents: see the cells of the upper side. Prosenchymatic elements. H. 10 μ, Lev. B. 7 μ. Outer walls somewhat thickened and having a granular cuticle. Cell contents: see the cells of the upper side. Guard-cells of

---

[1]) The development of the mucilage and the covering cellulose layer has not been examined, because in the fresh and in the alcoholic material the number of cells with mucilage was too small.

s t o m a t a. H. 7 μ, Lev. B. 9 μ, Lev. L. 18 μ. S u b s i d i a r y c e l l s hardly ever papilla-shaped.

Mesophyll.

P a l i s a d e c h l o r e n c h y m a. Consisting of 1 layer of cells, some cells with a transverse wall. Showing intercellular spaces.

C e l l s o f t h e a b o v e. H. 40—50 μ, Lev. 8 μ; polygonal prisms with strongly rounded edges or cylinders. Cell contents: chloroplasts containing starch grains, in chromic acid material these chloroplasts penta- or hexagonal tables paving the walls; alcaloid; in the uppermost part of some cells 1 larger or some smaller dark globules; in the divided cells often a simple crystal in one or in both partitions.

S p o n g y c h l o r e n c h y m a. Consisting of about 7 layers of stellate cells, branches always parallel to the surface of the leaf; cells of the uppermost 1—2 layers somewhat palisade-shaped. Corresponding with the curved line mentioned above 1—6 layers of polygonal, longitudinally elongated, collenchymatous chlorenchyma cells.

C e l l s o f t h e a b o v e. In the uppermost 1—2 layers H. 15 μ, Lev. 8 μ, the rest H. 15 μ, Lev. 30—50 μ; cells corresponding with the curved line 14 μ in diameter, smaller towards the epidermis. Cell contents: alcaloid; in some cells tannin; in leaves kept in potassium dichromate in many cells 1 or more dark brown globules; for the chloroplasts see the palisade chlorenchyma.

V e i n s.

Epidermis upper side. Epidermal cells of the midrib and the larger primary veins lying in longitudinal rows. Stomata wanting. Trichomes wanting.

C e l l s o f t h e a b o v e. H. 15 μ, Lev. B. 12 μ, Lev. L. 20—40 μ; quadrangular and mostly rectangular prisms. Inner walls slightly collenchymatous. See for the rest the epidermal cells of the intervenia.

Epidermis under side. Cells lying in longitudinal rows. Stomata wanting. Trichomes wanting.

C e l l s o f t h e a b o v e. H. 14 μ, Lev. B. 12—18 μ, Lev. L. 50—100 μ. Outer walls somewhat thickened, with a granular cuticle; inner walls a little collenchymatous; lateral walls pitted. Cell contents: in the drug in each cell a brown mass. See for the rest epidermal cells of upper side of intervenia.

Mesophyll.

C o l l e n c h y m a. At the upper side represented by a bundle of 5—6 layers of cells, the middle part of which filling up the projecting ridge of the midrib of the leaf, on either lateral side 1 layer of cells between the epidermis and a part of the palisade chlorenchyma projecting into the midrib; at the under side 3—4 layers of cells. The cells filling up the ridge arranged in longitudinal rows. On both sides of the leaf here and there idioblasts in the shape of crystal fibres.

C o l l e n c h y m a c e l l s o f t h e r i d g e. R. and T. 20 μ, close to the epidermis R. and T. 12 μ; polygonal prisms with a longitudinally directed axis. Walls somewhat collenchymatous. C e l l s b e t w e e n t h e e p i d e r m i s a n d t h e p a l i s a d e t i s s u e. R. 20 μ, T. 18 μ, L. 50 μ; rectangular prisms with a longitudinally directed axis. Walls collenchymatous; showing stratification. C e l l s o f t h e u n d e r s i d e. R. 15 μ, T. 18 μ, L. 40—190 μ, the longest cells in the outermost layer. Cell contents: near the epidermis in some cells alcaloid; in other cells tannin in globules; a few cells filled with a colourless granular mass. C r y s t a l f i b r e s. Divided in cells, each L. 15—20 μ, containing a simple crystal in the direction of the diagonal and surrounded by a crystal skin of cellulose connected with the cell wall.

C o l o u r l e s s p a r e n c h y m a. At the upper side in the middle part about 10 layers, at the under side 5—7 layers. Cells lying in longitudinal rows. Outermost 1—2 layers of the under side often flattened. On both sides idioblasts in the shape of crystal fibres, especially in the vicinity of the meristele.

C e l l s o f t h e a b o v e. At the upper side R. and T. 15—20 μ, L. 40—50 μ, outermost cells longer, up to 100 μ; at the under side R. and T. 20 μ, L. 50—70 μ; polygonal prisms with a longitudinally directed axis and rounded edges. Cell contents: chloroplasts containing starch grains, relatively numerous; here and there tannin in globules. C r y s t a l f i b r e s. See those of the collenchyma.

P a l i s a d e c h l o r e n c h y m a. Not present below the ridge; in the rest of the vein represented by 2 layers of cells.

C e l l s o f t h e a b o v e. H. 20 μ, Lev. B. 10 μ, Lev. L. 10 μ; cylindrical. Cell contents: chloroplasts containing starch grains; here and there tannin.

S c l e r e n c h y m a. Here and there an elliptical fibre between the colourless parenchyma.

S c l e r e n c h y m a f i b r e s. R. 18 μ, T. 12 μ. Walls thickened; showing stratification; lignified; pitted.

E n d o d e r m i s. Only to distinguish from the colourless parenchyma by its somewhat smaller and at the upper side somewhat shorter cells.

Meristele of the midrib. Gutter-shaped, the convex side turned downwards.

P e r i c y c l e. Always wanting in the middle part, sometimes over the whole of the upper side of the meristele. Consisting of sclerenchyma fibres and common parenchyma. Sclerenchyma fibres at the upper side in about 5 layers, this number decreasing at the lateral sides to 1—2; at the under side in bundles thick 1—3 layers, alternating with numerous bands of parenchyma cells showing intercellular spaces.

S c l e r e n c h y m a f i b r e s. R. and T. 10—15 μ, L. 500—2000 μ; polygonal. Walls thickened; middle lamella conspicuous; lignified. P a r e n c h y m a c e l l s. R. 17 μ, T. 20 μ, L. 60—100 μ, often smaller; polygonal prisms with a longitudinally directed axis and strongly rounded edges. Walls often thickened, then lignified; pitted. Cell contents: some chloroplasts containing starch grains; in leaves kept in potassium dichromate sometimes brown globules.

C o m p o u n d v a s c u l a r b u n d l e. A single one; containing 15—20 simple bundles, arranged in a curved plane. Simple bundles having the shape in a cross section of thin radially directed lamellae; collateral; open.

P h l o e m. Here and there somewhat more developed in a tangential direction than the accompanying xylem and sometimes fusing together with the phloem of an adjacent simple bundle; consisting of polygonal elements.

E l e m e n t s o f t h e a b o v e. R. and T. 5 μ, at the outside larger e. g. R. and T. 6—8 μ. Walls a little collenchymatous. Cell contents: sometimes starch grains.

C a m b i u m. Not clearly to distinguish.

X y l e m. Consisting of radial bands of vessels, 1—2 in a tangential and 6—7 in a radial direction; spiral and reticulate vessels, the innermost ones scalari-form.

V e s s e l s. Undermost ones R. and T. 12—14 μ, uppermost ones much smaller, L. of articulations 250—700 μ, those of the spiral vessels the longest; transverse walls showing a circular perforation. Walls lignified. In the undermost vessels walls sometimes much more thickened; middle lamella conspicuous.

M e d u l l a r y c o m m i s s u r e s. The phloem parts here and there wanting in consequence of the more extensive development of the phloem bundles mentioned above; corresponding with the parenchyma bands of the pericycle; uni-to biseriate. In the phloem part idioblasts in the shape of crystal fibres.

C e l l s o f t h e a b o v e. R. 8—15 μ, T. 4—8 μ, L. 60—100 μ; polygonal prisms with a longitudinally directed axis and rounded edges. Walls often thickened; lignified; pitted. Cell contents: chloroplasts containing starch grains; alcaloid; in leaves kept in potassium dichromate brown globules. **C r y s t a l f i b r e s.** L. of constituent cells e. g. 8 μ; see for the rest those of the collenchyma. **C e l l s o f t h e x y l e m p a r t** R. 10 μ, T. 6 μ, L. 30—50 μ; penta- or hexagonal prisms with a longitudinally directed axis. Walls lignified; pitted. Cell contents: see the phloem part.

M e d u l l a. Represented by 3—4 layers of somewhat thick-walled parenchyma cells.

C e l l s o f t h e a b o v e. R. and T. 10 μ, L. 80—120 μ; polygonal prisms with a longi-tudinally directed axis. Walls somewhat thickened; lignified; pitted. Cell contents: chloroplasts containing starch grains; alcaloid; in material kept in potassium dichromate brown globules.

Meristeles of the primary veins and the veinlets. All, except the smallest ones, showing a pericycle consisting of sclerenchyma fibres. Round this peri-cycle here and there a crystal fibre with simple crystals, like those of the collenchyma of the midrib. At the upper side these fibres often adjoining the epidermis.

November 1902.   J; M; L.

FOLIA DIGITALIS.
Digitalis Leaves. Foxglove Leaves.

The dried leaves of Digitalis purpurea, Linn. Sp. Pl. 621, collected from plants commencing to flower.

Macroscopic characters.

Simple, petiolate, up to 30 c.M. in length. Blade ovate-oblong; feather-veined; midrib, veins and veinlets strongly prominent on the under side; apex acute; base obtuse, tapering into the somewhat alate petiole; margin slightly crenate, in the drug seeming serrate, on each tooth a white tip. Surface, especially of the under side and margin, densely and softly pubescent; hairs chiefly on the veins and veinlets. Taste bitter.

Anatomical characters.

LITERATURE. Bohny. Beitr. z. Kenntn. d. Digitalisblattes, &c. Diss. Zürich. 1906. 8. Collin. Sur la Digitale. Journ. Pharm. et Chem. Sér. 6. Bd. 22. 1905. 56. Deutsch. Arzneib. 5. Ausg. 1910. 229. Ellrodt. Üb. d. Vert. d. Gerbst. i. off. Blättern, Kräutern u. Blüten. Diss. Würzburg. 1903. 12. Flückiger. Pharmakogn. 1891. 671. Gilg. Pharmakogn. 1910. 319. Glaser. Mikr. Anal. d. Blattpulver. Verh. d. Phys.-Med. Ges. z. Würzburg. N. F. Bd. 34. 1901. 254. Greenish & Collin. Anat. Atl. o. veget. powders. 1904. 64. Hager. Pharm. Praxis.

Bd. 1. 1900. 1037. Hérail. Mat. Méd. 1912. 513. Karsten u. Oltmanns. Pharmakogn. 1909. 163. Koch. Mikr. Anal. d. Drogenpulver. Bd. 3. 1906. 99. Koch. Einf. i. d. mikr. Anal. d. Drogenpulver. 1906. 69. Koch u. Gilg. Pharmakogn. Praktik. 1907. 137. Kraemer. Botany a. Pharmacogn. 1910. 615, 727. Marmé. Pharmacogn. 1886. 193. Meyer. Wiss. Drogenk. Bd. 2. 1892. 202. Mitlacher. Toxik. od. Forens. wicht. Pfl. u. Drogen. 1904. 171. Moeller. Pharmakogn. 1889. 64. Moeller. Mikr.-pharm. Üb. 1901. 95. Moeller. Digitalis u. Verbascum. Pharm. Post. Jg. 37. 1904. 677, also Apoth. Zeitg. 1904. 953. Oudemans. Aanteek. o. d. Pharmac. neerl. 1854—56. 335. Oudemans. Pharmacogn. 1880. 264. Planchon et Collin. Drogues simples. T. 1. 1895. 548. Schneider. Powdered Veget. Drugs. 1902. 182. Tschirch. Angew. Pfl.⊢anat. 1889. 324. Tschirch u. Oesterle. Anat. Atl. 1900. 319. Taf. 74. Vogl. Anat. Atl. 1887. Taf. 12. Wigand. Pharmakogn. 1879. 225. MATERIAL. The drug. Leaves gathered July and October 1902 from the Botanic Garden at Groningen, fresh, in alcohol and also after fixation in chromic acid 1 per cent. REAGENTS. Water, glycerine, chloral hydrate, potash, potassium iodide iodine, iodine in chloral hydrate, phloroglucin and hydrochloric acid, iodine and sulphuric acid 66 per cent., concentrated sulphuric acid, Schulze's macerating mixture, cupric acetate and ferric acetate, potassium dichromate.

MICROGRAPHY of the blade.

Intervenia.

Epidermis. Upper side. Stomata very much varying in number on different leaves, e.g. 8 and sometimes up to 100 to the sq. m.M.; phaneroporous; lying generally in the level of the epidermis; mostly surrounded by 4 subsidiary cells. Water-pores one on each tooth. Trichomes in 3 kinds: 1. Conical hairs very much varying in number on different leaves, sometimes nearly or even wholly wanting; along the margins always more numerous; often implanted upon 2 somewhat prominent epidermal cells having no sinuous walls and somewhat imitating in arrangement and shape the guard-cells of a stoma; L. 100—300 μ, diameter at the base 20—30 μ; uniseriate; unbranched; 2- to 5-cellular; usually more or less curved; often 1 or more cells flattened. 2. Capitate hairs with multicellular stalk not numerous, often nearly or even wholly wanting. Head globular, unicellular. See for the rest the conical hairs; somewhat less in length. 3. Capitate hairs with unicellular stalk not numerous; the stalk L. 12 μ, diam. 10 μ, very rarely consisting of 2 cells. Head H. 18 μ, Lev. 20 and 30 μ; consisting of 1 but generally 2 cells, separated from each other by a vertical wall; each of these cells showing a small lateral knob.

E p i d e r m a l c e l l s p r o p e r. H. 20 μ, Lev. 30 and 50 μ; polygonal tables having straight to strongly sinuous side walls. Outer walls a little thickened, showing a cuticle. Cell contents: some small chloroplasts; by treatment with potassium dichromate a brown mass in the large vacuole, some smaller adventitious vacuoles remaining colourless; by treatment with potash coloured yellow; by treatment with potassium iodide iodine a dark yellow to brown mass, chiefly along the walls. G u a r d - c e l l s o f s t o m a t a. H. 10 μ, Lev. 10 and 25 μ. G u a r d - c e l l s o f w a t e r - p o r e s. Lev. 18 and 40 μ. C e l l s o f c o n i c a l h a i r s. Outer wall of the undermost cells generally showing a parallel longitudinal cuticular striation; that of the other cells showing more or less projecting cuticular dots or stripes. Contents of the undermost cells generally the same as those of the epidermal cells proper. C e l l s o f c a p i t a t e h a i r s w i t h u n i c e l l u l a r s t a l k. Walls comparatively thick. Cell contents: in the head often some chloroplasts and by treatment with potassium dichromate no precipitate; see for the rest the epidermal cells proper.

Under side. Stomata 160—200 to the sq. m.M. Trichomes the same as on the upper side, but more numerous, especially the conical hairs. See for the rest the upper side.

E p i d e r m a l   c e l l s   p r o p e r. H. 15 μ, Lev. 20—30 and 30—50 μ; tabular, having strongly sinuous side walls. See for the rest the upper side.

Mesophyll.

P a l i s a d e   c h l o r e n c h y m a. 1—3 layers of cells; intercellular spaces. C e l l s   o f   t h e   a b o v e. H. 30 μ, Lev. 15—20 μ; polygonal prisms with more or less rounded edges. Cell contents: very numerous chloroplasts, paving the walls, thick 2.5 μ, diameter 5 μ, generally hexagonal tables, filled with starch grains; by treatment with potassium dichromate in each cell a dark brown globe; by treatment with potash coloured yellow; by treatment with potassium iodide iodine a brown granular mass; by concentrated sulphuric acid first yellow, soon after green.

S p o n g y   c h l o r e n c h y m a. 3—4 layers of stellate cells; large intercellular spaces. See for the rest the palisade chlorenchyma.

V e i n s.

Epidermis.
Upper side. Stomata wanting. Trichomes the same as on the mesophyll; capitate hairs with unicellular stalk much more numerous, even on the smaller veins and implanted upon very small cells — e.g. T. 20 μ, L. 30 μ.

E p i d e r m a l   c e l l s   p r o p e r. R. 40 μ, T. 30—40 μ, L. 50—150 μ; polygonal prisms with a radially directed axis. Outer walls a little thickened, showing a longitudinal parallel cuticular striation; inner walls a little collenchymatous. Cell contents: some small chloroplasts; by treatment with potassium dichromate a brown mass, sometimes with one or more cavities; by treatment with potash coloured yellow; by treatment with potassium iodide iodine a granular dark yellow mass; by concentrated sulphuric acid first yellow, soon after green; in material fixed with chromic acid a citrine granular mass, sometimes having one or more cavities.

Under side. Stomata rare, see for the rest the intervenia. Trichomes the same as on the upper side of the mesophyll, but more numerous; the conical hairs and capitate hairs with multicellular stalk also larger, e.g. L. 700 μ and up to 30 μ diameter at the base.

E p i d e r m a l   c e l l s   p r o p e r. R. 25 μ, T. 20—30 μ, L. 70—150 μ; generally hexagonal prisms with a radially directed axis. See for the rest those of upper side. G u a r d- c e l l s. H. 12 μ, Lev. 12 and 35 μ.

Mesophyll.

C o l l e n c h y m a. On the upper side 1 layer of cells, on the under side 3—5 layers. Intercellular spaces.

C e l l s   o f   t h e   a b o v e. R. 40 μ, T. 35 μ, L. 100—200 μ; tetra- to hexagonal prisms with a longitudinally directed axis and rounded edges. Cell contents: numerous chloroplasts, chiefly on the lateral sides of the midrib, each containing some globular starch grains; see for the rest epidermal cells proper of veins.

C o l o u r l e s s   p a r e n c h y m a. On the upper side some 10, on the under side some 15 layers of cells; intercellular spaces.

C e l l s   o f   t h e   a b o v e. The largest cells R. and T. 75 μ, L. 100—250 μ, in the vicinity of epidermis and meristele the cells smaller; polygonal prisms with a longitudinally directed axis and rounded edges. Cells contents: some chloroplasts, each containing some globular starch grains; by treatment with potassium dichromate a yellow granular mass; in material fixed with chromic acid in some cells, principally in the outermost ones, a citrine granular mass along the walls.

E n d o d e r m i s. Developed as a complete starch-sheath consisting of 1 or 2 layers of cells.

C e l l s   o f   t h e   a b o v e. R. 50 μ, T. 60 μ, L. on the upper side 50—60 μ, on the under side 70—100 μ; polygonal prisms with a longitudinally directed axis and rounded edges. Cell contents: large compound starch grains, 10 μ in diam. and consisting e. g. of 7 component grains; by treatment with potassium dichromate a yellow mass.

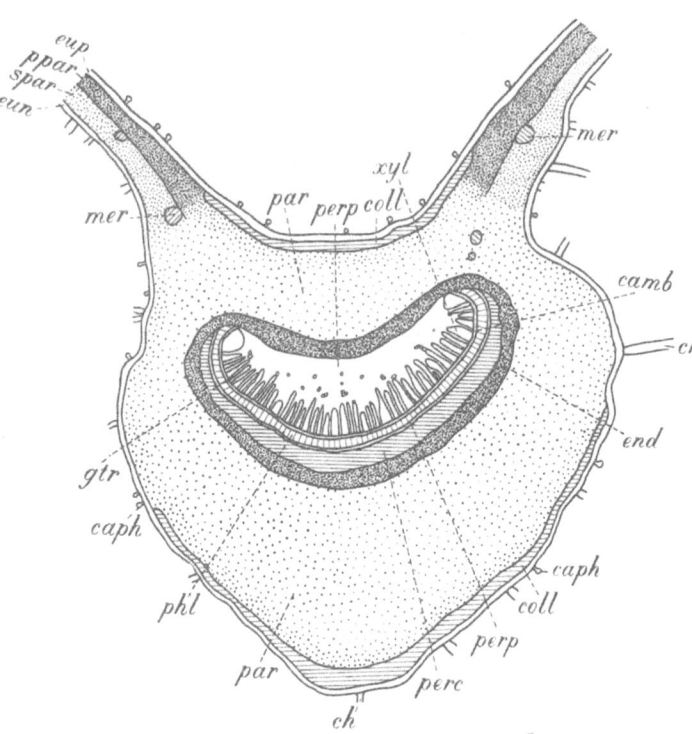

Fig. 27. *Digitalis purpurea*. Leaf, transverse section through the midrib and the neighbouring portions of the lamina. camb Cambium; caph Capitate hairs with unicellular stalk; ch Conical hairs; coll Collenchyma; end Starch-sheath; eun Epidermis under side; eup Epidermis upper side; gtr Medullary commissures; mer Small meristeles obliquely cut; par Colourless parenchyma; perc Pericyclic collenchyma; perp Pericyclic parenchyma; phl Phloem; ppar Palisade chlorenchyma; spar Spongy chlorenchyma; xyl Xylem.

Meristele. In the midrib flat gutter-shaped, the concave side turned upwards. **Pericycle.** On the upper side some layers of parenchyma cells. On the under side 6–8 layers of collenchyma fibres, becoming larger towards the endodermis and without intercellular spaces; moreover 1 or 2 layers of parenchyma cells, adjoining the phloem bundles.

P a r e n c h y m a   c e l l s of upper side. R. 30—35 μ, T. 35—40 μ, L. 70—120 μ; polygonal prisms with a longitudinally directed axis. Walls a little collenchymatous. Cell contents: by treatment with potassium dichromate a yellow mass. C o l l e n c h y m a   f i b r e s. R. 10—22 μ, T. 15—35 μ, L. 500—2500 μ; polygonal, a few having branched ends. Cell contents: by treatment with potassium dichromate in several fibres a yellow mass.

V a s c u l a r   b u n d l e. In the midrib a single compound one, containing numerous simple bundles. Simple bundles collateral, open.

P h l o e m. Bundles entirely consisting of sieve-tubes.

Sieve-tubes. R. and T. 6 μ, L. of articulations 100—160 μ; polygonal prisms with a longitudinally directed axis; transverse walls horizontal, each entirely occupied by a single sieve-plate.

Cambium. Consisting of some layers of cells.

Xylem. Bundles consisting of a single radial row of elements: annular, spiral, reticulate and pitted vessels, mixed with parenchyma cells between the first-named vessels.

Vessels. R. and T. up to 25 μ, L. of articulations 150—400 μ; polygonal prisms with more or less rounded edges; transverse walls with circular perforations. Walls lignified; pits elongated in a transverse direction. Parenchyma cells. R. 20—30 μ, T. 25—35, L. 100—200 μ; polygonal prisms with a longitudinally directed axis. Walls a little collenchymatous. Cell contents: by potassium dichromate a yellow mass.

Medullary commissures. Uni- or biseriate, consisting of parenchyma cells.

Cells of the above. R. 12 μ, T. 10 μ, L. 50—140 μ, polygonal prisms with a longitudinally directed axis. Cell contents: in the phloem parts by treatment with potassium dichromate a yellow to brown mass; in the xylem parts some chloroplasts, by treatment with potassium dichromate a dark brown mass, by treatment with potash coloured yellow, by treatment with potassium iodide iodine a brown granular mass, by treatment with concentrated sulphuric acid coloured yellow, soon after green, in material fixed with chromic acid a yellow granular mass.

Micrography of the powder.

Green parenchyma; 1—3 layers of palisade cells. Many 1- to 6- cellular conical hairs, having an obtuse apex. Capitate hairs, having an 1- or 2-cellular head, in much smaller number. Epidermal cells having sinuous lateral walls; stomata on the upper side less numerous. Spiral and reticulate vessels. Starch may be found in the green parenchyma and in the colourless parenchyma of the veins.

October 1902, December 1911.   J; M.

FOLIA EUCALYPTI.
Eucalyptus Leaves.
The dried leaves of Eucalyptus Globulus, Labill. Voy. I. 153. t. 13, collected from the older parts of the tree.

Macroscopic characters.

Simple, petiolate. Petiole up to 4 c.M. long, more or less twisted. Blade up to 25 c.M. long, up to 4 c.M. broad; pretty thick; stiff; elongated lanceolate-scythe-shaped; more or less unequal; feather-veined; numerous thin veins, diverging from the midrib at sharp angles, running into an undulating vein parallel and near to the margin; apex acute; base from obtuse to acute; margin entire, somewhat thickened. Surface of both sides the same, dead grayish green, glabrous, with innumerable fine dark cork patches.

The juvenile leaves 15 c.M. long, up to 9 c.M. broad, sessile, ovale, with auriculate base, are not officinal. Odour especially of bruised leaves aromatic; taste aromatic, somewhat bitter, first not, afterwards cooling.

Anatomical characters.

LITERATURE. De Bary. Vergl. Anat. 1877. 89. Brissi. Intorno all' anatomia delle foglie dell' Eucalyptus globulus, Lab. Milano. 1891. Engler u. Prantl. Pflanzenfam. III, 7. 58. Glaser. Mikr. Anal. d. Blattpulver v. Arzneipfl. Verh. d. Phys.-Med. Ges. z. Würzburg. N. F. Bd. 34. 1901. 256. Haberlandt. Phys. Pfl.-anat. 3. ed. 457. Hérail. Mat. Med. 1912. 341. Kraemer. Botany and Pharmacogn. 1910. 720. Kramer. Mikr.-pharmacogn. Beitr. z. Kenntn. v. Blättern u. Blüten. Diss. Würzburg, 1907. 16. Lutz. Die oblitoschizogenen Secretbehälter d. Myrtaceen. Diss. Bern. 1890. 20; also Bot. Centralbl. Bd. 64. 1895. 258. Lignier. Recherches sur l'anatomie comparée des Calycanthées, des Mélastomacées et des Myrtacées. Diss. Paris. 1887. 395—416. Marmé. Pharmacogn. 1886. 176. Moeller. Pharmacogn. 1889. 45. Planchon et Collin. Drogues simples. T. 2. 1896. 343. Schneider. Powd. Veget. Drugs. 1902. 188. Solereder. Syst. Anat. 1898. 398. Tschirch. Harze u. Harzbehälter. 1900. 367. MATERIAL. The drug; leaves gathered Oct. 1901 from the Botanic Garden at Groningen, fresh and in alcohol. REAGENTS. Water, glycerine, chloral hydrate, iodine in chloral hydrate, phloroglucin and hydrochloric acid, iodine and sulphuric acid 66 per cent., concentrated sulphuric acid, Schulze's macerating mixture, tannin reagents: cupric acetate and ferric acetate, potassium dichromate.

MICROGRAPHY.

Intervenia.

Epidermis. Upper side. Wanting on the cork-patches to be described afterwards. Stomata 40 to the sq. m.M., phaneroporous, with a large front chamber, the outer wall thickenings of the guard-cells being strongly developed and forming a broad circular ridge. Inner wall thickenings almost wanting,

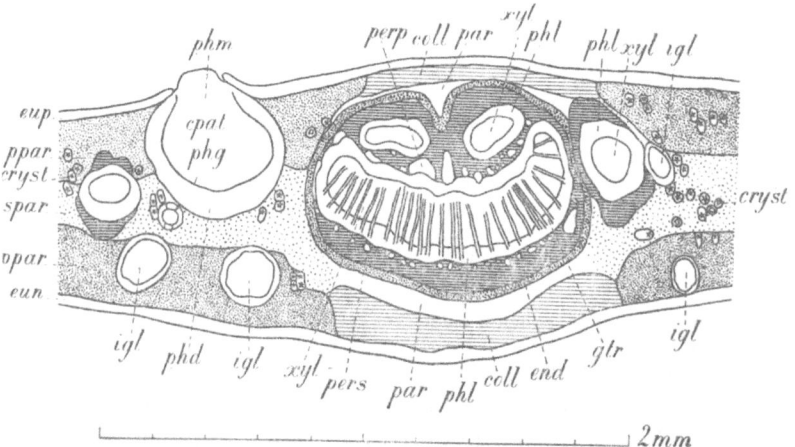

Fig. 28. *Eucalyptus globulus*. Leaf, transverse section through the midrib and the neighbouring portions of the lamina. coll Collenchyma; cpat Cork patch; cryst Crystal cells; end Starch-sheath; eun Epidermis under side; eup Epidermis upper side; gtr Medullary commissures not drawn out in some parts of the larger and in the smaller compound vascular bundles; igl Inner glands; par Colourless parenchyma; perp Pericyclic parenchyma; pers Pericyclic sclerenchyma; phd Phelloderm; phg Phellogen; phl Phloem; phm Phellem; ppar Palisade chlorenchyma; spar Spongy chlorenchyma; xyl Xylem.

back chamber united with the air chamber. Over each gland 2 half-circular cells, forming together a disc; separating wall somewhat undulating, often thickened in the middle part.

Epidermal cells proper. H. 30 μ, Lev. 17 and 20 μ, from tetra-to septagonal prisms. Outer walls strongly thickened, with a somewhat undulating outline; lateral walls cuneiform. Nearly the whole outer wall and a cuneiform part of the lateral walls cuticularized. Wax on the outer walls. Cell contents: some chloroplasts; tannin; sometimes anthocyanin. D i s c o v e r t h e g l a n d s. H. 18 μ, Lev. 35 μ. Outer walls of these cells somewhat thickened. G u a r d-c e l l s. Lev. 45 μ and 20 μ.

Under side. Stomata 72 to the sq. m.M.; see for the rest the upper side.

Mesophyll.

P a l i s a d e c h l o r e n c h y m a. On both sides; on the upper side 3—4, on the under 3 layers of cells; intercellular spaces, becoming larger in the inner layers. Some cells especially in the 2nd and 3rd layers containing a cluster crystal or a simple crystal.

C e l l s o f t h e a b o v e. Cells of upper side, outer layer H. 40—50 μ, Lev. 5 μ, inner layers H. 30 μ, Lev. 7—10 μ; of under side the corresponding dimensions H. 35 μ, Lev. 7—10 μ and H. 30 μ, Lev. 7—10 μ; cylinders or prisms with rounded edges. Cell contents: penta- to hexagonal tabular chloroplasts, paving the walls, 4—5 μ in diameter and 2 μ thick, containing 1 or 2 needle-shaped starch grains; tannin.

S p o n g y c h l o r e n c h y m a. 3—4 layers of cells with much larger intercellular spaces; numerous crystal cells.

C e l l s o f t h e a b o v e. H. 20 μ, Lev. 15 μ; see for the rest the palisade chlorenchyma.

I n t e r n a l g l a n d s. Only in the palisade chlorenchyma, the larger number in that of the upper side, impinging on the epidermis; H. 150—250 μ, Lev. 100—150 μ, usually pear-shaped; oblito-schizogenous (see Lutz); epithelium one or two layers of polygonal tabular cells, thick 8 μ, diameter 30—60 μ, with cuticularized walls. Contents of cavity: oil or a resinous mass.

C o r k p a t c h e s. Lying in the palisade chlorenchyma and protruding into the spongy chlorenchyma; diameter 250 μ, globular. Phellem, consisting of radial rows of many periderm cells, by far the largest part of the cork patches, breaking through the epidermis; phellogen 1 cell layer; phelloderm 5 radially arranged cell layers.

P e r i d e r m c e l l s. Polygonal discs, thick 10 μ, diameter 15 μ. Walls often brown, always lignified. Cell contents: tannin. P h e l l o d e r m c e l l s thick 5 μ. Walls thickened. Cell contents: tannin.

M a r g i n. Under the epidermis 8 layers of collenchyma cells, the 2 outer layers smaller than the rest; some cells containing a cluster crystal.

C o l l e n c h y m a c e l l s. H. 15—20 μ, Lev. 15—20 μ and 30—50 μ. Circular or elliptical cylinders. Walls thickened, pitted.

V e i n s.

Epidermis. Upper side. Cells in longitudinal rows.

E p i d e r m i s c e l l s p r o p e r. R. 30 μ, T. 18 μ, L. 30 μ; tetra- to hexa-, generally tetragonal prisms. Outer and especially lateral walls more thickened than in epidermis of intervenia; outer walls covered with a granular wax layer. See for the rest epidermis cells of intervenia.

Under side. The same as upper side.

Mesophyll.

C o l l e n c h y m a. On the upper side 6—7, on the under side 4—6 layers of cells.

Collenchyma cells. R. and T. 10—25 μ, L. of the outermost layer 30 μ, of the other layers 60—140 μ; cylindrical cells or fibres. Walls thickened; pitted; without intercellular spaces. Cell contents: some chloroplasts with needle-shaped starch grains; tannin; sometimes in the outermost layer anthocyanin.

Colourless parenchyma. On the upper side one, on the under side 4 layers of cells; some idioblasts containing a cluster crystal or a simple crystal, shorter and usually lying in rows.

Parenchyma cells. R. and T. 15—30 μ, L. 30—50 μ; circular or elliptical cylinders. Walls somewhat thickened. Intercellular spaces. Cell contents: some chloroplasts with needle-shaped starch grains; tannin.

Palisade chlorenchyma. Protruding somewhat into the veins.

Endodermis. Starch-sheath, 1 layer of cells.

Cells of the above. R. and T. 15—30 μ, L. 15—25 μ. Intercellular spaces wanting. Cell contents: many starch grains, usually compound, 3-adelphous. See for the rest colourless parenchyma.

Meristele. Somewhat cylindrical, slightly excavated on the upper side.

Pericycle. Chiefly consisting of some 5 or 6 layers of sclerotic prosenchyma, mixed with common parenchyma.

Sclerenchyma fibres. R. and T. 10—15 μ, L. 400—1100 μ; tetra- to hexagonal, sometimes with branched ends. Walls usually strongly thickened, sometimes only somewhat collenchymatous; showing stratification; mostly only the outer parts somewhat lignified; pitted. Parenchyma cells. R. and T. 4—10 μ, L. 40—60 μ; polygonal prisms. Cell contents: tannin.

Vascular bundles. In the midrib on the under side of the meristele a single compound one, gutter-shaped and containing many simple bundles; in the base of the midrib on the upper side 2 smaller more cylindrical compound vascular bundles. Simple bundles radially elongated in a cross section; bicollateral.

Phloem. Each bundle consisting in a cross section of about 30 elements; the innermost 5 layers radially arranged; sieve-tubes distinct; here and there crystal fibres.

Sieve-tubes. R. and T. 8—10 μ, L. of articulations 60—100 μ; polygonal prisms; partition walls oblique. Cell contents: sometimes callus plates; in a few instances a granular mass. Companion cells. R. and T. 3—4 μ; tri- to hexagonal prisms. Cell contents: a granular mass. Cambiform cells. R. and T. 4—10 μ, L. 40—60 μ. Crystal fibres. Polygonal; divided by cross walls in several cells, 16—20 μ in length. In each cell usually 1 simple crystal. In some cases one end of a crystal fibre without cross walls; side walls thickened and reticulated.

Xylem. Each bundle R. 150 μ; elements radially arranged.

Vessels. R. and T. 10—20 μ, L. of articulations 300—450 μ; spiral and reticulate vessels and pitted vessels. Walls thickened and lignified. Fibres. R. and T. 10 μ, L. 180—450 μ; tetra- to hexagonal. Walls thickened; lignified; reticulated or with bordered pits, in the outer part of the bundle often with cross slit-like pits. Parenchyma cells. R. and T. 6—10μ, L. 50—130μ; tetra- to hexagonal prisms. Walls thickened; lignified.

Medullary commissures. Uniseriate, consisting of common parenchyma.

Parenchyma cells. R. 10 μ, T. 6 μ, L. 30 μ; tetra- to hexagonal prisms. Walls thickened; lignified; pitted. Cell contents: tannin.

October 1901, September 1911.   J; M.

FOLIA JABORANDI.

Pilocarpus. Jaborandi Leaves.

The dried leaflets of Pilocarpus Jaborandi, Holmes, Bot. Mag. Plate 7483 and Pharm. Journ. ser. 3.
Vol. V. p. 582.

LITERATURE. Biechele. Microsc. Prüf. der off. Drogen. 1904. 37. Flückiger. Pharma-kogn. 1891. 694. Flückiger & Hanbury. Pharmacographia. 1879. 113. Geiger. Beiträge zur pharmakognostischen and botanischen Kenntnis der Jaborandiblätter. Diss. Zürich. 1897. Gilg. Pharmakogn. 1910. 183. Glaser. Microsc. Anal. d. Blattpulver. 1901. 260. Greenish & Collin. Anat. Atlas o. veget. powders. 1904. 72. Plate XXX. Hager. Pharm. Praxis. Bd. II. 1902. 100. Hérail. Pharmacol. 1912. 750. Karsten u. Oltmanns. Pharmakogn. 1909. 178. Marmé. Pharmacogn. 1886. 192. Meyer. Wissensch. Drogenkunde. II. 1892. 228. Moeller. Pharmacogn. 1889. 42. Moeller. Mikr. Pharm. Üb. 1901. 86. Planchon et Collin. Drogues Simples. T. II. 1896. 621. Schneider. Powdered Veg. Drugs. 1902. 256. Tschirch. Angew. Anat. 1889. 182. Tunmann. Pfl.-mikrochemie. 1913. 308. Vogl. Anat. Atlas. 1887. 5. Wigand. Pharmakogn. 1879. 200. MATERIAL. The drug. REAGENTS. Water, glycerine, chloral hydrate, potash, iodine in chloral hydrate, phloroglucin and hydrochloric acid, iodine and sulphuric acid 66 per cent., concentrated sulphuric acid, Schulze's macerating mixture, iron acetate.

MICROGRAPHY.

Intervenia.

Epidermis upper side. Cells of different dimensions, those above the centre of several of the internal glands being smaller. Stomata wanting. Trichomata very rare and only apparent as scars; at the base surrounded by about 6 radially arranged cells of smaller dimensions than the other epidermal cells and sometimes having a tangential partition-wall. The base of the hair 18 μ in diameter.

Epidermal cells proper. H. 25 μ, Lev. B. 30 μ, Lev. L. 30—40 μ. The smaller cells above the centre of glands Lev. 15 μ. Outer walls thickened; colouring yellow in potash, consisting of an outer cuticularized layer covered with a cuticle and an equally thick inner cellulose layer, the limit of both parts notched, above the lateral walls the cuticularized layer entering wedge-wise into the cellulose layer; cuticular striation in groups of parallel undulated lines, direction of the striae in the different groups not always the same. Cell contents: sometimes a small strip of a brown mass close to the outer or inner wall. Cells round the hair bases e. g. 18 μ in diameter; the cuticular striation of those cells radiating from the hair. Cells of trichomata. Lateral walls very thick and cuticularized.

Epidermis under side. Stomata about 20 to the square m.M., width 20 μ, length 22 μ, lying in the same level with the epidermal cells and surrounded by a ring of mostly 4 small subsidiary cells. File hairs. Only a single specimen notic-ed; unicellular; at the base 15 μ in diameter, 500 μ in length; conical; for the rest see trichomata of the upper side.

Epidermal cells proper. H. 25 μ, Lev. B. 30 μ, Lev. L. 35—50 μ; polygonal tables with somewhat sinuous or rounded lateral walls. Outer walls as thick as those of the upper side, though with much thinner cuticularized part; cuticular striation quite or nearly wanting. For lateral walls and cell contents see cells of the upper side. Subsidiary cells. E. g. 8 by 25 μ; cuticular striation radiating from the stoma.

Mesophyll.

Palisade chlorenchyma. Consisting of 1 layer of cells with intercellular spaces. Here and there a cell with a transverse partition-wall and then mostly having somewhat larger level dimensions; often in one or in both of the daughter cells a cluster crystal.

Cells of the above. H. 20—30 μ, Lev. B. 12 μ, Lev. L. 15 μ; polygonal prisms with strongly rounded edges or elliptical cylinders. Cell contents: chloroplasts, especially at the upper- and undermost ends.

Spongy chlorenchyma. Consisting of about 10 layers of cells, the lar-

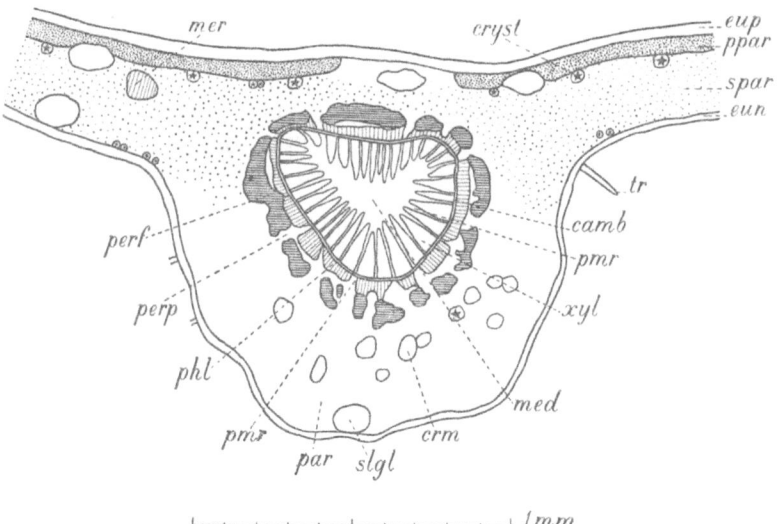

Fig. 29. *Pilocarpus Jaborandi.* Leaf, transverse section. camb. Cambium; crm Cells containing a reddish brown mass; cryst Crystal cells; eun Epidermis under side; eup Epidermis upper side; med Medulla; mer Meristele; par Colourless parenchyma; perf Pericyclic sclerenchyma fibres; perp Pericyclic parenchyma; phl Phloem; pmr Medullary commissures; ppar Palisade chlorenchyma; slgl Schizo-lysigenous glands; spar Spongy chlorenchyma; tr Trichomes; xyl Xylem.

gest in the middlemost layers, somewhat smaller towards the palisade tissue and still smaller in the direction of the epidermis of the under side; showing large intercellular spaces. Cells more or less stellate, lying in horizontal planes. Crystal idioblasts. In some of the cells round each air-cavity cluster crystals of 12 μ'in diameter; moreover close to the palisade tissue cluster crystals of e. g. 18 μ in diameter.

Cells of spongy chlorenchyma proper. In the middlemost layers e. g. H. 25 μ, Lev. B. and L. 40 μ; close to the palisade tissue H. 15 μ, Lev. B. 18 μ, Lev. L. 20 μ. Cell contents: chloroplasts; in some cells a brownish red mass; cells adjoining the glands often filled with a reddish brown mass.

Schizo-lysigenous glands. [1]) Lying close to the epidermis of the upper and the under side. H. 60—100 μ, Lev. B. and L. 80—150 μ; ellipsoids. Contents: along the periphery a granular mass, varying from yellow to brown; moreover remnants of cells.

---

[1]) See W. Sieck. Die Schizo-lysigenen Secretbehälter. Diss. Bern. 1895. 43 or Pringsheim's Jahrbücher. Bd. 27. 1895. 239.

V e i n s.

Epidermis upper side. Composed of about 4 longitudinal rows of cells. Stomata wanting. Trichomata here and there apparent, scars; the base surrounded by a ring of 4—6 radially arranged cells of smaller dimensions than the other epidermal cells and often apparently formed by divisions of the surrounding cells.

E p i d e r m a l   c e l l s   p r o p e r. H. 20 μ, Lev. B. 15—20 μ, Lev. L. 30—50 μ: quadrangular, and mostly rectangular prisms. See for the rest those of the intervenia. Cells surrounding the bases of the trichomata: length 20 μ, width 15 μ; see for the rest those of the intervenia.

Epidermis under side. Composed of cells lying in longitudinal rows. Stomata wanting. File hairs comparatively numerous; at the base 12—18 μ in diameter, length 150—350 μ; conical, often curved; unicellular. Walls except the inner ones strongly thickened and showing a cuticle.

E p i d e r m a l   c e l l s   p r o p e r. H. 20 μ, Lev. B. 20 μ, Lev. L. 40—70 μ; quadrangular, mostly rectangular prisms. Outer walls strongly thickened; consisting of a cuticularized part covered with a cuticle and a much thicker inner part; cuticular striation wanting. See for the rest the cells of the upper side.

Mesophyll.

C o l o u r l e s s   p a r e n c h y m a. Consisting of about 7 layers of cells at the upper side and about 10 layers at the under side; cells lying in longitudinal rows, in the middle part somewhat elongated. Intercellular spaces. Cluster crystal idioblasts of smaller dimensions sometimes adjoining one another and then lying in a longitudinal row. On both sides of the leaf one innermost layer differring from the others with regard to the dimensions of the cells (endodermis?).

C e l l s   o f   t h e   a b o v e. At the upper side R. 18 μ, T. 25 μ, L. 30—80 μ; in the innermost layer of this side R. 12 μ, T. 20 μ, L. 45—90 μ; polygonal prisms with a longitudinally directed axis and rounded edges. Walls somewhat pitted. Cell contents: some chloroplasts; often a reddish brown mass, especially in the outer layers and the innermost one (endodermis?). At the under side in the outer layers R. 20 μ, T. 15 μ; tetragonal prisms; the rest R. 20 μ, T. 30—130 μ. Elongated cells of the middle part R. and T. 60 μ. Cells of the innermost layer R. 12 μ, T. 20 μ, L. 45—90 μ. Walls somewhat collenchymatous; pitted. Cell contents: in the largest cells always a reddish brown mass. See for the rest cells of the upper side. C r y s t a l   i d i o b l a s t s e.g. R. 15 μ, T. and L. 20 μ.

P a l i s a d e   c h l o r e n c h y m a. Only found at the lateral parts of the upper side as continuations of the palisade chlorenchyma of the intervenia. Cells less high than those of the intervenia.

Meristele. In the shape of a semi-elliptical cylinder, the flat side turned upwards.

P e r i c y c l e. Consisting of bundles of sclerenchyma fibres, alternating with bands of common parenchyma. The sclerenchyma bundles mostly thick 4 layers of fibres in a radial direction. At the upper side of the meristele 3 bundles, the middlemost consisting of 4 layers of fibres, the 2 other ones thinner. Bundles at the lateral sides containing up to 40 elements, at the under side about 15.

Parenchyma bands at the upper side always uniseriate, at the under side uni-
to triseriate. Intercellular spaces wanting.

S c l e r e n c h y m a f i b r e s. R. 10 μ, T. 12 μ; polygonal. Walls thickened; middle
lamella distinct; nearly exclusively the middle lamella distinctly lignified; pitted.
C o m m o n p a r e n c h y m a c e l l s. R. and T. 15—18 μ, L. 70—120 μ; polygonal
prisms with a longitudinally directed axis. Walls a little thickened; pitted. Cell contents:
mostly a brown mass.

V a s c u l a r b u n d l e s. About 37 simple bundles; collateral; open; radial
arrangement of the constituting elements rather strongly developed.

P h l o e m. The outer part consisting of 3 layers of parenchyma cells, the
middle part of very small elements and the inner part of some layers of radially
arranged elements. In a certain number of vascular bundles only consisting
of secondary parenchyma and in other cases the phloem bundles of adjoining
vascular bundles united together.

P a r e n c h y m a c e l l s o f t h e o u t e r p a r t. R. 8 μ, T. 10 μ, L. 75 μ; polygonal
prisms with a longitudinally directed axis. S m a l l e l e m e n t s i n t h e m i d d l e. R. and
T. 4—5 μ; polygonal. Walls somewhat thickened. E l e m e n t s o f t h e i n n e r p a r t.
R. 4 μ, T. 10 μ, L. 50—100 μ, corresponding in length with the elements of the cambium;
mostly quadrangular with a longitudinally directed axis.

C a m b i u m. Clearly to be seen as 1 or 2 layers of elements.

X y l e m. Radially elongated in a cross section; the bundles at the under side
of the stele the largest, consisting of 8—10 radially arranged elements. Often 2
radial rows of the outer part corresponding with 1 of the inner part. The outer
part of the bundles consisting of polygonal prismatic cells and a few libriform
fibres; the inner part composed of pitted vessels and of spiral and annular
vessels on the side of the medulla.

V e s s e l s. R. and T. 10—15 μ, L. of articulations 200—1100 μ, the annular and spiral
vessels having the smallest diameter and the longest articulations, pitted vessels L. of articulations 200—500 μ; polygonal prisms, annular and spiral vessels with rounded edges.
Walls lignified. Contents: here and there sphaerocrystals of hesperidine. C e l l s o f
t h e o u t e r p a r t o f t h e b u n d l e s. R. and T. 10 μ, L. 50—100 μ; polygonal prisms.
Walls strongly thickened; middle lamella clearly to be distinguished; lignified; pitted.

M e d u l l a r y c o m m i s s u r e s. Uni- to triseriate; consisting of common
parenchyma cells; much more numerous in the xylem than in the phloem; all
bands in the phloem corresponding with bands in the xylem and with parenchyma bands of the pericycle. Crystal idioblasts divided in partitions by
transverse walls, each partition containing a cluster crystal; often some of these
cells adjoining each other in a longitudinal row.

M e d u l l a. Mostly showing a central cavity. Consisting of common parenchyma cells, the largest cells towards the centre; round the cavity cells
elongated. Intercellular spaces.

C e l l s o f t h e a b o v e. R. 5—12 μ, T. 8—30 μ, L. 30—100; polygonal prisms in
medulla with central cavity, nearly cylindrical in medulla without central cavity. Walls
sometimes a little thickened; lignified; pitted. Cell contents: sometimes a brown mass.

At the top of the midrib the meristele much reduced, only consisting of 2 bundles of pericyclic sclerenchyma fibres with a very few vascular bundles in the middle. Meristele of veinlets a cylindrical bundle of pericyclic sclerenchyma fibres enclosing hardly any vascular bundles.

November 1902.   J; M; L.

FOLIA LAUROCERASI.
Cherry-Laurel Leaves.
The fresh leaves of Prunus Laurocerasus, Linn. Sp. Pl. 474, full-grown and collected in autumn.

Macroscopic characters.

Simple, on a short petiole. Blade coriaceous, oblong; feather-veined; apex somewhat acuminate and recurved; base acute; margin slightly revolute, superficially and distantly serrate. Surface glabrous. On the under side on either side of the base of the midrib one or two circular, shallow, glandular depressions. Inodorous, but emitting when bruised an odour resembling that of bitter almonds; taste bitter, astringent, somewhat aromatic.

Anatomical characters.

LITERATURE. De Bary. Vergl. Anat. 1877. 102, 392. Flückiger. Pharmakogn. 1891. 765. Flückiger & Hanbury. Pharmacographia. 1879. 255. Guignard. Sur la localisation dans les Amandes et le Laurier-Cérise des principes qui fournissent l'acide cyanhydrique. Journ. d. Bot. T. 4. 1890. 6. Kramer. Mikrosk.-pharmacogn. Beitr. z. Kenntn. v. Blättern u. Blüten. Diss. Würzburg. 1907. 22. Marmé. Pharmacogn. 1886. 179. Molisch. Mikrochemie d. Pfl. 1913. 172, 290. Oudemans. Pharmacogn. 1880. 309. Pecke. Mikroch. Nachweis d. Cyanwasserstoffsäure in Prunus Laurocerasus. Ber. Wiener Akad. Bd. CXXI. 1912. 33. Planchon et Collin. Drogues simples. T. 2. 423. Tunmann. Pfl.-microchemie. 1913. 360, 341. Virchow. Üb. Bau u. Nervatur d. Blattzähne und Blattspitzen. Diss. Bern. 1895. 57. Wigand. Pharmakogn. 1879. 198. MATERIAL. Leaves gathered October 1901 from the Botanic Garden at Groningen, fresh and in alcohol. REAGENTS. Water, glycerine, chloral hydrate, iodine in chloral hydrate, phloroglucin and hydrochloric acid, iodine and sulphuric acid 66 per cent., concentrated sulphuric acid, ferric acetate, Schulze's macerating mixture, Millon's reagent.

MICROGRAPHY.
Intervenia.

Epidermis. Upper side. Stomata wanting; on each tooth of the serrate margin a water-pore.

E p i d e r m a l   c e l l s   p r o p e r. H. 25 µ, Lev. 35 and 45 µ; tabular, lateral walls sinuous. Outer walls thickened, showing 3 layers: 1. a cuticle; 2. cuticular layers containing crystal sand in their innermost parts; 3. a stratified layer showing a pure cellulose reaction. Lateral and inner walls somewhat thickened; stratified; pitted. Cell contents: some little chloroplasts; sometimes tannin. G u a r d - c e l l s   o f   w a t e r-p o r e s. Lev. 15 and 35 µ.

Under side. Stomata 180 to the sq. m.M., phaneroporous, in the level of the epidermis, nearly always surrounded by 4 subsidiary cells.

E p i d e r m a l   c e l l s   p r o p e r. H. 25 µ, Lev. 25 and 35 µ; lateral walls sometimes a little sinuous; see for the rest the upper side. G u a r d - c e l l s   o f   s t o m a t a. Lev. 12 µ and 35 µ; outer and inner walls very thick.

Mesophyll.

P a l i s a d e   c h l o r e n c h y m a. From 2—3 layers of cells; the innermost

12

layer consisting of somewhat shorter and thicker cells; intercellular spaces. Some often much shorter idioblasts, especially of the uppermost layer, containing a cluster crystal, sometimes a simple crystal, sometimes both.

C e l l s   o f   t h e   a b o v e. H. 35—55 μ, Lev. 15—20 μ; tetra- to hexagonal prisms with rounded edges, or cylinders. Walls pitted. Cell contents: poly-, mostly penta- or hexagonal tabular chloroplasts, thick 2 μ, diameter 4—6 μ, paving the walls, containing each some needle-shaped or ellipsoidal starch grains; globular granules, 6 μ in diameter; tannin in many cells. **S p o n g y   c h l o - r e n c h y m a.** From 6—7 layers of stellate cells; collecting cells to be distinguished; large intercellular spaces. Some idioblasts containing each a short bar-shaped simple crystal.

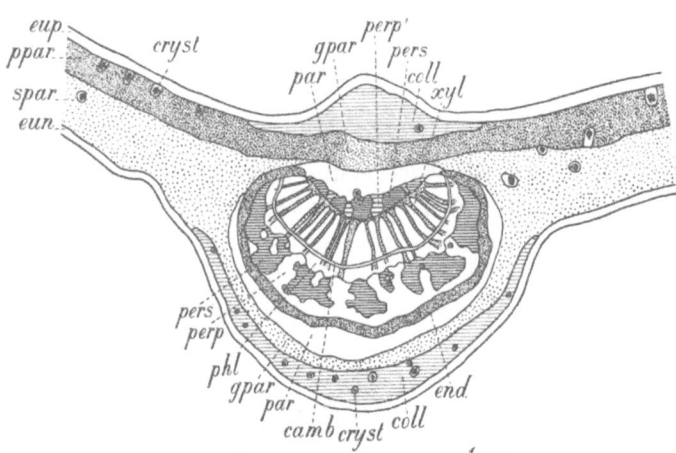

Fig. 30. *Prunus Laurocerasus.* Leaf, transverse section through the midrib and the neighbouring portions of the lamina. camb Cambium; coll Collenchyma; cryst Crystal cells; end Endodermis; eun Epidermis under side; eup Epidermis upper side; gpar Chlorenchyma; gtr Medullary commissures; par Colourless parenchyma; perp Pericyclic parenchyma; perp' Pericyclic somewhat collenchymatous parenchyma; pers Pericyclic sclerenchyma; phl Phloem; ppar Palisade chlorenchyma; spar Spongy chlorenchyma; xyl Xylem.

C e l l s   o f   t h e   a b o v e. H. 20—25 μ, Lev. 25—50 μ. Walls pitted. Cell contents: chloroplasts containing globular, sometimes compound starch grains; tannin in many cells.

V e i n s.

Epidermis. Upper side. Stomata wanting.

E p i d e r m a l   c e l l s   p r o p e r. H. 30 μ, T. 25 μ, L. 40 μ; polygonal prisms. Cuticular layers protruding wedge-like into the lateral walls; inner walls collenchymatous, see for the rest the intervenia.

Under side. Stomata wanting; cells often in longitudinal rows.

E p i d e r m a l   c e l l s   p r o p e r. R. 28 μ, T. 25 μ, L. 35 μ; tetra- to hexa-, mostly tetragonal prisms. Outer walls very much thickened; wedge-like parts of cuticular layers, protruding into the lateral walls, very distinct; two layers of crystal sand to be distinguished in the cuticular layers, the innermost accompanying the margin of the wedge-like parts; inner walls collenchymatous. See for the rest the intervenia.

Mesophyll.

C o l l e n c h y m a. In the midrib 2 flat bundles; one on the upper, the other on the under side. The upper bundle in the middle thick 10 layers of cells, at the margins 1 layer; that on the under side in the middle 4, at the margins 2 layers. Here and there, especially in the undermost bundle, crystal cells in longitudinal rows of 4—5 cells, containing simple bar-shaped or clustered crystals, or both.

Collenchyma cells proper. R. and T. 30 μ, L. 80—200 μ; polygonal prisms. Walls very thick; middle lamellae very distinct; stratified; after treatment with potassium iodide iodine and sulphuric acid 66 per cent. showing a very fine radial blue striation; showing pit canals. Cell contents: very few chloroplasts with starch grains; tannin in some cells. Crystal cells. L. 35 μ; separating walls tolerably thin.

Chlorenchyma. In the midrib on the upper side a layer of the same height as the palisade chlorenchyma in the adjoining intervenia and forming a continuation of this layer; 3—5 layers of cells. On the under side a much thinner layer continuous with the spongy parenchyma of the adjoining intervenia; 4 layers of cells. In both layers intercellular spaces and here and there crystal cells, the same as in the collenchyma.

Cells of the above. R. and T. 20—30 μ, L. 50—150 μ; circular or elliptical cylinders having a longitudinally directed axis. Walls on the margin of the veins a little, in the middle part much thickened; stratified; pitted. Cell contents: chloroplasts containing starch grains and diminishing in number towards the middle part of the vein.

Colourless parenchyma. On the upper side 3—4, on the under side 3—6 layers of cells; intercellular spaces. Here and there longitudinal rows of some cells containing generally each a clustered, sometimes a simple crystal.

Parenchyma cells. R. and T. 30—50 μ, L. 60—100 μ; circular or elliptical cylinders or polygonal prims having rounded edges. Walls a little thickened; pitted. Cell contents: here and there some starch; often tannin. Crystal cells. L. 12—35 μ.

Endodermis. In the midrib one layer of cells, completely surrounding the meristele; the cells on all sides containing emulsin and thus forming an emulsin-sheath; the cells on the under and lateral sides containing at the same time much starch and thus forming an incomplete starch-sheath as shown in the figure; the cells without starch on the upper side containing often tannin.

Cells of the above. R. and T. 30—50 μ, L. 40 μ; cells of the incomplete starch-sheath containing many pale chloroplasts with starch grains, sometimes compound up to 4-adelphous.

Meristele. In the midrib slightly gutter-shaped, the concave side turned upwards.

Pericycle. Wanting only at the margins of the gutter-shaped compound vascular bundle. Consisting of bundles of sclerenchyma fibres and between these common parenchyma. All cells containing emulsin and being crystal cells. On the upper side generally 5 bundles, consisting in a transverse section of 10—25 fibres; on the under side generally 7—8 irregularly shaped bundles, consisting in a cross section of 35—50 fibres.

Sclerenchyma fibres. R. and T. 15—25 μ, L. 1000—2000 μ, those of the upper side somewhat thinner; tetra- to hexagonal. Walls very thick; stratified. Parenchyma cells, those on the upper side R. and T. 15—25 μ, L. 150—200 μ, polygonal prisms with rounded edges and circular or elliptical cylinders. Walls thickened, sometimes collenchymatous; stratified; showing pit canals; sometimes with intercellular spaces. Cell contents: chloroplasts containing starch grains. Parenchyma cells of under side: see colourless parenchyma of mesophyll.

Vascular bundle. A single compound one, gutter-shaped and containing numerous simple bundles. Simple bundles collateral, open.

P h l o e m. Bundles consisting in a radial direction of 6—7 layers of elements, showing radial arrangement towards the cambium.

C a m b i u m. Two or 3 layers of elements.

X y l e m. Bundles consisting of 1—6 layers of elements in a tangential direction, showing radial arrangement towards the cambium. Spiral vessels only in the outer part; vessels with bordered pits; tracheid fibres; parenchyma cells not numerous.

V e s s e l s. R. 18—22 μ, T. 12—18 μ, L. of articulations 150—500 μ. Walls stratified; lignified. T r a c h e i d f i b r e s in all respects resembling articulations of pitted vessels. P a r e n c h y m a c e l l s. R. 25 μ, T. 20 μ, L. 100—150 μ.

M e d u l l a r y c o m m i s s u r e s. Uniseriate, consisting of common parenchyma.

P a r e n c h y m a c e l l s. R. 16 μ, T. 8—10 μ, L. 70 μ, those of the xylem parts somewhat smaller than those of the phloem parts; mostly hexagonal prisms. Walls thickened; generally lignified, except between the older xylem parts; reticulated. Cell contents: chloroplasts containing starch grains, sometimes compound, up to 4-adelphous; often tannin.

Micrography of circular glandular depressions before mentioned. Epidermal epithelium consisting of palisade parenchyma with thickened outer walls. The cells at the margin of the gland having moreover the outer ⅓ part of the lateral walls thickened. Some palisade cells especially at the margin of the gland divided by a transverse wall in 2 cells. In this case eventually occurring thickenings of the lateral walls, limited to the outer cell. Intercellular spaces to be distinguished. Below the epithelium common parenchyma, consisting of 12—14 layers of cells often without intercellular spaces. The cells of this parenchyma containing much starch, and especially the outer 1 or 2 layers being crystal cells; these crystal cells often divided in 2 cells.

E p i t h e l i u m c e l l s. Lev. 10 by 15 μ; tetra- to hexagonal, having somewhat curved lateral walls. Cell contents: now and then some starch; often a granular brown mass, especially in old leaves. C o m m o n p a r e n c h y m a c e l l s also often containing a brown mass. C r y s t a l c e l l s containing each a simple or a cluster crystal, sometimes both.

October 1901, October 1911.   J; M.

FOLIA MENTHAE PIPERITAE.
Peppermint.

The dried leaves of Mentha piperita, Linn. Sp. Pl. 576, collected when the plant is in flower.

Macroscopic characters.

Simple, petiolate. Petiole up to 15 m.M. long. Blade up to 7 c.M. long, oblong to oblong ovate, sometimes more lanceolate; feather-veined; apex acute; base from obtuse to acute; margin somewhat coarsely serrate, with somewhat sharp teeth pointing somewhat outwards. Surface often slightly pubescent on the under side of midrib and veins; globular, yellow, glandular hairs always to be seen, when magnified 50 times, in large numbers especially on the under side;

each glandular hair implanted in a shallow hollow. Odour strongly aromatic; taste aromatic, followed by a cooling sensation. Peppermint should not be kept longer than one year.

Anatomical characters.

LITERATURE. Deutsch. Arzneib. 5. Ausg. 1910. 235. Ellrodt. Üb. d. Verteil. d. Gerbst. in off. Blättern, Kräutern u. Blüten. Diss. Würzburg. 1903. 15. Gilg. Pharmacogn. 1910. 300. Glaser. Mikr. Anal. d. Blattpulver. Verh. d. Phys.-Med. Ges. z. Würzburg. N. F. Bd. 34. 1901. 267. Hager. Pharm. Praxis. Bd. 2. 1902. 372. Hérail. Mat. Méd. 1912. 274. Karsten u. Oltmanns. Pharmakogn. 1909. 172. Koch. Mikr. Anal. d. Drogenpulver. Bd. 3. 1906. 107. Koch u. Gilg. Pharmakogn. Praktik. 1907. 152. Kraemer. Botany a. Pharmacogn. 1910. 729. Marmé. Pharmacogn. 1886. 195. Meyer. Wissensch. Drogenkunde. Bd. 2. 1892. 213. Oudemans. Aanteek. o. d. Pharmac. neerl. 1854—56. 287. Oudemans. Pharmacogn. 1880. 255. Planchon et Collin. Drogues simples. T. 1. 1895. 500. Schneider. Powdered Veget. Drugs. 1902. 241. Tschirch. Angew. Pfl.-anat. 1889. 120, 180, 250, 462, 463. Tschirch u. Oesterle. Anat. Atl. 1900. 73. Taf. 19. Vogl. Anat. Atl. 1887. Taf. 13. MATERIAL. The drug. Leaves gathered Sept. 1901 from the Botanic Garden at Groningen; full-grown, fresh and in alcohol; young — 1 c.M. long — fresh. REAGENTS. Water, glycerine, chloral hydrate, iodine in chloral hydrate, phloroglucin and hydrochloric acid, iodine and sulphuric acid 66 per cent., concentrated sulphuric acid, Schulze's macerating mixture.

MICROGRAPHY.

Intervenia.

Epidermis. Upper side. Stomata phaneroporous, lying somewhat above the level of the epidermis, generally surrounded by only 2 subsidiary cells. Trichomes in 3 kinds. 1. Conical hairs 30 μ thick at the base, sometimes up to 60 μ if inserted in more than 1 epidermal cell, uniseriate, consisting of 1—4 usually 3—4 cells. 2. Capitate hairs somewhat more numerous, on young leaves very numerous; L. 30 μ; stalk usually, head always unicellular; head 15 μ in diam., without cuticular bladder. 3. Glandular hairs of the Labiate-type, on young leaves growing in number from the top to the base; surrounding epidermal cells radially arranged; stalk unicellular; head generally consisting of 8 cells, now and then some of these having divided.

Epidermal cells proper. H. 25 μ, Lev. 35 and 55 μ; tabular, lateral walls sinuous. Outer walls somewhat thickened, having a thin cuticle; lateral walls pitted. Cell contents: some small chloroplasts. Guard-cells of stomata. Lev. 30 μ and 10 μ. Cells of conical hairs. Outer walls a little thickened, having a cuticle and longitudinal cuticular striation. Cells of Labiate hairs. Stalk cells H. 8 μ, diameter 25 μ; circular tables. Cell contents: often some chloroplasts. Head cells H. 12 μ. Cuticular bladder H. 10 μ, diameter 60 by 65 μ; contents of the bladder a green clustered crystalline mass, now and then also some colourless granules.

Under side. Stomata 180 to the sq. m M. Trichomes the same as on the upper side, but generally more numerous, especially the glandular hairs.

Epidermal cells proper. H. 12 μ, Lev. 20 and 40 μ; side walls somewhat more undulated than on the upper side. See for the rest the upper side. Guard-cells. Lev. 27 and 9 μ.

Mesophyll.

Palisade chlorenchyma. One layer of cells, with intercellular spaces.

Cells of the above. H. 30 μ, Lev. 20 and 20 μ; poly-, mostly penta- or hexagonal prisms having rounded edges or cylinders. Cell contents: poly-, generally tetra- to hexagonal tabular chloroplasts, thick 2 μ, diameter 4—5 μ, paving the walls, containing 1 or 2 needle-shaped starch grains; globular colourless granules, 8 μ in diameter and of unknown origin.

Spongy chlorenchyma. Five layers of cells.

Cells of the above. Form more or less stellate. Cell contents: globular granules somewhat smaller; see for the rest palisade chlorenchyma.

Veins.

Epidermis. Upper side. Consisting of elongated cells. Stomata rare, having 2 subsidiary cells. Trichomes in 2 kinds; often inserted in the end of an epidermal cell. Conical hairs much more numerous than on the intervenia, also often longer — up to 12 cells — and broader. Capitate glandular hairs the same as on the mesophyll. Hairs of the Labiate type wanting.

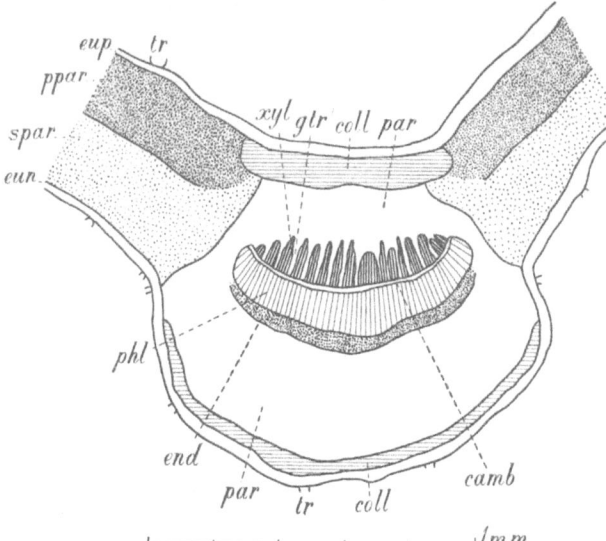

Epidermal cells proper. R. 20 μ, T. 15 μ, L. 90 μ; tetra- to hexagonal prisms with a radially directed axis. Outer walls thickened, having a thin cuticle and longitudinal parallel cuticular striation; with stratification; lateral walls pitted; inner walls collenchymatous. Cell contents: some little chloroplasts.

Fig. 31. *Mentha piperita.* Leaf, transverse section through the midrib and the neighbouring portions of the lamina. camb Cambium; coll Collenchyma; end Starch-sheath; eun Epidermis under side; eup Epidermis upper side; gtr Medullary commissures; par Colourless parenchyma; phl Phloem; ppar Palisade chlorenchyma; spar Syongy chlorenchyma; tr Trichomes; xyl Xylem.

Under side. In general the same as on the upper side; stomata and conical hairs more numerous.

Epidermal cells proper. R. 18 μ, T. 15 μ, L. 80 μ.

Mesophyll.

Collenchyma. In the midrib on the upper side 2—3, on the under side 1 layer of cells. Cells in longitudinal rows; intercellular spaces.

Collenchyma cells. R. 20 μ, T. 20 μ, L. 170 μ; cylinders. Walls thickened; pitted. Cell contents: a few chloroplasts.

Colourless parenchyma. In the midrib on the upper 3, on the under side 5 layers of cells; intercellular spaces.

Parenchyma cells. On the upper side the outmost layer R. 8 μ, T. 15 μ, the other layers R. and T. 20 μ, L. 150 μ; on the under side R. and T. 30—45 μ, L. 75—150 μ; tetra- to hexagonal prisms. Cell contents: a few chloroplasts.

E n d o d e r m i s. Starch-sheath. One layer of cells, to be distinguished in the midrib only on the under side.

C e l l s   o f   t h e   a b o v e. R. and T. 30 μ, L. 50—100 μ; polygonal prisms with rounded edges. Cell contents: chloroplasts filled with globular starch grains.

Meristele. In the midrib somewhat gutter-shaped, the concave side turned upwards.

P e r i c y c l e. Not sharply defined, therefore not drawn in the figure. On the upper side consisting of colourless parenchyma, sometimes mixed with a few sclerenchyma fibres; on the under side somewhat collenchymatous parenchyma.

P a r e n c h y m a   c e l l s   of upper side. R. and T. 6—8 μ, L. 100—120 μ; tetra- to hexagonal prisms. Walls a little thickened, stratified. Cell contents: some chloroplasts containing needle-shaped starch grains; sometimes green sphaero-crystals. S c l e r e n c h y m a   f i b r e s. R. and T. 10 μ, L. 800 μ; tetra- to hexagonal. Walls thickened and stratified. P a r e n c h y m a   c e l l s   of under side. R. 10 μ, T. 12 μ, L. 250 μ; tetra- to hexagonal prisms. Walls somewhat collenchymatous. Cell contents: some chloroplasts, each with 1 or 2 needle-shaped starch grains; sometimes green sphaero-crystals.

V a s c u l a r   b u n d l e. In the midrib a single compound one, containing about 16 simple bundles, arranged in a slightly curved plane. Simple bundles collateral, open.

P h l o e m. From 5—7 layers of elements, radially arranged towards the cambium; sieve-tubes distinct.

S i e v e-t u b e s. R. and T. 4—6 μ, L. of articulations 65—80 μ; tetra- to hexagonal prisms. Sieve-plates distinct, often with callus-plates.

C a m b i u m. 2 or 3 layers of elements, R. 4 μ, T. 10 μ, L. 70 μ.

X y l e m. Each bundle consisting in a radial direction of some 10 layers, in a tangential of 1—3 layers of elements. Annular vessels, often flattened; spiral and pitted vessels; tracheid fibres.

A n n u l a r   a n d   s p i r a l   v e s s e l s. R. 8 μ, T. 10 μ, L. of articulations 500 μ. Walls lignified. P i t t e d   v e s s e l s. L. of articulations 100—300 μ; now and then resembling scalariform or reticulate vessels; partition walls very oblique with a small perforation. Walls thickened; showing stratification; lignified. T r a c h e i d   f i b r e s. R. 8 μ, T. 10 μ, L. 200—300 μ; tetra- to haxagonal. Walls thickened; showing stratification; lignified; with slit-like bordered pits. Contents: sometimes green sphaero-crystals.

M e d u l l a r y   c o m m i s s u r e s. Ordinary parenchyma.

P a r e n c h y m a   c e l l s   of the xylem parts. R. 8 μ, T. 10 μ, L. 65—130 μ; tetra- to hexagonal prisms. Walls thickened; showing stratification; lignified. Cell contents: sometimes green sphaero-crystals.

September 1901, September 1911.   J; M.

FOLIA SALVIAE.
Salvia Leaves. Sage Leaves.
The dried leaves of Salvia officinalis, Linn. Sp. Pl. 23.

Macroscopic characters.

Simple, petiolate. Petiole half the length of the blade; somewhat gutter-shaped, the concave side turned upwards. Blade generally somewhat 5 c.M. in length;

oblong lanceolate or ovate; feather-veined, veins protruding on the under side; intervenia between the veinlets arched upwards; apex acute; base obtuse; margin finely crenate. Surface on both sides, sometimes only on the under side grayish pubescent; the hairs on the upper side exclusively on the arched intervenia, those on the under side especially on the prominent nerves; glandular hairs especially on the under side resembling fine, white, sparkling dots. Odour especially of bruised leaves aromatic; taste bitterish.

Anatomical characters.

LITERATURE. Deutsch. Arzneib. 5. Ausg. 1910. 235. Ellrodt. Üb. d. Verteil. d. Gerbst.. in off. Blättern, Kräutern u. Blüten. Diss. Würzburg. 1903. 16. Gilg. Pharmakogn. 1910. 293. Glaser. Mikr. Anal. d. Blattpulver. Verh. d. Phys.-Med. Ges. z. Würzburg. N. F. Bd. 34. 1901. 273. Greenish & Collin. Anat. Atl. o. veget. powders. 1904. 80. Hérail. Mat. Méd. 1912. 299. Karsten u. Oltmanns. Pharmakogn. 1909. 171. Koch. Mikr. Anal. d. Drogenpulver. Bd. 3. 1906. 125. Koch. Einf. i. d. mikr. Anal. d. Drogenpulver. 1906. 79. Koch u. Gilg. Pharmakogn. Praktik. 1907. 153. Kraemer. Botany and Pharmacogn. 1910. 730. Marmé. Pharmacogn. 1886. 201. Meyer. Wiss. Drogenk. Bd. 2. 1892. 215. Moeller. Mikr.-pharm. Üb. 1901. 107, 110. Oudemans. Aanteek. o. d. Pharmac. neerl. 1854—1856. 290. Oudemans. Pharmacogn. 1880. 253. Planchon et Collin. Drogues simples. T. 1. 1895. 507. Solereder. Zur mikr. Pulver-analyse d. Folia Salviae. Arch. d. Pharmacie. Bd. 249. 1911. 123. Tschirch. Angew. Pfl.-anat. 1889. 434. Tschirch. Harze u. Harzbeh. 1900. 381, 382. Vogl. Anat. Atl. 1887. Taf. 14. Vogl. Veget. Nahr. u. Genussm. 1899. 346. MATERIAL. The drug. Leaves gathered Sept. 1901 from the Botanic Garden at Groningen; fresh and in alcohol. REAGENTS. Water, glycerine, chloral hydrate, iodine in chloral hydrate, phloroglucin and hydrochloric acid, iodine and sulphuric acid 66 per cent., concentrated sulphuric acid, Schulze's macerating mixture.

MICROGRAPHY.

Intervenia.

Epidermis. Upper side. Stomata phaneroporous, lying in the level of the epidermis, generally surrounded by only 2 subsidiary cells. Trichomes in 3 kinds. 1. Conical file hairs long 250—350 μ; numerous; to be found only on the summits of the arched intervenia; thick at the base 10 μ; uniseriate; articulate; consisting of 2—3 cells; often crooked. 2. Simple capitate glandular hairs not numerous; L. 25—45 μ; stalk 1- to 3-cellular, thick 5 μ; head unicellular, 15 μ in diameter, showing a cuticular bladder. 3. Glandular hairs of the Labiate type. Stalk H. 7 μ, diameter 20 μ; unicellular; head 50 μ in diameter, 8-cellular; cuticular bladder 60 μ in diameter, generally opaque.

Epidermal cells proper. H. 15 μ, Lev. 20 and 30 μ; tabular, lateral walls very slightly sinuous. Outer walls showing a parallel undulating cuticular striation; lateral walls pitted. Cell contents: some chloroplasts. Guard-cells of stomata. Lev. 20 μ and 8 μ. Cells of conical hairs. Outer walls thickened, with a cuticle showing small warts. Cell contents: some small granules.

Under side. Stomata more numerous than on the upper side. Conical hairs less numerous; glandular hairs of both kinds more numerous than on the upper side. See for the rest the upper side.

Epidermal cells proper. H. 10 μ, Lev. 15 and 28 μ; tabular; lateral walls sinuous. Outer walls having a cuticle. Cell contents: some chloroplasts.

Mesophyll.

P a l i s a d e c h l o r e n c h y m a. From 2—3 layers of cells; intercellular spaces.

C e l l s o f t h e a b o v e. H. 40 μ; polygonal prisms with rounded edges or cylinders. Cell contents: poly-, mostly penta- or hexagonal tabular chloroplasts, thick 2 μ, diameter 5 μ, paving the walls, each containing 1 or 2 needle-shaped starch grains; globular granules, 5 μ in diameter.

S p o n g y c h l o r e n c h y m a. Three layers of cells; large intercellular spaces.

C e l l s o f t h e a b o v e. Shape more or less stellate. See for the rest the palisade chlorenchyma.

V e i n s.

Epidermis. Upper side. Cells elongated; arranged in longitudinal rows. Stomata wanting. Conical hairs and glandular hairs of the Labiate type, especially the latter, less numerous than on the intervenia. On all veins and veinlets simple capitate glandular hairs of an other type than those on the intervenia; L. 25 μ; stalk unicellular, thick 5 μ; head 15 μ in diameter, globular, divided by vertical walls in cells numbering up to 4, without cuticular bladder. Contents of head cells a granular mass.

E p i d e r m a l c e l l s p r o p e r. R. and T. 20 μ, L. 40 μ; those bearing a capitate glandular hair often shorter, up to 10 μ; tetra- to hexa-, generally tetragonal tables. Outer walls a little thickened; showing longitudinal parallel cuticular striation. Inner walls collenchymatous; lateral walls pitted. Cell contents: some chloroplasts.

Under side. In most respects the same as epidermis of upper side. Trichomata the same as on epidermis of intervenia; conical hairs very numerous on all veins and veinlets, often somewhat longer.

E p i d e r m a l c e l l s p r o p e r. R. and T. 15 μ, L. 20—60 μ. For the rest see upper side.

Mesophyll.

C o l l e n c h y m a. In the midrib on the upper side 1 or 2, on the under side 2 layers of cells; intercellular spaces.

C o l l e n c h y m a c e l l s. On the upper side R. 25 μ, T. 30 μ, L. 150 μ; on the under side R. 15 μ, T. 18 μ, L. 75 μ; tetra- to hexagonal prisms with rounded edges. Cell contents: some chloroplasts.

C o l o u r l e s s p a r e n c h y m a. In the midrib on both sides 4—5 layers of cells; small intercellular spaces.

C e l l s o f t h e a b o v e. R. 45 μ, T. 50 μ, L. 100 μ; penta- to heptagonal prisms. Walls pitted. Cell contents of the cells on the under side: some chloroplasts.

S p o n g y c h l o r e n c h y m a. Somewhat protruding on both sides into the midrib, continuing that of the intervenia.

E n d o d e r m i s. Starch-sheath. One layer of cells, to be distinguished in the midrib only on the under side.

C e l l s o f t h e a b o v e. R. 20 μ, T. 30 μ, L. 35 μ; polygonal prisms. Walls pitted. Cell contents: chloroplasts, each filled with a globular starch grain.

Meristele. In the midrib very slightly gutter-shaped, the concave side turned upwards.

P e r i c y c l e. Not sharply defined, therefore not drawn in the figure. On both sides somewhat collenchymatous parenchyma, on the under side mixed with very long fibres.

P a r e n c h y m a c e l l s. R. and T. 10—12 μ, L. on the upper side 100—130 μ, on the under side 50 μ; penta- to hexagonal prisms. Cell contents: some chloroplasts contain-ing starch. **F i b r e s.** R. and T. 10 μ, L. 500—2000μ, sometimes septate. Walls not lignified, showing a pure cellulose reaction.

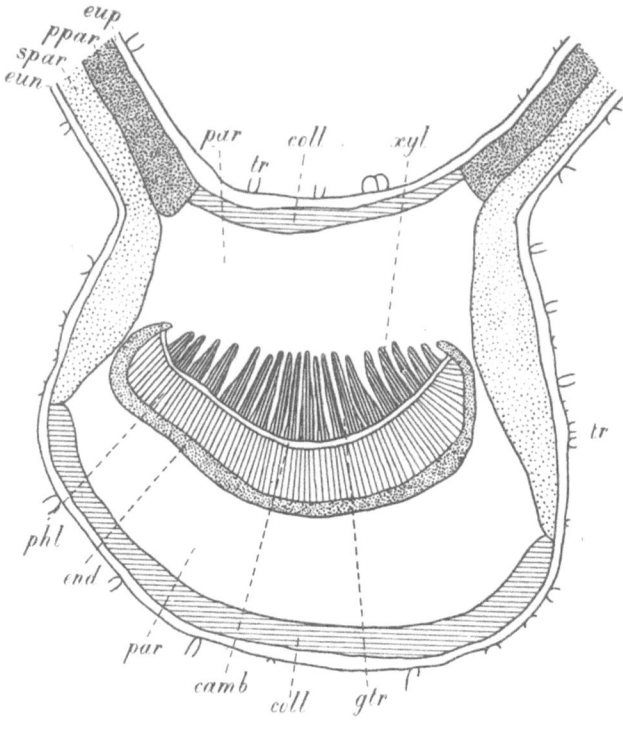

V a s c u l a r b u n d l e. In the midrib a single compound one, contain-ing about 19 simple bundles. Simple bundles collateral, open.

P h l o e m. Radially ar-ranged towards the cam-bium; sieve-tubes and cambiform distinct.

S i e v e - t u b e s. R. and T. 3—4 μ, L. of articulat-ions 90 μ; polygonal prisms. Sieve-plates often with callus-plates.

C a m b i u m. Some layers of elements. R. 3 μ, T. 9 μ, L. 35 μ; tetragonal prisms.

X y l e m. Each bundle consisting in a radial direction of some 10 layers, in a tangential direction of 1 to a few

Fig. 32. *Salvia officinalis*. Leaf, transverse section through the midrib and the neighbouring portions of the lamina. camb Cambium; coll Collenchyma; end Starch-sheath; eun Epidermis under side; eup Epidermis upper side; gtr Medullary commissures, not drawn out in the phloem; par Colourless parenchyma; phl Phloem; ppar Palisade chlorenchyma; spar Spongy chlorenchyma; tr Trichomes; xyl Xylem.

layers of elements. Annular and spiral vessels, those at the upper side often flattened; reticulate and pitted vessels.

A n n u l a r a n d s p i r a l v e s s e l s, the flattened ones R. and T. 3—4 μ, the other ones R. 14 μ, T. 15 μ, L. of articulations 200—300 μ. **R e t i c u l a t e a n d p i t t e d v e s s e l s.** R. 8 μ, T. 9 μ, L. of articulations 150—250 μ.

M e d u l l a r y c o m m i s s u r e s. Ordinary parenchyma.

P a r e n c h y m a c e l l s. Between the xylem bundles R. 6 μ, T. 9 μ, L. 20—35 μ, be-tween the phloem bundles somewhat larger; tetra- to hexagonal prisms. Walls pitted. Cell contents: some chloroplasts.

October 1901, October 1911.

J; M.

FOLIA SENNAE.

Est Indian Senna. Tinnivelly Senna.

The dried leaflets of the pinnate leaf of cultivated Cassia angustifolia, Vahl, Symb. Bot. I. 29.

Macroscopic characters.

Sessile, on an average length 2.5 c.M.; stiff, flat; lanceolate; feather-veined; apex acute; base on one side acute on the other somewhat obtuse; margin entire. Surface seeming glabrous, but in fact having, especially on the under side, many short adpressed hairs, distinctly to be seen when magnified 50 times; colour palegreen. Odour slightly aromatic; taste at first mucilaginous-sweet, afterwards somewhat bitter, more or less pungent. Senna indica should not be coloured yellow or brownish.

Anatomical characters.

LITERATURE. Deutsch. Arzneib. 5. Ausg. 1910. 236. Gilg. Pharmakogn. 1910. 153. Glaser. Mikr. Anal. d. Blattpulver. Verh. d. Phys.-Med. Ges. z. Würzburg. N. F. Bd. 34. 1901. 275. Hager. Pharm. Praxis. Bd. 2. 1902. 884. Hérail. Mat. Méd. 1912. 592. Karsten u. Oltmanns. Pharmakogn. 1909. 182. Koch. Mikr. Anal. d. Drogenpulver. Bd. 3. 1906. 135. Koch. Einf. i. d. mikr. Anal. d. Drogenpulver. 1906. 74. Koch u. Gilg. Pharmakogn. Praktik. 1907. 143. Kraemer. Botany a. Pharmacogn. 1910. 608 and 721. Marmé. Pharmacogn. 1886. 181. Meyer. Wiss. Drogenk. Bd. 2. 1892. 233. Meyer. Mikr. Unters. v. Pflanzenpulv. 1901. 207. Moeller. Pharmacogn. 1889. 56. Moeller. Mikr.-pharm. Üb. 1901. 93. Oudemans. Pharmacogn. 1880. 314. Planchon et Collin. Drogues simples. T. 2. 1896. 455. Tschirch. Angew. Pfl.-anat. 1889. 185. Tschirch u. Oesterle. Anat. Atl. 1900. 25. Taf. 7. Vogl. Anat. Atl. 1887. Taf. 19 & 20. Walliczek. Stud. ü. d. Membranschleime veget. Org. Pringsheim's Jahrb. Bd. 25. 1893. 234; the same as Dissertation Bern. 1893. 26. Wigand. Pharmakogn. 1879. 194. MATERIAL. The drug. REAGENTS. Water, glycerine, potash, iodine in chloral hydrate, phloroglucin and hydrochloric acid, iodine and sulphuric acid 66 per cent., concentrated sulphuric acid, Schulze's macerating mixture, ferric acetate.

MICROGRAPHY OF THE LEAFLET.

Intervenia.

Epidermis. Upper side. Stomata 180 to the sq. m.M., phaneroporous, almost always surrounded by only 2 subsidiary cells; these and the guard-cells obviously having originated from one cell. Guard-cells 2—3 times less in height than the subsidiary cells and attached to the lower parts of these; stomata thus lying lower than the surface of the epidermis. Subsidiary cells having $^1/_2$ of the height of the epidermal cells proper. Trichomes: file hairs; not numerous; the adjoining 5—7 epidermal cells radiately surrounding the base, shorter than the height of the epidermal cells; L. 50—250 μ, diameter at the base 10—15 μ; conical, bent towards the top of the leaf; unicellular.

Epidermal cells proper. H. 30 μ, Lev. 25 and 25—40 μ; polygonal prisms having somewhat curved lateral walls. Outer walls thickened; stratified; having a cuticle and being covered with wax particles; lateral walls very thin. Inner walls of almost all cells, except those of the margin, very considerably thickened, thick 15—25 μ; the thin uppermost and innermost parts of these walls consisting of cellulose, the rest of a mucilaginous substance, being somewhat folded and granular when examined in alcohol [1]).

---

[1]) These curious thickenings have led many investigators to the opinion that the epidermal cells were divided.

Cell contents of many cells: one or more yellow globes, becoming dark red on addition of potash. G u a r d - c e l l s. H. 8 μ, Lev. 17 and 20 μ. S u b s i d i a r y c e l l s. H. 15—20 μ, Lev. 22 and 30 μ; thickenings of inner walls wanting. F i l e h a i r s. Walls strongly thickened; rough on the outside by cuticularisation; the parts between the epidermal cells also cuticularized.

Under side. Stomata somewhat, file hairs much more numerous; see for the rest the upper side.

Mesophyll.

P a l i s a d e c h l o r e n c h y m a. One layer of cells on either side of the blade, those on the upper side having 2 times the height of those on the under

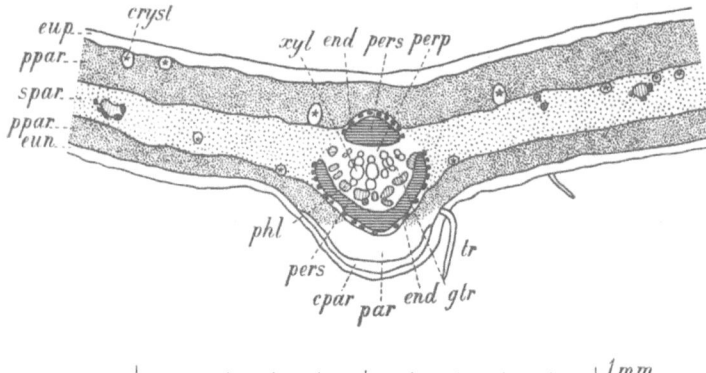

side. In the upper layer here and there 1 or 2 somewhat wider cells divided by 1 or 2 transverse walls in 2 or 3 cells; 1 or more of these containing a cluster crystal. Intercellular spaces.

Fig. 33. *Cassia angustifolia*. Leaf, transverse section through the midrib and the neighbouring portions of the lamina. cpar Collenchymatous part of colourless parenchyma; cryst Crystal cells; end Crystal-sheath; eun Epidermis under side; eup Epidermis upper side; gtr Medullary commissures; par Colourless parenchyma; perp Pericyclic parenchyma; pers Pericyclic sclerenchyma; phl Phloem; spar Spongy chlorenchyma; tr Trichomes; xyl Xylem.

C e l l s o f t h e a b o v e. H. of upper layer 80 μ, of under layer 40 μ, Lev. of both 12 μ;

circular cylinders. Cell contents: chloroplasts, in some cells, especially near to the meristeles, containing each some needle-shaped to cylindrical starch grains. By addition of iodine in chloral hydrate some cells showing a brown mass. In potash walls and contents of all cells reddish brown.

S p o n g y c h l o r e n c h y m a. Generally 4 layers of cells; some cells especially of the upper- and undermost layer containing a cluster crystal. Large intercellular spaces.

C e l l s o f t h e a b o v e. H. 15 μ, Lev. 20 and 20 μ; globes or polyhydra with strongly rounded edges. Cell contents: chloroplasts, in the undermost layers mostly containing each some needle-shaped to cylindrical starch grains. By addition of potash walls and contents of all cells reddish brown. At the margin of the blade some cells filled with a red mass.

M a r g i n. Generally below the epidermis one layer of colourless parenchyma cells. Walls a little thickened. Contents: sometimes a cluster crystal.

V e i n s.

Epidermis. Upper side. In all respects the same as that of the intervenia. Under side. Stomata very rare. File hairs the same as on the intervenia. Epidermal cells proper arranged in longitudinal rows.

E p i d e r m a l c e l l s p r o p e r. R. 12 μ, T. 9 μ. L. 20—40 μ; rectangular prisms. Outer walls thickened; stratified; showing a cuticle and wax particles. Here and there on the sides of the protruding part of the midrib cells having the same thickening of inner wall as described for the epidermal cells of the upper side of the intervenia. Cell contents and walls yellow with transitions to red by addition of potash.

Mesophyll.

P a l i s a d e   c h l o r e n c h y m a. The layer of the upper side of the intervenia continued without interruption in the midrib; the cells being somewhat less in height.

C o l o u r l e s s   p a r e n c h y m a. Only on the under side some 7 layers of cells in a radial direction; cells in longitudinal rows, some of these cells being divided by transverse walls in several smaller cells, each containing a simple crystal.

C e l l s   o f   t h e   a b o v e. R. and T. 12 μ, L. 50—100 μ; polygonal prisms with rounded edges. Walls of the outermost 1 or 2 layers collenchymatous; sometimes pitted. Cell contents: in the outermost layers and in the cells close to the palisade tissue of the under side of intervenia here and there chloroplasts with starch grains; all cells containing a mass coloured yellow with transitions to red.

E n d o d e r m i s. In the midrib, with the exception of a lateral longitudinal strip on either side, developed as a crystal-sheath of 1 layer of cells. These cells in longitudinal rows and here and there crystals wanting. In smaller veins the upper part of the crystal-sheath wanting; in the veinlets the under part also wanting.

C e l l s   o f   t h e   a b o v e. R. 10 μ, T. 12 μ, L. 10—15 μ; polygonal prisms.

Meristele. In the midrib somewhat irregularly cylindrical.

P e r i c y c l e. Consisting in the peripheral part of 2 bundles of sclerenchyma fibres, towards the centre of the meristele of some common parenchyma. The sclerenchyma bundles laterally extending as far as the crystal-sheath; the upper one semi-cylindrical, the undermost gutter-shaped. In smaller veins the upper bundle, in the veinlets both bundles wanting. No intercellular spaces. Some layers of common parenchyma on the under side here and there with crystal cells, filling up the space between sclerenchyma and vascular bundle.

S c l e r e n c h y m a   f i b r e s. R. and T. 8—12 μ, L. 250—500 μ; polygonal, often having branched ends. Walls thickened; somewhat yellow; lignified, especially the middle-lamellae; showing slit-like simple pits. P a r e n c h y m a   c e l l s. R. 8 μ, T. 10 μ; polygonal prisms with a longitudinally directed axis. Cell contents: often some starch grains; here and there a red mass.

V a s c u l a r   b u n d l e. In the midrib a single compound one, containing about 8 or 10 simple bundles. Simple bundles collateral.

P h l o e m. Bundles consisting of some 10 elements in a transverse section. Elements. R. and T. 5 μ; polygonal; walls somewhat collenchymatous; contents often a red mass.

X y l e m. Bundles consisting of radially arranged spiral, reticulate and pitted vessels.

V e s s e l s. The largest R. 20 μ, T. 18 μ, lying in the middle of the radial rows. Articulations of the pitted vessels L. 60—180 μ. All vessels being polygonal prisms with strongly rounded edges. Diaphragms of the pitted vessels with circular perforations. Walls lignified.

M e d u l l a r y　c o m m i s s u r e s. Parenchyma.

C e l l s　o f　t h e　a b o v e. R. 10 μ, T. 8 μ; polygonal prisms with a longitudinally directed axis. Cell contents: some starch grains.

Micrography of the powder.

Chlorenchyma partly consisting of very elongated palisade cells; in the spongy chlorenchyma some clustered crystals. Epidermis of upper and under side the same: penta- to hexagonal cells, often showing a very much thickened colourless inner wall and thereby seeming divided in 2 cells; stomata mostly between two subsidiary cells smaller than the epidermal cells. Unicellular, thick-walled, acute file hairs. Bundles of thick-walled long fibres, belonging to the pericyclic sheath of the larger meristeles, surrounded by a layer of short crystal cells, each containing a simple crystal. Spiral vessels, pitted vessels. Much starch in the chlorenchyma.

October 1902, November 1911.　　　　　　　　　　　　　　　　J; M.

FOLIA STRAMONII.
Stramonium Leaves.

The dried leaves of Datura Stramonium, Linn. Sp. Pl. 179, collected when the plant is in flower from cultivated plants and carefully dried so that the green colour is preserved.

Macroscopic characters.

Simple, petiolate. Petiole up to 10 c.M. in length, having a narrow groove on the upper side. Blade up to 20 c.M. in length and 15 c.M. in breadth; ovate, pinnately lobed, having acute teeth and obtuse arched incisions; apex acute; base obtuse to truncate; on each side of the midrib 3—5 veins; midrib and veins strongly prominent on the under side. Surface of the young leaves slightly pubescent, chiefly on the veins and intervenia of the upper side and also on the veins of the under side, afterwards glabrous. Odour of the fresh leaves nauseous; taste bitter and brinish.

Anatomical characters.

LITERATURE. Deutsch. Arzneib. 5. Ausg. 1910. 237. Ellrodt. Üb. d. Vert. d. Gerbst. i. off. Blättern, Kräutern u. Blüten. Diss. Würzburg. 1903. 16. Fedde. Vergl. Anat. d. Solanaceen. Diss. Breslau. 1896. Feldhaus. Quant. Unters. d. Verteilung des Alcaloides in den Org. von D. Stramonium. Diss. Marburg. 1903. Flückiger. Pharmakogn. 1891. 707. Gilg. Pharmakogn. 1910. 313. Glaser. Mikr. Anal. d. Blattpulver. Verh. d. phys.-med. Ges. z. Würzburg. N. F. Bd. 34. 1901. 278. Greenish & Collin. Anat. Atl. o. veget. powders. 1904. 86. Hager. Pharm. Praxis. Bd. 1. 1900. 1013. Herail. Mat. Méd. 2. ed. 1912. 686. Karsten u. Oltmanns. Pharmakogn. 1909. 166. Koch. Mikr. Anal. d. Drogenpulver. Bd. 3. 1906. 147. Koch. Einf. i. d. mikr. Anal. d. Drogenpulver. 1906. 86. Koch u. Gilg. Pharmakogn. Praktik. 1907. 147. Kraemer. The crystals in Datura Stramonium. Bull. of the Torrey Bot. Club. 1900. Kraemer. Bot. a. Pharmacogn. 1910. 208, 622, 723, 728. Marmé. Pharmacogn. 1886. 188. Meyer. Wiss. Drogenk. Bd. 2. 1892. 198. Mitlacher. Toxik. od. Forens. wicht. Pfl. u. Drogen. 1904. 153. Moeller. Pharmakogn. 1889. 69. Moeller. Mikr.-pharm. Üb. 1901. 105. Molle. Rech. de microchémie comparée sur la localisation des alcaloides dans les Solanacées. Mémoires de l'Académie des Sciences de Belgiques. 1896. T. LIII. Molisch. Mikrochemie d. Pfl. 1913. 257. Oudemans. Aanteek. o. d. Pharmac. Neerl. 1854—56. 324. Oudemans. Pharmacogn. 1880. 242. Planchon et Collin. Drogues simples.

T. 1. 1895. 588. Schneider. Powd. Veget. Drugs. 1902. 302. Tschirch. Angew. Pfl.-anat. 1889. 436. Tschirch u. Oesterle. Anat. Atl. 1900. 283. Taf. 65. Tunmann. Pflanzenmikrochemie. 1913. 325. Vogl. Anat. Atl. Taf. 11. Wigand. Pharmakogn. 1879. 214. MATERIAL. Leaves gathered Sept. 1903 from the Botanic Garden at Groningen, fresh, in alcohol and after fixation in chromic acid 1 per cent. REAGENTS. Water, glycerine, chloral hydrate, iodine in chloral hydrate, phloroglucin and hydrochloric acid, iodine and sulphuric acid 66 per cent., concentrated sulphuric acid, Schulze's macerating mixture, potassium dichromate, cupric acetate and ferric acetate.

MICROGRAPHY OF THE BLADE.

Intervenia.

Epidermis. Upper side. Stomata 60 to the sq. m.M.; phaneroporous; lying in the level of the epidermis; generally surrounded by 3 subsidiary cells, often somewhat projecting under the guard-cells. On some leaves here and there a stoma missing one of the guard-cells but having a well developed porus; sometimes both guard-cells flattened, only the porus being distinct. Trichomes not numerous, on old leaves rare; in 2 kinds: 1. Conical file hairs, generally inserted above the separating wall of 2 larger epidermal cells, without sinuous lateral walls; length 130—400 μ, diameter at the base 25—30 μ; uniseriate; unbranched; consisting of 1—4, mostly 3 cells becoming shorter towards the apex. 2. Ordinary capitate hairs; stalk unicellular; head multicellular, dilated, sometimes consisting of some polyhedral cells at the base, covered by a layer of some prismatic cells with vertically directed axis; cuticular bladder wanting.

E p i d e r m a l c e l l s p r o p e r. H. 25 μ, Lev. e. g. 30 by 50 μ, 40 by 50 μ, 40 by 60 μ; polygonal tables having not very sinuous lateral walls. Outer walls somewhat thickened, having a cuticle. Cell contents: some small chloroplasts; by potassium dichromate a red-brown mass, only partly filling the cell. G u a r d - c e l l s o f s t o m a t a. H. 9 μ, Lev. 9 and 25 μ. Cell contents: by the action of potassium dichromate a red-brown mass, especially at the ends of the cells. C e l l s o f f i l e h a i r s. Outer walls showing longitudinally elongated, cuticular warts. Cell contents: by the action of potassium dichromate a red-brown mass or globules. H e a d c e l l s o f c a p i t a t e h a i r s. Contents of material fixed in alcohol a yellow granular mass.

Under side. Stomata 80 to the sq. m.M., generally surrounded by 4 subsidiary cells. Trichomes less numerous than on the upper side. See for the rest epidermis of upper side.

E p i d e r m a l c e l l s p r o p e r. H. 20 μ, Lev. e. g. 40 by 40 μ, 40 by 60 μ, 30 by 80 μ; polygonal tables having very sinuous lateral walls. Outer walls somewhat thickened, with a cuticle. Cell contents: some chloroplasts, each containing some very small needle-shaped starch grains. G u a r d - c e l l s o f s t o m a t a. H. 8 μ, Lev. 10 by 25 μ. Cell contents: by the action of potassium dichromate some small yellow-green globules. See for the rest guard-cells and trichomes of upper side.

Mesophyll.

P a l i s a d e c h l o r e n c h y m a. One layer of cells; intercellular spaces. C e l l s o f t h e a b o v e. H. 120—150 μ, Lev. 15 by 20 or 15 by 15 μ; polygonal prisms with strongly rounded edges or cylinders. Cell contents: many penta- or hexagonal tabular chloroplasts, paving the walls, diameter 4 μ, thick 2 μ, each containing some needle-shaped starch grains.

S p o n g y c h l o r e n c h y m a. 4—5 layers of cells; the uppermost layer more or less distinctly developed as collecting cells; the other layers consisting of stellate cells, e.g. those of the layer adjoining the epidermis of the under side often having 4 branches in the level of the blade; large intercellular spaces. C e l l s o f t h e a b o v e. H. 20 μ, Lev. 25 by 25 μ; collecting cells H. 35 μ. Cell contents: the same as in the palisade chlorenchyma; in most collecting cells besides the chloroplasts a cluster crystal in the centre — e. g. 20 μ in diameter.

V e i n s.

Epidermis. Upper side. Epidermal cells very much elongated, fairly in longitudinal rows. Stomata wanting. Trichomes. Conical hairs numerous, see for the rest the intervenia.

E p i d e r m a l c e l l s p r o p e r. R. and T. 25 μ, L. 150— 300 μ; tetra- to hexa-, generally tetragonal tables with a radially directed axis. Outer walls strongly thickened, showing a longitudinal cuticular striation; inner walls strongly collenchymatously thickened, showing stratification. Cell contents: by the action of potassium dichromate a red-brown mass.

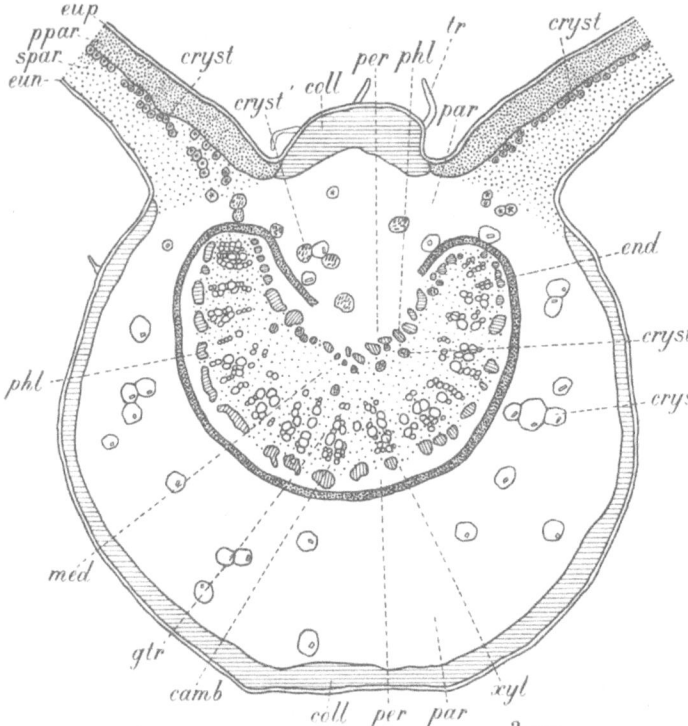

Fig. 34. *Datura Stramonium.* Leaf, transverse section through the midrib and the neighbouring portions of the lamina. camb Cambium; coll Collenchyma; cryst Crystal cells; cryst' Cells filled with crystal sand; end Starch-sheath; eun Epidermis under side; eup Epidermis upper side; gtr Medullary commissures; med Medulla; par Colourless parenchyma; per Pericycle; phl Phloem; ppar Palisade chlorenchyma; spar Spongy chlorenchyma; tr File hairs; xyl Xylem.

Under side. Epidermal cells elongated, fairly in longitudinal rows.

Stomata wanting. Trichomes: see those of upper side of intervenia.

E p i d e r m a l c e l l s p r o p e r. R. 20 μ, T. 25 μ, L. 70—160 μ; tetra- to hexa-, mostly tetragonal tables with a radially directed axis. Outer walls strongly thickened, showing a longitudinal cuticular striation; inner walls strongly collenchymatously thickened, showing stratification. Cell contents: by the action of potassium dichromate several dark yellow globules, often clustered together and seeming enclosed in a common wall.

Mesophyll.

C o l l e n c h y m a. On the upper side a somewhat flat bundle, 8 cells thick in a

radial direction; on the under side a collenchyma layer, thick 6 cells in the middle part, somewhat less towards the lateral sides; very small intercellular spaces.
C e l l s o f t h e a b o v e. R. and T. 25 μ, L. 200—350 μ; polygonal prisms with a longitudinally directed axis. Walls especially of the cells of the upper side strongly thickened, showing stratification.

C o l o u r l e s s p a r e n c h y m a. In the middle part of the midrib on the upper side 12, on the under side about 8 layers of cells; cells in longitudinal rows; intercellular spaces. Crystal idioblasts in 2 kinds: 1. entirely filled with crystal sand, consisting of small tetrahedral particles; these cells sometimes 2 or 3 joining upon each other in a longitudinal row and smaller — e. g. R. and T. 60 μ, L. 100 μ — than the adjoining parenchyma cells; 2. generally containing 1 simple crystal, sometimes a twin crystal or a cluster crystal.
P a r e n c h y m a c e l l s. On the upper side R. and T. 100 μ, L. 150—300 μ; on the under side R. and T. 140 μ, L. 250—350 μ; polygonal prisms with a longitudinally directed axis and rounded edges. Cell contents: some small chloroplasts, each containing some needle-shaped starch grains; for the rest in the crystal cells the before mentioned crystals.

E n d o d e r m i s. Developed in the midrib and veins as a nearly complete starch-sheath, consisting of 1 layer of cells and often interrupted on the upper and lateral sides. Cells in longitudinal rows. In the smaller veins and veinlets starch-sheath wanting.
C e l l s o f t h e a b o v e. R. and T. 60 μ, L. 70—130 μ; polygonal prisms with a longitudinally directed axis and rounded edges. Cell contents: some chloroplasts, entirely filled with some starch grains.

Meristele. In the midrib gutter-shaped, the concave side turned upwards.

P e r i c y c l e. In the midrib on all sides some layers of parenchyma cells.
C e l l s o f t h e a b o v e. On the upper side R. and T. 30 μ, on the under side R. and T. 50—80 μ; polygonal prisms with a longitudinally directed axis.

V a s c u l a r b u n d l e. In the midrib a single compound one, containing numerous (about 30) simple bundles. Simple bundles collateral[1]), open.

P h l o e m. Bundles chiefly consisting of sieve-tubes, further of companion cells and cambiform cells, the last not numerous and chiefly in the outer parts of the bundles.
S i e v e - t u b e s. R. and T. 10 μ; polygonal prisms. Sieve-plates very distinct, generally showing callus plates. Walls slightly thickened. Contents: close to each sieve-plate and always on the same side some small granules often along the longitudinal walls of the tube and becoming red-brown in iodine and chloral hydrate. C o m p a n i o n c e l l s. R. and T. 4—5 μ; polygonal prisms. Walls slightly thickened. Cell contents: in material fixed in alcohol a yellow granular mass, becoming red-brown in iodine and chloral hydrate. C a m b i f o r m c e l l s. R. and T. 10 μ; polygonal prisms. Walls slightly thickened.
C a m b i u m. Consisting of some layers of elements.
X y l e m. Bundles some elements broad, consisting of annular, spiral, reticu-

---

[1]) The vascular bundles in the leaf of *Datura* are generally called bicollateral, but without sufficient reason; in truth the phloem and xylem bundles lying on the under side of the meristele correspond with each other and constitute simple collateral vascular bundles, separated by a considerable layer of medullary parenchyma from the upper phloem bundles, which correspond neither in number nor in place with the collateral vascular bundles. See moreover *Folia Belladonnae.*

late vessels and parenchyma cells; the largest vessels lying in the middle parts of the radial rows.

V e s s e l s. R. 20—40 μ, T. 20—35 μ; polygonal prisms, those in the upper parts of the bundles with strongly rounded edges. Walls lignified. P a r e n c h y m a c e l l s. R. and T. 15—20 μ, L. 50—100 μ; polygonal prisms with a longitidunally directed axis.

M e d u l l a r y c o m m i s s u r e s. Consisting of parenchyma, bi- or triseriate, without cambium; intercellular spaces wanting.

C e l l s o f t h e a b o v e. R. and T. 30 μ, L. up to 130 μ.; polygonal prisms with a longitudinally directed axis. Cell contents: somewhat more chloroplasts than in the colourless parenchyma of mesophyll, each chloroplast containing some needle-shaped starch grains.

M e d u l l a. Consisting of some layers of parenchyma cells, and containing on the upper side a layer of phloem bundles in all respects resembling those of the simple vascular bundles; longitudinal rows of the parenchyma cells containing coarse crystal sand.

P a r e n c h y m a c e l l s. R. and T. 30 μ, L. up to 130 μ; polygonal prisms with a longitudinally directed axis. Cell contents: the crystal sand of the crystal cells generally consisting of tetrahedral particles, up to 10 μ in diameter; see for the rest the parenchyma cells of the medullary commissures.

Micrography of the powder.

Much parenchyma coloured bright green and remains of same. Numerous crystals, generally cluster crystals, sometimes also more simple crystals. Many small parts of the leaf, in chloral hydrate easily showing the complete structure of the leaf in a transverse section or in the level of the blade: cells of the upper epidermis having not very sinuous, those of the under epidermis having very sinuous lateral walls; many stomata on both sides; layer of high palisade chlorenchyma cells; some layers of spongy chlorenchyma, the uppermost containing in almost all cells one of the crystals mentioned before; veinlets chiefly consisting of spiral vessels, forming a net-work with many dead ends in the meshes. Pieces of veins showing oblong epidermal cells, sometimes also collenchyma. File hairs consisting of 1—4 cells or parts of these, and less numerous capitate hairs having an unicellular stalk and multicellular head. A few larger reticulate vessels. In the chloroplasts of the parenchyma sometimes very thin needle-shaped starch grains.

September 1903, January 1912. J; M.

FOLIA TARAXACI.
Taraxacum Leaves.
The fresh leaves of Taraxacum officinale, (Weber, in) Wigg. Prim. Fl. Holsat. 56.

LITERATURE. Erdmann-König. Allg. Warenk. 1895. 222. Hager. Pharm. Praxis. Bd. 2. 1902. 1014. Koch. Mikr. Anal. d. Drogenpulver. Bd. 2. 1903. 243. Marmé. Pharmacogn. 1886. 205. Moeller. Mikr. Pharm. Ueb. 1901. 97. Planchon et Collin. Drogues simples. T. II. 1896. 16. MATERIAL. Leaves gathered in May 1902 in the Botanic Garden at Groningen, fresh and in alcohol, after having been fixed in 0.5 % chromic acid. REAGENTS. Water, glycerine, chloral hydrate, iodine in chloral hydrate, phloroglucin and hydrochloric acid, iodine and sulphuric acid 66 per cent., concentrated sulphuric acid, Schulze's macerating mixture, potassium dichromate, iron acetate.

MICROGRAPHY.

Intervenia.

Epidermis. Upper side. Stomata 30 to the sq. m.M., in the same level with the epidermal cells proper. Hairs conical, usually strongly curved, articulate, consisting of 4—9 cells, uniseriate at the base often 4-seriate, often the uppermost cells more or less flattened.

Epidermal cells proper. H. 12 μ, Lev. 10—20 by 20—60 μ. Tabular, lateral walls strongly sinuous. Outer walls with a cuticle. Cell contents: some small chloroplasts; no tannin reaction by iron acetate in living cells, a brown precipitate by the action of potassium dichromate. Guard-cells of stomata. Lev. 18—20 by 18—28 μ.

Under side. Stomata 40 to the sq. m.M., in the same level with the epidermal cells proper. Trichomes, see upper side.

Mesophyll.

Palisade chlorenchyma. From 2—3 layers of cells, with intercellular spaces.

Cells of the above. H. 20—25 μ, Lev. 12 μ; cylinders. Cell contents: many tabular polygonal chloroplasts paving the walls, 3 μ in diameter and 1.5 μ thick.

Spongy chlorenchyma. 4—5 layers of cells; large intercellular spaces separated by series of 1 cell broad.

Cells of the above. Diameter 15—20 μ; globular, ellipsoidal or slightly stellate. Cell contents: see palisade chlorenchyma.

Veins.

Epidermis. Upper side. Epidermal cells proper arranged more or less in longitudinal rows. Stomata arranged in longitudinal rows consisting of a few longitudinally directed stomata, these rows placed irregularly; lying in the same level with the epidermal cells proper. Hairs long up to 600 μ, generally longer than those on the intervenia, consisting of 15 cells; the walls of the undermost cells often somewhat thickened. See for the rest intervenia upper side.

Epidermal cells proper. R. 20 μ, T. 18 μ, L. 100—250 μ; tetra- or penta-, mostly tetragonal tables. Outer walls thick 8 μ, with a cuticle and stratification; inner walls also thickened. Cell contents: no tannin reaction by iron acetate in living cells, a brown precipitate by the action of potassium dichromate; sometimes anthocyanin. Guard-cells of stomata. L. 30 μ, T. 20 μ.

Under side. Stomata less numerous than on the upper side. For the rest see upper side.

Epidermal cells proper. R. 18 μ, T. 13 μ, L. 100—250 μ.

Mesophyll.

Collenchyma. In the midrib adjoining the epidermis 2 flat bundles, the one on the upper, the other on the under side; the upper bundle thick 1 layer of cells, the undermost bundle 1 or 2. Under the stomata these bundles wanting. On places near the stomata and with these in the same longitudinal rows, here and there between these bundles and the epidermis intercellular spaces of elliptical shape in a tangential section. Intercellular spaces also between these bundles and the outermost layer of the colourless parenchyma.

C e l l s o f t h e a b o v e. R. 20 μ, T. 25 μ, L. 170 μ; tetragonal prisms with convex inner and outer walls and a longitudinally directed axis. Outer and inner walls thickened and with stratification. Cell contents: tannin and some small chloroplasts.

C o l o u r l e s s p a r e n c h y m a. Mostly torn in the middle part of the midrib; with intercellular spaces in the outer parts, these in the outermost parts having a cross diameter of up to 51 μ.

C e l l s o f t h e a b o v e. R. and T. 30—60 μ, L. 100—160 μ; circular or elliptical cylin-

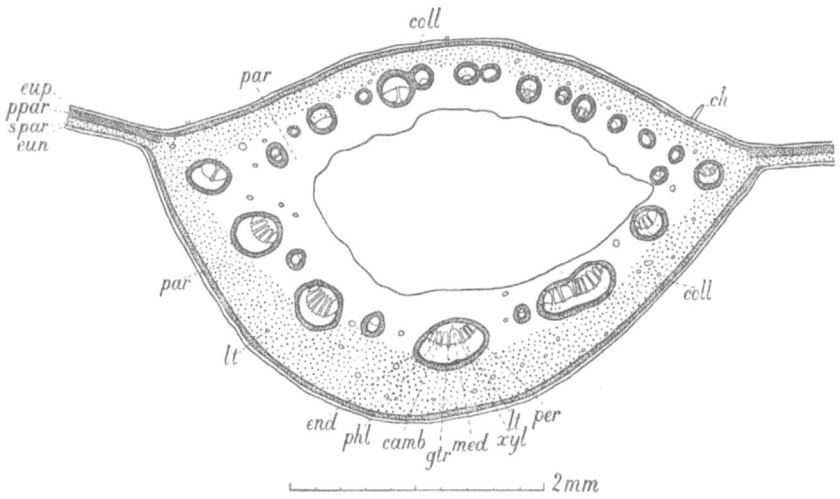

Fig. 35. *Taraxacum officinale.* Leaf, transverse section through the midrib and the neighbouring portions of the lamina. camb Cambium; ch Conical hairs; coll Collenchyma; end Starch-sheath; eun Epidermis under side; eup Epidermis upper side; gtr Medullary commissures; lt Latex tubes; med Medulla; par Colourless parenchyma; per Pericycle; phl Phloem; ppar Palisade chlorenchyma; spar Spongy chlorenchyma; xyl Xylem.

ders or polygonal prisms with strongly rounded edges, all these with a longitudinally directed axis. Cell contents: chloroplasts, especially in the outermost layers.

E n d o d e r m i s. Developed as a starchs-heath and surrounding all meristeles lying in a ring, even the smallest one.

C e l l s o f s t a r c h - s h e a t h o f t h e l a r g e s t m e r i s t e l e s. On the outer side R. 22 μ, T. 30 μ, L. 40—80 μ, on the inner side of the meristeles R. 18 μ, T. 20 μ, L. 100 μ; rectangular prisms with a longitudinally directed axis. Cell contents: chloroplasts quite filled with a large starch grain, adjoining one of the transverse walls, turned in all cells to the same side.

Meristeles. Numerous at the base of the midrib and diminishing in number towards the top, one by one bending outward into the primary nerves, some 25 at about one third of the length of the midrib from the base. Longitudinally arranged in the plane of an elliptical cylinder in the middle part of the colourless parenchyma. Thinner and thicker ones placed without fixed order, but the thickest generally at the under side of the leaf.

P e r i c y c l e. Developed only at the outside of the meristeles and extending over somewhat more than one half of their circumference; consisting of fibres, common parenchyma and latex vessels, the fibres mostly at the circumference and the latex vessels in one row and often anastomosing.

F i b r e s. R. and T. 10 μ, L. 500—800 μ; polygonal. Walls a little thickened. P a r e n-c h y m a c e l l s. R. and T. 3—6 μ; polygonal prisms with a longitudinally directed axis. Walls somewhat collenchymatous. L a t e x v e s s e l s. R. and T. 10 μ and less; polygonal prisms with rounded edges and longitudinally directed axis. Contents: a granular mass.

V a s c u l a r b u n d l e. In each meristele mostly a single compound one containing 2—10 simple bundles, the latter radially elongated in a transverse section; in a few meristeles only a single simple vascular bundle. Simple bundles collateral, open.

P h l o e m. With sieve-tubes to be distinghuished only in the inner portion. Latex vessels, like those of the pericycle, here and there between the other elements.

S i e v e-t u b e s. R. and T. 3—5 μ in diameter, L. of articulations 90—120 μ; polygonal prisms with a longitudinally directed axis. Walls somewhat collenchymatous.

C a m b i u m. To be distinguished.

X y l e m. Bundles consisting of 1 or more radial rows of annular and spiral vessels, up to 6 in number in a radial direction, the 3 nearest to the cambium not full grown, moreover some xylem parenchyma.

V e s s e l s. R. 22 μ, T. 20 μ and less, L. of the articulations of the vessels nearest to the cambium 130—200 μ; rectangular prisms with a longitudinally directed axis and rounded edges. Walls somewhat thickened, those of the vessels farthest from the cambium lignified.

M e d u l l a r y c o m m i s s u r e s. Uni- or biseriate, consisting of common parenchyma.

C e l l s o f t h e a b o v e. R. 5—8 μ, T. 4—5 μ; rectangular prisms with a longitudinally directed axis. Walls here and there somewhat lignified. Cell contents: some small chloroplasts.

M e d u l l a. Developed between the inner parts of the xylem bundles and the starch-sheath, consisting of about 3 layers of common parenchyma cells in a radial direction.

C e l l s o f t h e a b o v e. R. and T. 8 μ; polygonal prisms with a longitudinally directed axis. Walls somewhat thickened; intercellular spaces wanting.

May 1901.   J; M.

FOLIA TRIFOLII FIBRINI.

Folia Menyanthidis. Buchbean.

The dried leaves of Menyanthes trifoliata, Linn. Sp. Pl. 145, collected when the plant is in flower, rinsed with water and rapidly dried.

Macroscopic characters.

Ternate. Sheath 6—8 c.M. in length, cylindrical, mostly wanting in the drug. Petiole about 10 c.M. in length, terete. Leaflets up to 8 c.M. in length, 4 c.M. in breadth; sessile; obovate to elliptical; feather-veined; margin coarseley and superficially crenated, each incision showing at the bottom a small crescent-shaped brownish wart (emissarium with water-pores). Surface glabrous. Inodorous; taste very bitter.

Anatomical characters.

LITERATURE. De Bary. Vergl. Anat. 1877. 129. Deutsch. Arzneib. 5. Ausg. 1910. 238. Ellrodt. Üb. d. Vert. d. Gerbst. in off. Blättern, Kräutern u. Blüten. Diss. Würzburg. 1903. 17. Flückiger. Pharmakogn. 1891. 679. Gilg. Pharmacogn. 1910. 278. Glaser. Mikr. Anat. d. Blattpulver. Verh. d. Phys. Med. Ges. z. Würzburg. N. F. Bd. 34. 1901. 279. Karsten u. Oltmanns. Pharmakogn. 1909. 181. Koch. Mikr. Anal. d. Drogenpulver. Bd. 3. 1906. 157. Marmé. Pharmacogn. 1886. 203. Meyer. Wiss. Drogenk. Bd. 2. 1892. 223. Moeller. Pharmakogn. 1889. 59. Moeller. Mikr.-pharm. Üb. 1901. 89. Oudemans. Pharmacogn. 1880. 273. Perrot. Anat. comp. d. Gentiacées. Ann. sc. nat. Sér. 8. T. 7. 1898. 159, 160. Planchon et Collin. Drogues simples. T. 1. 1895. 653. Tschirch. Angew. Pfl.-anat. 1889. 322. Vogl. Anat. Atl. 1887. Taf. 7. Wigand. Pharmakogn. 1879. 213. MATERIAL. Leaves collected in the Botanic Garden at Groningen May 1902, fresh, in alcohol and after fixation in chromic acid $^1/_2$ per cent. REAGENTS. Water, glycerine, chloral hydrate, iodine in chloral hydrate, phloroglucin and hydrochloric acid, iodine and sulphuric acid 66 per cent., concentrated sulphuric acid, Schulze's macerating mixture, cupric acetate and ferric acetate, potassium dichromate.

MICROGRAPHY OF THE LEAFLET.

Intervenia.

Epidermis. Upperside. Stomata 28 to the sq. m.M., phaneroporous; guard-cells lying in the same level with the epidermis; surrounded by 4—6 subsidiary cells. E p i d e r m a l c e l l s p r o p e r. H. 20 µ, Lev. 25—40 by 50 µ; polygonal tables having

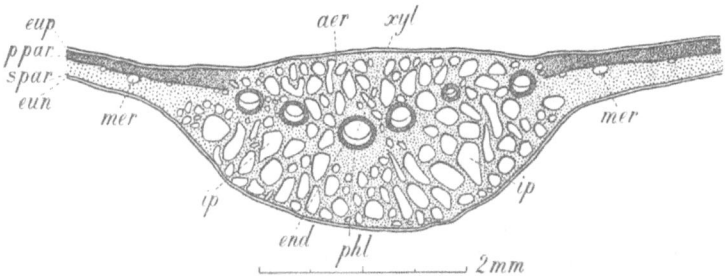

their inner walls curved towards the palisade chlorenchyma. Outer walls showing a parallel undulating cuticular striation, on the subsidiary cells directed perpendicularly to the circumference of the stoma. Cell contents: some chloroplasts; tannin;

Fig. 36. *Menyanthes trifoliata.* Leaf, transverse section through the midrib and the neighbouring portions of the lamina. aer Aerenchyma; end Starchsheath, protective sheath; eun Epidermis under side; eup Epidermis upper side; ip Intercellular passages; mer Thin meristeles; phl Phloem; ppar Palisade chlorenchyma; spar Spongy chlorenchyma; xyl Xylem.

in potassium dichromate 1 large vacuole showing a brown precipitate, or more smaller vacuoles remaining colourless. G u a r d-c e l l s. H. 10 µ, Lev. 15 by 40 µ. Cell contents: tannin.

Under side. Stomata 80 to the sq. m.M. See for the rest the upper side. E p i d e r m a l c e l l s p r o p e r. H. 20 µ, Lev. 20—30 by 40 µ; polygonal tables, having somewhat sinuous lateral walls. See for the rest the upper side. G u a r d-c e l l s. H. 10 µ, Lev. 12 by 35 µ. Cell contents: tannin.

Mesophyll.

P a l i s a d e c h l o r e n c h y m a. From 2—3 layers of cells, having intercellular spaces.

C e l l s   o f   t h e   a b o v e. H. 25 μ, Lev. 15 by 20 μ; circular or elliptical cylinders. Cell contents: many tabular polygonal chloroplasts paving the walls, 3 by 4 μ in diameter, thick 2 μ, each containing some needle-shaped starch grains.

S p o n g y   c h l o r e n c h y m a. Some 8 layers of stellate cells, having 3—6 rays. Intercellular spaces very large, especially in the neighbourhood of the epidermis.

C e l l s   o f   t h e   a b o v e. H. 20 μ, Lev. 40 by 50 μ. Cell contents: the same as those of the palisade cells.

M a r g i n. In each incision of the crenated margin an emissarium, consisting of an epidermis with some water-pores, an epithema formed by a small-celled parenchyma and the extremity of a xylem bundle with numerous free ends of spiral vessels.

V e i n s.

Epidermis. Upper side. Epidermal cells arranged in more or less distinct longitudinal rows. Stomata lying more or less in transverse bands and always longitudinally directed; 12 to the sq. m.M.; phaneroporous; guard-cells lying in the same level with the epidermis; surrounded by 4—8 subsidiary cells.

E p i d e r m a l   c e l l s   p r o p e r. R. 20 μ, T. 22 μ, L. 50—100 μ; tetra- to hexa-, generally tetragonal tables; the ends of the cells often narrower than the middle part. Outer walls showing a longitudinal parallel cuticular striation. Cell contents: some chloroplasts; tannin in the larger vacuoles, but wanting in some smaller adventitious ones. G u a r d- c e l l s. R. 10 μ, T. 15 μ, L. 40 μ.

Under side. In almost all respects the same as that of upper side.

E p i d e r m a l   c e l l s   p r o p e r. R. 18 μ, T. 25 μ, L. 75—125 μ. G u a r d-c e l l s. R. 10 μ, T. 15 μ, L. 50 μ.

Mesophyll.

A e r e n c h y m a. Strongly developed in the midrib, especially on the under side. On both sides below the epidermis, but not below the stomata, 1 layer of cells without intercellular spaces. Large and very long intercellular passages of aerenchyma separated from each other by diaphragms, consisting of 1 layer of cells; these cells having between them smaller intercellular spaces, uniting the adjoining chambers.

C e l l s   o f   t h e   a b o v e. R. and T. 25—40 μ, L. 50—120 μ; tetra- to hexagonal prisms with a longitudinally directed axis and having the walls adjoining the intercellular passages curved into these. The cells surrounding the air chambers of stomata smaller and more or less stellate. Outer walls of the cells belonging to the layer below the epidermis somewhat thickened. Walls adjoining intercellular passages showing in iodine and sulphuric acid 66 per cent. a thin outermost layer remaining colourless and coming off. Cell contents: chloroplasts, each containing some needle-shaped starch grains; in the neighbourhood of the air chambers of stomata chloroplasts more numerous; in the outermost layer below the epidermis some tannin.

E n d o d e r m i s. In the midrib each meristele, up to 6 in number, having a separate starch-sheath generally consisting of 1 layer of cells. Sometimes these cells divided by 1 or 2 partition walls extending in the direction of the circumference of the meristele. Often the mesophyll cells adjoining the starch-sheath obviously being sister cells of those of the starch-sheath, and having been separated from them by cell division. In several leaves the starch-sheath at the same time developed as a protective sheath.

C e l l s   o f   t h e   a b o v e. R. 20 μ, T. 25 μ, L. 50—60 μ; polygonal prisms with a longitudinally directed axis and rounded edges. The casparian dots lignified. Cell contents: chloroplasts containing large globular starch grains, all heaped together in one side of the cells.

Meristeles. In the base of the midrib 4—6 in number, thin cylindrical, running parallel and anastomosing by means of small cross bundles. The lateral ones successively bending outwards into the lateral veins, thus leaving only one meristele extending towards the top of the leaf and there ending in an emissarium.

P e r i c y c l e. Developed only on the upper and under side of the meristele in the form of 2 flat bundles. The upper one consisting of common parenchyma and thin-walled prosenchyma; the under one only of prosenchyma having very slightly thickened walls. Not indicated in the figure.

P a r e n c h y m a   c e l l s. R. 7 μ, T. 8 μ. Cell contents: chloroplasts. F i b r e s. R. and T. 10—20 μ, L. 250—900 μ; polygonal.

V a s c u l a r   b u n d l e. In each meristele a single compound one, containing in the largest meristeles 5—8 simple bundles. Simple bundles collateral.

P h l o e m. Bundles consisting of sieve-tubes and parenchyma cells.

E l e m e n t s   o f   t h e   a b o v e. R. and T. 4—8 μ; polygonal prisms. Contents: often tannin.

X y l e m. Bundles consisting of 1 element in a tangential, 6 in a radial direction; spiral vessels and parenchyma cells. The youngest elements not yet full grown and only these containing tannin.

V e s s e l s. R. 18 μ, T. 15 μ, L. of articulations of the youngest vessels 70—100 μ. Walls lignified. P a r e n c h y m a   c e l l s. R. 7 μ, T. 8 μ; tetragonal prisms.

M e d u l l a r y   c o m m i s s u r e s. Uniseriate; consisting of parenchyma cells.

May 1902, November 1911.   J; M.

FOLIA UVAE URSI.
Bearberry Leaves.
The dried leaves of Arctostaphylos Uva-ursi, Spreng. Syst. II. 287.

Macroscopic characters.

Simple, short stalked. Petiole long 2—5 m.M., in young leaves his margins finely pubescent. Blades stiff; coriaceous; up to 2.5 c.M. in length; oval-, oblong- or lanceolate-obovate; feather-veined; veinlets finely reticulate, on the upper surface deeply impressed, on the under surface brown; apex obtuse, sometimes somewhat reflexed and hence seeming emarginate, sometimes acute; base acute; margin entire, somewhat reflexed. Surface glabrous, but in a young state the margin sometimes also the midrib finely pubescent; colour green, not brown or red. Without odour; taste astringent.

Anatomical characters.

LITERATURE. Deutsch. Arzneib. 5. Ausg. 1910. 239. Ellrodt. Üb. d. Verteil. d. Gerbst. in off. Blättern, Kräutern u. Blüten. Diss. Würzburg. 1903. 17. Gilg. Pharmakogn. 1910. 263. Glaser. Mikr. Anal. d. Blattpulv. v. Arzneipfl. Verh. d. Phys.-Med. Ges. z. Würzburg. N. F. Bd. 34. 1901. 281. Greenish & Collin. Anat. Atl. o. veget. powders. 1904. 52. Hérail. Mat. Méd. 1912. 560. Karsten u. Oltmanns. Pharmakogn. 1909. 154. Koch u. Gilg. Pharmakogn. Praktik. 1907. 141. Marmé. Pharmacogn. 1886. 184. Meyer. Wiss. Drogenk. T. 2. 1892. 220. Oudemans. Aant. o. d. Pharmac. Neerl. 1854—56. 346. Oudemans. Pharmacogn. 1880. 284. Planchon et Collin. Drogues simples. T. 1. 1895. 780. Schneider. Powd. Veget. Drugs. 1902. 312. Tunmann. Ueber Folia Uvae Ursi und den mikrochemischen Nachweis des Arbutin. Pharm. Centralhalle. 47. 1906. 945—947. Vogl. Anat. Atl. 1887. Taf. 6 & 7. MATERIAL. The drug; leaves full grown and young, gathered April 1901 from the Botanic Garden at Groningen, fresh and in alcohol. REAGENTS. Water, glycerine, chloral hydrate, potassium iodide iodine, phloroglucin and hydrochloric acid, iodine and sulphuric acid 66 per cent., concentrated sulphuric acid, iron acetate, Schulze's macerating mixture.

MICROGRAPHY.

Intervenia.

Epidermis. Upper side. Stomata wanting. File hairs only at the margin, usually some cells distant from each other, directed to the apex of the leaf, from 50—320 μ in length, 15 μ thick, articulate, uniseriate, 2-celled, in young leaves sometimes branched, in the drug the top cell generally wanting.

Epidermal cells proper. H. 30 μ, Lev. 35 μ; penta- or hexagonal tables. Outer walls very thick, entirely coloured brown by iodine and sulphuric acid 66 per cent. Cell contents: chloroplasts; tannin; sometimes in groups of cells a light yellow transparent granular mass. Cells of the file hairs. Outer walls thickened; cell cavities small without contents.

Under side. Stomata 120 to the sq. m.M., in a level with the epidermal cells.

Epidermal cells proper. H. 30 μ, Lev. 25 by 30 μ. See for the rest those of the upper side. Guard-cells. Lev. 35 by 10 μ.

Mesophyll.

Palisade chlorenchyma. From 3 to 4 layers of cells; with intercellular spaces, becoming larger in the inner layers.

Cells of the above. H. 45 μ, Lev. 15 by 15 μ; hexagonal prisms with rounded edges; cell walls pitted. Cell contents: chloroplasts containing starch; tannin; in some cells transparent light yellow globular bodies resembling resin or oil, sometimes accompanied by a dark brown granular mass; other cells quite filled with bodies resembling cluster crystals or sphaero-crystals with a darker centre.

Spongy chlorenchyma. Six layers of cells, with large intercellular spaces.

Cells of the above. H. from 20 to 35 μ, Lev. 20 by 20 μ. Globes, ellipsoids or cylinders, the layer of cells adjoining the under epidermis consisting of tables. See for the rest the palisade chlorenchyma.

Margin. Under the epidermis 1—2 layers of collenchyma cells.

Collenchyma cells. Polyhedra, diameter from 15—30 μ, walls pitted.

Veins.

Epidermis. Upper side. Cells in longitudinal rows.

E p i d e r m a l   c e l l s   p r o p e r. R. 30 μ, T. 20—30 μ, L. 20—40 μ. Outer walls strongly thickened, the other walls also thickened and pitted; see for the rest those of epidermis of mesophyll. Cell contents: a light yellow, transparent, granular mass.

Under side. File hairs in small numbers. See for the rest the upper side.

E p i d e r m a l   c e l l s   p r o p e r. R. 30 μ, T. 20—25 μ, L. 25—35 μ.

Mesophyll.

C o l l e n c h y m a. In the midrib 8 or 9 layers of cells on the upper, 10 on the under side; longitudinal rows somewhat radially arranged; intercellular spaces only between the more thin-walled cells of the 5 innermost layers of the under side. Some cells, especially on the under side, divided by

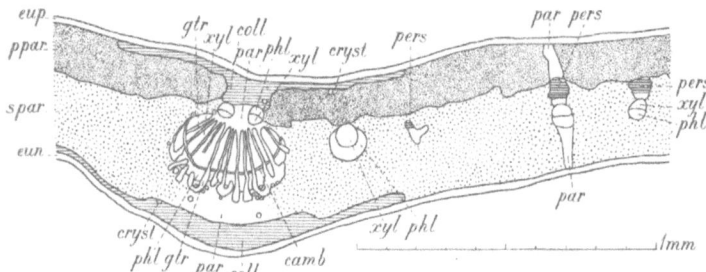

Fig. 37. *Arctostaphylos Uva-ursi.* Leaf, transverse section through the midrib and the neighbouring portions of the lamina. camb Cambium; coll Collenchyma; cryst Crystal cells; eun Epidermis under side; eup Epidermis upper side; gtr Medullary commissures; par Colourless parenchyma; pers Pericyclic sclerenchyma fibres; phl Phloem; ppar Palisade chlorenchyma; spar Spongy chlorenchyma; xyl Xylem.

2—3 thin cross walls, each division containing a simple crystal or a sphaerocrystal; sometimes a certain number of these cells in a longitudinal row.

C o l l e n c h y m a   c e l l s. R. 15 μ, T. 20 μ, L. 50—120 μ; from tetra- to hexagonal prisms with a longitudinally directed axis. Walls strongly thickened and pitted. Cell contents: some chloroplasts with starch grains; in many cells tannin.

C o l o u r l e s s   p a r e n c h y m a. On the under side only.

P a r e n c h y m a   c e l l s. R. 18 μ, T. 20 μ; polygonal prisms; walls sometimes a little thickened.

E n d o d e r m i s. Not to be distinguished.

Meristele.

P e r i c y c l e. Bundles of sclerenchyma, consisting in a cross section of 1—10 fibres, to be found only in the smaller meristeles and on their upper side; generally the thickest bundles in the smallest meristeles. Some colourless parenchyma between these bundles and the vascular bundle.

S c l e r e n c h y m a   f i b r e s. R. and T. 10—12 μ, L. 200—800 μ. From tetra- to hexagonal, sometimes at the ends bent rectangularly or branched. Walls very thick; showing stratification; lignified; with slit-like pit canals.

V a s c u l a r   b u n d l e. In the midrib a single compound one, containing about 14 simple bundles; moreover in the base of the midrib towards the upper side of the leaf 2 smaller bundles. Simple bundles in a cross section radially elongated, collateral, open.

P h l o e m. Sometimes the elements in radial rows; sieve-tubes to be distinguished; elements generally thinner than those of the medullary commissures.

C a m b i u m. To be distinguished now and then in the larger bundles.

X y l e m. Consisting of 10—15 elements in a radial, of 2—4 elements in a tangential direction; radial arrangement. Spiral vessels and reticulate vessels only in the part turned towards the upper side of the leaf; pitted vessels; tracheid fibres.

S p i r a l a n d r e t i c u l a t e v e s s e l s. R. and T. 10—12 μ; articulations L. 350—600 μ. Walls thickened, showing stratification, lignified. P i t t e d v e s s e l s. R. and T. 10—12 μ, the outer ones somewhat smaller; articulations L. 120—300 μ. Transverse walls oblique and with a single perforation. Walls thickened, showing stratification, lignified. T r a c h e i d f i b r e s resembling the vessels with bordered pits.

M e d u l l a r y c o m m i s s u r e s. Uniseriate, consisting of parenchyma.

P a r e n c h y m a c e l l s. R. 10 μ, T. 5 μ, L. 35—40 μ; tetragonal prisms. Walls a little thickened, lignified and pitted. Cell contents: some chloroplasts containing starch grains; tannin.

April 1901, September 1911.   J; M.

FRUCTUS ANETHI.
Dill Fruit.
The dried ripe fruit of Peucedanum graveolens, Benth. et Hooker f. Gen. I. 919.

LITERATURE. Berg. Anat. Atl. 1865. Pl. 43. Bortsch. Beitr. z. Anat. u. Entw. d. Umbellif. Fr. Diss. Breslau. 1882. Colignon. Canaux sécrét. d. l. Ombellif. 1874. Flückiger & Hanbury. Pharmacogr.1879. 327. Kayser. Beitr. z. Kenntn. d. Entw. gesch. d. Samen. Pringsheim's Jahrb. 1893. 79. Lange. Entw. d. Oelbeh. i. d. Fr. d. Umbellif. Schr. d. phys. oek. Ges. Königsberg. Jahrg. 25. 1884. 27. Meyer. Entst. d. Scheidew. i. d. secretf. plasmafr. Intercellularraume d. Vittae d. Umbellif. Bot. Zeit. 1889. 341. Oudemans. Pharmacogn. 1881. 388. Planchon et Collin. Drogues simples. T. II. 1896. 262. Pfeiffer. Meth. d. mikrosk. Anat. ruh. Umbellif. Fr. Mikrokosmos. 1918/19. 8. Rompel. Kryst. v. Caoxal. i. d. Fruchtw. d. Umbellif. u. ihre Verw. f. d. Syst. Ber. Wien. Akad. Bd. 104. Abt. 1. 1895. 417. Styger. Beitr. z. Anat. d. Umbellif. Fr.Schweiz. Apoth. Zeit. LVII. 1919. 3. Tanfani. Morfol. e. Istolog. d. fr. e. d. seme d. Apiacee. N. G. B. J. Vol. XXIII. 1891. 451; report Bot. Jahrb. 1891. 598. Tschirch. Harze u. Harzbeh. 1900. 366, 367. Tunmann. Resinog. Schicht d. Secretbeh. d. Umbellif. Ber. d. pharm. Ges. Bd. XVII. 1907. 456. Wilke. Anat. Bez. d. Gerbst. z. d. Sekretbeh. d. Pfl. Diss. Halle. 1883. v. Wisselingh. Vittae d. Umbellif. Verh. d. Kon. Acad. v. Wetensch. Amsterdam. 1894. (2de sect.) T. IV. No. 1. v. Wisselingh. Bijdr. II, Zaadhuid. Umbellif. Pharm. Weekbl. 1918. 1530. MATERIAL. The drug. REAGENTS. Glycerine, potash 50 per cent., chloral hydrate, phloroglucin and hydrochloric acid, Schulze's macerating mixture.

MICROGRAPHY.
P e r i c a r p.

Wall. Thick about 40 μ in the vallecular parts of a soaked seed.

E p i d e r m i s o u t e r s i d e. Showing a re-entering part above the commissure, the cells of this part practically in longitudinal rows. In the vallecular parts cells polygonal tables; on the ribs longitudinally arranged rectangular tables. Stomata rare.

E p i d e r m a l c e l l s p r o p e r. R. 8 μ, T. 18 μ, L. 25 μ. Outer walls thickened, especially on the marginal ribs; showing a cuticle with longitudinal striation. Cell contents: here and there a crystal. C e l l s o f t h e r e-e n t e r i n g p a r t. R. 14 μ, T. 12 μ, L. 6—10 μ; tetragonal prisms.

Ground tissue.

Valleculae. On the outer side of the vittae some layers of cells, cells of the layer contiguous to the vittae larger than the others. On the lateral sides of the vittae about 7 layers in a radial direction of small common parenchyma cells, more or less flattened especially the innermost ones; cells of the innermost layer longitudinally arranged and somewhat bulging out into the inner epidermis.

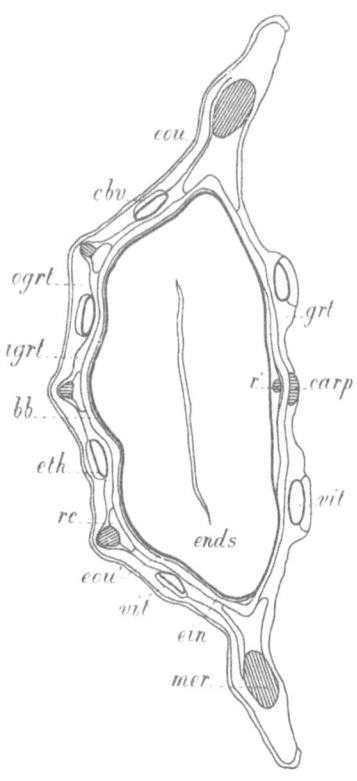

Fig. 38. *Peucedanum graveolens*. Fruit, transverse section. bb Brown band of flattened tissue; carp Carpophore; cbv Cells bordering the vittae; ein Epidermis inner side; ends Endosperm; eou Epidermis outer side; eou' Epidermis outer side of seed-coat; eth Epithelium; grt Ground tissue of commissure; igrt Inner part of the ground tissue; mer Meristeles; ogrt Outer part of the ground tissue; r' Meristele of raphe; rc Reticulate cells; vit Vittae.

Cells of the layer bordering the outer side of the vittae. Thick 20 μ, wide 22 μ, L. 30—50 μ; hexagonal tables. Walls somewhat collenchymatous; brown. On the lateral sides of the vittae. Thick 12 μ, wide 35—60 μ, L. 18 μ; tetra- to hexagonal tables. Walls in the innermost cell layers yellow to brown. Cell contents: in the outermost layers some chloroplasts.

V i t t a e. 1 in each vallecula; running from the base of the mericarp to the top of the stylopode; in the shape of an elliptical or semi-elliptical cylinder, in the latter case the flat side turned outwards, pointed at both ends, at the top somewhat more so; both ends closed; R. 35 μ, T. 180 μ, inclusive of the epithelium; schizogenous [1]); showing numerous septa; in the middle part those septa at a distance of 200—800 μ from each other, towards the ends 200—400 μ. Septa 4 μ thick in their middle part; towards the circumference the thickness increasing and becoming up to 70 μ; the upper and lower surface covered by a thin reddish brown lamella continuous with the corresponding lamella of the epithelium cells; in macerated material this lamella often folded. In the thick part of the septa generally cavities, e. g. 50 μ in diameter; these cavities surrounded by the same reddish brown lamella. In the top of a vitta once noticed a uniformly very thick septum, thick 40 μ, with numerous large cavities. Epithelium of 1 layer of flattened cells; walls on the side of the cavity with a rather indistinct reddish brown lamella like that mentioned for the septa and continuous with it.

Cavity containing a light yellow to colourless oil.

E p i t h e l i u m c e l l s. Thick 4 μ, wide 20 μ, L. 20 μ; polygonal tables. Walls varying from yellow to brown; in macerated material the walls turned towards the

---

[1]) T a n f a n i. l. c. 598.

ground tissue very granular; for the walls bordering the cavity of the vittae see vittae.
Ribs. At the inner side of the meristele and somewhat extending tangentially on both sides reticulate cells, like those at the inner side of the meristele in Fructus Foeniculi. These cells very distinct in the marginal ribs. In the marginal ribs between the band of sclerenchyma fibres of the meristele and the outer epidermis radially directed prismatic cells with thickened and pitted walls.
M e r i s t e l e s. Structure like that of the meristeles of Fructus Foeniculi. Sclerenchyma fibres of the band showing lignified walls.
P h l o e m hardly discernible.
X y l e m containing spiral vessels, especially at the lateral parts of the inner side of the meristeles; vessels showing lignification.
E p i d e r m i s i n n e r s i d e. Shape of the mother cells fairly well to be distinguished, those cells giving rise to longitudinal rows of component cells by 8—11 transverse partition walls; sometimes walls and partition walls sinuous. Component cells varying in size especially with regard to the tangential dimension, in case of a large tangential dimension the shape of the mother cells becoming indistinct.
C o m p o n e n t c e l l s. R. 14 μ, T. 35 μ, L. 40—60 μ, or larger; tetragonal tables. Walls yellowish, in potash 50 per cent. more intensely yellow.

Commissure.

E p i d e r m i s of both faces. Essentially the same as the inner epidermis of the wall of the pericarp.
G r o u n d t i s s u e. Only the layers bordering the epidermis present on the mericarps. V i t t a e. In each mericarp 1 vitta on either side of the carpophore; somewhat larger than those of the wall of the pericarp; for the rest see there.
C a r p o p h o r e. Composed of 2 strongly tangentially elongated bundles, adjoining the epidermis; consisting of fibres with strongly thickened lignified and pitted walls.
S e e d 1).

Seed-coat.

Entirely produced by the integument.
E p i d e r m i s o u t e r s i d e. Cells generally only distinct in the middle part of the dorsal side; for the rest flattened.
C e l l s o f t h e a b o v e. R. 8 μ, T. 25 μ, L. 15 μ; rectangular. Walls yellow to brown.
G r o u n d t i s s u e. Represented by a thin brown band of entirely flattened parenchyma, along the commissure the outermost part less flattened. In the raphe part small prismatic elements.
M e r i s t e l e of raphe. Containing spiral vessels with lignification of the wall.
E p i d e r m i s i n n e r s i d e. Quite flattened.
Nucleus of the seed. See Fructus Foeniculi. Distribution of the aleurone grains as in Fructus Coriandri.

March 1903.   J; M; L.

---

1) As far as I know the development of the ovule has not been investigated. See the ovule of *Fructus Foeniculi*, *Anisi* and *Coriandri*.

FRUCTUS ANISI.
Anisum. Anise Fruit. Anise.
The ripe fruits of Pimpinella Anisum, Linn. Sp. Pl. 264.

Macroscopic characters.

Diakenes with a pedicel, the mericarps mostly remaining connected. Pedicel long about 3 m.M. Fruit long up to 5 m.M., thick up to 3 m.M.; ovoid-conical, somewhat flattened in a direction at right angles to the commissure; crowned by the knob-shaped bifurcous stylopode and 2 short styles; carpophore divided; each mericarp showing 5 filiform, little projecting primary ribs — 2 marginal ones and 3 dorsal ones, the latter nearer to each other — and 4 shallow flat valleculae. Vittae on the convex surface not visible on the outside; on the flat commissural plane of each mericarp mostly 2 wide brown vittae conspicuous. Surface greenish greyish brown, the ribs somewhat coloured paler; the whole fruit showing short white hairs directed upwards. In a transverse section up to 50 vittae, regularly distributed in 1 ring in the wall of the mericarp of each akene, but in the commissure larger and less numerous. Endosperm orthosperm. Odour strongly aromatic; taste sweetish and aromatic.

Anatomical characters.

LITERATURE. Attema. De zaadhuid d. Angiosp. en Gymnosp. Diss. Groningen. 1901. 113. Berg. Anat. Atl. 1865. Pl. 42. Benecke. Mikrosk. Drogenprakt. 1912. 84. Biechele. Mikr. Prüf. d. off. Drogen. 1904. 42. Bortsch. Beitr. z. Anat. u. Entw. d. Umbellif. Fr. Diss. Breslau. 1882. Colignon. Canaux sécrét. d. l. Ombellif. 1874. Flückiger. Pharmakogn. 1891. 945. Flückiger & Hanbury. Pharmacogr. 1879. 310. Gilg. Pharmakogn. 1910. 247. Greenish & Collin. Anat. Atl. of veget. Powders. 1904. 142. Hager. Pharm. Praxis. Bd. I. 1900. 313. Hérail. Mat. Med. 1912. 321. Karsten u. Oltmanns. Pharmakogn. 1909. 285. Kayser. Beitr. z. Kenntn. d. Entw. gesch. d. Samen. Pringsheim's Jahrb. 1893. 79. Koch. Mikr. Anal. d. Drogenpulv. Bd. IV. 1908. 75, Pl. VII. Koch. Einf. i. d. mikr. Anal. d. Drogenpulver. 1906. 159. Koch u. Gilg. Pharmakogn. Prakt. 1907. 246. Kraemer. Botany a. Pharmacogn. 1910. 561. Lange. Entw. d. Oelbeh. in d. Fr. d. Umbellif. Schr. d. phys. oek. Ges. Königsberg. Jhrg. 25. 1884. 27. Marmé. Pharmakogn. 1886. 316. Meyer. Wiss. Drogenk. Bd. II. 1891. 443. Meyer. Mikr. Unters. v. Pflanzenpulv. 1901. 32. Meyer. Entst. d.Scheidew. i. d. sekretführ. plasmafr. Interc.-raume d. Vittae d. Umbellif. Bot. Ztg. 1889. 341. Mitlacher. Toxik. od. Forens. wicht. Pfl. u. Drogen. 1904. 126. Moeller. Pharmakogn. 1889. 126, 129. Moeller. Mikr. pharm. Üb. 1901. 181. Oudemans. Aant. o. d. Pharmacop. neerl. 1854—56. 352. Oudemans. Pharmacogn. 1881. 381. Planchon et Collin. Drogues simples. T. II. 1896. 249, 263. Pfeiffer. Meth. d. mikr. Anat. ruh. Umbellif. Fr. Microkosmos. 1918/19. 8. Rompel. Kryst. v. Caoxal. i. d. Fruchtw. d. Umbellif. u. ihre Verw. f. d. Syst. Ber. Wien. Akad. Bd. 104. Abt. 1. 1895. 417. Schneider. Powdered veget. Drugs. 1902. 118. Styger. Beitr. z. Anat. d. Umbellif.-fr. Schweiz. Apoth. Zeit. LVII. 1919. 3. Tanfani. Morfol. ed. Istolog. d. frutto e. d. seme d. Apiacee. N. G. B. J. Vol. XXIII. 1891. 451; report Bot. Jahresb. 1891. 598. Tschirch. Angew. Pfl.-Anat. 1889. 260, 495. Tschirch. Pharmakogn. Bd. 2. 1911. 1184. Tschirch. Harze u. Harzbeh. 1900. 366, 367. Tschirch u. Oesterle. Anat. Atl. 1900. Pl. XIV. Tunmann. Resinog. Schicht d. Secretbeh. d. Umbellif. Ber. d. pharm. Ges. Bd. 17. 1907. 456. Vogl. Veget. Nahr. u. Genussm. 1899. 411. Wigand. Pharmakogn. 1879. 274. Wilke. Anat. Bezieh. d. Gerbst. z. d. Secretbeh. d. Pfl. Diss. Halle. 1883. 9. v. Wisselingh. Vittae d. Umbellif. Verh. d. Kon. Ak. v. Wetensch. Amsterdam. 1894. (2de sect.) T. IV. No. 1. v. Wisselingh. Bijdr. II, Zaadhuid. Umbellif. Pharmac. Weekbl. 1918. 1530. MATERIAL. The drug. Fruits in alcohol. REAGENTS. Water, glycerine, potash 50 per cent., chloral hydrate, iodine in chloral hydrate, phloroglucin and hydrochloric acid, iodine and sulphuric acid 66 per cent., concentrated sulphuric acid.

MICROGRAPHY.

P e r i c a r p. Thick about 40 μ in the vallecular parts of an imbibed seed.
Wall.

E p i d e r m i s   o u t e r   s i d e. Above the commissure mostly bending rather far inward; sometimes both mericarps having a rib in common corresponding to the commissure. Stomata not numerous; lying in the same level as the epidermis cells; surrounded by 4—6 subsidiary cells. File hairs numerous; sometimes placed on a protuberance formed by some epidermal cells; only in a very few cases separated by a transverse partition wall from the cells bearing them; sclerotic; mostly more or less conical, widening at the base and extending over the epidermal cells; generally bending towards the top of the fruit.

E p i d e r m a l   c e l l s   p r o p e r. R. 5 μ, T. 18 μ, L. 20 μ; polygonal tables. Outer walls showing cuticular striation. S t o m a t a. 10 by 20 μ. H a i r s. Diameter 10—25 μ, length 25—150 μ. Outer walls strongly thickened; showing a cuticle with oblong longitudinally arranged walls.

G r o u n d   t i s s u e. Consisting of 6—8 layers of more or less flattened common parenchyma cells, practically arranged in longitudinal rows. Cells at the inner side of the vittae often strongly tangentially elongated.

C e l l s   o f   t h e   a b o v e. R. 10 μ, T. 15 μ; polygonal prisms with a longitudinally directed axis. Cell contents: some starch grains.

V i t t a e. In each mericarp 40—50, arranged in a single ring; 1 vitta at the inner side of each meristele. Only the large vittae running from the base of the fruit as far as the top; nearly all the shorter ones originating at the base and ending at a shorter or larger distance from the top thin and finely tapering. Sometimes bifurcating, a few times 1 of the branches anastomosing with an other vitta. Cylindrical, often showing short strictures and widenings; pointed at both ends, the top more pointed than the base; closed at both ends; R. 30—60 μ, T. 50—100 μ; schizogenous, see Fructus Foeniculi; showing numerous septa, at 50—300 μ in distance from each other. Septa very thin; towards the circumference thickness equally increasing at the upper and the under side, becoming up to 60 μ; thereby in a tangential section sometimes looking very thick. In the thick parts of the septa minute cavities. For the rest of the septa see Fructus Foeniculi. Epithelium of 1 layer of cells in a transverse section. Cell walls on the side of the cavity thickened; brown; covered with a very thin brown lamella; in iodine and sulphuric acid 66 per cent. the part immediately bordering the cavity colouring brown, the rest weakly blue. [1]) See moreover for this lamella

---

[1]) v. W i s s e l i n g h l. c. 10, comes essentially to the same result; according to him this cell wall consists of cellulose, except the outermost lamella bordering the cavity, the layer next to this consisting of cellulose and vittine.

Fructus Foeniculi. Contents of the cavity: a pale yellow to colourless oil; in old fruits sometimes a resinous mass.

E p i t h e l i u m   c e l l s. Thick 7 μ, wide 14 μ, L. 30 μ; polygonal tables with more or less sinuous longitudinal walls. Walls colourless; containing cellulose. For the walls bordering the cavity of the vittae, see the vittae.

M e r i s t e l e s. In each rib 1 meristele, often showing a structure like that of the meristeles of Fructus Foeniculi; often again the phloem apparently consisting of 1 crescent-shaped bundle half encircling the xylem. The sclerenchyma fibres described for Fructus Foeniculi here only represented by some longitudinal rows of prismatic sclereids. S c l e r e i d s. Diameter 8 μ, L. 75—100 μ. Walls thickened; pitted.

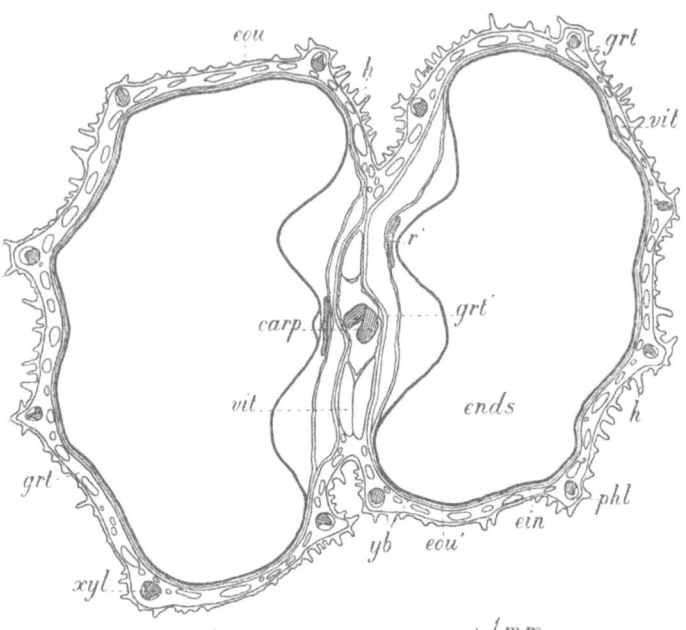

E p i d e r m i s   i n n e r   s i d e. See for this the epidermis inner side of Fructus Foeniculi. Mother cells not distinct; component cells larger.

Fig. 39. *Pimpinella anisum*. Fruit, transverse section. carp Carpophore; ein Epidermis inner side; ends Endosperm; eou Epidermis outer side; eou' Epidermis outer side seed-coat; grt Ground tissue; grt' Parenchyma of the ground tissue resembling the epidermis; h Hairs; phl Phloem; r' Meristele of raphe; vit Vittae; xyl Xylem; yb Yellow band of flattened ground tissue.

C e l l s   o f   t h e   a b o v e. Component cells R. 10 μ, T. 80 μ, L. 10 μ; transverse walls somewhat sinuous. Outer walls showing a cuticle; all the walls colouring yellow in iodine and sulphuric acid 66 per cent. and persisting in concentrated sulphuric acid.

Commissure.

E p i d e r m i s of both faces. On the whole similar to the inner epidermis of the wall of the pericarp. In the places corresponding to the carpophore cells arranged in longitudinal rows; component cells showing a more irregular shape.

C e l l s   o f   t h e   p l a c e s   c o r r e s p o n d i n g   t o   t h e   c a r p o p h o r e. T. 15 μ, L. 15—30 μ; tetragonal prisms with a longitudinally directed axis. Walls thickened; lignified; showing slit-like pits.

G r o u n d   t i s s u e. Cells strongly flattened or torn. Between the epidermis of both faces and the carpophore 1—2 layers of parenchyma cells resembling the epidermal cells of the corresponding places. V i t t a e. On both sides of the commissure 2 very large vittae; now and then one of these vittae wanting. In the case of the mericarps having a marginal rib in common besides the large vittae some smaller ones. R. 30 μ, T. 250 μ. For the rest see the vittae of the wall.

C a r p o p h o r e. Consisting of 2 bundles of fibres ending in the stylopode with 2 cudgel-shaped parts; those parts partially consisting of parenchyma cells like those lying between the epidermis and the carpophore.

F i b r e s. Diameter 8 μ; polygonal. Walls thickened; lignified.

S e e d.

Ovule. Closely resembling that of Fructus Foeniculi. In each loculus 2 ovules, 1 of which aborting. Anatropous; pendulous; with 1 integument [1]).

Ripe seed.

S e e d - c o a t. Entirely produced by the integument.

Epidermis outer side. Not so closely joining the pericarp as in Fructus Foeniculi.

C e l l s o f t h e a b o v e. R. 12 μ, T. 35 μ, L. 20 μ; those corresponding in place to the commissure R. 25 μ, T. 16 μ, L. 20 μ; polygonal tables. Walls not colouring blue in iodine and sulphuric acid 66 per cent., persisting in concentrated sulphuric acid; outer walls with a cuticle.

Ground tissue. Represented by a thin yellow band of entirely flattened tissue; here and there the outermost cell layer still discernible. In the raphe part near the meristele the outermost layers not flattened.

C e l l s o f t h e a b o v e. In a cross section 20 by 25 μ to 30 by 45 μ; polygonal. Cell contents: starch grains, especially in the outermost cells.

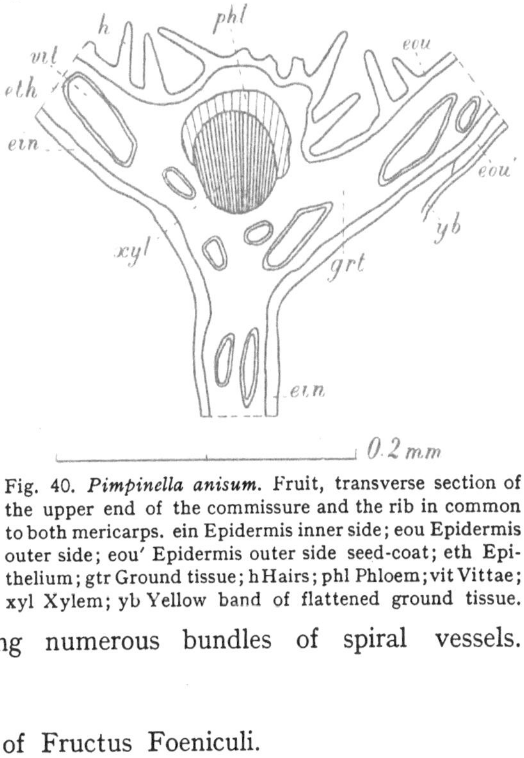

Meristele of raphe. Strongly tangentially elongated; containing numerous bundles of spiral vessels.

Epidermis inner side. Flattened.

N u c l e u s of the seed.

Endosperm. See the endosperm of Fructus Foeniculi.

Embryo. Only present in the uppermost part of the endosperm; long about 1100 μ. R a d i c l e. All the elements in longitudinal rows; showing a central procambial bundle, also with longitudinally arranged elements; calyptra discernible; cell contents like those of the endosperm. C o t y l e d o n s. Not quite covering each other; elements arranged in longitudinal rows; the central pro-

Fig. 40. *Pimpinella anisum*. Fruit, transverse section of the upper end of the commissure and the rib in common to both mericarps. ein Epidermis inner side; eou Epidermis outer side; eou′ Epidermis outer side seed-coat; eth Epithelium; gtr Ground tissue; h Hairs; phl Phloem; vit Vittae; xyl Xylem; yb Yellow band of flattened ground tissue.

---

[1]) After T a n f a n i. l. c. 599, and T s c h i r c h and O e s t e r l e. l. c. 51.

cambial bundle of the radicle dividing into 2 and giving off 1 bundle for each cotyledon; cell contents like those of the endosperm.

P e d i c e l of the fruit. Epidermis. Cells longitudinally more elongated then in the pericarp; only here and there a hair; outer walls showing longitudinal cuticular striation. In the centre a meristele containing strongly flattened phloem and a xylem showing annular and spiral vessels, polygonal fibres and polygonal cells with thickened lignified and pitted walls. Here and there a schizogenous gland.

Micrography of the powder.

File hairs mostly 1-, sometimes 2-celled; conical with an obtuse top; curved; with ovate warts; colourless. Parts of vittae yellowish brown, often paved with thin-walled polygonal cells. Inner epidermis of the pericarp formed of low strongly tangentially elongated cells; often still connected with the vittae. Colourless parenchyma of the endosperm: in many cells several small crystal clusters with a central air-bubble. Outer epidermis of the fruit formed of polygonal cells showing cuticular striation; stomata; many hairs. Epidermis of the pedicel with more oblong cells, stronger cuticular striation and few hairs. Annular and spiral vessels; bundles of thick-walled pitted vessels. Sclereids short, not very thick-walled, showing slit-like pits. A few parenchyma cells quite filled with starch; moreover a very few scattered smal starch grains.

March 1903.   J; M; L.

FRUCTUS CARUI.
Caraway Fruit.
The dried fruit of Carum Carvi, Linn. Sp. Pl. 263.

LITERATURE. Berg. Anat. Atl. 1865. Pl. 42. Benecke. Mikr. Drogenprakt. 1912. 82. Bortsch. Beitr. z. Anat. u. Entw. d. Umbellif. Fr. Diss. Breslau. 1882. Colignon. Canaux sé-crét. d. l. Ombellif. 1874. Flückiger. Pharmakogn. 1891. 941. Flückiger & Hanbury. Pharmacogr. 1879. 305. Gilg. Pharmacogn. 1910. 245. Greenish & Collin. Anat. Atl. of veget. Powders. 1904. 146. Hager. Pharm. Praxis. Bd. I. 1900. 660. Harold Matthews. Vittae of Caraway fruits. Pharmac. Journ. Ser. IV. 1898. No. 1446. Karsten u. Oltmans. Pharmacogn. 1909. 279. Kayser. Beitr. z. Kenntn. d. Entw. gesch. d. Samen. Pringsheim's Jahrb. 1893. 79. Koch. Mikr. Anal. d. Drogenpulv. Bd. IV. 1908. 91, Pl. IX. Kraemer. Botany a. Pharmacogn. 1910. 566, 772. Lange. Entwick. d. Oelbeh. i.d. Früchten d. Umbellif. Schr. d. phys.-oekon. Ges. Königsberg. Jhrg. 25. 1884. 27. Marmé. Pharmacogn. 1886. 313. Meyer. Wiss. Drogenk. Bd. II. 1891. 440. Meyer. Entst. d. Scheidew.i.d. sekr.-führ. plasm.-fr. Intercellularraume d. Vittae d. Umbellif. Bot. Zeit. 47. 1889. 341. Mitlacher. Tox. od. Forens. wicht. Pfl. u. Drogen. 1904. 131. Moeller. Mikr. pharm. Üb. 1901. 178. Oudemans. Pharmacogn. 1881. 380. Planchon et Collin. Drogues simples. T. II. 1896. 249, 260. Pfeiffer. Meth. d. mikr. Anat. ruh. Umbellif. Fr. Mikrokosmos. 1918/19. 8. Rompel. Kryst. v. Caoxal. i. d. Fruchtw. d. Umbellif. u. ihre Verw. f. d. Syst. Ber. Wien. Akad. Bd. 104. Abt. 1. 1895. 417. Schneider. Powdered veg. Drugs. 147. Styger. Beitr. z. Anat. d. Umbellif. Fr. Schweiz.Apoth. Zeit. LVII. 1919. 3. Tanfani. Morfol. ed Istol. d. fr. e. d. seme d.Apiacee. N. G. B. J. Vol. XXIII. 1891. 451; report Bot. Jahresber. 1891. 598. Tschirch. Harze u. Harzbeh. 1900. 366, 367. Tschirch. Pharmakogn. Bd. II. 1911. 1091. Tunmann. Resinog. Schicht d. Sekretbeh. d. Umbellif. Ber. d. pharm. Ges. Bd. XVII. 1907. 456. Vogl. Veget.

Nahr. u. Genussm. 1899. 407. Wigand. Pharmakogn. 1879. 274. Wilke. Anat. Bezieh. d. Gerbst. z. d. Sekretbeh. d. Pfl. Diss. Halle. 1883. v. Wisselingh. Vittae d. Umbellif. Verh. d. Kon. Academie v. Wetensch. Amsterdam. 1894. (2 de sect.) T. IV. No. 7. v. Wisselingh. Bijdr. II, Zaadhuid. Umbellif. Pharm. Weekbl. 1918. 1530. Zijlstra. Ueber Carum Carvi L. Rec. de Trav. Bot. néerl. T. XIII. 1916. 183. MATERIAL. The drug. REAGENTS. Water, glycerine, potash 50 per cent., chloral hydrate, Schulze's macerating mixture, osmic acid.

MICROGRAPHY.

Pericarp.

Wall. Thick 50—60 μ in the vallecular parts of a soaked seed.

Epidermis outer side. Re-entering very far into the commissure. Cells mostly in longitudinal rows. Stomata very rare.

Cells of the above. In the vallecular parts R. 10 μ, T. 30 μ, L. 30—45 μ; on the ribs R. 15 μ, T. 25 μ, L. 45 μ; rectangular tables, often with sinuous lateral walls. Outer walls thickened, especially on the ribs; with longitudinal cuticular striation and showing a peculiar figuration of it above the lateral walls; those walls cuneiform in their outer part, for the rest thin; no pits to be distinguished in a transverse section. Cell contents: here and there a simple crystal, rhombic or somewhat longitudinally elongated; often a mass soluble in chloral hydrate and colouring black in osmic acid.

Ground tissue. Chiefly consisting of flattened parenchyma cells. The tissue on the outside of the vittae and the meristeles flattened; between the vittae and the meristeles the outermost 4 layers flattened; the rest entirely flattened, except the tissue on the inner side of the vittae. Flattened cells brown; less strongly flattened cells sometimes containing small oil drops and some chloroplasts. Reticulate cells, often occurring in other fruits of the Umbelliferae, on the inner side of the meristeles wanting.

Cells on the outer side of the meristeles. R. 15 μ, T. 20 μ, L. 50—80 μ; prisms. Walls a little thickened.

Vittae. 1 in each vallecula. Running from the base of the fruit to the top of the stylopode; often in the shape of triangular prisms, 1 edge turned outwards, with rounded sides and edges and pointed ends, the top somewhat more pointed than the base, closed at both ends; R. 75 μ, T. 250—300 μ; schizogenous [1]); showing numerous septa, in the middle part of the vittae 400—500 μ in distance from each other, towards the ends the distance much smaller. Septa thick 3—4 μ in their middle part, towards the circumference much increasing in thickness; in those thick parts numerous smaller or larger cavities; sometimes a cavity extending from the upper to the under surface of the septum. Sometimes in a tangential section only this part of the septum discernible, the middle part not being distinguishable. Septa covered with a thin reddish brown lamella continuous with the reddish brown lamella of the epithelium cells and surrounding the cavities; this lamella sometimes folded in macerated material; the rest of the septa yellow. Epithelium of 1 layer of cells; those cells flattened. The walls on the side of the cavity strongly thickened and covered with a very thin brown lamella like that mentioned above for the septa; next to this lamella, towards

---

[1]) Tanfani. l. c. 598.

the lumen of the cell, a yellow to brown granular layer; the rest of the walls yellow to colourless. Cavity partly filled with a yellow to brown oil or a resinous mass not becoming black in osmic acid.

E p i t h e l i u m   c e l l s. Thick 4 μ, wide 20 μ, L. 25 μ. Lumen mostly not discernible or fissure-shaped. Walls yellow to colourless; for the walls bordering the cavity see vittae.

S c h i z o g e n o u s   g l a n d s. 1 on the outer edge of each meristele, close to it or separated from it by 1 or more flattened cells; a few times moreover a gland on one of the lateral sides of the meristele; continued into the stylopode and into the pedicel; in a transverse section elliptical; R. 15 μ, T. 20 μ; showing transverse septa resembling the septa of the vittae; sometimes showing strictures, now and then combined with a transverse septum. Septa covered with a reddish brown lamella like that mentioned for the vittae, continuous with the corresponding lamella of the lateral walls; small cavities wanting. Epithelium of 1 layer of cells. Cavity filled with a yellow to colourless oil or a resinous mass not becoming black in osmic acid.

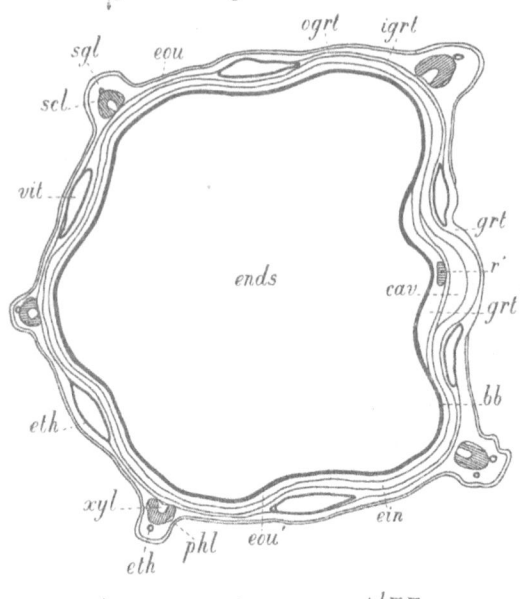

Fig. 41. *Carum Carvi*. Fruit, transverse section. bb Brown band of flattened tissue of seed-coat; cav Cavity; ein Epidermis inner side; ends Endosperm; eou Epidermis outer side; eou′ Epidermis outer side seed-coat; eth Epithelium; grt Ground tissue of commissure; grt′ Unflattened ground tissue of seed-coat; igrt Innermost part of ground tissue; ogrt Outermost 4 layers of ground tissue; phl Phloem; r′ Meristele of raphe; scl Sclerenchyma of meristele; sgl Schizogenous glands; vit Vittae; xyl Xylem.

E p i t h e l i u m   c e l l s. Rectangular in a cross section. Cell contents: a reddish brown mass.

M e r i s t e l e s. Triangular prisms, 1 edge turned outwards. Consisting of 2 closed **vascular bundles**, not clearly to be distinguished from the rest of the tissue and united by a strongly developed band of sclerenchyma fibres and very long prismatic sclereids; both polygonal and showing slit-like pit canals. Towards the top of the fruit the sclereids increasing in number and here along the margins of the meristeles shorter and more abundantly pitted.

P h l o e m. 1 bundle on each lateral side of the meristele, very indistinct. X y l e m. At the base in the middle part spiral vessels; walls thinner than those of the fibres.

E p i d e r m i s   i n n e r   s i d e. Mother cells not clearly discernable. Component cells arranged in longitudinal rows, the tapering ends of these fitting into each other.

C e l l s   o f   t h e   a b o v e. Component cells R. 20 μ, T. 40—100 μ, L. 10—20 μ; tetragonal prisms. Walls yellowish.

Commissure. Showing in its middle part a very slight longitudinal stricture.
E p i d e r m i s of both faces. On the whole similar to the inner epidermis of the wall of the pericarp. On the stricture cells R. 20 μ, T. 20 μ, L. 10—18 μ; rectangular prisms.
G r o u n d   t i s s u e. Consisting of flattened common parenchyma cells. Along the stricture and partly along the vittae some layers of flattened cells with thickened walls. **V i t t a e.** 2 on each side of the carpophore; somewhat smaller than those of the wall; for the rest see those of the wall.
C a r p o p h o r e. Consisting of 2 bundles of fibres and some vessels, at the top of the fruit surrounded by some layers of sclereids.
Stylopode. Consisting for the greater part of longitudinally arranged sclereids, like those of the upper part of the meristeles of the wall.
S e e d. [1]) Showing 5 projecting edges corresponding to the primary ribs of the pericarp.
Seed-coat. Layers produced by the integument.
E p i d e r m i s   o u t e r   s i d e. In the projecting ridges cells somewhat flattened; along the commissure cells more distinct; on the raphe cells smaller in a transverse section.
C e l l s   o f   t h e   a b o v e. R. 18 μ, T. 30 μ, L. 20 μ; on the raphe e. g. R. and T. 7 μ; polygonal tables. Walls yellow. Cell contents: yellow oil drops.
G r o u n d   t i s s u e. Represented by a thin brown band of entirely flattened tissue; along the commissure outermost layers not flattened and consisting of longitudinally arranged prismatic parenchyma cells without intercellular spaces. Moreover in the undermost part of the seed some layers of this unflattened parenchyma. In the raphe cells much smaller than the rest.
M e r i s t e l e of raphe. Apparently consisting of an amphivasal vascular bundle; the spiral vessels mostly surrounding some torn tissue.
E p i d e r m i s   i n n e r   s i d e. Flattened.
Nucleus of the seed. See Fructus Foeniculi. Aleurone grains apparently distributed in the same way as in Fructus Coriandri.

April 1903. J; M; L.

FRUCTUS COLOCYNTHIDIS.

Pulpa Colocynthidis. Colocynth Pulp. Colocynthis. Colocinth.
The peeled fruits of Citrullus Colocynthis, Schrad. in Linnaea, XII. (1838) 414.

Macroscopic characters.

Diameter up to 8 c.M.; about globular, sometimes somewhat shrivelled, light, spongy, dry, easily to be compressed. External surface somewhat angular from being peeled, dull, white or yellowish white, here and there still exhibiting

---

[1]) As far as I know the development of the ovule has not been investigated, conf. the ovule of *Fructus Foeniculi, Anisi* and *Coriandri*.

fragments of the peel; peel thick hardly 1 m.M., dull orange-brown. Gourd fruit composed of 3 carpels, with 3 complete septa, so actually 3-locular; the edges of the carpels however from the centre of the fruit bending outwards so as to form in each loculus 2 thick perpendicular plates extending to the circumference and there bending away from each other, constituting the placentae. In consequence seemingly 6 loculi, in which the seeds in 2 or 3 perpendicular rows horizontally attached to the placentae. All the empty spaces of the loculi, also those between the seeds, filled with a very spongy tissue; the outermost layer of the peeled fruit somewhat less spongy, more fleecy. The plates mentioned above generally strongly diverging in drying, especially in the uppermost part of the fruit, in consequence the fruit showing internally 3 longitudinally running fissures, these joining in the middle, narrowing cuneiformly towards the exterior, in their middle part 0.5—1 c.M. By this the inner part of the fruit easily separating into 3 cuneiform parts, each corresponding to 2 halves of contiguous carpels, placentae also to be made conspicuous by carefully peeling the fruit. Placentae often grown together, but generally only adhering to the wall of the pericarp in the upper part of the fruit. The somewhat concave walls of the fissures and the adjacent outwards bending parts of the placentae covered with a tighter more shining tissue, in which horizontal veins; these veins about parallel, little branched, slightly depressed and pale brownish yellow.

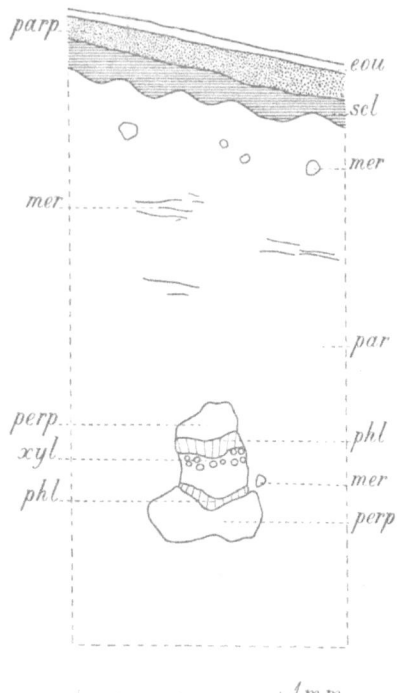

Fig. 42. *Citrullus Colocynthidis*. Fruit, transverse section. eou Epidermis outer side; mer Meristeles; par Common parenchyma forming the pulp; parp Parenchyma containing plastids; perp Pericyclic parenchyma; phl Phloem; scl Sclereids; xyl Xylem.

Seeds up to 300 in number, long 7.5 m.M., wide 5 m.M., thick 1—2 m.M., oval with a somewhat pointed top, flat with rounded edges. External seed-coat bony hard, thick about 0.5 m.M.; hilum a little below the top of the seed, oblong; micropyle on the top; near the edges 2 fissures, directed towards the top, 1.5 m.M. long; surface dull, paler or darker greyish brown. Internal seed-coat fleecy, white, somewhat shining. No albumen. Odour slight; taste intensely bitter, already perceptible by the dust thrown out by the fruits when being prepared.

Anatomical characters.

LITERATURE. Attema. De Zaadhuid d. Angiosp. en Gymnosp. Diss. Groningen. 1901. 101. Berg. Anat. Atl. 1865. 89. Benecke. Mikrosk. Drogenprakt. 1912. 85. Biechele. Mikr. Prüf.

d. off. Drogen. 1904. 43. Flückiger. Pharmakogn. 1891. 885. Flückiger & Hanbury. Pharmacographia. 1879. 296. Gilg. Pharmakogn. 1910. 340. Greenish & Collin. Anat. Atl. of veget. Powders. 1904. 156. Hérail. Pharmacogr. 1912. 535. Karsten u. Oltmanns. Pharmakogn. 1909. 303. Koch. Mikr. Anal. d. Drogenpulver. Bd. IV. 1908. 99, Pl. X. Koch u. Gilg. Pharmakogn. Prakt. 1907. 256. Kraemer. Botany a. Pharmacogn. 1910. 743. Luerssen. Syst. Bot. Bd. II. 1882. 1079. Mc. Alpine. The Fibro-vasc. syst. of the Quince fr. Proc. Linn. Soc. N. S. Wales. XXXVII. 1912. 689. Meyer. Wiss. Drogenk. Bd. II. 1892. 427. Meyer. Mikr. Unters. v. Pflanzenpulv. 1901. 38. Mitlacher. Toxik. od. Forens. wicht. Pfl. u. Drogen. 1904. 183. Moeller. Mikr. pharm. Üb. 1910. 174. Oudemans. Aant. o. d. Pharmac. neerl. 1854—56. 438. Oudemans. Pharmacogn. 1880. 359. Planchon et Collin. Drogues simples. T. II. 1896. 300. Schneider. Powdered veget. Drugs. 1902. 169, 170. Tschirch. Pharmakogn. Bd. II. 1911. 1606. Tunmann. Pflanzenmicrochemie. 1913. 338. MATERIAL. The drug. Dry peels. Nearly ripe fresh fruit, collected in the Botanic Garden at Groningen, in alcohol. Young fruits about 1 c.M. in length, collected in the Botanic Garden at Groningen, in alcohol. REAGENTS. Glycerine, glycerine jelly, boiling water, potassium iodide iodine, iodine and sulphuric acid 66 per cent., phloroglucin and hydrochloric acid, concentrated sulphuric acid, Schulze's macerating mixture.

MICROGRAPHY.

Pericarp.

Epidermis outer side. Stomata 15—20 to the square m.M.; somewhat cryptoporous with a circular vestibulum. Each stoma surrounded by 2 concentric rings of 5—6 much smaller epidermal cells. Trichomes in 3 kinds; present only on the young fruit.

Epidermal cells proper. R. 30 μ, T. 19 μ, L. 18 μ; penta- to hexagonal prisms with a radially directed axis. Outer walls rather strongly thickened; radial and transverse walls also thickened, but becoming thinner towards the inner part; outermost half of the outer walls and the outer part of the middle lamella in the thicker outer parts of the radial and transverse walls cuticularized; radial and transverse walls pitted. Cell contents: yellowish brown granular. Stomata 20 by 10 μ.

Ground tissue. 3 parts to be distinguished, the outer 2 forming together with the epidermis the peel of the fruit. 1. Parenchyma containing plastids, about 8 layers of cells; triangular intercellular spaces running into all directions. 2. Sclereids in 5—10 layers, gradually passing into the underlying parenchyma; triangular intercellular spaces extending into all directions. 3. Large celled common parenchyma constituting the pulp of the fruit; very strongly developed intercellular spaces of various shapes, the cells having only small circular or elliptical surfaces of contact; those surfaces surrounded by conspicuous connecting frames.

Cells of parenchyma containing plastids. R. 20 μ, T. 40 μ, L. 35 μ. Walls very superficially pitted. Sclereids. R. 32 μ, T. 24 μ, L. 30 μ; very irregular polyhedra often with rounded edges. Walls thickened; light yellow; lignified; showing pit canals. Cells of large celled common parenchyma. R. 60 μ, T. 60—75 μ, L. 65 μ, towards the interior the dimensions increasing to 250 μ; globes or ellipsoids. Walls pitted.

Meristeles. Of various size; extending in all directions in the innermost part of the sclereid layer and in the pulp; consisting of a single vascular bundle, surrounded by a pericyclic sheath of parenchyma, on the outer and inner sides several cell layers thick, on the radial sides a few layers; this parenchyma with intercellular spaces, the latter extending in all directions. **V a s c u l a r b u n d l e s** of the larger meristeles bicollateral, of the smaller ones collateral; closed; the phloem and xylem part separated from each other by a parenchyma like that of the sheath. **P h l o e m** consisting of sieve-tubes with companion cells and cambiform cells without intercellular spaces. **X y l e m** consisting of annular and spiral vessels and parenchyma without intercellular spaces.

C e l l s   o f   s h e a t h. R. 25 μ, T. 30 μ, L. 30—50 μ; tetra- to hexagonal prisms. Walls somewhat collenchymatously thickened; pitted. S i e v e - t u b e s. R. and T. 9—12 μ, L. of articulations 130—180 μ; tetra- to hexagonal prisms; transverse walls sometimes oblique, always showing distinct sieve-plates. Small callus masses. C o m p a n i o n c e l l s. Thick about 5 μ; tri- or tetragonal prisms. Cell contents: a granular yellow mass. C a m b i f o r m   c e l l s. R. and T. 12 μ, L. 70—100 μ; tetra- to hexagonal prisms, axis parallel to the longitudinal direction of the meristeles. Walls pitted. X y l e m   v e s s e l s. Diameter 25 μ. Walls showing lignification. X y l e m   p a r e n c h y m a. R. and T. 16 μ, L. 70 μ; tetra- to hexagonal prisms, axis parallel to the longitudinal direction of the meristeles. Walls pitted.

April 1901.   J; M; L.

FRUCTUS CONII.
Conium. Conium Fruit.
The dried, full grown, unripe fruits of Conium maculatum, Linn. Sp. Pl. 243.

LITERATURE. Anema. Zetel d. alcaloiden b. enk. narc. Pl. Diss. Utrecht. 1892. 56. Attema. Zaadhuid d. Angiosp. en Gymnosp. Diss. Groningen. 1901. 112. Barth. Stud. üb. d. Nachw. v. Alkaloiden in pharmac. verw. Drogen. Bot. Centrbl. Bd. 75. 1898. 292. Berg. Anat. Atl. 1865. Pl. 42. Bortsch. Beitr. z. Anat. u. Entw. d. Umbellif. Fr. Diss. Breslau. 1882. Clautriau. Localis. et signific. d. Alcaloides d. quelques graines. 1894. Flückiger & Hanbury. Pharmacogr. 1879. 299. Gilg. Pharmacogn. 1910. 244. Hager. Pharm. Prax. Bd. 1. 1900. 946. Hérail. Mat. Med. 1912. 706. Karsten u. Oltmanns. Pharmacogn. 1909. 190. Kayser. Beitr. z. Kenntn. d. Entw.-gesch. d. Samen. Pringsheim's Jahrb. 1893. 79. Kraemer. Botany a. Pharmacogn. 1910. 568, 719. Marmé. Pharmacogn. 1886. 325. Meyer. Entstehung d. Scheidew. i. d. secretführ. plasmafr. Intercellularraume d. Vittae d. Umbellif. Bot. Ztg. 47. 1889. 341. Mitlacher. Toxik. od forens. wicht. Pfl. u. Drogen. 1904. 122. Molisch. Microchemie d. Pfl. 1913. 255. Planchon et Collin. Drogues simples. T. II. 1896. 249, 272. Pfeiffer. Meth. d. mikr. Anat. ruh. Umbelliferenfr. Mikrokosmos. 1918/19. 8. Rompel. Kryst. v. Caoxal. i. d. Fruchtw. d. Umbellif. u. ihr. Verwert. f. d. Syst. Ber. Wiener Akad. Bd. 104. Abt. 1. 1895. 417. Schneider. Powdered veget. Drugs. 1902. 170. Styger. Beitr. z. Anat. d. Umbellif. Fr. Schweiz. Apoth. Zeit. Bd. 57. 1919. 3. Tanfani. Morfol. ed Istolog. d. frutto e. d. seme d. Apiacee. N. G. B. J. T. XXIII. 1891. 451; report. Bot. Jahresber. 1891. 598. Tschirch. Angew. Pfl. Anat. 1889. 473. Tschirch u. Oesterle. Anat. Atlas. 1900. Pl. 37. Tunmann. Pfl.-mikrochemie 1913. 315. v. Wisselingh. Bijdr. II. Umbelliferae. Pharmac. Weekbl. 1918. 1530. MATERIAL. The drug; flowers and unripe fruits (up to 1.3 m.M. thick and 4 m.M. long) in alcohol, collected in the Botanic Garden at Groningen. REAGENTS. Water, glycerine, potash 50 per cent., chloral hydrate, potassium iodide iodine, iodine in chloral hydrate, phloroglucin and hydrochloric acid, iodine and sulphuric acid 66 per cent., concentrated sulphuric acid, Schulze's macerating mixture, osmic acid, oil of cloves, origanum oil, potassium iodide and cadmium iodide.

MICROGRAPHY.

P e r i c a r p. Showing on both sides a longitudinal furrow corresponding in place to the commissure. Depth of the furrow over $^1/_3$ of the total breadth of the fruit.

Wall. Thick about 70 µ in a fruit soaked in water.

E p i d e r m i s o u t e r s i d e. In the drug mostly flattened in a radial direction. In young fruits the cells becoming much smaller, especially in a radial direction, towards the bottom of the furrow; in the drug cells hardly discernible at the bottom of the furrow. Stomata rare; lying in the same level as the surrounding epidermal cells. Trichomes. In the valeculae and on the lateral sides of the ribs here and there conical papillae, diameter 60 µ, length 80 µ; a few times in the same places conical pluricellular hairs, somewhat larger than the papillae.

E p i d e r m a l c e l l s p r o p e r. R. 20 µ, T. 30 µ, L. 40 µ; polygonal. Outer walls thickened, with longitudinal cuticular striation; inner walls somewhat thickened. Cell contents: especially on the ribs smaller or larger, globular, semi-globular or more irregular sphaero-crystals, often yellow, dissolving in potash with a yellow colour, not soluble in alcohol, chloral hydrate and glycerine, colouring yellow in iodine.

G r o u n d t i s s u e.

Valleculae. Consisting of common parenchyma in about 8 practically tangentially arranged cell layers; especially the outermost layers containing chlorophyll. Cells flattened. Cells of the innermost layer mostly in longitudinal rows; here and there the shape of the very large mother cells giving rise to the longitudinal rows by transverse partition walls still discernible. Cells strongly flattened, the cavity all but disappearing. V i t t a e. In the drug in a cross section only on treatment with hot chloral hydrate or potash here and there indistinctly appearing. In the wall of the ovary distinct; very numerous; in a cross section showing one layer of 4—6 epithelium cells and a small cavity. In older material not much growing out; in the thickest specimen of the alcoholic material epithelium cells on the radial side thick 5 µ, on the tangential side 10 µ, cavity R. 10 µ, T. 20 µ.

C e l l s o f t h e a b o v e, except of the innermost layer. R. 8 µ, T. 17 µ, L. 25 µ; mostly tetragonal prisms with a longitudinally directed axis. Cell contents: chloroplasts, especially in the outermost part; simple and 2-to 3-adelphous starch grains; in the young fruit all the cells filled with starch grains; in the outermost layers here and there a sphaero-crystal like those of the epidermis. In the innermost layer component cells R. 15 µ, T. 60 µ, L. 20 µ. Inner tangential walls thickened, in the oldest alcoholic material still unthickened, radial walls cuneiformly thickened; thickened walls yellow; showing an innermost lignified lamella like that of the cells of the inner epidermis, but thinner and often very indistinct in several reagents; for the rest see walls of the inner epidermis; inner walls showing stratification and small pits. Cell contents: in the oldest alcoholic material cells filled with starch grains like those of the other layers of the ground tissue, only the grains somewhat smaller. E p i t h e l i u m c e l l s. Walls bordering the cavity somewhat

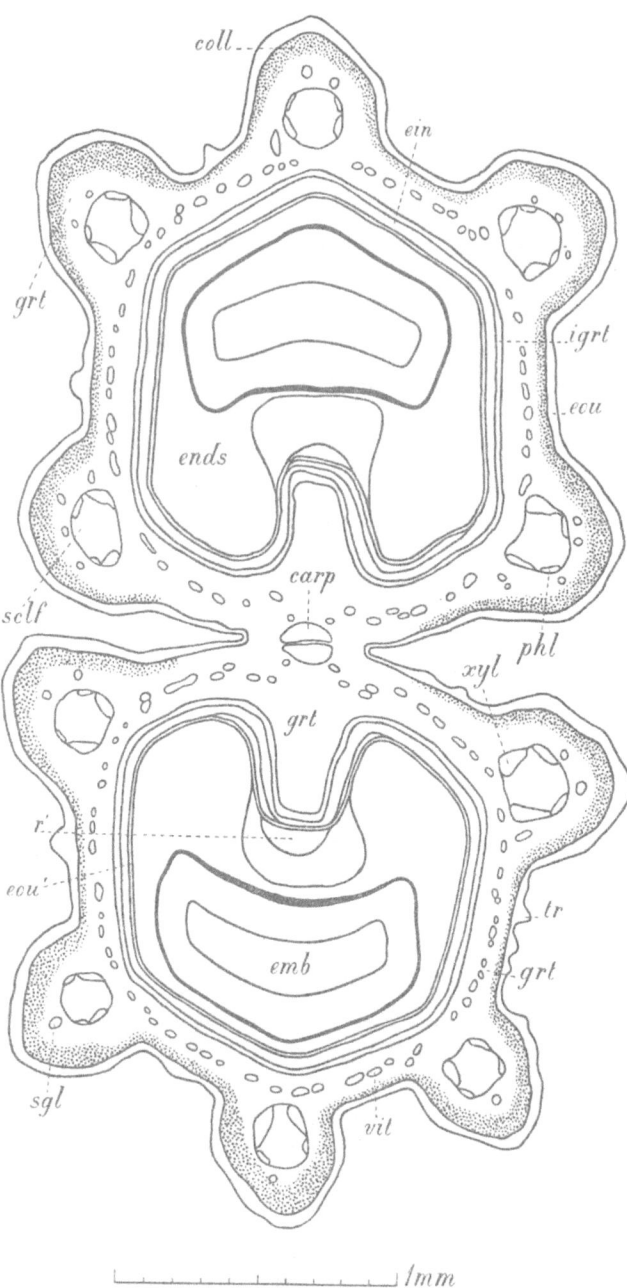

Fig. 43. *Conium maculatum*. Fruit, transverse section of the alcoholic material. carp Carpophore; coll Collenchyma; ein Epidermis inner side; emb Embryo; ends Endosperm; eou Epidermis outer side; eou′ Epidermis outer side seed-coat; grt Ground tissue; igrt Inner layer of the ground tissue; phl Phloem; r′ Meristele of raphe with surrounding ground tissue; sclf Sclerenchyma fibres; sgl Schizogenous glands; tr Trichomes; vit Vittae; xyl Xylem.

thickened in the young fruit. Cell contents: a greenish yellow mass without starch grains.

Ribs. In several ribs 1 or 2 outermost layers of collenchyma, the rest being chiefly made up of common parenchyma, see for this the vallecular part. Along the inner side and the inner part of the lateral sides of the meristeles cells often more or less fibre-shaped; in the same parts but somewhat farther from the meristeles cells polygonal prisms. Fibre-shaped and polygonal cells showing spiral or reticulate thickenings or very numerous pits. S c h i z o g e n o u s g l a n d s [1] mostly 1 on the outside of each meristele at a short distance from it; becoming visible in the drug on treatment with hot chloral hydrate or potash; these glands

[1] Similar glands corresponding to the vascular bundles are described for the stem in *Herba Conii*. These glands have not been considered as vittae, because in *Fructus Anethi, Anisi, Carui, Coriandri* and *Foeniculi* no normally developed vittae have been observed in the ground tissue of the ribs outside the meristeles. In *Fructus Carui* schizogenous glands occur in the same place.

often slightly developed and still showing in a transverse section their schizogenous way of formation from a single mother cell divided in 4 daughter cells; in the alcoholic material resembling vittae, but not tangentially elongated in a cross section and the epithelium cells not yet reduced.

C o l l e n c h y m a c e l l s. Cell contents: a yellow to green mass, in potash forming drops, in iodine and sulphuric acid 66 per cent. colouring reddish brown. C o m m o n p a-r e n c h y m a c e l l s. Polygonal. Cell contents: simple and compound starch grains, more numerous than in the vallecular parts; compound grains 2-or3-adelphous. F i b r e-s h a p e d c e l l s. Diameter 10 μ, L. 150 μ. Walls lignified. P o l y g o n a l c e l l s. R. 10 μ, T. 12—15 μ, L. 15—40 μ. Walls lignified. E p i t h e l i u m c e l l s. Walls bordering the cavity somewhat thickened.

M e r i s t e l e s. 1 in each rib; in the shape of triangular prisms with 1 edge turned outwards and with rounded sides and edges. Consisting of 2 collateral closed **vascular bundles** united by a band of sclerenchyma fibres, this band containing a few spiral vessels in its inner part corresponding with the xylem parts of the vascular bundles. P h l o e m lying on the lateral sides of the meristeles and turned outwards; radially elongated x y l e m lying along the inner edges of the meristeles.

V e s s e l s. Having a smaller diameter than the fibres. Spirals lignified. F i b r e s. Diameter 8—10 μ, L. 200—500 μ; polygonal. Walls thickened; middle lamella conspicuous; only the middle lamella lignified; showing not numerous slit-like pits.

E p i d e r m i s i n n e r s i d e. Cells mostly arranged in longitudinal rows. Shape of the very large mother cells, giving rise to the longitudinal rows by several, up to 14 transverse partition walls, still discernible. Component cells increasing in length towards the top and especially towards the base of the fruit. This epidermis sometimes called coniine layer, showing in its cells a dark gray precipitate when treated with potassium iodide and cadmium iodide.

C e l l s o f t h e a b o v e. Mother cells L. e. g. 220 μ; component cells R. 22 μ, T. 40—70 μ, L. 20 μ, at the base of the fruit L. up to 40 μ; tetragonal prisms. Outer and inner walls thickened; yellow; the innermost lamella of all the walls colouring somewhat red in phloroglucin and hydrochloric acid, brown in iodine and sulphuric acid 66 per cent. and persisting in concentrated sulphuric acid; the rest of the walls colouring yellow to brown in iodine and iodine in chloral hydrate and somewhat blue in iodine and sulphuric acid 66 per cent.; in boiling potash yellow globules emerging from the walls, in cold potash the whole wall colouring somewhat more yellow; outer and inner walls showing stratification; outer walls now and then with small pits. Cell contents: in each cell 1, sometimes 2 small prismatic crystals; in iodine and in iodine and chloral hydrate numerous small reddish brown drops or a dark brown mass; in potash yellow with numerous yellow drops; in alcohol very numerous small colourless oil drops.

Commissure. Projecting on both sides into the loculi. **E p i d e r m i s** of both faces. Mother cells of longitudinal rows not or hardly discernible.

C e l l s. R. 22 μ, T. 25—40 μ, L. 20 μ. See for the rest the inner epidermis of the wall.

G r o u n d t i s s u e. Consisting of common parenchyma. Round the carpophore and in the vicinity of the furrow cells large and polygonal, often strongly elongated in the direction of the plane of the commissure. V i t t a e. Numerous; see for the rest those of the ground tissue of the valleculae.

C a r p o p h o r e. Consisting of 2 bundles of sclerenchyma fibres in the shape of 2 semi-elliptical cylinders, the flat sides joining; in the young fruit separated by some layers of parenchyma cells.

Stylopode. Containing many sclereids with thickened walls and numerous pits.

S e e d. Quite filling up the loculus.

Ovule. According to Tschirch und Oesterle, Atlas, Pl. 37 pendulous, anatropous, with 1 integument of about 10 layers of cells.

Ripe seed.

S e e d-c o a t.

Entirely produced by the integument.

Epidermis outer side. Consisting of tabular [1]) cells with radially directed axis; in the concave part of the seed, turned towards the commissure, cells not to be distinguished.

C e l l s   o f   t h e   a b o v e. R. 4 μ, T. and L. 10 μ. Cell contents: some yellow granules colouring brown in potash.

Ground tissue. Quite resorbed on all the sides of the seed, except on the raphe part turned towards the commissure; there the ground tissue strongly developed and consisting of common parenchyma; cell layers bordering on the commissure and often those bordering on the endosperm flattened; those of the middle part polygonal, towards the endosperm increasing in size and often torn. Cells surrounding the meristele flattened.

C o m m o n   p a r e n c h y m a   c e l l s. R. 12 μ, T. 15 μ, L. 50 μ; polygonal prisms. Cell contents: here and there a dark granular mass lighting up under the polarization microscope, soluble in hydrochloric acid.

Meristele. Running longitudinally in the vicinity of the commissure and not quite reaching the chalaza; in the shape of a semi-elliptical cylinder, the flat side turned towards the commissure; consisting of a ring of spiral vessels, surrounding mostly some torn tissue.

Epidermis inner side. Crushed.

N u c l e u s of the seed.

Endosperm. Campylosperm. Furrow sometimes only in the middle, and not continued to the top and the base. E p i d e r m i s. Hardly to be distinguished from the rest of the endosperm. Outer walls showing a cuticle.

G r o u n d   t i s s u e. Cells arranged in radial rows.

C e l l s   o f   t h e   a b o v e. Walls colouring yellow in iodine, blue in iodine and sulphuric acid 66 per cent. Cell contents: 1. Cells with large aleurone grains, mostly containing 1 sometimes 2 or 3 cluster crystals; in a few cases moreover grains containing a single crystal and other grains without contents. The largest grains 10—13 by 13—15 μ; the cluster crystals 4—6 μ in diameter and having a small central air bubble; these crystals disappearing in potash. 2. Cells with smaller aleurone grains (5 μ in diameter) mostly quite filled by a globoid, often showing a small shining central spot. In all the cells oil as oil plasm.

Embryo. See Fructus Foeniculi.

April 1903.   J; M; L.

---

[1]) According to T s c h i r c h   und   O e s t e r l e of polygonal cells.

FRUCTUS CORIANDRI.

Coriandrum. Coriander Fruit. Coriander.

The ripe fruit of Coriandrum sativum, Linn. Sp. Pl. 256.

Macroscopic characters.

Diakenes, without a pedicel; the 2 mericarps closely covering; diameter up to 5 m.M., hard, globular, crowned by 5 very short calyx teeth, 2 of which sometimes somewhat larger and the conical stylopode sometimes bearing remains of the styles; each mericarp with 5 slightly projecting wavy primary ribs, every other rib in superposition with the lobes of the calyx (carinal ribs) and with 4 straight, somewhat more prominent secondary ribs; moreover the margins of the mericarps united on both sides of the fruit in a somewhat wavy rib. External surface pale yellowish brown, glabrous. Internally showing a lenticular cavity in which the carpophore, only united with the pericarp at the top and at the base. Vittae only in the commissure, 2 in each mericarp, in the shape of curved brown ridges easily to be detached. Endosperm coelosperm. Odour of the fresh seed nauseous, after bedbugs, but becoming by and by peculiarly aromatic in drying; taste aromatic, first sweetish, but pungent when chewed.

Anatomical characters.

LITERATURE. Berg. Anat. Atl. 1865. Pl. 41. Bortsch. Beitr. z. Anat. u. Entw. d. Umbellif. Fr. Diss. Breslau. 1882. Bochman. Beitr. z. Entw. gesch. off. Samen u. Früchte. Diss. Bern. 1901. 41. Colignon. Canaux sécrét. d. l. Ombellif. 1874. Flückiger. Pharmacogn. 1891. 953. Flückiger & Hanbury. Pharmacogr. 1879. 329. Greenish & Collin. Anat. Atl. of veget. powders. 1904. 150. Hager. Pharm. Praxis. Bd. I. 1900. 961. Hérail. Mat. Med. 1912. 260. Karsten u. Oltmanns. Pharmakogn. 1909. 278. Kayser. Beitr. z. Kenntn. d. Entw.-gesch. d. Samen. Pringsheim's Jahrb. 1893. 79. Kraemer. Bot. a. Pharmacogn. 1910. 772. Lange. Entwickel. d. Oelbeh. i. d. Früchten d. Umbellif. Schr. d. Phys. oek. Gesellsch. zu Königsberg. Jahrg. 25. 1884. 27. Marmé. Pharmakogn. 1886. 326. Meyer. Mikr. Unters. v. Pflanzenpulv. 1901. 38. Meyer. Entstehung d. Scheidew. i. d. secretführ. plasmafr. Intercellularraume d. Vittae d. Umbellif. Bot. Ztg. 47. 1889. 341. Moeller. Pharmakogn. 1889. 126, 133. Moeller. Mikr. pharm. Üb. 1901. 182. Oudemans. Aant. o. d. Pharmac. neerl. 1854—56. 376. Oudemans. Pharmacogn. 1881. 389. Perrot. Anat. du Fruit de Coriandre. Bull. Sc. pharmacol. Ann. 3. 1901. 385. Planchon et Collin. Drogues simples. T. II. 1896. 249, 256. Pfeiffer. Meth. d. mikrosk. Anat. ruh. Umbellif. Fr. Microkosmos. 1918/19. 8. Rompel. Kryst. v. Caoxal. i. d. Fruchtw. d. Umbellif. u. ihre Verwert. f. d. Syst. Ber. Wien. Akademie. Bd. 104. Abt. 1. 1895. 417. Schneider. Powder. veg. Drugs. 1902. 172. Styger. Beitr. z. Anat. d. Umbellif. Fr. Schweiz. Apoth. Zeit. LVII. 1919. 3. Tanfani. Morfol. ed Istolog. d. fr. e. d. seme d. Apiacee. N. G. B. J. Vol. XXIII. 1891. 451; report Bot. Jahresber. 1891. 598. Tschirch. Angew. Pfl. Anat. 1889. 44, 495. Tschirch. Harze u. Harzbeh. 1900. 366, 367. Tschirch. Pharmakogn. Bd. 2. 1911. 837. Tunmann. Resinog. Schicht d. Secretbeh. d. Umbellif. Ber. d. pharm. Ges. Bd. 17. 1907. 456. Vogl. Veget. Nahr.-u. Genussm. 1899. 419. Wigand. Pharmakogn. 1879. 278. v. Wisselingh. Vittae d. Umbellif. Verh. d. Kon. Acad. v. Wet. Amsterdam. 1894. (2de sect.) T. IV. No. 1. v. Wisselingh. Bijdr. II. Umbellif. Pharm. Weekbl. 1918. 1530. Wilke. Anat. Bez. des Gerbst. z. d. Secretbeh. d. Pfl. Diss. Halle. 1883. 9. MATERIAL. The drug. Young fruits in alcohol. REAGENTS. Water, glycerine, chloral hydrate, potassium iodide iodine, phloroglucin and hydrochloric acid, Schulze's macerating mixture, osmic acid, oil of cloves, origanum oil.

MICROGRAPHY.

Pericarp.

Wall. Thick about 200 μ in the vallecular parts of an imbibed seed.

Epidermis outer side. Stomata wanting. Some crystal idioblasts. Cells of the above. R. 10 μ, T. 15 μ, L. 20 μ; poly-, mostly tetragonal tables. Outer walls a little thickened. Cell contents: a yellow granular mass; sometimes a small prismatic crystal.

Ground tissue. To be divided in 3 parts. 1. Outer part. In the valleculae consisting of 4—10 layers of common parenchyma cells, in the outermost 1 or 2 layers collenchymatous and more or less flattened. In the primary and secondary ribs consisting of rectangular common parenchyma cells mostly arranged in tangential planes; the outermost cell layers more or less flattened. At the inside of this tissue some layers of prismatic cells, arranged in wavy longitudinal rows, hence

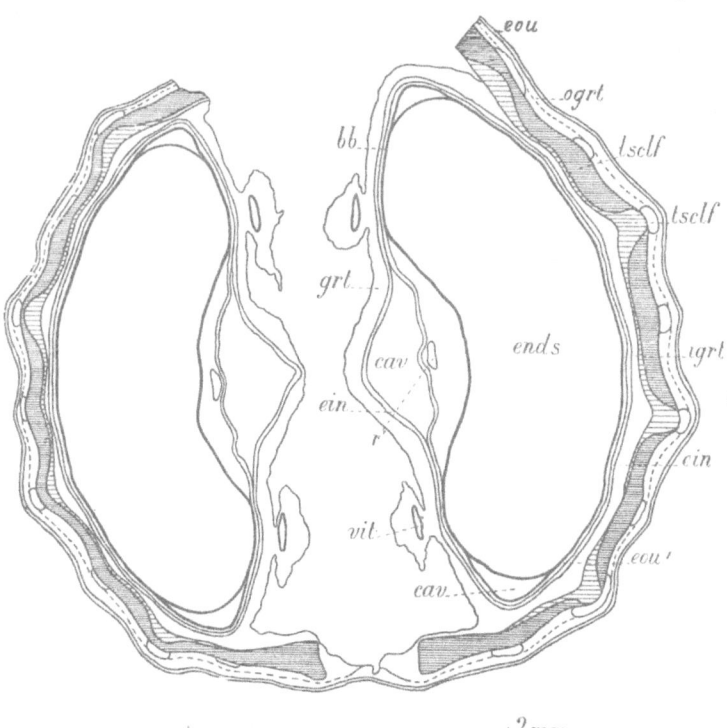

Fig. 44. *Coriandrum sativum.* Fruit, transverse section. bb Thin brown band of entirely flattened tissue; cav Cavity; ein Epidermis inner side; ends Endosperm; eou Epidermis outer side; eou' Epidermis outer side seed-coat; igrt Inner part of the ground tissue; lsclf Longitudinally directed sclerenchyma fibres; ogrt Outer part of the ground tissue; r' Meristele of raphe; tsclf Tangentially directed sclerenchyma fibres; vit Vittae.

the cells often more or less hook-shaped. V i t t a e.[1] In several fruits tangentially elongated cavities, being remains of glands formed at a very early stage and obliterated. In an ovary about 1 m.M. in diameter on the dorsal side of each mericarp 10—12 glands [2]), largest in the centre of the dorsal side and very

---

[1] B o c h m a n mentions 5.

[2] We have not called these glands vittae, though the walls of the epithelium cells bordering the cavity do not show the structure characteristic of such walls in real vittae, the latter however occurring at the same time in the commissure of this fruit.

small in the vicinity of the commissure. In a transverse section at the inside of 2 or 3 of the largest cavities 1 or 2 smaller ones. Not showing a distinct epithelium. C o l l e n c h y m a t o u s c e l l s. R. 10 μ, T. 25 μ; polygonal prisms with a longitudinally directed axis. C o m m o n p a r e n c h y m a c e l l s o f t h e r i b s. R. 10 μ, T. 25 μ, L. 30—50 μ; in the innermost layers R. often larger; rectangular prisms with a longitudinally directed axis. Cell contents: some chloroplasts; here and there a small oil drop, especially in the innermost layer. P r i s m a t i c c e l l s o f t h e l o n g i t u d i n a l r o w s. R. 25 μ, T. 30 μ, L. 60—100 μ; with a mostly longitudinally directed axis. Walls somewhat thickened; pitted.

2. Middle layer, showing shallow grooves corresponding to the valleculae; in the vallecular part thick 70 μ, in the primary ribs 100—120 μ. Consisting of sclerenchyma fibres; in the outermost 5—6 layers fibres directed longitudinally, in the innermost 1—3 layers tangentially; in a tangential section bundles of fibres very sinuous. In the primary ribs the number of longitudinally directed fibres increased; in the secondary ribs nearly all the fibres running tangentially. On both ends of the commissure the sclerenchyma layer interrupted by a smaller or larger band of parenchyma. The ends of the sclerenchyma layer joining on this parenchyma up to 200 μ thick and consisting only of longitudinally running fibres.

F i b r e s. R. and T. 10—12 μ, L. 80—350 μ; polygonal, often curved, a few times branched. Walls strongly thickened; middle lamella conspicuous; lignified, especially the middle lamella; showing pit canals.

3. Inner part, consisting of 4—5 layers of common parenchyma cells; often more or less flattened; innermost layer often bulging out towards the centre of the fruit. Intercellular spaces wanting.

C e l l s o f t h e a b o v e. R. 20 μ, T. and L. 50 by 80 μ; polygonal. Walls of the innermost 1 or 2 layers thickened; yellow; lignified; pitted. Cell contents: sometimes a small oil drop.

M e r i s t e l e s. In or near the primary ribs in the outer part of the sclerenchyma layer mostly some spiral vessels with lignified walls; diameter 3—5 μ.

E p i d e r m i s i n n e r s i d e. Cells arranged in longitudinal rows; mother cells of those rows not distinct (see Fructus Foeniculi); component cells often showing sinuous walls. Radial dimension of the cells varying, in accordance with the cells of the innermost layer of the ground tissue, bulging out. Stomata not noticed. [1])

E p i d e r m a l c e l l s p r o p e r. Component cells R. 6—10 μ, T. 70—170 μ, L. 10 μ. Walls sometimes brown; lignified.

Commissure. Near the carpophore showing a longitudinal stricture in its middle part.

E p i d e r m i s of both faces. See the inner epidermis of the wall.

G r o u n d t i s s u e. In the ripe fruit only along the epidermis of both faces some layers of parenchyma cells, resembling those of the inner part of the ground

---

[1]) B o c h m a n noticed a stoma here and there.

tissue of the wall. In the vicinity of the vittae those cell layers more numerous. Intercellular spaces wanting. The main part having disappeared, leaving a large cavity traversed by the carpophore. V i t t a e. 4 in each fruit, viz. in each mericarp 1 on either side of the carpophore; running from the centre of the base of the mericarp in a curved line towards the top, ending lower than the vittae in Fructus Foeniculi; spindle-shaped, the top somewhat more pointed than the base, closed at both ends; R. 50—70 μ, T. 250—400 μ; schizogenous [1]); showing an epithelium of 1 layer of cells; epithelium cells largest on the side turned towards the centre of the fruit. Walls of these cells on the side of the cavity strongly thickened; covered with a reddish brown lamella containing small bladders bulging out into the cavity of the vittae; this lamella thicker than in Fructus Foeniculi; next to this lamella towards the lumen of the cell a yellow to brown layer filled with very small granules [2]); lastly, immediately bordering the lumen of the cell, an equally thick pale yellow layer resembling the other walls of the epithelium cells. Transverse septa wanting in these vittae. Cavity of vittae in fresh fruits containing a colourless to pale yellow oil, in older fruits the oil becoming darker yellow. Moreover about 4 obliterated schizogenous glands like those of the wall of the mericarp.

P a r e n c h y m a c e l l s. R. 40 μ, T. 50 μ, L. 80—130 μ; polygonal. Walls showing somewhat thickened edges; sometimes brown. E p i t h e l i u m c e l l s. Thick 18 μ, wide 25 μ, long 30 μ; polygonal tables. Cell contents wanting.

C a r p o p h o r e. Consisting of 2 flat bundles of fibres surrounded by spiral vessels; those bundles arranged in the commissure in a transverse, not in a median plane.

V e s s e l s. E. g. R. and T. 12 μ, L. of articulations 230 μ.

S e e d.

Ovule. In each loculus 2 ovules, 1 of which abortive; anatropous; pendulous; with 1 integument. Nucellus soon quite resorbed. [3])

Ripe seed.

S e e d - c o a t.

Entirely produced by the integument.

Epidermis outer side.

C e l l s. R. 10 μ, T. 25 μ, L. 30 μ; polygonal tables. Walls brown to red. Cell contents: some small yellow granules.

Ground tissue. Represented by a thin brown band of entirely flattened tissue; in the raphe part the outermost layers not flattened, but neither filling up the cavity formed by the longitudinal stricture of the commissure within the epidermis of the seed-coat; cells mostly more or less torn. Round the meristele cells smaller.

---

[1]) T a n f a n i. Report Bot. Jahresb. 1891. 598.
[2]) According to v. W i s s e l i n g h. Vittae d. Umbellif. 11, the layer containing cellulose, the granules consisting of vittine.
) After B o c h m a n. l. c. 42.

Meristele. In the middle of the band of ground tissue just mentioned. In its outer part consisting of bundles of spiral vessels.

Epidermis inner side. Flattened.

N u c l e u s of the seed.

Endosperm. Coelosperm. Cells arranged in radial rows. In the centre the embryo; below the embryo a central cavity; below this again a tissue consisting of longitudinally arranged cells.

R a d i a l l y a r r a n g e d c e l l s. R. 25—40 μ, T. and L. 20 μ; polygonal prisms with a radially directed axis. L o n g i t u d i n a l l y a r r a n g e d c e l l s. R. 20 μ, T. and L. 25 μ; tetragonal prisms. Walls of all cells somewhat thickened; colouring yellow in iodine. Cell contents of all cells: numerous aleurone grains, ovoid or globular. In the greater part of the cells grains 4—8 μ in diameter; with 1 sometimes 2 or 3 cluster crystals, 2—4 μ in diameter, soluble in potash and showing a small central air-bubble; exceptionally a grain with a simple crystal. In the other cells aleurone grains globular and generally somewhat smaller, containing 1 globoid, 3 μ in diameter and often with a small central shining spot. Ground mass soluble in water. Oil as oil plasm.

Embryo. See Fructus Foeniculi.

March 1903.   J; M; L.

FRUCTUS CUBEBAE.
Cubebs.
The dry fruits, for the greater part unripe, of Piper Cubeba, Linn. f. Suppl. 90.

Macroscopic characters.

Globular monopyrenous drupes; diameter up to 6 m.M.; contracted at the base into a tail, long 6—10 m.M. and thick hardly 1 m.M. Surface sometimes even, mostly reticulately wrinkled, dull black or dark brown. Stone light brown, with a smooth internal surface. Seeds often little developed; funicle very short, attached at the base of the cavity of the fruit. Seed-coat wrinkled, dark brown; hilum at the base, large, circular. Nucleus of the seed consisting for the greater part of a solid perisperm; at its top a small conical endosperm with an embryo. Odour aromatic; taste aromatic and bitter, not pungent.

Anatomical characters.

LITERATURE. Benecke. Mikr. Drogenprakt. 1912. 81. Berg. Anat. Atl. 1865. 86. Biechele. Mikr. Prüf. d. off. Drogen. 1904. 32. Biermann. Ueber Bau und Entwicklungsgeschichte der Oelzellen. Diss. Bern. 1898. Deiveure. Recherches sur le cubebe et sur les Piperaceae etc. 1894. Flückiger. Pharmakogn. 1891. 925. Flückiger & Hanbury. Pharmacographia. 1879. 584. Gilg. Pharmakogn. 1910. 78. Greenish & Collin. Anat. Atl. veget. powders. 1904. 158. Hager. Pharm. Praxis. Bd. I. 1900. 972. Hérail. Mat. Med. 1912. 442. Karsten u. Oltmanns. Pharmakogn. 1909. 240. Koch. Mikr. Anal. d. Drogenpulver. Bd. IV. 1908. 107. Koch. Einf. i. d. mikr. Anal. d. Drogenpulver. 1906. 155. Koch u. Gilg. Pharmakogn. Praktik. 1907. 258. Kraemer. Botany and Pharmacogn. 1910. 571. Luerssen. Syst. Bot. Bd. II. 1882. 517. Marmé. Pharmacogn. 1886. 280. Meyer. Wiss. Drogenk. Bd. II. 1892. 405. Moeller. Pharmakogn. 1889. 137. Moeller. Mikr. Pharm. Ueb. 1900. 187. Molisch. Mikrochemie der Pflanzen. 1913. 256. Oudemans. Pharmacogn. 1880. 362. Planchon et Collin. Drogues simples. T. I. 1895. 44. Schneider. Powdered Veget. Drugs. 1902. 176. Tschirch. Angew. Pfl.-anat. 1889. 85 and 474. Tschirch und Oesterle. Anat. Atl. 1900. Taf. 77. Vogl.

Anat. Atl. 1887. Taf. 29. Wigand. Pharmakogn. 1879. 287. MATERIAL. The drug; the ripest fruits being least shrivelled. REAGENTS. Water, glycerine, potash, iodine in chloral hydrate, phloroglucin and hydrochloric acid, iodine and sulphuric acid 66 per cent., concentrated sulphuric acid, Schulze's macerating mixture, iron acetate, cupric sulphate and potash 50 per cent., nitric acid, Millon's reagent, molybdate of ammonia [1]) and concentrated sulphuric acid.

MICROGRAPHY.

Globular head.

Pericarp.

Epidermis outer side. At the top of the fruit often small papillae. Sometimes 2 cells showing the origin from a single mother cell.

Cells of the above. R. 12 μ, T. and L. 20 μ; polygonal tables; papillae R. 14 μ, T. 12 μ, L. 15 μ. Outer walls thickened, lateral walls somewhat thickened; outer walls colouring yellow in iodine and sulphuric acid 66 per cent.; showing a cuticle. Cell contents: in several cells 1 or more prismatic crystals; often a brown mass.

Ground tissue. To be divided into 4 parts. 1. An outer part, consisting of 3—5 layers of common parenchyma cells, intermixed with several sclereids, especially in the outermost layer; towards the top of the fruit sclereids more numerous, also in the second and third parenchyma layer of this part; cells practically arranged in tangential planes, except in the outermost layer. In the inner layers cells showing radial and

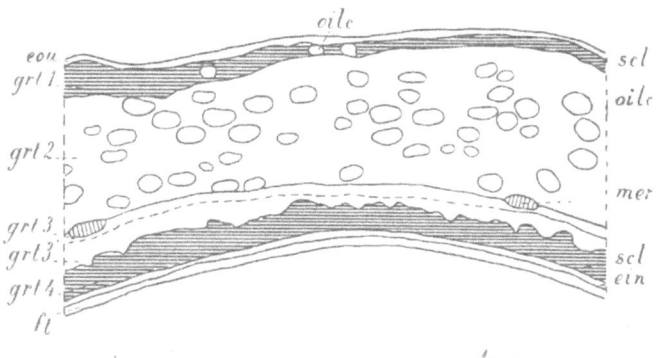

Fig. 45. *Piper Cubeba.* Dry ripe fruit, transverse section of the middle part of the pericarp. ein Epidermis inner side; eou Epidermis outer side; ft Layer of flattened tissue of unknown origin, not mentioned in the description; grt1 Ground tissue outer part; grt2 Ground tissue 2nd part; grt3 Ground tissue 3rd part, flattened portion; grt3′ Ground tissue 3rd, not flattened portion; grt4 Ground tissue 4th part; mer Meristeles; oilc Oil cells; scl Sclereids.

transverse partition walls; here and there an oil cell, see the second part. **Common parenchyma cells.** In the outermost layer R. 10 μ, T. and L. 12 μ; mostly polygonal prisms with a longitudinally directed axis; in the other 2—4 cell layers R. 11 μ, T. and L. 20—30 μ; polygonal. Walls somewhat thickened. Cells contents of the inner layers: a few starch grains. **Sclereids.** R. 25 μ, T. 18 μ, L. 20 μ; polygonal; the outermost layers polygonal prisms with a radially directed axis. Walls strongly thickened; yellowish; lignified, colouring blue on the addition of concentrated sulphuric acid if not previously treated with alcohol, see the oil cells of the second part; showing stratification and pit canals.

2. A part, consisting of 15 layers of common parenchyma cells, intermixed with numerous oil cells and at the top of the fruit containing sclereids often united

---

[1]) E. Schmidt. Ausführliches Lehrbuch der Pharmaceutischen Chemie. Bd. I. 4th Ed. 1898, 346.

in clusters. Showing intercellular spaces. Near the top of the fruit often a cell with an oil drop of the same kind as the oil of the oil cells.

C o m m o n p a r e n c h y m a c e l l s. R. 18 μ, T. and L. 25 μ; polygonal. Cell contents: many starch grains, mostly simple, a few compound, 2- to 3-adelphous; the simple grains globular or ellipsoidal, 3 μ in diameter, with a centric hilum; near the top of the fruit sometimes small granules persisting in potash. S c l e r e i d s. R. 30 μ, T. and L. 35 μ; polyhedra. See for the rest those of the outer part. O i l c e l l s [1]). R. 45 μ, T. and L. 80 μ; ellipsoidal. Walls thin; consisting of an outermost thin lamella persisting in concentrated sulphuric acid, not swelling in potash, and an innermost somewhat thicker lamella, swelling in potash. Cell contents: mostly 1 large yellow oil drop or a more irregular mass of oil not entirely filling the cell; in a transverse section of the pericarp small oil drops spread all through the section; no crystals [2]). On the addition of concentrated sulphuric acid the whole tissue colouring somewhat red, a few minutes after some parts of the oil drops becoming intensely cherry red; after 2—3 hours the red colour passing into violet and blue; this reaction only taking place in sections not previously treated with alcohol, solving the oil; during this reaction the sclereids first green, afterwards dark blue. On treatment first with molybdate of ammonia and afterwards with concentrated sulphuric acid about the same reaction as by treatment with concentrated sulphuric acid only; in the latter case the oil becoming directly blue. Oil soluble in alcohol as mentioned above, but often leaving a reddish brown mass along the walls of the cells.

3. A part, consisting of 5—7 layers of common parenchyma cells; the outermost layers flattened and containing meristeles; at those places the flattened layers bulging out into the preceding layer; flattened cells often brown; the unflattened layers always showing intercellular spaces. Unflattened layers containing a few small oil cells; near the top of the fruit numerous mostly tangentially elongated sclereids. At the top of the fruits this tissue bending outwards in the shape of a column traversing the other parts and passing into the tissue

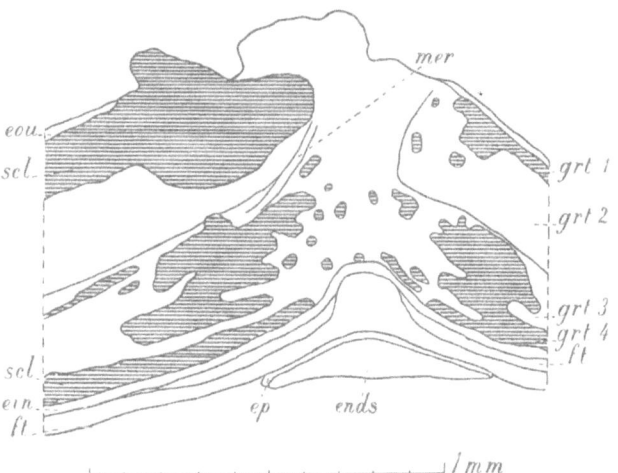

Fig. 46. *Piper Cubeba.* Ripe fruit, longitudinal section of the top. ein Epidermis inner side; ends Endosperm; eou Epidermis outer side; ep Epidermis of the endosperm; ft Layers of flattened tissue of unknown origin, not mentioned in the description; grt1 Ground tissue outer part; grt2 Ground tissue 2nd part; grt3 Ground tissue 3rd part; grt4 Ground tissue 4th part; mer Meristeles; scl Sclereids.

of the short stigmata. This prismatic part showing more or less radial arrangement of the cells; containing meristeles and radially elongated sclereids in the vicinity of the meristeles and near the margin.

---

[1]) See R. B i e r m a n n. Ueber Bau und Entwicklungsgeschichte der Oelzellen. Diss. Bern. 1898.
[2]) T s c h i r c h und O e s t e r l e mention crystals of cubebin.

C o m m o n p a r e n c h y m a c e l l s o f t h e u n f l a t t e n e d l a y e r s. R. 20 μ, T. and L. 25—40 μ; polygonal or prismatic. Walls yellowish; showing a pure cellulose reaction [1]). Cell contents: often an oil drop consisting of the same oil as that of the oil cells; sometimes, in a cell close to the following fourth part of the ground tissue, a granular dark mass. S c l e r e i d s. For the walls see sclereids of the outer part. O i l c e l l s. See those of the second part. C o m m o n p a r e n c h y m a c e l l s o f t h e u p p e r m o s t p r i s m a t i c p a r t. R. 30 μ, T. and L. 15 μ; polygonal prisms with a radially directed axis. Walls somewhat thickened. Cell contents: sometimes a few granules persisting in potash. S c l e r e i d s o f t h e p r i s m a t i c p a r t. R. 35—100 μ, T. and L. 10—20 μ.

4. An inner part, consisting of 1 sometimes of 2 or 3 layers of sclereids. At the top and at the base of the fruit sclereids smaller. At the base often intermixed with common parenchyma cells.

S c l e r e i d s. R. 50—100 μ and smaller in case of 2 or more layers of sclereids, T. and L. 15—25 μ; polygonal prisms with a radially directed axis. Walls more thickened than those of the sclereids of the outer part; for the rest see there.

M e r i s t e l e s. Running in the third part of the ground tissue, as well in the outer flattened layers as in the prismatic top; showing spiral vessels, those of the outer layers with lignification.

E p i d e r m i s i n n e r s i d e. [2]) A single layer of polygonal tabular common parenchyma cells; lateral walls sometimes a little rounded.

C e l l s o f t h e a b o v e. R. 10 μ, T. 25 μ, L. 60 μ. The characteristic thickenings, mentioned by T s c h i r c h u n d O e s t e r l e, l. c., not observed in this material.

T a i l.

E p i d e r m i s. Cells arranged in longitudinal rows.

C e l l s o f t h e a b o v e. R. 18 μ, T. 10 μ, L. 10—20 μ; mostly tetragonal prisms with a radially directed axis. See for the rest the epidermal cells of the outer side of the head.

G r o u n d t i s s u e. To be divided in 2 parts. 1. An outer part, consisting of 2 layers of longitudinally arranged common parenchyma cells, intermixed with sclereids.

C o m m o n p a r e n c h y m a c e l l s. R. 20 μ, T. 15 μ, L. 15—25 μ; polygonal prisms with a longitudinally directed axis. S c l e r e i d s. R. 25 μ, T. 15 μ, L. 40—70 μ; polygonal prisms with a longitudinally directed axis. Walls thickened, especially the outer ones. See moreover those of the head of the pericarp.

2. An inner part, consisting of about 10 layers of common parenchyma cells, intermixed with sclereids and a few oil cells; common parenchyma cells and sclereids arranged in longitudinal rows; common parenchyma showing intercellular spaces.

C o m m o n p a r e n c h y m a c e l l s. R. 25 μ, T. 30 μ, L. 70 μ; polygonal prisms with a longitudinally directed axis and rounded edges. Walls somewhat thickened. Cell contents: starch grains and a brown mass. S c l e r e i d s. R. and T. 25 μ, L. 100—150 μ; polygonal prisms with a longitudinally directed axis. Walls thickened; yellowish; lignified; with pit canals; showing stratification. O i l c e l l s. R. 25 μ, T. 30 μ, L. 50 μ. See for the rest those of the head of the fruit.

---

[1]) T s c h i r c h u n d O e s t e r l e mention these walls as lignified.
[2]) See T s c h i r c h u n d O e s t e r l e. p. 232.

E n d o d e r m i s. Not to be distinguished.

S t e l e.

Pericycle. Consisting of sclerenchyma fibres.

Vascular bundles. 8 in a ring; xylem containing spiral vessels with lignification.

Medullary commissures. Consisting of sclerenchyma fibres.

Medulla. The outer part consisting of sclerenchyma fibres, the inner part of common polygonal parenchyma cells, often somewhat flattened or torn, and showing in the centre a vascular bundle, containing spiral vessels with lignification, and a bundle of sclerenchyma fibres on two sides.

S c l e r e n c h y m a f i b r e s o f s t e l e. R. and T. 18 μ. Walls thickened; lignified, especially the middle lamellae; showing stratification and pit canals.

Seed. In a great many fruits only present as a shrivelled mass; in ripe fruits the well developed seed mostly attached to the bottom of the cavity of the pericarp; developed from an atropous ovule, showing in Piper nigrum a single integument, at the base sometimes indications of an outer one, see Tschirch und Oesterle, l.c. 103—106. Funicle short and thick; the outer part consisting of some layers of more or less flattened common parenchyma intermixed with some sclereids; the inner part consisting of common parenchyma cells without intercellular spaces, arranged in 6—8 transverse layers.

C o m m o n p a r e n c h y m a c e l l s o f t h e i n n e r p a r t. R. 10 μ, L. 18 μ; polygonal in a longitudinal section. Walls a little yellow. Cell contents: a mass coloured light red.

S e e d-c o a t. For the greater part only consisting of 2 epidermal layers and no ground tissue; often flattened into a brown mass without distinct structure; at the top of the seed those cells practically arranged in longitudinal rows and moreover between the 2 epidermal layers some layers of ground tissue; in the vicinity of the micropyle the cells of the inner epidermal layer papillary.

C e l l s o f o u t e r e p i d e r m a l l a y e r. R. 10 μ, T. 30—40 μ, L. 60—120 μ; strongly elongated polygonal tables with a radially directed axis; at the top of the seed-coat R. 10 μ, T. 30 μ, L. 50 μ; rectangular tables with a radially directed axis. Walls, outer ones strongly thickened, lateral ones thickened, inner ones thin; all walls red till brown, lignified; lateral walls with pit canals. C e l l s o f i n n e r e p i d e r m a l l a y e r. R. 10 μ, T. 12—20 μ, L. 35—40 μ; polygonal tables with a radially directed axis; the papillary cells R. 35 μ, L. 15 μ.

N u c l e u s of the seed.

Albumen.

Perisperm.

E p i d e r m i s.

C e l l s o f t h e a b o v e. R. 10 μ, T. and L. 12 μ; polygonal prisms. Outer walls thickened; in iodine and sulphuric acid 66 per cent. colouring yellow and in phloroglucin and hydrochloric acid mostly red; showing stratification. Cell contents: some globular starch grains; the rest of each cell filled with small aleurone grains, biuret reaction, in Millon's reagent and in nitric acid protein reaction very distinct.

G r o u n d t i s s u e. Consisting of common parenchyma cells; the cells of the outermost layer being smaller, for the rest decreasing in size and becoming more isodiametric towards the centre; intermixed with very numerous oil idioblasts especially in the outermost part. The outer layer of common parenchyma cells showing starch and aleurone grains like the epidermal cells; in a few cell layers towards the interior the aleurone grains decreasing in number and only present along the walls; the parenchyma cells of the inner part without aleurone grains. Starch grains in all cells, except those of the outermost layer, very numerous and sometimes coloured greenish yellow; in many cases quite filling up the cell as a lump composed of polyhedral grains; in other cells some simple grains and 1 or more polyadelphous compound grains, the latter e.g. 20 μ in diameter; in other cells again many simple grains, 2—5 μ in diameter.

Common parenchyma cells of the outermost layer. R. 12 μ, T. and L. 25 μ; polyhedral. Cells of the outer layers of the rest. R. 70 μ, T. and L. 40 μ; polyhedral with strongly rounded edges. Walls sometimes brown. Cell contents: see above. Oil cells. R. 100 μ, T. and L. 70 μ; ellipsoidal with a radially directed longest axis. See for the rest those of the pericarp.

Micrography of the powder. Sclereids oblong, very thick-walled and light yellow, forming in 1 or 2 layers the stone and very firmly cohering. Layer of sclereids under the brown polygonal epidermal cells and closely united to them; these sclereids not forming a continuous layer, but interchanging with common parenchyma cells, thus forming a kind of mosaic against the epidermis; isodiametrical, polygonal and half the size of those of the stone. Thin bright reddish brown seed-coat, often showing but little cell structure. Oblong polyhedral common parenchyma cells of the perisperm, quite filled with starch. Common parenchyma of the pulp, containing oil cells with a brownish wall. A few spiral vessels. Bundles of long sclerenchyma fibres of the tail not permitted to occur to a large number. Much oil (in chloral hydrate). Many starch grains from the parenchyma of the perisperm and the pulp; those of the perisperm penta- or hexagonal, with a central hilum, many of them 6—7 μ in diameter, sometimes compound, 2- or 3-adelphous. Numerous smaller starch grains, some originating from the pulp.

December 1902.   J; M; L.

FRUCTUS FOENICULI.
Foeniculum. Fennel. Fennel Fruit.
The ripe fruits of Foeniculum vulgare, Mill. Gard. Dict. ed. VIII. n. 1.

Macroscopic characters.

Diakenes with or without a pedicel; the mericarps not closely cohering, often separated. Pedicel long up to 1 m.M. Fruit long up to 1 c.M., thick up to 4 m.M., cylindrical, somewhat curved, crowned by the conical bifurcous stylopode on which 2 short curved stigmata; carpophore divided; each mericarp with 5 strongly projecting keel-shaped primary ribs — the 2 marginal ones somewhat wider, the 3 dorsal ones somewhat narrower and nearer to each other — and with 4 deep valleculae. Vittae 1 in each vallecula, 2 on the flat commissural face of each mericarp; all clearly visible from the outside of the mericarp. Surface smooth, greenish brownish yellow; vittae dark brown. Endosperm orthosperm. Odour and taste strongly aromatic.

Anatomical characters.

LITERATURE. Attema. De zaadhuid d. Angiosp. en Gymnosp. Diss. Groningen. 1901. 113. Berg. Anat. Atl. 1865. Pl. 43. Benecke. Mikrosk. Drogenprakt. 1912. 83. Bortsch. Beitr. z. Anat. u. Entw. d. Umbellif. Fr. Diss. Breslau. 1882. Colignon. Canaux sécrét. d. l. Ombellif. 1874. Flückiger. Pharmakogn. 1891. 948. Flückiger & Hanbury. Pharmacogr. 1879. 308. Gilg. Pharmakogn. 1910. 251. Greenish & Collin. Anat. Atl. of veget. Powders. 1904. 154. Hager. Pharm. Praxis. Bd. I. 1900. 1163. Hérail. Mat. Med. 1912. 328. Karsten u. Oltmanns. Pharmakogn. 1909. 283. Kayser. Beitr. z. Kenntn. d. Entw.-gesch. d. Samen. Pringsheim's Jahrb. 1893. 79. Koch. Mikr. Anal. d. Drogenpulver. Bd. IV. 1908. 117. Pl. XII. Koch. Einf. i. d. mikr. Anal. d. Drogenpulver. 1906. 163. Koch u. Gilg. Pharmakogn. Praktik. 1907. 251. Kraemer. Botany a. Pharmacogn. 1910. 564, 740, 173. Lange. Entw. d. Oelbeh. i. d. Früchten d. Umbellif. Schr. d. Phys. oekon. Gesellsch. zu Königsberg. Jahrg. 25. 1884. 27. Marmé. Pharmakogn. 1886. 318. Meyer. Wiss. Drogenk. Bd. II. 1892. 37. Meyer. Entstehung d. Scheidew. i. d. secretf. plasmafr. Intercellularraume d. Vittae d. Umbellif. Bot. Ztg. 47. 1889. 341. Meyer. Grundl. u. Meth. f. Unters. v. Pfl.-pulv. 1901. 32, 38. Moeller. Pharmacogn. 1889. 126, 131. Moeller. Mikr. pharm. Üb. 1901. 178. Oudemans. Pharmacogn. 1881. 386. Planchon et Collin. Drogues simples. T. II. 1896. 249, 267. Pfeiffer. Meth. d. mikr. Anat. ruh. Umbellif. Fr. Mikrokosmos. 1918/19. 8. Rompel. Kryst. v. Caoxal. i. d. Fruchtw. d. Umbellif. u. ihre Verw. f. d. Syst. Ber. Wien. Akad. Bd. 104. Abt. 1. 1895. 417. Schneider. Powd. veg. Drugs. 1902. 192. Styger. Beitr. z. Anat. d. Umbellif. Fr. Schw. Apoth. Zeit. Bd. LVII. 1919. 3. Tanfani. Morfol. ed. Istol. d. fr. e. d. seme d. Apiacee. N.G.B.J. Vol. XXIII. 1891. 451; report Bot. Jahresber. 1891. 598. Tschirch. Angew. Pfl. Anat. 1889. 44, 157, 484, 494, 495. Tschirch. Pharmakogn. Bd. 2. 1911. 1194. Tschirch u. Oesterle. Anat. Atl. 1900. 52. Pl. XIV. Tunmann. Resinog. Schicht. d. Sekretbeh. d. Umbellif. Ber. d. pharm. Gesellsch. Bd. 17. 1907. 456. Vogl. Veget. Nahr. u. Genussm. 1899. 407. Wigand. Pharmakogn. 1879. 276. Wilke. Anat. Bez. d. Gerbst. z. d. Secretbeh. d. Pflanze. Diss. Halle. 1883. 9. v. Wisselingh. Vittae d. Umbellif. Verh. d. Kon. Acad. v. Wet. Amsterdam. 1894. (2de sect.) T. IV. No. 1. v. Wisselingh. Bijdr. II, Zaadhuid. Umbellif. Pharm. Weekbl. 1918. 1530. MATERIAL. The drug. REAGENTS. Water, glycerine, potash 50 per cent., chloral hydrate, iodine in chloral hydrate, phloroglucin and hydrochloric acid, iodine and sulphuric acid 66 per cent., concenrated sulphuric acid, Schulze's macerating mixture, osmic acid, oil of cloves, origanum oil.

MICROGRAPHY.

Pericarp.

Wall. In a soaked seed thick about 200 μ in the vallecular parts.

Epidermis outer side. Somewhat curving inwards above the place of the future line of dehiscence. The origin of 2 cells from 1 mother cell often distinct in the valleculae. Stomata very rare, lying in the same level as the epidermal cells, with 2 subsidiary cells.

Epidermal cells proper. In the valleculae R. 10 μ, T. 20 μ, L. 25 μ; on the ribs R. 10 μ, T. 15 μ, L. 30 μ; polygonal tables. Outer walls somewhat thickened; with a cuticle. Cell contents: some small granules. Stomata. 15 by 20 μ. Subsidiary cells 22 μ by 22 μ.

Ground tissue. Chiefly consisting of common parenchyma.

Valleculae. On the lateral sides of the vittae the parenchyma 8—9 cells thick in a radial direction; in the 2 outermost layers, especially in the second layer, cells much smaller than the rest and often flattened; cells largest in the middle part and tangentially elongated in the vicinity of the meristele. On the outside of the vittae some radial rows of periderm-like cells, in transverse and radial

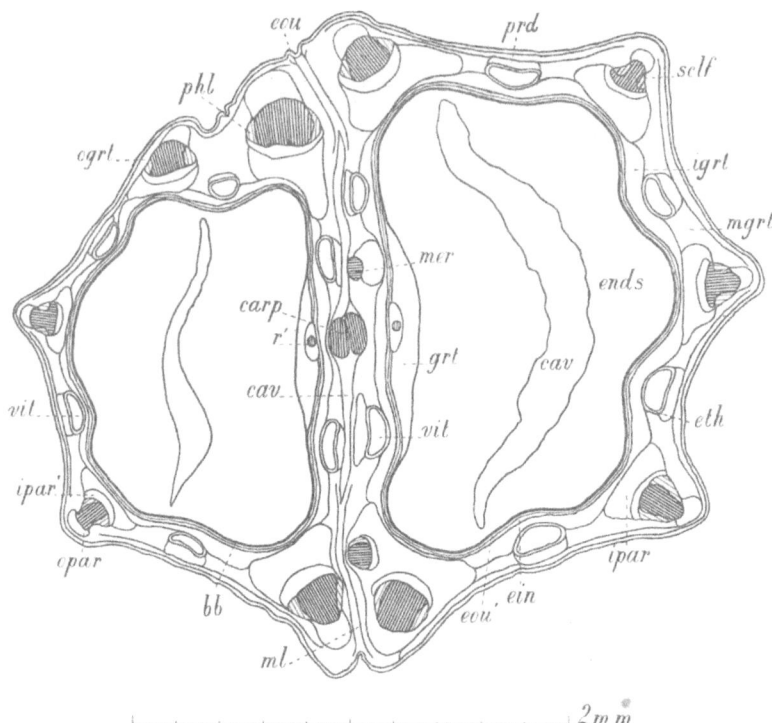

Fig. 47. *Foeniculum vulgare.* Fruit, transverse section. bb Brown band of flattened ground tissue; carp Carpophore; cav Cavity; ein Epidermis inner side; ends Endosperm; eou Epidermis outer side; eou' Epidermis outer side seed-coat; eth Epithelium; grt Ground tissue; igrt Inner part of the ground tissue; ipar Parenchyma on the inner side of the meristeles; ipar' The same parenchyma with smaller diameter; mer Meristele; mgrt Middle part of the ground tissue; ml 2 middlemost layers between which the dehiscence takes place; ogrt Outer part of the ground tissue; opar Parenchyma on the outside of the meristeles; phl Phloem; prd Periderm-like cells outside the vittae; r' Meristele of raphe; sclf Sclerenchyma band and xylem; vit Vittae.

sections in radial rows, in a tangential section not conspicuously in longitudinal rows; this radial arrangement wanting on the outer side of the thickest vittae; without intercellular spaces. At the inner side of the vittae 3 layers of cells, strongly tangentially elongated, especially those of the innermost layer. V i t t a e. 1 in each vallecula, sometimes accompanied by a small one. Running from the base of the fruit to the base of the stylopode; elliptical or semi-elliptical cylinders, the flat side turned outwards; pointed on both ends, the upper end somewhat more so; closed at both ends; R. 60 μ, T. 140 μ; schizogenous; [1]) Showing numerous transverse septa, in the middle part of the vitta 400—600 μ in distance from each other, at the ends 150—200 μ. Septa thick 5 μ in their middle part, towards the circumference of the vitta strongly increasing in thickness, especially at their under side, and measuring up to 100 μ; in those thick parts numerous small cavities bordering on each other like the cells of a tissue and each surrounded by a reddish brown lamella like that mentioned below for the epithelium cells. Transverse septa on their upper and under surface covered by the same reddish brown lamella continuously running over the septa and the walls of the epithelium cells, sometimes folded in macerated material; for the rest yellow and consisting of vittine, containing more pectine than the outer reddish brown lamella and no cellulose. [2]) Epithelium of 1 layer of cells without intercellular spaces and not corresponding in place to the radial rows of ground tissue mentioned above outside the vittae. Walls of these cells on the side of the cavity strongly thickened, covered with a very thin reddish brown lamella continuous with the layer of the septa mentioned above; yellow to brown [3]). Contents of the cavity and of the small cavities in the septa: a pale yellow to colourless oil; in old fruits sometimes a resinous mass; the oil colouring red in concentrated sulphuric acid.

C o m m o n   p a r e n c h y m a   c e l l s   o n   t h e   l a t e r a l   s i d e s   o f   t h e   v i t t a e. In the middlemost cell layers R. 25 μ, T. 40 μ, L. 40—60 μ; the rest smaller; in the second layer R. 5 μ, T. 10 μ; polyhedral or prismatic with a longitudinally directed axis; in the second layer mostly rectangular prisms. Walls of the innermost 1 or 2 cell layers yellow to brown; persisting in concentrated sulphuric acid, with potash 20 per cent. no yellow globules, in osmic acid not colouring black. Cell contents: in the outermost layers some chloroplasts, in the second layer a yellow mass. P e r i d e r m - l i k e   c e l l s   o n   t h e   o u t s i d e   o f   t h e   v i t t a e. R. 11 μ, T. 18 μ, L. 40 μ; hexagonal tables with a radially directed axis. Walls with strongly thickened edges; brown; persisting in concentrated sulphuric acid, not swelling neither colouring brown in potash 50 per cent., not becoming black in osmic acid. In earlier stages of development these reactions not appearing. P a r e n c h y m a   c e l l s   a t   t h e   i n n e r   s i d e   o f   t h e   v i t t a e. Walls brown; persisting in concentrated sulphuric acid, not swelling neither becoming brown in potash 50 per cent., not becoming black in osmic acid. E p i t h e l i u m   c e l l s. Thick 8 μ, wide 20 μ, long 35 μ; polygonal tables. Walls, except the thickenings bordering the cavity and described above yellow [4]); in the drug these walls remaining yellow to brown in iodine and sulphuric acid 66 per cent. Cell contents: a dirty yellow granular mass.

Ribs. On the outside of the meristele first polygonal reticulate parenchyma cells; between this tissue and the phloem some minute parenchyma cells. On the inside of the meristeles some layers of prismatic parenchyma cells passing

---

[1]) See Tanfani. l.c. 599.
[2]) C. f. v. Wisselingh. l.c. 9, 22.
[3]) According to v. W i s s e l i n g h. l. c. 10, fig. 6, in these walls 4 parts to be distinguished: 1. the reddish brown lamella, bordering on the cavity, consisting of vittine and some pectose and no cellulose; 2. a thin lamella of cellulose; 3. a lamella of vittine and some pectose and no cellulose; 4. a lamella of cellulose. The layers 2, 3, 4 of each cell separated from the corresponding layers of the adjacent cell by the cellulose layer of the lateral walls.
[4]) According to v. W i s s e l i n g h containing cellulose.

into fibres; without intercellular spaces; the prismatic cells often tangentially elongated.

Parenchyma cells of the outside. Polygonal cells R. 18 μ, T. 20 μ, L. 50—120 μ; prisms with a longitudinally directed axis. Walls showing reticulate thickenings; yellow; lignified. Small prismatic cells. Walls; see polygonal cells. Parenchyma cells of the inner side. R. 25 μ, T. 35 μ, L. 50—80 μ. Walls like those of the parenchyma cells of the outer side. Fibres. R. 10 μ, T. 10 μ, L. 300—400 μ. Walls showing reticulate thickenings, often reminding of a tracheid fibre; lignified.

Meristeles. 1 in each rib, sometimes accompanied by a small one. In the 3 dorsal ribs in the shape of a semi-elliptical cylinder, in a transverse section the longest axis radially directed; the others having a somewhat more irregular shape. Ending in the stylopode with a cudgel-shaped part bending outwards. Consisting of 2 collateral closed vascular bundles united together by a band of sclerenchyma fibres without intercellular spaces; this band distinct between the 2 phloem parts, but hardly to be distinguished from the xylem parts.

Phloem. Lying on the lateral sides of the meristele; triangular in a transverse section; elements often more or less flattened. In Roman Fennel often in each bundle a schizogenous, often tetragonal duct, e.g. 7 μ in diameter in a transverse section, surrounded by 4 or more epithelium cells and apparently connected with the corresponding ducts of the stem. Xylem consisting of a few spiral vessels in the innermost part of the meristele.

Sclerenchyma fibres. R. and T. 8 μ, L. 200—400 μ; polygonal. Walls thickened; lignified; along the circumference showing reticulate thickenings, the others slit-like pits. Elements of phloem. Sometimes filled with a yellow mass. Xylem vessels. Diameter 4—6 μ. Walls showing lignification.

Epidermis inner side. The shape of the mother cells, giving rise to longitudinal rows of cells by 6—10 transverse partition walls, clearly to be distinguished. These mother cells again arranged in fairly longitudinal rows; mostly very elongated hexagonal tables; here and there a mother cell of a more irregular shape, showing partition walls differently directed, but always parallel to each other.

Cells of the above. Component cells R. 12 μ, T. 40 μ, L. 45 μ. Outer walls a little thickened; outer and inner walls yellowish; somewhat lignified, persisting in concentrated sulphuric acid, colouring yellow in potash 50 per cent., when boiled producing yellow drops; cuticle not distinct, the whole wall colouring yellowish brown in iodine and sulphuric acid 66 per cent.

Commissure.

Epidermis of both faces. See the inner epidermis of the wall.

Ground tissue. Thick 24 layers of parenchyma cells. The 2 middlemost layers, uniting the 2 re-entering parts of the wall of the pericarp mentioned above, consisting of tetragonal prismatic cells; the dehiscense afterwards taking place between these 2 layers. Cells of the other layers polygonal. Round the carpophore cells not reticulate. Moreover some radial rows of somewhat collenchymatous rectangular cells, corresponding in place to the raphe. Vittae mostly 4, in each mericarp 1 on either side of the carpophore; sometimes 5—8; see for the rest the vittae of the wall.

P o l y g o n a l   c e l l s. R. 30 μ, T. 30 μ, L. 40—60 μ. T e t r a g o n a l   p r i s m a t i c c e l l s. R., T. and L. 15 μ. Cell contents: often some small granules. C o l l e n c h y m a- t o u s   c e l l s. R. and T. 12 μ; rectangular.

M e r i s t e l e s. Mostly 1 or 2 small meristeles on each side of the carpophore in the plane of dehiscence. See meristeles of the wall.

C a r p o p h o r e. Consisting of 2 bundles of sclerenchyma fibres in the shape of semi-elliptical cylinders, the flat sides joining; lying in the plane of dehiscence of the fruit.

S e e d. Quite filling up the loculus; very closely joining on the pericarp, making the epidermis of the seed-coat look like a part of the pericarp.

Ovule. [1]) In each loculus 2 ovules, 1 of which aborting; anatropous; pendu- lous; with 1 integument of 20 layers of cells. Nucellus consisting of only 1 layer of cells; quite resorbed by the time of the ripeness of the ovule.

Ripe seed.

S e e d - c o a t.

Entirely produced by the integument.

Epidermis outer side. Cells arranged in longitudinal rows; rectangular tables; at the base of the seed in a radial section often papilla-like; in the places corre- sponding to the raphe cells in a transverse section much smaller.

C e l l s   o f   t h e   a b o v e. R. 10 μ, T. 12 μ, L. 25—40 μ. Walls, especially the inner walls brown; persisting in concentrated sulphuric acid, in potash 50 per cent. no yellow bladders, in osmic acid not colouring black; outer walls showing a cuticle.

Ground tissue. Represented by a thin brown band of flattened tissue; in the raphe part outermost cell layers not flattened.

C e l l s   o f   t h e   a b o v e. R. and T. 16 μ, L. 50 μ; in the raphe part R. and T. 8 μ; polygonal prisms with a longitudinally directed axis.

Meristele of raphe. Consisting of an amphivasal vascular bundle. Phloem mostly torn. Xylem composed of spiral vessels.

Epidermis inner side. Flattened.

N u c l e u s of the seed.

Endosperm orthosperm.

E p i d e r m i s. Outer walls of the cells somewhat thickened; for the rest see the ground tissue.

G r o u n d   t i s s u e. Cells arranged in radial rows. In the centre at the top the embryo; below this a central cavity and below this again a tissue consisting of cells differing from the other endosperm cells and differently arranged.

C e l l s   o f   t h e   a b o v e. Radially arranged cells R. 30 μ, T. and L. 20 μ; polygonal prisms with a longitudinally directed axis; cells of the tissue below the cavity R. 20 μ, T. 25 μ, L. 35 μ; rectangular prisms with a longitudinally directed axis. Walls somewhat thickened; colouring yellow in iodine, not or hardly colouring blue in iodine and sulphuric acid 66 per cent. Cell contents: numerous aleurone grains, 5 μ in diameter or smaller. The

---

[1]) After K a y s e r. l. c. 79, and T a n f a n i. l. c. 599.

greater part of the grains containing crystals, mostly cluster crystals with a small central air-bubble, a few times a simple crystal; the other grains containing 1 or more globoids. See for the distribution of the aleurone grains in the several cells Fructus Coriandri. Ground mass of the grains soluble in water. Oil as oil plasm. Cell nucleus conspicuous.

Embryo. Much varying in length, from 500 to 1500 μ. The 2 cotyledons distinct; showing procambial bundles. Once noticed an embryo with 3 cotelydons. Cell contents: aleurone grains and oil.

Micrography of the powder.

A great many oil drops (in chloral hydrate). Parenchyma cells with dark brown, somewhat thickened walls. Colourless parenchyma of the endosperm, somewhat thick-walled; in many cells several small cluster crystals with a central air-bubble. Inner epidermis of the wall of the mericarp composed of low cells, strongly elongated in an all but tangential direction, often connected with the vittae. Oblong colourless parenchyma cells, strongly and coarsely reticulate. Parts of vittae yellowish brown, sometimes paved with thin-walled polygonal cells. Bundles of annular and spiral vessels and thick-walled fibres. Epidermis consisting of polygonal colourless cells; stomata.

February 1903.   J; M; L.

FRUCTUS PIMENTAE.
Pimenta. Pimento.
The dried, full grown, unripe fruit of Pimenta officinalis, Lindl. Coll. Bot. sub t. 19.

LITERATURE. Erdmann-König's. Allg. Warenkunde. 1895. 322. Flückiger. Pharmakogn. 1891. 957. Flückiger & Hanbury. Pharmacogr. 1879. 287. Gilg. Pharmakogn. 1910. 238. Greenish & Collin. Anat. Atl. of veget. Powders. 1904. 162. Hager. Pharm. Prax. Bd. 2. 1902. 627. Ingerman. Mikr. d. voorn. Handelsw. 1910. 158. Kraemer. Bot. a. Pharmacogn. 1910. 755. Lutz. Oblito-schizog. Sekretbeh. d. Myrtaceen. Bot. Centralbl. Bd. 64. 1895. 114. Also Diss Bern. 1895. Marmé. Pharmacogn. 1886. 330. Moeller. Mikr. d. Nahr. u. Genussm. 1886. 254. Moeller. Pharmakogn. 1886. 142. Moeller. Mikr.-pharm. Üb. 1901. 193. Oudemans. Pharmacogn. 1880. 393. Planchon et Collin. Drogues simples. T. II. 1896. 331. Schimper. Mikr. Unters. d. veget. Nahr. u. Genussm. 1900. 97. Schneider. Powdered veget. Drugs. 1902. 257. Solereder. Bemerkungsw. anat. Vorkommn. Archiv d. Pharm. Bd. 245. 1907. 410. Tschirch. Angew. Pfl. Anat. 1889. 165, 487. Tschirch. Harze u. Harzbeh 1900. 367. Tschirch. Pharmacogn. II. 1911. 1239. Tunmann. Unters. üb. d. Sekretbeh. (Drüsen) ein. Myrtaceen spez. üb. ihr. Entleerungsapparat. Arch. d. Pharmac. Bd. 248. 1910. 23. Vogl. Veget. Nahr. u. Genussm. 1899. 426. MATERIAL. The drug. REAGENTS. Water, glycerine, potash 50 per cent., chloral hydrate, potassium iodide iodine, iodine in chloral hydrate, phloroglucin and hydrochloric acid, iodine and sulphuric acid 66 per cent., concentrated sulphuric acid, Schulze's macerating mixture, osmic acid, iron acetate, ammonia.

MICROGRAPHY.
Pericarp.
Wall.

Epidermis outer side. Cells more or less flattened; those above the glands somewhat larger than the other ones. Moreover above the centre of each gland a perforated cover consisting of 2 cells, resembling a stoma, lying in the same level as the epidermal cells, T. 25 μ, L. 20 μ; the whole cover elliptical; perforations circular. Sometimes below this cover a similar one with or without a perforation and joining the epithelium of the gland [1]). Trichomes. Only the scars corresponding to them to be seen, circular or elliptical, e.g. 8 μ by 10 μ. Walls somewhat thickened [2]); unicellular trichomes, long 75—220 μ, more or less curved and twisted, thick-walled.

E p i d e r m a l c e l l s p r o p e r. R. 10 μ, T. 6—8 μ, L. 8—10 μ; above the glands R., T. and L. 10 μ; polygonal prisms. Outer walls thickened, with a thick cuticularized part; lateral walls showing a cuneiform cuticularized part; all the walls brown in potash 33 per cent., in phloroglucin and hydrochloric acid reddish. Cell contents: here and there a red mass. Walls and contents infiltrated with oil, for the greater part consisting of eugenol; therefore colouring black in iron acetate and in osmic acid 1 per cent. The same phenomenon in all the cell walls of the wall of the pericarp. C e l l s o f c o v e r. Walls slightly lignified; the outer ones covered with a cuticle continued inwards along the perforation. Cell contents: a granular mass.

G r o u n d t i s s u e. Consisting of 3 parts.

1. An outer part formed of common parenchyma containing numerous closely approximated glands and idioblasts containing a cluster crystal, exceptionnally a simple crystal. The cluster crystals, not the simple ones, dissolving in potash. Between the epidermis and the glands 1—3 layers of common parenchyma cells; the cells of those outer layers smaller than the rest; no intercellular spaces. G l a n d s oblito-schizogenous [3]); diameter 100—180 μ; globular, sometimes a little radially elongated. The sclerenchyma sheath not well developed. Epithelium of 1 or 2 layers of cells; resinogenous layer not distinct. Cavity filled with a yellow to reddish brown mass, colouring black in iron acetate and in osmic acid 1 per cent. In macerated material the glands with their epithelium present in entire.

C o m m o n p a r e n c h y m a c e l l s. In the outermost layers R., T. and L. 10 μ, for the rest R. 12—15 μ, T. and L. 20—35 μ. Walls dark brown; lignified. Cell contents: here and there a brown mass; in the inner layers often starch grains. Walls and contents infiltrated like those of the cells of the outer epidermis. E p i t h e l i u m c e l l s. Thick 8 μ, wide 25 μ, long 35 μ; polygonal with rounded sides. Walls yellow to brown; persisting in concentrated sulphuric acid; lignified.

2. A middle part constituting the bulk of the ground tissue, consisting of common parenchyma, sclerenchyma and crystal idioblasts. The parenchyma more or less flattened, towards the interior forming a flattened layer. Sometimes a cell divided in 4 by 2 partition walls at right angles to each other, each partition filled with a cluster crystal. Showing intercellular spaces. Sclereids isolated or in groups; towards the interior the groups larger and more numerous. Idioblasts containing cluster crystals, rarely simple crystals.

C o m m o n p a r e n c h y m a c e l l s. R. 15 μ, T. 30 μ, L. 35 μ; towards the interior T. and L. larger; polygonal with rounded edges. Walls lignified. Cell contents: here and

[1]) L u t z. l. c. 262.
[2]) According to V o g l. l. c. 426.
[3]) According to L u t z. l. c. 263.

there a starch grain. Walls and contents infiltrated like those of the outer epidermis. S c l e r e i d s. The smallest 20 by 30 by 30 μ, the largest 60 μ in diameter; polygonal with smaller and larger branches, the isolated ones with strongly rounded edges. Walls strongly thickened, sometimes thick about 15 μ; showing very fine stratification; ligni-fied, especially the middle lamella, in potash 50 per cent. often sea-green; showing much branched pit canals. Cell contents: several cells filled with a brown granular mass.

3. An inner part composed of a narrow band of flattened tissue.

C e l l s   o f   t h e   a b o v e. Often more or less black in iron acetate and in osmic acid. Cell contents: now and then some starch grains.

M e r i s t e l e s. In the middle part of the ground tissue; much branched; not all running in the same plane. The larger meri-steles in the shape of semi-elliptical, strong-ly elongated cylinders, the flat side turned outwards. P e r i c y c l e. On the outside of the meristeles sometimes a single sclerenchyma fibre, sometimes also on other sides a single crystal fibre containing simple crystals. V a s-c u l a r   b u n d l e s   o f the larger meristeles

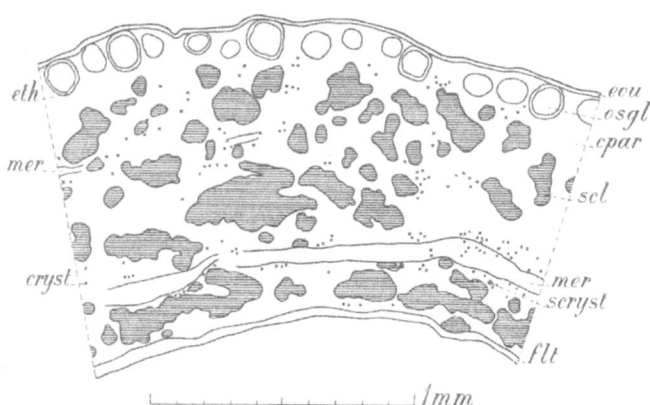

Fig. 48. *Pimenta officinalis.* Fruit, longitudinal section. cpar Common parenchyma; cryst Cells with cluster crystals; eou Epidermis outer side; eth Epithelium; flt Flattened tissue of the inner part of the ground tissue; mer Meristeles; osgl Oblito-schizogenous glands; scl Sclereids; scryst Cells with a simple crystal.

amphicribral, those of the smaller meristeles rather indistinct. Phloem represent-ed by a flattened brown mass. Xylem consisting of radial strips of 1 vessel wide, intermixed with radial strips of parenchyma also 1 cell wide.

E p i d e r m i s   i n n e r   s i d e. Flattened. Cells showing a cuticle; often colour-ing black in iron acetate.

Septum. General structure the same as that of the wall of the pericarp. Con-sisting of flattened parenchyma, sometimes containing starch grains; here and there a small group of sclereids; numerous cluster crystal idioblasts, some simple crystals; here and there a meristele.

S e e d. [1]

Seed-coat.

L a y e r s produced by the **o u t e r   i n t e g u m e n t.**

Epidermis outer side. Mostly more or less flattened. Cells polygonal tables, some-times more or less fibre-shaped.

---

[1] Ovule according to E n g l e r   u n d   P r a n t l, Die natürlichen Pflanzenfamilien, Teil III, Abt. 7, 57 and B e n t h a m   e t   H o o k e r, Genera Plantarum, Vol. I, 690 anatropous or campylotro-pous.

Cells of the above. R. 8 μ, T. 9—20 μ, L. 60—80 μ. Outer walls somewhat thickened.

Ground tissue. Consisting of some layers of strongly flattened cells; cells of these layers containing a reddish brown mass; the innermost 1 or 2 layers showing thinner cell walls than the rest. Sometimes moreover long fibre-like elements occasionally containing a cluster crystal. Walls of all these elements colouring brown in potash. In the uppermost part of the seed the seed-coat much thicker and protruding between radicle and cotyledons; in this place, going from the exterior towards the interior, the ground tissue showing the following parts: 1. next to the epidermis some layers of cells similar to those of the rest of the ground tissue; 2. several layers of polygonal thin-walled common parenchyma cells, sometimes containing cluster

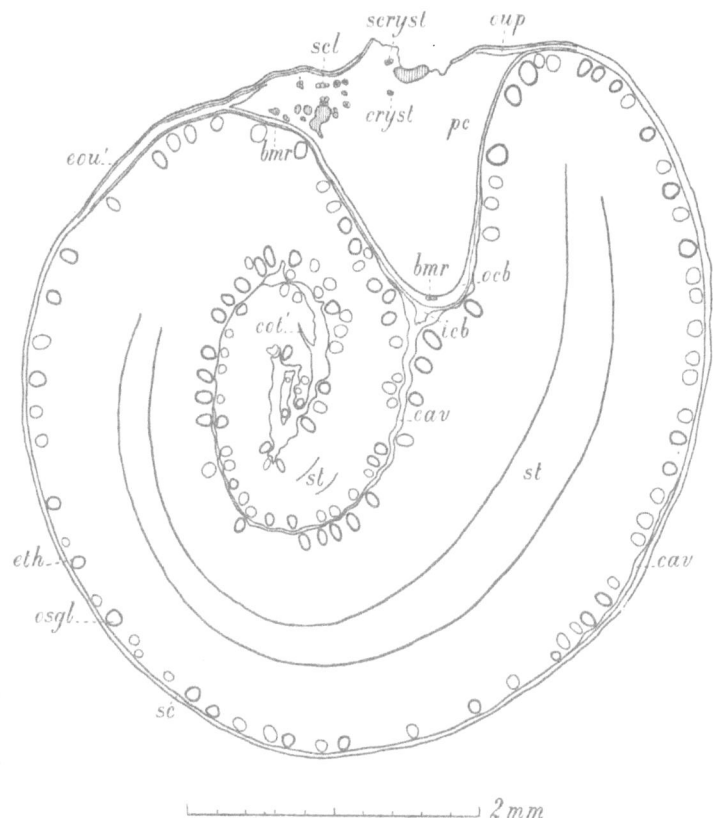

Fig. 49. *Pimenta officinalis*. Seed, longitudinal section. bmr Branched meristele of raphe; cav Cavity; cot' Cut cotyledons; cryst Cells with cluster crystals; eth Epithelium; eou' Epidermis outer side seed-coat; icb Innermost band of colourless parenchyma; ocb Outermost band of colourless parenchyma; osgl Oblito-schizogenous glands; oup Outermost part of ground tissue; pc Second part of ground tissue consisting of polygonal parenchyma cells; sc Seed-coat; scl Sclerenchymatous elements; scryst Cells with a simple crystal; st Stele.

crystals or simple crystals; contents of these cells reddish brown, colouring black in osmic acid and blue in iron acetate; cells R. 20 μ, T. 100 μ; 3. a thin band of colourless tissue, the cells of which indistinguishable; 4. another narrow band of colourless tissue. Between the other parts of the embryo the seed-coat protruding very little or not at all.

Meristele of raphe. Running in the first thin colourless band in the thick part of the seed-coat mentioned above. Course of the meristele not distinct. Meristele branched; showing on its outer side sclerenchymatous elements; containing spiral vessels, 5 µ in diameter, with lignification.

Epidermis inner side. Not discernible, perhaps represented by a cuticle in the vicinity of the embryo and in the thick part of the seed-coat on the inner side of the meristele of the raphe.

L a y e r s produced by the **i n n e r   i n t e g u m e n t** and the **e n d o s p e r m.** Insignificant.

Nucleus of the seed. Albumen wanting.

E m b r y o. With very small cotyledons, sometimes the latter wanting. Nearly quite consisting of radicle and plumula.

Epidermis. Cells arranged in longitudinal rows. Above each gland a mostly perforated cover consisting of 2 cells as described for the epidermis of the wall of the pericarp; 25 by 25 µ. Cells of cover in a transverse section very shallow; under each cell another cell, both apparently originating from the same cell by a tangential partition wall; this second cell showing the resinogenous layer of the gland and consequently belonging to the secreting cells.

Cortex. Cells arranged in longitudinal rows; rows of shorter and longer cells next to each other. Small intercellular spaces.

C e l l s   o f   t h e   a b o v e. Short cells R. and T. 45 µ, L. 35 µ; long cells R. and T. 25 µ, L. 50 µ. Walls bordering the intercellular spaces slightly thickened; somewhat brownish red; in phloroglucin and hydrochloric acid or in hydrochloric acid only colouring red, in ammonia decolorating, in concentrated sulphuric acid nearly not colouring red, in osmic acid black, in iron acetate blue, in potash brown. Cell contents: starch grains, simple or compound; the latter 2- to 4-adelphous and diameter e. g. 11 µ; simple grains globular, diameter 8 µ or smaller; oil; in several embryo's in many cells in the parts of the embryo towards the cotyledons a red globule, not colouring blue in iron acetate, in phloroglucin and hydrochloric acid or in hydrochloric acid only colouring more intensely red.

G l a n d s. Arranged in a layer close to the epidermis; less numerous than in the pericarp; oblito-schizogenous [1]); R. and L. 50 µ; ellipsoidal. Sclerenchyma sheath not developed; secreting layer not clearly to be distinguished. Resinogenous layer very conspicuous; yellowish brown; forming caps strongly bulging out into the cavity of the gland, about 10 µ thick; sometimes connection between those caps and secreting cells distinct, e.g. in the cells immediately lying under the epidermis (see epidermis); showing colourless, often radially elongated cavities, extending from the margin towards the centre of the gland; in those cavities no oil to be seen; on the cavity side of the resinogenous layer a very thin lamella persisting in concentrated sulphuric acid; the whole layer colouring intensely black in osmic acid, very dark blue in iron acetate, reddish brown in potash, yellow in concentrated sulphuric acid and in phloroglucin and hydrochloric acid or in hydrochloric acid only. Contents of the cavity of the gland a yellow to reddish brown mass, colouring black in iron acetate and in osmic acid 1 per cent.

---

[1]) According to L u t z. l. c. 263.

Stele. Thin; consisting of a thin ring of small parenchyma cells intermixed with bundles of minute elements; at the inner side of the ring here and there some fibres with spiral thickenings. M e d u l l a consisting of parenchyma cells like those of the cortex, only somewhat smaller.

S m a l l   p a r e n c h y m a   c e l l s. Diameter 10 μ. Cell contents: like those of the cortical parenchyma cells; in the cells of the ring sometimes red globules like those described for the parenchyma of the cortex. M i n u t e   e l e m e n t s. Diameter 3—4 μ. F i b r e s. Diameter 5 μ. Walls strongly thickened and lignified.

April 1903.   J; M; L.

FRUCTUS PIPERIS NIGRI.
Piper Nigrum. Black Pepper. Piper. Pepper.
The dried unripe fruits of Piper nigrum, Linn. Sp. Pl. 28.

LITERATURE. Attema. De zaadhuid der Angiosp. en Gymnosp. Diss. Groningen. 1901. 169. Biermann. Ueber Bau u. Entwicklungsgesch. d. Oelzellen. Diss. Bern. 1898. 50—53. Copper. Beitr. z. Entwicklungsgesch. der Samen u. Früchte offizin. Pflanzen. Diss. Bern. 1908. Deiveure. Recherches sur le cubebe et sur les Piperaceae etc. 1894. Erdmann-König. Allg. Warenk. 1895. 317. Flückiger. Pharmakogn. 1891. 913. Flückiger & Hanbury. Pharmacographia. 1879. 576.* Gilg. Pharmakogr. 1910. 80. Greenish & Collin. Anat. Atl. o. veget. powders. 1904. 160. Hager. Pharm. Praxis. Bd. 2. 1902. 634. Hérail. Mat. Med. 1912. 427. Ingerman. Mikrosk. der voornaamste Handelswaren. 1910. 149. Ironside. The anatomical structure of the New-Zealand Piperaceae. Trans. N. Zealand Inst. XLIV. 339—348. Johnson. Studies in the development of the Piperaceae. Bot. Centrbl. Bd. 117. 290. Karsten u. Oltmanns. Pharmakogn. 1909. 242. Luerssen. Syst. Bot. Bd. 2. 1882. 517. Marmé. Pharmacogn. 1886. 282. Meunichet. Ueber eine Verfälschung des Pfeffers etc. Journal de Pharmacie. XLVII. 1901. No. 15. Moeller. Mikr. d. Nahr. u. Genussm. 1886. 226. Moeller. Pharmakogn. 1889. 134. Moeller. Mikr.-pharm. Ueb. 1901. 184. Molisch. Mikrochemie d. Pflanze. 1913. 256. Oudemans. Pharmacogn. 1880. 361. Planchon et Collin. Drogues simples. T. I. 1895. 407. Schimper. Mikr. Unters. d. veget. Nahr. u. Genussm. 1900. 80. Schneider. Powdered Veget. Drugs. 1902. 258. v. Tieghem. Sur les canaux à mucilage des Piperacées. Ann. d. sc. nat. Série 9. T. 7. 1908. 117—127. Tschirch. Angew. Pfl. Anat. 1889. 77. Tschirch u. Oesterle. Anat. Atl. 1900. 103. Taf. 25. Tunmann. Mikrochemie. 1913. 283. Vogl. Veget. Nahr. u. Genussm. 1899. 391. Wigand. Pharmakogn. 1879. 285. MATERIAL. The drug. REAGENTS. Water, glycerine, potash, iodine in chloral hydrate, phloroglucin and hydrochloric acid, iodine and sulphuric acid 66 per cent., concentrated sulphuric acid, Schulze's macerating mixture, iron acetate, copper sulphate and potash 50 per cent., nitric acid, Millon's reagent, molybdate of ammonia [1]) and concentrated sulphuric acid.

MICROGRAPHY.
P e r i c a r p.

Epidermis outer side. Stomata not observed [2]).

C e l l s   o f   t h e   a b o v e. R. 20 μ, T. and L. 10—20 μ; polygonal prisms. Outer walls thickened; showing a cuticle. Cell contents: a brown mass.

Ground tissue. To be divided into 4 parts:

1. An outer part consisting of 3 layers of sclereids, mixed with a very few common parenchyma cells, especially in the innermost layer.

---

[1]) See footnote p. 226.
[2]) T s c h i r c h has seem them, Anat. Atl. p. 104 and Fig. 12.

Sclereids. R. 25—80 μ, T. and L. 20 μ; polyhedra or polygonal prisms. Walls thickened; lignified, lignification diminishing towards the cell cavity; showing pit canals. Cell contents: a red-brown mass. Parenchyma cells. R. 20 μ, T. and L. 15 μ. Cell contents: a brown mass.

2. A mostly strongly flattened part consisting of about 10 layers of common parenchyma cells, intermixed at the top of the pericarp with numerous, at the base with a few sclereids and here and there with an also flattened oil cell. Parenchyma cells. R. 40 μ, T. and L. 50 μ; polyhedra. Walls somewhat brown. Cell contents: many very small starch grains, often united into a brown mass. Sclereids. Somewhat longitudinally elongated. See for the rest those of the first part. Oil cells. R. 45 μ, T. and L. 50 μ; ellipsoidal; see for the rest those of the following part.

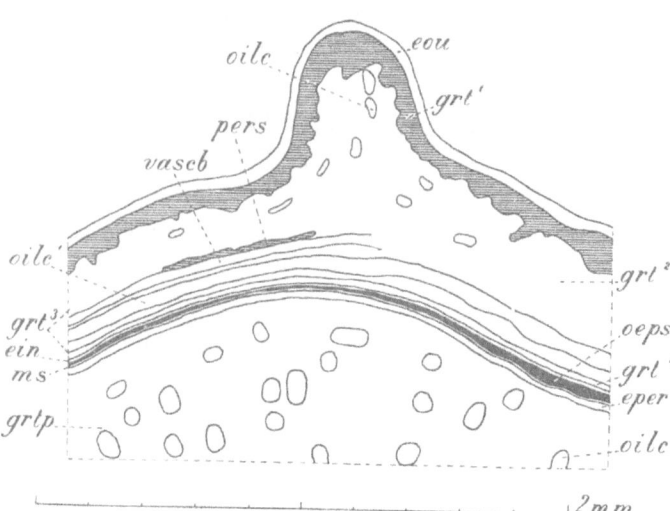

Fig. 50. *Piper nigrum*. Unripe fruit, transverse section of the middle part, after boiling in potash 50 per cent. ein Epidermis inner side of the pericarp; eou Epidermis outer side of the pericarp; eper Epidermis of perisperm; grt1 Ground tissue outer part, consisting of sclereids; grt2 Ground tissue 2nd part; grt3 Ground tissue 3rd part; grt4 Ground tissue 4th part; grtp Ground tissue of perisperm; ms Membrane of seed-coat; oeps Outer epidermal layer of the seed-coat; oilc Oil cells; oilc' A fairly complete continuous layer of oil cells; pers Pericyclic sclerenchyma of the meristeles; vascb Vascular bundles of meristeles of the ground tissue of pericarp.

3. An often more or less flattened part consisting of about 10 often concentric layers of common parenchyma cells, containing meristeles and oil cells in the outer part. Those meristeles separated by 2 innermost layers of parenchyma cells from the fourth part. Oil cells in groups of 1—4 cells in number, thick only 1 cell in a radial direction or sometimes forming a fairly complete continuous layer. Oil cells less numerous at the top and the base of the pericarp.

Parenchyma cells. 1. Of the outer 8 cell layers R. 10 μ, T. 20—35 μ, L. 40—75 μ; polygonal tables with a radially directed axis. Walls pitted; of cells close to the groups of oil cells colouring red in phloroglucin and hydrochloric acid. Cell contents: small starch grains. 2. Of the innermost cell layers R. 12 μ, T. and L. 20 μ; polygonal tables. Walls pitted. Oil cells[1]). R. 45 μ, T. and L. 60—90 μ; polyhedra, often with rounded sides. Walls often colouring red in phloroglucin and hydrochloric acid; the outermost lamella persisting in concentrated sulphuric acid (probably suberized). Cell contents: one large oil drop or a more irregular mass of oil, often containing some smaller globules and some small needle-shaped crystals; in concentrated sulphuric acid oil colouring at first red,

---

[1]) See Biermann, l. c.

afterwards brown and darker; in nitric acid contents colouring yellow and showing evolution of gas; on treatment first with molybdate of ammonia followed by concentrated sulphuric acid about the same reaction as by treatment with concentrated sulphuric acid only.

4. An inner part consisting of a single layer of cells; at the top and at the base of the pericarp not clearly discernible.

C e l l s o f t h e a b o v e. R., T. and L. 20 μ; polygonal prisms with a radially directed axis. Inner walls and inner parts of radial walls thickened; lignified; showing stratification and pit canals.

Meristeles. Occurring in the outer part of the third part of the ground tissue; showing a pericycle, consisting of 1—5 sclereids in a cross section; vascular bundles collateral.

S c l e r e i d s. R. 15 μ, T. 20 μ, L. 50—150 μ; polygonal prisms with a longitudinally directed axis and rounded edges. Walls thickened; lignified; showing stratification and pit canals. S p i r a l v e s s e l s showing lignification.

Epidermis inner side. Most clearly discernible only at the top of the pericarp, less so for the rest; consisting of cells, rectangular in cross and radial sections; mostly flattened.

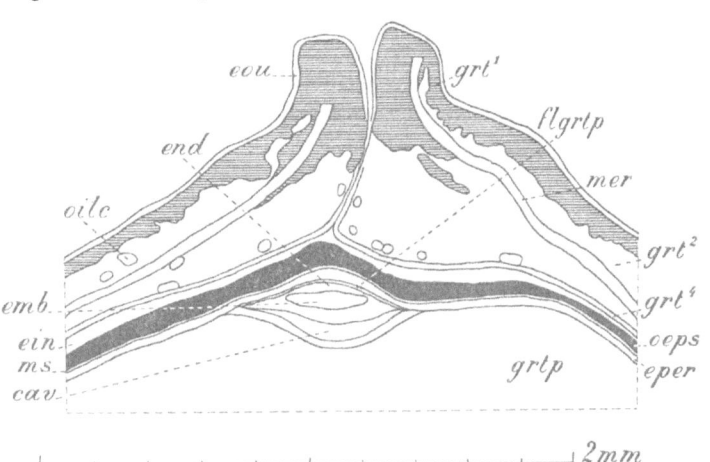

Fig. 51. *Piper nigrum.* Unripe fruit, longitudinal section of the top. cav Cavity; ein Epidermis inner side of the pericarp; emb Embryo; end Endosperm; eou Epidermis outer side of the pericarp; eper Epidermis of perisperm; flgrtp Flattened layer of ground tissue of perisperm; grt1 Ground tissue outer part, consisting of sclereids; grt2 Ground tissue 2nd part; grt4 Ground tissue 4th part; grtp Ground tissue of perisperm; mer Meristeles of the ground tissue of pericarp; ms Membrane of seed-coat; oeps Outer epidermal layer of the seed-coat; oilc Oil cells.

C e l l s o f t h e a b o v e. At the top of the pericarp R. 35 μ, T. 50—90 μ, L. 60—90 μ. Tangential walls turned towards the inside of the pericarp and parts of the radial and transverse walls turned the same way, both strongly thickened; coloured red-brown. For the rest R. 8 μ, T. and L. 20—40 μ. Walls not thickened. Walls and cell contents somewhat brown.

S e e d. [1])

―――――――――――

[1]) I have followed T s c h i r c h u. O e s t e r l e l. c. in the interpretation of the several cell layers constituting the fruit and seed because T s c h i r c h u. O e s t e r l e have examined their development and because I myself had no material for such an investigation at my disposition. Nevertheless I cannot suppress some doubts about the correctness of their interpretation, raised by the following observations. Between the 2 cell layers interpreted, according to T s c h i r c h u. O e s t e r l e, as innermost cell layers of ground tissue of pericarp and inner epidermis of same, there appeared in iodine and sulphuric acid 66 per cent. a brown lamella, exhibiting incurvations corresponding with the lateral walls of the layer considered as epidermis; moreover there appeared here and there a thin brown

Ovule. Atropous; with a single integument; showing at the base sometimes the remainders of a second integument, extending up to one third of the height of the ovule.

Ripe seed.

S e e d-c o a t. Consisting in the first place of a cell layer, only to be distinguished with difficulty, corresponding with the outer epidermis of the integument, and secondly of an uniform transparent membrane, corresponding with the flattened ground tissue and inner epidermis of the integument.
C e l l s o f o u t e r e p i d e r m a l l a y e r. R. 8 μ; those at the top of the seed R. 40 μ, T. 8 μ, L. 10 μ; those at the base of the seed radially also much more elongated. Cell contents: a red-brown mass.
M e m b r a n e. ¹) Resembling exactly thickened outer walls of the epidermal cells of the perisperm; colouring yellow in iodine and sulphuric acid 66 per cent.
N u c l e u s of seed.

Albumen.

Perisperm. Forming the bulk of the seed; showing a central cavity extending into a narrow canal and leading towards the top of the perisperm.
E p i d e r m i s and some flattened cell layers of ground tissue extending over the endosperm.
E p i d e r m a l c e l l s. R. 15 μ, T. 20 μ, L. 30 μ; polygonal prisms. Cell contents: some small globular starch grains; the rest of the cell cavity filled up with small aleurone grains.
G r o u n d t i s s u e. Consisting of common parenchyma cells; with the exception of the 3—4 outermost cell layers intermixed with oil cells, larger in size than the surrounding parenchyma cells.
P a r e n c h y m a c e l l s. Of the outermost 4 cell layers R. 15 μ, T. and L. 20 μ; of the rest R. 80 μ, T. and L. 40 μ; polyhedra. Cell contents: 1. of the outermost 4 cell layers some small starch grains, the rest of the cell cavity filled up with aleurone grains; 2. of the inner part of the perisperm very numerous small starch grains, sometimes clustered together into a single mass or sometimes into some masses, measuring e. g. 15 by 18 μ and separated from one another by thin layers of protoplasm; these layers and the layer of protoplasm covering the walls colouring red in Millon's reagent; 3. of cells in the vicinity of the central cavity of the perisperm the starch grains less clustered together. O i l c e l l s. R. 90 μ, T. and L. 60 μ; polyhedra, often with rounded sides. Oil coloured somewhat yellow. See for the rest those of ground tissue of pericarp.
Endosperm. Very small, lying at the top of the perisperm. Embryo sometimes discernible.

January 1903.   J; M; v. E. d. W.

---

lamella projecting from the brown lamella mentioned above into the lateral walls of the cells belonging to the layer interpreted as the innermost layer of the ground tissue. Further investigations concerning the development of the fruit will show whether my observations are due to the exsistance of 2 cuticular formations in this place. In this case the cell layer considered by T s c h i r c h u. O e s t e r l e as the innermost layer of ground tissue of the pericarp would appear to be the inner epidermis of the pericarp, whilst the layer considered by these authors as the inner epidermis would appear to be the outer epidermis of the seed-coat.
¹) According to T s c h i r c h und O e s t e r l e, after treatment with Schulze's macerating mixture, appearing to consist of 3 layers of flattened cells.

FRUCTUS SAMBUCI RECENTES.
Fresh Elder Fruit.

The fresh ripe fruits of Sambucus nigra, Linn. Sp. Pl. 269.

Macroscopic characters.

Globular or ovate compound drupes, long 5—7 m.M., purplish black, shining, crowned with the calyx teeth and the scars of the stigmata. Stones 3, rarely 2 in number, oblong, flattened laterally, yellowish brown.

Anatomical characters.

LITERATURE. Bochmann. Beitr. z. Entw.-gesch. off. Samen u. Früchte. Diss. Bern. 1901. 58. Hager. Pharm. Praxis. Bd. II. 1902. 801. Pampe. Z. Kenntn. d. Baues u. d. Entw. säft. Früchte. Diss. Halle-Wittenberg. 1884. Tschirch u. Oesterle. Anat. Atl. 1900. 44. Pl. XII. Tschirch. Pharmakogn. Bd. II. 1911. 53. MATERIAL. Unripe fruits, gathered in the Botanic Garden at Groningen, in alcohol. Ripe fruits, gathered October 1901, fresh and in alcohol. REAGENTS. Water, glycerine, chloral hydrate, iodine in chloral hydrate, phloroglucin and hydrochloric acid, iodine and sulphuric acid 66 per cent., concentrated sulphuric acid, Schulze's macerating mixture.

MICROGRAPHY.

Pericarp.

Peel.

Epidermis. Stomata very rare; lying in the same level as the epidermal cells.

Epidermal cells proper. R. 35 μ, T. 120 μ, L. 70 μ; tabular. Walls somewhat thickened; outer walls showing coarse longitudinal cuticular striation, striae sometimes at 20 μ distance from each other and independent of the circumference of the cells; lateral and inner walls pitted. Cell contents: chloroplasts containing starch grains; cell sap red to brown, sometimes nearly black. Stomata. 50 by 50 μ. Guard-cells. Containing colourless cell sap.

Ground tissue of the outermost part of the pericarp. Consisting of a few not very fleshy layers of parenchyma cells.

Cells of the above. Tabular. Walls somewhat thickened; pitted. Cell contents: many chloroplasts containing starch grains; cell sap often coloured like that of the epidermal cells.

Pulp.

Middle part of the ground tissue of the pericarp. Forming the bulk of the pulp and consisting of common parenchyma with intercellular spaces; cells smaller in the layers adjoining the stone; here and there a crystal idioblast especially in the neighbourhood of the stone.

Cells of the principal part. 125 by 150 μ; globular, ellipsoidal, pear-shaped or polyhedra with rounded edges, sometimes more irregular. Cell contents: like those of the cells forming the inner part of the peel, chloroplasts less numerous. Cells adjoining the stone. R. 15 μ, T. 20 μ, L. 25 μ; tetragonal prisms. Cell contents: like those of the preceding cells. Crystal idioblasts. Filled with crystal sand.

Meristeles of pericarp. Extending in a longitudinal direction through the pulp, 6 to 10 in number in the peripheral part of the pulp and 3 in the central part.

Stone.

E l e m e n t s forming the inner part of the **g r o u n d   t i s s u e** of the pericarp. Sclereids in a single outer layer forming groups of more radially elongated elements intermixed with groups of less elongated ones.

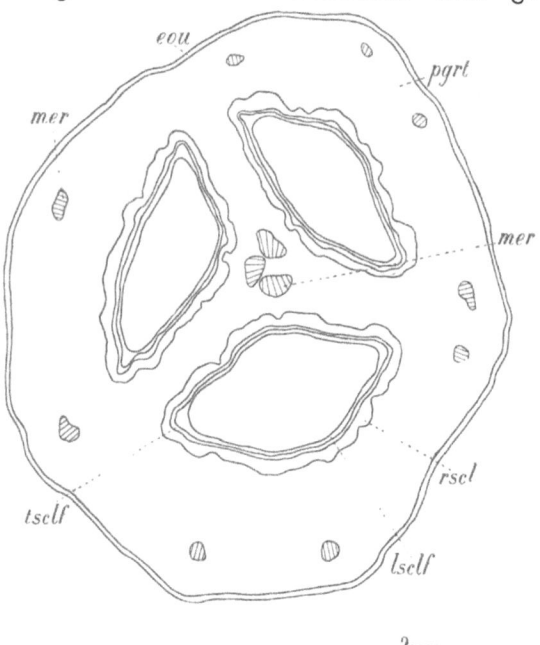

Sclerenchyma fibres in 1 or 2 layers, transversely directed and often arranged in longitudinal rows.

S c l e r e i d s. R. 20—100 μ, T. 20 μ, L. 35 μ; tetra- to hexagonal prisms with a radially directed axis. Walls very thick; yellow; somewhat lignified; lumen branched. Cell contents: in unripe fruits in each cell a heap of small crystals. S c l e r e n c h y m a f i b r e s. R. and L. 15—20 μ, T. 250—400 μ; tetra- to hexagonal, often with branched ends. Walls very thick; especially the outer part showing stratification; outer part yellow; lignified; showing pit canals; lumen sometimes branched.

E p i d e r m i s of pericarp inner side. Sclerenchyma fibres in 1 or 2 layers, longitudinally directed and often arranged in transverse rows.

Fig. 52. *Sambucus nigra*. Unripe fruit, transverse section. eou Epidermis outer side; lsclf Longitudinally directed sclerenchyma fibres of the stone; mer Meristeles of pericarp; pgrt Parenchymatous ground tissue forming the pulp and a part of the peel; rscl Radially elongated sclereids of the stone; tsclf Transversely directed sclerenchyma fibres of the stone.

F i b r e s o f t h e a b o v e. R. and T. 15—20 μ, L. 250—400 μ. See for the rest the preceding fibres.

October 1901.

J; M; L.

GALLAE.

Galla. Galls. Nutgall.

Excrescences, resulting from the action of a gall-wasp, Cynips Gallae tinctoriae Oliv., on the young branches of Quercus lusitanica, Lam. Encyc. I. 719, Quercus infectoria, Oliv. Voy. Atlas, tt. 14, 15.

LITERATURE. Berg. Anat. Atl. 1865. Tafel XXXIX. 98. Benecke. Mikr. Drogenprakt. 1912. 89. Beyerinck. Beobachtungen ü. d. ersten Entwicklungsphasen einiger Cynipidengallen. Amsterdam. 153. Flückiger. Pharmakogn. 1891. 263. Flückiger & Hanbury. Pharmacographia. 1879. 595. Gilg. Pharmakogn. 1910. 84. Greenish & Collin. Anat. Atl. o. veget. powders. 1904. 286. Hager. Pharm. Praxis. Bd. I. 1900. 1194. Hartwig. Ueber Gerbstoffkugeln u. Ligninkörper i. d. Nahrungsschicht d. Infectoriagalle. Ber. d. d. bot. Ges. 1885. Hérail. Mat. Med. 1912. 487. Karsten u. Oltmanns. Pharmakogn. 1909. 313. Luerssen. Syst. Bot. Bd. 2. 1882. 500. Marmé. Pharmacogn. 1886. 118. Moeller. Mikr. pharm. Ueb. 1900. 326. Oudemans. Aanteek. o. d. Pharmac. neerl. 1854—56. 103. Oudemans. Pharmacogn. 1880. 443. Planchon et Collin. Drogues simples. T. I. 1895. 261. Tschirch. Angew. Pfl. Anat. 1889. 104, 126, 451. Tunmann. Pfl.-microchemie. 1913.

259. Wigand. Pharmakogn. 1879. 321. MATERIAL. The drug; galls 2 c.M. in diameter. REAGENTS. Water, glycerine, safranin, potash, iodine in chloral hydrate, phloroglucin and hydrochloric acid, iodine and sulphuric acid 66 per cent., concentrated sulphuric acid, Schulze's macerating mixture, iron acetate, iron chloride, copper sulphate and potash.

MICROGRAPHY.

The whole body of the gall consisting of ground tissue, containing some meristeles; the outer and inner epidermis having disappeared.

Ground tissue. Developed in 3 different parts.

I. The outer and main part of the gall. Consisting of about 110 layers of parenchyma cells; again to be divided into 3 parts: a. 5—10 outer layers of smaller cells, intermixed with a few simple crystal or cluster crystal idioblasts, each showing a single cluster crystal; b. 40 middle layers of larger cells, also intermixed with crystal idioblasts; c. 50—70 inner layers also consisting of larger cells, but more or less arranged in radial rows; crystal idioblasts more numerous and also more or less arranged in radial rows.

In the superficial knobs and ridges of the gall the outer layers consisting of common parenchyma cells, resembling those of the 5—10 outer cell layers mentioned above; the more central part consisting of cylindrical common parenchyma cells with a longitudinally directed axis.

Parenchyma cells. a. Of outer part R. 8 μ, T. 15 μ, L. 20 μ; polygonal tables or more or less flattened polyhedra. Walls thickened; of the outermost 1—2 layers suberized and of a few cells also lignified; of the other layers colouring blue in potassium iodide iodine only; pitted; intercellular spaces wanting. b. Of middle part, e. g. 60—70 μ in diameter, or R. 45 μ, T. 100 μ, L. 70 μ; polyhedra with rounded edges, often tangentially elongated. Walls a little thickened; pitted; showing intercellular spaces. c. Of inner part R. 150 μ, T. 70 μ, L. 80 μ; polygonal prisms with rounded edges or cylinders with a radially directed axis. Contents of the cells in the 3 parts: a transparent colourless crystalline mass of tannin in almost all cells; chloroplasts, especially in the outermost cell layers; a dark brown mass in some cells.

Meristeles. Especially in the outer part of the ground tissue, not wanting

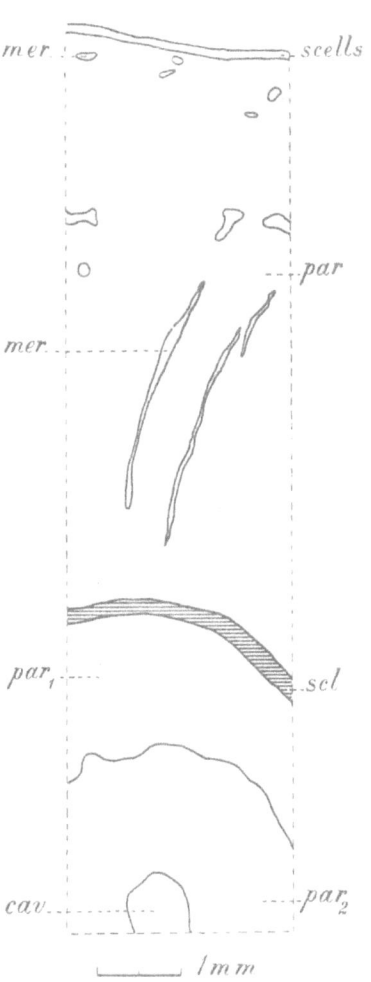

Fig. 53. *Gallae*. Transverse section. cav Cavity; mer Meristeles; par Parenchyma cells; par1 Parenchyma containing much starch; par2 Parenchyma containing less starch; scells 1—2 outermost layers of parenchyma cells with suberized walls; scl Sclereids.

in the knobs and ridges. Pericycle sometimes represented by some sclerenchymatic elements, bounding the phloem.

II. S c l e r e i d s. In 3—5 layers; often wanting in galls without central cavity; crystal idioblasts here and there.

S c l e r e i d s. R. 70 μ, T. 35 μ, L. 40 μ; polyhedra. Walls thickened; yellow; lignified; showing pit canals; intercellular spaces wanting.

III. C o m m o n p a r e n c h y m a c e l l s. Chiefly starch cells; surrounding in a variable number of layers the cavity containing the insect; some inner cell layers of this tissue containing much less starch; cluster crystal idioblasts here and there.

P a r e n c h y m a c e l l s. R. 70—120 μ, T. 40 μ, L. 45 μ; polyhedra, towards the centre more radially elongated. Walls yellow to brown; showing intercellular spaces. Cell contents: simple starch grains, 20 μ in diameter or 15 by 20 μ, globular or ellipsoidal, with a centric or excentric slit-like or cross slit-like hilum, showing stratification; several cells near the central cavity showing a brown globule; these globules about 20 μ in diameter, surrounded by a membrane with the same characters as that sometimes surrounding the starch grains, insoluble in water and alcohol, in potash colouring darker brown, in osmic acid blacker like the surrounding tissue, persisting in concentrated sulphuric acid but often with a dark coloured central part, in iron acetate and iron chloride colouring black after a long time; in other cells a more irregular mass between the persisting starch grains; sometimes a single cell containing some smaller globules lying in a row close to each other; in a more or less large number of cells close to the cavity egg-shaped masses, measuring 15 by 35 μ, colouring yellow in potassium iodide iodine and in iodine and sulphuric acid 66 per cent., in the latter reagent also swelling up somewhat, in phloroglucin and hydrochloric acid colouring red, in safranin red like the walls, in potash yellow, in concentrated sulphuric acid black, in iron acetate and iron chloride not colouring black, in copper sulphate and potash not colouring blue, showing double refraction, in Schulze's macerating mixture a thin striation as a spiral line discernible; in other cells smaller more irregular colourless or light yellow masses, lying close to the wall and often filling up the whole cell [1]).

January 1901.   J; M; v. E. d. W.

GLANDULAE LUPULI.
Lupulinum. Lupulin.
Glandular hairs occurring on the scales of the conical fruits of Humulus Lupulus, Linn. Sp. Pl. 1028.

Macroscopic characters.
Coarsely granular, unequal, brownish yellow powder, consisting of glandular hairs described beneeth. Odour aromatic; taste aromatically bitter.

Anatomical characters.
LITERATURE. Berg. Anat. Atl. 1865. 97. Tafel XXXIX. Bigelow. Glands in the Hop tree. Proceed. Jowa Acad. of Sc. II. 1895. 13. Erdmann-König. Allg. Warenk. 1895. 309. Flückiger. Pharmakogn. 1891. 254. Flückiger & Hanbury. Pharmacographia. 1879. 551. Gilg. Pharmakogn. 1910. 92. Greenish & Collin. Anat. Atl. veget. powders. 1904. 282. Hager. Pharm. Praxis. Bd. 2. 1902. 311. Holzner u. Lermer. Hopfen. Zeitschr. ges. Brauwesen. Jhrg. 1892—95. Hérail. Mat. Med. 464. Ingerman. Microskopie d. Voorn. Handelsw. 1910. 177. Karsten u. Oltmanns. Pharmakogn. 1909. 309. Kraemer. Botany a.

---

[1]) See B e y e r i n c k and H a r t w i g ll. cc.

Pharmacogn. 1910. 715. Luerssen. Syst. Bot. Bd. 2. 1882. 527. Marmé. Pharmacogn. 1886. 289. Meyer. Wiss. Drogenk. Bd. 2. 1892. 458. Moeller. Pharmacogn. 1889. 354. Moeller. Mikr. pharm. Ueb. 1900. 42. Oudemans. Pharmacogn. 1880. 438. Planchon et Collin. Drogues simples. T. 2. 1896. 297. Tschirch. Angew. Pfl.-anat. 1889. 466. Vogl. Anat. Atl. 1887. Taf. 59. 119. Wiesner. Rohstoffe. 1900. 818. Wigand. Pharmakogn. 1879. 315. MATERIAL. Glandular hairs gathered in October 1901 from the Botanic Garden at Groningen, fresh and in alcohol. REAGENTS. Water, glycerine, chloral hydrate, iodine in chloral hydrate, phloroglucin and hydrochloric acid, iodine and sulphuric acid 66 per cent., concentrated sulphuric acid.

MICROGRAPHY.

G l a n d u l a r   h a i r s. Of the labiate type; stalk very small; head consisting of a single hollow about hemispherical layer of parenchyma cells, their common cuticle on the upper side being raised bluntly dome-like, high 70—90 μ, thick 150 —250 μ; the cuticular bladder sometimes showing the marks of cells. Contents of the bladder of the gland: a yellow substance, soluble in water.

C e l l s   o f   t h e   a b o v e. High 12 μ, wide 15 by 22 μ; tetra- to hexagonal tables. Walls a little thickened; lateral walls with a clearly discernible middle lamella, showing stratification in the secondary layers.

July 1901.   J; M; v. E. d. W.

HERBA ABSINTHII.

Summitates Absinthii. Absinthium.

The dried leaves and flowering tops of Artemisia Absinthium, Linn. Sp. Pl. 848, gathered just before flowering.

Macroscopic characters.

Inflorescence mixed: consisting of nodding heads either isolated or in pairs, united in racemes and these again in panicles. Bracts at the bases of all branches, also 1 or 2 close under each head; the higher ones lanceolate to linear, the lower ones up to bipinnatisect. Branches and bracts greyish tomentose by the presense of longitudinally directed balance hairs; between these numerous oblong more or less spherical glandular hairs. Head up to 5 m.M. in diameter, almost globular. Involucre composed of imbricate bracts, the outer ones lanceolate, the inner ones oval; all showing a more or less membranaceous margin and hairs like the rest of the plant. General receptacle very small, solid, covered with white hairs. Flowers epigynous; pappus wanting. Ray-flowers not numerous, thin, female; corolla tubular and now and then somewhat zygomorphous, with an entire or at utmost 3-toothed limb. Disk-flowers numerous, hermaphrodite, heterodistylus; corolla tubular, with a 5-toothed limb, the lower part green, the upper part dirty yellow. Odour aromatic; taste aromatic, very bitter.

Anatomical characters.

LITERATURE. Le Blois. Can. sécrét. e. Poches sécrétr. Ann. d. Sc. nat. Série 7. Botanique. T. VI. 1887. 274. Col. Appareil sécrét. int. d. C. Thèse Paris. 1903. Tiré apart du Journ. de bot. 1903. 252. Dafert u. Mihlauz. Unters. üb. d. Kohleähnl. Masse d. Komp. (Chem. Teil) Denkschr. Wien. Akad. Bd. 87. 1912. 143. Daniel. Rech. anat. s. l. Bractées d. l'Involucre d. Comp. Ann. d. Sc. nat. Série 7. T. XI. 1890. 17. Ellrodt. Vert. d. Gerbst. i.

off. Blätt. Kräut. u. Blüten. Diss. Würzburg. 1913. 20. Flückiger. Pharmakogn. 1891. 682. Feuilloux. Contrib. à l'étude de l'app. tecteur et glandulaire des Comp. Diss. Paris. 1901. 24. Grimm. Beitr. z. vergl. Anat. d. Compositenblätter. Diss. Kiel. 1904. Hager. Pharm. Praxis. Bd. 1. 1900. 408. Hanausek. Unters. üb. d. Kohleähnl. Masse d. Komp. Anz. k. Akad. Wiss. Bd. XLVII. 1910. 388. Hérail. Mat. Med. 1912. 295. Heineck. Beitr. z. Kenntn. d. fein. Baues d. Fruchtschale d. Comp. Diss. Giessen. 1890. Karsten u. Oltmanns. Pharmakogn. 1909. 195. Koch. Mikr. Anal. d. Drogenpulver. Bd. 3. 1906. 18. Koch. Pharmacogn. Atlas. Bd. 2. 1914. 133. Koch u. Gilg. Pharmacogn. Praktik.1907. 187. Kraemer. Botany a. Pharmacogn. 1910. 396. Lavialle. Recherches s. l. développe-ment d. l'ovaire en fruit ch. 1. Comp. Ann. d. Sc. nat. Série 9. T. XV. 1912. 39—151. Luerssen. Syst. Bot. Bd. 2. 1882. 1135. Marmé. Pharmacogn. 1886. 236. Moeller. Phar-macogn. 1889. 213. Müller. Beiträge z. Kenntn. d. Baues u. d. Inhaltsstoffe d. Compositen-blätter. Diss. Göttingen. 1912. Oudemans. Pharmacogn. 1880. 274. Planchon et Collin. Drogues simples. Bd. 2. 1896. 53. Schneider. Powdered Veget. Drugs. 1902. 105. Small. Flor. Anat. of som. Comp. Linn. Soc. Journ. Bot. XLIII. 1917. 517. Solereder. Syst. Anat. d. Dicot. 1899. 515. v. Tieghem. Can. sécrét. d. Plantes. Ann. d. Sc. nat. Série 5. Bo-tanique. T. XVI. 1872. 97. v. Tiegh-em. Sec. Mém. s. l. Can. sécrét. d. Plantes. Ann. d. Sc. nat. Série 7. Botanique. T. 1. 1885. 6. Tschirch. Angew. Pfl. Anat. 1889.262.Tschirch. Harze u. Harzbeh. 1900. 343. Vesque. Anat. d. Tissus. Nouv. Arch. du Museum d'Hist. nat. Sér. 2. T. IV. 1881. 11. Virchow. Ueber Bau u. Ner-vatur d. Blattzäh-ne u. Blattspitzen. Diss. Bern. 1895. 16. Vogl. Anat. Atl. 1887. Tafel 3, 7. Vuillemin. Val. d. Caract. anat. (tige des Compo-sées). Paris. 1884. Wigand. Pharmakogn. 1879. 234.

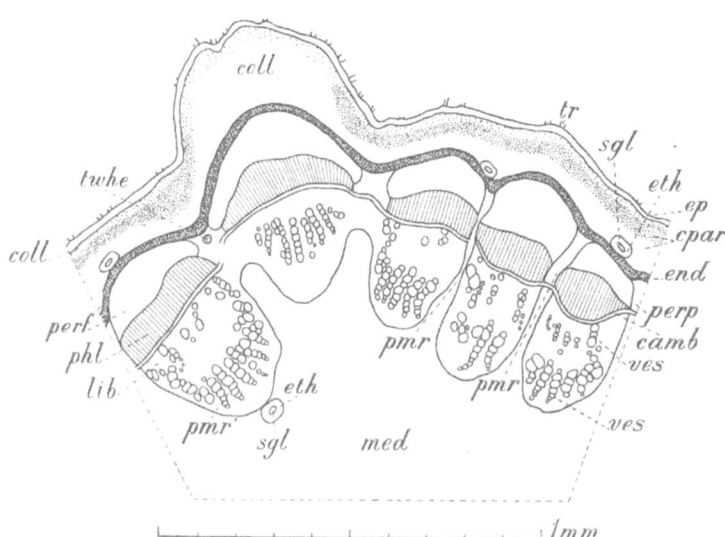

Fig. 54. *Artemisia Absinthium.* Stem, transverse section of a part of the stem thick 0.22 c.M. camb Cambium; coll Collenchyma; cpar Common parenchyma; end Endodermis; ep Epidermis; eth Epithelium; lib Libriform fibres; med Medulla; perf Pericyclic sclerenchyma fibres; perp Pericyclic parenchyma; phl Phloem; pmr Medullary commisuress between the compound vascular bundles; pmr' Medullary commissures within the compound vascular bundles; sgl Schizogenous glands; twhe Group of thick-walled elements; tr Trichomes; ves Vessels.

Stem.

MATERIAL. Stems gathered in July and September 1902 in the Botanic Garden at Groningen, fresh, in alcohol and fixed with chromic acid 1 per cent. These stems were examined at different distances from the growing top; in these parts they were thick 0.07, 0.22 (chiefly used) and 0.4 c.M. REAGENTS. Water, glycerine, choral hydrate, potash, iodine in chloral hydrate, phloroglucin and hydrochloric acid, iodine and sulphuric acid 66 per cent., concentrated sulphuric acid, Schulze's macerating mixture, copper acetate and iron acetate, potassium dichromate.

MICROGRAPHY.

Epidermis. Stomata only present in the furrows of the stem; not numerous; longitudinally directed; the subsidiary cells rather strongly curved outwards, thus raising the guard-cells above the other epidermal cells. Trichomes of two kinds: glandular hairs only present in the furrows of the stem; not numerous. See for the rest the trichomes of the leaf.

E p i d e r m a l c e l l s p r o p e r. On the ribs of the stem R. 18 μ, T. 10 μ, L. 30—80 μ; fibre-shaped; in the furrows of the stem R. 15 μ, T. 15 μ, L. 30—50 μ; polygonal tables, lateral walls often somewhat curved outwards. Outer and inner walls thickened, the outer ones with a cuticle. Cell contents: often some small chloroplasts; in a few cells a granular mass; in the thinner parts of the stem here and there anthocyanin; in chromic acid material in many cells a somewhat yellow globule, or a more distinctly yellow granular globule. S t o m a t a. R. 10 μ, T. 20 μ, L. 30 μ; outer walls of the subsidiary cells with a cuticular striation perpendicular to the outline of the stoma. B a s a l c e l l o f b a l a n c e h a i r s mostly T. 15 μ, L. 30 μ.

Cortex.

C o l l e n c h y m a. In the ribs of the stem consisting of 8—10 layers of cells and constituting the whole of the cortex; in the parts corresponding with the furrows of the stem forming the greater part of the outermost cell layer; showing intercellular spaces.

C e l l s o f t h e a b o v e. Those in the ribs somewhat flattened; those corresponding with the furrows of the stem R. and T. 10—15 μ, L. 50—100 μ; tetragonal prisms with a longitudinally directed axis and somewhat rounded edges. Cell contents: especially in the second and third cell layer of the ribs chloroplasts with some needle-shaped starch grains; here and there sometimes inulin.

C o m m o n p a r e n c h y m a. In the furrows of the stem in the parts corresponding with groups of vascular bundles consisting of about 5 cell layers, in the parts corresponding with the larger medullary commissures consisting of 8 cell layers.

C e l l s o f t h e a b o v e. R. and T. 10—15 μ, L. 50—100 μ; cells of innermost layers mostly the shortest; polygonal prisms with a longitudinally directed axis and somewhat rounded edges. Cell contents: the same as those of collenchyma.

G l a n d s. Not numerous; lying in the parenchyma, contiguous to the endodermis and corresponding in place either with the middle or with a lateral part of a group of vascular bundles; resembling those of the leaf [1]).

E n d o d e r m i s. In the parts of the stem thick 0.07 c.M. developed as a distinct starch-sheath; thick 1 cell layer, here and there 2 cell layers with radially arranged cells.

C e l l s o f t h e a b o v e. In all parts of the stem R. 14 μ, T. 22 μ, L. 30 μ; polygonal prisms with a longitudinally directed axis. Walls, inner walls always lignified, radial and cross walls mostly, outer walls sometimes lignified. Cell contents: in the thinnest parts of the stem in all cells some chloroplasts with 1 or more globular starch grains; in the thicker parts these occurring only here and there.

Stele.

---

[1]) See moreover *Rhizoma Arnicae*.

Pericycle. In the parts corresponding in place with the compound vascular bundles consisting of a layer of sclerenchyma fibres, thick 20 elements in a radial direction and mixed with these fibres here and there some elements with thin lignified walls. In the parts corresponding with the larger medullary commissures between the compound vascular bundles consisting of common parenchyma thick 2—3 cell layers; in the thickest parts of the stem here moreover a small group of thick-walled elements, not radially arranged, the cross dimensions of these elements smaller than those of the surrounding parenchyma cells.

Parenchyma cells. R. 15 μ, T. 12 μ, L. 150—300 μ; polygonal prisms with a longitudinally directed axis and rounded edges. Walls thickened; lignified; with slit-like pits; showing intercellular spaces. Cell contents: sometimes a sphaero-crystal of inulin. Sclerenchyma fibres. R. and T. 8 μ, L. considerable; polygonal. Walls strongly thickened; lignified, especially the clearly discernible middle lamella; with slit-like pit canals; showing stratification.

Vascular bundles. Larger or smaller compound bundles; 3 or more corresponding with each furrow of the stem; only 1 corresponding with each rib and lying at a somewhat larger distance from the axis of the stem; containing 4—10 simple bundles. Simple bundles collateral; open.

Phloem.

Primary phloem. Exarch. Fascicular secondary phloem only in the thickest parts of the stem; consisting of radially arranged elements. Elements of primary and secondary phloem. R. and T. 4—10 μ; polygonal prisms with a longitudinally directed axis.

Fascicular cambium. Already present in the parts of the stem thick 0.07 c.M.

Xylem.

Fascicular secondary xylem. Containing here and there a group of some pitted vessels, only 1 element broad, and close to those some parenchyma cells; for the rest consisting of libriform fibres arranged in radial rows. Vessels. R. and T. 15 μ, L. of articulations 200—250 μ; polygonal prisms; transverse walls with a circular perforation. Walls thickened; lignified; with slit-like bordered pits, 2-sided when adjoining one another, 1-sided when adjoining a parenchyma cell. Parenchyma cells. R. and T. 10 by 6 μ; polygonal prisms. Walls thickened; lignified. Libriform fibres. R. and T. 8 μ, L. 150—300 μ; tetra- to hexagonal. Walls thickened; lignified, especially the clearly discernible middle lamella; pitted.

Primary xylem. Endarch; consisting of radially arranged vessels, inwardly diminishing in size; the outer ones pitted, spiral vessels in the middle, annular vessels in the inner part.

Pitted vessels. R. 12 μ, T. 20 μ, L. of articulations 200—400 μ; tetragonal prisms with strongly rounded edges; transverse walls with circular perforations. Walls thickened and lignified. Spiral vessels. R. and T. 17 μ. Annular vessels. R. 5 μ, T. 10 μ.

Medullary commissures.

Between the compound vascular bundles. Broader than those within the compound bundles; in all their parts chiefly consisting of somewhat thick-walled elongated parenchyma cells; only some interfascicular secondary phloem and cambium being developed, in the thickest parts of the stem the cambium sometimes developed into adult parenchyma cells.

P a r e n c h y m a c e l l s. R. 15 μ, T. 10 μ, L. 200—300 μ; polygonal prisms with a longitudinally directed axis. Walls thickened; lignified. Cell contents: sometimes a sphaerocrystal of inulin.

W i t h i n t h e c o m p o u n d b u n d l e s. Narrower than those between them; only consisting of radially arranged parenchyma cells.

C e l l s o f t h e a b o v e. R. and T. 8 μ; polygonal prisms with a longitudinally directed axis.

M e d u l l a. Consisting of parenchyma cells, augmenting in size towards the centre and in this respect divided in 3 parts: outer layers, middle layers and the central part. Here and there cluster crystal idioblasts and in the middle layers some schizogenous glands, resembling in all respects those of the cortex.

P a r e n c h y m a c e l l s. Cells of the outer layers R. and T. 10 μ, L. 70—100 μ; polygonal prisms with a longitudinally directed axis. Walls somewhat thickened. Cells of some adjoining more central layers R. and T. 20 μ, L. up to 140 μ; of the same shape, but with rounded edges. Walls often somewhat collenchymatous; often somewhat lignified; with slit-like pits; showing intercellular spaces. Cells of the central part R. and T. 25—60 μ, L. 70—100 μ. For the rest like the preceding cells.

L e a f.

MATERIAL. Leaves, gathered in July and September 1902 in the Botanic Garden at Groningen, fresh, in alcohol and fixed with ½ per cent. chromic acid. REAGENTS. The same as for the stem, potash excepted.

MICROGRAPHY.

I n t e r v e n i a.

Epidermis upper side. Stomata very rare; Lev. 18 by 35 μ; see for the rest those of under side. Water-pores: some on each tooth of the margin of the leaf; Lev. 20 by 30 μ. Trichomes of two kinds: 1. Glandular hairs. Of the composite type, 3—4 stories high, there flat sides directed transversely to the veins; the uppermost story of cells with a cuticular bladder; the whole hair long 24—32 μ, Lev. 30 by 40 μ; cuticular bladder H. 30 μ, Lev. 30 by 40 μ. 2. Balance hairs. Consisting of a 1-seriate stalk long 2—6 cells and bearing a longitudinally directed cross-beam, consisting of a single fibre-shaped cell; stalk longer or shorter according to the inequalities of the surface of the leaf, the cross-beams thus all being brought up to the same level; stalk thick 10 μ, the cells of short stalks long 15 μ, those of long stalks 10 μ; beam cell thick 15 μ, long 300—600 μ.

E p i d e r m a l c e l l s p r o p e r. H. 15 μ, Lev. B. 20—30 μ, Lev. L. 40 μ; polygonal, with strongly sinuous lateral walls. Outer walls a little thickened, with a cuticle. Cell contents: some small chloroplasts; in potassium dichromate material the large vacuole filled with a brown mass and one (8 μ in diameter) sometimes some still smaller colourless vacuoles; in chromic acid material the large vacuole filled with a light brown granular mass, the smaller vacuoles coloured yellow. B e a m c e l l s o f b a l a n c e h a i r s. Walls without a cuticle. Cell contents: see epidermal cells proper.

Epidermis under side. Stomata 20 to the sq. m.M.; phaneroporous; surrounded by 5—6 subsidiary epidermal cells mostly curved somewhat outwards, thus

raising the stoma somewhat above the epidermis; air-cavity distinct and large.
Trichomes. See those of upper side, but glandular hairs here more numerous.
E p i d e r m a l   c e l l s   p r o p e r. See those of upper side. Lateral walls more sinuous.
Cell contents: not much discernible. S t o m a t a. H. 8 μ, Lev. 20—30 μ.

Mesophyll.

P a l i s a d e   c h l o r e n c h y m a. Present as well at the upper as at the un-

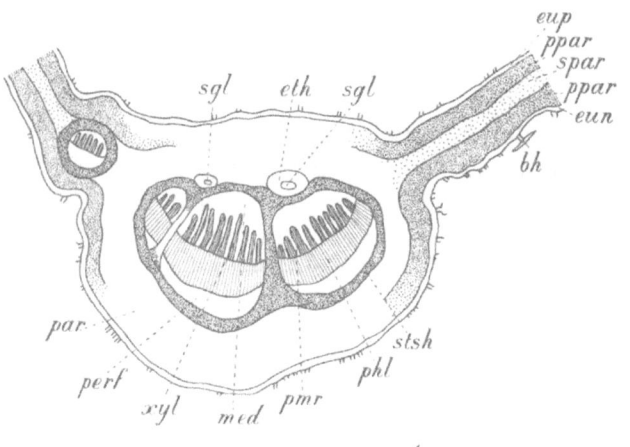

der side of the leaf; consisting on both sides of
one cell layer, the cells at
the under side less regu-
larly arranged; showing
intercellular spaces, those
of the under layer mostly
rather large.

C e l l s   o f   t h e   a b o v e.
Those of the upper layer Lev.
B. and L. 15 μ, those of the
under layer Lev. B. and L.
10 μ; H. of all cells 40 μ; cylindrical; in some cells, especially of the upper layer,
a cross wall. Cell contents:
many penta- or hexagonal
tabular chloroplasts, paving
the wall, 5 μ in diameter,
thick 2 μ, containing some
needle-shaped starch grains.

Fig. 55. *Artemisia Absinthium*. Leaf, transverse section through
the midrib and the neighbouring portions of the lamina. bb Balance hairs; eun Epidermis under side; eup Epidermis upper side;
eth Epithelium; med Medulla; par Colourless parenchyma; perf
Pericyclic sclerenchyma fibres; phl Phloem; pmr Medullary commissures; ppar Palisade chlorenchyma; sgl Schizogenous glands;
spar Spongy chlorenchyma; stsh Starch-sheath; xyl Xylem.

S p o n g y   c h l o r e n c h y m a. Constituting the middle part of the mesophyll; mostly consisting of 3 cell layers; showing intercellular spaces.
C e l l s   o f   t h e   a b o v e. H. 20 μ, Lev. B. and L. 15—20 μ; globular, ellipsoidal or
cylindrical with the axis perpendicular to the surface of the leaf. Cell contents: see those
of palisade chlorenchyma.

M a r g i n. That of the lobes showing under the epidermis a single layer of
palisade chlorenchyma, uniting together the same layers at the upper and under
side of the leaf. The margin nearest to the midrib, in the deepest part of the incisions, showing in the same way the union of the collenchyma layers of upper
and under side; this tissue here being 1 or 2 cell layers thicker.

V e i n s.

Epidermis upper side. Stomata wanting. Trichomes of two kinds. 1. Glandular hairs. Very rare; see for the rest those of mesophyll. 2. Balance hairs. Mosty
with a basal cell smaller than the other epidermal cells; stalk mostly 2—3 cells
high; see for the rest those of mesophyll.

E p i d e r m a l   c e l l s   p r o p e r. R. 15 μ, T. 12 μ, L. 30—130 μ; polygonal prisms
with the axis perpendicular to the surface of the leaf. Walls, outer and inner ones thickened; outer walls with a parallel longitudinal cuticular striation. Cell contents: in potassium

dichromate material the vacuole filled with a yellow to brown mass; in chromic acid material each cell with some small colourless vacuoles and for the rest filled with a yellow to greenish granular mass.

Epidermis under side. Stomata wanting. Trichomes. See those of mesophyll; glandular hairs not numerous.

E p i d e r m a l c e l l s p r o p e r. See those of upper side, but outer and inner walls a little more thickened; the precipitates in potassium dichromate and chromic acid material here less distinct.

Mesophyll.

P a l i s a d e c h l o r e n c h y m a. Towards the top of the midrib and in the primary veins, especially at the upper side of the leaf, forming a continuous layer with that of the intervenia; at the base of the midrib only occurring in the lateral parts of the under side.

C o l o u r l e s s p a r e n c h y m a. In the basal half of the midrib at the upper and under side about 7 cell layers; cells arranged in longitudinal rows.

C e l l s o f c o l o u r l e s s p a r e n c h y m a. R. 35 μ, T. 35 μ, L. 60—100 μ; polygonal prisms with a longitudinally directed axis; the cells of the under layers somewhat larger than those of the upper ones; the outermost 1 or 2 layers of both sides having smaller radial and tangential dimensions and larger longitudinal dimensions, moreover collenchymatously thickened walls. Cell contents: here and there some small chloroplasts and sometimes inulin.

G l a n d s. Only in the basal part of the midrib; lying above and in contact with the group of meristeles 2 larger or smaller schizogenous glands; measuring R. 10 μ, T. 30 μ; epithelium consisting of 1 layer of cells, each divided in an inner and an outer one by a partition wall, the cells smaller than those of the surrounding colourless parenchyma and often containing some chloroplasts; cavity filled with colourless oil [1]).

E n d o d e r m i s. Developed as a starch-sheath in the midrib; starch mostly wanting in the primary veins and veinlets.

C e l l s o f t h e a b o v e. R. 30 μ, T. 30 μ, L. 40—60 μ; polygonal prisms with a longitudinally directed axis. Cell contents: some globular starch grains. In alcoholic material in the veinlets the walls often brown and the contents showing some brown globules.

Meristeles. In the basal half of the midrib 3 in number; closely united together and forming a semi-cylinder with the flat side turned upwards; the several meristeles somewhat varying in dimensions and more or less clearly separated from each other.

P e r i c y c l e. Developed only at the under side of the meristele; consisting of a layer of sclerenchyma fibres.

F i b r e s o f t h e a b o v e. R. and T. 10 μ, L. 300 μ; polygonal. Walls somewhat thickened, lignified.

C o m p o u n d v a s c u l a r b u n d l e. A single one in each meristele; containing 10 simple bundles, sometimes a smaller number; arranged in a semi-

---

[1]) See v. T i e g h e m, L e B l o i s and T s c h i r c h, Harze u. Harzbeh. ll. cc.

circle. Simple vascular bundles having the shape of thin radially directed lamellae; collateral; open.

Phloem. Exarch; consisting of sieve-tubes and cambiform cells, not always clearly to be distinguished from one another.

E l e m e n t s   o f   t h e   a b o v e. R. and T. 4—8 μ, L. about 100 μ; polygonal prisms with a longitudinally directed axis. Walls collenchymatously thickened; intercellular spaces wanting. Contents: here and there some small chloroplasts.

Cambium. Mostly not clearly discernible.

Xylem. Endarch; chiefly consisting of radially arranged pitted, reticulate and spiral vessels.

R e t i c u l a t e   a n d   p i t t e d   v e s s e l s. R. and T. 10—15 μ, L. of articulations 100 —300 μ; rectangular with rounded edges; pitted vessels showing transverse walls with a circular perforation. S p i r a l   v e s s e l s. R. and T. 10 μ, L. of articulations 200—300 μ.

M e d u l l a r y   c o m m i s s u r e s. Narrow and consisting of common parenchyma.

C e l l s   o f   t h e   a b o v e. R. 10 μ, T. 7 μ, L. 80—100 μ; polygonal prisms with a longitudinally directed axis. Cell contents: some very small chloroplasts.

M e d u l l a. Consisting of some layers of common parenchyma cells.

C e l l s   o f   t h e   a b o v e. R. 8 μ, T. 12 μ, L. 90—150 μ; polygonal prisms with a longitudinally directed axis. Walls somewhat collenchymatously thickened and pitted; intercellular spaces wanting.

F l o w e r - h e a d s.

MATERIAL. Gathered in July and September 1902 from the Botanic Garden at Groningen, fresh and in alcohol. REAGENTS. Water, glycerine, chloral hydrate, phloroglucin and hydrochloric acid, potassium dichromate, copper acetate and iron acetate.

MICROGRAPHY.

R e c e p t a c l e.

E p i d e r m i s. Stomata wanting. Capitate hairs numerous; stalk consisting of some small cells; head consisting of 1 very large spindle-shaped cell and liable to be separated from the stalk.

C e l l s   o f   s t a l k. Thick 10 μ, long 20 μ. H e a d - c e l l. Thick 25 μ, long 700— 1000 μ.

G r o u n d   t i s s u e. Consisting of common parenchyma, the outermost cell layer having smaller cells than the rest; containing some glands with an epithelium of 1 or 2 cell layers; in the central part some large intercellular spaces. Here and there some inulin in the cells. **M e r i s t e l e s** numerous.

P e d i c e l s   o f   t h e   f l o r e t s. Consisting of common parenchyma.

B r a c t s   o f   t h e   i n v o l u c l e. The 5 bracts of the outer whorl somewhat differing in anatomical structure from those of the inner whorl; the latter chiefly used for the following description.

E p i d e r m i s. Inner side. Consisting of rather long fibres, diminishing in size towards the margin, those in the vicinity of the margin directed perpendicularly to it; the margin thick only 1 cell layer, these cells often prolonged into a 1-cellular often flattened hair; stomata present only on the bracts of the outer whorl, longitudinally directed, lying in the same level with the epidermal cells, Lev. B. 20 μ, Lev. L. 35 μ.

E p i d e r m a l   f i b r e s. In the vicinity of the midrib H. 18 μ, Lev. B. 25 μ, Lev. L. 150—300 μ. Walls a little lignified; outer walls thickened, with a cuticle, showing stratification; lateral walls on several places thickened and pitted. Cell contents: those of the bracts of the outer whorl some small chloroplasts; in potassium dichromate material all cells filled with a brown granular mass, showing colourless cavities.

Outer side. Stomata. See those of the inner side; sometimes also present on the bracts of the inner whorl. Trichomes of two kinds: 1. Glandular hairs. See those of the leaf. 2. Balance hairs. Especially on the bracts of the outer whorl; longitudinally directed.

E p i d e r m a l   c e l l s   p r o p e r. H. 15 μ, Lev. B. 15 μ, Lev. L. 50 μ; tetra- to hexagonal prisms. Walls a little lignified; outer walls with a cuticle. Cell contents: see those of inner side.

M e s o p h y l l. In the bracts of the inner whorl on both sides adjacent to the midrib a bundle of sclerenchyma fibres thick some elements; for the rest chlorenchyma containing a few chloroplasts. In the bracts of the outer whorl these bundles of sclerenchyma wanting; palisade chlorenchyma 1 cell layer at the upper, mostly 2 at the under side; these cells arranged longitudinally along the midrib; between the palisade layers some chlorenchyma not distinctly characterized as spongy tissue.

S c l e r e n c h y m a   f i b r e s. R. and T. 10 μ, L. 150 μ; polygonal. Walls strongly thickened; with a clearly discernible middle lamella; showing stratification; lignified and pitted. C e l l s   o f   p a l i s a d e   c h l o r e n c h y m a. H. 50 μ, Lev. 12 μ; cylinders or rectangular prisms with rounded edges.

M e r i s t e l e. A single one in a midrib, only with some small branches. In the bracts of the outer whorl branches numerous. Pericycle consisting of prismatic parenchyma cells, 20 μ in diameter in a cross section.

R a y-f l o w e r.
Corolla. See the disk-flower. Pistil. See the disk-flower.

D i s k-f l o w e r.

Corolla.
Epidermis. Inner side. Stomata and trichomes wanting.

E p i d e r m a l   c e l l s   p r o p e r. In the basal half of the corolla H. 15 μ, Lev. B. and Lev. L. 25 μ; polygonal tables; in the apical half H. 10 μ, Lev. B. 10 μ, Lev. L. 50 μ; tetra- to hexagonal prisms. Outer walls a little thickened; with a cuticular striation. Cell contents: in the basal half chloroplasts and a cluster crystal; in the apical half some small yellow to green plastids and some cells filled with very small globules. In potassium dichromate material all cells filled with a brown granular mass, showing colourless cavities.

Outer side. Stomata wanting. Glandular hairs especially on the teeth; see for the rest those of the leaf.

E p i d e r m a l   c e l l s   p r o p e r. In the basal half of the corolla H. 20 μ, Lev. B. and Lev. L. 20 μ; tetra- to hexagonal tables; arranged in longitudinal rows; for the rest not much different from those of the inner side.

Mesophyll. Present only in the basal half of the corolla; consisting of common parenchyma, showing between the veins large intercellular cavities.

Meristeles. 5 in number; not reaching beyond the basal half of the corolla.
Stamen.

F i l a m e n t. Cells of the outer half of the uppermost part Lev. B. and Lev. L.
10 μ; rectangular. Walls thickened and lignified.

A n t h e r. Connective without a meristele, but with a well developed part
extending above the anther.

Cells of extending part. Lev. B. 10 μ, Lev. L. up to 100 μ; fibre-shaped.
Walls somewhat thickened and lignified.

Pollen grains. 18 μ in diameter; slightly tetrahedral, almost globular. Exine
having 3 pores and showing a rather distinct layer of rod-shaped corpuscules.
Pistil.

O v a r y.

Wall.

Epidermis. Outer side. Showing some longitudinal rows of flat but not projecting cells; in the other parts of the epidermis sometimes cluster crystal idioblasts; at the base a ring of some rectangular cells with somewhat thickened
and lignified walls; the scars of the corolla also colouring red in phloroglucin
and hydrochloric acid.

Inner side.

Cells of the above. Tetragonal tables, showing radial and cross walls. Walls
thickened, especially the innermost parts; yellow.

Mesophyll. Mostly wanting; sometimes developed as 1 layer of small cells and
some cells in the vicinity of the meristeles.

Meristeles. 3 in number on one side of the ovary; lying close to each other.

Conducting tissue. Forming a bundle in the wall opposite to the meristeles and
almost touching the ovule; consisting of thick-walled cells.

Ovule. Inserted at the base of the ovarial cavity; anatropous. Integument
flattened, with the exception of 1 or 2 outer cell layers; these cells sometimes
developed as cluster crystal idioblasts. Endosperm flattened, with the exception
of 1 or 2 outer cell layers. Embryo in a rather developed stage; the epidermis
showing in bichromate of potash material the brown precipitate mentioned above.

S t y l e. The swollen part of the base covered with rectangular epidermis cells,
showing thickened and lignified walls.

S t i g m a. On the inner and lateral sides 2 bands of papillae and long hairs
on the top; in each stigma 1 schizogenous gland.

September 1902.   J; M; v. E. d. W.

HERBA ACONITI RECENS.
The aerial parts of the flowering plant of Aconitum Napellus, Linn. Sp. Pl. 532.

LITERATURE. Barth. Stud. üb. d. Nachw. v. Alcaloiden in pharm. verw. Drogen. Bot.
Centrbl. 1898. Bd. 75. 338. Berg. Anat. Atl. 1865. Erréra, Maistriau e. Clautriau. Prem.
Rech. s. l. Local. e. l. Signif. d. Alcaloides d. l. Plantes. Journ. d. Méd. d. Chir. e. d.
Pharm. publ. p. l. Soc. royale d. Sc. méd. e. nat. d. Bruxelles. 1887. 108. Flückiger

Pharmakogn. 1891. 692. Goffart. Rech. s. l'Anat. d. feuilles d. l. Renonculac. Arch. d. l'Inst. bot. d. l'Université d. Liège. T. III. 1901. Goris. Anat. Untersch. d. wicht. Aconitenarten. Bull. d. Sc. pharm. 1901. n° 4. Haberlandt. Vergl. Anat. d. assimil. Gewebesyst. d. Pfl. Pringsheim's Jahrbücher. 1882. 96. Pl. III. fig. 4,5,10. v. d. Linden. Rech. microchim. s. l. prés. d. alcaloides e. d. glycosides d. l. fam. d. Renonculac. Rec. d. l'Inst. Bot. d. Bruxelles. T. V. 1902. Lonay. Struct. anat. d. pericarpe e. d. spermoderme ch. l. Renonculac. Liège. Arch. Inst. bot. 4. 1907. Marié. Rech. s. l. struct. d. Renonculac. Ann. d. Sc. nat. Sér. 6. Bot. T. XX. 1885. 5. Mitlacher. Tox. od. forens. wicht. Pfl. u. Drogen. 1904. 73. Molisch. Mikrochemie d. Pfl. 1913. 269. Oudemans. Pharmacogn. 1880. 292. Planchon e. Collin. Drogues simples. T. II. 1896. 905. Schneider. Powdered veget. Drugs. 1902. 107. Tschirch. Angew. Pfl. Anat. 1889. 38, 244. Vogl. Anat. Atl. 1887. Pl. XVI. Tunmann. Pfl. Mikrochemie. 1913. 285. REAGENTS. Water, glycerine, chloral hydrate, potassium iodide iodine (stem and leaf, not for the flower), iodine in chloral hydrate, phloroglucin and hydrochloric acid (stem and leaf), hydrochloric acid (flower), iodine and sulphuric acid 66 per cent. (stem and leaf), concentrated sulphuric acid (stem and leaf), Schulze's macerating mixture (stem and leaf), potassium dichromate, copper acetate and iron acetate.

S t e m.

MATERIAL. Several stems, up to 0.6 c.M. thick, gathered June 1903 in the Botanic Garden at Groningen, fresh, fixed with chromic acid 1 per cent. and in alcohol.

MICROGRAPHY.

Epidermis. Cells lying fairly well in longitudinal rows. Stomata rare; phaneroporous; mostly lying somewhat above the level of the epidermal cells; usually surrounded by 4 subsidiary cells; in some cases the stoma more or less flattened. See also the stomata of the under side of the leaf. Trichomes more numerous than on the leaf; thick 12 μ at the base and 350—400 μ long; the part between the epidermal cells rectangular, in a tangential section 8 by 10 μ. See for the rest the trichomes of the leaf.

E p i d e r m a l c e l l s p r o p e r. R. 30 μ, T. 25 μ, L. 50—140 μ; tetra- to hexa-, mostly tetragonal prisms with a longitudinally directed axis. Outer walls thickened; showing longitudinal cuticular striation; in iodine and sulphuric acid the cuticularized part colouring yellow to brown; this part thicker above the lateral walls; lateral walls slightly thickened, pitted; inner walls collenchymatous. Cell contents: some small chloroplasts; in material treated with potassium dichromate all the cells filled with a granular mass varying from yellow to brown, these contents apparently by thin membranes divided in globular partitions (vacuoles?); nucleus conspicuous; when treated with potassium iodide iodine no alcaloid to be found. S t o m a t a. T. 30 μ, L. 40 μ. When treated with potassium dichromate no precipitate in the guard-cells. Alcaloid wanting.

Cortex.

C o l l e n c h y m a. Consisting of 1 layer of cells, pretty well arranged in longitudinal rows. Intercellular spaces wanting.

C e l l s o f t h e a b o v e. R. 25 μ, T. 35 μ, L. 70—120 μ; mostly tetragonal tables with a radially directed axis and sometimes more or less rounded sides. Walls thickened collenchymatously; in alcoholic material mostly yellow to brown; lateral walls slightly pitted. Cell contents: several chloroplasts; in material treated with potassium dichromate colourless vacuoles apparently divided by membranes in smaller ones of e. g. 15 μ in diameter; in these vacuoles clusters of very dark brown grains apparently surrounded by a thin wall; alcaloid wanting.

Common parenchyma. Mostly 3 layers of cells lying in longitudinal rows; in the small ribs of the stem some layers more. Showing intercellular spaces, especially between the collenchyma and the parenchyma.

Cells of the above. R. and T. 30—35 μ, L. 80—170 μ; polygonal prisms with a longitudinally directed axis and rounded edges. Walls of the outermost layers in the ribs sometimes collenchymatous; those of the rest often slightly pitted. Cell contents: in the thinner stems chloroplasts paving the walls and containing some needle-shaped starch grains; in material treated with potassium dichromate granular clusters similar to those described for the collenchyma but of a smaller size; alcaloid wanting.

Endodermis. Discernible only in the middle part of a stem of about 2 m.M. thick. Developed as a starch-sheath; consisting of 1, sometimes of 2 layers of cells lying in longitudinal rows. Cells of the above. R. 25 μ, T. 40 μ, L. 150—250 μ; polygonal prisms with a longitudinally directed axis and strongly rounded edges. Walls often thickened, especially the inner walls and the innermost halves of the radial and

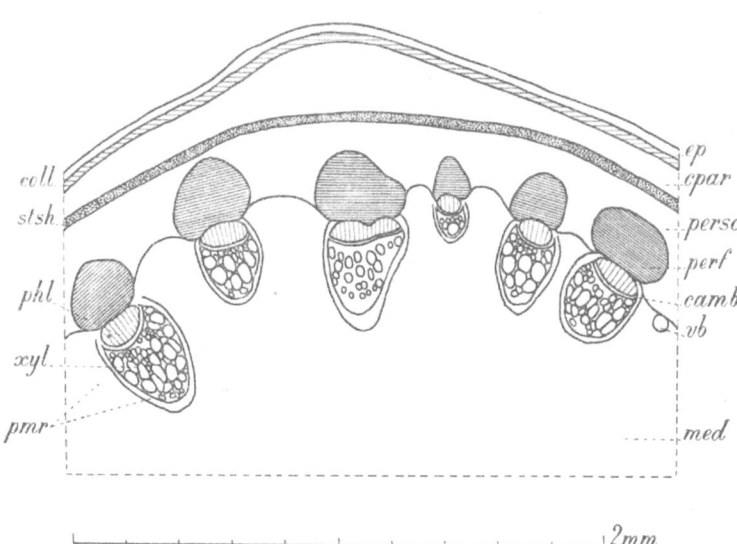

Fig. 56. *Aconitum Napellus*. Stem, transverse section. camb Cambium; coll Collenchyma; cpar Common parenchyma; ep Epidermis; med Medulla; perf Pericyclic sclerenchyma fibres; persc Pericyclic sclerenchyma cells; phl Phloem; pmr Medullary commissures; stsh Starch-sheath; vb Small vascular bundle.

transverse walls; the thicker parts lignified and pitted. Cell contents: chloroplasts, each containing a compound starch grain, or perhaps simple grains filling up the chloroplasts; in material treated with potassium dichromate some minute clusters similar to those of the collenchyma; alcaloid wanting.

Stele.

Pericycle. Consisting of sclereids in several layers, e. g. 12 in the thickest stem; in these layers longitudinal bundles of sclerenchyma fibres, every bundle of fibres corresponding in place to a vascular bundle. Number of layers of cells and fibres increasing from the top towards the base of the stem. Bundles of fibres here and there adjoining the endodermis. Sclereids arranged in longitudinal rows and showing no intercellular spaces.

Sclereids. R. and T. 20—30 μ, L. 200—700 μ, the outermost cells the shortest; polygonal, axis longitudinally directed, edges strongly rounded. Walls strongly thickened; showing stratification; middle lamella slightly yellow; lignified, especially the middle lamella; showing numerous pit canals. Cell contents: chloroplasts containing starch grains, especially in stems showing a distinct starch-sheath the grains resembling those of the medullary commissures, but smaller; in material treated with potassium dichromate clusters like those of the cortical parenchyma, but smaller; alcaloid wanting. S c l e r e n-c h y m a   f i b r e s. R. 15 μ, T. 20 μ, L. 1, up to over 2 m.M.; polygonal. Walls strongly thickened; showing stratification; middle lamella somewhat yellow; lignified, especially the middle lamella; pit canals not numerous.

V a s c u l a r   b u n d l e s. Simple; number of bundles increasing from the top of the stem towards the base, e. g. in a top thick 2 m.M. 11 bundles and in a base thick 2.4 m.M. 21 bundles. Arranged in a ring; collateral; open. Bundles strongly varying in size; between the large bundles very small ones often abutting on a single thick-walled cell of the medulla. These small bundles apparently developed not secondarily, but in the punctum vegetationis [1]; somewhat larger bundles already showing a cambium; the largest bundles showing a small quantity of secondary tissue.

Phloem. Only a small innermost part radially arranged. Consisting of sieve-tubes with companion cells and parenchyma cells. In a transverse section companion cells not discernible in the radially arranged part. Parenchyma cells lying in 1 or more rows at the outer and lateral sides of the phloem.

S i e v e-t u b e s. R. and T. 10 by 20 μ, or R. and T. 13 μ, L. of articulations 200—360 μ; polygonal prisms. Sieve-tubes of the radial rows R. 5 μ, T. 10 μ, tetragonal; transverse walls horizontal or oblique with distinct sieve-plates; callus plates wanting. Walls in chromic acid, in iodine in chloral hydrate and in sulphuric acid (66 per cent.) strongly swelling up, mostly only the lamella surrounding the cavity recognisable after the swelling [2]. Cell contents: numerous minute starch grains in the vicinity of the sieve-plates, colouring blue in iodine in chloral hydrate and in potassium iodide iodine, fresh material treated with potassium iodide iodine producing in these places, but often somewhat farther from the sieve-plates, a brownish red precipitate (alcaloid) [3]. C o m p a n i o n   c e l l s. R. and T. 6—8 μ; mostly tetragonal prisms; nearly always the transverse walls of these cells corresponding with the sieve-plates. Walls: see those of the sieve-tubes. Cell contents: a granular mass; for alcaloid see the sieve-tubes. P a r e n c h y m a   c e l l s. R. and T. 12 μ, L. 40—100 μ; polygonal prisms. Walls sometimes slightly thickened; pitted.

Cambium. Consisting of 1—2 layers of elements.

E l e m e n t s   o f   t h e   a b o v e. R. 4 μ, T. 10 μ; tetragonal in a transverse section.

Xylem. Consisting of 3 parts: 1°. a small outermost part chiefly composed of pitted vessels, for the rest of fibres; the elements arranged in radial rows corresponding to the radial rows of the phloem and the cambium; intercellular spaces wanting; 2°. a middle part constituting the bulk of the xylem, composed of

---

[1] From the facts mentioned above it follows that there is a certain periodical development of the punctum vegetationis, the structure of the flowers rendering this still more probable. M a r i é does not mention this for annual stems like those of *Aconitum*, but only treats of the increase of the vascular bundles in *Clematis* and *Astragenus* (l. c. p. 17 and 163, and in his historical and literary notes p. 9.) On p. 163 he says: Mais plus tard il va se former dans les espaces interfasciculaires et aux dépens des assises profondes du péricycle de nouveaux faisceaux également isolés et de même forme que les précédents.

[2] See G o f f a r t. l. c. 67 and 162.

[3] This result agrees not with that of E r r é r a, C l a u t r i a u et M a i s t r i a u l. c. 111. „L'aconitine se rencontre principalement autour des faisceaux, au voisinage du liber. La quantité est faible dans le parenchyme, et il existe une très légère accumulation vers les couches sous-épidermiques''.

pitted vessels intermixed with some parenchyma cells without intercellular spaces; this part wanting in a certain number of vascular bundles, e. g. in the middle part of a thinner stem in 5 bundles, in the top and in the base of such a stem in a smaller number; 3°. a smaller or larger innermost primary xylem composed of spiral and reticulate vessels and parenchyma, the vessels often more or less flattened.

Pitted vessels. Of the outermost part R. 15 μ, T. 12 μ; rectangular prisms; transverse walls showing a circular perforation and bordered pits. Walls somewhat thickened; lignified. Of the middle part R. and T. 45—55 μ, L. of articulations 300—550 μ; polygonal prisms with rounded edges, transverse walls with circular perforation. Walls somewhat thickened; lignified; showing bordered pits at the places of contact of 2 vessels. Spiral and reticulate vessels. R. and T. 15—20 μ; in the bundles deprived of the middle part of the xylem R. 35 μ, T. 25 μ; polygonal prisms with somewhat rounded edges. Walls lignified. Parenchyma cells. Of the middle part R. and T. 15—20 μ, L. 125—170 μ. Walls lignified; showing perhaps sometimes unilateral bordered pits. In the primary xylem R. and T. 15 μ; polygonal prisms. Walls often slightly collenchymatous. Cell contents: sometimes some small starch grains.

Medullary commissures. Consisting of common parenchyma cells, arranged in longitudinal rows. Round the vascular bundles 1 layer of much smaller cells probably more or less flattened. Intercellular spaces.

Cells of the above. R. 65 μ, T. 50 μ, L. 80—150 μ; polygonal prisms with a longitudinally directed axis and rounded edges; round the vascular bundles R. 30 μ, T. 25 μ, L. 40—80 μ. Walls of the outermost layers thickened, towards the centre gradually less so; the thickened walls stratified and lignified; showing pits and pit canals. Cell contents: chloroplasts, nearly quite filled with simple and compound, up to 6-adelphous starch grains; in the innermost cells the largest simple grains 6—8 μ in diameter, the largest compound grains 7—8 μ in diameter; alcaloid wanting.

Medulla. Often showing a central lysigenous cavity. Intercellular spaces sometimes very large.

Cells of the above. R. and T. 50 μ, L. 80—130 μ; polygonal prisms with a longitudinally directed axis and rounded edges. Cell contents: chloroplasts containing starch grains similar to those of the medullary commissures, mostly only occurring along the margin; alcaloid wanting.

Leaf.

MATERIAL. Leaves, gathered June 1903, in the Botanic Garden at Groningen, fresh, fixed with chromic acid 1 per cent., and in alcohol.

MICROGRAPHY.

Intervenia.

Epidermis upper side. Cells at the base of the leaf often lying in longitudinal rows; along the margin also in longitudinal rows and resembling those of the veins. Stomata wanting. Water-pores 1, sometimes 2 quite at the top of every tooth. Trichomes not numerous, only occurring at a short distance from the veins and occasionally along the margin; inserted between the epidermis cells; at the base 20 μ in diameter, 250 μ long; consisting mostly of 1 sometimes of 2 cells; conical; curved.

Epidermal cells proper. At the base of the leaf H. and Lev. B. 40 μ, Lev. L.

50—90 μ; mostly rectangular tables with somewhat sinuous lateral walls. For the rest H. 35 μ, Lev. B. 50 μ, Lev. L. 90 μ; polygonal, lateral walls sinuous. Close to the tops of of the lobes dimensions much smaller; lateral walls not or hardly sinuous. Outer walls slightly thickened and showing a thin cuticle, along the margin cuticular striation. Cell contents: some minute chloroplasts; alcaloid (potassium iodide iodine); in potassium dichromate nearly all the cells containing an evenly granular yellowish brown mass, the remaining cells numerous brown granules; in chromic acid the nucleus showing up very clearly. W a t e r p o r e s. 30—40 μ; guard-cells dead[1]). C e l l s o f t r i c h o m e s. Walls slightly thickened; with a cuticle with warty markings.

Epidermis under side. At the base of the leaf cells more or less arranged in longitudinal rows. Stomata about 80 to the sq. m.M.; phaneroporous; mostly lying somewhat above the level of the epidermal cells; usually surrounded by 4 subsidiary cells. Several stomata more or less compressed during the development; in some stomata the full-grown mother cell not divided. Trichomes less numerous than at the upper side; see for the rest those of the upper side.

E p i d e r m a l c e l l s p r o p e r. At the base of the leaf Lev. B. 40 μ, Lev. L. 100 μ; polygonal tables; the rest H. 20—30 μ, Lev. 80 μ; polygonal tables with strongly sinuous lateral walls. Outer walls slightly thickened, exhibiting a thin cuticle, at the base of the leaf longitudinal cuticular striation. Cell contents: chloroplasts, more numerous than at the upper side; in chromic acid the nucleus showing up clearly. S t o m a t a. H. 20 μ, Lev. B. 35 μ, Lev. L. 40 μ. In potassium dichromate several of the subsidiary cells containing some brown granules. T r i c h o m e s. See those of the upper side.

Mesophyll.

P a l i s a d e c h l o r e n c h y m a. Consisting of 1 layer of cells. In the uppermost half of the leaf most of the cells divided in 2 parts by a tranverse wall, each part similar to a palisade cell; the innermost one of these 2 cells becoming larger at some distance from the margin, now and then somewhat indented. Showing intercellular spaces.

C e l l s o f t h e a b o v e. Lev. 20 μ; of the divided cells the outermost one Lev. 20 μ, the innermost one H. 60—70 μ, Lev. 20—30 μ; cylindrical. Cell contents: numerous chloroplasts, 1—2 μ thick and 4—6 μ in diameter, mostly penta- to hexagonal tables, containing mostly 1, sometimes 2 minute needle-shaped starch grains.

S p o n g y c h l o r e n c h y m a. At the base of the leaf 7—8, towards the top 5 layers of cells; the uppermost layer developed as collecting cells. Cells showing numerous branches. Intercellular spaces very large, especially in the undermost layers.

C e l l s o f t h e a b o v e. H. 20 μ, Lev. B. 40 μ, Lev. L. 180 μ; or in a 4-branched cell H. 20 μ, Lev. 70 μ; in the uppermost layer (collecting cells) H. 25 μ, Lev. 35 μ. Cell contents: in potassium dichromate in each cell some small dark brown granules of about 3 μ in diameter; see for the rest contents of the palisade tissue, the chloroplasts here somewhat smaller. In the cell layer adjoining the colourless parenchyma of the veins the starch grains in the chloroplasts larger than in the rest of the spongy tissue.

M a r g i n. On the whole similar to the intervenia, a few times 1 or 2 layers of collenchyma.

---

[1]) See D e B a r y. Vergleichende Anatomie. 1877. 54 and 55.

Collenchyma cells. 20 μ in diameter. Walls collenchymatous. Cell contents: some chloroplasts.

Veins.

Epidermis upper side. Only above the smallest veinlets the epidermis resembling that of the intervenia. For the rest cells arranged very regularly in longitudinal rows, at the base of the leaf longitudinal arrangement less clear. Stomata wanting. Trichomes, see those of the upper side of the intervenia.

Epidermal cells proper. At the base R. 40 μ, T. 20 μ, L. 60 —125 μ; mostly tetragonal prisms; in the middle part longer tetragonal prisms; at about 1 c.M. in distance from the top of the leaf T. 18 μ, L. 125 μ. Outer

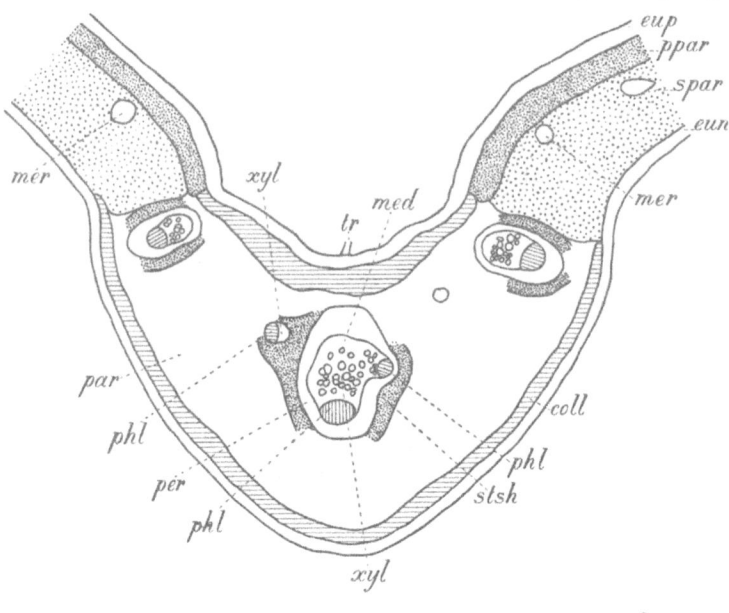

Fig. 57. *Aconitum Napellus*. Leaf, transverse section through the midrib and the neighbouring portions of the lamina. coll Collenchyma; eun Epidermis under side; eup Epidermis upper side; med Medulla; mer Small meristeles; par Colourless parenchyma; per Pericycle; phl Phloem; ppar Palisade chlorenchyma; spar Spongy chlorenchyma; stsh Starch-sheath; tr Trichomes; xyl Xylem.

walls slightly thickened and showing very fine longitudinal cuticular striation; inner walls thickened collenchymatously. Cell contents: see those of the upper side of the intervenia.

Epidermis under side. Cells arranged very regularly in longitudinal rows; at the base of the leaf longitudinal arrangement less clear. Stomata wanting. Trichomes more numerous than at the upper side; see for the rest those of the upper side.

Epidermal cells proper. At the base of the leaf R. and T. 30 μ, L. 70—130 μ; polygonal tables; in the middle part tetragonal tables; at about 1 c.M. in distance from the top R. 35 μ, T. 25 μ, L. 110 μ. Outer walls thickened, showing very fine longitudinal cuticular striation; inner walls collenchymatous. Cell contents: see those of the under side of the intervenia.

Mesophyll. More particularly that of the basal part of the midrib.

Collenchyma. 1 layer of cells at the upper and 1 at the lower side; at the base of the leaf over a short distance on both sides 2 layers. Small intercellular spaces.

C e l l s o f t h e a b o v e. At the upper side in the outermost layer R. 30 μ, T. 20 μ; in the innermost layer R. 25 μ, T. 30 μ, L. 180—300 μ; at the under side R. 35 μ, T. 30 μ, L. 150—325 μ; polygonal prisms with a longitudinally directed axis and somewhat rounded edges. Walls thickened collenchymatously, near.the top of the leaf walls hardly thickened. Cell contents: some small chloroplasts; in potassium dichromate some small dark brown granules; in chromic acid material the nucleus conspicuous.

C o l o u r l e s s p a r e n c h y m a. At the upper side 1—2, at the under side 3 layers of cells. Showing intercellular spaces.

C e l l s o f t h e a b o v e. At the upper side R. and T. 20—30 μ, L. 150—250 μ; at the lower side in the outermost layer R. 70 μ, T. 60 μ; for the rest R. and T. 45 μ, L. 200—300 μ; polygonal prisms with a longitudinally directed axis and rounded edges. Walls of the cells at the under side somewhat collenchymatous. Cell contents: see those of the collenchyma; besides some small starch grains.

P a l i s a d e c h l o r e n c h y m a. Covering only the smallest meristeles.

E n d o d e r m i s. Developed as a starch-sheath. Consisting of 1 layer of cells, showing very small intercellular spaces between themselves and the surrounding tissue. In the largest meristeles of the midribs and in the larger primary veins, corresponding with the divisions of the leaf, only present at both lateral sides. For the rest quite surrounding the smaller and even the smallest meristeles; here however containing less starch, though always larger grains than the adjacent mesophyll.

C e l l s o f t h e a b o v e R. and T. 30 μ, L. 100—150 μ; polygonal prisms with a longitudinally directed axis and slightly rounded edges. Cell contents: spurious chloroplasts quite filled with generally compound, up to 6-adelphous starch grains; 6-adelphous grains e. g. 12 by 18 μ, 3-adelphous grains e. g. 6 by 10 μ, etc.

Meristele.

P e r i c y c l e. Consisting of common parenchyma and thin-walled fibres. Only present round the larger meristeles. In the base of the midrib at the upper and under side of the larger meristeles 5, at either lateral side 3 layers of elements.

P a r e n c h y m a c e l l s. R. and T. 12 μ, in the outermost layer R. and T. larger, L. 300 μ; polygonal prisms with a longitudinally directed axis. Walls, especially at the edges thickened collenchymatously. Cell contents: some chloroplasts, especially at the lateral sides of the xylem; in potassium dichromate some small dark brown granules. F i b r e s. L. 800 μ; see for the rest the parenchyma cells.

V a s c u l a r b u n d l e. A single one, not clearly differentiated either as compound or simple but represented by a phloem and a xylem part not separated from one another by a distinct cambium.

P h l o e m. The elements nearly quite arranged in radial rows; sieve-tubes and companion cells distinct in a longitudinal section. The rows pretty well corresponding with those of the xylem. Intercellular spaces wanting.

S i e v e - t u b e s o f t h e r a d i a l r o w s. R. 5 μ, T. 8 μ, the other ones R. and T. 6 μ, L. of articulations of all e. g. 60 μ; sieve-plates on the transverse walls distinct; tetragonal. Walls slightly thickened. Cell contents: near the sieve-plates a little alcaloid (fresh material, potassium iodide iodine); in alcoholic material no such reaction. C o m p a n i o n c e l l s with granular contents.

X y l e m. The undermost part consisting of elements arranged in radial row; the uppermost and largest part of numerous vessels, mostly spiral, rarely reticulate vessels, intermixed with common parenchyma. In the smaller bundles the radially arranged elements wanting.

V e s s e l s. R. and T. 15 μ; polygonal prisms with rounded edges. Walls lignified. E l e m e n t s o f t h e u n d e r m o s t p a r t. At the base of the radial rows R. 5 μ, T. 10 μ; tetra- to hexagonal.

M e d u l l a. Represented by a longitudinal strip of common parenchyma cells between the upper side of the xylem and the pericycle.

C e l l s o f t h e a b o v e. R. and T. 10 μ; polygonal prisms with a longitudinally directed axis. Cell contents: some chloroplasts; in the chromic acid material nucleus conspicuous.

F l o w e r.

MATERIAL. Flowers, gathered June 1903 in the Botanic Garden at Groningen, fresh and in alcohol.

MICROGRAPHY.

C a l y x.

Helmet-shaped sepal.

E p i d e r m i s. Inner side. In the upper part of the sepal cells practically lying in longitudinal rows; stomata wanting; hairs especially along the margin and near or above the meristeles, diameter at the base 20 μ, length 160—700 μ, conical, unicellular.

E p i d e r m a l c e l l s p r o p e r. Close to the base of the sepal Lev. B. 25 μ, Lev. L. 60—130 μ; polygonal tables; farther towards the middle part Lev. B. 25 μ, Lev. L. 30—80 μ; mostly rectangular tables; in the middle part and the uppermost half of the sepal H. 22 μ, Lev. B. 30 μ, Lev. L. 40 μ; polygonal tables with sinuous lateral walls; axis of all the cells at right angles with the surface of the sepal. Outer walls a little thickened; with longitudinal cuticular striation; lateral walls often pitted. Cell contents: nucleus mostly distinct and surrounded by some chloroplasts containing starch grains [1]); in material treated with potassium dichromate some brown granules united in a cluster. C e l l s o f h a i r s. Walls more or less thickened, in the part between the epidermal cells not thickened; showing a cuticle with warty markings.

Outer side. In the undermost half of the sepal the cells practically in longitudinal rows. Stomata in the middle part 20 to the square m.M., increasing in number towards the top, at the base nearly wanting; phaneroporous; lying in the same level with the epidermal cells. Hairs very numerous; regularly spread over the surface of the sepal; see for the rest those of the inner side.

E p i d e r m a l c e l l s p r o p e r. In the undermost half of the sepal Lev. B. 20 μ, Lev. L. 60 μ; mostly rectangular prisms, for the rest polygonal tables; in the middle part e. g. H. 30 μ, Lev. B. 30 μ, Lev. L. 45 μ; axis of all the cells at right angles with the surface of the sepal. Outer walls showing a cuticle often with longitudinal striation. Cell contents: anthocyanin [2]); in material treated with potassium dichromate cells containing a yellow to brown granular mass, sometimes still a colourless vacuole. See for the rest the inner

---

[1]) For alcaloid see E r r é r a, M a i s t r i a u e t C l a u t r i a u, l. c. 111.

[2]) Alcaloid to a greater quantity than in the inner epidermis and in the parenchyma of the mesophyll, according to E r r é r a etc., l. c. 111.

epidermis. S t o m a t a. In the middle part Lev. 30 by 35 µ, near the top Lev. 40 by 50 µ. G u a r d-c e l l s. Contents: numerous chloroplasts containing starch grains; anthocyanin wanting. H a i r s. See those of the inner side.

M e s o p h y l l.

S p o n g y c h l o r e n c h y m a. In the middle part of the sepal consisting of about 6 layers of cells; cells practically arranged in planes parallel to the outer and inner surface of the sepal.

C e l l s o f t h e a b o v e. At the base of the sepal e. g. Lev. B. 25 µ, Lev. L. 190 µ, in the middle part e. g. H. 20 µ, Lev. B. 50 µ, Lev. L. 70 µ; showing numerous branches. Cell contents: especially in the outermost layers chloroplasts, containing some needle-shaped starch grains and paving the walls; nuclei mostly very large [1]); in material treated with potassium dichromate, especially in the outermost cell layers, some very dark brown granules.

E n d o d e r m i s. Developed as a starch-sheath; consisting of 1—2 layers of cells. Starch grains in the chloroplasts larger than in the spongy chlorenchyma.

M e r i s t e l e s.

Anastomosing and showing blind ends. Sieve-tubes containing minute starch grains. In material treated with potassium dichromate in many cells brown granules.

Lateral sepals.

E p i d e r m i s. Inner side. Cells of the colourless part in the bud hidden under the helmet somewhat more longitudinally elongated than those of the other parts, along the margin of the sepal a strip of cells with exceedingly sinuous lateral walls; see for the rest those of the helmet; stomata wanting; hairs spread all over the sepal, along the margin the longest and to the greatest number, on the part lying under the helmet the longest hairs up to 3 m.M. in length and at their base up to 50 µ in diameter.

Outer side. Along the margin lateral cell walls sinuous, see for the rest the epidermal cells of the helmet; stomata wanting on the part covered by the helmet and along the margin on the strip with strongly undulated lateral cell walls, for the rest distributed like those of the helmet, but somewhat more numerous, in the middle part of the sepal about 30 to the sq. m.M., near the top about 60; hairs along the margin very long, for the rest see those of the helmet. M e s o-p h y l l: see that of the helmet. M e r i s t e l e s: see those of the helmet.

Undermost sepals. Stomata more numerous than on the helmet; for the rest see the helmet.

C o r o l l a.

Nectaries.

E p i d e r m i s. Inner side. Stomata wanting; hairs wanting at the top of the nectary, for the rest not numerous, see those of the helmet.

E p i d e r m a l c e l l s p r o p e r. At the top, up to about 1.2 m.M. in distance from the margin H. 20 µ, Lev. 15 µ; polygonal prisms. Outer walls thin and covered with a cuticle, lateral walls very thin. Cell contents: nucleus very large and distinct; some very small

---

[1]) For alcaloid see E r r é r a etc., l. c. 111.

chloroplasts containing minute starch grains [1]). Covering the ramifications of the meristeles H. 24 µ, Lev. 25 µ. Outer walls thickened and showing a cuticle; lateral walls pitted. Cell contents: nucleus very large and distinct.

Outer side. Cells covering the meristeles and those covering the base of the nectary in longitudinal rows. Lateral cell walls sinuous, except at the base. Stomata wanting. Hairs not numerous, the greater number on the furrows; for the rest see those of the helmet.

E p i d e r m a l c e l l s p r o p e r. At the top of the nectary H. 30 µ, Lev. B. 25 µ, Lev. L. 40 µ; on the outer side cells larger than on the inner side; farther towards the base H. and Lev. B. 30 µ, Lev. L. 130—300 µ; above the meristeles and at the base rectangular prisms. Outer walls thickened; with longitudinal cuticular striation. Cell contents: mostly anthocyanin, especially at the top; nucleus conspicuous; in material treated with potassium dichromate some very large granules, varying from very dark brown to dark blue.

M e s o p h y l l.

C o l o u r l e s s p a r e n c h y m a. At the top at the outer side 10 layers of cells, at the inner side 3 layers. Cells increasing in size towards the base of the nectary and there showing smaller or larger branches. Intercellular spaces. In the lip 1—2 layers of cells.

C o l o u r l e s s p a r e n c h y m a c e l l s. At the top of the nectary R., T. and L. 20 µ; polygonal with rounded edges. Cell contents: some chloroplasts containing needleshaped starch grains; in material treated with potassium dichromate some small brown granules. Cells of base R. and T. 50 µ, L. up to 200 µ; cylinders or polygonal prisms with a longitudinal directed axis and strongly rounded edges.

E n d o d e r m i s. Developed more or less as a starch-sheath. Consisting of 2 layers of cells.

C e l l c o n t e n t s: Chloroplasts containing needle-shaped or compound starch grains.

M e r i s t e l e s. 3 in number. The 2 lateral meristeles bending forward at a right angle at about 3 m.M. from the top; a smaller branch parallel to the middle bundle pursuing its course towards the top; the deviating branch emitting several branches upwards and at its end 1 downwards into the lip. Ends mostly anastomosing. Containing a pericycle of common parenchyma and a vascular bundle. In the sieve-tubes very small starch grains.

S t a m e n.

Filament. Epidermis, mesophyll and meristele. See those of the base of the nectary [2]).

Anther. C o n n e c t i v e containing a fibrovasal bundle. T h e c a e: fibrous layer conspicuous; middle layer still extant; tapetum no more to be distinguished; pollen here and there still united in tetrades, grains smooth.

P i s t i l.

Ovary [3]). Without complete fusion of the margins of the carpel. Carpel contain-

---

[1]) „Dans les pétales modifiés, allant vers la forme staminale, la quantité d'alcaloide diminue et il se localise principalement à leur base." E r r é r a etc. l. c. 112.

[2]) Les étaraines renferment aussi de l'aconitine, principalement à la base et dans la partie axiale; dans les anthères on n'en trouve pas. E r r é r a etc. l. c. 112.

[3]) „Toutes les parties de l'ovaire et de l'ovule contiennent de l'aconitine, qui a une tendance dans ces organes à se masser dans les cellules sous-épidermiques. L'ovule en renferme beaucoup, ainsi que le pistil. E r r é r a etc. l. c. 112.

ing a mesophyll consisting of about 6 layers of common parenchyma cells. Meristeles, without a starch-sheath, scattered here and there; the one of the mid-rib being thicker than the others. The meristele of the raphe not reaching quite to the chalaza. Ovules anatropous; with 2 integuments, the outer one consist-ing of about 4 layers of cells with small starch grains and near the raphe of about 12, the inner one consisting of 2 layers of cells and overtopping the outer in-tegument.

Style. With a longitudinal furrow. Epidermis with thickened outer walls and longitudinal cuticular striation. Mesophyll consisting of common parenchyma cells with large compound starch grains. Meristele in the centre. Conducting tissue in the mesophyll opposite to the furrow.

Stigma. Papillae not discernible.

July 1903.   J; M; L.

HERBA BELLADONNAE RECENS.

The fresh aerial parts of Atropa Belladonna, Linn. Sp. Pl. 181, gathered during the flowering season.

LITERATURE. Anema. Zetel d. Alcaloiden. Diss. Utrecht. 1892. Benecke. Mikr. Dro-genprakt. 1912. 54. Ellrodt. Verteil. d. Gerbstoffs in offizinellen Blättern, Kräutern u. Blüten. Diss. Würzburg. 1903. 11. Fedde. Vergl. Anat. d. Solanaceen. Diss. Breslau. 1896. Flückiger. Pharmakogn. 1891. 702. Gilg. Pharmakogn. 1910. 303. Glaser. Mikr. Anal. d. Blattpulver. Verh. d. Phys. Med. Ges. z. Würzburg. N. F. Bd.. XXXIV. 251, 285. Greenish & Collin. Anat. Atl. o. Veget. Powders. 1904. 54. Hager. Pharm. Praxis. Bd. I. 1900. 467. Hérail. Pharmacogr. 1912. 678. Karsten u. Oltmanns. Pharmakogn. 1909. 166. Koch. Mikr. Anal. d. Drogenpulver. Bd. III. 1906. 91. Pl. VIII. Koch. Einf. i. d. mikr. Anal. d. Drogenpulver. 1906. 82. Koch u. Gilg. Pharmakogn. Praktik. 1907. 147, 149. Kraemer. Botany a. Pharmacogn. 1910. 620. Marmé. Pharmacogn. 1886. 186. Meyer. Wiss. Drogenk. Bd. II. 1892. 271. Mitlacher. Tox. od. forens. wicht. Pfl. u. Drogen. 1904. 140. Moeller. Mikr. Pharm. Ueb. 1901.101. Moeller. Pharmakogn. 1889. 67. Molisch. Mikrochemie der Pfl. 1913. 257. Oudemans. Aant. o. d. Pharm. neerl. 1854—56. 331. Oudemans. Pharmacogn. 1880. 254. Planchon e. Collin. Drogues simples. T. I. 1895. 575. Sievers. Verd. d. Alcaloïden i.d. versch. Deelen d. Atropa Belladonna, L. Am. J. Pharm. 1914. 97—112. Tschirch u. Oesterle. Anat. Atl. 1900. 329. Pl. 76. Tunmann. Pflan-zenmikrochemie. 1913. 325. Unger. Folia Belladonnae. Apoth. Ztg. XXVII. 1912. 763. Vogl. Anat. Atl. 1887. 8, 9. de Wèvre. Local. d. l'Atropine. Bull. d. Séances d.1. Soc. belge d. Micr. Oct. 1887. Wigand. Pharmakogn. 1879. 215. Wittlin. Bildung d. Kalk-Oxal. Taschen. Bot. Centralbl. Bd. 67. 1886. 101. and Diss. Bern. 1896. 21. REAGENTS. Water, sugar and water (flower), glycerine, chloral hydrate, potassium iodide iodine (leaf), iodine and sulphuric acid 66 per cent., iodine in chloral hydrate (flower), phlo-roglucin and hydrochloric acid, concentrated sulphuric acid, Schulze's macerating mixture (leaf).

Stem.

MATERIAL. Stems 6 c.M. thick, collected June 1901 in the Botanic Garden at Gro-ningen, fresh, in alcohol and fixed with chromic acid 1 per cent.

MICROGRAPHY.

Epidermis. Stomata 5 to 10 to the sq. m.M.; phaneroporous; lying in the same level with the epidermal cells or somewhat lower; for the rest see those of the leaf. Trichomes. Capitate hairs like those of the leaf; sometimes very long. Glandular hairs wanting.

E p i d e r m a l c e l l s p r o p e r. R. 18 μ, T. 25 μ, L. 25—50 μ; tetra- to hexagonal tables with a radially directed axis. Outer walls somewhat thickened; with a cuticle showing parallel longitudinal striation; for the rest walls cuticularized (in iodine and sulphuric acid 66 per cent. colouring greenish blue). Cell contents: some small chloroplasts.

Cortex.

C o m m o n p a r e n c h y m a. Consisting of 7 layers of cells intermixed with idioblasts. Sometimes a cell divided in 2 or more cells by transverse partition-walls. Cells of the outermost 3 layers much smaller in diameter than the rest; sometimes the cells of the outermost layer having the same length as the epidermal cells. Showing intercellular spaces. Idioblasts in the shape of crystal sand cells not numerous; sometimes some of them forming a longitudinal row.

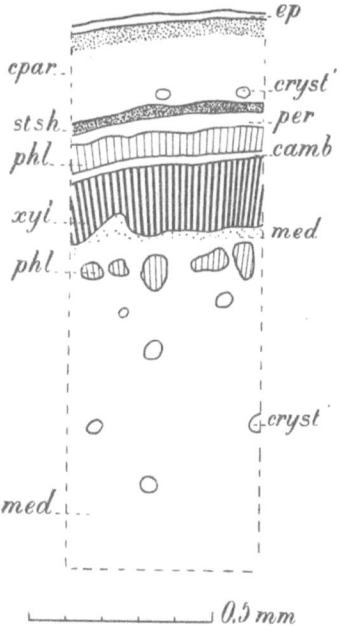

C o m m o n p a r e n c h y m a c e l l s. The largest R. and T. 40 μ, L. 150—300 μ; in the outermost layer strictly tetragonal prisms, for the rest tetra- to hexagonal prisms with a longitudinally directed axis. Radial and transverse walls collenchymatous, especially characteristic for the 4th, 5th and 6th layer (c. f. colourless parenchyma of the midrib of the leaf); thickened parts stratified, pitted. Cell contents: in the outermost 2—3 layers many small chloroplasts, containing some crescent-shaped starch grains; farther towards the inner part a few minute chloroplasts; in the outermost layer, on the side of the stem turned towards the light the vacuole coloured red, the chloroplasts paving the walls; in some cells a yellow granular mass. I d i o b l a s t s. L. 60 μ; for the rest similar to the surrounding parenchyma. Crystal sand, see the leaf.

Fig. 58. *Atropa Belladonna*. Stem, transverse section. camb Cambium; cpar Common parenchyma; cryst' Crystal sand cells; ep Epidermis; med Medulla; per Pericycle; phl Phloem; stsh Starch-sheath; xyl Xylem.

E n d o d e r m i s. Developed as a starch-sheath; consisting of 1, sometimes of 2 layers of cells. C e l l s o f t h e a b o v e. R. 35 μ, T. 60 μ, L. 70 μ; elliptical cylinders, surfaces of contact often flattened; sometimes divided in 2 cells by a transverse wall; often at the upper or at the under side a corner cut off by a subsequently formed wall. Walls pitted, at least the longitudinal ones. Cell contents: spurious chloroplasts united in a cluster, sometimes to be found on one side of the cell, mostly quite filled with simple or up to 4 compound starch grains.

Stele.

P e r i c y c l e. Consisting of 1—3 layers of common parenchyma cells. Here and there small bundles, thick 1—4 sclerenchyma fibres, imbedded in the parenchyma.

. C o m m o n p a r e n c h y m a c e l l s. R. 24 μ, T. 32 μ, L. 160 μ; elliptical cylinders like the endodermis cells. Walls slightly thickened; pitted . S c l e r e n c h y m a f i b r e s. R. 20 μ, T. 30 μ, very long; one whole fibre noticed 1 m.M. long, for the rest fragments of fibres long up to 3.3 m.M.; sometimes perhaps divided by transverse walls in partitions 300 μ long, this fact however not being corroborated by maceration; tetra- to hexagonal, mostly with rounded edges and with rather obtuse ends. Walls somewhat thickened; not lignified; with a highly refractive inner mucilaginous layer (Gelatinous layer), sometimes letting loose and then looking like flattened tissue; secondary layers showing stratification; cellulose reaction more or less strong in the outer secondary layers: yellowish green, sometimes quite blue; the innermost tertiary layer consisting of pure cellulose. Cell contents: sometimes in a transverse section fibres seen to be filled with a yellow granular mass.

P h l o e m. Forming a continuous ring of 6—8 elements thick. The outermost part not radially arranged, probably representing the primary phloem. Primary phloem consisting of sieve-tubes sometimes having companion cells, cambiform cells showing intercellular spaces and crystal sand idioblasts sometimes forming longitudinal rows of e. g. 6 cells. Secondary phloem consisting of sieve-tubes sometimes having companion cells, bast parenchyma without intercellular spaces and crystal idioblasts like those mentioned above; for the medullary rays see below.

S i e v e-t u b e s. In the primary part R. and T. 12 μ, in the secondary part R. 12 μ, T. 18 μ, L. of articulations 130 μ; tetra- to hexagonal prisms; sieve-plates on the transverse walls conspicuous, showing callus. C o m p a n i o n c e l l s. L. corresponding to that of the articulations of the sieve-tubes; tri- or tetragonal prisms. Cell contents often granular. C a m b i f o r m c e l l s. R. and T. 15 μ, L. 80—150 μ; tetra- to hexagonal prisms with a longitudinally directed axis. Cell contents: sometimes some chloroplasts containing globular or crescent-shaped starch grains. B a s t p a r e n c h y m a c e l l s. R. 15 μ, T. 18 μ, L. 100—120 μ, corresponding to the length of the elements of the cambium. Cell contents: see those of the cambiform cells. C r y s t a l s a n d i d i o b l a s t s. Shape and dimensions like those of the parenchyma cells. Contents: see the leaf.

C a m b i u m. Consisting of 2—3 layers of elements.

E l e m e n t s o f t h e a b o v e. T. 18 μ, L. 100—120 μ. Contents: sometimes some yellow granules.

X y l e m. Forming a continuous ring of 12—14 elements thick. Secondary xylem taking up the outermost 8 layers; radially arranged; consisting of pitted vessels, libriform fibres occasionally divided in partitions by transverse walls and wood parenchyma; fibres radially arranged. Medullary rays see below. No intercellular spaces. Primary xylem consisting of detached spiral vessels, common parenchyma cells in longitudinal rows and an occasional crystal sand idioblast. No intercellular spaces.

P i t t e d v e s s e l s. R. and T. 50 μ, L. of articulations 250—330 μ; varying from polygonal prisms to circular or elliptical cylinders. Walls somewhat thickened; showing stratification; lignified; showing circular or slit-like bordered pits. S p i r a l v e s s e l s. R. and T. 25 μ. Walls lignified. L i b r i f o r m f i b r e s. R. 18 μ, T. 20 μ, L. 100—400 μ; tetragonal with acute, sometimes branched ends. Walls thickened; middle lamella sometimes lignified, the rest of the walls consisting of cellulose; generally not pitted, sometimes showing cross-wise slit-like pits. W o o d p a r e n c h y m a c e l l s. R. 15 μ, T. 20 μ, L. 100—140 μ; of the cells with unpitted walls lying towards the interior and less radially arranged L. 150—200 μ; sometimes the cells of the same radial row increasing in length towards the outside and even attaining twice the length of the innermost cells;

tetragonal prisms with a longitudinally directed axis. Walls thickened; middle lamella sometimes lignified, the rest of the walls consisting of cellulose; sometimes unpitted, in other
cases with pits of various sizes, sometimes one pit taking up the whole breadth of the cell,
in other cases with one-sided bordered pits, in other cases again with reticulate thickenings.
P a r e n c h y m a c e l l s o f t h e p r i m a r y x y l e m. R. and T. 20 μ, L. 80 μ; pen-
ta- or hexagonal prisms. Cell contents: often chloroplasts containing starch. C r y s t a l
s a n d i d i o b l a s t s. Shape and dimensions like those of the surrounding cells. Contents: see the leaf.

M e d u l l a r y r a y s. Mostly 1- to 2-seriate; consisting of parenchyma cells, the
cell walls bearing the same character as those of the surrounding tissue. In a
transverse section cells radially elongated and increasing in size towards the
outside.

C e l l s o f t h e a b o v e. R. 20 μ, T. 18 μ. See for the rest wood parenchyma cells.

M e d u l l a. Consisting of common parenchyma containing crystal sand idioblasts and a very few idioblasts with a simple crystal, sometimes forming longitudinal rows, and phloem bundles with a few adjacent sclerenchyma fibres.
Phloem bundles separated by uni- to biseriate bands of parenchyma; lying at
different distances from the xylem ring; sometimes radially elongated; containing 20—35 elements in a transverse section; elements similar to those of the
outer primary phloem.

P a r e n c h y m a c e l l s. R. and T. 90 μ, L. 75—150 μ; polygonal prisms with a longitudinally directed axis and sometimes with rounded edges. Walls pitted, transversely
slit-like pits on the longitudinal walls. Cell contents: sometimes minute chloroplasts.
P a r e n c h y m a c e l l s b e t w e e n t h e p h l o e m b u n d l e s. R. and T. smaller
than those of the other parenchyma cells; for the rest similar to them. E l e m e n t s o f
t h e p h l o e m b u n d l e s. See those of the primary phloem. S c l e r e n c h y m a
f i b r e s. Quite similar to those of the pericycle. C r y s t a l s a n d i d i o b l a s t s.
R. and T. 40 μ, L. 90 μ. Cell contents: see those of the leaf.

L e a f.

Macroscopic characters.

Simple, petiolate, up to 2 d.M. long. Blade when dried very thin and brittle;
oblong; feather-veined; apex somewhat taper-pointed; base acute and tapering into the petiole; margin entire. Surface: midrib, at the lower surface also
the primary veins and the smaller veins, moreover the margin sparingly hairy,
at all events the young leaves; dried leaves, when seen through the magnifying
glass, showing minute white dots marking the places of the crystal sand cells.
Odour of the fresh rubbed leaf somewhat narcotic; taste bitterish.

Anatomical characters.

MATERIAL. Leaves, gathered June 1901 in the Botanic Garden at Groningen; full
grown leaves fresh and in alcohol; young leaves, 1 c.M. long, fresh.

MICROGRAPHY.

I n t e r v e n i a.

Epidermis upper side. Stomate about 60 to the sq. m.M.; phaneroporous;
lying in the same level as the epidermal cells; generally surrounded by 3 subsidiary cells. Trichomes not numerous, the greater number along the margin

of the leaf; occurring in 2 kinds: 1°. Common capitate hairs, forming the majority of the trichomes; articulate, consisting of 1 row of 4—5 cells; the apical cell nearly always dilated into a globular head; stalk cells often flattened; at the margin, especially of the young leaf, some of these hairs having a single small branch also with a globular head. 2°. Glandular capitate hairs, stalk unicellular; head pluricellular, generally consisting of 3 stories of 4 cells each.

E p i d e r m a l c e l l s p r o p e r. H. 20 μ, Lev. B. 30 μ, Lev. L. 50 μ; tabular, lateral walls slightly sinuous. Outer walls with a thin cuticle forming an undulated striation; yellowish. Cell contents: some chloroplasts. S t o m a t a. Lev. 20 μ by 30 μ. C e l l s o f c o m m o n c a p i t a t e h a i r s decreasing in length towards the top of the hair. Walls thin; quite cuticularized. Cell contents: some chloroplasts. C e l l s o f g l a n d u - l a r h a i r s. Walls quite cuticularized; no cuticular bladder. Cell contents: some chloroplasts.

Epidermis under side. Stomata about 150 to the sq. m.M. See for the rest those of the upper side. Trichomes like those of the upper side, but more numerous.

E p i d e r m a l c e l l s p r o p e r. H. 20 μ, Lev. B. 35 μ, Lev. L. 60 μ; lateral walls more sinuous than at the upper side; see for the rest the epidermal cells of the upper side.

Mesophyll.

P a l i s a d e c h l o r e n c h y m a. Consisting of 1—2 layers of cells; the innermost one less regularly arranged than the other. Showing intercellular spaces.

C e l l s o f t h e a b o v e. H. 40—50 μ, Lev. 20 μ, in the innermost layer H. much smaller; polygonal prisms or cylinders. Cell contents: many mostly pentagonal tabular chloroplasts, 5 μ in diameter and 3 μ thick, paving the walls, containing some cylindrical or needle-shaped starch grains.

S p o n g y c h l o r e n c h y m a. Consisting of 5—7 layers of stellate, 6- to 8-radiate cells. Showing large intercellular spaces.

C e l l s o f t h e a b o v e. H. 30 μ, Lev. 50 μ. Cell contents: see those of the palisade chlorenchyma.

I d i o b l a s t s. Crystal cells occurring in the undermost layer of the palisade tissue and the uppermost layer of the spongy tissue.

C e l l s o f t h e a b o v e. 25 μ in diameter; globular or polyhedral; quite filled with crystal sand consisting of tetrahedral particles with curved sides and sharp angles.

V e i n s.

Epidermis upper side. Stomata about 10 to the sq. m.M.; see for the rest those of the intervenia. Trichomes like those of the intervenia, but more numerous especially the glandular hairs.

E p i d e r m a l c e l l s p r o p e r. R. 25 μ, T. 25 μ, L. 70 μ; generally tetra- to hexagonal tables, axis at right angles with the surface of the leaf. Outer walls slightly thickened; showing a thin cuticle and longitudinal cuticular striation; yellowish; inner walls collenchymatous. Cell contents: some chloroplasts.

Epidermis under side. Stomata 5 to the sq. m.M.; see for the rest those of the upper side. Trichomes like those of the upper side, but much more numerous.

E p i d e r m a l c e l l s p r o p e r. H. and Lev. B. 30 μ, Lev. L. 130 μ. Some cells with finely granular, light brown, transparent contents. See for the rest the upper side.

18

Mesophyll.

C o l o u r l e s s p a r e n c h y m a. In the midrib about 12 layers of cells at the upper and at the under side. Cells arranged in longitudinal rows. Showing intercellular spaces.

C e l l s o f t h e a b o v e. R. and T. 50 µ, L. 100—250 µ; cylinders or polygonal prisms with a longitudinally directed axis. Only the parts of the walls adjoining the inter-

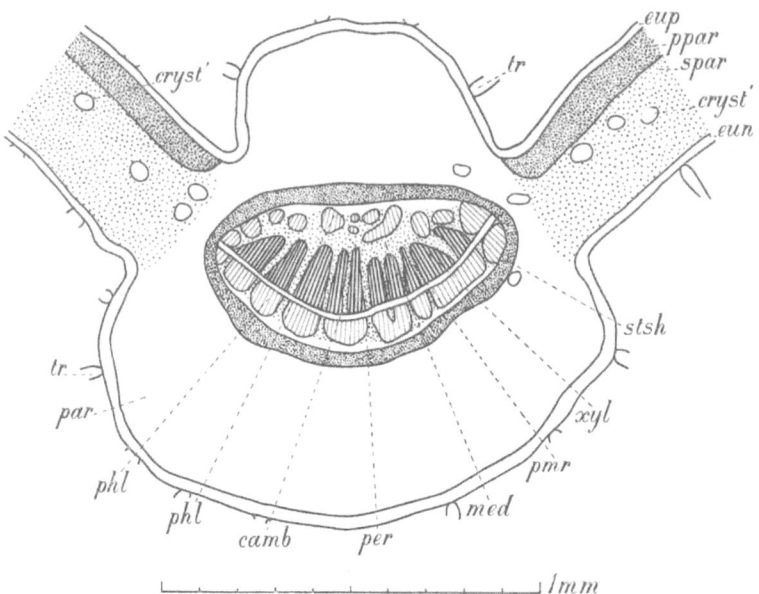

Fig. 59. *Atropa Belladonna.* Leaf, transverse section through the midrib and the neighbouring portions of the lamina. camb Cambium; cryst' Cells filled with crystal sand; eun Epidermis under side; eup Epidermis upper side; med Medulla; par Colourless parenchyma; per Pericycle; phl Phloem; ppar Palisade chlorenchyma; prm Medullary commissures; spar Spongy parenchyma; stsh Starch-sheath; tr Trichomes; xyl Xylem.

cellular spaces thickened; walls of the 5 outer layers of the upper side and of all the layers of the under side somewhat collenchymatous. Cell contents: a few chloroplasts, containing cylindrical and needle-shaped starch grains.

I d i o b l a s t s. Crystal cells, often 7 or less arranged in a longitudinal row.

C e l l s o f t h e a b o v e. R. 40 µ, T. 40 µ, L. 45 µ; shape like that of the parenchyma cells. Contents: crystal sand; see for this the intervenia.

E n d o d e r m i s. In the midrib and the primary veins developed as a starch-sheath; in the smaller veinlets wanting.

C e l l s o f t h e a b o v e. R. and T. 30 µ, L. 70 µ; polygonal prisms with a longitudinally directed axis and rounded edges. Cell contents: simple globular starch grains, all on one side of the cell.

Meristele. In the shape of a semi-elliptical cylinder, the flat side turned upwards.

Pericycle. In the larger meristeles developed as sclerenchyma; in the smaller veinlets wanting.

Vascular bundle. A single compound one containing about 8 simple bundles, arranged on the undermost part of a cylinder surface.

Simple bundles. Radially elongated in a cross section; collateral [1]); open.

Phloem. Structure essentially the same as that of the medullary phloem bundles.

Cambium. Consisting of some layers of radially arranged elements.

Xylem. Bundles somewhat cuneiform; containing spiral vessels.

Vessels. R. and T. 30 μ, L. of articulations 200 μ. Walls lignified.

Medullary commissures. Narrow; consisting of common parenchyma.

Cells of the above. R. and T. 15 μ, L. 100 μ; cylinders or polygonal prisms with a longitudinally directed axis. Walls slightly thickened.

Medulla. Developed only at the upper side of the meristele; in the shape of a flat elliptical band; consisting of common parenchyma and about 10 phloem bundles lying at some distance from the xylem of the simple vascular bundles and not corresponding in place with it. Phloem bundles consisting of about 30 elements in a cross section; sieve-tubes and companion-cells.

Common parenchyma cells. R. and T. 20 μ, L. 80—90 μ; tetra- to hexagonal prisms with slightly rounded edges and a longitudinally directed axis. Sieve-tubes. R. and T. 10 μ, L. of articulations 130—170 μ; tetra- to hexagonal prisms; sieve-plates on the transverse walls conspicuous; often showing callus plates; walls slightly thickened. Companion-cells. R. and T. 3 μ, L. corresponding to that of the articulations of the sieve-tubes; tri- or tetragonal prisms; cell contents granular.

Flower.

MATERIAL. Flowers, collected in the Botanic Garden at Groningen June 1901, fresh and in alcohol.

MICROGRAPHY.

Calyx.

Intervenia.

Epidermis inner side. Stomata about 10 to the sq. m.M. Trichomes numerous; occurring in 2 kinds: 1°. Common capitate hairs like those of the leaf, most numerous at the base of the calyx; 2°. Glandular capitate hairs, stalk consisting of 1 row of cells, followed by some shorter cells containing much protoplasm; head formed by the apical cell swollen in a globule or by some more cells swollen in tables.

Epidermal cells proper. At the base of the calyx H. 20 μ, Lev. B. 50 μ, Lev. L. 70 μ; farther towards the top cells smaller and more rounded; at the base lateral walls a little sinuous. Outer walls somewhat thickened. Cell contents: some chloroplasts.

Epidermis outer side. Stomata about 125 to the sq. m.M.; mostly surrounded by 3 subsidiary cells. See for the rest the epidermis of the inner side.

Epidermal cells proper. H. 25 μ, Lev. B. 30 μ, Lev. L. 45 μ; some cells show-

---

[1]) See *Folia Stramonii*, foot note.

ing slightly sinuous lateral walls. See for the rest the inner side. S t o m a t a. Lev. 25 by 40 μ.

M e s o p h y l l. Spongy chlorenchyma consisting of 5—6 layers of cells; essentially similar to the chlorenchyma of the leaf.

Veins.

E p i d e r m i s. Stomata occurring. Cells somewhat narrower than the epidermal cells of the intervenia, with a longitudinally directed axis; see for the rest the epidermis of the intervenia.

M e s o p h y l l.

Common parenchyma. Consisting of 4 layers of cells at the inner and 5 layers at the outer side. Large intercellular spaces.

C e l l s o f t h e a b o v e. R. and T. 30 μ; tetragonal prisms or cylinders with a longitudinally directed axis. Cell contents: some small chloroplasts.

Endodermis. Not discernible.

M e r i s t e l e. Constituted of a pericycle of sometimes sclerotic cells, a single vascular bundle and some medullary parenchyma containing a phloem bundle like those of the leaf.

V a s c u l a r b u n d l e. Collateral [1]); closed. Phloem substantially the same as the phloem of the vascular bundles of the leaf. Xylem consisting of spiral vessels.

V e s s e l s. R. and T. 10 μ. Walls lignified.

C o r o l l a.

Intervenia.

E p i d e r m i s. Inner side. Stomata wanting; hairs only found in a horizontal zone 2 m.M. wide and at 2 m.M. distance from the base of the corolla, 500—700 μ long, conical, consisting of 1 row of 3—5 cells.

E p i d e r m a l c e l l s p r o p e r. At a few m.M. from the base of the corolla H. 20 μ; at the base tetra- to hexagonal tables with somewhat sinuous lateral walls and slightly collenchymatous; farther towards the top isodiametrical, hexagonal, walls not undulated; still farther towards the top of the corolla papilla-like. Outer walls showing a thin striated cuticle. Cell contents: in the yellow part of the corolla numerous small yellow plastids.

C e l l s o f h a i r s. 40 μ wide, 160 μ long. Walls showing parallel longitudinal cuticular striation.

Outer side. Epidermal cells longest in the middle of the corolla. Stomata rare. Trichomes like those of the calyx, wanting along the margin of the corolla.

E p i d e r m a l c e l l s p r o p e r. In the middle part of the corolla H. 20 μ, Lev. B. 28 μ, Lev. L. 60 μ; smaller towards the top and towards the base; at the base irregular tables with sinuous lateral walls; farther towards the top rectangular tables with a longitudinally directed axis and sinuous lateral walls; still farther towards the top stellate tables. Outer walls showing a thin striated cuticle.

M e s o p h y l l. Common parenchyma containing spherical or ellipsoidal crystal sand idioblasts. In the brown part of the corolla anthocyanin cells, especially in the layer adjoining the epidermis; the other cells containing chloroplasts.

---

[1]) See *Folia Stramonii*, foot note.

In the yellow part some yellow to yellowish green plastids, being larger than the yellow plastids of the epidermis. In the colourless cells plastids wanting.

Veins.

Epidermis, cells T. 30 μ, L. 100 μ. Mesophyll consisting of cylindrical common parenchyma cells; in the outermost layers of the red veins anthocyanin cells; endodermis wanting. Meristeles substantially the same as those of the calyx but less developed, the medullary phloem sometimes wanting.

S t a m e n.

Filament.

E p i d e r m i s. Showing distinct longitudinal cuticular striation and numerous deep undulations of the lateral cell walls. Stomata wanting. Hairs in places corresponding in height to the places of the hairs of the inner side of the corolla; only occurring on the lateral sides; for the rest see the hairs of the corolla.

M e s o p h y l l.

Common parenchyma. Showing large intercellular spaces.

C e l l s o f t h e a b o v e. R. and T. 30 μ, L. 200 μ; cylinders with a longitudinally directed axis. Cell contents: starch grains, sometimes 3-adelphous.

M e r i s t e l e. Consisting only of a single amphicribral vascular bundle.

Anther.

C o n n e c t i v e. Containing a mesophyll composed of common parenchyma; cells showing large intercellular spaces, R. and T. 30 μ, L. 30—45 μ, spherical or ellipsoidal. Moreover a single meristele containing annular vessels with lignified rings.

T h e c a e.

Epidermis. Cells transversely elongated, showing papilla-like swellings, above the future fissures smaller, isodiametrical, hexagonal. Stomata about 10 to the sq. m.M.

E p i d e r m a l c e l l s p r o p e r. R. 25 μ, T. 35 μ, L. 50 μ. Walls yellow; lateral walls pitted. Cell contents: of the common epidermal cells a yellow finely granular mass; of the cells above the fissures 1 yellow lump surrounded by a ring of yellow plastids. S t o m a t a Lev. 25 by 30 μ.

Fibrous layer. Wanting at the places of the future fissures; for the rest consisting of 2 layers of transversely elongated reticulate cells, mounting to 4 layers in the vicinity of the connective.

C e l l s o f t h e a b o v e. H. 25—30 μ, T. 55 μ, L. 22 μ; rectangular; each cell with several horse-shoe-shaped thickenings grown more or less together, each horse-shoe covering the outer and side walls; inner walls not thickened; thickenings lignified.

Middle layer and tapetum. Only represented by remainders of flat cells.

Septa of the thecae. Consisting of common parenchyma.

Pollen. Grains yellow; pentagonal or globular; having 3, a few times 4 pores. Extine showing radial striation; intine somewhat thickened under the pores.

P i s t i l.

Ovary.

O u t e r   w a l l.

Epidermis. Inner and outer side consisting of polygonal cells filled with starch.
Mesophyll. Common parenchyma containing crystal sand idioblasts. The inner
and outer parenchyma layer, joining the epidermis, without intercellular spaces;
the other layers showing intercellular spaces.

C e l l s   o f   t h e   a b o v e. Longitudinally elongated polyhedra; in the outer and the
inner layer shape similar to that of the epidermal cells. Cell contents: starch grains, a
great many of them compound and up to 3-adelphous.

Meristeles. Containing distinct spiral vessels.

S e p t a   a n d   p l a c e n t a e. The parenchyma somewhat spongy. Cell contents: see the outer wall.

O v u l e. Sometimes distinctly anatropous; integument showing a long micropyle.

Style.

E p i d e r m i s. Half-way the height of the style cells R. and T. 18 μ, L. 150
μ; from the base upwards becoming much longer and somewhat wider; tetra-to
hexagonal prisms with a radially directed axis. Outer walls somewhat thickened;
showing longitudinal cuticular striation. Cell contents: some starch grains,
sometimes 3-compound.

M e s o p h y l l. Common parenchyma, showing intercellular spaces.

C e l l s   o f   t h e   a b o v e. R. and T. 20 μ, L. 150 μ; cylinders. Cell contents: starch
grains, some of which 3-compound.

M e r i s t e l e s. 2, viz. 1 on each side of the conducting tissue.

C o n d u c t i n g   t i s s u e. In the shape of an elliptical cylinder, thick 80 by
200 μ; consisting of cylindrical cells, 7 μ in diameter and having thickened walls.
Afterwards 1, sometimes 2 pollen canals, perhaps containing some remainders
of cells and surrounded by many small cells.

Stigma. Not much developed; papillae isodiametrical polyhedra, measuring 40
by 50 μ.

June 1901.   J; M; L.

HERBA CANNABIS INDICAE.

Cannabis Indica. Indian Hemp.

The dried flowering tops of the female plant of Cannabis sativa, Linn. Sp. Pl. 1027, grown in the East
Indies, from which the resin has not been removed.

Macroscopic characters.

The larger leaves and branches being removed as much as possible, the several
parts mostly more ore less adhering to one another and flattened, long 2—10
c.M., brittle, dirty green. A few small palmate leaves with linear-lanceolate leaflets present. Inflorescence consisting of compact only partly developed dichotomous cymes clustered together. Linear-lanceolate bracts, each with two shorter lanceolate stipules; apparently in the xil aof each of the latter a single flower,

surrounded by an other bract forming a sheath, developing together with the fruit. Flowers in all stages of development, female, hypogynous; perigon bell-shaped without limb, closely surrounding the ovary up to half its height, membranaceous; pistil with two filiform stigmata. The surface of branches, leaves and bracts showing many short adpressed hairs and less easily discernible, globular, glandular hairs. Odour slight, narcotic; taste slightly bitter.

Anatomical characters.

LITERATURE. Attema. De zaadhuid der Angiosp. en Gymnosp. Diss. Groningen. 1901. 176. de Bary. Vergl. Anat. 1899. 865. Berg. Anat. Atl. 1865. 86. Erdmann-König. Allgem. Warenkunde. 1895. 397. Flückiger. Pharmakogn. 1891. 749. Gilg. Pharmakogn. 1910. 93. Greenish & Collin. Anat. Atl. o. veget. powders. 1904. 100. Guérin. Cellules à mucilage chez les Urticées. Bull. Soc. bot. d. France. T. LVII. 399—406. Hager. Pharm. Praxis. Bd. I. 1900. 590. Hérail. Mat. Med. 1912. 462. Ingerman. Microsk. d. voorn. Handelswaren. 1910. 88. Karsten u. Oltmanns. Pharmakogn. 1909. 184. Kraemer. Botany a. Pharmakogn. 1910. 255. Luerssen. Syst. Bot. Bd. 2. 1882. 528. Marmé. Pharmacogn. 1886. 212. Mitlacher. Toxik. od. Forens. wicht. Pfl. u. Drogen. 1904. 164. Moreau. Étude sur le Hachich. Paris. 1904. Molisch. Über das Verhalten der Zystolithen gegen Silber- und andere Metallsalze. Ber. d.d. bot. Ges. Bd. XXXVI. 1918. 477. Netolitzky. Ein Kennzeichen der Cannabis-Frucht. Arch. Chem. Mikrosk. Bd. 5. 1912. 2509, 2510. Moeller. Pharmakogn. 1889. 205. Moeller. Mikr. pharm. Üb.1900. 210. Oudemans. Pharmacogn. 1880. 366. Oudemans. Aanteek. o. d. Pharmac. neerl. 1854—56. 112. Parmentier. Recherches sur l'influence d'un mouvement continu regulier imprimé à une plante en végétation normale. Revue gen. Bot. T. 22. 1910. 137. Planchon et Collin. Drogues simples. T. I. 1895. 299. Schneider. Powdered Veget. Drugs. 1902. 144. Schorn. Ueber Schleimzellen bei Urticaceen und über Schleimcystolithen von Girardinia palmata. Ber. Wiener Akad. Bd. 116. Abth. 1. 1906. 393—410. Solereder. Syst. Anat. 1899. 865. Solereder. Ergänzungsband. 1908. 296. Tschirch. Angew. Pfl.-anat. 1889. 44, 113, 158, 163, 286, 293, 464. Tschirch. Pharmakogn. Bd. 2. Abt. 2. 1917. 558. Tschirch u. Oesterle. Anat. Atl. 1900. Tafel 15. Tschirch. Die Harze und die Harzbehälter. 1900. 381. Vogl. Anat. Atl. 1887. Tafel 2. Wiesner. Rohstoffe. II. 1900. 300, 520. Wigand. Pharmakogn. 1879. 236. Winton. Anatomie des Hanfsamens. Zeitsch. Unters. Nahr. u. Genussmittel. 1904.385. REAGENTS. Water, glyceryne, chloral hydrate, iodine in chloral hydrate, phloroglucin and hydrochloric acid, iodine and sulphuric acid 66 per cent., concentrated sulphuric acid, potassium dichromate, copper acetate and iron acetate.

Stem.

MATERIAL. Stems gathered in July and September 1903 in the Botanic Garden at Groningen, fresh and in alcohol. These stems were examined at different distances from the growing top; in these parts they were thick 0.25, 0.6 (chiefly used) and 1 c.M. The drug.

MICROGRAPHY.

Epidermis. Cells on the ribs here and there arranged in longitudinal rows. Stomata wanting. Trichomes of three kinds: 1 Conical cystolith hairs. In all principal respects like those of the leaf, only showing the following differences: total length 150—500 μ; on the ribs of the stem those with strongly enlarged bases the most numerous; the bases mostly not penetrating into the cortex but partly lying above the level of the epidermis, the largest bases T. up to 110 μ, L up to 140 μ, the smaller ones e. g. T. and L. 15 μ; the upper part of all hairs curved nearly always towards the top of the stem; in very numerous hairs cystolith wanting; outer walls with warty cuticular markings; in alcoholic material surrounding each hair a crystal mass, dissolving in glycerine and not showing double refraction. 2. Capitate hairs. Only present in the furrows of the stem; see for the rest those of the leaf. 3. Glandular hairs. Not numerous, to be found only in the furrows; see for the rest those of the leaf.

E p i d e r m a l   c e l l s   p r o p e r. R. 12 μ, T. 18 μ, L. 20 μ, those on the ribs of the stem R. 12 μ, T. 12 μ, L. 15—35 μ; tetragonal tables with a radially directed axis. Walls,

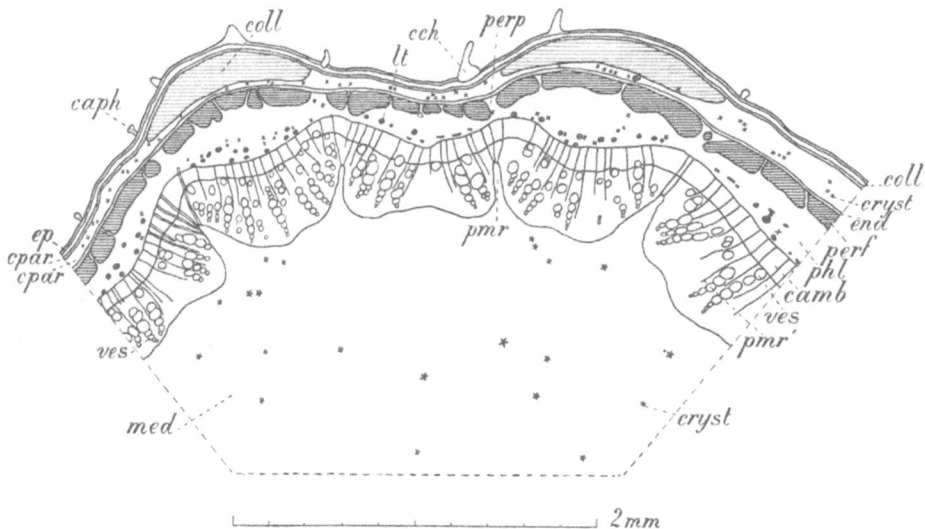

Fig. 60. *Cannabis indica*. Stem, transverse section of a part of the stem thick 0.6 c.M. camb Cambium; caph Capitate hairs; cch Conical cystolith hairs; coll Collenchyma; cpar Common parenchyma; cryst Cells with cluster crystals; end Endodermis; ep Epidermis; lt Latex tubes; med Medulla; perf Pericyclic sclerenchyma fibres; perp Pericyclic parenchyma; phl Phloem; pmr Medullary commissures between the compound vascular bundles; pmr′ Medullary commissures within the compound vascular bundles; ves Vessels.

outer ones a little thickened, with a cuticle, inner walls somewhat collenchymatous, lateral walls somewhat pitted. Cell contents: some relatively large chloroplasts, each containing some small needle-shaped starch grains; in alcoholic material in almost every cell 1 or 2 crystal masses resembling cluster crystals, dissolving in glycerine, not showing double refraction; in material treated with potassium dichromate a yellowish brown homogeneous mass; in chloral hydrate or iodine in chloral hydrate along the outer walls numerous needle-shaped crystals.

Cortex.

C o l l e n c h y m a. In the parts corresponding with the furrows of the stem forming only the outermost cell layer; in the ribs consisting of about 6 cell layers and constituting the whole of the cortex with the exception of the second and 1 or 2 innermost layers; the cells of the outermost layer arranged in longitudinal rows.

C e l l s   o f   t h e   a b o v e. Those corresponding with the furrows of the stem R. 20 μ, T. 25 μ, L. 40—100 μ; tetragonal prisms with a longitudinally directed axis. Walls, especially the outer walls collenchymatous. Cell contents: some chloroplasts, each containing some small needle-shaped starch grains. Cells of outermost layer in the ribs R. 15 μ, T. 18

μ, L. 35—100 μ; tetragonal prisms with a longitudinally directed axis. Walls and contents: see the cells of the furrows, but here more chloroplasts. Inner cells of ribs R. and T. 20—30 μ; polygonal prisms with a longitudinally directed axis and here and there a rounded edge. Walls with very typical collenchymatous thickenings. Here and there with a small intercellular space. Cell contents: here and there a small chloroplast containing some small

needle-shaped starch grains; in alcoholic material here and there close to a wall a globular mass resembling a sphaero-crystal, dissolving in glycerine, not showing double refraction.

Common parenchyma. In the parts corresponding with the furrows of the stem consisting of 4 cell layers; in the ribs constituting the second and the innermost 1 or 2 cell layers; intercellular spaces. Cluster crystal idioblasts here and there in the 2 inner cell layers; often some arranged in longitudinal rows.

Common parenchyma cells. Those corresponding with the furrows of the stem R. 10—12 μ, T. 12—15 μ, L. 25—50 μ; those corresponding with the ribs R. 18 μ, T. 25 μ, L. 40—80 μ; polygonal prisms with often strongly rounded edges. Cell contents: especially of the cells of the outermost layer many chloroplasts, each with some needle-shaped starch grains. Cluster crystal idioblasts. Each containing one cluster crystal, in a single instance a prismatic simple crystal.

Endodermis. In the parts of the stem thick 0.6 c.M. developed as a distinct starch-sheath, thick 1 layer of cells; the cells showing intercellular spaces only towards the common parenchyma of the cortex.

Cells of the above. Those corresponding with the furrows of the stem R. 20 μ, T. 25—45 μ, L. 20—35 μ; mostly rectangular prisms with a longitudinally directed axis. Those corresponding with the ribs R. 20—30μ, T. 30—40 μ, L. 25—80 μ; sometimes polygonal prisms with a longitudinally directed axis. Those

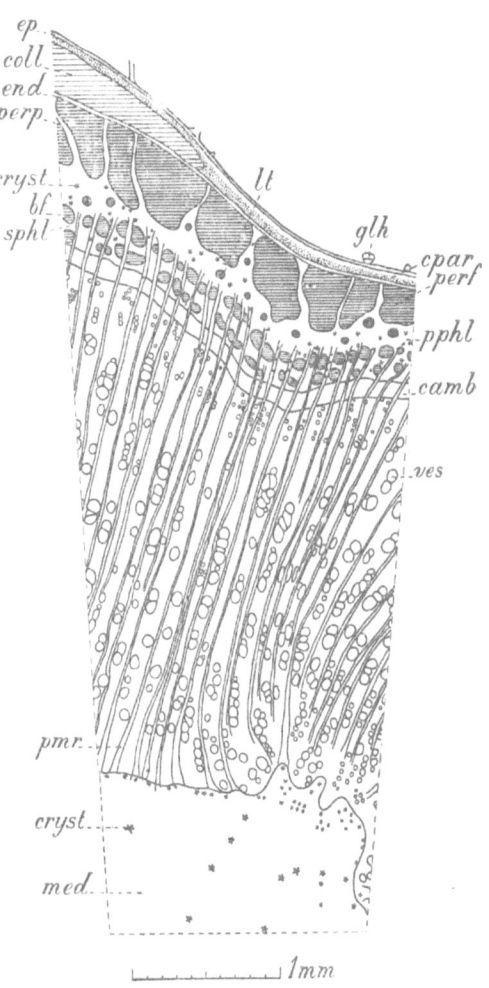

Fig. 61. *Cannabis indica.* Stem, transverse section of a part of the stem thick 1 c.M. bf Bast fibres; camb Cambium; coll Collenchyma; cpar Common parenchyma; cryst Çells with cluster crystals; end Endodermis; ep Epidermis; glh Glandular hairs; lt Latex tubes; med Medulla; perf Pericyclic sclerenchyma fibres; perp Pericyclic parenchyma; phl Primary phloem; pmr Medullary commissures; sphl Secondary phloem; ves Vessels.

corresponding with the medullary commissures R. and T. 20—25 μ; always polygonal prisms with a longitudinally directed axis. Cell contents of all cells: see the common parenchyma of the cortex, only the starch grains here somewhat larger and in the alcoholic material the chloroplasts, containing these starch grains, often lying at the bottom of the cells.

Stele.

P e r y c y c l e. In the parts corresponding with the ribs of the stem and the compound vascular bundles consisting of bundles of sclerenchyma fibres; these bundles 3—4 layers thick in a radial direction and separated from one another by 1-seriate strips of common parenchyma; in the parts corresponding with the furrows of the stem and the compound vascular bundles having the same structure but the bundles of sclerenchyma fibres thick 1—2 layers in a radial direction; in the parts between the ribs and the furrows, and corresponding with the medullary commissures between the compound vascular bundles, consisting only of common parenchyma thick some cell layers in a tangential direction. S c l e r e n c h y m a f i b r e s. In the parts corresponding with the ribs R. 20—25 μ, T. 25 μ; in the parts corresponding with the furrows if forming only a single layer R. 30 μ, T. 15—30 μ, if forming 2 layers R. 20 μ, T. 25 μ; L. of all fibres at least up to 4 m.M. [1] Walls strongly thickened only in the thickest parts of the stem; the inner layers often separated from the rest of the wall; the whole wall colouring blue by treatment with iodine and sulphuric acid 66 per cent. in the part thick 0.6 c.M. (gelatinous layer), the middle lamellae lignified in the part thick 1 c.M., the other layers sometimes slightly lignified; pit canals in thicker material; intercellular spaces wanting. P a r e n c h y m a c e l l s b e-t w e e n t h e f i b r e s. R. 30 μ, T. 15 μ; polygonal prisms. Cell contents: chloroplasts, each with some needle-shaped starch grains. P a r e n c h y m a c e l l s c o r r e s p o n d-i n g w i t h m e d u l l a r y c o m m i s s u r e s b e t w e e n c o m p o u n d b u n d l e s. R. 25—30 μ, T. 25—40 μ; polygonal prisms with somewhat rounded edges; showing intercellular spaces. Cell contents: chloroplasts, each with some needle-shaped starch grains.

C o m p o u n d v a s c u l a r b u n d l e s. Corresponding in place to the ribs and furrows of the stem, containing each about 10 simple bundles; the medullary commissures between the compound bundles being at least in the thinnest parts of the stem broader than those within them. In the compound bundles corresponding to the furrows of the stem the middlemost simple bundles only represented by primary phloem; the xylem being substituted by common parenchyma. Simple vascular bundles collateral; open.

Phloem.

P r i m a r y p h l o e m. Exarch; forming a continuous layer together with the primary phloem parts of the medullary commissures lying within the compound bundles; without radial arrangement of the constituting elements; corresponding with the ribs somewhat thicker in a radial direction than in the furrows, this being chiefly due to somewhat larger radial dimensions of the elements; consisting of about 8 layers of elements: sieve-tubes in a large number,

---

[1] According to T s c h i r c h. Angew. Pfl. Anat. Bd. I. 288 measures of these fibres are the following: thick 16—50 μ, length 5—55 m.M.; other authors mention a length of 22 m.M. It should be kept in mind, that the part, used for ditermining the measure mentioned in the text, was only 0.6 c.M. thick. The so called „Verschiebungen" mentioned by T s c h i r c h, p. 293 were not observed by me; see for this subject T i n e T a m m e s. Der Flachsstengel. 1907. p. 234—240.

companion cells, cambiform cells, somewhat collenchymatously thickened fibres and latex tubes either isolated or in groups of 2 or 3 together. Cluster crystal idioblasts, here and there arranged in longitudinal rows, mixed with the cambiform cells.

S i e v e-t u b e s. R. and T. 15—20 μ, L. of articulations 140 μ; polygonal prisms. Transverse walls oblique with mostly very distinct sieve-plates, callus plates always at the same side of the transverse walls. Contents: sometimes in the vicinity of the callus plates numerous small granules. C o m p a n i o n c e l l s. L. 70 μ, 2 cells corresponding with each articulation of a sieve-tube. Cell contents: a granular mass. C a m b i f o r m c e l l s o f t h e o u t e r m o s t p a r t. R. and T. 15—20 μ; polygonal prisms with a longitudinally directed axis and somewhat rounded edges; showing intercellular spaces. Cell contents: some chloroplasts, each containing some needle-shaped starch grains. C a m b i f o r m c e l l s o f t h e i n n e r p a r t. R. 10—20 μ, T. 15—25 μ, L. 20—60 μ; polygonal prisms with a longitudinally directed axis and somewhat rounded edges; showing intercellular spaces. Cell contents: especially of the innermost cells chloroplasts, each with some needle-shaped starch grains. F i b r e s. R. and T. 15—20 μ; polgygonal; intercellular spaces wanting. L a t e x t u b e s. E. g. R. 10 μ by T. 15 μ, R. 25 μ by T. 30 μ, R. 30 μ by T. 40 μ; unbranched and without anastomoses; polygonal prisms with rounded edges, transverse walls wanting. Walls sometimes a little thickened. Contents: a brown granular mass like that in the tubes of the leaf.

F a s c i c u l a r s e c o n d a r y p h l o e m. Still wanting in the part of the stem thick 0.6 c.M., present in the part thick 1 c.M.; constituted by tangential layers of equal thickness, alternately consisting of bast fibres only and a complex of cribal and parenchymatical elements; moreover a few secondary medullary rays developed here and there.

F a s c i c u l a r c a m b i u m. Joining the interfascicular cambium in all parts of the stem and both together forming a continuous layer of tissue; 8—10 layers of radially arranged elements; in the elements of the outermost layers often a radial partition wall. In the material gathered in September the elements being full grown.

E l e m e n t s o f t h e a b o v e. T. 15—20 μ, L. 40—60 μ; rectangular in a cross section and hexagonal in a radial section.

Xylem.

F a s c i c u l a r s e c o n d a r y x y l e m. Already present in the parts of the stem thick 0.25 c.M., slightly developed in the parts thick 0.6 c.M. and containing only 1 or 2 radially arranged vessels surrounded by radially arranged libriform fibres, in the parts thick 1 c.M. vessels more numerous, isolated and in groups consisting of some vessels, moreover a few secondary medullary rays developed here and there.

V e s s e l s. R. 40 μ, T. 30 μ, L. of articulations 200—1000 μ; polygonal prisms; transverse walls with a single circular perforation; sometimes an articulation with one pointed end showing near the top a large lateral perforation. Walls a little thickened; lignified; with bordered pits or network thickenings when adjoining one another, with network thickenings when adjoining cells of medullary rays. L i b r i f o r m f i b r e s. R. and T. 15—20 μ, L. up to 800 μ; polygonal, sometimes with branched ends. Walls somewhat thickened; lignified, in iodine and sulphuric acid 66 per cent. the innermost thin lamellae blue; not or naerly not pitted. Cell contents: mostly some protoplasm.

Primary xylem. Endarch; consisting of radially arranged vessels, inwardly diminishing in size and the innermost, constituting the protoxylem, often flattened.

Vessels. R. 45 μ, T. 30 μ, L. of articulations 550 μ; polygonal prisms with strongly rounded edges and transverse walls with a circular perforation. Walls a little thickened, with annular spiral and reticulate thickenings, spirals 1 or more in number and dextrors; lignified. Contents: sometimes a yellow-brown mass quite filling a vessel.

Medullary commissures. In consequence of the formation of secondary tissues grown out into large medullary rays.

Between the compound bundles.

Primary phloem part. Consisting of common parenchyma.

Cells of the above. R. and T. 20 μ; polygonal, with rounded edges; showing intercellular spaces. Cell contents: chloroplasts, each with some needle-shaped starch grains.

Interfascicular secondary phloem. Still wanting in the part of the stem thick 0.6 c.M.; in the part thick 1 c.M. developed and in all respects like the fascicular secondary phloem.

Interfascicular cambium. See the fascicular cambium.

Interfascicular secondary xylem. In all respects like the fascicular secondary xylem, but the tangential dimensions somewhat smaller.

Primary xylem part. In the part of the stem thick 0.6 c.M. having the same tangential dimensions as the corresponding secondary xylem; in the part thick 1 c.M. much less wide; consisting of common parenchyma; the cells of the outermost part radially elongated.

Cells of the above. Those of the outermost part R. 25 μ, T. 12 μ; those of the inner part 15—30 μ in diameter; polygonal with somewhat rounded edges. Walls mostly somewhat thickened; lignified; with numerous pits; showing small intercellular spaces. Cell contents: some small needle-shaped starch grains.

Within the compound bundles.

Primary phloem part. Constituting together with the primary phloem of the vascular bundles the continuous layer of primary phloem mentioned above. For the rest consisting of common parenchyma and uni- sometimes biseriate.

Elements of cambium. T. 10 μ. Contents: some small starch grains. Parenchyma cells corresponding with the secondary xylem. R. 15—25 μ, T. 10 μ, L. 50—110 μ. Walls somewhat thickened; lignified; mostly with numerous pits and reticulate thickenings when adjoining a vessel. Cell contents: some very small starch grains especially in the outermost part of the ray. Parenchyma cells corresponding with the primary xylem. R. 10—15 μ, T. 15—25 μ, L. 50—100 μ; polygonal prisms with a longitudinally directed axis; intercellular spaces mostly wanting. Cell contents: some small chloroplasts, each with some needle-shaped starch grains.

Medulla. A thin outer part consisting of very small common parenchyma cells; the central and principal part, showing in the middle a lysigenous cavity, consisting of much larger longitudinally arranged parenchyma cells. Those of the outermost part with somewhat thickened walls. Cluster crystal idioblasts in both parts often in short longitudinal rows. Intercellular spaces in both parts.

C e l l s   o f   o u t e r   p a r t. R. and T. 10—30 μ; polygonal prisms with a longitudinally directed axis. Cell contents: some small chloroplasts, each with some needl-shaped starch grains. C e l l s   o f   c e n t r a l   p a r t. R. and T. 130 by 150 μ, L. 100—120 μ; polygonal prisms with a longitudinally directed axis and somewhat rounded edges. Walls lignified, especially in the parts of the stem thick 1 c.M.; pitted. Cell contents: in the alcoholic material a globular mass adjoining the walls, resembling a sphaero-crystal, dissolving in glycerine, not showing double refraction.

L e a f.

MATERIAL. The drug. Leaves gathered July and September 1903 in the Botanic Garden at Groningen, fresh, fixed with chomic acid 1 per cent. and in alcohol.

MICROGRAPHY.

I n t e r v e n i a.

Epidermis upper side. The cells surrounding the large conical hairs radially arranged around them. Stomata wanting. Trichomes of two kinds: 1. Conical cystolith hairs. Very numerous; 1-cellular, in a very few cases 2-cellular; showing a strongly enlarged globular or ellipsoidal base penetrating into the mesophyll, sometimes even reaching to the spongy chlorenchyma, and a mostly smaller conical top, curved towards the top of the leaf and in the vicinity of the teeth of the leaf and on those teeth themselves towards the tops of the teeth. Base of the hair measuring H. 40—75 μ, Lev. B. 40—75 μ, Lev. L. 45—90 μ; top e. g. L. 60 μ, along the margin the hairs longer like those of the under side of the leaves. 2. Capitate hairs. Mostly uniseriate and consisting of a 1-cellular stalk and a 1-cellular head, sometimes biseriate in both parts. Stalk long 6 μ, thick 10 μ; head long e. g. 12 μ, thick 16 μ; head containing a green granular mass. Cuticular bladder mostly wanting, sometimes present in developed state.

E p i d e r m a l   c e l l s   p r o p e r. H. 18 μ, Lev. B. 25 μ, Lev. L. 30 μ; polygonal tables with the axis perpendicular to the surface of the leaf and lateral walls somewhat coarsely sinuous. The cells, radially arranged around the larger hairs, often curving upwards along the hairs and without sinuous walls. Walls, outer walls a little thickened, those on the margin of the leaf very strongly; with a cuticle forming an undulated parallel striation, this striation on the cells surrounding the large conical hairs directed at right angles to the outline of the hair; lateral walls sometimes pitted. Cell contents: some chloroplasts, containing a few short needle-shaped starch grains; in material treated with potassium dichromate a yellow homogeneous mass; in chloral hydrate or iodine in chloral hydrate numerous needle-shaped crystals along the outer walls. C e l l s   o f   c o n i c a l   c y s t o l i t h   h a i r s. Walls somewhat thickened; lateral walls pitted; attached to an excrescence of the wall a cystolith. These cystoliths somewhat above the level of the epidermal cells, at the convex side of the hair, more ore less filling the base of the hair; showing concentric stratifications around the excrescence after solution of the carbonate of lime by means of hydrochloric acid and attended by the formation of bubbles of carbon dioxyde, these stratifications also conspicuous in material fixed with chromic acid; appearing to consist of a granular mass after treatment with hydrochloric acid or chromic acid, this same mass seeming to fill up the conical top of the hair. The excrescence, bearing the cystolith, showing numerous holes in material fixed in chromic acid. Cellulose reaction appearing in all parts of the wall with the exception of the cuticle and in a few cases the granular mass filling the top, the latter parts being than coloured brown in iodine and suphuric acid 66 per cent.

Cell contents: in the base of the hair protoplasm, nucleus and some small chloroplasts; also in the base conspicuous in material treated with potassium dichromate often a yellow homogeneous mass. Round about the conical hairs numerous needle-shaped or prismatic crystals, appearing in chloral hydrate or iodine in chloral hydrate.

Epidermis under side. Stomata numerous; phaneroporous; lying above the level of the epidermal cells; the adjacent cells sometimes strongly curved out-wards and thus raising the stoma as it were on a stalk, e. g. in the neigh-bourhood of the midrib; the air-cavity not large. Trichomes of three kinds: 1. Conical cysto-lith hairs. More numer-ous than those of the upper side; the enlarged base lying for the great-er part above the level of the leaf, not much penetrating into the mesophyll, smaller than on the upper side, e. g. H. 30 μ, Lev. B. 30 μ, Lev. L. 40 μ; here and there a hair without enlarged base, the cys-tolith wanting at the same time; the conical tops of the hairs much longer than on the upper

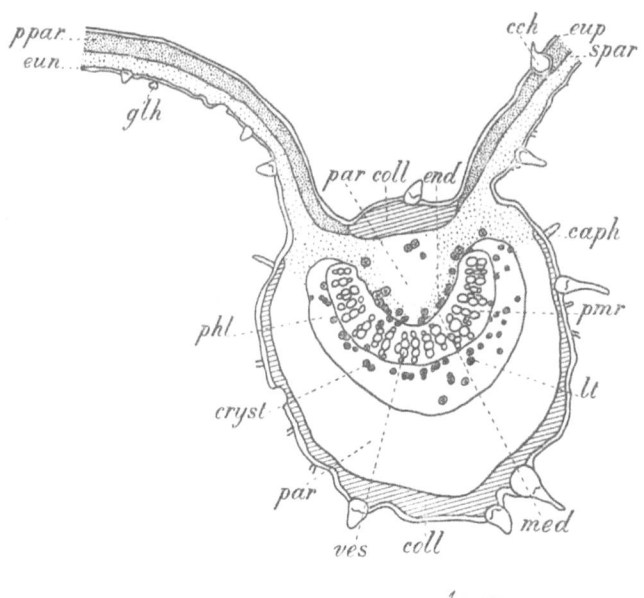

Fig. 62. *Cannabis indica*. Leaf, transverse section through the midrib and the neighbouring portions of the lamina. caph Capitate hairs; cch Conical cystolith hairs; coll Collenchyma; cryst Cells with cluster crystals; end Cluster crystal sheath; eun Epidermis under side; eup Epidermis upper side; glh Glandular hairs; lt Latex tubes; med Medulla; par Colourless parenchyma; phl Phloem; pmr Medullary commissures; ppar Palisade chlorenchyma; spar Spongy chlorenchyma; ves Vessels.

side, L. 90—180 μ; see for the rest these hairs of upper side. 2. Capitate hairs. See upper side. 3. Glandular hairs. Not numerous in material gathered from the Botanic Garden at Groningen; belonging to the labiate type; stalk mostly consisting of 1 tabular cell; head consisting of 8 or more cells, cuticular bladder large and filled with a greenish often somewhat brown mass; measures of head e. g. H. 35 μ, Lev. 50 by 50 μ.

Epidermal cells proper. H. 10 μ, Lev. B. 10 μ, Lev. L. 18 μ; polygonal tables with sinuous lateral walls. Outer walls slightly thickened; with a cuticle. Cell contents: some chloroplasts, containing some short needle-shaped starch grains. Stomata. H. 8 μ, Lev. 14—18 by 18—20 μ.

Mesophyll.

Palisade chlorenchyma. A single cell layer; sometimes the under-most ends of these cells bending in groups towards one another; a few crystal idioblasts containing a small cluster crystal; intercellular spaces. C h l o r e n c h y m a c e l l s. H. 50 μ, Lev. 8 by 8 μ; polygonal prisms with strongly rounded edges or cylinders. Cell contents: many tabular penta- and hexagonal chloroplasts, having a diameter of 8 μ and thick 2 μ, paving the walls and containing some cylindrical or needle-shaped starch grains.

S p o n g y c h l o r e n c h y m a. Three cell layers, thick together about 35 μ; the uppermost layer showing more or less the development of collecting cells; cells more or less stellate with large intercellular spaces; crystal idioblasts more numerous than in the palisade chlorenchyma and resembling in all respects those of the latter tissue.

C h l o r e n c h y m a c e l l s. Cell contents: like those of the palisade chlorenchyma.

V e i n s.

Epidermis upper side. Stomata wanting. Trichomes of two kinds: 1. Conical cystolith hairs. Like those of the intervenia but with a smaller base and a much longer conical top. 2. Capitate hairs. See those of the intervenia.

E p i d e r m a l c e l l s p r o p e r. H. 20 μ, Lev. B. 20 μ, Lev. L. 30—50 μ; tetra- to polygonal prisms, lateral walls not sinuous even on the smaller veinlets. Walls, outer walls a little thickened and with a cuticle forming a longitudinal striation; inner walls collenchymatous. Cell contents: some chloroplasts containing some short needle-shaped starch grains; in material traeted with chromic acid a yellow granular mass; in material treated with potassium dichromate a yellow homogeneous mass.

Epidermis under side. Cells surrounding the cystolith hairs often smaller, e. g. Lev. 10 by 10 μ, and curved upwards against the hair. Stomata wanting. Trichomes of three kinds: 1. Conical cystolith hairs. Partly with bases as strongly enlarged as on the upper side of the intervenia, but not penetrating into the mesophyll, partly without or nearly without enlarged bases like the hairs on the under side of the intervenia; the largest hairs often implanted on a protuberance of the leaf; the conical top often very long, up to 500 μ. 2. Capitate hairs. See those of upper epidermis of intervenia. 3. Glandular hairs. See those of under epidermis of intervenia.

E p i d e r m a l c e l l s p r o p e r. H. 16 μ, Lev. B. 10 μ, Lev. L. 20 μ, polygonal prisms. Walls, outer walls somewhat thickened, with a cuticle forming a longitudinal striation; inner walls collenchymatous. Cell contents: some chloroplasts containing some short needle-shaped starch grains.

Mesophyll.

C o l l e n c h y m a. At the upper side in the middle part about 6, at the under side 2—3 cell layers; cells arranged in longitudinal rows.

C e l l s o f t h e a b o v e. R. and T. 20 μ, L. 100—270 μ, polygonal prisms with a longitudinally directed axis. Walls strongly collenchymatous, especially the cells of the upper side. Cell contents: some chloroplasts containing some short needle-shaped starch grains; these grains being larger than those of the epidermis of the veins.

C o l o u r l e s s p a r e n c h y m a. At the upper side in the middle part about 9, at the under side about 7 cell layers; cells arranged in longitudinal rows. Cluster crystal idioblasts especially in the upper part; often arranged in longitudinal rows, forming part of rows of colourless parenchyma cells.

Cells of colourless common parenchyma. At the upper side cells of the outermost layer R. and T. 20 by 25 μ; those of the middle cell layers R. and T. 40 by 40 μ, L. 50—100 μ; those of the remaining cell layers R. and T. 50 by 60 μ, L. 110—200 μ; polygonal prisms with a longitudinally directed axis and somewhat rounded edges. Cell contents: in the upper and lateral parts of the parenchyma many chloroplasts, each with some short needle-shaped starch grains; for the rest only a few chloroplasts, each with some short needle-shaped starch grains. Cluster crystal idioblasts. R. and T. 15—20 μ, L. 20—30 μ; resembling in shape the cells of colourless parenchyma. Cell contents: a single cluster crystal.

Endodermis. Developed at the under side in the form of a starch-sheath; each cell containing some chloroplasts, quite filled with some starch grains adjoining one another. At the upper side developed as a fairly complete cluster crystal-sheath, consisting of cells arranged in longitudinal rows.

Cells of starch-sheath. At the upper side R. and T. 20 by 25 μ, at the under side R. and T. 25 μ, L. 60—100 μ. Cells of crystal-sheath. R. and T. 15—20 μ, L. 20—30 μ; resembling in shape the cells of colourless parenchyma. Cell contents: a single cluster crystal.

Meristele. Distinctly gutter-shaped, the concave side turned upwards.

Pericycle. Consisting of some layers of thin-walled fibres, without intercellular spaces.

Fibres. R. and T. 15—20 μ, L. very long. Walls not thickened; not lignified.

Compound vascular bundle. A single one containing about 20 simple bundles, arranged in a semi circle. Simple bundles having the shape of thin radially directed lamellae; collateral; open.

Phloem. Exarch; not clearly to be distinguished from the adjoining medullary commissures; consisting of sieve-tubes, cambiform cells and in the middle part containing cluster crystal idioblasts in longitudinal rows, moreover longitudinally directed latex tubes.

Sieve-tubes. R. and T. 15 μ, L. of articulations e. g. 110 μ; transverse walls oblique with very distinct sieve-plates and callus plates very distinct all at the same side of the transverse walls. Cambiform cells. Near the sieve-tubes R. and T. 10 μ, L. 70—90 μ; those of the under part R. and T. 15—20 μ, L. up to 100 μ. Cell contents: here and there a few chloroplasts, each containing some small needle-shaped starch grains. Cluster crystal idioblasts. R. and T. 15—20 μ, L. 10—20 μ; polygonal prisms with a longitudinally directed axis. Cell contents: a single cluster crystal. Latex tubes. R. and T. 18 μ, L. very considerable; polygonal prisms with a longitudinally directed axis; not branched; without transverse walls. Contents: a brown granular mass, mostly along the walls, insoluble in alcohol, remaining yellow in chromic acid material and in concentrated sulphuric acid.

Cambium. More ore less developed.

Xylem. Endarch; forming the principal part of the vascular bundle and chiefly consisting of a single radial row of spiral, reticulate and pitted vessels, a double row only occurring in the vicinity of the cambium and in the bundles lying at both ends of the semi circle.

Vessels. The largest ones in the middle part of the xylem, e. g. R. 25 μ, T. 22 μ; polygonal prisms with rounded edges; the undermost ones with oblique transverse walls, showing a single large perforation. Walls lignified.

Medullary commissures. Consisting of common parenchyma; uni-to tri-, mostly biseriate.

Parenchyma cells of the parts corresponding with the xylem. R. 8 μ, T. 10 μ, L. 60—80 μ; of those parts corresponding with the phloem. R. and T. 15 μ, L. 40—70 μ.

Medulla. Consisting of common parenchyma without intercellular spaces, moreover cluster crystal idioblasts in longitudinal rows.

Parenchyma cells. R. 10 μ, T. 10 μ, L. 90 μ; polygonal prisms with a longitudinally directed axis. Cell contents: some small chloroplasts, each containing some small needle-shaped starch grains. Cluster crystal idioblasts. L. up to 20 μ; moreover of the same shape as the parenchyma cells.

Inflorescence, flower.

MATERIAL. Gathered in July and September 1903 from the Botanic Garden at Groningen, in alcohol. REAGENTS. Glycerine and chloral hydrate.

MICROGRAPHY.

Large bract. In many respects showing the structure of the leaf.

Stipules of the above.

Epidermis. Upper side. Stomata and trichomes wanting. Under side. Stomata wanting; trichomes of the same three kinds as on the leaf, viz. conical cystolith hairs, enlarged base mostly wanting, only occurring along the margin of the top of the stipule; cystolith mostly wanting, even in the hairs with enlarged base; for the rest see the leaf.

Epidermal cells proper. Those of the under side smaller than those of the upper side; both with somewhat sinuous lateral walls.

Mesophyll. Chlorenchyma. Only present in the middle part of the stipule; decreasing towards the margins; the margins themselves only consisting of 2 epidermal layers. Cluster crystal idioblasts here and there, especially in the uppermost cell layer. Intercellular spaces wanting.

Chlorenchyma cells. Those of the uppermost layer polygonal tables with the axis perpendicular to the surface of the stipule.

Sheath-like bract.

Epidermis. Inner side. Stomata wanting. Trichomes: only conical hairs without a cystolith; present only here and there in young material, especially along the margin; without enlarged base; thick at the base 5—8 μ, long 50—150 μ.

Epidermal cells proper. H 8 μ, Lev. B. 10—20 μ, Lev. L. 25—35 μ; polygonal tables with somewhat sinuous lateral walls. Outer walls a little thickened.

Outer side. The cells surrounding the conical hairs often radially arranged around them. Stomata wanting. Trichomes of three kinds. 1. Conical hairs without a cystolith, numerous, especially on the veins of the material in a more developed stage; mostly curved towards the top of the bract, those on the veins perpendicular to its surface; unicellular; mostly with an enlarged base especially on the veins, not penetrating into the mesophyll; the largest hairs with enlarged base long 500 μ, thick at the base Lev. B. 140 μ, Lev. L. 180 μ; the hairs without enlarged base long 50—100 μ, thick at the base 20—40 μ. 2. Capitate hairs. See those of the leaf, but biseriate more numerous. 3. Glandular hairs. In young inflorescences resembling those of the leaf. In more developed inflorescences these hairs placed on conical excrescences; these excrescences thick at the base 150 μ, long 400 μ, consisting of often longitudinally arranged epidermal cells and now and then containing some prismatic mesophyll cells. Hairs very numerous, especially on the veins of material in a more developed stage; showing the labiate type; long without the single basal cell 50 μ; stalk biseriate, a single cell in height; head wide 80 μ, consisting of 8 or more cells, with a cuticular bladder.

E p i d e r m a l c e l l s p r o p e r. H. 8 μ, Lev. B. 10—15 μ, Lev. L. 20—30 μ; polygonal tables with somewhat sinuous lateral walls. Outer walls a little thickened. E p i d e r m a l c e l l s f o r m i n g t h e e x c r e s c e n c e. Wide 10 μ, long 30 μ; tetragonal tables. Cell contents: some small granules. C e l l s o f c o n i c a l h a i r s. Walls somewhat thickened. Cell contents: the protoplasm and nucleus mostly clearly discernible, also in the conical part. C e l l s o f g l a n d u l a r h a i r s. Contents: of the basal and stalk cells in alcoholic material a yellow-green granular mass, of the bladder entirely dissolved.

M e s o p h y l l. Consisting of about 4 layers of parenchyma cells; in the veins of some more layers. Cluster crystal idioblasts constituting nearly the whole of the innermost layers; also occurring here and there in the outer ones. Intercellular spaces wanting.

P a r e n c h y m a c e l l s. In the innermost layer H. 6 μ, Lev. B. 10 μ, Lev. L. 12 μ; poly- often hexagonal tables. Walls thin.

F e m a l e f l o w e r.

Perigon. Surrounding the under half of the ovary and growing up with it for a long time; consisting of small parenchyma cells; epidermis without trichomes.

Pistil.

O v a r y. Epidermis of wall without trichomes; in the uppermost part of the ground tissue of the wall numerous cluster crystal idioblasts; 2 meristeles running on opposite sides of the wall. Ovule inserted somewhat askance at the top of the ovarial cavity; semi-anatropous; with 2 integuments, the outer integument with a large exostomium leaving the upper part of the ovule uncovered.

S t y l e a n d s t i g m a. 2 cylindrical styles, each continued into a cylindrical stigma of the same length as the style; these organs in various stages of development; in the adult state a stigma with a pointed end. Style. Epidermal cells often in longitudinal rows, square tables, papillae wanting; ground tissue consisting of longitudinally elongated parenchyma cells; meristeles wanting. Stigma. Epidermal cells longitudinally elongated, rectangular; in the adult state the uppermost ends grown out into papillae and curved outwards; for the rest like the style.

September 1903.   J; M; v. E. d. W.

HERBA CARDUI BENEDICTI.

Cnicus. Thistle.

The aerial parts of Carbenia benedicta, Adans. Fam. II. 116, gathered in the beginning of the flowering season.

Macroscopic characters.

Annual plant, up to 6 c.M. in height. Stem erect, herbaceous, upwardly corymbously branched, angular; surface showing long white hairs and short glandular hairs with a small round white head, consequently the surface viscous, green; the ribs red. Leaf arrangement scattered; the uppermost leaves closely approximated, forming a cover wrapping up the head. Leaf long up to 3 d.M., simple, sessile, at the base of the stem with a triangular petiole. Blade lanceolate-pinnatisect with lobes at right angles to the midrib; blade of the uppermost leaves cordate and lobed; apex of leaf and tops of lobes ending acutely in a sharp bristle; incisions of the lobes rounded; base blunt, gradually passing into the stem, in the uppermost leaves eared, in the undermost ones acute, passing into the petiole; margin coarsely and irregularly toothed; hairs like those of the stem. Inflorescence: heads high 3 c.M., thick 2 c.M., 1 on the top of each branch. Receptacle flat, white, scaly, densely beset with bristle-like paleae. Involucre about spherical, constituted of several ranks of bracts; bracts green, the outer ones ovate, shorter, and having at the top a long taper-pointed prickle, the inner ones longer and narrower with a dark red pinnatisect prickle curved outwards at a right angle; all the prickles densely beset with hairs forming a cob-web-like covering. Flower long up to 3 c.M. Ray-flowers 4—6, neutral, only consisting of a corolla inserted on a rudimentary ovary; this corolla tubular, with 3 linear yellow lobes. Disk-flowers numerous, complete, epigynous. Pappus in 3 circles: brim with 10 teeth, 10 long and 10 short white bristles. Corolla gamopetalous, pentamerous, tubular, somewhat zygomorphic with 1 long lobe, 2 middling and 2 short lobes, all curved towards the axis of the head; tube white, lobes yellow; glandular hairs on the tube and the lobes. Stamens 5, inserted on the corolla, syngenesious; anthers yellow with purplish brown margin; upper part of the anthers with a long, flat, crescent-shaped continuation; those continuations forming a curved tube. Pistil compound, composed of 2 carpels, with 1 long style and 2 short stigmata; ovary with ribs, white, containing 1 ovule. Taste bitter.

Anatomical characters.

LITERATURE. Le Blois. Can. sécrét. e. Poches sécrétr. Ann. d. Sc. nat. Bot. Sér. 7. T. VI. 1887. 274. Dafert u. Mihlauz. Unters. üb. d. Kohle-ähnl. Masse d. Komp. (Chem. Teil) Denkschr. Wien. Akad. Bd. 87. 1912. 143. Daniel. Rech. anat. s. l. Bractées d. l'Involucre d. Comp. Ann. d. Sc. nat. Sér. 7. T. XI. 1890. 17. Ellrodt. Vert. d. Gerbst. i. off. Blätt., Kräut. u. Blüten. Diss. Würzburg. 1913. 20. Feuilloux. Appareil tect. e. glandul. d. Comp. Thèse Paris. 1901—02. Flückiger. Pharmakogn. 1891. 680. Gerdts. Bau u. Entw. d. Komp. frucht. Diss. Bern. 1905. 65. Gilg. Pharmakogn. 1910. 359. Grimm. Beitr. z. vergl. Anat. d. Compositenblätter. Diss. Kiel. 1904. Hager. Pharm. Praxis. Bd. I. 1900. 864. Heineck. Beitr. z. Kenntn. d. fein. Baues d. Fruchtschale d. Comp. Koch. Mikr. Anal. d. Drogenpulver. Bd. III. 1906. 27. Pl. II. Koch. Pharmakogn. Atlas. Bd. 1914.

141. Pl. XIX. Lavialle. Develop. de l'Ovaire en Fruit ch. l. Comp. Ann. d. Sc. nat. Bot. Sér. 9. T. XV. 1912. 39. Marmé. Pharmakogn. 1886. 234. Müller. Bau. u. Inhaltst. d. Comp.-blätter. Diss. Göttingen. 1912. Oudemans. Pharmacogn. 1880. 277. Rosenthaler u. Stadler. Anat. v. Cnicus benedictus L. Arch. Pharm. Bd. 246. 1908. 436. Small. Flor. Anat. of som. Comp. Linn. Soc. Journ. Bot. XLIII. 1917. 517. Solereder. Syst. Anat. d. Dicot. 1899. 515. v. Tieghem. Can. sécrét. d. Pl. Ann. d. Sc. Sér. 5. Bot. T. XVI. 1872. 97. v. Tieghem. Sec. Mém. s. l. Can. sécrét. d. Pl. Ann. d. Sc. Bot. Sér. 7. T. I. 1885. 6. Tschirch. Harze u. Harzbeh. 1900. e. g. 343. Vesque. Anat. d. Tissus. Nouv. Arch. du Museum d'Hist. nat. Sér. 2. T. IV. 1881. 11. Vuillemin. Val. d. Caract. anat. 1884. v. Wisselingh. Bijdr. Kennis Zaadh. I. Comp. Pharm. Weekbl. 1918. 864. REAGENTS. Water, glycerine, chloral hydrate, potash (stem), iodine in chloral hydrate (stem and leaf), phloroglucin and hydrochloric acid, iodine and sulphuric acid 66 per cent (stem and leaf), concentrated sulphuric acid (stem and leaf), Schulze's macerating mixture (stem and leaf), copper acetate and iron acetate (stem and leaf), potassium dichromate.

Stem.

MATERIAL. Stems gathered July and October 1902 in the Botanic Garden at Groningen, thick 0.3 c.M. and 0.7 c.M., fresh, in alcohol and fixed with chromic acid 1 per cent. The drug.

MICROGRAPHY.

Epidermis. On the edges of the stem cells in longitudinal rows and mostly tetragonal tables, for the rest often showing a more irregular shape. The cells bearing hairs much larger than the rest. Stomata 5 to the sq. m.M., wanting on the edges; phaneroporous, lying in the same level as the epidermal cells; generally surrounded by 4 subsidiary cells much smaller than the epidermal cells proper and partly covered by the stoma. Trichomes represented by glandular and conical hairs; conical hairs larger than

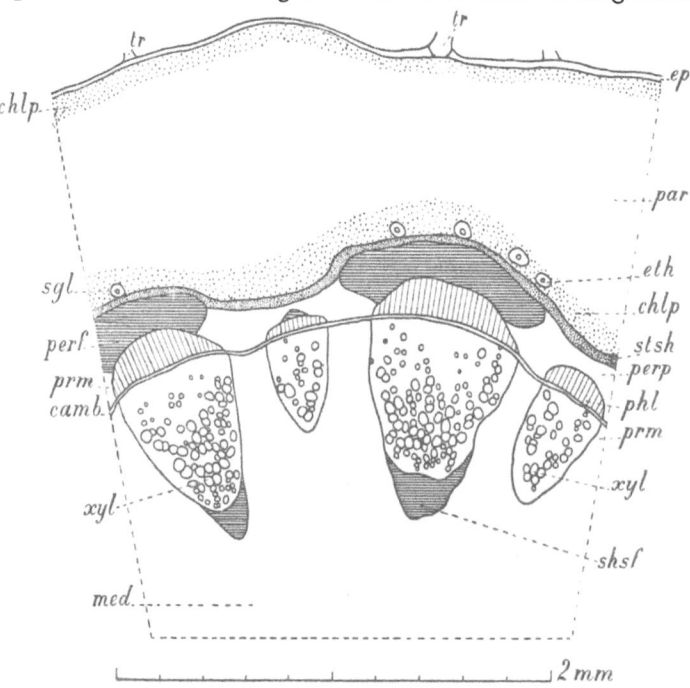

Fig. 63. *Carbenia benedicta*. Stem, transverse section. camb Cambium; chlp Common parenchyma containing chloroplasts; ep Epidermis; eth Epithelium; med Medulla; par Colourless parenchyma; perf Pericyclic sclerenchyma fibres; perp Pericyclic parenchyma; phl Phloem; prm Medullary commissures; sgl Schizogenous glands; shsf Sheath of sclerenchyma fibres; stsh Starch-sheath; tr Trichomes; xyl Xylem.

those of the leaf. See for the rest trichomes of the leaf.

Epidermal cells proper. On the edges of the stem R. 20 μ, T. 22 μ, L. 150—

250 μ; for the rest T. up to 30 μ, L. mostly not over 150 μ; inner walls somewhat convex. Outer walls a little thickened; with parallel longitudinal cuticular striation. Cell contents: some chloroplasts; in potassium dichromate the large vacuole filled with a dark yellow to brown mass showing some mostly small colourless vacuoles; in some cells very numerous dark brown granular globules: on the edges and the top of the stem in all the cells anthocyanin. S t o m a t a. Lev. 20 by 30 μ. G u a r d-c e l l s. Contents: in potassium dichromate some yellow globules.

Cortex.

C o m m o n p a r e n c h y m a. Consisting of about 12 layers; the outer and inner 1 or 2 cell layers containing chloroplasts, the middle part colourless. Schizogenous glands lying in the innermost 1 or 2 layers of the common parenchyma; groups of 1—4 corresponding in place to the vascular bundles of the stele; in a transverse section surrounded by 4 or more epithelium cells. Cavity of a large gland e.g. R. 8 μ, T. 16 μ; filled with a red oil. See moreover Rhizoma Arnicae.

E n d o d e r m i s. Developed as a starch-sheath; not very distinct. Consisting of 1 layer of cells in places corresponding to the vascular bundles, for the rest 1 or 2 layers.

C e l l s o f t h e o u t e r m o s t 1 o r 2 g r e e n l a y e r s. R. 25 μ, T. 30 μ, L. 60—120 μ; mostly tetragonal prisms with a longitudinally directed axis and rounded edges. At the edges of the stem walls collenchymatously thickened. Cell contents: chloroplasts, containing small starch grains; in potassium dichromate the large vacuole filled with a pale brown mass, sometimes containing colourless vacuoles. C e l l s o f t h e c o l o u r l e s s p a r e n c h y m a. In the middlemost layers the largest cells R. 100 μ, T. 150 μ, L. 140—270 μ; all the cells with longitudinally directed axis and rounded edges. C e l l s o f t h e i n n e r m o s t 1 o r 2 g r e e n l a y e r s. R. 45 μ, T. 50 μ, L. 70 μ. Cell contents: chloroplasts; when treated with potassium dichromate the large vacuole filled with a pale brown mass like that of the outer green cell layers. E p i t h e l i-u m c e l l s. Thick 8 μ, wide 7 μ, L. 100 μ; tetragonal prisms. Walls bordering the cavity mostly persisting in concentrated sulphuric acid. Cell contents: some chloroplasts. C e l l s o f e n d o d e r m i s. In places corresponding to the vascular bundles R. 20 μ, T. 30 μ, L. 60—100 μ; mostly tetragonal prims with a longitudinally directed axis. Contents of all the cells: some chloroplasts, containing a few globular to needle-shaped starch grains; on treatment with potassium dichromate quite the same precipitate as in the outer green cell layers.

Stele.

P e r i c y c l e. Consisting of 3 kinds of tissue. 1°. The parts corresponding in place to the fibrovasal bundles consisting of larger or smaller bundles of sclerenchyma fibres intermixed with some thin-walled elements, 1 or 2 layers of common parenchyma cells on the side of the endodermis being added to the smaller bundles. 2°. The parts corresponding in place to the detached bundles of secondary phloem consisting of common parenchyma like that of the parts mentioned first. 3°. The parts corresponding to the medullary commissures consisting of common parenchyma with intercellular spaces.

S c l e r e n c h y m a f i b r e s. R. 12 μ, T. 15 μ, L. 500—1000 μ; polygonal. Walls strongly thickened; middle lamella conspicuous; lignified, especially the middle lamella; pitted.

P a r e n c h y m a   c e l l s mentioned under 1° and 2°. R. 10 μ, T. 18 μ, L. 100—200 μ; polygonal prisms with a longitudinally directed axis and strongly rounded edges. Walls thickened; lignified; pitted. C o m m o n   p a r e n c h y m a   c e l l s mentioned under 3°. R. 30 μ, T. 40 μ, L. 80—120 μ; polygonal prisms with a longitudinally directed axis and somewhat rounded edges. Cell contents: here and there chloroplasts; in potassium dichromate quite the same precipitate as in the outer green cortical layers.

Larger fibrovasal bundles, smaller vascular bundles and detached phloem bundles arranged in a ring.

F i b r o v a s a l   b u n d l e s. Sheath only present at the inner side; varying in thickness; consisting of sclerenchyma fibres.

V a s c u l a r   b u n d l e s. Simple; collateral; open. Vascular bundles without sclerenchyma sheath also simple; collateral; open.

Phloem. Primary and secondary phloem. Not clearly to be distinguished from one another; in a transverse section mostly only the elements of the inner part in radial rows. Consisting of sieve-tubes with companion cells and bast parenchyma, not to be distinguished from one another in a radial section.

Cambium. Even in the thinnest stems forming part of a continuous cambium ring, composed of alternating fascicular and interfascicular portions.

Xylem. Secondary xylem consisting of pitted vessels, most numerous in the inner part, intermixed with libriform fibres with irregular or branched ends and often some vessel tracheids adjoining the vessels. Primary xylem containing annular and spiral vessels.

F i b r e s   o f   s c l e r e n c h y m a - s h e a t h. R. and T. 10 μ, L. 1000 μ; polygonal. Walls strongly thickened; middle lamella conspicuous; lignified; pitted. S i e v e - t u b e s. R. and T. 15—20 μ, or much smaller, L. of articulations 80—140 μ; tetragonal to polygonal; transverse walls sometimes oblique, sieve-plates on them conspicuous. Walls sometimes a little collenchymatous. Contents: numerous small chloroplasts containing small starch grains in the callus near the sieve-plates; callus mostly on one, and in a radial section always on the same side of the sieve-plates. C o m p a n i o n   c e l l s. R. and T. 3—5 μ, L. of 1 cell or of 2 cells together corresponding to L. of the articulations of the sieve-tubes. Cell contents: granular. B a s t   p a r e n c h y m a   c e l l s. R. and T. 8—10 μ, L. 60—100 μ; polygonal prisms with a longitudinally directed axis. Walls a little collenchymatous. E l e m e n t s   o f   f a s c i c u l a r   c a m b i u m. R. 5 μ, T. 8 μ; tetragonal prisms. P i t t e d   v e s s e l s. R. 35 μ, T. 22 μ, in the inner part R. up to 50 μ, T. up to 35 μ, L. of articulations 120—300 μ; polygonal prisms with strongly rounded edges; transverse walls showing a circular perforation. Walls thickened; lignified; walls adjoining other vessels or vessel tracheids showing bordered pits. V e s s e l   t r a c h e i d s. R. and T. 12—20 μ, L. 150—300 μ; polygonal. For the rest see the vessels. L i b r i f o r m   f i b r e s. R. 9—15 μ, T. 8 μ, L. 250—300 μ; polygonal. Walls thickened; middle lamella conspicuous; lignified, especially the middle lamella; pitted. Cell contents: sometimes sphaero-crystals of inulin.

M e d u l l a r y   c o m m i s s u r e s. Mostly some cells wide; the elements radially arranged in the parts corresponding to the secondary xylem; consisting of common parenchyma.

C e l l s   o f   t h e   p a r t s   c o r r e s p o n d i n g   t o   t h e   p r i m a r y   p h l o e m   a n d   x y l e m. R. 30—50 μ, T. 25 μ, L. 100—200 μ; polygonal prisms with a longitudinally directed axis and somewhat rounded edges. Walls a little thickened; lignified; pitted. Cell contents: some chloroplasts; sometimes sphaero-crystals of inulin; in potassium dichromate quite the same precipitate as in the outermost green layers of the cortex. C e l l s o f t h e p a r t s c o r r e s p o n d i n g t o t h e s e c o n d a r y p h l o e m. R. 22 μ, T. 25 μ, L. 50—80 μ; mostly tetragonal prisms with a longitudinally directed axis. Cell contents: some chloroplasts; in potassium dichromate the same precipitate as in the outer green cortical layers. C e l l s o f t h e p a r t s c o r r e s p o n d i n g t o t h e s e c o n d a r y x y l e m. R. 20—30 μ, T. 15 μ, L. 50—80 μ; mostly rectangular prisms with a longitudinally directed axis. Walls thickened; lignified; pitted. Cell contents: some chloroplasts; sometimes sphaero-crystals of inulin; in potassium dichromate quite the same precipitate as in the outer green layers of the cortex.

M e d u l l a. Consisting of common parenchyma. In the outermost part cells smallest. Showing a central cavity.

C e l l s o f t h e a b o v e. R. and T. 40—110 μ, L. 50—150 μ; polygonal prisms with a longitudinally directed axis and somewhat rounded edges. Walls of the outermost cells somewhat thickened; pitted. Cell contents: sometimes sphaero-crystals of inulin.

L e a f.

MATERIAL. Leaves gathered July and September 1902 in the Botanic Garden at Groningen, fresh, in alcohol and fixed with chromic acid 1 per cent. The drug.

MICROGRAPHY.

I n t e r v e n i a.

Epidermis upper side. Cells with strongly sinuous lateral walls. Cells bearing hairs, especially those bearing long hairs, much larger than the other and with smooth lateral walls. Stomata 20 to the sq. m.M.; phaneroporous; the surrounding epidermal cells somewhat raised and partly covered by the stomata; generally surrounded by 4 subsidiary cells. Trichomes very numerous; in 3 kinds. 1°. Glandular hairs of a somewhat modified Compositae type: stalk consisting of 1 or 2 stories of 2 cells each, forming together an elliptical cylinder mostly directed transversely; head composed of some stories each consisting of a single tabular cell and covered by a cuticular bladder. 2°. Capitate hairs. Consisting of 1 row of 6—10 cells, the apical cell somewhat larger in diameter than the other; conical; diameter at the base 30—60 μ, length up to 300 μ. 3°. Conical hairs. Consisting of 1 row of 10—20 cells; the uppermost cells more or less flattened; on the uppermost leaves these hairs often showing thread-like continuations, some m.M. in length and forming the cob-web-like covering of the plant; diameter at the base 120 μ, length up to 3 m.M.

E p i d e r m a l c e l l s p r o p e r. H. 30 μ, Lev. 25 by 60 μ; cells bearing hairs e. g. H. 80 μ, Lev. 90 μ; polygonal prisms. Outer walls a little thickened; with a cuticle. Cell contents: some small chloroplasts; when treated with potassium dichromate a yellow sometimes granular mass, sometimes showing colourless globules; a few times many globules with a brown outline. S t o m a t a. H. 10 μ, Lev. 18 by 25 μ. C e l l s o f g l a n-d u l a r h a i r s. Stalk high 45 μ, wide 30 by 35 μ; head high 5 μ, wide 30 by 45 μ; cuticular bladder high 35 μ, wide 50 μ. C e l l s o f c a p i t a t e h a i r s. Contents: often some minute chloroplasts; in potassium dichromate sometimes a yellow mass. C e l l s o f c o n i c a l h a i r s. Outer walls a little thickened. Cell contents: here and there some small chloroplasts.

Epidermis under side. Stomata 30 to the sq. m.M. See for the rest those of the upper side. Trichomes. See those of the upper side.

E p i d e r m a l c e l l s p r o p e r. See those of the upper side. S t o m a t a. H. 10 μ, Lev. 18 by 20 μ. See for the rest those of the upper side.

Mesophyll. Constituted of about 6 layers of cells, mostly developed as palisade

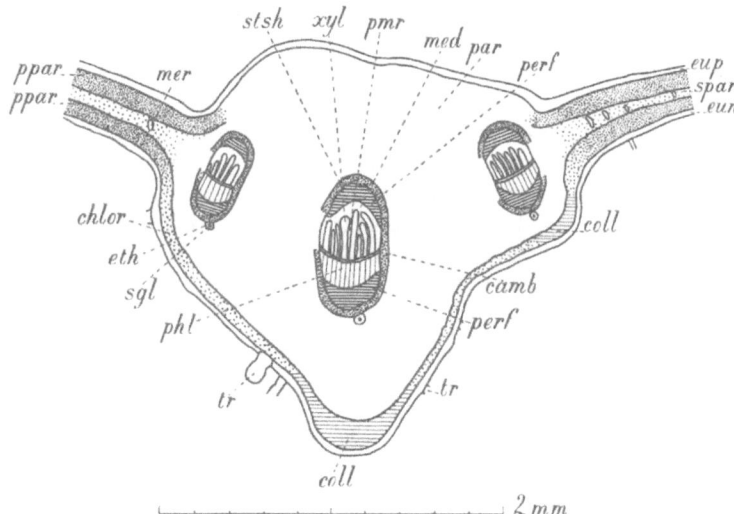

Fig. 64. *Carbenia benedicta*. Leaf, transverse section through the midrib and the neighbouring portions of the lamina. camb Cambium; chlor Chlorenchyma; coll Collenchyma; eth Epithelium; eun Epidermis under side; eup Epidermis upper side; med Medulla; mer Small meristeles; par Colourless parenchyma; perf Pericyclic sclerenchyma fibres; phl Phloem; ppar Palisade chlorenchyma; prm Medullary commissures; sgl Schizogenous glands; spar Spongy chlorenchyma; stsh Starch-sheath; tr Trichomes; xyl Xylem.

chlorenchyma; the cells of the 3rd and 4th layer only forming spongy chlorenchyma. Intercellular spaces largest in the undermost cell layers.

C e l l s o f t h e a b o v e. In the upper 2 layers H. 45 μ, Lev. 15 by 20 μ; in the undermost 2 layers H. 30 μ, Lev. 15 by 20 μ; in the middlemost layers Lev. e. g. 20 μ, in the middlemost layers polyhedra with rounded edges, globes, ellipsoides, sometimes branched cells. Contents of all the cells: chloroplasts, paving the walls, mostly hexagonal, tabular, 4 μ in diameter, 1. 5 μ thick, containing some minute needle-shaped starch grains.

V e i n s.

Epidermis. Upper side. Stomata wanting; trichomes: see those of the intervenia.

E p i d e r m a l c e l l s p r o p e r. H. 22 μ, Lev. B. 22 μ, Lev. L. 100—250; penta- or hexagonal tables. Outer and inner walls a little thickened; outer walls showing a cuticle. Cell contents: here and there anthocyanin; for the rest see the intervenia, on treatment with potassium dichromate the precipitate more clearly to be seen.

Under side. Stomata 3 to the square m.M.; surrounded by 4 much smaller subsidiary cells. Trichomes. See those of the intervenia.

E p i d e r m a l c e l l s p r o p e r. See those of the upper side. S t o m a t a. Lev. 20 by 25 μ. S u b s i d i a r y c e l l s. Cell contents: in potassium dichromate a yellow mass with colourless vacuoles. See for the rest the intervenia.

Mesophyll.

C o l l e n c h y m a. A few bundles adjoining the epidermis of the under side in places corresponding to the meristeles. C h l o r e n c h y m a. In bands adjoining the lower epidermis and quite as thick as the collenchyma bundles. C o l o u r l e s s p a r e n c h y m a. In the middle part of the midrib consisting of about 25 layers of cells. In the vicinity of the epidermis and of the meristeles cells much smaller. G l a n d s. Lying in the colourless parenchyma; 1 or 2 close to the under or lateral sides of each meristele; schizogenous; surrounded by an epithelium of mostly 4 cells [1]). The cavity filled with a reddish brown oil. E n d o d e r m i s. Developed as a starch-sheath. Consisting of 1 or 2 layers of cells.

C e l l s o f c h l o r e n c h y m a. Cells shorter than the colourless parenchyma cells and the edges more rounded. Cell contents: many chloroplasts, containing needle-shaped starch grains; a few times inulin. C e l l s o f c o l o u r l e s s p a r e n c h y m a. R. and T. 70—100 μ, L. 150—250 μ; polygonal prisms with a longitudinally directed axis and somewhat rounded edges. Cell contents: here and there chloroplasts, a few times inulin. E p i t h e l i u m c e l l s. Thick and wide 10 μ. Cell contents: some small chloroplasts. E n d o d e r m i s c e l l s. At the upper side R. and T. 30 μ, L. 40 μ; at the under side R. and T. 25 μ, L. 60 μ. Cell contents: some chloroplasts, each containing a globular starch grain.

Meristeles. In the midrib parallel to one another; at the base somewhat larger in number; number decreasing upwardly in consequence of the outer meristeles ending blind, thus in the middle part of the midrib 3 and in a small part of the top of the midrib only 1. Outer meristeles giving off the lateral ones; moreover longitudinal ones connected by transverse meristeles. Longitudinal meristeles in the shape of elliptical cylinders; each consisting of a pericycle, a single compound vascular bundle, containing about 7 simple bundles, and some medullary tissue.

P e r i c y c l e. Developed as a sheath of sclerenchyma fibres; only at the under and upper side of the meristeles and wanting at the lateral sides; under side consisting of more layers of fibres than the upper side. Here and there intercellular spaces.

F i b r e s o f t h e a b o v e. R. and T. 15 μ, L. 300—600 μ; polygonal, here and there with rounded edges. Walls thickened; middle lamella distinct; lignified, especially the middle lamella; pitted. Contents: here and there in the outermost fibres a dark green mass.

S i m p l e v a s c u l a r b u n d l e s. Collateral; open; in the shape of radially elongated narrow plates.

P h l o e m. Consisting of sieve-tubes, companion cells and cambiform cells. S i e v e - t u b e s. R. and T. 8—10 μ, L. of articulations 45—90 μ; sieve-plates on the transverse walls distinct. Contents: in the callus near a sieve-plate chloroplasts containing small starch grains, mostly lying on one side of the sieve-plate and always on the same side. C o m p a n i o n c e l l s. R. and T. 3 μ, L. corresponding to that of the articulations of sieve-tubes; polygonal prisms. Cell contents: granular. C a m b i f o r m c e l l s. R. and T. 8 μ, L. 45—80 μ; polygonal prisms with a longitudinally directed axis.

C a m b i u m. Mostly represented by some layers of radially arranged elements.

---

[1]) See v a n T i e g h e m, L e B l o i s, T s c h i r c h, Harze u. Harzbeh. 343.

X y l e m. Consisting of pitted, scalariform, spiral and annular vessels; the pitted vessels lying undermost. Uppermost vessels sometimes flattened; vessels of the middle part the largest.

V e s s e l s. In the middle part R. 30 μ, T. 25 μ, L. of articulations 100—225 μ; pitted vessels having the largest articulations; tetra- to hexagonal prisms with strongly rounded edges; transverse walls with a circular perforation. Walls lignified.

M e d u l l a r y   c o m m i s s u r e s. Not to be distinguished in the phloem and cambium part. Consisting in the xylem part of parenchyma cells.

C e l l s   o f   t h e   a b o v e. R. and T. 10 μ, L. 30—60 μ; polygonal prisms with a longitudinally directed axis. Walls mostly a little thickened; lignified; pitted.

M e d u l l a. Between the xylem and the upper part of the pericycle; consisting of common parenchyma. Intercellular spaces wanting.

C e l l s   o f   t h e   a b o v e. R. and T. 12 μ, L. 100—120 μ; polygonal prisms with a longitudinally directed axis.

F l o w e r.

MATERIAL. Flowers gathered July 1902 in the Botanic Garden at Groningen, fresh and in alcohol.

MICROGRAPHY.

R e c e p t a c l e.

E p i d e r m i s. Covered with paleae; these consisting of rectangular tabular cells or slightly twisted fibres. At the top of the paleae cells sometimes more or less turned off from the axis.

E l e m e n t s   o f   p a l e a e. Wide 14 μ, long 500 μ. Lateral walls a little thickened and pitted; outer walls with a cuticle; all the walls somewhat lignified.

M e s o p h y l l. Consisting of common colourless parenchyma, at the base showing large intercellular spaces.

C e l l s   o f   t h e   a b o v e. At the base walls mostly thickened and pitted. Cell contents: often sphaero-crystals of inulin.

M e r i s t e l e s. Extending in all directions.

I n v o l u c r e.

Foliaceous part of the bract.

E p i d e r m i s. Inner side. Consisting of fibres; stomata wanting.

F i b r e s   o f   t h e   a b o v e. H. 14 μ, Lev. B. 10 μ, Lev. L. 300—500 μ. Walls thickened; a little lignified; lateral walls pitted.

Outer side. For the greater part consisting of thick-walled cells, especially the lateral walls thickened. Thin-walled cells in longitudinal strips. Stomata only occurring in the strips of thin-walled cells; phaneroporous; lying in the same level as the epidermal cells.

E p i d e r m a l   c e l l s   p r o p e r. H. 8 μ, Lev. B. 16 μ, Lev. L. 70—100 μ; tetra-to hexagonal tables. Thickened walls lignified; pitted. Cell contents: often a crystal; in the thin-walled cells chloroplasts. St o m a t a. Lev. 15—20 μ.

M e s o p h y l l. Consisting of chlorenchyma. The layer adjoining the inner epidermis composed of strongly rounded or somewhat fibre-shaped cells; showing very large intercellular spaces. In the middle part the cells larger and polygonal prisms with rounded edges; the intercellular spaces much smaller. Towards the outer epidermis cells smaller. The layers adjoining the outer epidermis consisting again of prismatic or fibre-shaped cells.

C e l l s   o f   t h e   a b o v e. In the layers adjoining the inner epidermis H. 10 μ, Lev. B. 10 μ, Lev. L. 300—500 μ. Walls somewhat thickened; lignified; pitted. Larger cells of the inner part H. 50 μ, Lev. B. 40 μ, Lev. L. 150 μ. Cell contents: chloroplasts; sometimes sphaero-crystals of inulin. In the layers adjoining the outer epidermis H. and Lev. B. 10 μ, Lev. L. 200—300 μ. Walls thickened; lignified; pitted.

M e r i s t e l e s. Showing a pericyclic sclerenchyma-sheath at the outer side. Xylem containing spiral vessels with lignification of the spiral thickenings.

Prickle of the bract.

E p i d e r m i s. Inner side. Cells decreasing in size towards the top of the prickle, quite at the top larger again and fibre-shaped; trichomes in 2 kinds, forming the cob-web-like covering, 1°. conical hairs, unicellular, diameter at the base 10 μ, length 150 μ, 2°. pluricellular hairs, consisting of one row of some cylindrical cells at the base, and 1 very long but narrower cell at the top.

E p i d e r m a l   c e l l s   p r o p e r. R. 25 μ, T. 15 μ, L. 45 μ; polygonal prisms. Outer walls a little thickened; walls somewhat lignified. Cell contents: some minute chloroplasts; anthocyanin. C e l l s   o f   u n i c e l l u l a r   h a i r s. Walls thickened; lignified. P l u r i-c e l l u l a r   h a i r s. Cylindrical cells at the base thick 15 μ, long 70 μ.

Outer side. At the top of the prickle cells larger than the rest and fibre-shaped. Only pluricellular hairs, see for these the inner side.

E p i d e r m a l   c e l l s   p r o p e r. H. 10 μ, Lev. B. 12 μ, Lev. L. 60—80 μ; tetra- to hexagonal tables. Tops of the cells somewhat turned outwards. Outer walls thickened; showing stratification; lateral walls pitted; all the walls somewhat lignified. Cell contents: some minute chloroplasts; anthocyanin.

M e s o p h y l l. The same as that of the foliaceous part, only all the elements much smaller; the layer at the upper side with smaller intercellular spaces; all the walls lignified. Schizogenous glands close to the outside of the meristeles; showing an epithelium of very thin-walled cells; cavity filled with a colourless mass.

M e r i s t e l e s. Containing a pericyclic sclerenchyma-sheath. Walls of sclerenchyma and xylem elements lignified.

D i s k-f l o w e r.

Pappus. Showing 3 circles of different structure.

1°. Brim showing 10 teeth.

Epidermis. Inner side.

C e l l s. H. 10 μ, Lev. B. 15 μ, Lev. L. 25 μ. Cell contents: often sphaero-crystals of inulin.

Outer side. Hairs conical; unicellular; diameter at the base 3μ, length 70 μ. E p i d e r m a l   c e l l s   p r o p e r. H. 18 μ, Lev. B. 10 μ, Lev. L. 45 μ. Cell contents: often sphaero-crystals of inulin.

Mesophyll. Consisting of common parenchyma.

C e l l s polygonal. Cell contents: often sphaero-crystals of inulin.

Meristeles. Wanting.

2°. Long bristles. 10 in number; in a transverse section semi-circular, the convex side turned outwards.

Epidermis. Inner side. Consisting of longitudinally directed fibres, the top of which somewhat curved outwards, especially towards the top of the bristle; hairs not numerous, unicellular, conical, diameter at the base 40 μ, length 100 μ. E p i d e r m a l f i b r e s. H. 25 μ, Lev. B. 10 μ; polygonal .Walls thickened; lignified; showing stratification; pitted.

Outer side. For the epidermal fibres see the inner side. Trichomes in 2 kinds. 1°. Conical hairs like those of the inner side. 2°. Pluricellular hairs only occurring on the undermost half of the bristles; directed towards the top; cudgel-shaped; consisting of 2 rows of 10—15 cells; thick at the base 18 by 40 μ, length 100—200 μ, Cell contents: chloroplasts.

Mesophyll. Consisting of parenchyma with intercellular spaces.
C e l l s o f t h e a b o v e. 10 μ in diameter in a cross section; polygonal prisms with a longitudinally directed axis and rounded edges. Walls thickened; lignified.

Meristele. 1 in the undermost half of the bristle. Xylem containing spiral vessels with lignification.

3°. Short bristles. Essentially the same as the undermost half of the long bristles. Especially the pluricellular hairs more numerous. Meristeles wanting.

Corolla.

Epidermis. Inner side. Cells arranged in longitudinal rows; along the margin of the lobes some cells papilla-form, the longest at the top; on the incisions cells much smaller and forming a projecting ridge.
C e l l s o f t h e a b o v e. At some m.M. in distance from the top of the corolla H. 15 μ, Lev. B. 12 μ, Lev. L. 110 μ; at the base H. 15 μ, Lev. B. 15 μ, Lev. L. 25 μ; at the top tetragonal, the rest tetra- to hexagonal tables; at the top lateral walls sinuous. Outer walls a little thickened; in some heads in all the flowers at the base of the corolla transverse walls thickened, those walls joining in a tangential section the longitudinal ones with a triangular collenchymatous swelling. Cell contents: some small yellow chromoplasts in the yellow part of the corolla; in some cells inulin; at the base of the corolla several cells of the epidermis and the mesophyll forming together a hemisphere filled with inulin.

Outer side. Cells see those of the inner side. Hairs not numerous; in the expanded flower generally wanting. Mostly some glandular hairs of the Compositae type at the top and the base of the corolla; consisting of about 10 stories of 2 cells each, becoming larger towards the top of the hair. At the top thick 30 μ, L. 100 μ. With a cuticular bladder.
E p i d e r m a l c e l l s p r o p e r. At the top H. 10 μ, Lev. B. 10 μ, Lev. L. 60—100 μ; at the base H. 8 μ, Lev. 15 μ, Lev. L. 30 μ. See for the rest those of the inner side.

Mesophyll. Consisting of common parenchyma; at the top of the corolla cells stellate, for the rest cylindrical or prismatic with rounded edges and a longitudinally directed axis. Cell contents: often sphaero-crystals of inulin; in the cells surrounding the meristeles a yellow mass. Glands, showing an epithelium, close to the outside of the meristeles.

Meristeles. 5 in number; bifurcating just below the incisions of the lobes; both parts running along the margins of the lobes and ending in the top. Spiral vessels showing lignification.

Stamen.

Filament.

Epidermis. Cells in longitudinal rows. At the upper part of the filament, adjacent to the anther, the cells of the outer side showing strongly collenchymatously thickened walls and also arrangment in longitudinal rows (so-called articulation). Hairs only on the outer side of the articulation; 1-to 2-cellular; conical; diameter at the base 10 μ, length 50—100 μ.

Epidermal cells proper. T. 20 μ; rectangular prisms; in the vicinity of the articulation T. 8 μ, L. 20 μ; rectangular. Walls a little thickened; lateral walls pitted. Cells on the articulation T. 8 μ, L. 25 μ; rectangular. Walls yellow; lignified; pitted.

Mesophyll. Consisting of longitudinally arranged common parenchyma cells, sometimes containing sphaero-crystals of inulin. Meristele. 1 small meristele.

Anther.

Connective.

Epidermis. After the dehiscence the outermost parts of the anthers joined together by the epidermis. [1])

Cells containing here and there a very small chloroplast.

Mesophyll. Adjacent to the epidermis of the outer side some layers of sclerenchyma fibres, forming at the top of the anther the only elements of the mesophyll; the lower part containing moreover some common parenchyma.

Sclerenchyma fibres. R. and T. 8 μ; polygonal. Walls strongly thickened; middle lamella distinct; with pit canals.

Meristele. 1 in the common parenchyma of the lower part of the anther.

Thecae.

Epidermis. Cells in longitudinal rows.

Cells of the above. R. 10 μ, T. 12 μ, L. 20—40 μ; tetra- to hexagonal. Outer walls a little thickened. Cell contents: red to dark blue anthocyanin; for the rest here and there some minute yellow chloroplasts.

Fibrous layer. 1 layer of cells, arranged in longitudinal rows.

Cells of the above. R. 25 μ, T. 10 μ, L. 15—20 μ; R. larger in the vicinity of the connective; rectangular prisms. Only the reticulate thickenings of the walls a little lignified. Cell contents: some small yellow chloroplasts.

Pollen. Grains mostly ovate; diameter 30 μ, length 40 μ; with 3 lateral pores. Extine relatively thick; pitted.

Pistil.

Ovary.

Wall.

Epidermis. Outer side. Cellls arranged in longitudinal rows; conical hairs, unicellular, diameter at the base 4 μ, length 50—90 μ.

Epidermal cells proper. R. and T. 10 μ, L. 70—100 μ; tetra- to hexa-, mostly tetragonal tables. Cell contents: often sphaero-crystals of inulin; also hemispheres of inulin as described for the epidermis of the corolla.

Inner side.

---

[1]) Tschirch. „Sind die Antheren der Compositen verklebt oder verwachsen?" Flora. Bd. 93. 1904. 51, comes to a different conclusion.

C e l l s. R. 10 μ, T. 15 μ, L. 50 μ; polygonal tables. Walls mostly collenchymatously thickened. Cell contents: often a prismatic crystal tapering on both ends, those crystals but rarely to be found in the ripe fruit; sometimes inulin, see for this the inner epidermis.

M e s o p h y l l. Consisting of common parenchyma, containing in a transverse section a single schizogenous gland in each rib at the outer side of the meristele.

C e l l s o f t h e a b o v e. R. and T. 15 μ, L. 70—90 μ; polygonal prisms with a longitudinally directed axis. In the ripe fruit walls thickened. Cell contents: here and there inulin; in the innermost cells often crystals as described for the inner epidermis.

M e r i s t e l e s. In each rib the rudiments of 2 meristeles.

C o n d u c t i n g t i s s u e. The 2 bands lying on the flat sides of the ovary, diametrically opposed; the ovule lying immediately against them; consisting of small thick-walled cells.

Ovule. Anatropous; attached to the base of the ovarial cavity.

I n t e g u m e n t. [1])

Epidermis. Outer side.

C e l l s. R. 15—70 μ, T. 7 μ, L. 7 μ, the radial dimension strongly increasing towards the top of the ovule; prisms with a radially directed axis. In the ripe seed walls strongly thickened. Cell contents: much protoplasm; sometimes sphaero-crystals of inulin.

Inner side. More or less obliterated.

Mesophyll. Consisting of common parenchyma cells arranged in longitudinal rows; cells of the middle part larger than those of the inner part and reduced. Cells polygonal prisms with a longitudinally directed axis. Raphe containing a single meristele; continued through the chalaza. [1])

E n d o s p e r m. Slightly developed.

Nectary. Irregularly annular; high about 400 μ; attached on the ovary; at some distance surrounding the style.

Epidermis. Inner and outer side consisting of small polygonal cells of 8 μ in diameter; stomata resembling water-pores, numerous, especially on the margin, 25 μ in diameter. Mesophyll at the base about 10 layers of common parenchyma cells. Meristeles not distinguishable.

S t y l e.

E p i d e r m i s. Cells arranged in longitudinal rows.

C e l l s o f t h e a b o v e. R. 20 μ, T. 15 μ, L. 50 μ; tetragonal prisms. Outer walls thickened; showing a cuticle; transverse walls thickened in heads showing the thickening of the transverse walls in the epidermal cells of the corolla mentioned above. Cell contents: some small yellow chromoplasts; sometimes sphaero-crystals of inulin.

M e s o p h y l l. Consisting of common parenchyma; showing intercellular spaces.

C e l l s o f t h e a b o v e. R. and T. 15 μ; cylinders. Cell contents: some minute yellow chromoplasts; in some cells sphaero-crystals of inulin.

M e r i s t e l e s. 1 meristele on each side of the central conducting tissue. On the outer side of the meristeles often a schizogenous gland. Xylem containing spiral vessels with lignification.

---

[1]) See G u i g n a r d. Recherches sur le dévelopement de la graine. Journ. de Bot. Tome VII. 295.

Conducting tissue. Forming an elliptical cylinder in the centre of the style; consisting of thick-walled tissue. A few times the style 3-compound with 3 stigmata; in this case the conducting tissue trigonal and containing a pollen-canal.

Cells of conducting tissue. R. and T. 40—60 μ. Walls strongly thickened; middle lamella not distinguishable.

Stigmata.

Epidermis. Outer side. Above the meristeles some longitudinally elongated cells; for the rest polygonal papilla-form cells.

Cells of the above. R. and T. 7 μ.

Inner side. Cells bearing pointed papillae; at the base here and there a strip of much longer conical papillae.

Conical papillae. Diameter at the base 15 μ, length 80 μ.

Meristele. Lying in the middle part of the mesophyll, nearly reaching the top.

Ray-flower.

Many glandular hairs on the base of the 3 lobes; see for the rest the disk-flower.

July and October 1902.   J; M; L.

HERBA CENTAURII.

Herba Centaurii minoris.

The dried aerial parts of Erythraea Centaurium, Pers. Syn. I. 283, gathered when just flowering.

Macroscopic characters.

Stem high up to 5 d.M., herbaceous, erect, mostly unbranched, sometimes strongly branched at the base, tetra- to hexagonal with narrow wings at the edges; surface smooth. Arrangement of the leaves decussate. Leaves simple, sessile. Limb long up to 3 c.M.; oblong, in the higher leaves narrower, up to linear; curve-veined with 3 or 5 veins; those veins showing a depression at the upper side of the leaf and protruding on the under side; apex and base acute; margin entire; surface smooth. Inflorescence mixed; dichotomous cyme, often passing into small true scorpioid cymes, these all united together in a loose compound cyme; bracts lanceolate to linear; bracteoles present. Flower long about 1.5 c.M., complete, actinomorphous, hypogynous. Calyx gamosepalous, 5-sect, tubular, green, remaining. Corolla gamopetalous, with a 5-sect limb, salver-shaped; teeth oval, red with a somewhat lighter coloured spot at the base, contorted before and after the flowering period. Stamens 5, inserted on the throat of the corolla; anthers innate, dehiscent with longitudinal slits introrse, after the flowering period contorted like a cork-screw. Pistil compound, consisting of 2 carpels, with 1 cylindrical style and 2 spatulate stigmata; ovary cylindrical, incompletely bilocular, with axile placentae and numerous ovules. Odour wanting; taste bitter.

Anatomical characters.

LITERATURE. Benecke. Mikr. Drogenprakt. 1912. 62. Ellrodt. Vert. d. Gerbst. i. off.

Blätt., Kräut. u. Blüten. Diss. Würzburg. 1913. 20. Flückiger. Pharmakogn. 1891. 676. Gilg. Pharmakogn. 1910. 272. Hager. Pharm. Praxis. Bd. 1. 1900. 684. Hérail. Pharmacol. 1912. 539. Karsten u. Oltmanns. Pharmakogn. 1909. 192. Koch. Mikr. Anal. d. Drogenpulver. Bd. 3. 1906. 35. Koch. Pharmakogn. Atlas. Bd. II. 1914. 147. Kraemer. Botany a. Pharmacogn. 1910. 362. Luerssen. Syst. Bot. Bd. 2. 1882. 1053. Marmé. Pharmacogn. 1886. 228. Meyer. Drogenk. Bd. 1891. 33. Moeller. Pharmakogn. 1889. 212. Oudemans. Aanteek. o.d. Pharmac. neerl. 1854—56. Oudemans. Pharmacogn. 1880. 268. Perrot. Anatomie comparée d. Gentianacées. Ann. d. Sc. nat. Bot. Serie 8. T. 7. 1898. 105. Planchon et Collin. Drogues simples. T. 1. 1895. 650.

Stem.

MATERIAL. The drug. Stems, gathered in September 1901 from the Botanic Garden at Groningen, fresh and in alcohol. These sems were examined at the top and in the middle part. REAGENTS. Water, sugar solution 3 per cent., glycerine, chloral hydrate, phloroglucin and hydrochloric acid, iodine and sulphuric acid 66 per cent., concentrated sulphuric acid, Schulze's macerating mixture.

MICROGRAPHY.

Epidermis. On the wings of the stem somewhat collenchymatous. Stomata lying somewhat above the level of the epidermal cells; generally with only 3 subsidiary cells. Inner walls of the guard-cells more than usually thickened.

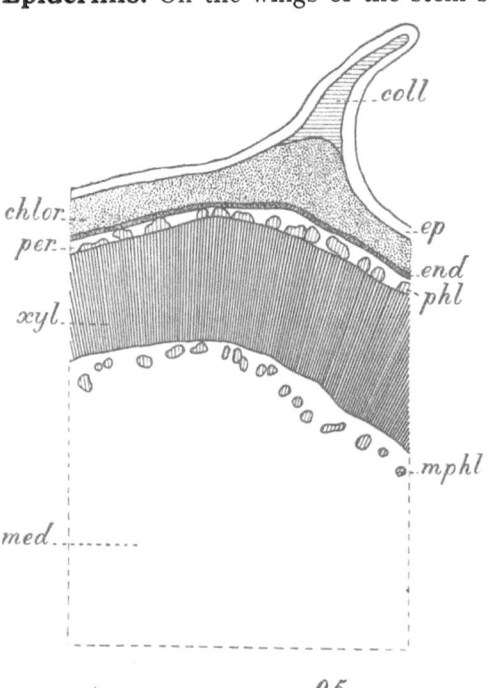

Fig. 65. *Erythraea Centaurium.* Stem, transverse section. chlor Chlorenchyma; coll Collenchyma; end Endodermis; ep Epidermis; med Medulla; mphl Medullary phloem; per Pericycle; phl Phloem; xyl Xylem.

E p i d e r m a l c e l l s p r o p e r. R. and T. 25 μ, L. 150—200 μ; tetra- to hexagonal prisms, with a radially directed axis. Outer walls thickened and consisting of 3 parts: 1. the cuticle giving rise to a cuticular stiation; 2. the middle layer showing a pure cellulose reaction; 3. the inner less refringent layer in iodine and sulphuric acid 66 per cent. not colouring blue, but much swelling up and showing very distinct stratification; inner walls collenchymatous. Cell contents: small chloroplasts. S t o m a t a. long 33 μ, wide 22 μ.

Cortex.

C h l o r e n c h y m a. Consisting of 5 cell layers and constituting the bulk of the cortex, with the exception of the outer parts of the wings; these consisting of more or less flattened collenchyma.

C h l o r e n c h y m a c e l l s. R. 15 μ, T. 15—20 μ, L. 50—100 μ; circular or elliptical cylinders with a longitudinally directed axis; showing intercellular spaces. Cell contents: some chloroplasts, each with oval starch grains.

E n d o d e r m i s. Developed as a protective-sheath and in the top of the stem developed as a distinct starch-sheath; generally consisting of 1, sometimes 2 layers of cells; in the middle part of the stem in the case of 2 cell layers the outer layer flattened in a radial direction.

Cells of the above. R. 15 μ, T. 35 μ, L. 65 μ; elliptical cylinders with a longitudinally directed axis. Radial walls persisting in concentrated sulphuric acid and colouring red with phloroglucin and hydrochloric acid. Cell contents: only in the top of the stem some globular starch grains, mostly united in a cluster.

Stele.

Pericycle. Consisting of 1 layer of common parenchyma cells; sometimes wanting in parts corresponding in place to the primary phloem bundles.

Cells of the above. R. 10 μ, T. 15 μ, L. 120 μ; elliptical cylinders or polygonal prisms with rounded edges and a longitudinally directed axis. Walls a little thickened. Cell contents: some small chloroplasts.

Vascular bundles.

Primary phloem. Exarch; in bundles of 5—15 elements in a cross section; consisting of sieve-tubes and cambiform cells. Between these bundles the phloem parts of the medullary commissures, consisting of common parenchyma, like that of the pericycle.

Sieve-tubes. R. and T. 4 by 4 to 4 by 7 μ, L. of articulations 30—90 μ; polygonal, often pentagonal prisms; distinct sieve-plates on the transverse walls; callus plates always at the same side of the transverse walls. Cambiform cells. R. and T. 4 by 4 to 4 by 7 μ, L. 125 μ; polygonal, often pentagonal.

Secondary phloem. Here and there developed as 1 layer of elements.

Cambium. Also here and there developed as 1 layer of elements.

Secondary xylem. Forming a continuous band; thick about 20 elements; consisting of tracheid fibres and medullary rays.

Tracheid fibres. R. 11 μ, T. 15 μ, L. 500—1200 μ; generally tetra-, sometimes hexagonal and a few times with bifurcated ends. Walls very thick; lignified, especially the middle lamellae. Cells of medullary rays. R. and T. 18 μ, L. 150—250 μ; hexagonal prisms, with a longitudinally directed axis. Walls very thick; lignified. Cell contents: chloroplasts.

Primary xylem. Endarch; in bundles consisting of annular and spiral vessels. Between these bundles the xylem parts of the medullary commissures, consisting of parenchyma.

Vessels. R. and T. 6—8 μ. Walls lignified. Parenchyma cells. R. and T. 10 μ, L. 100—150 μ; poly- often hexagonal prisms with a longitudinally directed axis. Walls sometimes thickened and pitted. Cell contents: some small chloroplasts.

Medulla. Consisting of parenchyma and containing in its outer part phloem bundles like those of the primary phloem, but not corresponding in place to the latter and separated by parenchyma from the primary xylem. Parenchyma showing intercellular spaces and sometimes torn in its central part.

Parenchyma cells. R. 40 μ, T. 40—50 μ, L. 100—200 μ; circular or elliptical cylinders, with a longitudinally directed axis. In the outer cell layers walls sometimes thickened and pitted.

Leaf.

MATERIAL. Leaves, gathered in September 1901 from the Botanic Garden at Groningen, fresh and in alcohol. REAGENTS. Water, sugar solution 3 per cent., chloral hydrate, iodine and sulphuric acid 66 per cent., iodine in chloral hydrate, phloroglucin and hydrochloric acid, concentrated sulphuric acid, Schulze's macerating mixture.

MICROGRAPHY.

Intervenia.

Epidermis. Upper side. Stomata 80—100 to the sq. m.M.; phaneroporous; often with only 3 subsidiary cells; lying somewhat above the level of the epidermal cells; the subsidiary cells lying sometimes more or less under the guard-cells; long 25 μ, wide 20 μ.

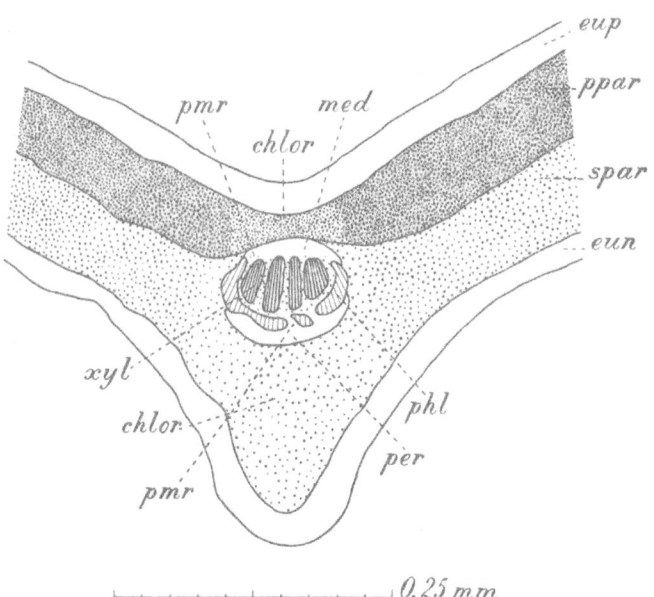

Fig. 66. *Erythraea Centaurium*. Leaf, transverse section through the midrib and the neighbouring portions of the lamina. chlor Chlorenchyma; eun Epidermis under side; eup Epidermis upper side; med Medulla; per Pericycle; phl Phloem; pmr Medullary commissures; ppar Palisade chlorenchyma; spar Spongy chlorenchyma; xyl Xylem.

Epidermal cells proper. E. g. H. 20 μ, Lev. B. 50 μ, Lev. L. 80 μ, but very different in size; polygonal, often rectangular with sinuous lateral walls. Outer walls somewhat thickened, with cuticular striation. Cell contents: some small chloroplasts.

Under side. Stomata 100—120 to the sq. m.M.; see for the rest those of upper side.

Epidermal cells proper. E. g. H. 20 μ, Lev. B. 30 μ, Lev. L. 80 μ; still more different in size than those on the upper side; see for the rest upper side.

Mesophyll.

Palisade chlorenchyma. Consisting at the upper side of the leaf of mostly 2, sometimes 1 layer of cells; not having the typical shape, especially those of the second layer.

Cells of the above. H. 40—60 μ, Lev. 20—25 μ; cylinders or prisms with rounded edges. Cell contents: chloroplasts, each containing some needle-shaped starch grains; on the bottom of the majority of the cells a prismatic crystal or a group of clustered prismatic crystals; here and there a yellow to green mass.

Spongy chlorenchyma. Consisting at the under side of the leaf of 4—6 cell layers and containing in the upper part many small meristeles; showing large intercellular spaces.

Cells of the above. H. and Lev. B. 25—30 μ, Lev. L. up to 50 μ; more or less branched, sometimes ellipsoidal or spherical. Cell contents: a few chloroplasts, each with some needle-shaped starch grains; here and there a yellow to green mass.

Veins.

Epidermis. Upper side. Cells generally arranged in longitudinal rows; stomata at places showing epidermal cells with somewhat more sinuous lateral walls.

Epidermal cells proper. H. 25 μ, Lev. B. 20 μ, Lev. L. 150 μ; rectangular, a few times pentagonal, with lateral walls much less sinuous than those of the cells of the intervenia. Walls, outer walls thickened, with longitudinal cuticular striation; inner walls somewhat collenchymatously thickened; lateral walls pitted. Cell contents: a few small chloroplasts.

Under side. Cells mostly arranged in longitudinal rows. Stomata wanting.

Epidermal cells proper. H. 25 μ, Lev. B. 20 μ, Lev. L. 90—150 μ; tetragonal, sometimes penta- or hexagonal. See for the rest those of upper side.

Mesophyll.

Chlorenchyma. Consisting above the meristele of 3 cell layers, under the meristele of 7—8 cell layers; the undermost layers somewhat collenchymatous; the middle layers sometimes somewhat flattened; the uppermost layers containing more chlorophyll.

Cells of the above. Above the meristele R. 14—15 μ, T. 15—18 μ; elliptical or circular cylinders with a longitudinally directed axis; showing intercellular spaces. Under the meristele R. and T. 20—30 μ, L. 50—120, increasing in length towards the meristele; cylinders with a longitudinally directed axis; showing intercellular spaces. Cell contents: chloroplasts, each with some starch grains.

Endodermis. Not discernible.

Meristele. Having the shape of an elliptical cylinder.

Pericycle. Only present at the under side; consisting of a few layers of collenchyma cells; showing small intercellular spaces.

Cells of the above. R. and T. 10—12 μ, L. 80—180 μ; penta- or hexagonal.

Vascular bundle. Single compound one, containing 4 simple bundles; the 2 lateral ones being somewhat broader. Simple bundles having in a cross section the shape of radially directed lamellae; collateral; open.

Phloem. Exarch; thick 5—8 elements; the phloem of the lateral simple bundles extending over there radial sides turned towards the periphery of the meristele; consisting of sieve-tubes and cambiform cells.

Sieve-tubes. R. and T. 3—4 μ, L. of articulations 60—70 μ. Cambiform cells. L. 80—100 μ.

Xylem. Endarch; consisting of a single row of elements, near the cambium of 2 rows; annular, spiral and pitted vessels.

Pitted vessels. R. and T. 8—9 μ, L. of articulations 230 μ; tetra- to hexagonal, especially towards the cambium. Walls strongly thickened; coloured brown.

Medullary commissures. Uniseriate; consisting of chlorenchyma; the phloem part showing larger cells.

Cells of the above. R. and T. 5—10 μ, L. 50—90 μ; hexagonal.

Medulla. Consisting of some layers of chlorenchyma cells.

Cells of the above. R. and T. 8—10 μ, L. 80—180 μ.

Flower.

MATERIAL. Flowers, gathered in September 1901 from the Botanic Garden at Groningen, fresh and in alcohol.

MICROGRAPHY.

Calyx.

Central green part of the lobes.

Epidermis. Inner side. Cells fairly well arranged in longitudinal rows, rather elongated, but diminishing in length from the base towards the top of the lobes, at the base and at the top themselves the cells much shorter. Stomata about 20 to the sq. m.M.; long 25 µ, wide 20 µ.

Epidermal cells proper. Measures on the middle part of the lobes H. 14 µ, Lev. B. 18 µ, Lev. L. 150 µ; rectangular, those at the base and on the top with sinuous lateral walls; at the base, especially above the midrib, lateral walls often thickened and the cells somewhat macerated and separated from each other. Outer walls with longitudinal cuticular striation. Cell contents: some granules.

Outer side. See the upper side; at the top of the lobes the cells being of a less regular shape. Stomata more numerous, 80 to the sq. m.M.; long 25 µ, wide 20 µ.

Epidermal cells proper. Measures on the middle part of the lobes H. 18 µ, Lev. B. 15 µ, Lev. L. 100 µ. See for the rest inner side.

Mesophyll. Consisting of chlorenchyma; the cells increasing in size towards the under epidermis.

Chlorenchyma cells. Lev. 15—20 µ; globular or ellipsoidal. Cell contents: chloroplasts.

Meristeles. Not numerous; containing some spiral vessels.

Membranaceous margin.

Epidermis. Inner side. Consisting of rather long cells, often arranged in longitudinal rows and diminishing in length towards the top of the lobes.

Cells of the above. H. 15 µ, Lev. B. 12 µ, Lev. L. 150—200 µ; mostly tetragonal, sometimes penta- or hexagonal, with somewhat sinuous lateral walls. Outer walls somewhat thickened; with longitudinal cuticular striation. Cell contents: some granules.

Outer side. See the inner side.

Cells of the above. Measures on the middle part of the margin H. 15 µ, Lev. B. 10 µ, Lev. L. 150—180 µ. Lateral walls less sinuous than those at the upper side. See for the rest those of upper side.

Mesophyll. Wanting.

Corolla.

Epidermis inner side. Of the tube. Cells nearly always arranged in longitudinal rows.

Cells of the above. H. 10 µ, Lev. B. 12 µ, Lev. L. 80—120 µ, at the base of the tube Lev. L. often 150 µ; mostly rectangular, sometimes penta- or hexagonal at the end of the longitudinal rows. Outer walls with a longitudinal cuticular striation.

Of the throat.

Cells of the above. H. 20 µ, Lev. B. 15 µ, Lev. L. 25 µ; mostly hexagonal, somewhat papilliform, often with rounded edges at the longitudinal sides. Outer walls with radial cuticular striation. Cell contents: a few small granules along the wall and in the centre a small vacuole containing anthocyanin.

Epidermis outer side. Of the tube. See the inner side; cell contents here a few granules.

Of the throat. See the inner side.

C e l l s o f t h e a b o v e. Here H. 30 μ, Lev. B. 15 μ, Lev. L. 60 μ; less papillate, with somewhat sinuous lateral walls and partly with a transverse cuticular striation. Cell contents: more clearly discernible.

M e s o p h y l l.

Of the tube. Consisting of somewhat flattened and thick-walled parenchyma; showing large intercellular spaces. Of the lobes wanting.

S t a m i n a.

Filament.

Epidermis. Cells mostly arranged in longitudinal rows. Stomata wanting.

E p i d e r m a l c e l l s p r o p e r. T. 6 μ, L. 100—150 μ, towards the anther L. somewhat diminishing; tetragonal, sometimes pentagonal with sinuous lateral walls. Outer walls with longitudinal cuticular striation. Cell contents: many small granules.

Mesophyll. Consisting of cylindrical parenchyma cells.

Meristele. A single central one.

Anther.

C o n n e c t i v e.

Epidermis.

C e l l s o f t h e a b o v e. On the outer side R. 15 μ, T. 20 μ, L. 15—80 μ; on the inner side somewhat smaller; the shorter cells looking like partitions of longer ones; rectangular with somewhat sinuous lateral walls. Outer walls thickened and with a thin cuticle. Cell contents: a yellow mass, not quite filling up the vacuole; for the rest starch grains.

Mesophyll. Consisting of common parenchyma and on the outside of the connective, adjoining the epidermis, sometimes elongated spiral cells. Mixed with the common parenchyma a few idioblasts each containing a prismatic crystal. In the flowering period the common parenchyma somewhat flattened.

Meristele. Only a single one.

T h e c a e.

E p i d e r m i s. Cells somewhat papillate; on the line of the dehiscence of the unripe theca 2 longitudinal rows of cells, having smaller transverse dimensions.

C e l l s o f t h e a b o v e. R. 25 μ, T. and L. 15—20 μ; polygonal, often hexagonal. Walls in the vicinity of the line of dehiscence collenchymatous. Outer walls with a cuticle, but a very thin one on the somewhat collenchymatous cells. Cell contents: starch grains, wanting in the collenchymatous cells. See for the rest the connective.

F i b r o u s l a y e r. Also present under the line of dehiscence; consisting of 1 layer of cells, some more in the vicinity of the connective.

C e l l s o f t h e a b o v e. R. 5—15 μ, increasing towards the connective, T. 15 μ, L. 20—60 μ, increasing towards the connective; rectangular prisms with a radially directed axis. Walls reticulate thickened near the line of dehiscence, towards the connective the cells showing various less regular thickenings.

P o l l e n. Grains 15—25 μ in diameter; oblong, sometimes tending to globular or tetrahedral; showing 3 pores and a smooth surface. Cell contents: starch grains.

P i s t i l.

Ovary.

W a l l.

E p i d e r m i s. Outer side. Cells mostly lying in longitudinal rows, with a cuneiform inner part protruding into the mesophyll.

C e l l s   o f   t h e   a b o v e. R. and T. 20 μ, L. 100—150 μ; tetra- to hexagonal prisms with a radially directed axis. Outer walls somewhat thickened, with parallel longitudinal cuticular striation. Cell contents: small chloroplasts.

Inner side. Consisting of tangential directed fibres, arranged in transverse and longitudinal rows.

F i b r e s. R. 9 μ, T. 150 μ, L. 5 μ. Walls, especially in more developed ovaries, somewhat thickened.

M e s o p h y l l. Thick 4—5 layers of elements; chiefly consisting of common parenchyma; the outer cell layer, consisting of somewhat radially elongated cells, protruding with there cuneiform outer part between the inner parts of the epidermal cells; the inner layer consisting of fibres similar to those of the inner epidermis, but directed at an angle to them. Here and there an idioblast containing prismatic crystals. M e r i s t e l e s. 3 in each carpel.

P a r e n c h y m a   c e l l s. Prisms or more irregular polyhedra. Cell contents: chloroplasts containing starch.

S e p t a   a n d   p l a c e n t a e.

Epidermis. Like that of the inner epidermis of the wall. Mesophyll like that of the wall but the middle layers consisting of smaller somewhat collenchymatous cells and the layer of fibres, adjoining the epidermis, wanting in the vicinity of the placentae and replaced by 1 or 2 layers of sclerenchyma cells; these cells somewhat elongated in a direction perpendicular to the epidermis.

Style.

Epidermis. Cells arranged in longitudinal rows and somewhat papillate.

C e l l s   o f   t h e   a b o v e. R. 20 μ, T. 7 μ, L. 120—240 μ; rectangular. Outer walls with cuticular striation. Cell contents: some small granules.

Mesophyll. Consisting of common parenchyma.

Meristeles. 2 in number; adjoining the narrow sides of the conductive tissue.

Conductive tissue. Forming an elliptical cylinder in the centre of the style; consisting of somewhat thick-walled elements.

C e l l s   o f   m e s o p h y l l. R. and T. 8—9 μ; polygonal prisms with a longitudinally directed axis and rounded edges; showing intercellular spaces. Cell contents: starch grains.
E l e m e n t s   o f   c o n d u c t i v e   t i s s u e. R. and T. 4 μ; polygonal in a cross section.

Stigmata.

Epidermis. Consisting of irregularly shaped cells with sinuous lateral walls, wide 20 μ, long 50—100 μ. At the top of each stigma a crest consisting of conical fairly elongated papillae.

September 1901.   J; M; v. E. d. W.

HERBA CONII.

Conii Folia. Conium Leaves. Conium.

The fresh leaves and young branches of Conium maculatum, Linn. Sp. Pl. 243, collected when the fruit begins to form.

LITERATURE. Anema. De zetel d. alcaloiden b. enk. nark. planten. Diss. Utrecht. 1892. 56. Barth. Stud. üb. d. microchem. Nachw. von Alcaloiden in pharm. verw. Drogen. Bot. Centrbl. 1898. Bd. 75. 292. Flückiger. Pharmacogn. 1891. 697. Funk. Beitr. z. Kenntn. d. mechan. Gew. syst. i. Stengel u. Blatt d. Umbellif. 1912. 81. Gilg. Pharmacogn. 1910. 243. Greenish & Collin. Anat. Atl. o. veget. powders. 1904. 66. Hager. Pharm. Praxis. Bd. I. 1900. 945. Hoaz. Comp. o. the Stem anat. o. the Cohort Umbellif. Ann. o. Bot. T. 29. 1915.55. Karsten u. Oltmanns. Pharmacogn. 1909. 189. Koch. Mikr. Anal. d. Drogenpulver. Bd. III. 1906. 43. Pl. IV. Koch u. Gilg. Pharmakogn. Praktik. 1907. 185. Kraemer. Botany a. Pharmacogn. 1910. 719. Marmé. Pharmacogn. 1886. 218. Martel. Note s. l'Anat. d. l. fleur d. Ombellif. Bot. Centrbl. T. 101. 226. Mitlacher. Tox.od. forens. wicht. Pfl. u. Drogen. 1904. 117. Moeller. Mikr. pharm. Üb. 1901. 213. Molisch. Microchemie d. Pfl. 1913. 255. Nestel. Beitr. z. Kenntn. d. Stengel. u. Blattanat. d. Umbellif. Diss. Zürich. 1905. 126. Oudemans. Pharmacogn. 1880. 306. Perrot e. Morel. Quelques Remarques sur l'Anat. d. Ombellif. Bull. Soc. bot. France. LX. 99, 141. Planchon e. Collin. Drogues simples. T. II. 1896. 218. Tschirch. Angew. Pfl. Anat. 1889. 169, 246. Tschirch u. Oesterle. Anat. Atl. 1900. Pl. 36. Tunmann. Krist. i. Herba Conii. Pharm. Zeit. Bd. L. 100. 1905. 1055. Tunmann. Pflanzenmicrochemie. 1913. 317. Vogl. Anat. Atl. 1887. Pl. 18, 19. Wigand. Pharmacogn. 1879. 211.

Stem.

MATERIAL. Stems of 0.15—1.3 c.M. and especially of 0.4—0.5 c.M. in diameter, gathered June 1903 in the Botanic Garden at Groningen; fresh, fixed with chromic acid 1 per cent. and in alcohol. REAGENTS. See those of the leaf.

MICROGRAPHY.

Epidermis. Stomata wanting in places corresponding to the collenchyma bundles mentioned below, for the rest rare, about 4 to the square m.M.; phaneroporous; lying in the same level with the epidermal cells; surrounded by 4—5 subsidiary cells.

Epidermis cells proper. R. 20 μ, T. 20—30 μ, L. 30—40 μ; polygonal tables with a radially directed axis. Outer walls somewhat thickened; showing a cuticle, here and there longitudinal cuticular striation. Cell contents: nucleus mostly distinct; some small chloroplasts; alcaloid (for material treated with potassium iodide iodine see also Anema l.c. 58); sometimes a colourless rarely a yellow oil-drop, containing numerous minute more strongly refringent globules and becoming brown in potassium iodide iodine and sometimes persisting in the alcoholic material; in several cells anthocyanin; in material treated with chromic acid often needle-shaped, occasionally branched crystals soluble in glycerine; in alcoholic material in some cells sphaero-crystals, see for these the epidermal cells of the upper side of the leaf. Stomata. R. 20 μ, T. 18 μ, L. 30 μ.

Cortex.

Consisting of collenchyma, chlorenchyma, colourless parenchyma containing glands and an endodermis. In the thicker stems the parenchyma cells regularly increasing in size towards the endodermis.

Collenchyma. In larger and smaller flat longitudinal bundles of fibres. The innermost fibres here and there somewhat flattened. Intercellular spaces

wanting. The larger bundles separated from the epidermis by a single layer of
common parenchyma, in the thinnest specimens adjoining the epidermis and
then corresponding with the most projecting ribs. Each larger bundle corresponding with a larger vascular bundle and having a larger tangential dimension
than this vascular bundle, so as to project over it on both sides; in a stem of

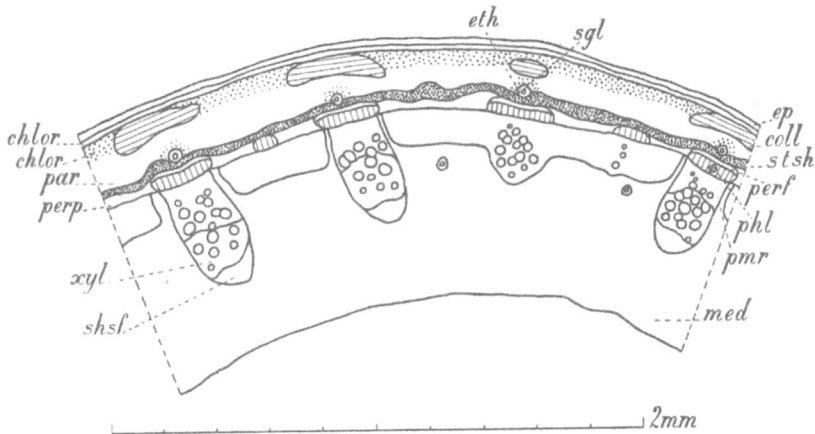

Fig. 67. *Conium maculatum*. Stem, transverse section. chlor Chlorenchyma; coll Collenchyma; ep Epidermis; eth Epithelium; med Medulla; par Colourless parenchyma;
perf Pericyclic fibres; perp Pericyclic parenchyma; phl Phloem; pmr Medullary commissures; sgl Schizogenous glands; shsf Sheath of sclerenchyma fibres; stsh Starch-
sheath; xyl Xylem.

4.5 m.M. in diameter 12 layers of fibres thick. Smaller collenchyma bundles
especially in the thicker stems, corresponding to the smaller vascular bundles
and farther removed from the epidermis than the larger collenchyma bundles.
C o l l e n c h y m a f i b r e s. R. 5 μ, T. 10 μ; polygonal. Walls collenchymatous. Cell
contents: some minute chloroplasts.
C h l o r e n c h y m a. 4 layers of cells; the walls paved with chloroplasts as
described for the palisade tissue of the leaf, in the outermost layer chloroplasts
less numerous. The cells of the outermost layer arranged in longitudinal rows in
parts corresponding to the vascular bundles. Showing intercellular spaces. Under and near a stoma cells of the outermost layer mostly somewhat branched;
at the lower side of a stoma the branches of some cells joining each other, so as
to divide the air cavity into some parts. In the vicinity of these large intercellular spaces generally moreover some smaller ones.
C h l o r e n c h y m a c e l l s. Of the outermost layer R. 20 μ, T. 30 μ, L. 35 μ; mostly
rectangular tables with a radially directed axis and rounded edges; a more irregular cell
R. 20 μ, T. 40 μ, L. 70 μ; in the inner 3 layers R. 18 μ, T. 20 μ, L. 40—50 μ; mostly
polygonal prisms with a longitudinally directed axis, strongly rounded edges and often a
tendency to ramification. Cell contents: numerous chloroplasts containing minute starch
grains; in some cells anthocyanin, sphaero-crystals like those of the epidermal cells.
C o l o u r l e s s p a r e n c h y m a. 3 layers of cells; 1 or 2 innermost layers
often flattened in parts corresponding to the smaller vascular bundles and de-

prived of collenchyma. Cells surrounding the glands smaller than the other. Showing intercellular spaces. S c h i z o g e n o u s g l a n d s. 1, or rarely 2, corresponding to each larger or smaller vascular bundle; similar to those of the colourless parenchyma of the midrib of the leaf; the cavity filled with a colourless oil.

C e l l s o f c o l o u r l e s s p a r e n c h y m a. R. 40 μ, T. 50 μ; polygonal prisms with a longitudinally directed axis and somewhat rounded edges; cells round the glands R. and T. 10—20 μ, L. 70—110 μ. Cell contents: here and there a small chloroplast; occasionally an oil-drop as described for the epidermal cells; sphaero-crystals like those of the epidermal cells.

E n d o d e r m i s. Distinct only in a specimen of 0.45 c.M. in diameter; developed as a starch-sheath; consisting of 1, sometimes of 2 layers of cells. By way of exception the epithelium cells of the glands of the cortex interrupting the endodermis and adjoining the pericycle.

C e l l s o f t h e a b o v e. R. and T. 25—40 μ, L. 25—60 μ; polygonal prisms with a longitudinally directed axis and rounded edges. Cell contents: chloroplasts containing simple and compound starch grains.

Stele.

P e r i c y c l e. Consisting of thin-walled fibres and common parenchyma. The fibres in some layers along the phloem bundles, e. g. 4 layers along the largest bundles; intercellular spaces wanting. The parenchyma part of the pericycle made up of 1 or 2 layers of cells.

F i b r e s. R. 5 μ, T. 10 μ; polygonal. C o m m o n p a r e n c h y m a c e l l s. R. 14 —25 μ, T. 18—20 μ; polygonal prisms with a longitudinally directed axis and rounded edges. Cell contents: often some chloroplasts containing small starch grains.

V a s c u l a r b u n d l e s. Number increasing towards the base in consequence of the periodical development of the punctum vegetationis [1]); arranged in a ring; simple, but most in places of the circumference showing a decided tendency to fuse with one another. Mostly 3 kinds of bundles, showing well marked differences: 1°. larger collateral open bundles, in their primary part developed as fibrovasal bundles, each corresponding in place to a larger cortical collenchyma bundle; 2°. smaller bundles of the same kind, especially occurring in the thicker stems, each corresponding to a smaller cortical collenchyma bundle; 3°. very small fibrovasal bundles, consisting either of phloem and xylem, the latter showing a thick sheath of sclerenchyma fibres, or of phloem and in stead of the xylem only a large bundle of sclerenchyma fibres, or again only of a large bundle of sclerenchyma fibres.

P h l o e m. Relatively much more developed in the smaller bundles than in the larger ones. Of a larger tangential dimension than the xylem, so as to project on both sides. The elements of the innermost part arranged in radial rows. Consisting of very large sieve-tubes with companion cells, distinct in the outermost part, and common parenchyma containing glands. Glands not always

---

[1]) C.f. *Herba Aconiti*, stem.

distinct, always having a single layer of epithelium cells consisting of 4 cells in a cross section. See for the rest the glands of the leaf.

S i e v e - t u b e s. The largest ones in the outermost parts e. g. R. and T. up to 15 μ; polygonal prisms. Sieve-plates on the transverse walls conspicuous. Contents: close to the sieve-plates very numerous small starch grains; perhaps some alcaloid. C o m- p a n i o n   c e l l s. R. and T. 3—5 μ, mostly tetragonal prisms. Cell contents: a granular mass, colouring yellow in iodine and chloral hydrate. C o m m o n   p a r e n c h y m a c e l l s. In a cross section smaller than the sieve-tubes. Cell contents: some chloroplasts, especially in the inner part of the tissue.

F a s c i c u l a r   c a m b i um. Sometimes distinct.

X y l e m. Of the larger vascular bundles consisting of 3 parts: 1° the outermost secondary part, consisting of libriform fibres and an occasional pitted vessel; arranged in radial rows corresponding to the rows of the phloem; intercellular spaces wanting; 2° a middle part composed of libriform fibres and numerous larger pitted vessels, probably surrounded by tracheid fibres; intercellular spaces wanting; 3° primary xylem endarch, composed chiefly of annular and spiral vessels, intermixed with xylem parenchyma and surrounded on its lateral and inner sides by a sheath of sclerenchyma fibres; intercellular spaces wanting. Xylem of the smaller vascular bundles similar to that of the larger ones, only sometimes the inner sclerenchyma-sheath wanting. Of the small fibrovasal bundles mostly only consisting of libriform fibres and some vessels, and sur- rounded on its lateral and inner sides by a sheath of sclerenchyma fibres, thick 16 elements in a radial direction, those of the outer part being radially arranged; in some cases only the sclerenchyma-sheath present.

V e s s e l s. In the outermost part R. and T. 15 μ, L. of articulations 190—250 μ; in the middle part R. and T. 20—35 μ, L. of articulations 190—300 μ; polygonal prisms with strongly rounded edges; transverse walls with a circular perforation. Walls lignified. In the primary xylem R. and T. 15—30 μ. L i b r i f o r m   f i b r e s. In the outermost part R. 6 μ, T. 9 μ; tetra- to hexagonal; in the middle part R. and T. 10 μ; polygonal. Walls thickened; lignified; pitted. F i b r e s   o f   t h e   p r i m a r y   x y l e m. R. and T. 10 μ, L. very great; polygonal. Walls thickened; lignified; pitted sparingly. X y l e m   p a- r e n c h y m a   c e l l s. R. 10—15 μ, T. 10 μ, L. 60—80 μ; polygonal prisms with a longitudinally directed axis. Cell contents: chloroplasts. F i b r e s   o f   s c l e r e n c h y- m a-s h e a t h. R. and T. 10—12 μ; polygonal. Walls thickened; lignified; pitted sparingly. Here and there on the outer side 2 fibres probably developed out of one fibre by the for- mation of a tangential wall, then the outermost fibre often showing unthickened walls. Contents: in several fibres an oil-drop like those of the epidermis.

M e d u l l a r y   c o m m i s s u r e s. Consisting of common parenchyma; 3- to 4- seriate. Interfascicular cambium wanting. Small intercellular spaces.

C e l l s   o f   t h e   a b o v e. Outermost cells R. 8 μ, T. 10 μ; tetra- to hexagonal prisms, sometimes more or less fibre-shaped; innermost cells R. and T. 18 μ, L. 80 μ; tetra- to hexagonal prisms with slightly rounded edges; in both cases axis longitudinally directed. Walls thickened; lignified; pitted. Cell contents: some small chloroplasts; in several cells an oil-drop like those of the epidermis.

M e d u l l a. Consisting of common parenchyma and glands; projecting be- tween the xylem of those vascular bundles farthest extending towards the centre;

showing a very large lysigenous central cavity, except in the vicinity of the inflorescence; round the cavity some layers of flattened parenchyma cells with small intercellular spaces. G l a n d s. Small glands lying close to the vascular bundles and alternating with them; farther towards the centre here and there a gland; mostly showing an epithelium consisting of 5 cells in a cross section. See for the rest the glands of the cortex.

C o m m o n p a r e n c h y m a c e l l s. In the outermost border R. and T. 25 μ, L. up to 120 μ; farther towards the centre R. 40 μ, T. 45 μ, L. 50—75 μ; polygonal prisms with a longitudinally directed axis and somewhat rounded sides. Cell contents: in several cells an oil-drop like those of the epidermis cells.

L e a f.

MATERIAL. Leaves collected June 1903 in the Botanic Garden at Groningen; fresh, fixed with chromic acid 1 per cent. and in alcohol. REAGENTS. Water, glycerine, chloral hydrate, potassium iodide iodine, iodine in chloral hydrate, phloroglucin and hydrochloric acid, iodine and sulphuric acid 66 per cent., concentrated sulphuric acid, Schulze's macerating mixture, potassium dichromate, copper acetate and iron acetate, glacial acetic acid.

MICROGRAPHY of the leaf-blade.

I n t e r v e n i a.

Epidermis upper side. Along the margin of the leaf the cells lying in about 5 longitudinal rows, near the tops of the teeth the cells very small. Stomata rare; towards the tops of the teeth more numerous and sometimes more or less flattened; phaneroporous; lying in the same level as the epidermal cells; with a very large air cavity, sometimes filled with sphaero-crystals like those of the epidermal cells; generally surrounded by 3 or 4 subsidiary cells, sometimes being smaller than the other epidermal cells. Water-pores numerous near the tops of the teeth. Trichomes wanting.

E p i d e r m a l c e l l s p r o p e r. H. 20 μ, Lev. B. 35 μ, Lev. L. 70 μ, or e. g. Lev. B. 50 μ, Lev. L. 60 μ; polygonal tables with sinuous lateral walls; cells of the innermost rows along the margin H. 20 μ, Lev. B. 20 μ, Lev. L. 70—100 μ; cells of all these rows rectangular prisms with slightly sinuous lateral walls; near the tops of the teeth e. g. Lev. 20 μ and smaller. Outer walls somewhat thickened; showing a cuticle and in the vicinity of a stoma showing cuticular striation at right angles with the outline of the stoma; along the margin the cuticular striation longitudinal. Cell contents: alcaloid [1]), some small chloroplasts; in the material treated with potassium dichromate a granular yellow precipitate, sometimes enclosing a colourless spherical cavity; in the alcoholic material, but not in the fresh and the chromic acid material in many cells sphaero-crystals; the component needle-shaped crystals distinct, not soluble in cold nor in hot water, when dissolved in potash yielding a yellow liquid, not soluble in cold nor in boiling glacial acetic acid even in a test-tube [2]); not soluble in iodine in chloral hydrate nor in phloroglucin and hydro-

---

[1]) T s c h i r c h u n d O e s t e r l e and A n e m a mention alcaloid in the epidermis without making a difference between the upper and the lower side of the leaf. T s c h i r c h u n d O e s t e r l e use for reagents vanadium suphuric acid, phosphomolybdic acid, potassium iodide iodine, and potassium bismuth iodide; A n e m a and myself only used potassium iodide iodine.

[2]) Hesperidine? see Z i m m e r m a n n. Die bot. Mikrotechnik. 1892. 91.

chloric acid. S t o m a t a. H. 9 μ, Lev. B. 10 μ, Lev. L. 12 μ. G u a r d-c e l l s. Containing alcaloid. W a t e r-p o r e s. Lev. B. 15 μ, Lev. L. 20 μ or smaller.

Epidermis under side. Along the margin about 5 longitudinal rows of cells. Stomata exceedingly numerous, nearly every epidermal cell adjoining a stoma, 140 to the square m.M.; see for the air cavities the stomata of the upper side; wanting along the margin; lying in the same level as the other epidermal cells; closed in by 3—4 subsidiary cells. Trichomes wanting.

E p i d e r m a l   c e l l s   p r o p e r. H. 15 μ, Lev. B. 25 μ, Lev. L. 60 μ; polygonal tables with highly sinuous lateral walls; cells of the longitudinal rows along the margin H. 20 μ, Lev. B. 20 μ, Lev. L. 80—110 μ; rectangular prisms; cells of the innermost ones of these rows with sinuous lateral walls. Outer walls showing cuticular striation at right angles with the outline of the stoma, along the margin the striation longitudinal. Cell contents: alcaloid wanting; in alcoholic material sphaerocrystals like those of the upper side. S t o m a t a. H. 8 μ, Lev. B. 20 μ, Lev. L. 30 μ.

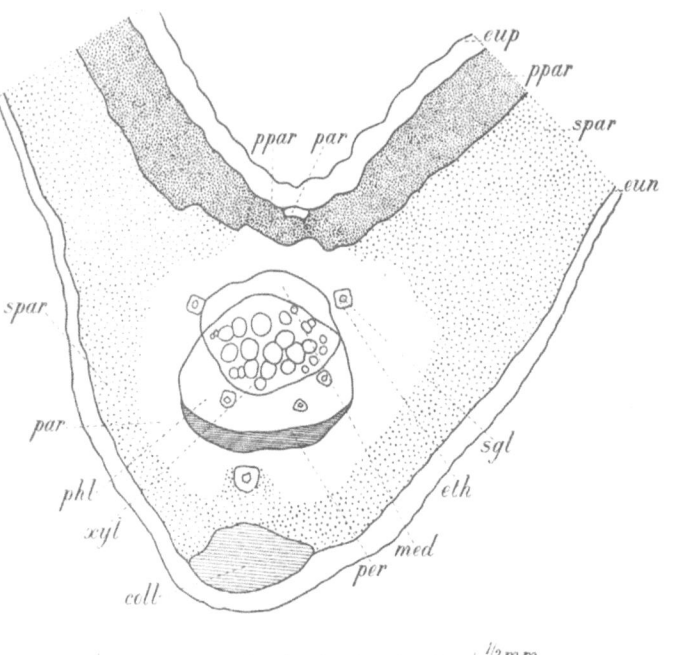

Fig. 68. *Conium maculatum*. Leaf, transverse section through the midrib and the neighbouring portions of the lamina. coll Collenchyma; eth Epithelium; eun Epidermis under side; eup Epidermis upper side; med Medulla; par Colourless parenchyma; per Pericycle; phl Phloem; ppar Palisade chlorenchyma; sgl Schizogenous glands; spar Spongy chlorenchyma; xyl Xylem.

Mesophyll.

P a l i s a d e   c h l o r e n c h y m a. Consisting of 1 layer of cells; showing intercellular spaces.

C e l l s   o f   t h e   a b o v e. H. 75 μ, Lev. 10—15 μ, in the middle part of the intervenia often smaller; polygonal prisms with strongly rounded edges or circular and elliptical cylinders. Cell contents: a great number of chloroplasts, thick 3 μ and 5—6 μ in diameter, mostly penta- or hexagonal tables, paving the walls and containing some needle-shaped starch grains; here and there a cell having in its uppermost part a sphaerocrystal like those of the epidermal cells.

S p o n g y   c h l o r e n c h y m a. Consisting of about 5 layers of stellate cells, the uppermost layer pretty well developed as collecting cells; the cells of the undermost layer having their branches strongly elongated in the direction perpendicular to the epidermis. Showing large intercellular spaces, especially in the undermost layer.

Cells of the above. Lev. e. g. 20 μ, in the undermost layer H. 30 μ, Lev. 10 μ.
Cell contents: see those of the palisade tissue, the chloroplasts here somewhat smaller.

Margin. Palisade tissue occurring everywhere in the margin. At the bases
of the teeth containing a midrib or larger primary veins bundles of collenchyma similar to those lying under the meristeles. In the white uppermost
small tops of the teeth palisade tissue wanting; in these places an epithema of
longitudinally elongated colourless parenchyma cells.

Veins.

Epidermis upper side. Cells lying in longitudinal rows. Stomata not numerous, lying in the same level as the epidermal cells. Trichomes wanting.

Epidermal cells proper. R. 30 μ, T. 30 μ, L. 70—200 μ; rectangular prisms;
more towards the top of the leaf lateral walls more or less sinuous, the rectangular shape
however preserved. Outer walls somewhat thickened; showing longitudinal cuticular
striation. Cell contents: see those of the epidermal cells of the intervenia; here and there
a cell containing anthocyanin, especially in the vicinity of a stoma. Stomata. Lev.
B. 20 μ, Lev. L. 30 μ.

Epidermis under side. Cells lying in longitudinal rows; those covering the
collenchyma bundles, mentioned below, different from the other. Stomata wanting in the parts of the epidermis covering the collenchyma bundles, for the
rest relatively numerous; lying in the same level as the epidermal cells; showing
a very large conical air-cavity, H. 40 μ, diameter of inwards turned base 35 μ;
generally closed in by 3 smaller subsidiary cells. Trichomes wanting.

Epidermal cells proper. R. 25 μ, T. 30 μ, L. 100—180 μ; rectangular prisms;
farther towards the top lateral walls more or less sinuous, but the rectangular shape preserved; cells covering the collenchyma bundles R. and T. 20 μ; L. 300—350 μ. Outer
walls more strongly thickened than those of the upper side, inner walls of the cells having
no collenchyma underneath also slightly collenchymatous; outer walls showing longitudinal cuticular striation. Cell contents: see those of the intervenia. Stomata. R.
10 μ, T. 20 μ, L. 35 μ.

Mesophyll. Consisting of collenchyma, colourless parenchyma, palisade chlorenchyma, spongy chlorenchyma, inner glands and an endodermis.

Collenchyma. In the larger veins a bundle of collenchyma fibres adjoining the epidermis of the under side and filling up its curve. At the upper side,
sometimes also adjoining the epidermis, a bundle consisting in a cross section
of some small elements having slightly collenchymatous walls, those bundles
corresponding only with the largest meristeles. Intercellular spaces wanting.

Fibres. R. and T. 10 μ, polygonal; no transverse walls noticed. Walls strongly collenchymatously thickened; middle lamella pretty distinct.

Colourless parenchyma. Represented by 2 layers of cells surrounding the larger meristeles, at the base of the leaf often 3 layers. Cells lying in
longitudinal rows; those surrounding the glands smaller than the rest. Intercellular spaces.

Colourless parenchyma cells. R. and T. 30 μ, L. 70—150 μ; polygonal
prisms with a longitudinally directed axis and rounded edges. Cell contents: some chloroplasts containing a few small starch grains, especially in the cells surrounding the glands;

sphaero-crystals similar to those of the epidermal cells of the upper side of the intervenia.

Palisade chlorenchyma. Only present at the upper side and continuing the palisade tissue of the intervenia; fully developed in the base of the midrib; the cells of the innermost layers often isodiametrical in a transverse section; various transitory forms between palisade chlorenchyma and common colourless parenchyma along the midrib; less developed towards the top of the midrib, sometimes even wanting there.

Spongy chlorenchyma. Represented by 2—3 layers of cells at the lower side of the larger veins, interrupted by the collenchyma bundles; cells more or less stellate. Showing large intercellular spaces.

Cells of the above. R. and T. 20 μ, L. 30—70 μ. Cell contents: chloroplasts, containing some needle-shaped starch grains.

Glands. Schizogenous; 1 gland below every meristele, with the exception of the smallest ones; in the undermost half of the leaf 1 or 2 glands on either lateral side of the meristele at the level of the xylem bundles; cavity R. and T. 10 μ, polygonal, probably filled with a colourless oil. The glands below the meristele larger than the other ones. Epithelium of glands below the meristeles consisting of 1 layer of cells, 5 or more in a cross section, of the other glands 4 cells in a cross section.

Epithelium cells. Of the glands lying below the meristeles. Thick 10 μ, wide 15 μ; polygonal prisms with a longitudinally directed axis. Inner walls of the cells borderingthe cavity somewhat thickened; swelling in potassium iodide iodine, more strongly so in iodine and sulphuric acid 66 per cent., the lamella of the walls bordering the cavity colouring blue, the rest hardly or not, the outermost very thin lamella bordering the cavity colouring brown; in concentrated sulphuric acid this lamella not to be found. Cell contents: a granular protoplasm, especially close to the wall adjoining the cavity.

Endodermis. Consisting of smaller cells than the colourless parenchyma; partly developed as a starch-sheath, for the rest as common parenchyma; the starch cells at the lower side in longitudinal rows. Starch grains simple and compound.

Meristele. Irregularly cylindrical.

Pericycle. Outside the phloem bundles consisting of up to 4 layers of thin-walled fibres; intercellular spaces wanting. The parts between the bundles of fibres consisting of 1 or 2 layers of common parenchyma cells.

Fibres of the above. R. and T. 12 μ; polygonal. Walls a little collenchymatous.

Vascular bundle. A single compound one, containing a number of simple bundles not to be determined.

Simple bundles collateral, open.

Phloem. Gutter-shaped; lying under the xylem and half-way encircling it. Constituted of 3 layers; the outer layer consisting of sieve-tubes and companion cells and containing up to 2 glands; the middle layer of bundles of small elements intermixed with parenchyma cells and containing up to 3 glands alternating with those of the outer layer; the inner layer of 2—3 layers of parenchyma cells. Intercellular spaces wanting throughout the phloem. Glands with an epithelium of 1 layer of cells, consisting in a cross section of 4, mostly of 3 cells [1]); cavity R. and T. 5—6 μ, resp. tetra- or trigonal.

S i e v e - t u b e s. R. and T. 10 μ; polygonal prisms, sieve-plates on the transverse walls distinct. Contents: minute starch grains and alcaloid near the sieve-plates. C o m- p a n i o n c e l l s. R. and T. 5 μ; tetragonal prisms. Cell contents granular. S m a l l e l e- m e n t s o f t h e m i d d l e p a r t. R. and T. 5 μ; polygonal prisms with a longi- tudinally directed axis. Walls somewhat thicker than those of the other elements. P a r e n- c h y m a c e l l s. R. and T. 6—9 μ, L. 35—70 μ; polygonal prisms with a longitudinally directed axis. Cell contents: often chloroplasts. E p i t h e l i u m c e l l s. E. g. thick 6 μ, wide 9 μ; polygonal prisms with a longitudinally directed axis. Walls thinner than those of the other elements.

C a m b i u m. Not to be distinguished.

X y l e m. Forming an elliptical cylinder. Consisting of spiral and reticulate vessels, interchanging with parenchyma cells. Intercellular spaces wanting.

V e s s e l s. The largest R. 25 μ, T. 22 μ, L. of articulations of reticulate vessels e. g. 400—500 μ; polygonal prisms with strongly rounded edges, transverse walls having a circular perforation. Walls lignified.

M e d u l l a r y c o m m i s s u r e s. Not to be distinguished.

M e d u l l a. Represented by a band of very long, perhaps fibrous parenchyma cells, encircling the upper half of the xylem and on both sides joining the phloem. C e l l s o f t h e a b o v e. R. and T. 10 μ, L. very great; polygonal with a longitudinal- ly directed axis. Walls somewhat collenchymatous.

P e t i o l e. Mesophyll: between the epidermis and the collenchyma bundles 1 layer of parenchyma cells containing chloroplasts; spongy chlorenchyma 3 layers of cells lying at the under side; colourless parenchyma composed of about 3 layers at the upper side and about 10 layers at the under side; in the colour- less parenchyma here and there a small gland.

July 1903.   J; M; L.

HERBA HYOSCYAMI.
Hyoscyamus. Henbane.

The aerial parts of Hyoscyamus niger, Linn. Sp. Pl. 179, collected from the flowering plant.

Macroscopic characters of the leaf.
Simple; the longest ones petiolate, the higher ones sessile, the highest semi- amplexicaul. Blade long 1—2 d.M., wide up to 1 d.M.; thin; when fresh flaccid and viscous; oblong or ovate pinnatilobate; apex acute. Surface pale green; hairy, especially on the veins and the margin with common and glandular hairs.

Odour narcotic; taste bitter and somewhat pungent.

Anatomical characters.
LITERATURE. Benecke. Mikr. Drogenprakt. 1912. 54. Flückiger. Pharmakogn. 1891. 710. Gilg. Pharmakogn. 1910. 306. Greenish & Collin. Anat. Atl. o. powd. veget. Drugs. 1904. 68. Hager. Pharm. Praxis. Bd. II. 1902. 93. Hérail. Mat. Méd. 1912. 682. Kar-

---

[1]) T s c h i r c h and his disciples always mention 4 cells.

sten u. Oltmanns. Pharmakogn. 1909. 167. Koch. Mikr. Anal. d. Drogenpulver. Bd. III. 1906. 53. Pl. V. Kraemer. Botany a. Pharmacogn. 1910. 720, 728, 755. Kuntz. A Hyoscyamus niger alkaloidatartalmànak szövetrendszerbeli eloszlása. Bot. közlem. XVII. $^1/_3$. 1—16. 1918 (With a German summary). Luerssen. Syst. Bot. Bd. II. 1882. 987. Marmé. Pharmakogn. 1886. 424. Mitlacher. Tox. od. Forens. wicht. Pfl. u. Drogen. 1904. 149. Moeller. Mikr. pharm. Üb. 1901. 104. Moeller. Pharmakogn. 1889. 70. Molisch. Mikrochem. d. Pfl. 1913. 255. Oudemans. Pharmacogn. 1880. 243. Planchon e. Collin. Drogues simples. T. I. 1895. 583. Schneider. Powdered veget. Drugs. 215. Tschirch u. Oesterle. Anat. Atl. 1900. 168. Pl. XXXIX. Tunmann. Pfl.-mikrochemie. 1913. 325. Vogl. Anat. Atl. Pl. IX. X. Wigand. Pharmakogn. 1879. 214. REAGENTS. Water, glycerine, chloral hydrate (not for the leaf), iodine in chloral hydrate, iodine and sulphuric acid 66 per cent., phloroglucin and hydrochloric acid, concentrated sulphuric acid.

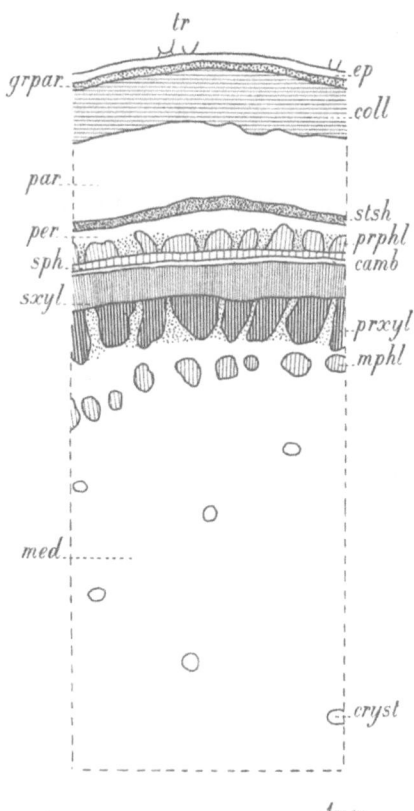

Fig. 69. *Hyoscyamus niger*. Stem, transverse section. camb Cambium; coll Collenchyma; cryst Crystal cells; ep Epidermis; grpar Outermost green layer of common parenchyma; med Medulla; mphl Medullary phloem; par Colourless parenchyma; per Pericycle; prphl Primary phloem; prxyl Primary xylem; sph Secondary phloem; stsh Starch-sheath; sxyl Secondary xylem; tr Trichomes.

Stem.

MATERIAL. The drug. Stems, gathered July 1901 in the Botanic Garden at Groningen, fresh and in alcohol.

MICROGRAPHY.

Epidermis. Epidermal cells bearing hairs mostly somewhat wider and shorter than the rest. Stomata rare; phaneroporous; lying in the same level as the epidermal cells. Trichomes. See those of the leaf; conical hairs the most numerous.

Epidermal cells proper. R. 35 μ, T. 28 μ, L. 100—200 μ; tetra- to hexagonal prisms with a longitudinally directed axis. Outer and inner walls thickened; outer walls with cuticular striation, at their inner side showing a notched outline, sometimes also to be seen on other walls, under the polarization microscope no crystals to be seen. Cell contents: a few small chloroplasts; highly refractive granules and irregular fragments. Stomata. 20 by 35 μ. Cells of trichomes. Contents: plastids, not numerous, generally pale green; green colour more prominent in the glandular hairs.

Cortex. Consisting of: 1 outermost green layer of common parenchyma; 5 layers of collenchyma, cells and fibres intermixed, showing intercellular spaces; 2—4 layers of somewhat collenchymatous colourless parenchyma cells, showing intercellular spaces; an endodermis, developed as a starch-sheath and consisting of 1—2, sometimes of 3 layers of cells.

Cells of outermost green parenchyma layer. Tetragonal prisms

with a longitudinally directed axis. Cell contents: many chloroplasts. C o l l e n c h y m a-
t i c e l e m e n t s. R. 40 μ, T. 35 μ, L. 300; tetra- to octogonal, cells and fibres, sometimes divided by 1 or 2 transverse partition walls. Thickenings of the walls stratified. Cell
contents: a few chloroplasts; in the outermost collenchymatic layers often small granules
along the walls, not distinctly appearing as crystals under the polarization microscope.
C o l o u r l e s s p a r e n c h y m a c e l l s. R. and T. somewhat larger than those of
the collenchyma cells, L. 100—200 μ; tetra- to octogonal prisms with a longitudinally directed axis and strongly rounded edges. Cell contents: here and there a chloroplast. C e l l s
o f e n d o d e r m i s. R. 30 μ, T. 35 μ, in the additional 1 or 2 layers sometimes somewhat
smaller, L. 50—75 μ; ellipsoidal cylinders with a longitudinally directed axis, surfaces of
contact somewhat flattened. Cell contents: some spurious chloroplasts, not containing
much starch, generally heaped up on one side of the cavity; starch grains sometimes
compound, 3-adelphous.

Stele.

P e r i c y c l e. Consisting of 1—2 layers of common parenchyma cells; on the
outside of the primary phloem small bundles of sclerenchymatic elements.

C o m m o n p a r e n c h y m a c e l l s. R. 22 μ, T. 30 μ, L. 125 μ; tetra- to hexagonal
prisms with a longitudinally directed axis and mostly with strongly rounded edges.
S c l e r e n c h y m a t i c e l e m e n t s. R. 12 μ, L. 120 μ.

Vascular bundles.

Phloem.

P r i m a r y p h l o e m. Enarch; lying in bundles containing smaller bundles
of sievetubes and companion cells, smaller bundles 10—15 elements in a
transverse section, separated by bands of cambiform cells, those bands 1—3 cells
wide and showing intercellular spaces.

S e c o n d a r y p h l o e m. Forming a continuous ring.

S i e v e-t u b e s. R. and T. 10—15 μ, L. of articulations 100 μ; tetra- to hexagonal
prisms; transverse walls often very oblique; sieve-plates and callus conspicuous. Walls
often somewhat thickened. C o m p a n i o n c e l l s. R. and T. 3—4 μ; tri- to tetragonal
prisms. C a m b i f o r m c e l l s. R. 15 μ, T. 20 μ, L. 100 μ; tetra- to hexagonal prisms
with more or less rounded edges. Contents: sometimes a small simple crystal.

Cambium. Consisting of 1—2 layers of elements.

E l e m e n t s o f t h e a b o v e. R. 4 μ, T. 12 μ, L. 80 μ; tetragonal prisms.

Xylem.

S e c o n d a r y x y l e m. Forming a continuous ring. Vessels pitted; articulations short; transverse walls with 1 large circular perforation. Libriform
fibres showing slit-like pits. Wood parenchyma cells without intercellular
spaces. Medullary rays uni- to triseriate, many cells high, sometimes blended
together in a longitudinal direction, consisting of common parenchyma.

W o o d p a r e n c h y m a c e l l s. R. 12 μ, T. 18 μ, L. 70—100 μ; penta- or hexagonal
prisms. C e l l s o f m e d u l l a r y r a y s. Axis longitudinally directed. Walls pitted.
In the young stem vessels with obliquely placed transverse walls; wood parenchyma
cells and cells of medulary rays containing chloroplasts.

P r i m a r y x y l e m. Endarch; lying in bundles, 15 elements thick, always
corresponding in place to the bundles of primary phloem. Outermost and largest
part of the bundles consisting of radial bands of annular spiral and reticulate

vessels with lignification and some parenchyma cells, alternating with bands made up of parenchyma cells only (medullary commissures); innermost part of the bundles consisting of parenchyma with occasionally a vessel. Sometimes vessels and parenchyma cells intermixed without distinct arrangement.

Medulla. Consisting of common parenchyma without intercellular spaces. In its outermost part containing a ring of phloem bundles, on the whole similar to the fascicular phloem; however the cribral bundles containing more elements, the elements themselves larger; the sclerenchymatic elements on the the inside of the phloem bundles like those of the pericyclic sclerenchyma of the fascicular phloem more numerous, lying in bundles thick 15—25 elements. Common parenchyma cells. R. and T. 60—70 μ; spherical or ellipsoidal. Sclerenchymatic elements of the phloem bundles. R. and T. 20—25 μ; tetra- to hexagonal.

Leaf.

MATERIAL. The drug. Leaves collected in the Botanic Garden at Groningen July 1901; fresh, in alcohol and fixed with chromic acid.

MICROGRAPHY.

Intervenia.

Epidermis upper side. Stomata 125 to the sq. m.M.; phaneroporous; lying above the level of the epidermal cells; generally surrounded by 3—4 subsidiary cells, 1 of which often smaller than the epidermal cells proper and lying on one of the longitudinal sides of the stoma; the lower parts of the subsidiary cells partly covered by the stoma. Water-pores on the apex of the leaf; not numerous; perhaps also on the tops of the lobes; much larger than the stomata. Trichomes in 2 kinds, inserted on the middle of the epidermal cells. 1°. Conical hairs rather numerous, more so along the margin; consisting of 1 row of 2—4 cells; long 100—300 μ. 2°. Glandular hairs rare; capitate; length about 100—300 μ, much varying; stalk consisting of 1 row of 2—4 cells; head showing various sizes and structures: either a smaller head consisting of 1 row of some cells, the undermost ones of which tabular and the topmost one spherical; or a larger head consisting of 2—3 rows of some cells. Cuticular bladder wanting.

Epidermal cells proper. H. 30 μ, Lev. 45 by 65 μ; lateral walls sinuous. Outer walls showing a thin cuticle. Cell contents: chloroplasts. Stomata. Lev. 22 by 28 μ. Conical hairs. Cell contents: plastids, sometimes green, not numerous. Glandular hairs. Cell contents of the head: mostly a yellowish green granular mass; of the stalk: plastids, sometimes green, not numerous.

Epidermis under side. On the whole similar to that of the upper side. Stomata somewhat more numerous. Trichomes more numerous, especially those with the larger head; the long conical hairs mentioned for the upper side of the veins here much more numerous.

Mesophyll.

Palisade chlorenchyma. At the upper side of the leaf; consisting of

1 layer of cells. Often the lower ends of some cells bending towards each other so as to join on the same collecting cell of the underlying spongy chlorenchyma, this collecting cell nearly always containing a crystal [1]). Showing large intercellular spaces.

Cells of the above. H. 60 μ, Lev. 16 μ; cylinders. Cell contents: chloroplasts in the shape of tetra- to hexagonal tables, lying with their flat sides along the cell wall; thick 2 μ, wide 3 by 4 μ.

Spongy chlorenchyma. Consisting of 3, sometimes of 4 layers of cells. In the uppermost layer cells spherical, ellipsoidal, or ellipsoidal-polygonal; longest axis at right angles to the surface of the leaf. The other 2 or 3 layers showing the common shape of spongy parenchyma cells; in the undermost layer cells often somewhat palisade-like. Large intercellular spaces. Idioblasts, containing a single crystal or a cluster crystal, occurring in the upper-most layer; numerous (collecting cells).

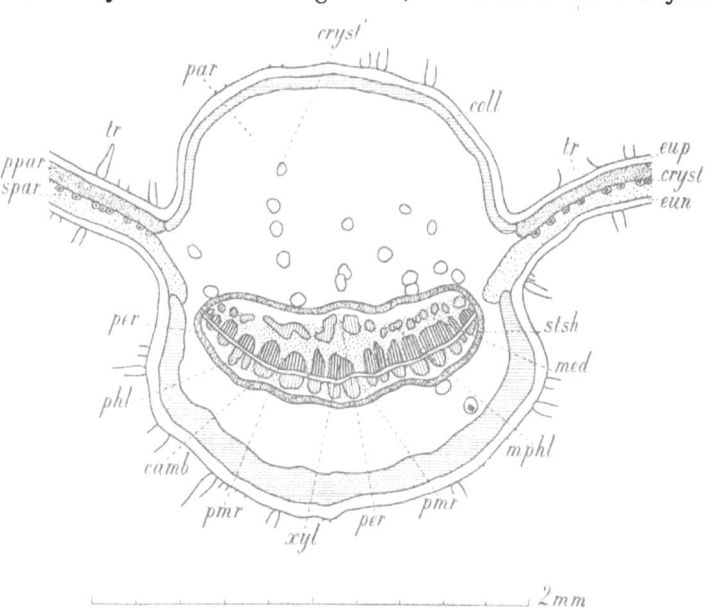

Fig. 70. *Hyoscyamus niger*. Leaf, transverse section through the midrib and the neighbouring portions of the lamina. camb Cambium; coll Collenchyma; cryst Crystal cells; cryst′ Cells containing crystal sand or a simple crystal; eun Epidermis under side; eup Epidermis upper side; med Medulla; mphl Medullary phloem; par Colourless parenchyma; per Pericycle; phl Phloem; ppar Palisade parenchyma; prm Medullary commissures; spar Spongy parenchyma; stsh Starch-sheath; tr Trichomes; xyl Xylem.

Cells of spongy chlorenchyma. H. 28 μ, Lev. 24 by 30 μ; in the undermost layer H. 50, Lev. 20 by 30 μ. Cell contents: in the uppermost layer chloroplasts resembling those of the palisade chlorenchyma. Idioblasts. Crystal prisms, hexahedra, rhomboids, or twin-crystals in the shape of crosses, the clusters not very complex.

Veins.

Epidermis upper side. Cells bearing hairs differing in shape from the other epidermal cells: higher, wider and shorter and showing rounded edges; the hair inserted in the middle. This shape of the hair-bearing cells also to be found in the primary veins, gradually disappearing in the smaller ones. Stomata

---

[1]) See Haberlandt. Vergleichende Anatomie des assimilatorischen Gewebesystemes der Pflanzen. Jahrbücher für wiss. Botanik. Bd. 13. 74.

wanting on the midrib and the primary veins. Trichomes like those of the intervenia. Conical hairs more numerous; longer; some of them consisting of up to 10 cells and then having a length of about 5 m.M. Glandular hairs much more numerous than on the intervenia, yet less numerous than the conical hairs; sometimes those with the smaller head much elongated.

E p i d e r m i s c e l l s p r o p e r. H. 45 μ, Lev. B. 30 μ, Lev. L. 200—300 μ; tetra-, mostly penta- or hexagonal. Outer walls somewhat thickened; showing cuticular striation; inner walls distinctly thickened; lateral walls pitted. Cell contents: some chloroplasts, especially above the lateral parts of the midrib.

Epidermis under side. On the whole similar to that of the upper side. Stomata and trichomes somewhat more numerous, especially the glandular hairs with the larger head; the long hairs of the upper side much more numerous.

Mesophyll.

C o l l e n c h y m a. Consisting of 1—2 layers of cells at the upper and of 1—3 layers at the under side. At the under intercellular spaces larger than at the upper side and in the colourless parenchyma.

C e l l s o f t h e a b o v e. R. and T. 40 μ, L. 150—300 μ, sometimes longer. Cell contents: at the under side in the outermost layer cells filled with chloroplasts, especially at the lateral sides of the midrib; for the rest a few plastids; in many cells small granules along the walls.

C o l o u r l e s s p a r e n c h y m a. At the upper side about 20 cell layers, at the under side about 4. Showing intercellular spaces. Idioblasts containing crystal sand, sometimes a simple prismatic crystal.

C o l o u r l e s s p a r e n c h y m a c e l l s. R. and T. 60 μ, L. 100—300 μ; cylinders with a longitudinally directed axis. I d i o b l a s t s. L. 120—130 μ.

E n d o d e r m i s. Developed as a starch-sheath. Consisting of 1—2 layers of common parenchyma cells.

C e l l s o f t h e a b o v e. At the upper side R. 28 μ, T. 35 μ, L. 80—100 μ, sometimes 40—50 μ; at the under side R. and T. somewhat larger; elliptical cylinders with a longitudinally directed axis, surfaces of contact flattened. Cell contents: little starch, often collected on one side of the cell; chloroplasts.

Meristele. Gutter-shaped, the concave side turned upwards.

P e r i c y c l e. Nearly continuous; least developed at the under side of the meristele; consisting of thick-walled elements, partly—perhaps all—being fibres.

S c l e r e n c h y m a f i b r e s. R. and T. 20—25 μ, L. 900—1400 μ; tetra- to hexagonal. Walls not much thickened; giving a very distinct cellulose reaction.

V a s c u l a r b u n d l e. A single compound one, containing about 15 simple bundles arranged in a curved plane. Simple bundles irregularly circular or elliptical cylinders; collateral; open.

P h l o e m Structure similar to that of the medullary phloem.

C a m b i u m. Consisting of some layers of elements. Towards the top of the midrib disappearing.

X y l e m. Consisting of 7—8 layers of elements. The uppermost 3—4 layers consisting of parenchyma and rare small vessels; the rest of radial sheets of 1—2 vessels wide. Vessels annular, spiral or reticulate.

Vessels. R. and T. 15—30 μ. Walls with lignification. Parenchyma cells. R. and T. 15—20 μ, largest in the uppermost layers, L. 60 μ. Cell contents: a few chloroplasts.

Medullary commissures. In the phloem part continuous with those of the xylem part; in the xylem part uni- or biseriate; in the phloem part cells larger than in the xylem part and showing intercellular spaces. The interfascicular cambium similar to the fascicular.

Cells of the above. In the phloem part with rounded edges. Contents of both parts: chloroplasts.

Medulla. Consisting of common parenchyma without intercellular spaces and bundles of phloem more or less corresponding in number and place to the simple vascular bundles. Those bundles largest in the middle part, very small on the lateral sides of the medulla and separated from each other by some parenchyma cells.

Common parenchyma cells. R. and T. 15—20 μ, L. 80 μ; tetra- to hexagonal prisms. Sieve-tubes. R. and T. 5—6 μ, L. of articulations 140 μ; walls somewhat thickened. Cell contents: a granular mass.

Flower.

MATERIAL. Collected July 1901 in the Botanic Garden at Groningen, fresh and in alcohol. Material for the examination of the pistil collected 4 weeks afterwards.

MICROGRAPHY.

Calyx.

Epidermis inner side. At the base cells rectangular with smooth lateral walls; upwards hexagonal and lateral walls sinuous; farther upwards the regular shape disappearing and the sinuosities much deeper, up to half the diameter of the cell; still farther towards the top of the calyx cells much smaller again; above the principal meristeles the rectangular shape with sinuous lateral walls always preserved and cells larger. Stomata only in places showing sinuous lateral cell walls; number gradually increasing towards the top of the calyx, in a lobe 100 to the sq. m.M., wanting above the principal meristeles; lying in the same level as the epidermal cells; often surrounded by 3 subsidiary cells. Trichomes only in places showing sinuous lateral cell walls, most numerous above the meristeles and along the margin; on the whole similar to those of the leaf; the long conical hairs on the margin of the lobes.

Epidermal cells proper. Lev. 45 by 70 μ. Outer walls somewhat thickened; showing a thin cuticle; attached to the outer walls or within short prismatic to rhomboid crystals. Stomata. Lev. 22 by 25 μ. Trichomes. Walls showing crystals like those of the epidermal cells proper.

Epidermis outer side. Cells like those of the inner side, but somewhat smaller. Stomata like those of the inner side, more numerous. Trichomes like those of the inner side; more numerous; on the lower part of the calyx numerous long conical hairs like those occurring on the veins of the leaf.

Mesophyll.

In the intervenia consisting of 4—5 cell layers. Generally the inner layer consisting of palisade chlorenchyma with large intercellular spaces. Sometimes all the layers spongy chlorenchyma with very large intercellular spaces. These cells often stellate; containing chloroplasts with a single rods-haped or ellipsoidal granule. Crystal idioblasts, corresponding in place to the meristeles, spherical or ellipsoidal; generally containing a cluster crystal, sometimes a prismatic or rhombic crystal.

In the veins on the inner side 3 and on the outer side 4 layers of common parenchyma cells, showing distinct intercellular spaces.

Cells of the above. R. and T. 25 μ, L. 70—150 μ; cylindrical with a longitudinally directed axis.

Meristeles. Essentially the same as those of the leaf. Pericyclic sclerenchyma-sheath wanting. Cambium thick 10—15 layers of radially arranged elements, in the smaller veins less developed. Xylem. Vessels all developed as spiral vessels, with lignification.

Elements of cambium. R. 3 μ, T. 6 μ; tetra- to hexagonal prisms.

Corolla.

Epidermis inner side. At the base of the corolla cells lying in rather regular longitudinal rows and rectangular; now and then the rows shifting and then cells tri- to pentagonal; somewhat farther upwards the regular arrangement less clear and the cells tetra- to hexagonal and nearly isodiametrical. At $^1/_3$ of the length of the corolla from the base lateral walls becoming sinuous; farther towards the top cells much larger and sinuosities deeper; near the top of the corolla sinuosities angular and some folds of the lateral walls, those folds only thickened at the top. Above the midrib cells longitudinally elongated and modified shape of cells appearing much farther upwards. Stomata rare. Trichomes wanting.

Epidermal cells proper. At $\frac{1}{3}$ of the length of the corolla from the base H. 22 μ, Lev. 20 by 25 μ. Outer walls with faint cuticular striation of widely dispersed lines. Cell contents: in the purple part of the corolla here and there anthocyanin. Stomata. Lev. 16 by 20 μ.

Epidermis outer side. Practically the same as the inner epidermis; cells deeper and wider on the more projecting midrib. Trichomes here present in places with sinuous lateral cell walls; in 2 kinds. 1°. Conical hairs, mostly 1—4 cells long. 2°. Glandular hairs, in the lower part of the corolla nearly exclusively on the veins; sometimes having a short stalk of 1—2 cells; head generally well developed.

Mesophyll.

In the intervenia mostly consisting of 4—5 layers of spongy parenchyma; cells mostly stellate. Cell contents: here and there yellowish green chloroplasts.

In the veins common parenchyma; on the inner side some layers of cells, at the outer side about 4 layers. Showing intercellular spaces.

Cells of the above. At the inner side R. and T. 10—15 μ, at the outer side R. and T. 30 μ; cylinders. Cell contents: in the reddish purple part of the veins at the inner and the outer side in the outermost cell layer here and there anthocyanin, especially on the lateral sides of the midrib.

Endodermis. Developed as a starch-sheath.

Meristeles. Practically the same as those of the calyx. Cambium wanting.

S t a m e n.

Filament.

Epidermis. Length of cells increasing upwards. Only in the middle part lateral walls sinuous. Stomata wanting. Trichomes occurring on the insertion in the corolla and on the larger under part of the filament, on all the stamens up to the same level; in 2 kinds, like those of the calyx moreover a 3rd kind; capitate hairs showing a long apical cell; the conical hairs here also forming the majority; length of all the hairs about 500 µ.

E p i d e r m a l   c e l l s   p r o p e r. R. 60 µ, T. 18 µ, L. 80—100 µ; tetra- to hexagonal with a longitudinally directed axis; sometimes showing a tangential partition wall. Outer and inner walls somewhat thickened.

Mesophyll. Consisting of 5 layers of common parenchyma cells.

C e l l s   o f   t h e   a b o v e. R. and T. 35 µ; tetra- to octogonal prisms with rounded edges. Cell contents: in a few cells a simple crystal, sometimes a twin crystal.

Endodermis. Developed as a starch-sheath.

Meristele.

Vascular bundle amphicribral. Phloem with distinct sieve-tubes. Xylem containing spiral vessels with lignification.

Anther.

C o n n e c t i v e. Epidermis. See Thecae.

Mesophyll. Consisting of common parenchyma showing intercellular spaces.

C e l l s   o f   t h e   a b o v e. R. 20 µ, T. 50 µ, L. 25 µ; ellipsoidal polyhedra; longest dimension transversely directed axis.

Meristele.

Xylem. Containing spiral vessels with lignification.

T h e c a e. Epidermis. In the vicinity of the fissure cells without regular arrangement and showing an irregular shape, sometimes with slightly sinuous lateral walls. Above the fissure cells much smaller and nearly isodiametrical. Stomata about 40 to the sq. m.M.; sometimes lying above the level of the epidermal cells.

E p i d e r m a l   c e l l s   p r o p e r. Near the fissure R. 22 µ, T. 30 µ, L. 80 µ; penta- or hexagonal; axis at right angles to the fissure; above the fissure T. 10 µ, L. 13 µ. Outer walls showing cuticular striation. Cell contents: except above the fissure a dark red to purple liquid. S t o m a t a. Lev. 20 by 25 µ.

Fibrous layer. Consisting of 1 layer of cells, near the connective of 2—3 layers.

C e l l s   o f   t h e   a b o v e. R. 30 µ, T. 20 µ, L. 70 µ, near the fissure T. greater, L. smaller; tetra- to hexagonal prisms with their longest dimension at right angles to the fissure, radially directed axis and more or less rounded edges. Thickenings of the walls slightly lignified.

Middle layer and tapetum. Generally already having disappeared. Middle layer sometimes still to be distinguished.

Pistil.

Ovary.

W a l l. Epidermis outer side. Cells tetra- or hexagonal, containing much protoplasm with a large nucleus. Inner side. Cells polygonal, containing much protoplasm with a large nucleus. Stomata numerous.

Mesophyll. About 7 layers of cells; without intercellular spaces. Inner and outer cell layer resembling the epidermis and closely joining it.

C e l l s o f t h e a b o v e. Polyhedra. Containing much protoplasm with a large nucleus. Meristeles. Showing distinct spiral vessels.

S e p t u m a n d p l a c e n t a e.

Epidermis, see the inner epidermis of the wall. Mesophyll consisting of somewhat spongy parenchyma with large intercellular spaces; cells mostly ellipsoidal.

Style.

E p i d e r m i s. Cells showing finely sinuous longitudinal lateral walls. Stomata rare. Trichomes, especially at the base of the style, like those of the calyx; wanting on the upper part.

E p i d e r m a l c e l l s p r o p e r. In the middle part of the style R. 20 μ, T. 12 μ, L. 80—100 μ, near the base L. smaller; mostly tetra-, sometimes penta- or hexagonal prisms with a radially directed axis. Outer walls a little thickened; showing a thin cuticle. Cell contents: mostly starch grains, some of them compound, the compound ones especially lying against the inner walls; some cells quite filled with anthocyanin. S t o m a t a. Lev. 30 by 35 μ.

M e s o p h y l l. Consisting of 12—13 layers of common parenchyma cells. Cells of the innermost layers much smaller in diameter and somewhat flattened.

C e l l s o f t h e a b o v e. R. 18 μ, L. 100—150 μ. Cell contents: starch grains, some of which compound with up to 3 component grains.

M e r i s t e l e s. 1 meristele on each side of the central conducting tissue; reaching into the stigmata; containing 1 vascular bundle. Vascular bundle collateral; xylem showing distinct spiral vessels with lignification.

C o n d u c t i n g t i s s u e. In the shape of a circular or elliptical cylinder. Cells cylindrical with a small diameter.

Stigma. Beset with papillae, not numerous, long, conical, unicellular.

July 1901.   J; M; L.

HERBA SABINAE.

Sabina. Savin.

The dried youngest tops of Juniperus Sabina, Linn. Sp. Pl. 1039, gathered in spring.

Macroscopic characters.

Branchlets with very short internodes. Arrangement of the leaves decussate or whorled, the whorl formed by 3 leaves. Leaves long 2—8 m.M.; sessile; either rhomboid and imbricated or more needle-shaped and standing out; on each leaf at the outside an oblong gland in a shallow groove.

Anatomical characters.

LITERATURE. De Bary. Vergl. Anat. 1877. 171. Flückiger. Pharmakogn. 1891. 743. Flückiger & Hanbury. Pharmacographia. 1879. 626. Geyler. Ueber den Gefäszbündelverlauf in den Laubblattregionen d. Coniferen. Jahrb. f. Wissenschaftl. Botanik. Bd. 6. 65—69. Gilg. Pharmakogn. 1910. 27. Hager. Pharm. Praxis. Bd. 2. 1902. 763. Hérail. Mat. Méd. 1912. 247. Kraemer. Botany a. Pharmacogn. 1910. 682. Leth. A comp. morph. a. anat. examination of Ramuli Sabinae a. substitutes. Archiv. Pharm. 13 (63). 1906. 381—388. Luerssen. Syst. Bot. Bd. 2. 1882. 96. Mahlert. Anat. der Laubblätter der Coniferen. Bot. Centrlbl. Bd. 24. 1885. 54. Marmé. Pharmacogn. 1886. 207. Moeller. Anat. d. Baumrinden. 1882. 14. Moeller. Pharmakogn. 1889. 204. Moeller. Mikr. pharm. Ueb. 1901. 208. Molisch. Mikrochem. d. Pfl. 1913. 150—153. Oudemans. Aanteek. o.d. Pharmac. neerl. 1854—56. 92. Oudemans. Pharmacogn. 1880. 239. Planchon et Collin. Drogues simples. T. 1. 1895. 82. Tschirch. Angew. Pfl. Anat. 1889. 318. Tschirch. Harze u. Harzbeh. 1900. Tunmann. Pfl.-microchemie. 1913. 234. Wigand. Pharmakogn. 1879. 239. MATERIAL. The drug. Tops, gathered in July 1901 in the Botanic Garden at Groningen, fresh and in alcohol. REAGENTS. Glycerine, chloral hydrate, iodine in chloral hydrate, iodine and sulphuric acid 66 per cent., phloroglucin and hydrochloric acid, Schulze's macerating mixture.

MICROGRAPHY.

Stem.

Epidermis. During the first year quite surrounded by and grown together with the bases of the leaves; therefore showing no epidermis of its own.

Cortex.

C h l o r e n c h y m a. Thick some layers of cells.

C e l l s o f t h e a b o v e. R. and T. 12 μ, L. 40 μ; cylinders. Cell contents: chloroplasts and in several cells starch.

E n d o d e r m i s. In no stage of development of the stem clearly discernible.

Stele.

P e r i c y c l e. In shoots of the first year chiefly consisting of thin-walled elements and only in the places corresponding with the phloem of the vascular bundles showing 1 or 2 sclerenchyma fibres with brown contents; in the second year sclerenchyma fibres much increased in number and forming 2 or 3 almost continuous tangential layers; in shoots having developed secondary cork tissue consisting of bundles of sclerenchyma fibres alternating with thin-walled elements.

F i b r e s o f t h e a b o v e. R. 4 μ, T. 9 μ. Walls colouring yellow in iodine and sulphuric acid 66 per cent.

V a s c u l a r b u n d l e s. Each leaf-trace descending through 2 internodes and bifurcating in the middle part of the lower one, the 2 branches uniting with the leaf-traces of the 2 leaves inserted in the lower internode. Thus in the upper half of each internode 4, in the lower half 6 vascular bundles in a transverse section. In branches with a whorled arrangement of the leaves the number of vascular bundles corresponding with the number of leaves in the whorls[1]). Collateral, open. P r i m a r y p h l o e m. Exarch; forming a thick bundle

---

[1]) See G e y l e r l. c.

of elements without radial arrangement and here and there containing some chloroplasts.

Secondary phloem. Showing radial arrangement; in young shoots consisting only of 6 to 7 layers of elements. Afterwards showing a regularly reiterated succession of the following tangential layers: 1. one layer of bast fibres with walls lignified, especially in their inner part, and radial walls somewhat thickened and practically showing no crystals in polarized light; 2. one layer of sieve-tubes often containing starch; 3. one layer of bast parenchyma fibres often containing starch; 4. one layer of sieve-tubes often containing starch.

Cambium. Not clearly discernible in very young shoots.

Secondary xylem. In young shoots consisting only of 5—6 layers radially arranged tracheid fibres and a few uniseriate secondary medullary rays 1—8 cells in hight.

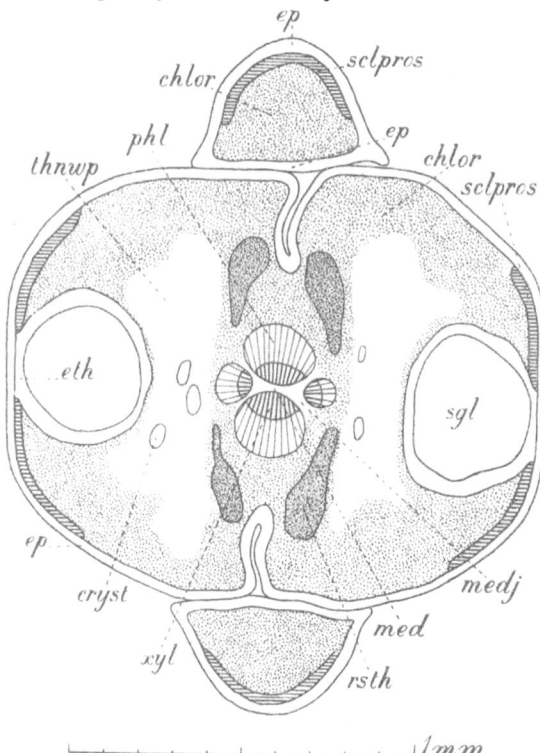

Fig. 71. *Juniperus Sabina*. Transverse section of a young top of the stem with 2 leaves. chlor Chlorenchyma; cryst Cells with simple crystals; ep Epidermis; eth Epithelium; med Medulla; medj Medullary commissures; phl Phloem; rsth Parenchyma with cross rod-shaped thickenings of the walls; sclpros Sclerenchymatic prosenchyma; sgl Schizogenous glands; thnwp Thin-walled parenchyma cells; xyl Xylem.

Tracheid fibres. R. 5 μ, T. 6 μ, L. 400 μ; tetra- to hexagonal. Walls thickened; middle lamella in the shape of a dark coloured line; lignified. Cells of medullary rays. R. 10 μ, T. 7 μ, L. 20 μ; prisms with a longitudinally directed axis; showing intercellular spaces. Cell contents: chloroplasts and starch.

Primary xylem. Endarch; consisting of about 3 layers of spiral vessels, fairly well radially arranged.

Vessels. R. and T. 3—4 μ.

Medullary commissures. Consisting of parenchyma; pluriseriate; extending in length through the whole internode.

Cells of the above. R. and T. 5—9 μ, L. 40 μ; tetra- to hexagonal prisms. Walls somewhat thickened. Cell contents: chloroplasts containing starch, and here and there a brown mass entirely filling a cell.

Medulla. Almost wanting; see for the rest medullary commissures.

Leaf.

Epidermis upper side. Cells arranged in longitudinal rows; those in the vicinity of the stomata smaller than the rest. Stomata 100 to the sq. m.M.; but only covering one third part of the upper surface of the leaf in the vicinity of the midrib; longitudinally directed and often some of them lying close to each other in longitudinal rows; cryptoporous, the subsidiary cells nearly quite covering the guard-cells and the latter lying beneeth the surface of the surrounding epidermal cells.

Epidermal cells proper. H. 12 μ, Lev. B. 16 μ, Lev. L. 40—80 μ; tetra- to hexagonal, in the vicinity of the stomata hexagonal. Walls, outer ones strongly thickened, with a very thick cuticle and 1, sometimes 2 curved layers of minute crystals; lateral ones also strongly thickened and pitted; lateral and inner walls showing only a feeble cellulose reaction. Cell contents: small starch grains; above the midrib in material gathered in the Botanic Garden and treated with chloral hydrate in a few cells some yellow globules; in the drug soaked in water a yellow to brown mass, coloured somewhat darker than the chloroplasts of the mesophyll. Guard-cells of stomata. Walls thickened and lignified with the exception of the following thin and not lignified parts connected with one another: 1. a strip parallel to the surface of the leaf and adjoining the porus; 2. a strip parallel to the surface of the leaf running along the base of the outer lateral wall; 3. a curved strip at each end of the guard-cell connecting together the strips mentioned sub 1 and 2.

Epidermis under side. Resembling that of the upper side. Stomata present only at the base at the margins of the leaves; see for the rest those of upper side.

Epidermall cells proper. See those of upper side, but H. and Lev. B. larger, Lev. L. smaller.

Mesophyll.

Sclerenchymatic prosenchyma. 1, sometimes 2 layers of fibres, adjoining the epidermis of the under side; wanting between the glands and the epidermis of the under side and under the strips of stomata.

Fibres of the above. H. 8 μ, Lev. B. 11 μ, Lev. L. 400—800 μ; sometimes tri- to hexagonal, mostly tetragonal. Walls very much thickened, cavity sometimes in the shape of a fissure; middle lamella thin and lignified; secondary layers strongly refringent, folded, showing stratification and a pure cellulose reaction (gelatinous layer).

Chlorenchyma. Forming the bulk of the mesophyll; consisting of 2 layers of palisade chlorenchyma cells adjoining the prosenchyma and for the rest of chlorenchyma cells having a more irregular shape. On the inner side of the gland a tissue of some large thin-walled parenchyma cells without or nearly without chlorophyll. Intercellular spaces increasing towards the interior. Crystal idioblasts mixed with these large thin-walled cells; containing small prismatic or rhombic crystals.

Chlorenchyma cells. Measures of palisade cells H. 40 μ, Lev. 20 μ. Walls of both kinds of chlorenchyma cells containing either in their middle or in their inner part small prismatic crystals, already visible without the use of polarized light. Cell contents: chloroplasts containing starch grains, mostly of a considerable size.

Parenchyma with cross bar-like thickenings of the walls. In the upper part of the leaf forming several cell layers surrounding the meristele, towards the base forming 2 lateral bands not adjacent to the meristele and spreading towards the margin of the leaf.

Cells of the above. R. and T. 10—20 μ, L. 25—40 μ; often prisms or cylinders, often of a more irregular shape. Walls showing inner bar-like thickenings of the most various shapes, sometimes connected with the thickening part of the bordered pits; walls and bars lignified, showing also a yellowish green cellulose reaction.

Schizogenous glands. One in the middle part of the base of each leaf; adjacent to the under epidermis or separated from it by a single cell layer; R. 220 μ, T. 160 μ, L. 1300—2500 μ, varying in length according to the length of the leaf; ellipsoidal. Epithelium thick mostly 2 layers of cells, sometimes 3 or 4; inner cells somewhat smaller; cells of innermost layer sometimes protruding into the cavity or obliterated. Cavity in the drug sometimes quite filled with irregular fragments, probably proceeding from the cracking of the contents during the process of drying; this mass showing in some cases an internal cavity.

Cells of epithelium. Thick 10 μ, wide 25—50 μ, long 40—100 μ; disks, paving the cavity. Walls of the outermost cell layer colouring red in phloroglucin and hydrochloric acid in the part of the epithelium adjoining the epidermis. Cell contents: chloroplasts containing starch; in material treated with chloral hydrate also large globules or ellipsoidal bodies, 35—70 μ in diameter.

Endodermis. Often developed as a starch-sheath and consisting of 1 cell layer.

Cells of the above. R. 7 μ, T. 9 μ; cylindrical. Cell contents: starch, quite filling up the cells.

Meristele. Single and unbranched, in many respects resembling the stele of the stem.

Vascular bundles. Simple; collateral; open.

Phloem. Exarch; at the base of the leaf elements of the inner half radially arranged.

Xylem. Endarch; principally consisting of spiral vessels with lignified walls; in a cross section nearly always radially arranged.

Micrography of the powder. Chlorenchyma. Epidermal cells oblong, with thickened and pitted lateral walls. Stomata on both sides of the leaf; guard-cells with partly lignified walls, thereby in chloral hydrate the guard-cells seeming separated at there ends, the cuneiform space between them containing a lignified rod with a bifurcating end. Adjacent to the meristele of the leaf common parenchyma cells (in chloral hydrate) with very conspicuous, rod-like, sometimes branched thickenings of the walls. Thick-walled fibres, mostly remaining connected to the epidermis. Very thin spiral vessels. Thin tracheid fibres with small bordered pits. Much starch in the parenchyma, almost entirely filling up the chloroplasts.

Odour, especially after rubbing, strongly terebinthinate; taste bitter and pungent.

June 1901.   J; M; v. E. d. W.

LICHEN ISLANDICUS.
Cetraria. Iceland Moss.

Cetraria islandica, Ach. Lich. univ. 512, a Lichen, occurring in several varieties.

Macroscopic characters.

Thallus long up to 15 c.M., broad up to 1.5 c.M. and thick up to 0.5 m.M., fruticose, ribbon-like, crookedly wrinkled, branches irregularly stiff; the lower part rolled into a tube, channelled or tubular, the lobes flattened; the margins quite fringed with short unflexible more or less darkly coloured cilia. The upper surface olive green - chestnutbrown; the under surface paler, grey or whitish, with white somewhat depressed spots; showing when damp numerous less transparent, irregularly roundish spots. Apothecia to be found now and then in the drug towards the upper end of the plant, on the upper side of the apices of broader rounded lobes sometimes several together; up to 1.5 c.M. in diameter, but also much smaller, disk-shaped, sessile, scarcely elevated above the thallus, dark chestnutbrown. Often several individual plants more or less grown together or entangled by the cilia; in a dry state cartilaginous and brittle; when damp soft and somewhat coriaceous.

Taste bitter.

Cetraria should be freed from pine leaves, mosses and other lichens, frequently mixed with it.

Anatomical characters.

LITERATURE. Berg. Anat. Atl. 1863. 3. Taf. 2. Cohn. Krypt. Flora v. Schlesien. Bd. 2. Hälfte 2. 1879. 62. Deutsch. Arzneib. 5. Ausg. 1910. 300. Flückiger. Pharmakogn. 1891. 307. Flückiger & Hanbury. Pharmacographia. 1879. 738. Gilg. Pharmakogn. 1910 .11. Hérail. Mat. Méd. 1912. 88. Karsten u. Oltmanns. Pharmakogn. 1909. 5. Koch u. Gilg. Pharmakogn. Praktik. 1907. 22. Luerssen. Syst. Bot. Bd. 1. 1879. 176, 179, 223. Marmé. Pharmacogn. 1886. 12. Moeller. Pharmakogn. 1889. 29. Oudemans. Aanteek. o.d. Pharmac. neerl. 1854—56. 7. Oudemans. Pharmacogn. 1880. 8. Tschirch. Pharmakogn. Bd. 2. 1911. 266. Wiesner. Rohstoffe. Bd. 1. 1900. 670. Wigand. Pharmakogn. 1879. 29. MATERIAL. The drug. REAGENTS. Water, glycerine, potassium iodide iodine, iodine and sulphuric acid 66 per cent.

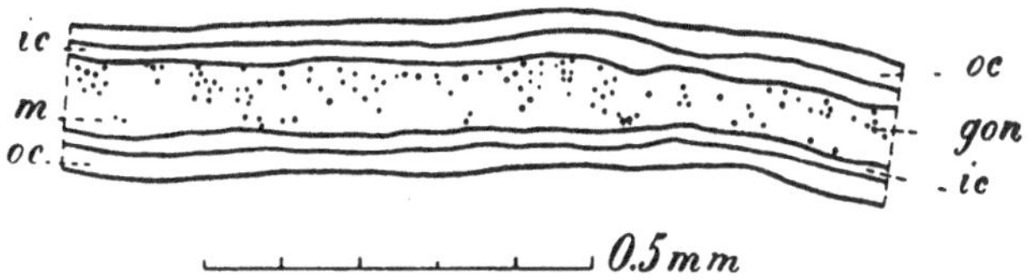

Fig. 72. *Cetraria*. Thallus, transverse section. gon Gonidia; ic Inner cortical layer; m Medullary layer; oc Outer cortical layer.

MICROGRAPHY.

T h a l l u s heteromerous.

Cortical layer. Thick 35 μ; covering the thallus on all sides; thicker at the margins of the lobes than elsewhere, thicker on the convex than on the concave side in undulated parts of the thallus; wanting on the under side at the places corresponding with the depressed white spots, mentioned under macroscopic characters, divided in 2 separate layers.

Outer layer. Thick 20 μ, consisting of 3—4 cell layers of a pseudo-parenchyma, coloured brown in its outer parts, especially on the upper side of the thallus.

C e l l s o f t h e a b o v e. Isodiametric; thin-walled; the walls being sometimes coloured blue in potassium iodide iodine, according to the strength of the iodine solution. Cell contents: some granular matter, covering the walls.

Inner layer. Thick 15 μ; consisting of very closely woven thin hyphae, showing in dry material many intercellular spaces and sometimes coloured blue in potassium iodide iodine, according to the strength of the iodine solution.

Medullary layer. Thick 50 μ, much thicker on the naked places corresponding with the depressed white spots; distinctly separated from the cortical layer; consisting of very loosely woven hyphae principally running in a longitudinal direction; showing numerous large mostly longitudinal intercellular spaces.

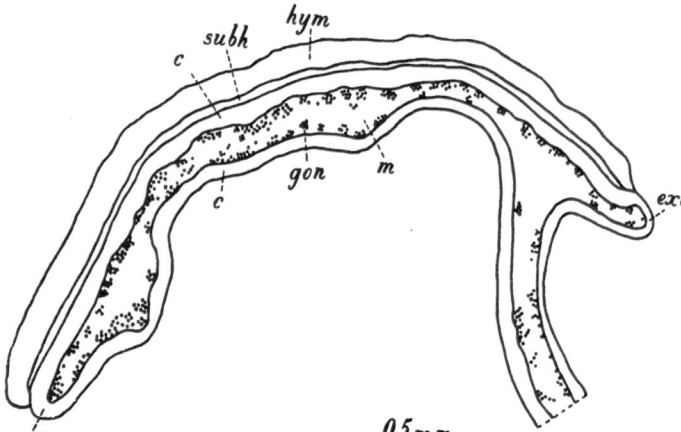

Fig. 73. *Cetraria*. Apothecium, longitudinal section. c Cortical layer; exc Excipulum; gon Gonidia; hym Hymenium; m Medullary layer; subh Subhymenial layer.

Hyphae of the above. Thick 3—4 μ, no transverse septa to be distinguished. Walls very thick, almost quite closing up the lumina; sometimes showing a feeble blue colour in potassium iodide iodine.

G o n i d i a. Cystococcus humicola Näg., scattered through the medullary layer in very variable numbers, often more numerous on the upper side of the thallus, and here and there arranged in a definite gonidial layer; unicellular, having a diameter of 6—8 μ, spherical. Walls thin, containing cellulose. Contents: granular, green, becoming dark brown in potassium iodide iodine,

A p o t h e c i a. Covering an almost normally developed cortical layer of the underlying thallus.

Excipulum. Showing the same structure as the thallus, containing many gonidia.

Subhymenial layer. Thick 25 μ; consisting of hyphae, interwoven in all directions but somewhat less loosely than those of the medullary layer; showing numerous intercellular spaces.

Hyphae of the above. Somewhat thinner than those of the medullary layer. Walls becoming more or less blue in potassium iodide iodine, according to the strength of the iodine solution.

Hymenium. Consisting of asci and paraphyses.

Asci. Long 35 μ, somewhat club-shaped, thick in their upper part 5—8 μ. Walls thin; the upper part — long 5 to 10 μ — coloured brown; in potassium iodide iodine the walls colourless, brown or blue according to the strength of the iodine solution.

Spores. 8 in number, arranged in a single row, sometimes 2 beside each other, ellipsoidal to spherical; the walls becoming dark blue to black in potassium iodide iodine.

Paraphyses. Thinner than the asci and surpassing them somewhat in length, cylindrical, coloured brown in their upper part. Walls becoming yellow to brown in potassium iodide iodine, blue in a feeble solution.

C i l i a. Long e. g. 300 μ; consisting only of a cortical layer, without medullary layer or gonidia; each containing a spermogonium in the top.

Spermogonium. Consisting of a layer — thick 20 μ — of radially arranged very thin sterigmata, surrounding on all sides a cavity of 40—45 μ in length and 20 μ in breadth; this cavity corresponding at the top by a narrow tubular aperture with the outer air and containing many small spermatia.

September 1911.   J; M.

LIGNUM GUAIACI.
Guaiacum Wood.

The heart-wood of Guaiacum officinale, Linn. Sp. Pl. 381, or of Guaiacum sanctum,
Linn. Sp. Pl. 382.

LITERATURE. de Bary. Vergl. Anat. 1877. 486, 524. Beauverie. Le Bois. II. 1905. 951. Berg. Anat. Atl. 1865. Tafel XXVII. 53—54. Benecke. Mikr. Drogenprakt. 1912. 33. Boulger. Wood. 1908. 202, 311. Erdmann-König. Allg. Warenk. 1895. 387. Flückiger. Pharmakogn. 1891. 485. Flückiger & Hanbury. Pharmacographia. 1879. 100. Gilg. Pharmakogn. 1910. 180. Greenish & Collin. Anat. Atl. veget. powders. 1904. 172. Gurnik. Kernholzbildung. Diss. Bern. 1915. 9,42, 50 a. 53. Hager. Pharm. Praxis. Bd. I. 1900. 1260. R. Hartig. Unterscheidungsmerkmale der wichtigeren in Deutschland wachsenden Hölzer. 1898. 36. Th. Hartig. Beitr. z. vergl. Anat. d. Holzpflanzen. Bot. Zgt. Jhrg. 17. 1859. 100 a. 109. Hérail. Mat. Méd. 1912. 368. v. Höhnel. Ueber stockwerkartig aufgebaute Holzkörper. Ber. Wiener Akad. Bd. 89. Abth. 1. 1884. 40. Karsten u. Oltmanns. Pharmakogn. 1909. 112. Koch. Mikr. Anal. d. Drogenpulver. Bd. I. 1901. 144. Taf. XII. Koch. Pharmakogn. Atlas. Bd. I. 1911. 65. Taf. XII. Koch u. Gilg. Pharmakogn. Praktik. 1907. 57. Kraemer. Botany a. Pharmacogn. 1910. 668. Luerssen. Syst. Bot. Bd. II. 1882. 679. Marmé. Pharmacogn. 1886. 106. Meyer. Wiss. Drogenk. Bd. 2. 1892. 167. Meyer. Mikr. Unters. v. Pflanzenpulv. 1901. 154. Moeller. Vergl. Anat. d. Holzes. Denkschr. Wiener Akad. Bd. 36. 1876. 297. Moeller. Pharmakogn. 1889. 257. Moeller. Mikr. pharm. Ueb. 1900. 227. Nördlinger. Querschnitte. Bd. VI. 1874. 25. Oudemans. Aanteek. o. d. Pharmac. neerl. 1854—56. 536. Oudemans. Pharmacogn. 1880. 132. Pätzoldt. Harz u. Holz von Guaiacum. Diss. Straszburg. 1902. Planchon et Collin. Drogues simples. T. II. 1896. 629. Präel. Vergl. Unters. über Schutz- und Kern-Holz der Laubbäume. Pringsheim's Jahrbücher. Bd. 19. 1888. 32. Record. Economic Woods of the United States. 1912. 95 (G. Sanctum). Sargent. The Woods of the United States. 1885. 7. (G. Sanctum). Solereder. Holzstructur. Diss. München. 1885. 90. Solereder. Syst. Anat. 1899. 191. Stone. Timbers of Commerce. 1904. 18. Tjaden. Hout. 1919. 33. Tschirch. Pharmakogn. Bd. 2. Abt. 2. 1917. 1535. Wiesner. Rohstoffe. Bd. II. 1918. 628. MATERIAL. A disc belonging to a stem or branch, thick 13 c.M., from commercial source and thus leaving undetermined the origin of this specimen, either from G. officinale or from G. sanctum. REAGENTS. Water, alcohol 96 per cent., glycerine, potassium iodide iodine, phloroglucin and hydrochloric acid, iodine and sulphuric acid 66 per cent., concentrated sulphuric acid, Schulze's macerating mixture, crystalized phenol and oil of cloves ([1]), ammonia, hydrochloric acid.

MICROGRAPHY.

G r o w t h   r i n g s. Thick 0.5—2.5 m.M.; often clearly to be distinguished by means of a zone, containing rather larger and more numerous vessels, in the inner part of the rings. The transverse dimensions of the libriform fibres and of the wood parenchyma fibres often somewhat larger in the inner part of the rings than in the outer; the walls of the libriform fibres in the outer part in a few cases somewhat thicker. The boundary of a ring sometimes showing in several places a very varying distinctness. **S t o r i e d   a r r a n g e m e n t** distinct. The articulations of the vessels, the libriform fibres, wood parenchyma fibres and the medullary rays all arranged in stories, high 70—110 μ; in a very few cases 2 articulations of a vessel together corresponding in length with a single story. **L o n g i t u d i n a l   a r r a n g e m e n t**. The vessels, libriform fibres and wood parenchyma fibres in a longitudinal direction bending hither and thither and forming short waves, especially in a tangential plane and showing inclinations of at most 45° to the longitudinal direction. This phenomenon very variable in different concentric zones. **T r a c h e a l   s y s t e m** consisting of vessels. Vessels usually more numerous in the inner part of the growth rings than in the outer, for the rest rather regularly distributed; 10—20, often about 18 to the square m.M. of the transverse section; solitary and in groups. The groups very scarce and nearly always consisting of 2 vessels arranged in a radial row. Vessels nearly always joining upon medullary rays either on one side or on both sides, both cases being of equal frequency; for the rest of their surface joining with about equal parts upon libriform fibres and wood parenchyma fibres. **L i b r i f o r m   s y s t e m** consisting of non-septate libriform fibres. Non-septate libriform fibres forming the bulk of the wood; usually not arranged in radial rows. **P a r e n c h y m a t i c   s y s t e m** consisting of wood parenchyma and medullary rays.

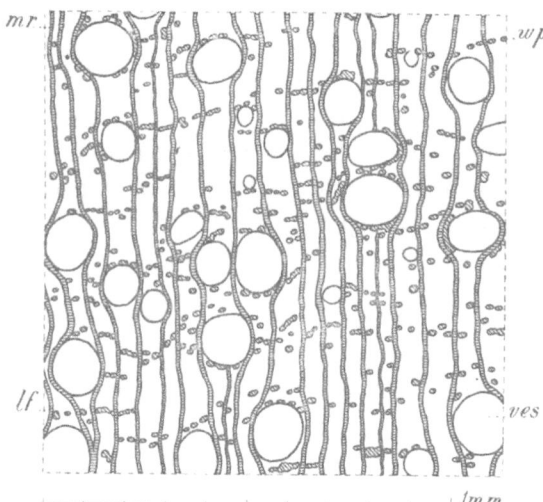

Fig. 74. *Guaiacum officinale*. Wood, transverse section. lf Libriform fibres; mr Medullary rays; ves Vessels.; wp Wood parenchyma.

---

[1]) For discovering siliceous bodies according to K ü s t e r, Die anat. Charaktere d. Chrysobalaneen, ins besondere ihre Kieselablagerungen, Bot. Centrlbl. Bd. 69, 1897, 50.

Wood parenchyma rather scarce; paratracheal, metatracheal and scattered among the libriform fibres. The paratracheal parenchyma scarce, surrounding in a single cell layer about one half of the surface of the vessel, not joining upon medullary rays. The metatracheal parenchymatous layers rather scarce, very slightly extending in a tangential direction, nearly always 1 cell thick. Wood parenchyma scattered among the libriform fibres usually more abundant than both other kinds of parenchyma, passing by numerous intermediate stages into the short metatracheal layers, very often occurring along a medullary ray. All or nearly all wood parenchyma consisting of wood parenchyma fibres; substitute fibres perhaps present in a very small number, scattered among the wood parenchyma fibres and resembling them in most respects. The wood parenchyma fibres usually with 1 transverse wall, sometimes fibres with 2 or 3. Rather often conjugated. **Medullary rays** 2—5, mostly 3 layers of libriform ·fibres distant from each other; usually uni- sometimes biseriate, the latter usually only in their middle part; 3—6, very often 4 or 5 cells high; in a very few cases in a radial direction interrupted by a vessel. All cells procumbent; the cells of the upper- and undermost radial row sometimes somewhat shorter in a radial and somewhat longer in a longitudinal direction; the cells adjoining wood parenchyma cells often in that part somewhat wider in a tangential direction.

V e s s e l s, the solitary ones R. 20—110 μ, T. 15—100 μ; those of the groups e. g. R. 8—110 μ, T. 25—85 μ; L. of articulations 70—110 μ. Elliptical or circular cylinders, flattened when joining one another. Transverse walls placed about horizontally; with a single circular or elliptical perforation; the remaining rings thin, narrow and with margins smooth or resembling those of bordered pits. Walls thick 5 or 6 μ between 2 vessels joining one another, for the rest 3—5 μ; yellow, brownish yellow or brownish greenish yellow; lignified and in potassium iodide iodine colouring greenish blue; — with very numerous bordered pits between 2 vessels; the borders small; these pits only observed by me in transverse sections; — with numerous uni- or bilateral bordered pits between vessels and libriform fibres; — with very numerous slit-like unilateral bordered pits between vessels and wood parenchyma cells or cells of medullary rays; the borders elliptical and placed transversely or pentagonal and hexagonal with rounded edges, 1½ by 2 μ; the canals slit-like, transversely directed or slanting and the inner aperture smaller than the diameter of the border. Contents: a yellow or brownish or greenish mass, usually partly or entirely filling the vessels, soluble in ammonia, not colouring blue in potassium iodide iodine. N o n - s e p t a t e l i b r i f o r m f i b r e s. R. 8—12 μ, T. 7—12 μ, L. 375—500 μ; pentagonal to dekagonal, the thin ends usually not sharply separated from the thicker midlle part. Walls thick 2—4, mostly 3 or 4 μ; yellow, brownish yellow or brownish greenish yellow; lignified and in potassium iodide iodine colouring greenish blue; — with numerous uni- or bilateral bordered pits between libriform fibres and vessels; — with numerous slit-like pit canals between libriform fibres reciprocally or between libriform fibres and wood parenchyma cells or cells of medullary rays; these pit canals about equally numerous in the ends as in the middle parts and somewhat more numerous in the radial sides than in the tangential ones; the inner aperture slanting; in a few cases these pit canals becoming wider like bordered pits in the neighbourhood of the thin middle lamella. Intercellular spaces wanting. Contents: sometimes a yellow to brown mass, colouring greenish blue in potassium iodide iodine. W o o d p a r e n c h y m a c e l l s. R. 7—20 μ,

T. 8—15 μ, L. 25—45 μ; the cells bordering on vessels often somewhat tabular in shape, e. g. deep 6 μ, wide 12 μ. All cells tetragonal to octogonal prisms with a longitudinally directed axis. The wood parenchyma fibres composed in the following manner: L. 45 + 45 μ, 45 + 50 μ, 25 + 25 + 30 μ; their ends in radial sections neither acute nor quite obtuse, on tangential sections fairly acute. Walls thick 1—2 mostly 1 μ; in cells bordering upon vessels the longitudinal walls, placed perpendicularly to the surface of these vessels, often much thicker; sometimes showing small projecting parts on their inner side; yellow, brownish yellow or brownish greenish yellow; lignified and in potassium iodide iodine colouring greenish blue; — with unilateral bordered pits between wood parenchyma cells and vessels; see the description of the vessels; — with pit canals between these cells and libriform fibres; see the description of the libriform fibres; — with simple pits between these cells reciprocally and between these cells and cells of medullary rays; pits rather numerous in the transverse and radial walls, in the latter usually not in groups, in the tangential walls scarce. Intercellular spaces wanting. Contents: often a yellow or brownish or greenish brown mass; this mass often divided in crystal-shaped parts, usually showing no double refraction, not soluble in hydrochloric acid, not completely soluble in alcohol 96 per cent., colouring greenish blue in potassium iodide iodine. P r o c u m b e n t c e l l s o f m e d u l l a r y r a y s. R. 25—75 μ, T. 5—10 μ, L. 8—15 μ; tetragonal to octogonal prisms with a radially directed axis and rounded radial edges. Walls thick 1—1$\frac{1}{2}$ μ, the tangential ones not much thicker; yellow, brownish yellow or brownish greenish yellow; lignified and in potassium iodide iodine colouring greenish blue; — see for the pits cell description of the wood parenchyma cells; the simple pits here rather numerous in tangential walls. Intercellular spaces only in a radial direction between these cells reciprocally and between these cells and libriform fibres or cells of wood parenchyma. Cell contents: see those of cells of wood parenchyma.

January 1921.   J; M.

LIGNUM QUASSIAE.
Quassia Wood.
The wood of the trunk and branches of Picraena excelsa, Lindl. Fl. Med. 208.

Macroscopic characters.
Occurring in logs long 15 c.M., obtained from stems thick up to 40 c.M.; showing a bast, thick 1 c.M., cohering to the wood, externally grey yellowish brown and exhibiting rhytidoma. Wood whitish or slightly yellowish, of an uniform structure, light, tough, splitting easily, often marked with bluish black spots. Layers like growth rings, but mostly not closed ones, appearing on a well smoothed transverse section of the wood; moreover fine medullary rays and many dots corresponding with groups of vessels. The wood generally used in the shape of chips. Odour wanting; taste very bitter.

Anatomical characters.
LITERATURE. Berg. Anat. Atl. 1865. Tafel XXVI, 51—52. Benecke. Mikr. Drogenprakt. 1912. 33. Flückiger. Pharmakogn. 1891. 497. Flückiger & Hanbury. Pharmacographia. 1879. 131. Gilg. Pharmakogn. 1910. 190, 192. Greenish & Collin. Anat. Atl. veget. powders. 1904. 174. Hager. Pharm. Praxis. Bd. 2. 1902. 709. Hérail. Mat. Méd. 1912. 545, 546. v. Höhnel. Ueber stockwerkartig aufgebaute Holzkörper. Ber. Wiener Akad. Bd. 89. Abth. 1. 1884. 41. Janssonius. Mikrographie des Holzes. Bd. II. 1908. 72 Simarubeae). Karsten u. Oltmanns. Pharmakogn. 1909. 108, 110. Koch. Mikr. Anal.

d. Drogenpulver. Bd. I. 1901. 151, Taf. XIII. Koch. Pharmakogn. Atlas. Bd. I. 1911. 71, Taf. XIII. Koch u. Gilg. Pharmakogn. Praktik. 1907. 59, 64. Kraemer. Botany a. Pharmacogn. 1910. 545. Luerssen. Syst. Bot. Bd. II. 1882. 695. Marmé. Pharmacogn. 1886. 109. Meyer. Wiss. Drogenk. Bd. 2. 1892. 162, 165. Meyer. Mikr. Unters. v. Pflanzenpulv. 1901. 154. Moeller. Pharmakogn. 1889. 259. Moeller. Mikr. pharm. Ueb. 1900. 228. Oudemans. Aanteek. o. d. Pharmac. neerl. 1854—56. 526. Oudemans. Pharmacogn. 1880. 129. Planchon et Collin. Drogues simples. T. II. 1896. 635, 637. Solereder. Syst. Anat. 1899. 210. Stone. The Timbers of Commerce. 1904. 29. Stone. Les Bois utiles de la Guyane Française. Ann. du Musée Colonial de Marseille. Année 25. 3e serie, 4e Vol. 1916. 127. Stone & Freeman. The Timbers of British Guiana. 1914. 81 (P. officinalis). Wiesner. Rohstoffe. Bd. II. 1918. 633. Wigand. Pharmakogn. 1879. 134. MATERIAL. A specimen of the wood thick 8 c.M., wide 11 c.M., long 18 c.M. REAGENTS. Water, alcohol 96 per cent., glycerine, phloroglucin and hydrochloric acid, iodine and sulphuric acid 66 per cent., concentrated sulphuric acid, Schulze's macerating mixture.

MICROGRAPHY.

G r o w t h   r i n g s. Probably always wanting; perhaps here and there indicated by the following phenomena: the metatracheal parenchymatous layers of the kind much extended in a tangential direction (see below) arranged at distances in a radial direction corresponding with the thickness of many growth rings, viz 3—8 m.M. These metatracheal layers mostly single, but now and then accompanied by a more irregular metatracheal layer of the same kind at a radial distance of 300—700 μ; the latter layer occasionally fusing with the principal ones. In the neighbourhood of the inner boundary of the parenchymatous layers the transverse diameter of the libriform fibres becoming outwards somewhat larger and the fibre walls in a very few cases somewhat thicker. S t o r i e d   a r r a n g e m e n t distinct. The articulations of the vessels, the libriform fibres, wood parenchyma fibres, crystal fibres and the medullary rays all arranged in stories, high 250—350 often 300 μ. In the libriform fibres the pits in those parts of the radial walls, corresponding with the boundary of 2 stories still more numerous than elsewhere. T r a c h e a l   s y s t e m consisting, of vessels. Vessels sometimes varying in number in the different growth rings; often more or less distinctly arranged in tangential layers, for the rest regularly distributed; about 7 to the square m.M. of the transverse section; solitary and in groups. The groups usually consisting of 2—5, very often of 2 or 3 vessels in a radial row; sometimes of 2 or 3 vessels in a tangential row and the latter groups often breaking off a medullary ray in a radial direction; sometimes showing a more irregular arrangement of the vessels. Isolated vessels and groups of vessels rather often forming larger complexes; the tangential dimensions of these complexes often as large or larger than the radial dimensions. Vessels very often joining upon medullary rays and much more often on one than on both radial sides; for the rest of their surface joining upon wood parenchyma and libriform fibres and upon the latter usually only for a very small part. L i b r i f o r m   s y s t e m consisting of non-septate libriform fibres. Non-septate libriform fibres forming the bulk of the wood, at best rather regularly arranged in radial rows. The thin ends of the fibres usually running along the radial sides of the thicker middle parts of the fibres belonging to the stories immediately above and below and thus in transverse sections usually a more or less regular radial row of thicker fibres alternating with a more or less distinct radial strip of thinner ones. The ends of fibres reaching the upper- or undermost cell row of medullary rays sometimes bending rectangularly. **Parenchymatic system** consisting of wood parenchyma and medullary rays. **Wood parenchyma** rather abundant; paratracheal, metatracheal and scattered among the libriform fibres. Paratracheal wood parenchyma forming 1 to some cell layers around those parts of the vessels not joining upon medullary rays and metatracheal wood parenchyma; sometimes rather abundant on the radial sides of the vessels and thus passing into metatracheal wood parenchyma. Metatracheal wood parenchyma rather abundant; distinctly in 2 kinds: 1. layers much extended in a tangential direction, already mentioned under growth rings, showing only a few blind ends and interruptions in a tangential direction, in radial direction 3—8 m.M. in distance from each other, up to 15 cell layers thick; 2. layers with a usually insignificant tangential extension, only in a very few cases bifurcating in a tangential direction, less far apart from one another in a radial direction than those of the first kind, usually about 5 cells thick. In both kinds of layers the elements in transverse sections in radial rows, corresponding with those of the libriform fibres. Vessels occurring in both kinds of metatracheal layers. Wood parenchyma scattered among the libriform fibres very scarce. All wood parenchyma composed of wood parenchyma fibres; these fibres, except the crystal fibres, nearly always with 3 transverse walls and the cells in adjoining fibres often fairly well forming transverse layers. Crystal fibres rather regularly distributed among the other wood parenchyma fibres, but somewhat more numerous among the fibres joining upon the libriform fibres; sometimes arranged in longitudinal rows of various length; usually partly or entirely divided in short cells; each cell usually containing a single simple crystal. Sometimes cells of the same length as the other wood parenchyma

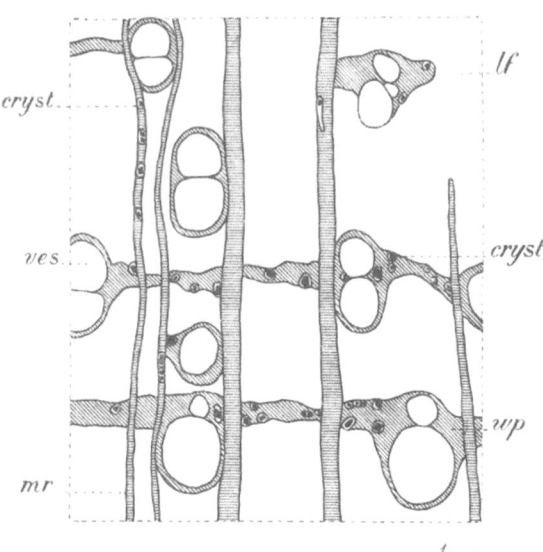

Fig. 75. *Picraena excelsa*. Wood, transverse section. cryst Cells with simple crystals; lf Libriform fibres; mr Medullary rays; ves Vessels; wp Wood parenchyma.

cells, each containing 1—8 simple crystals; these crystals usually arranged in a longitudinal row. Transverse dimensions of cells in the neighbourhood of vessels often larger than elsewhere. **Medullary rays** in tangential direction distant from one another 2—7, often about 4 radial rows of libriform fibres; uni- to tri-, usually biseriate; 5—13, usually 10—13 cells high; the uniseriate ones usually lower than the other and the triseriate ones usually only so in their middle part. All cells procumbent; cells of the upper- and undermost radial row often somewhat shorter in a radial and sometimes somewhat longer in a longitudinal direction; cells corresponding in place with the metatracheal parenchymatous layers often shorter in a radial and wider in a tangential direction. 1—4 simple crystals occurring in a much varying number of cells; these cells sometimes divided by tangential partition walls, each partition containing a single simple crystal. Incipient medullary rays sometimes observed in radial sections; the beginning only 2 cells high. These cells often arbitrary in shape, often rounded at their upper and lower ends and often each containing 1—5 simple crystals; adjacent walls of these cells often thicker than elsewhere.

V e s s e l s. The solitary ones R. 75—140 μ, T. 60—110 μ; those of the groups R. 35—130 μ, T. 50—120 μ; L. of articulations 250—325 μ. Elliptical or circular cylinders, flattened when joining one another. Transverse walls placed about horizontally, a small upper and under part often very obliquely; with a single circular or elliptical perforation; the remaining rings narrow and with a smooth margin. Walls thick 3 or 4 μ between 2 vessels joining one another, for the rest 2 or 2.5 μ; sometimes somewhat yellow; lignified; — with very numerous combined slit-like bordered pits between 2 vessels; the borders circular, elliptical or hexagonal with rounded angles, about 2 μ in diameter; the canals slit-like, directed transversally and the inner aperture much wider than the diameter of the border; — with very numerous combined slit-like unilateral bordered pits between vessels and wood parenchyma cells or cells of medullary rays; these pits for the rest almost similar to the bilateral ones. Contents: sometimes here and there a dirty yellow or yellow-brown granular mass, not soluble in alcohol 96 per cent. N o n-s e p t a t e l i b r i f o r m f i b r e s. R. and T. 15—18 μ, L. 750—900 μ; pentagonal to dekagonal, the mostly thin ends rather often more or less sharply separated from the thicker middle part, long about 350 μ. Walls thick $1^{1}/_{2}$ μ; sometimes somewhat yellow; lignified, in iodine and sulphuric acid 66 per cent. the inner secondary layers colouring blue; — with numerous slit-like simple pits or bordered pits with very small borders between 2 libriform fibres; these pits nearly exclusively in the radial sides of the fibres, rather often not all placed in a longitudinal row and in a longitudinal direction often 2 to 6 μ in distance from each other; the inner apertures of the canals in both kinds of pits very wide and directed rather longitudinally; — with numerous slit-like simple pits or unilateral bordered pits between libriform fibres and wood parecnhyma cells or cells of medullary rays; these pits somewhat more numerous than those between 2 libriform fibres joining one another, but for the rest quite similar to them. Intercellular spaces and contents wanting. W o o d p a r e n c h y m a c e l l s. R. 10—30 μ, T. 15—25 μ, L. 45—165 μ; those in the neighbourhood of the vessels e. g. R. 25—40 μ, T. 20—35 μ; those bordering on vessels often tabular in shape, e. g. deep 6—15 μ, wide 22—30 μ; the crystal cells L. 15—100 often 20—40 μ and the transverse diameters often somewhat smaller than those of the other parenchyma cells. Nearly all cells tetragonal to octogonal prisms with a longitudinally directed axis; the ends of the wood parenchyma fibres on tangential sections roof-shaped, on radial obtuse or very slightly tapering. The wood parenchyma fibres composed in the following manner: L. 60 + 50 + 70 + 85 µ, 80 + 75 + 65 + 75 µ, 70 + 80 + 75 + 80 µ, 90 + 75 + 45 + 80 µ, 85 + 80 + 130 µ. Walls thick 1 µ, the radial ones often distinctly thicker than the tangential, the longitudinal walls of the cells bordering on vessels and placed perpendicularly to the surface of these vessels often much thicker; sometimes somewhat yellow; lignified, in iodine and sulphuric acid 66 per cent. coloured somewhat blue; — with unilateral bordered pits between wood parenchyma cells and vessels; see the description of the vessels; — with numerous pit canals or unilateral bordered pits between wood parenchyma cells and libriform fibres; see the description of the libriform fibres; — with simple pits between these cells joining one another and between these cells and cells of medullary rays; the pits very small, circular or elliptical, numerous in the transverse and radial walls and also in the roof-shaped end walls of the parenchyma fibres, in the radial walls usually in groups, in the tangential scarce. Intercellular spaces wanting. Cell contents: sometimes a yellow mass. In the crystal cells each of the simple crystals surrounded by a rather thick lignified crystal skin; this skin cohering with the cell wall; the crystals in a few cases with a central cavity. Procumbent cells of medullary rays. R. 60—150 µ, T. 8—16 µ, L. 10—30 µ; tetragonal to octogonal prisms with a radially directed axis and rounded radial edges. Walls thick 1 µ, the tangential ones somewhat thicker; sometimes somewhat yellow; lignified, in iodine and sulphuric acid 66 per cent. colouring somewhat blue; — see for the pits the cell description of the wood parenchyma cells; the simple pits here numerous in the tangential walls, rather numerous in the transverse and radial ones and here often arranged in radial rows and usually directed towards the intercellular spaces. Intercellular spaces only present in a radial direction between these cells joining one another and between these cells and libriform fibres or cells of wood parenchyma. Cell contents: sometimes a yellow mass. In the crystal cells each of the simple crystals surrounded by a rather thick lignified crystal skin; this skin cohering with the cell wall.

Mikrography of the powder. Libriform fibres often united into bundles; each fibre thick on an average 16—20 µ; their walls not very thick, colourless and exhibiting slit-like simple pits. Cells of medullary rays often crossing in rows the bundles of fibres; their tangential walls bead-shaped by the consequence of numerous pit canals. Fragments of vessels with short articulations and bordered pits very small, much crowded, sometimes somewhat slit-like in a transverse direction. Starch in a very slight quantity. Sometimes a fungus mycelium branched, blackish brown or purplish.

January 1921.   J; M.

LIGNUM SANTALINUM.
Rasura Ligni Santalini rubri. Red Sanders Wood. Red Sandal-Wood.
The heart-wood of Pterocarpus santalinus, Linn. f. Suppl. 318.

Macroscopic characters.

A powder red, fine, light, flockly, partly cohering into woolly aggregates.

Anatomical characters.

LITERATURE. De Bary. Vergl. Anat. 1877. 148, 502 a. 523. Beauverie. Le Bois. II. 1905. 977. Engler u. Prantl. III. 3. 1894. 78. Erdmann-König. Allg. Warenk. 1895. 518. Flückiger. Pharmakogn. 1891. 500. Flückiger & Hanbury. Pharmacographia. 1879. 199. Foxworthy. Indo-Malayan woods. The Philippine Journ. of Science. C. Botany. Vol.

IV. 1909. 467. Gamble. Ind. Timbers. 1902. 259. Greenish & Collin. Anat. Atl. veget. powders. 1904. 178. Hager. Pharm. Praxis. Bd. 2. 1902. 820. v. Höhnel. Ueber stockwerkartig aufgebaute Holzkörper. Ber. Wiener Akad. Bd. 89. Abth. 1. 1884. 34. Jaensch. Anat. einiger Leg. Hölzer. Ber. d.d. bot. Ges. Bd. II. 1884. 279. Janssonius. Mikrographie d. Holzes. Bd. III. 1918. 64. (here a description of the wood of P. indicus and the complete summary of literature concerning the anatomical structure of Pterocarpus species. Jentsch. Der Urwald Kameruns. Beih. z. Tropenflanzer. Bd. XII. 1911. 171. Taf. IV. Karsten u. Oltmanns. Pharmakogn. 1909. 115. Krah. Ueber die Verteilung d. par. Elemente im Xylem etc. Diss. Berlin. 1883.20. Kraemer. Botany a. Pharmacogn. 1910. 547. Luerssen. Syst. Bot. Bd. II. 1882. 885. Marmé. Pharmacogn. 1886. 112. Moeller. Vergl. Anat. d. Holzes. Denkschr. Wiener Akad. Bd. 36. 1876. 409. Moeller. Mikr. d. Nahr. u. Genuszm. 1886. 259. Moeller. Pharmakogn. 1889. 262. Moeller. Mikr. pharm. Ueb. 1900. 223. Perrot et Gérard. Recherches sur les bois de diff. espèces de Légumineuses africaines. 1907. 122. The same as Diss. Paris. 1907 of Gérard. Piccioli. I caratteri anatomici per conoscere i principali legnami adoperati in Italia. 1906. 68. Planchon et Collin. Drogues simples. T. II. 1896. 532. Prael. Vergl. Unters. über Schutz- u. Kern-Holz d. Laubbaüme. Pringsheim's Jahrbücher. Bd. 19. 1888. 20. Saupe. Der anat. Bau d. Holzes der Leguminosen u. sein syst. Werth. Flora. Bd. 70. 1887. 263, 313, 314 u. 315. Tschirch. u. Oesterle. Anat. Atl. 1900. Taf. 27. 113. Vogl. Veget. Nahr. u. Genuszm. 1899. 561. Wiesner. Rohstoffe. Bd. II. 1918. 601. Wittlin. Ueber die Bildung der Kalkoxalat-taschen. Bot. Centrbl. Bd. 67. 1896. 72. MATERIAL. A piece of the heart-wood, 5 c.M. thick, 13 c.M. wide and 16 c.M. long. REAGENTS. Water, alcohol 96 per cent., glycerine, chloral hydrate, phloroglucin and hydrochloric acid, iodine and sulphuric acid 66 per cent., iron acetate, Schulze's macerating mixture.

MICROGRAPHY.

Growth rings never clearly discernible, perhaps indicated here and there by the following phenomena: tangential layers thick 1 or more m.M. containing more numerous vessels and a greater number of the layers of metatracheal wood parenchyma, to be mentioned below. Those layers moreover being more regular and somewhat thicker. In the neighbourhood of the inner boundary of these layers the vessels sometimes somewhat smaller, somewhat scarcer and the metatracheal wood parenchyma often less distinct. Storied arrangement distinct. The articulations of the vessels, the libriform fibres, wood parenchyma fibres, substitute fibres, crystal fibres and the medullary rays all arranged in stories, high 200—300 µ; sometimes 2 articulations of a vessel together corresponding in length with 1 single story. Tracheal system consisting of vessels. Vessels rather regularly distributed, except in the layers mentioned above under growth rings; 2—4 to the square m.M. of the transverse section; nearly always to be found in the metatracheal parenchymatous layers, to be mentioned below; solitary and in groups. These groups less numerous than the isolated vessels and mostly consisting of 2—5, often 4 or 5 vessels, arranged in a radial row; sometimes some additional small vessels along the radial sides of the group. Vessels nearly always joining upon medullary rays on both radial sides; for the rest of their surface joining upon wood parenchyma. Libriform system consisting of non-septate libriform fibres. Non-septate libriform fibres forming the bulk of the wood; arranged in radial rows. The thin ends of the fibres running along the radial sides of the much thicker middle parts of the fibres of the upwards and underwards following stories and thus in transverse sections one radial row of thciker fibres alternating with usually two radial row of thinner ones. The fibres joining upon the parenchymatous layers sometimes with somewhat thinner walls. **Parenchymatic system** consisting of wood parenchyma and medullary rays. **Wood parenchyma** rather abundant; paratracheal, metatracheal and scattered among the libriform fibres. The paratracheal parenchyma scarce, surrounding by 1 to several cell layers that part of the surface of the vessels not joining upon the medullary rays and the metatracheal wood parenchyma. Metatracheal parenchyma abundant. The metatracheal parenchymatous layers rather often with blind ends; interrupted or 2 layers anastomosing; here and there sometimes somewhat sinuous;

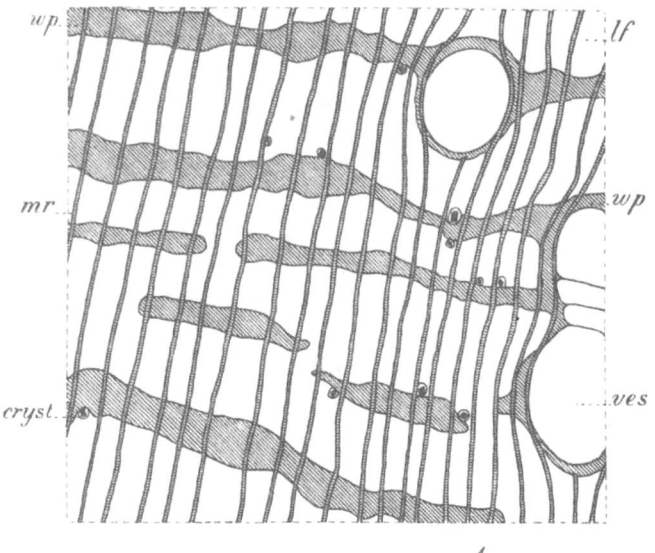

Fig. 76. *Pterocarpus santalinus*. Wood, transverse section. cryst Cells with simple crystals; lf Libriform fibres; mr Medullary rays; ves Vessels; wp Wood parenchyma.

separated from each other in a radial direction by 7—25 layers of libriform fibres; in a radial direction 1—7, often 3 or 4 elements thick. In the layers the elements in a transverse section in radial rows, corresponding with those of the libriform fibres. The wood parenchyma scattered among the libriform fibres very scarce; only sometimes here and there a wood parenchyma fibre joining upon a medullary ray. Nearly all wood parenchyma consisting of wood parenchyma fibres; substitute fibres only present in a very small number, scattered between the wood parenchyma fibres and for the rest resembling those fibres in all respects. These wood parenchyma fibres, except the crystal fibres, usually with 1 transverse wall; sometimes fibres with 2 or 3 transverse walls often joining upon vessels or scattered between the fibres with a single transverse wall and in a fibre with 2 transverse walls often 1 of the end cells of about the same length as the cells in a fibre with a single transverse wall The cells of the wood parenchyma fibres usually fairly well arranged in radial rows but not in transverse layers. The crystal fibres sometimes arranged in longer and shorter longitudinal rows; usually divided in about 8—10 short cells, each cell containing a single simple crystal; in a few cases such a fibre only partly divided into a corresponding number of such cells. The crystal fibres usually only on the inner and outer side of the metatracheal parenchymatous layers and scattered between the libriform fibres. **Medullary rays** 1—4, mostly 2 radial rows of libriform fibres in distance from each other; usually uniseriate, sometimes biseriate, but this only in their middle part; 2—12, very often 7—9 cells in high. All cells procumbent; the cells of the upper- and undermost radial row often somewhat shorter in a radial and somewhat longer in a longitudinal direction; the cells corresponding with the metatracheal parenchymatous layers often shorter in a radial and wider in a tangential direction.

V e s s e l s. The solitary ones R. 250—500 μ, T. 200—400 μ; those of the groups R. 70—350 μ, T. 100—350 μ; L. of articulations 120—300 μ. Elliptical or circular cylinders, flattened when joining one another. Transverse walls usually placed about horizontally; with a single circular or elliptical perforation, the remaining rings narrow and with smooth margin. Walls thick 5–12 μ between 2 vessels joining upon each other, for the rest about 5 μ; the innermost layer of the walls appearing sometimes somewhat swollen and sometimes more or less loosened from the rest of the wall; red, red-brown or dark red; lignified; — with numerous sometimes combined slit-like bordered pits between 2 vessels; sieve structure often rather distinct; the borders tetra- to hexa- often hexagonal with more or less rounded edges or elliptical, 3 by 5 μ, 4 by 6 μ; the canals slit-like, placed horizontally and the inner aperture usually smaller than the diameter of the border; — with numerous slit-like unilateral bordered pits between vessels and wood parenchyma cells or cells of medullary rays; these pits for the rest almost similar to the bilateral ones, the borders sometimes with a larger transverse diameter. Contents: often a yellow red, red, dark red or red-brown mass; this mass occurring especially in the neighbourhood of the transverse walls, here and there quite filling the vessels, entirely dissolving in chloral hydrate, partly in alcohol 96 per cent., colouring redder in phloroglucin and hydrochloric acid.

N o n - s e p t a t e l i b r i f o r m f i b r e s. R. 20—30 μ, T. 20—25 μ; L. 700—1300 μ; pentagonal to decagonal, the mostly thin ends usually more or less abruptly passing into the much thicker middle part, long about 250—350 μ. Walls thick 3, sometimes 4 μ; red, dark red or red-brown and the middle lamella often darker red than the rest; lignified, in iodine and sulphuric acid 66 per cent. the middle lamella becoming often more distinct; — with rather numerous slit-like simple pits between 2 libriform fibres, also between libriform fibres and wood parenchyma cells or cells of medullary rays; these pits nearly exclusively in the radial walls of the middle part, more or less distinctly arranged in a longitudinal row and in this row 5—20 often 8 μ in distance from each other, sometimes combined; the inner aperture very long in a surface view and usually placed rather vertically. Intercellular spaces wanting. Contents: often a red, dark red or red-brown mass; this mass often filling the thin ends of the fibres, here and there filling the middle part and sometimes resembling a thick transverse wall. W o o d p a r e n c h y m a c e l l s. R. 15—25 μ, T. 15—30 μ, L. 50—140 μ; the cells bordering on vessels often tabular, e. g. deep 10—15 μ, wide 20—30 μ; the crystal cells L. 15—35 μ. All cells tetragonal to octogonal prisms with a longitudinally directed axis; the ends of the wood parenchyma fibres in tangential sections roof-shaped, in radial sections obtuse. The wood parenchyma fibres partitioned as follows: L. 130 + 120 μ, 90 + 90 μ, 80 + 60 + 60 + 70 μ, 75 + 50 + 50 + 60 μ etc. Walls thick 1 μ; red, dark red or red-brown; lignified; — with unilateral bordered pits between wood parenchyma cells and vessels; see the cell description of the vessels; — with simple pits between these cells and libriform fibres; see the cell description of the libriform fibres; — with simple pits between 2 wood parenchyma cells, also between these cells and substitute fibres or cells of medullary rays; the pits rather numerous in the transverse and the radial walls and also in the roof-shaped end walls of the parenchyma fibres, in the radial walls usually not in groups, in the tangential walls scarce, often small, circular or elliptical. Intercellular spaces wanting. Cell contents: especially in the neighbourhood of the vessels usually red or red-brown globes or masses; moreover in the crystal cells a single simple crystal surrounded by a rather thick lignified crystal skin, cohering with the cell wall. P r o c u m b e n t c e l l s o f m e d u l l a r y r a y s. R. 45—100 μ, T. 8—15 μ, L. 12—20 μ; tetragonal to octogonal prisms with a radially directed axis and rounded radial edges. Walls thick 1—1$^1/_2$ μ; red, dark red or red-brown; lignified; —see for the pits the cell description of the wood parenchyma cells; the simple pits here rather numerous in all walls. Intercellular spaces only in a radial direction between these cells reciprocally, also between them and libriform fibres or substitute fibres or wood parenchyma cells. Cell contents: an abundant red or red-brown mass or numerous red or red-brown globes, often entirely filling the cells.

Micrography of the powder. The walls of all elements brown-red, moreover in the cell cavities often brown-red spheres or otherwise shaped masses of gum resin. Isolated libriform fibres, many of these 800 μ in length, some slender in all their parts, others with slender ends only and a swollen middle part more or less regularly cylindrical; sometimes bundles of fibre fragments, traversed by medullary rays. Brown-red masses, irregularly spherical and consisting of the cohering contents of the cells forming the crystal fibres; large, simple, polyhedral, colourless crystals more or less distinct in these masses; these crystals not seldom isolated. Fragments of the walls of vessels with many slit-like bordered pits crowded together. Parenchyma cells oblong, with circular pits in the radial walls. Starch wanting. The colouring matter extracted in chloral hydrate, not in water.

Odour and taste wanting.

January 1921.   J; M.

LYCOPODIUM.

Pollen Lycopodii. Pulvis Lycopodii. Semen Lycopodii. Sporae Lycopodii.
The spores of Lycopodium clavatum, Linn. Sp. Pl. 1101 et ed. 2. 1564.

Macroscopic characters.

Pale yellow, most mobile powder; soft to the touch; sticking to the fingers; floating upon water, but sinking on being boiled with it and burning quickly, almost without smoke, when thrown into a flame.

Inodorous, tasteless.

Anatomical characters.

LITERATURE. Berg. Anat. Atl. 1865. 97. Taf. 49. Deutsch. Arzneib. 5. Ausg. 1910. 326 Flückiger & Hanbury. Pharmacographia. 1879. 732. Flückiger. Pharmakogn. 1891. 251. Gilg. Pharmakogn. 1910. 19. Greenish & Collin. Anat. Atl. o. veget. powders. 1904. 280. Pl. 125. Hager. Handb. d. Pharmaceut. Praxis. Bd. 2. 1902. 314. Hérail. Mat. Méd. 1912. 124. Ingerman. Mikrosk. d. voorn. handelsw. 1910. 176. Luerssen. Med. Pharmac. Botanik. Bd. 1. 1879. 636. Karsten u Oltmanns. Pharmakogn. 1909. 12. Koch u. Gilg. Pharmakogn.

Praktikum. 1907. 18. Koch. Die mikrosk. Analyse der Drogenpulver. Bd. 4. 1908. 177. Kraemer. Botany a. Pharmacogn. 1910. 694, 749. Marmé. Pharmacogn. 1886. 347. Oudemans. Aanteekeningen op de Pharmacopoea Neerlandica. 1854—56. 22. Oudemans. Pharmacogn. 1880. 15. Planchon et Collin. Les drogues simples. Bd. 1. 1895. 54. Tschirch. Handb. d. Pharmakogn. T. 2. 1910. 477. Vogl. Anat. Atl. 1887. Taf. 58. MATERIAL. The drug. REAGENTS. Water, chloral hydrate, iodine in chloral hydrate.

MICROGRAPHY. Spores sphaero-tetrahedral; largest diameter about 30 μ. Exosporium marked with elevated ridges, strongly developed and grooved along the ribs of the trilateral pyramid and covering for the rest nearly the whole surface with a fine reticulum, showing penta- or hexagonal meshes; the spores thereby under a low magnifying power seeming to be beset with short projections. No starch-reaction.

November 1901, July 1911.   J; M.

PETALA RHOEADOS.
Red-Poppy Petals.
The dried petals of Papaver Rhoeas, Linn. Sp. Pl. 507.

Macroscopic characters.
Long up to 5 c.M., wide up to 9 c.M.; claw wanting; limb transversely oblong, often at the base with a smaller or larger incision, and consequently kidney-shaped; margin entire; when fresh scarlet, often with a purplish-black spot of various shape at the base; when dried dirty-purple, thin, very much shriveled up, often sticking together.

When fresh odour slightly narcotic, when dried nearly odourless; taste somewhat mucilaginous, more or less bitter.

Anatomical characters.
LITERATURE. Flückiger. Pharmakogn. 1891. 781. Marmé. Pharmacogn. 1886. 248. Meyer. Wiss. Drogenk. 1891. 276, 277. Oudemans. Pharmacogn. 1880. 331. Tschirch. Angew. Pfl. Anat. 1889. 39. MATERIAL. Petals, gathered July 1902 in the Botanic Garden at Groningen, fresh and in alcohol. REAGENTS. Water, glycerine, chloral hydrate, iodine in chloral hydrate, iodine and sulphuric acid 66 per cent., phloroglucin and hydrochloric acid, concentrated sulphuric acid, copper acetate and iron acetate, potassium dichromate.

MICROGRAPHY.

E p i d e r m i s u p p e r s i d e. Consisting of elongated cells showing a transverse undulation of their inner and lateral longitudinal walls, the folds at right angles with the longitudinal axis of the cell and being continued over the 3 walls. Stomata rare, somewhat more numerous towards the base; phaneroporous; lying in the same level as the epidermal cells.

E p i d e r m a l c e l l s p r o p e r. H. 25 μ, Lev. B. 15 μ, Lev. L. 60—150 μ; rectangular prisms; at the base of the petal Lev. B. 22 μ, Lev. L. 50 μ; polygonal prisms, only the inner walls sinuous. Outer walls slightly thickened and showing a cuticle. Cell contents: nucleus conspicuous with some small starch grains surrounding it; in the red part of the petal in nearly all the cells only 1 vacuole containing a red liquid (anthocyanin), soluble in alcohol and chloral hydrate, colouring blue and afterwards yellow in potash, and becoming red again on the addition of hydrochloric acid; in some cells, especially in the undermost half of the red part of the petal, here and there small colourless vacuoles lying in the projecting parts of the folds of the lateral walls; in the purplish-black spot of the petal the vacuoles containing a violet to dark blue liquid (anthocyanin) colouring red in hydrochloric acid and becoming blue again on the addition of potash; in many cells some very dark globules united in a cluster, these globules showing up very clearly in material treated with copper acetate; in alcoholic material the colouring matter of the cells heaped up in the nucleus and the surrounding starch grains, sometimes also in the walls; in potassium dichromate in many cells of the purplish part numerous minute dark brown granules along the walls, in other cells a larger vacuole containing a brownish red precipitate, the smaller ones remaining colourless. S t o m a t a. H. 12 μ, Lev. B. 15 μ, Lev. L. 20 μ. The vacuoles of the guard-cells containing no coloured liquid.

E p i d e r m i s   u n d e r   s i d e. Stomata somewhat more numerous. For the rest see the upper side.

M e s o p h y l l.

Spongy parenchyma consisting of about 12 layers of cells at the base of the petal, of 1—2 at the top. Cells elongated and distinctly stellate, except at the base of the petal. Large intercellular spaces.

C e l l s   o f   t h e   a b o v e. In the middle part of the petal H. 20 μ, Lev. B. 20 μ, Lev. L. 100 μ; at the base H. 30 μ, Lev. B. 30 μ, Lev. L. 200 μ. Walls at the base of the petal somewhat sinuous.

M e r i s t e l e s. Containing only a single simple vascular bundle; medulla wanting.

Pericycle. Mostly consisting of 1—2 layers of cylindrical common parenchyma cells.

Vascular bundles. Anastomosing throughout the petal, blind ends only at its margin; sometimes a bundle connecting 2 longitudinal branches showing no spiral vessels over part of its length. Collateral; closed.

P h l o e m. Consisting of bundles of smaller elements and of common parenchyma cells.

C o m m o n   p a r e n c h y m a   c e l l s. R. and T. 6 μ. S m a l l e r   e l e m e n t s. R. and T. 3 μ.

X y l e m. Consisting of spiral vessels and common parenchyma cells.

V e s s e l s. R. and T. 10 μ. Walls lignified. C o m m o n   p a r e n c h y m a   c e l l s. R. and T. 6 μ; prismatic.

June 1902. J; M; L.

PETALA ROSAE GALLICAE.
Rosa Gallica. Red-rose Petals. Red Rose.
The dried unexpanded petals of Rosa gallica, Linn. Sp. Pl. 492, gathered from double flowers.

Macroscopic characters.
Long up to 3.5 c.M.; claw small, when fresh for the greater part white, when dried yellow; limb obovate or nearly orbicular, the apex mostly with a smaller or larger incision; margin entire; dark red, velvety, more or less shriveled. Odour agreeable; taste a little astringent and bitterish.

Anatomical characters.

LITERATURE. Ellrodt. Verteil. d. Gerbstoffes i. off. Blättern, Kräutern u. Blüten. Diss. Würzburg. 1903. 27. Flückiger. Pharmacogn. 1891. 784. Karsten u. Oltmanns. Pharmakogn. 1909. 201. Kramer. Micr.-pharm. Beitr. z. Kenntn. v. Blättern u. Blüten. Diss. Würzburg. 1907. 51. Meyer. Wiss. Drogenk. Bd. II. 1892. 338. Oudemans. Pharmacogn. 1881. 341. Schneider. Powdered veget. Drugs. 1902. 276. Tschirch. Pharmacogn. Bd. II. 1911. 797. Vogl. Veget. Nahr. u. Genussm. 1899. 263. Wiesner. Rohstoffe. 1900. 650. MATERIAL. The drug; petals gathered November 1901 in the Botanic Garden at Groningen, fresh, fixed with chromic acid $\frac{1}{2}$ per cent. and in alcohol. REAGENTS. Water, glycerine, chloral hydrate, iodine in choral hydrate, phloroglucin and hydrochloric acid, iodine and sulphuric acid 66 per cent. concentrated sulphuric acid, Schulze's macerating mixture.

MICROGRAPHY.

Intervenia.

E p i d e r m i s   u p p e r   s i d e. Cells grown out into conical papillae increasing in size towards the top of the petal. Stomata wanting. Hairs only along the margin, about 200 μ in distance from each other; curved; unicellular.

E p i d e r m a l   c e l l s   p r o p e r. In the middle part of the petal H. 28 μ, Lev. B. 18 μ, Lev. L. 20 μ; poly- nearly always hexagonal prisms. Outer walls of each cell showing a radiating cuticular striation. Cell contents: starch grains; tannin; a red colouring matter (anthocyanin). H a i r s. 8 μ in diameter. Walls somewhat thickened; cavity small.

E p i d e r m i s   u n d e r   s i d e. Consisting of cells with sinuous lateral walls; in the upper part of the petal the lateral walls more strongly sinuous and moreover provided with folds projecting into the cell and sometimes enlarged at the top. Stomata very rare and small. Cuticular striation at the base of the petal consisting of parallel undulating lines; towards the centre of the petal consisting of ridges dividing the surfaces into small striated areas; farther towards the top consisting of circular ridges surrounded by striation, the enclosed areas showing no striation; round the stomata radiating from them.

E p i d e r m a l   c e l l s   p r o p e r. In the middle part of the petal H. 35 μ, Lev. B. 20 μ, Lev. L. 35 μ; tetra- to hexagonal prisms. Outer walls somewhat thickened; showing a cuticle; lateral walls pitted. Cell contents: see the cells of the upper side, but the starch grains less numerous and smaller.

M e s o p h y l l. In the middle part of the petal 10—12 cell layers of spongy parenchyma; the uppermost and the undermost cell layer, especially the first one, with small intercellular spaces and somewhat resembling the epidermis.

C e l l s   o f   t h e   a b o v e. H. 20—30 μ, Lev. B. 20—30 μ, Lev. L. 50—100 μ; stellate. Cell contents: in the outermost layers starch grains and tannin; the latter sometimes also in cells of the other layers.

Veins.

E p i d e r m i s   u p p e r   s i d e. Cells on the larger veins somewhat elongated.

E p i d e r m i s   u n d e r   s i d e. On the larger veins cells somewhat elongated and less wide; lateral walls not sinuous and without folds.

M e s o p h y l l.

Common parenchyma. At the upper side 4, at the under side 2 cell layers. Cells with folded longitudinal walls. Showing intercellular spaces.

C e l l s o f t h e a b o v e. R. and T. 25 μ, L. 60—150 μ, cells of the outermost layer the shortest; cylindrical. Cell contents: starch grains and tannin.

Endodermis. Not discernible.

M e r i s t e l e s. Much branched and especially at the margin of the petal with blind ends; containing a single simple vascular bundle surrounded by a pericycle; medulla wanting.

Pericycle. Represented by a single layer of parenchyma cells, resembling those of the mesophyll.

Vascular bundles. Collateral; closed.

P h l o e m. Consisting of bundles of sieve-tubes intermixed with cambiform cells.

S i e v e - t u b e s. R. and T. 2—3 μ, articulations pretty long; sieve-plates on the transverse walls distinct. C a m b i f o r m c e l l s. R. and T. 5—10 μ; tetra- to septagonal prisms. Cell contents: often tannin.

X y l e m. Composed of some bundles of spiral vessels containing 2—5 vessels in a transverse section, and common parenchyma cells and fibres.

S p i r a l v e s s e l s. R. and T. 5—8 μ. Middle lamella lignified. P a r e n c h y m a c e l l s and f i b r e s. See the cambiform cells.

November 1901.   J; M; L.

RADIX ACONITI.

Aconitum. Aconite Root. Aconite.

The tubers of Aconitum Napellus, Linn. Sp. Pl. 532, collected about the flowering season.

Macroscopic characters.

Tubercles. A mother and a daughter tuber at their thick bases being connected together by a very short branch; in the commercial ware usually detached for the greater part; this ware consisting of a mixture of both tubers, of which only the daughter tubers should be used. Tubers up to 11 c.M. in length, at the base thick up to 3 c.M.; hard; rather brittle; when broken throwing out dust; slenderly conical, the apex always broken off. Surface mostly showing coarse longitudinal wrinkles, dull dark greyish brown, at the base usually exhibiting the scar of the short connecting -branch, moreover on the whole tuber and especially on the thicker half scars or remains of thin branches, both ones sometimes more or less arranged in horizontal rows. Transverse fraction dull, uneven, mealy, white or more or less brown. In a transverse section a very thin, dark outermost layer; secondary phloem and xylem constituting nearly the whole tuber and developed about equally; cambium stellate, often 5- to 7-radiate, sometimes coloured somewhat darker. Mother tubers having at the base a remnant of an often hollow stem; the tuber often somewhat lighter in weight, externally and especially internally of a darker colour, internally often torn or hollow. Daughter tubers having at the base a dark obtuse bud, surrounded by leaf-sheaths or the remains corresponding to them; the tuber heavier, especially internally of a paler colour, internally solid; surface less wrinkled, sometimes even.

Odour characteristic; taste at first sweetish, soon pungent and astringent, joined to a sensation of numbness in the mouth.

Anatomical characters.

LITERATURE. Barth. Stud. üb. d. Nachweis v. Alcaloiden in pharm. verw. Drogen. Bot. Centrlbl. 1898. Bd. 75. 338. Berg. Anat. Atl. 1865. 58. Pl. XXIV. Erréra, Maistriau a. Clautriau. Prem. Recherches s. l. localis. e.l. signif. d. alcaloides d. l. Plantes. Journ. d. Méd. 1887. 110. Flückiger. Pharmakogn. 1891. 481. Flückiger & Hanbury. Pharmacographia. 1879. 8. Gilg. Pharmakogn. 1910. 104. Greenish & Collin. Anat. Atl. o. veget. powders. 1904. 216. Hager. Pharm. Praxis. Bd. I. 1900. 153. Hérail. Mat. Méd. 1912. 702. Hartwich. Aconitumkn. Abnormitäten. Bot. Centlbl. 1897. 70, 114. Koch. Pharmakogn. Atl. Bd. 2. 1914. 101. Pl. XIV. Kraemer. Botany a. Pharmacogn. 1910. 771. Luerssen. Syst. Bot. Bd. II. 597. Marié. Struct. d. Renonculacées. Ann. d. Sc. nat. Sér. 6. Bot. T. XX. 1885. 142. Marmé. Pharmacogn. 1886. 54. Meyer. Wiss. Drogenk. Bd. I. 219. Meyer. Aconitum Napellus u. s. wicht. Verwandten. Archiv d. Pharm. 1881. 171. Mitlacher. Tox. od. Forens. wicht. Pfl. u. Drogen. 1904. 70. Moeller. Leitfaden Mikr.-pharm. Üb. 1901. 318. Moeller. Pharmacogn. 1889. 312. Oudemans. Pharmacogn. 1880. 122. Planchon e. Collin. Drogues simples. T. II. 1896. 906. Schneider. Powdered veget. drugs. 1902. 108. Tschirch. Angew. Pfl. Anat. 1889. 414. Tunmann. Pfl. Mikrochemie. 1913.285. MATERIAL. Tubers gathered October 1903 in the Botanic Garden at Groningen, fresh and in alcohol; the thickest parts in a cross section 2.2 by 1.8 c.M. The drug. REAGENTS. Water, glycerine, potash, chloral hydrate, potassium iodide iodine, phloroglucin and hydrochloric acid, iodine and sulphuric acid 66 per cent., concentrated sulphuric acid, potassium dichromate, copper acetate and iron acetate.

MICROGRAPHY.

Tangential expansion necessary for keeping up with the increase in thickness of the root, due to the tangential growth of the cortical parenchyma, the endodermis cells (up to 8 radial partition-walls), pericyclic cells and the parenchymatic elements of the secondary phloem; moreover the primary phloem bundles tangentially elongated.

Epidermis. Thrown off; perhaps here and there a cell being left. Mostly still extant on the thinner parts of the apices, e. g. measuring 0.5 c.M. in diameter; in these parts the lateral walls mostly already quite disjoined and together with the outer walls forming a single dome-shaped wall. Cells lying in longitudinal rows. Root-hairs very numerous on parts of about 0.05 c.M. in diameter.

Epidermal cells proper. R. 20 μ. T. 25 μ. L. 120—230 μ; rectangular tables with a radially directed axis, sometimes having a transverse partition wall. Outer and lateral walls somewhat thickened; brown, except the innermost lamella; lateral walls with stratification; sometimes pitted. Root-hairs. e. g. 18 μ in diameter at the base, 350 μ in length; conical. Walls not thickened; yellow.

Cortex proper.

Primary cortex. Consisting of common parenchyma and idioblasts developed as sclereids. Composed of 10—20 layers of cells lying in longitudinal rows,

and often more or
less flattened; in
parts of the tuber
of 0.2 c.M. in di-
ameter in cross sec-
tion the innermost
layers in radial
rows. The outer-
most layer of cells
resembling the epi-
dermis; sometimes
the lateral and ou-
ter cell walls dome-
shaped like those
of the epidermis.
In parts of 0.5 c.M.
in diameter and
somewhat thinner
this layer showing
no alteration. Of-
ten 1 or 2 layers,
a few times the
whole cortical pa-
renchyma, thrown
off in thicker tu-
bers as well as in
small ones of about
6.5 m.M. in diam-
eter. The outermost
2 layers mostly
without intercellu-
lar spaces, the o-
ther layers showing
them. Sclereids.
Not to be distin-
guished in the
thinner parts.
Common paren-
chyma cells. In
the outermost 1 or 2
layers R. 30 μ, T. 60

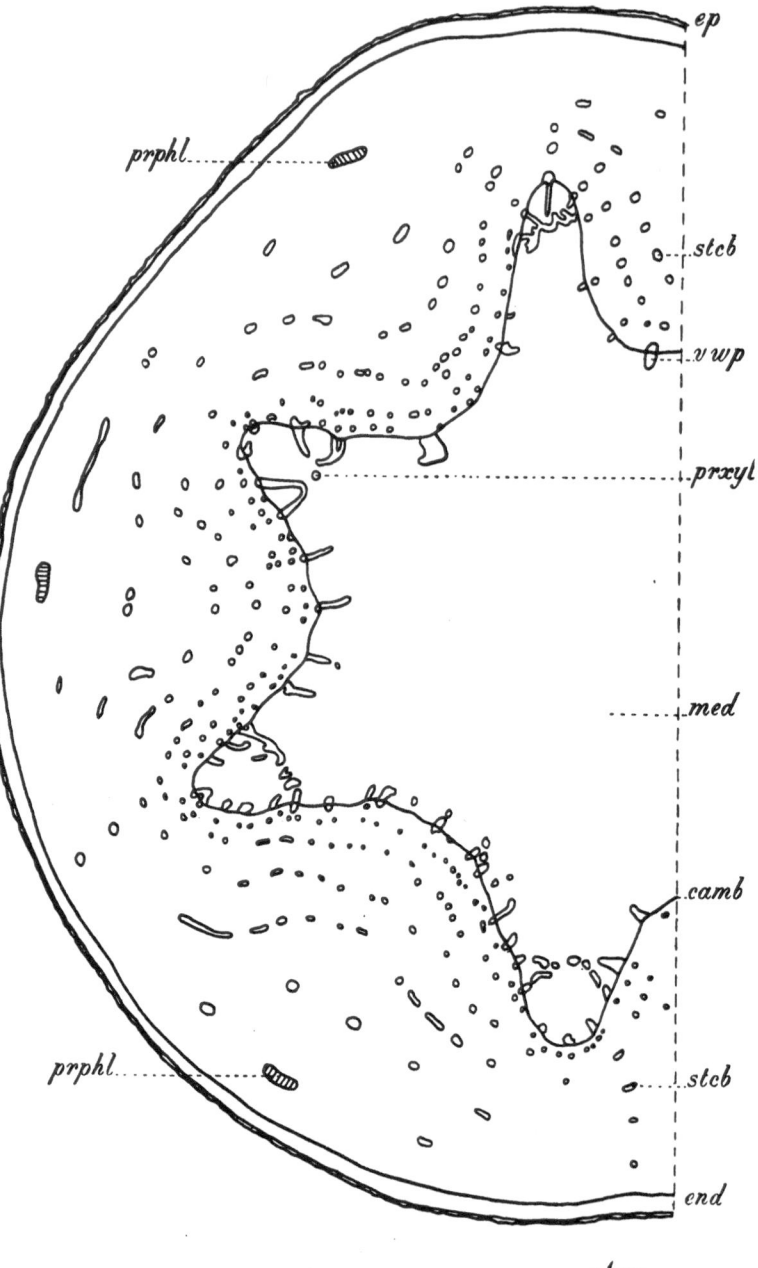

Fig. 77. *Aconitum Napellus*. Tuber, transverse section. camb Cambium; end Protective-sheath; ep Epidermis; med Medulla; prphl Primary phloem; prxyl Primary xylem; stcb Bundles of sieve-tubes, companion cells and elements of bast
parenchyma; vwp Bundles of vessels also containing wood parenchyma elements.

—100 μ, L. 70 μ; for the rest R. 35 μ, T. 125—230 μ, L. 70 μ; in the innermost layers

sometimes smaller again; polygonal prisms with a longitudinally directed axis and rounded edges, often tangentially strongly elongated, in a tangential section often hexagonal;
often showing a radial partition-wall. Walls often a little thickened, especially the outer
walls of the outermost layer; of the outermost layer or layers and sometimes of the innermost layer walls brown, brown walls of the outermost layer persisting in concentrated sulphuric acid at any rate much longer than the other walls; pitted. Cell contents: in the drug
here and there some starch grains resembling those of the pericycle; in the thinner parts
of the tuber the starch grains numerous, however in the outermost layers of these parts
less numerous and smaller than in the rest; some alcaloid [1]), in fresh material treated with
potassium iodide iodine a slightly reddish brown granular mass in all the cells, alcoholic
material not showing this peculiarity; in the thickest parts of the specimens treated
with potassium dichromate a yellow mass. S c l e r e i d s. Dimensions and shape like
those of the parenchyma cells. Walls strongly thickened, sometimes the outer or the inner
tangential walls not thickened; showing stratification; lignified, the innermost lamella
colouring blue in iodine andsu lphuric acid 66 per cent., showing pit canals. Cell contents:
see those of the parenchyma cells.

E n d o d e r m i s. Developed as a protective-sheath; in the thicker parts conspicuous as well as in the thinner parts. Consisting of 1 layer of cells, in places of
union of the 2 tubers sometimes 2 layers; often more or less flattened, except
in places deprived of the cortex. Cells showing a great number, e. g. 8, of radial
partition-walls; in parts of 0.18 c.M. in diameter 3—4 radial walls already
developed.

C e l l s   o f   t h e   a b o v e. Dimensions of the constituent cells formed by the development of radial walls R. 20 μ, T. 30 μ, L. 30—75 μ; rectangular prisms with a longitudinally directed axis. Walls often very thin; mostly brown; the thinner walls persisting
unaltered in concentrated sulphuric acid, the thicker walls consisting of an inner and an
outer cellulose layer and a middlemost layer persisting in concentrated sulphuric acid [2]).
Cell contents: alcaloid, see for this the cortical parenchyma cells; only in the thinner parts
of the roots some small starch grains.

S t e l e. In the examined specimens 4- to 8-arch; according to Hartwich up to
10-arch.

P e r i c y c l e. Consisting of common parenchyma and iodioblasts in the shape
of sclereids, occurring in the outermost layers; in parts of 0.09 and 0.18 c.M.
in diameter 1—3 layers of cells; in the thicker parts of tubers of e. g. 0.5 c.M.
in diameter about 8 layers of cells; in the still thicker parts of the same roots
20—25 layers. According to Meyer in the spring the outermost layers already
often thrown off. Only in a radial section the cells of the outermost 10 layers
lying in radial rows. In a transverse section often only the outermost 2 layers
showing radial rows, the cells having small radial dimensions. Intercellular
spaces in the outermost layers.

C o m m o n   p a r e n c h y m a   c e l l s. In the outermost layer R. 20 μ, T. 50—75 μ, L.
60 μ; for the rest R. 50 μ, T. 100—200 μ, L. 60 μ; in the outermost layer often rectangular

---

[1]) See E r r é r a, l. c. 111.
[2]) According to H e y e r. Archiv d. Pharmacie. 257, the walls are pitted; these pits seeming to be
closed only by cork lamella and middle lamella.

prisms, for the rest polygonal prisms with a longitudinally directed axis and strongly rounded edges, and often having a radial and a transverse partition-wall. Walls of the outermost layers a few times brown; all the walls pitted. Cell contents: simple and compound starch grains, filling all the cells; simple grains up to 12 μ in diameter, globular; compound grains more numerous than the simple ones, mostly 8—12 μ in diameter or e. g. 5 by 12 μ, 2-to 7-adelphous, in sulphuric acid the hilum visible and showing e. g. 3 or more slits; alcaloid [1]), in fresh material treated with potassium iodide iodine in all the cells a reddish brown granular mass between the starch grains, alcoholic material not showing this phenomenon. Sclereids. Dimensions and shape like those of the parenchyma cells. See for the rest the cortical sclereids.

Phloem.

Primary phloem. Consisting of very distinct bundles of sieves-tubes, corresponding in place to the re-entering edges of the cambium, larger than the bundles of the secondary phloem, in a cross section tangentially elongated.

Secondary phloem. Consisting of elements of the cribral and the parenchymatic system. Elements of the cribral system. Sieve-tubes and companion cells united in bundles, also containing small bast parenchyma cells and consisting of 10—25 elements in a cross section. Bundles pretty well lying in concentric zones and in radial rows, the radial rows mostly corresponding with the radial rows of the bundles of xylem vessels. Companion cells here and there to be seen in a cross section, not discernible in a longitudinal section. Elements of the parenchymatic system. Bast parenchyma composed of substitute fibres and some bast parenchyma fibres divided in 2 cells; in the innermost part elements lying in radial rows; showing intercellular spaces; the bast parenchyma elements in the bundles of sieve-tubes mostly lying in radial rows on the outer side of the bundles, each row probably developed out of 1 element, without intercellular spaces. Medullary rays hardly discernible.

Sieve-tubes. R. and T. 10—15 μ, L. of articulations 100 μ; polygonal prisms; sieve-plates on the transverse walls and showing minute pores. Companion cells. Smaller than the sieve-tubes; in a cross section mostly tri- or tetragonal. Elements of bast parenchyma. E. g. R. and T. 70 μ, L. 120 μ; near the cambium e. g. R. 25—35 μ, T. 30 μ, L. 100 μ; near the pericycle resembling the cells of the pericycle; in the interior part rectangular prisms with a longitudinally directed axis and somewhat rounded edges; here and there an element with a radial partition-wall. Walls pitted. Contents: see the parenchyma cells of the pericycle. Elements of bast parenchyma in the bundles of sieve-tubes. E. g. R. 8 μ, T. 16 μ, L. about 100 μ. Contents: often a large nucleus; some starch grains; the outer cells filled with small simple and compound starch grains; alcaloid, see the parenchyma cells of the pericycle.

Cambium. Consisting of some layers of fibres showing longitudinal projecting and reentering parts, hence stellate in a cross section; the number of the projecting parts corresponding to that of the radii of the root, in the thicker part of the tuber a greater number of projecting parts; in the vicinity of a branch the cambium bending still farther outwards.

---

[1]) See Erréra, l. c. 111.

F i b r e s o f t h e a b o v e. R. 25—35 μ, L. about 100 μ; in places corresponding with the radial rows of the phloem T. about 10 μ; tetragonal prisms with a longitudinally directed axis and tapering ends.

X y l e m.

Secondary xylem. Not much developed. Consisting of elements of the tracheal and the parenchymatic system. Elements of the t r a c h e a l s y s t e m. Bundles of reticulate and pitted vessels [1]), also containing elements of wood parenchyma, lying in the vicinity of the cambium and often in radial rows; in the projecting wedges of the xylem not far from the top a bundle of vessels starting from either side of the wedge and joining on the primary xylem lying in the middle of the wedge (see figure 77); in the tops of the wedges bundles of secondary vessels often wanting. Intercellular spaces wanting. Elements of t h e p a r e n c h y m a t i c s y s t e m. Wood parenchyma, about 10 layers of elements arranged in radial rows and showing small intercellular spaces. Wood parenchyma elements in the bundles of vessels mostly lying along the periphery of the bundle and showing no intercellular spaces.

V e s s e l s. E. g. R. and T. 20—30 μ, L. of articulations 60—200 μ; polygonal prisms; transverse walls with a circular perforation. Walls somewhat thickened; lignified; showing slit-like bordered pits, the slits running horizontally. E l e m e n t s o f w o o d p a r e n-c h y m a. E. g. R. and T. 45 μ, L. 100 μ, farther from the cambium especially the radial dimensions larger; tetragonal prisms with a longitudinally directed axis and somewhat rounded edges. Walls pitted. Contents: see the bast parenchyma. W o o d p a r e n-c h y m a e l e m e n t s i n t h e b u n d l e s o f v e s s e l s. E. g. R. and T. 25—30 μ, or smaller, L. 75—120 μ; poly-, often tetragonal prisms with a longitudinally directed axis. Contents: starch grains, not numerous and smaller than those of the rest of the wood parenchyma; alcaloid, see the bast parenchyma.

Primary xylem. Bundles of spiral and reticulate vessels, lying in the centre of the wedge at the juncture of 2 bundles of secondary xylem.

V e s s e l s. R. and T. about 10 μ. Walls somewhat thickened; lignified.

M e d u l l a. Consisting of common parenchyma; according to Meyer in tubers having borne a stem sclerenchyma in the medulla as well as in the cortex and in the pericycle. Showing intercellular spaces.

C o m m o n p a r e n c h y m a c e l l s. E. g. R. 120 μ, T. 130 μ, L. 140 μ; polyhedra with rounded edges. Walls pitted. Cell contents: see the bast parenchyma; in fresh material and in material treated with iodine numerous small vacuoles to be distinguished in the cells.

Micrography of the powder. Starch grains, for the greater part detached, but also contained in parenchyma cells; grains often 8—15 μ in diameter, simple, about spherical, with centric hilum, but also a great many compound 2-to 4-adelphous grains. Common parenchyma cells thin-walled, colourless; those from the outermost part of the tuber having brown walls. Sclerenchyma cells of greatly varying shape, shorter or larger prismatical, also more rounded;

---

[1]) See for the arrangement of these bundles cribral system of secondary phloem.

walls not very strongly thickened, pale yellowish, with distinct pit canals; contents of the spacious cell cavities sometimes brown. Xylem vessels isolated or in bundles, often having slit-like bordered pits, sometimes spiral vessels.

October 1903. J; M; L.

RADIX ALTHAEAE.

Althaea.

The secondary roots 1 or 2 years old or the primary root 1 year old of Althaea officinalis, Linn. Sp. Pl. ed. I. 686, deprived of branches, fibrils and outermost layer of the bark, immediately after being dug out.

Macroscopic characters.

Cylindrical or sub-conical, sometimes slightly enlarged at the base, 10—20 c.M. in length, up to 2.5 c.M. thick, straight or slightly curved, somewhat angular from being peeled, and showing shallow wide longitudinal furrows from drying up. Many roots halved lengthwise; the section then somewhat concave, or with a strong lengthwise crest in the middle. Surface even, somewhat fibrous, dirty white, showing a great number of transversely elongated brown scars of branches. When broken throwing out dust; transverse fracture dirty white, in the bark strongly fibrous, within the cambium uneven, mealy. Root easily to be cut across, the smooth transverse section brighter white; bark about 1 m.M. thick; cambium light brown; extremely fine radial lines in bark and xylem; similar concentric rings in the bark.

Odour faint; taste flat, mucilaginous.

Anatomical characters.

LITERATURE. Berg. Anat. Atl. 1865. 19. Pl. XI. Benecke. Microsc. Drogenprakt. 1912. 13. Biechele. Mikr. Prüf. d. off. Drogen. 1904. 52. Flückiger. Pharmakogn. 1891. 373. Flückiger & Hanbury. Pharmacographia. 1879. 92. Gilg. Pharmakogn. 1910. 217. Hager. Pharm. Prax. Bd. I. 1900. 230. Hartwich. Über die Schleimzellen von Althaea officinalis. L. Pharm. Centralhalle. 1891. 586. Hérail. Mat. Méd. 1912. 99. Karsten u. Oltmanns. Pharmakogn. 1909. 76. Koch u. Gilg. Pharmakogn. Prakt. 1907. 100. Koch. Mikr. Anal. d. Drogenpulver. Bd. II. 1903. 144. Pl. XII. Koch. Einf. i.·d. mikr. Anal. d. Drogenpulver. 1906. 44. Koch. Pharmakogn. Atl. Bd. I. 1914. 3. Pl. I. Kraemer. Botany a. Pharmacogn. 1910. 754. Luerssen. Syst. Bot. Bd. II. 1882. 665. Marmé. Pharmacogn. 1886. 25. Meyer. Wiss. Drogenk. Bd. I. 1891. 228. Moeller. Pharmakogn. 1889. 300. Moeller. Mikr. Pharm. Üb. 1901. 34. Oudemans. Aant. o. d. Pharmac. neerl. 1854—56. 446. Oudemans. Pharmacogn. 1889. 67. Planchon et Collin. Drogues simples. T. I. 1896. 703. Schneider. Powdered veget. drugs. 1902. 111. Tschirch. Pharmakogn. Bd. II. Abt. I. 1911. 349. Tschirch u. Oesterle. Anat. Atl. 1900. 125. Pl. 30. Vogl. Anat. Atl. 1887. Pl. 51. Walliczek. Studien über den Membranschleim vegetativer Organe. Diss. Bern. 47. See also Pringsheim's Jahrbücher. 1893. 255. Wigand. Pharmakogn. 1879. 40. MATERIAL. The drug. Chromic acid material ($\frac{1}{2}$ per cent) and alcoholic material, prepared from fresh roots, collected in the Botanic Garden at Groningen, thick 1 c.M. REAGENTS. Water, glycerine, potash, phloroglucin and hydrochloric acid, iodine and sulphuric acid 66 per cent., concentrated sulphuric acid, Schulze's macerating mixture.

MICROGRAPHY.

Epidermis and cortex. Thrown off in the drug.

Stele.

Pericycle.

Secondary cork tissue.

Phellem. Developed as periderm; consisting of 8—12 layers of cells, lying in radial rows and also pretty regularly in horizontal rows. Cells regularly scaling off.

Cells of the above. R. 14 μ, T. 20—80 μ, L. 30—40 μ; tetra- to hexagonal tables, sometimes with curved sides. Walls sometimes granular; yellow; suberized, the innermost lamella lignified.

Phellogen. Consisting of 1 layer of cells.

Cells of the above. Walls often more yellow than those of the periderm cells; the outermost tangential walls suberized and lignified like those of the periderm cells.

Phelloderm. Consisting of 3—5 layers of cells.

Cells of the above. Shape like that of the periderm cells, but radial dimensions larger and edges more rounded. Walls varying from colourless to yellow. Cell contents: starch grains and sometimes a cluster crystal in a smaller cell.

Phloem.

Primary phloem. No more to be recognised.

Secondary phloem.

Rhytidoma. Slightly developed by formation of cork layers, 3 cells thick.

Secondary phloem proper. Consisting of about 65 layers of elements, somewhat flattened in the outer part, showing radial arrangement in the inner part. Composed of elements of the cribral system, of the system of bast fibres and of the parenchymatic system.

Cribal system. Sieve-tubes united with companion cells in bundles of 5—10 elements; bundles arranged in concentric rings, separated from each other by 5—7 layers of parenchyma cells; these bundles distinct only in the inner part.

Sieve-tubes. R. and T. 6—8 μ, L. of articulations 20 μ; polygonal prisms. Sieve-plates rarely to be distinguished. Companion cells. The same dimensions as the sieve-tubes.

System of bast fibres. Bast fibres in bundles of 3—25; bundles arranged in concentric rings alternating with those of the sieve bundles, the bundles of the same ring strongly anastomosing. Intercellular spaces wanting.

Bast fibres. R. and T. 9—12 μ, L. 550—700 μ; tri- to hexagonal, the ends often bifurcate, rarely trifurcate. Walls not very thick; nearly always exclusively the middle lamella lignified; showing cross-wise slit-like pits.

Parenchymatic system.

Bast parenchyma. Forming the main part of the secondary phloem, consisting of common parenchyma cells and idioblasts; radial and especially tangential dimensions of the cells decreasing towards the cambium; showing intercellular spaces. Idioblasts in 3 kinds: 1. crystal cells, sometimes in longitudinal rows;

cluster crystals 15 μ; 2. asparagine cells, single or in groups, containing sphaero-crystals; 3. mucilage cells.

B a s t   p a r e n c h y m a   c e l l s. In the outer part R. 40 μ, T. 65 μ, L. 60—90 μ; in the middle part R. 35 μ, T. 45 μ, L. 60—90 μ; in the vicinity of the cambium R. 35 μ, T. 30 μ, L. 60—90 μ; in the outer part elliptical cylinders or ellipsoids, for the rest tetra- to hexagonal prisms with strongly rounded edges; often a cell divided in 2 parts by a radial or transverse wall. Walls often yellow; striation showing up by the cellulose reaction; often lignified; pitted. Cell contents: mostly ellipsoidal starch grains, measuring 5—12 μ. M u c i l a g e c e l l s. Of the same shape as the surrounding cells, but somewhat larger and having more strongly rounded edges. Walls with stratification, especially distinct in alcohol; consisting of a cellulose layer and a thinner or thicker mucilage layer, nearly quite filling the cell and mostly leaving only a small cavity; without protoplasm. [1]

Medullary rays. Consisting of common parenchyma cells and idioblasts; distinct only in the inner part of the phloem; separated by 3—15, mostly 6 layers of cells; 1- to 3-mostly 2-seriate and several cells high. Cells with larger tangential dimensions towards the outer part. Idioblasts: mucilage cells.

C o m m o n   p a r e n c h y m a   c e l l s. R. 50 μ, T. 30 μ, L. 30—50 μ; circular or elliptical cylinders or tetra to hexagonal prisms with rounded edges, both with radially directed axis. See for the rest bast parenchyma cells, but more starch and a little oil. M u c i l a g e   c e l l s: see those of the bast parenchyma.

Cambium. Consisting of some layers of elements; the tangential and longitudinal dimensions corresponding to those of the adjoining cells and articulations of sieve-tubes.

Xylem.

S e c o n d a r y   x y l e m. Arranged in radial rows only in the vicinity of the cambium. Consisting of elements of the tracheal system, libriform system and the parenchymatic system.

T r a c h e a l   s y s t e m. Represented by vessels and tracheid fibres united in sinuous bundles, thick 1—5 elements; close to the cambium a ring of these bundles; in the centre

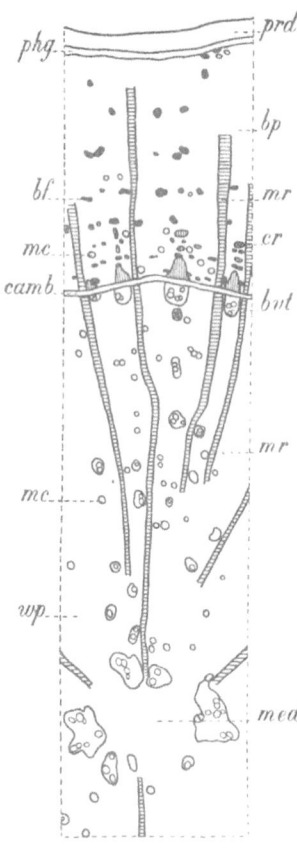

Fig. 78. *Althaea officinalis*. Root, transverse section. bf Bast fibres; bp Bast parenchyma; bvt Bundles of vessels and tracheid fibres; camb Cambium; cr Bundles of elements of cribal system; mc Mucilage cells; med Medulla; mr Medullary rays; phg Phellogen; prd Periderm; wp Wood parenchyma.

---

[1] According to W a l l i c z e k the mucilage is secreted by the protoplasm directly against the walls, the protoplasma gradually disappearing. Later on the mucilage may be partly dissolved again, the solution beginning from the centre. According to H a r t w i c h a thin layer of protoplasm is present between the secreted mucilage and the wall at first, the protoplasm disappearing later on or changing into mucilage.

of the xylem sometimes a large vessel surrounded by smaller vessels and fibres or parenchyma cells, this central bundle closed in by sometimes very large bundles of vessels and fibres. Vessels not numerous, scalariform and reticulate. Tracheid fibres without intercellular spaces.

V e s s e l s. R. and T. 25—30 μ, L. of articulations 150—200 μ; polygonal prisms, having transverse partition-walls with circular perforations. Walls thickened; yellow; lignified; with two-sided and one-sided bordered pits. Cell contents: here and there thyloses. T r a c h e i d   f i b r e s. R. and T. 14—18 μ, L. 90—180 μ; tetra- to hexagonal. Walls somewhat thickened; slightly yellow; the middle lamella lignified.

L i b r i f o r m   s y s t e m. Libriform fibres in the drug lying in bundles, more or less surrounding the bundles of vessels; in the alcoholic material not to be found. See for the rest the fibres of the secondary phloem.

P a r e n c h y m a t i c   s y s t e m. Consisting of wood parenchyma and medullary rays.

Wood parenchyma. Forming the main part of the secondary xylem; in the vicinity of the tracheal bundles the parenchyma cells sometimes fibre-shaped and having larger longitudinal and smaller radial and tangential dimensions; towards the interior the radial dimensions often much larger, the largest cells in the centre. Idioblasts: mucilage cells.

P a r e n c h y m a   c e l l s. In the vicinity of the cambium R. and T. 30 μ, L. 60—90 μ; farther towards the interior R. 40—60 μ, T. 40 μ, L. 70—150 μ; in the centre R. and T. 40—80 μ, L. 80—100 μ; near the cambium tetragonal prisms with rounded edges and a longitudinally directed axis, towards the interior the edges more strongly rounded and transverse walls occurring. M u c i l a g e   c e l l s. See the secondary phloem.

Medullary rays. See the secondary phloem; often the radial dimensions of the cells larger.

Medulla. Wanting or more or less developed.

Micrography of the powder. Starch, for the greater part detached grains, but also colourless common parenchyma cells wholly filled with them; grains oblong ellipsoids or more or less ovate, many of them 10—15 μ in length, wide 6—10 μ; hilum centric or eccentric ($\frac{1}{3}$), in the latter case often lying towards the somewhat broader end, often invisible as well as the stratification; not rarely an inner cleft filled with air replacing the hilum; some compound grains, 2- to 4-adelphous. Fragments of thick-walled fibres, not rarely in bundles; also tracheid fibres, thinner-walled and shorter. Fragments of vessels with slit-like bordered pits, often also in bundles. Cluster crystals sometimes still contained in common parenchyma cells. Mucilage cells, oblong or about spherical, with very thick secondary wall-layers, best to be seen in alcohol, oil of origanum and especially in oil of cloves; in oil of cloves often also numerous sphaerites, probably of asparagin, in different cells. The powder colouring yellow in strong potash.

March 1902.   J; M; L.

RADIX ARMORACIAE.

Horseradish Root.

The fresh root of Cochlearia Armoracia, Linn. Sp. Pl. t. 648, collected from cultivated plants late in the autumn and in early spring.

Macroscopic characters.

Roots mostly about 2 c.M. thick and over 30 c.M. in length, about cylindrical, at the crown often with semi-amplexicaul leaf-scars. Surface pale yellowish brown, rather smooth. Internally white, leathery-fleshy; fracture smooth. When intact odourless, when bruised smelling after oil of mustard; taste pungent, resembling that of mustard.

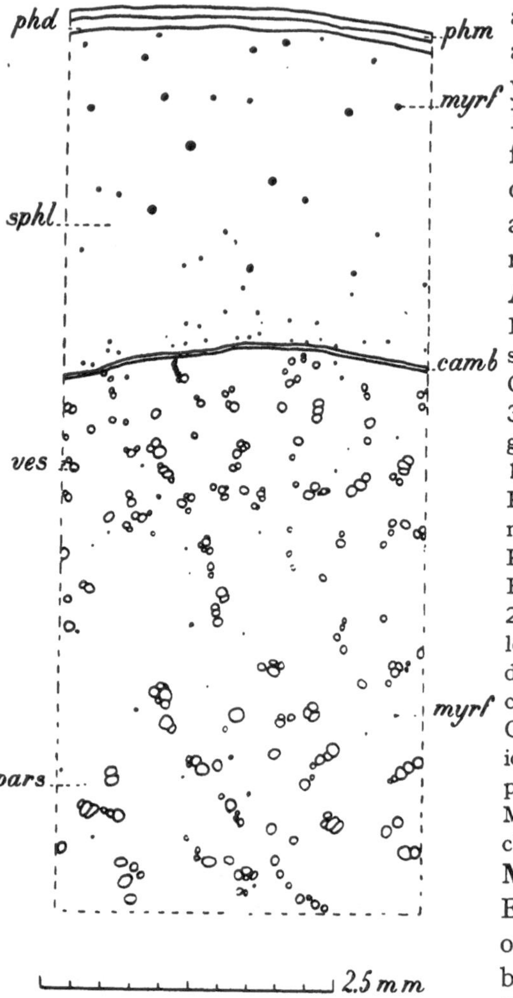

Fig. 79. *Cochlearia Armoracia.* Root, transverse section. camb Cambium; myrf Myrosin fibres; pars Elements of the parenchymatic system, except the myrosin fibres; phd Phelloderm; phm Phellem; sphl Secondary phloem; ves Vessels.

Anatomical characters.

LITERATURE. Guignard. Recherches s.l. localisation d. Principes actifs d. Crucifères. Journ. de Bot. T. IV. 1890. 385. Flückiger & Hanbury. Pharmacographia. 1879. 72. Hérail. Mat. Méd. 1912. 617. Oudemans. Aant. o. d. Pharm. neerl. 1854—56. 428. Oudemans. Pharmacogn. 1880. 60. Wigand. Pharmakogn. 1879. 42. MATERIAL. Fresh material, the same kept in alcohol, 2.5 c.M. in diameter and thinner, collected March 1901 in the Botanic Garden at Groningen; chromid acid (1 per cent) material. MEDIA and REAGENTS. Glycerine jelly, potassium iodide iodine, iodine and sulphuric acid 66 per cent., phloroglucin and hydrochloric acid. Myrosin reagents: concentrated hydrochloric acid, Millon's reagent.

MICROGRAPHY.

Epidermis and cortex soon thrown off by the secondary cork layers being developed in the pericycle.

S t e l e.

Pericycle. Secondary cork tissue. Initial-celled cork.

Phellem. Developed as periderm; 5—10 cell layers thick; showing radial arrangement.

P e r i d e r m c e l l s. R. 6 μ, T. 25 μ, L. 15 μ; tetra- to hexagonal prisms with a radially directed axis. Walls nearly colourless; lignified.

Phellogen. Consisting of 1 layer of elements.

Contents: starch.

Phelloderm. 15—25 layers; consisting of collenchyma, common parenchyma, sclerenchyma and idioblasts.

C o l l e n c h y m a. 6—8 cell layers thick; radial arrangement still to be distinguished, especially in a radial section, in roots with a diameter of 1.8 c.M. Intercellular spaces wanting.

C e l l s o f t h e a b o v e. R. 18 μ, T. 50 μ, L. 16 μ; polyhedra with a radially directed axis, sometimes divided in 2 parts by a radial wall. Walls rather strongly thickened; pitted. Cell contents: ellipsoidal starch grains.

C o m m o n p a r e n c h y m a. Somewhat radially arranged; showing large intercellular spaces extending to all sides.

C e l l s o f t h e a b o v e. R. 40 μ, T. 90 μ, L. 35 μ; tangentially elongated ellipsoids; sometimes divided in 2 or 3 cells by radial walls. In young roots, not over 1 c.M. thick, many parenchyma cells having collenchymatous thickenings on one of their common cell walls. Walls pitted. Cell contents: many simple and a few compound, 2- to 3-adelphous starch grains.

S c l e r e n c h y m a. Developed in roots of 1 c.M. thick, especially between collenchyma and parenchyma; mostly in ellipsoidal radial somewhat flattened clusters of cells, 3—6 elements in a radial, 6—12 elements in a tangential, 6—10 elements in a longitudinal direction. Intercellular spaces wanting.

C e l l s o f t h e a b o v e. R. 17 μ, T. 40 μ, L. 18 μ; tangentially elongated polyhedra. Walls strongly thickened, leaving a distinct cavity; showing stratification; lignified; having pit canals.

I d i o b l a s t s. Myrosin cells, here more numerous than in the parts lying farther towards the interior.

My r o s i n c e l l s. Similar in size to or somewhat larger than the common parenchyma cells. Contents: a yellowish brown granular mass; when heated in hydrochloric acid first colouring pink, afterwards purple; in Millon's reagent brownish red.

Rhytidoma like portions. Here and there developed in the collenchyma of the thicker roots by the formation of elliptical tertiary cork layers, measuring about 1 m.M. in a tangential direction. Tertiary cork layers with their edges closing in upon the secondary periderm; the tissue lying on the outside and consisting of 4—50 layers of collenchyma and periderm being thrown off.

Phloem.

Primary phloem. No more to be distinguished in the drug.

Secondary phloem. 110—120 layers of elements thick; showing storied structure; consisting of elements of the cribral and the parenchymatic system, both containing idioblasts.

C r i b r a l s y s t e m. Sieve-tubes united in anastomosing bundles of 2—12 tubes in a radial and 2—6 tubes in a tangential direction; elements in the bundles radially arranged. Here and there intercellular spaces.

S i e v e-t u b e s. R. 20 μ, T. 30 μ, L. of articulations 80—100 μ; tetra- to hexagonal prisms with a longitudinally directed axis. Sieve-plates often to be seen in a transverse and especially in a radial section. Walls somewhat thickened; pitted. Contents: starch, to a somewhat smaller amount than in the bast parenchyma.

P a r e n c h y m a t i c s y s t e m. Consisting of bast parenchyma and medullary rays.

Bast parenchyma. Forming the main part of the secondary phloem and consisting of substitute fibres, some bast parenchyma fibres composed of 2 cells and myrosin idioblasts of the same shape and size.

S u b s t i t u t e f i b r e s. R. 25 μ, T. 30 μ, L. 80—100 μ; longitudinal dimension like that of the articulations of sieve-tubes. Walls pitted. Contents: starch grains, sometimes simple, sometimes compound, 2- to 3-adelphous, strongly varying in size.

Myrosin fibres. Less numerous than the myrosin cells in the phelloderm. Shape and size like those of the surrounding fibres. For the rest see the myrosin cells of the phelloderm.

Medullary rays. Not to be distinguished from the rest of the parenchymatic system.

Cambium. 3—5 layers of elements thick.

E l e m e n t s o f t h e a b o v e. R. 8 μ, T. 25 μ, L. 80—100 μ; shape similar to that of the articulations of the sieve-tubes. Contents: small starch grains.

Xylem.

S e c o n d a r y x y l e m. Showing storied structure; consisting of elements of the tracheal system and the parenchymatic system.

T r a c h e a l s y s t e m. Consisting of bundles of vessels, surrounded by 2—3 layers of modified wood parenchyma elements. Bundles anastomosing in a tangential direction. In each bundle in a transverse section 1—30 vessels, radially arranged with regard to the centre of the bundle.

V e s s e l s. R. 20—40 μ, T. 30 μ, L. of articulations ·50—70 μ; hexagonal prisms with a longitudinally directed axis; transverse walls oblique and having 1 large perforation. Walls rather thick; yellow; lignified; showing reticulately arranged, closely approximated slit-like bordered pits. Contents: often a brown finely granular mass, dissolving in alcohol.

P a r e n c h y m a t i c s y s t e m. Consisting of wood parenchyma and medullary rays.

Wood parenchyma. Constituting the chief part of the secondary xylem; consisting of substitute fibres, some wood parenchyma fibres composed of 2 cells and myrosin idioblasts of the same shape and size. Somewhat modified in the vicinity of the tracheal bundles. The radial arrangement disappearing in the inner part.

S u b s t i t u t e f i b r e s. R. 28 μ, T. 32 μ, L. 70—100 μ; shape like that of the substitute fibres of the secondary phloem, in the inner part spherical or ellipsoidal. Walls and contents: see the secondary phloem. M o d i f i e d e l e m e n t s. R. 13 μ, T. 17 μ, L. 70—100 μ. Contents: small starch grains.

Myrosin fibres. Throughout the whole parenchymatic system; quite similar to those mentioned above.

Medullary rays. Only a few elements wide; in the drug not to be distinguished from the rest of the parenchymatic system.

Primary xylem. Hardly to be distinguished in the drug. Young roots 3-arch, the radii of the xylem joining in the centre.

March 1901.   J; M; L.

RADIX BELLADONNAE.
Belladonna Root.
The root of Atropa Belladonna, Linn. Sp. Pl. 181, collected in the autumn and dried.

LITERATURE. Berg. Anat. Atl. 1865. 23. Pl. XIII. Flückiger & Hanbury. Pharmacographia. 1879. 457. Gilg. Pharmacogn. 1910. 305. Greenish & Collin. Anat. Atl. of veget. powders. 1904. 218. Hager. Pharm. Prax. Bd. I. 1900. 408. Kraemer. Botany a. Pharmacogn. 1910. 446, 461, 463, 557. Luerssen. Syst. Bot. 1882. 982. Marmé. Pharmacogn. 1886. 42. Mitlacher. Toxik. od. Forens. wicht. Pfl. u. Drogen. 1904. 147. Moeller. Mikr. pharm. Üb. 1901. 304. Oudemans. Aant. o.d. Pharm. neerl. 1854—56. 330. Oudemans. Pharmacogn. 1880. 38. Planchon et Collin. Drogues simples. T. I. 1895. 578. Sievers. Verd. d. Alcaloïden i. d. versch. deelen der Atropa Belladonna L. Am. J. Pharm. 1914. 97—112. Schneider. Powdered veget. Drugs. 1902. 132. Stscherbatscheff. Beitr. z. Entw. gesch. ein. off. Pfl. Arch. d. Pharm. Bd. 245. 1907. 50.Tschirch u. Oesterle. Anat. Atl. 1900. 327. Pl. 76. Vogl. Anat. Atl. 1887. Pl. 48. Wigand. Pharmacogn. 1879. 68. MATERIAL. The drug. Fresh material, 0.8 m.M. thick, alcoholic material 7 m.M. thick; both from roots collected in the Botanic Garden at Groningen July 10th 1901. REAGENTS. Glycerine, iodine in chloral hydrate, iodine and sulphuric acid 66 per cent., phloroglucin and hydrochloric acid, concentrated sulphuric acid, diluted potash, Schulze's reagent.

MICROGRAPHY.
Epidermis. Thrown off by formation of cork in the pericycle.

Cortex. Here and there some thick-walled ellipsoidal brown cells, being the remains of the cortical parenchyma; for the rest thrown off in consequence of the cause mentioned above.

Stele.

Pericycle. Secondary cork tissue. Initial-celled cork. Often somewhat flattened in a radial direction; sometimes quite thrown off in the drug. Colourless in iodine and sulphuric acid, some cell walls slightly red in phloroglucin and hydrochloric acid.

Phellem. Developed as periderm; consisting of 6—8 radially arranged layers of cells.

Cells of the above. R. 15 μ, T. 45 μ, L. 35 μ; tetragonal tables with a radially directed axis. Walls yellowish brown; in concentrated sulphuric acid divided in 3 parts, the cork lamellae remaining more distinct than the middle lamella, some walls lignified.

Phellogen. Not clearly to be distinguished.

Phelloderm. Consisting of common parenchyma and idioblasts; up to 5 cell layers thick; radially arranged. Showing intercellular spaces.

Phelloderm cells. R. 30 μ, T. 45 μ, L. 35 μ; tetragonal tables with a radially directed axis and rounded edges. Walls yellowish blue in iodine and sulphuric acid 66 per cent. Cell contents: starch, or yellowish brown granules. Idioblasts. Cells with crystal sand, only lying in the phelloderm, quite similar to the phelloderm cells. Contents: see Folia Belladonnae.

Primary phloem. No more to be recognized in the drug.

Secondary phloem. About 35 layers thick. Consisting of elements of the cribral and the parenchymatic system, the parenchyma forming the main part. Radial arrangement only distinct in the inner part; the outer part here and there rather strongly flattened, especially in places having lost the cork tissue. The strips of phloem adjoining the outermost groups of xylem vessels nearly wholly consisting of entirely flattened tissue. Keratenchyma through the whole phloem. **Elements of the cribral system.** Sieve-tubes in small bundles close to the cambium, farther to the outside flattened. Companion cells perhaps present.

Sieve-tubes. R. 15 μ, T. 20 μ, L. of articulations 160 μ; tetragonal prisms; sieve-plates conspicuous, showing a callus. Companion cells. R. and T. 4 μ, L. 130 μ. **Elements of the parenchymatic system.**

Bast parenchyma. Forming the chief portion of the secondary phloem, and consisting of common parenchyma and idioblasts. The inner part radially arranged; with intercellular spaces. In the outer part intercellular spaces wanting.

Common parenchyma cells. In the outer part R. 40 μ, T. 120 μ, L. 70—100 μ; ellipsoidal polyhedra, often divided in 2 parts by a radial wall. Walls pitted. Cell contents: many simple and some compound, up to 3-adelphous starch grains, 10 μ in diameter, spherical, sometimes slightly angular or showing papilliform protuberances; sometimes a yellow mass. In the inner part R. 20 μ, T. 35 μ, L. 130—200 μ. Walls pitted. Cell contents: yellow mass wanting; see for the rest cells of the outer part. Idioblasts. Cells with crystal sand, sometimes lying in longitudinal rows, quite similar to the bast parenchyma cells, the radial dimensions often somewhat smaller. Contents: see Folia Belladonnae.

Medullary rays. In the vicinity of the cambium triseriate; more to the outside wider, the cells increasing in width.

Cambium. In the drug quite flattened, in alcoholic material 5—8 layers of elements.

Elements of the above. R. 10 μ, T. 20 μ, L. 70—130 μ.

Secondary xylem. Consisting of elements of the tracheal system, libriform fibres and elements of the parenchymatic system. **Elements of the tracheal system.** Bundles of vessels, sometimes running sinuously, nearly wholly constituting the outer border of the secondary xylem; irregularly scattered in the inner part. Bundles consisting of pitted, perhaps reticulate vessels, surrounded by tracheid and libriform fibres.

Vessels. In the outer part of the xylem R. 40 μ, T. 50 μ, more to the centre R. 80 μ, T. 90 μ, L. of articulations 120—200 μ; tetra- to hexagonal, sometimes with rounded edges. Transverse walls showing a large circular perforation. Walls thickened; lignified; the innermost layer showing a weak cellulose reaction, strongly dilating in sulphuric acid; showing bordered pits and slit-like bordered pits. Contents: vessels sometimes quite filled with yellowish green, irregular shaped granules. Tracheid fibres. R. 20 μ, T. 25 μ, L. 300—450 μ; tetra- to hexagonal; surface sometimes slightly sinuous. Walls thickened; yellow; stratified; chemical constitution like that of the vessels, the cellulose reaction being somewhat stronger; showing slit-like bordered pits and reticulate thickenings.

Libriform fibres. Mostly surrounding the tracheid fibres, sometimes adjoining the vessels. In 2 kinds: for the greater part showing the common form of fibres; those lying towards the outside of the bundles shuttle-shaped and forming a transition to the shuttle-shaped parenchyma cells.

F i b r e s. R. 15 μ, T. 20 μ, L. 300—400 μ; tetra- to hexagonal. Walls not strongly and very irregularly thickened; stratified; chemical constitution the same as that of the vessels; showing cross wise slit-like pits.

Elements of the parenchymatic system.

W o o d p a r e n c h y m a. Forming the main part of the secondary xylem and consisting of common parenchyma and idioblasts. Radially arranged; sometimes a little flattened in the drug. Sometimes some layers of parenchyma radiately arranged round the bundles of vessels, without intercellular spaces; sometimes the cells of this parenchyma shuttle-shaped (substitute fibres?). W o o d p a r e n c h y m a c e l l s. R. 22 μ, T. 30 μ, L. 120—200 μ; tetragonal prisms with a longitudinally directed axis. Walls slightly thickened; with some stratification; pitted. In the vicinity of the outer border of the xylem walls thickened; yellow; somewhat lignified; with numerous pits showing transitions to reticulate thickenings. Cell contents: see the bast parenchyma, starch grains somewhat smaller. P a r e n c h y m a c e l l s radiately arranged round the bundles of vessels. R. and T. 20 μ, L. 50 μ; tetragonal prisms with a longitudinally directed axis. Hardly any starch. S h u t t l e-s h a p e d p a r e n c h y-m a c e l l s. R. 40 μ, L. 100 μ, or R. 20 μ, L. 60 μ. I d i o b l a s t s. Cells with crystal sand. Similar to the surrounding cells, perhaps the radial dimension somewhat smaller. Contents: see Folia Belladonnae.

M e d u l l a r y r a y s. Triseriate.

C e l l s o f t h e a b o v e. R. 20 μ, L. 65 μ. In the parts between the outermost groups of vessels walls somewhat thickened; yellow; with numerous pits showing transitions to reticulate thickenings; walls turned towards the libriform fibres thickened.

Medulla. Wanting.

June 1901.   J; M; L.

RADIX HELENII.

The root with the caudex of Inula Helenium, Linn. Sp. Pl. 881, collected from plants 2—3 years old.

Macroscopic characters.

Up to 15 c.M. long, up to 3 c.M. thick; hard and brittle, in damp air soon flexible and tough; conical; having at the summit 1 or more short thicker caudices usually exhibiting several annular scars of dead stems; the lower portion cut off; deprived of the branches; mostly halved lengthwise. Surface of the outer side wrinkled longitudinally, yellowish-greyish-brown, darker in the slits; the surface of the longitudinal section hollow from drying up, in the middle showing a longitudinal embossed ridge, evenly yellowish-greyish-brown. In a transverse section cortex and phloem thin, the latter with dark rays; xylem thick with radially arranged groups of vessels; medulla mostly wanting; colour lighter or darker brown; dots, corresponding to resin glands, scattered over the whole section. Transverse fracture smooth, not fibrous. Thinner roots not cut and terete. Odour peculiarly aromatic; taste aromatic and bitter

Anatomical characters.

LITERATURE. Berg. Anat. Atl. 1865. 17. Pl. X. Le Blois. Les Canaux sécréteurs et les poches sécrétrices. Ann. d. Sciences. Sér. 7. Bot. T. VI. 1887. 274. Dye. Unterirdische Org. von Valeriana, Rheum u. Inula. Diss. Bern. 1901. Flückiger. Pharmakogn. 1891. 477. Flückiger & Hanbury. Pharmacographia. 1879. 381. Gilg. Pharmakogn. 1910. 343. Greenish & Collin. Anat. Atl. of veget. powders. 1904. 224. Hager. Pharm. Prax. Bd. II. 1902. 5. Kraemer. Botany and Pharmacogn. 1910. 398 (only a fig.). Luerssen. Syst. Bot. Bd. II. 1882. 101. Meyer. Wiss. Drogenk. Bd. I. 1891. 254. Oudemans. Aant. o.d. Pharmac. neerl. 1854—56. 155. Oudemans. Pharmacogn. 1880. 47. Planchon et Collin. Drogues simples. T. II. 1896. 34. Schneider. Powdered Veget. Drugs. 1902. 219. Solereder. Syst. Anat. d. Dicotyledonen. 1899. 519. Tschirch. Angew. Pflanzenanat. 1889. 116, 159, 282. 409, 411. Tschirch. Harze u. Harzbeh. 1900. 357, 361. v. Tieghem. Les canaux sécréteurs des plantes. Ann. d. Sciences. Sér. 5. Bot. T. XVI. 1872. 97. v. Tieghem. Sec. Mémoire sur les Canaux sécréteurs des Plantes. Ann. d. Sciences. Sér. 7. Bot. T. I. 1885. 6. Triebel. Ueber Oelbehälter in Wurzeln von Compositen. Nova Acta. Leop. Car. Ak. Naturf. Bd. I. no. 7. 1885. 34, Pl. I—VII. Wigand. Pharmacogn. 1879. 69. MATERIAL. The drug. Fresh roots, collected in the Botanic Garden at Groningen, thick 0.8 m.M. to 11 m.M. REAGENTS. Water, glycerine, potash, phloroglucin and hydrochloric acid, iodine and sulphuric acid 66 per cent., concentrated sulphuric acid, Schulze's macerating mixture.

MICROGRAPHY of the thinnest roots.

Epidermis. Thrown off.

Cortex.

Primary cork tissue. Nearly always consisting of 2 layers of cells; cells sometimes a little flattened. Intercellular spaces wanting.

Cells of the above. R. and T. 25 μ, L. 80—120 μ; tetra- to hexagonal prisms with a longitudinally directed axis; sometimes divided in 2 parts by a tangential or a transverse partition-wall. Walls varying from brown to red; suberized, often lignified. Cell contents: some granules along the walls.

Primary cortex proper. Consisting of common parenchyma, arranged in 4 successive different groups of layers.

1°. Some layers of pretty large very thin-walled cells, intermixed with bundles of smaller ones; generally flattened or torn to fragments.

2°. 2—3 layers of radially arranged cells. In a cross section the cells of these layers forming radial rows with the cells of the following layer. Showing intercellular spaces.

Cells of the above. R. 20 μ, T. 50 μ, L. 65 μ; elliptical cylinders or ellipsoids with a longitudinally directed axis; mostly divided in 2 parts by a radial wall. Walls yellow.

3°. 1 layer of cells showing incipient cork formation: 2 tangential and also 1 or 2 radial walls already having been formed [1]). Intercellular spaces wanting.

Cells of the above. L. 35—65 μ, depending on the presence or absence of a transverse wall; tetragonal prisms with a longitudinally directed axis. Walls persisting in sulphuric acid, those of the outermost layer sometimes lignified.

4°. 1 layer of cells, often showing tangential partition-walls. Cells in a cross

---

[1]) Tschirch. Angew. Pflanzenanat. 282.

section sometimes forming radial rows with those of the preceding layers. Intercellular spaces between this layer and the preceding and following ones.

C e l l s o f t h e a b o v e. R. 15 μ, T. 10—15 μ, L. 70 μ; tetragonal prisms with a longitudinally directed axis. Tangential walls mostly a little thickened; walls varying from yellow to brown.

E n d o d e r m i s. The parts corresponding with the bundles of primary xylem consisting of 1 layer of cells, sometimes having a tangential and a radial partition-wall; the parts corresponding to the bundles of primary phloem consisting of 2 layers of cells, the cells of the outermost of these 2 layers already again having formed 1 or 2 tangential walls. The latter parts each containing 1—5 long prismatical schizogenous resin glands, 15 μ in cross diameter; poly-, mostly tetragonal; cavity containing brown resin. Way of formation of glands still clearly to be traced.

E n d o d e r m i s c e l l s p r o p e r. In the parts corresponding to the xylem bundles R. 30 μ, T. 50—100 μ, depending on the presence or absence of radial walls, L. 55 μ; in the parts corresponding with the phloem bundles R. 25 μ, T. 40 μ, L. 35 μ; poly-, mostly tetragonal prisms with rounded edges and a longitudinally directed axis. In the parts corresponding to the xylem bundles in the innermost layer the radial walls partly lignified and suberized. Parts of the walls bordering on the cavities of the resin glands reddish brown; persisting in sulphuric acid.

S t e l e. 4-to 5-arch.

P e r i c y c l e. Sometimes consisting of 1 layer of cells.

See for the rest micrography of the drug.

MICROGRAPHY of the drug.

C o r t e x.

S e c o n d a r y c o r k t i s s u e. Initial-celled cork. More or less strongly developed.

P h e l l e m. Developed as periderm; 4—8 layers of cells lying in radial rows, often also pretty regularly arranged in tangential rows; many cells divided by numerous radial walls.

P e r i d e r m c e l l s. R. 16 μ, T. 25—80 μ, L. 20—35 μ, depending on the presence or absence of new transverse walls; tetragonal prisms with a longitudinally directed axis. Walls yellow, especially those of the innermost layers; suberized (potash). Cell contents: some granules.

P h e l l o g e n. Not clearly to be distinguished.

P h e l l o d e r m. Wanting.

Occasional formation of scaly rhytidoma; consequently in some roots the cork layers immediately adjoining the phloem.

E n d o d e r m i s. The parts containing schizogenous resin glands consisting of 3—4 layers of cells; the other parts often consisting of 2 layers of cells, tangential walls having been formed. Glands to the same number as in the thinnest roots. Intercellular spaces.

E n d o d e r m a l c e l l s p r o p e r. R. 30 μ, T. 50—100 μ, depending on the presence or absence of radial walls, L. 55 μ; ellipsoids or tetragonal prisms with strongly rounded edges and a longitudinally directed axis.

Stele.

Pericycle. Not clearly to be distinghuished.

Primary phloem. No more to be recognized.

Secondary phloem. Consisting of about 100 layers of elements, belonging to the cribral system and the parenchymatic system. **Cribral system.** Sieve-tubes and companion cells united with surrounding elements of the parenchymatic system in bundles of 5—30 elements in a cross section. Bundles lying in radial rows, the larger bundles on the outer side.

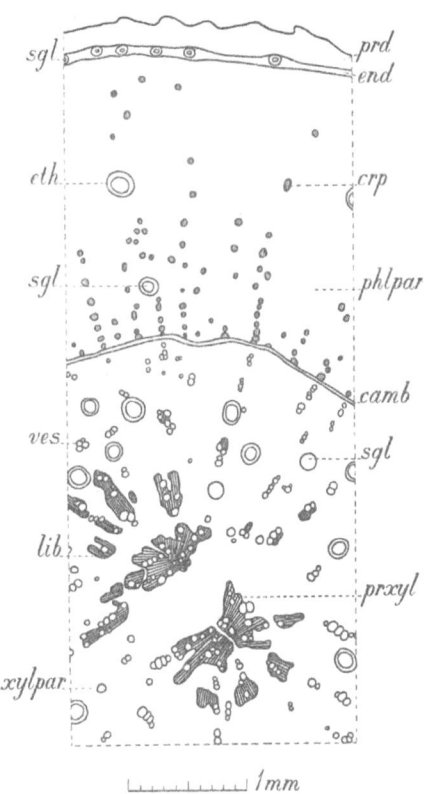

Fig. 80. *Inula Helenium*. Root, transverse section. camb Cambium; crp Bundles of elements of the cribral and parenchymatic system; end Endodermis; eth Epithelium; lib Libriform fibres; prd Periderm; phlpar Elements of the parenchymatic system of the secondary phloem; prxyl Primary xylem; sgl Schizogenous glands; ves Vessels; xylpar Elements of the parenchymatic system of the xylem.

Sieve-tubes. R. and T. 8—12 μ, L. of articulations 100—160 μ; polygonal prisms. Walls a little thickened; sieve-plates a few times distinct. Companion cells. R. and T. 10—30 μ, L. 100—160 μ; polygonal prisms. Walls somewhat thickened.

Parenchymatic system. Bast parenchyma consisting of common parenchyma and glands. The parenchyma showing an outer and an inner part, differing with regard to dimensions and contents of cells; radial arrangement hardly ever discernible. Glands occurring in both parts. Intercellular spaces often containing inulin.

Cells of the above. In the outer part R. 35 μ, T. 75 μ, L. 80 μ; tetragonal prisms with rounded edges or elliptical cylinders with a longitudinally directed axis; often a cell divided in 4 parts by a radial and a transverse wall. Cell contents: inulin in groups of cells. In the inner part R. and T. 30 μ, L. 150 μ; mostly tetragonal prisms with rounded edges and a longitudinally directed axis; often divided in 2 parts by a transverse wall. Cell contents: inulin along the walls of nearly every cell; in some cells a brown mass.

Schizogenous glands [1]). Lying in the bast parenchyma; showing an epithelium of some layers of cells without intercellular spaces. Glands in the outer part of the bast parenchyma. R. 200 μ, T. 230 μ, L. 250—400 μ; in the inner part smaller and less numerous; oval in a cross section. Contents of cavities: yellow resin, sometimes inulin.

---

[1]) See moreover v. Tieghem, Le Blois and Tschirch.

E p i t h e l i u m c e l l s. Thick 15 μ, wide 40—60 μ, L. 40 μ; tabular. Walls very thin. Cell contents: protoplasm.

Medullary rays. Generally not distinct; 1- to 3-seriate and separated from each other by 5—15 rows of cells. Small intercellular spaces.

C e l l s o f t h e a b o v e. R. and T. 40—45 μ, L. 35—70 μ; the wider cells in the outer part. Cell contents: see the parenchyma cells.

C a m b i u m. Consisting of some layers of elements.

C e l l s o f t h e a b o v e. R. 5—10 μ, T. 25 μ, L. 150 μ.

S e c o n d a r y x y l e m. Consisting of vessels, libriform fibres and elements of the parenchymatic system.

Elements of the tracheal system. Bundles of reticulate and pitted vessels, nearly always 1—3 vessels wide and 1—8 vessels thick.

V e s s e l s. R. 10—30 μ, T. 35 μ, L. of articulations 110—160 μ; polygonal prisms; transverse walls oblique, with 1 large perforation. Walls somewhat thickened; yellow; lignified; showing two- or one-sided bordered pits. C e l l s s u r r o u n d i n g t h e b u n d l e s. Walls a little thickened; pitted.

Libriform system. Libriform fibres sometimes lying close to a vessel.

Elements of the parenchymatic system.

Wood parenchyma. See the inner part of the bast parenchyma. Transverse walls more numerous. In the vicinity of the bundles of vessels the wood parenchyma fibres conspicuous. The inulin generally contained in groups of cells.

Glands. See those of the outer part of the secondary phloem.

Medullary rays. See those of the secondary phloem.

P r i m a r y x y l e m. Consisting usually of 4—5 bundles of vessels and libriform fibres. Bundles sometimes united together in the centre of the root. Each bundle containing only a few vessels.

V e s s e l s. 50 μ in diameter; polygonal. See for the rest those of the secondary xylem. L i b r i f o r m f i b r e s. R. and T. 12—18 μ; L. 250—350 μ; polygonal. Walls thickened; middle lamella conspicuous; lignified, especially the middle lamella; showing slit-like pits.

M e d u l l a r y c o m m i s s u r e s. Without intercellular spaces.

C e l l s o f t h e a b o v e. R. and T. 20—25 μ; L. 70 μ; polyhedra. Cell contents: inulin.

M e d u l l a. Wanting in roots with xylem bundles united in the centre. Without intercellular spaces.

C e l l s o f t h e a b o v e. See those of the medullary commissures.

Micrography of the powder.
Colourless parenchyma cells, spherical or more oblong, the latter often arranged in rows. A great many inulin sphaerites mostly shaped irregularly, sometimes showing radial striation, immediately dissolving in water when warmed. Reticulate vessels and vessels with slit-like bordered pits. Periderm cells square, wide, with thin light brown walls. Bundles of thick-walled libriform fibres with yellowish walls. Numerous nearly colourless bodies, measuring up to 60 μ, more or less obtuse-edged; crystaline, strongly refracting, looking like oily drops after being warmed in water. A few resin lumps, irregular, yellow to brown. Starch quite wanting.

November 1901.   J; M; L.

24

RADIX JALAPAE.
Jalapa. Jalap.
The turnip-shaped roots of Ipomoea Purga, Hayne, Arzneigew. XII. tt. 33. 34.

Macroscopic characters.

Generally thick 3—4 c.M.; hard; sinking down in a solution of sodium chloride, having a specific gravity of 1.2; pyriform, oblong or nearly spherical; entire, with incisions into pieces. Surface more or less wrinkled, with oblong transversely directed lenticels; colour dark, ashy-brown, especially in the wrinkles blackish. Internally massive or cracked; grayish white if containing starch grains not swollen up and then with a mealy fracture, or gray-brown if containing starch grains swollen up and then with a horny fracture; often both conditions occurring in the same root. Layer on the outside of the cambium thick $\frac{1}{2}$ up to 2 m.M.; dark coloured by the occurrence of many resin cells, the latter looking like dots under the magnifying glass. In the secondary xylem, forming the main part of the root, many cambia; the distribution of these principally to be recognized by the resin cells, occurring in the phloem layers formed by them. These cambia appearing in a cross section sometimes like large more or less regular circles, concentric with the original cambium, sometimes like small scattered circles, in the outer part of the xylem also now and then bow-shaped with the concave side turned outwards, sometimes like irregularly curved lines. Accordingly the cross section of different tubercules marbled in very different manners.

Odour peculiar; taste at first faint, afterwards pungent.

Anatomical characters.

LITERATURE.Berg. Anat. Atl. 1865. Taf. XXIII. Benecke. Mikr. Drogenprakt. 1912. 22. Biechele. Mikr. Prüf. d. off. Drogen. 1904. 106. Biermann. Ueber Bau u. Entwicklungs-gesch. d. Oelzellen u. d. Oelbildung in ihnen. Diss. Bern. 1897. 60. Erdmann-König. Allg. Warenk. 1895. 333. Florence A. Mc. Cormick. Notes on the Anatomy of the Young Tuber of Ipomoea Batatas Lam. Bot. Gazette. Vol. 61. 1916. 388. Flückiger & Hanbury. Pharmacographia. 1879. 443. Gilg. Pharmakogn. 1910. 288. Greenish & Collin. Anat. Atl. o. veget. powders. 1904. 242. Haase. Pharmakognotisch-chemische Untersuchung d. Ipomoea fisculosa, Mart. Diss. Strassburg. 1908. Hager. Pharm. Praxis. Bd. II. 1902. 102. Hérail. Mat. Méd. 1912. 524. Karsten u. Oltmanns. Pharmakogn. 1909. 99. Koch u. Gilg. Pharmakogn. Praktik. 1907. 89. Koch. Pharmakogn. Atlas. Bd. II. 1914. 109. Taf. XV. Koch. Mikr. Anal. d. Drogenpulver. Bd. 2. 1903. 103. Taf. X. Kraemer. Botany a. Pharmacogn. 1910. 452. Luerssen. Syst. Bot. Bd. 2. 1882. 958. Lloyd. Ascission in Mirabilis Jalapa. Bot. Gazette. Vol. 61. 1916. 213—230. Marmé. Pharmacogn. 1866. 59. Meyer. Wiss. Drogenk. Bd. I. 1891. 293. Moeller. Pharmakogn. 1889. 305. Moeller. Mikr. pharm. Ueb. 1900. 320. Oudemans. Aanteek. o. d. Pharmac. neerl. 1854—56. 312. Oude-mans. Pharmacogn. 1880. 28. Planchon et Collin. Drogues simples. T. I. 1895. 599. Schmitz. Sitzungsberichte der Naturforschende Gesellschaft zu Halle. Bot. Zeitung. 1875. 677. Schneider. Powdered Veget. Drugs. 1902. 224. Scott. On some points in the Anatomy of Ipomoea versicolor, Meissn. Ann. of Botany. Vol. 5. 1890. 173. Solereder. Syst. Anat. 1899. 647. Tschirch. Angew. Pfl. Anat. 1889. 6, 95, 241, 334, 472. Tschirch. Pharmakogn. Bd. II. Abt. 2. 1917. 1324. Vogl. Anat. Atl. 1887. 55. Wigand. Pharmakogn. 1879. 63. MATERIAL. The drug: tubercules, greatly varying in size, some of them thick about 33—40 m.M. Tubercules cultivated in the Botanic Garden at Groningen, gathered November 1901, thick 1, 3.7, 4.5, 6.5, 9, 12, 22, 30 and 55 m.M., fresh and in alcohol. REAGENTS. Water, glycerine, potash, iodine in chloral hydrate, phloroglucin and hydrochloric acid, iodine and sulphuric acid 66 per cent., concentrated sulphuric acid, Schulze's macerating mixture, copper acetate and iron acetate.

MICROGRAPHY.

E p i d e r m i s. Wanting in the drug. Only occurring in the thinnest material.
C e l l s o f t h e a b o v e. R. 10 μ, T. 15 μ. Walls lignified.

C o r t e x.

Exodermis. Wanting in the drug. Occurring only in the thinnest material and consisting of a single layer of parenchymatic cells. Idioblasts containing a yellow mass.
P a r e n c h y m a t i c c e l l s. R. 25 μ, T. 30 μ, L. 170 μ; tetra- to hexagonal prisms with a longitudinally directed axis. Walls lignified. I d i o b l a s t s. R. 25 μ, T. 30 μ, L. 60 μ; tetragonal prisms with strongly curved outer walls. Outer walls thickened; all walls yellow, suberized and lignified. Cell contents: a yellow mass.

Secondary cork tissue. Formed in the outmost cell layer of the primary cortex proper; its development beginning under the idioblasts of the exodermis, mentioned above. Lenticels surrounded by a slightly developed rhytidoma, consisting of a few scales. For the greater part remaining in the drug.

O r d i n a r y s e c o n d a r y c o r k t i s s u e.

P h e l l e m. Consisting of periderm; thrown off in large parts thick 5—8 cells; for the rest formed by 8—15 layers of cells, arranged radially and fairly well in longitudinal rows; outermost cell layers sometimes flattened.
P e r i d e r m c e l l s. R. 15 μ, T. 40—60 μ, L. 30 μ; tetra- to hexagonal, mostly tetragonal tables; sometimes the cells of the innermost layers with a radial partition wall. Walls yellow to brown; suberized; in the youngest and thinnest material also lignified. Cell contents: sometimes a brown mass.

P h e l l o g e n. In the thickest alcoholic material from the Garden very conspicuous; in the drug almost never discernible.

P h e l l o d e r m. Mostly wanting, a few times represented by 1 or 2 layers of cells.

L e n t i c e l s. Showing a more or less regular arrangement of cells in radial rows of about 25; cells often with 1 or more tangential partition walls and diminishing in size towards the outside.
C e l l s o f t h e a b o v e. R. 15—20 μ, T. and L. 35 μ; polygonal tables with strongly rounded edges. Walls of the cells of the outermost cell layers coloured brown; suberized; showing intercellular spaces.

Primary cortex proper. In the thinnest material from the Garden consisting of 4—5 cell layers, sometimes very conspicuously arranged in radial rows, but these rows not corresponding with those of the cork tissue; cells of the outermost cell layer sometimes showing a tangential partition wall; here and there a cell with some cross partition walls, each partition containing a cluster crystal. In the drug consisting of about 10 layers of cells or often less, without any radial arrangement. Cluster crystal idioblasts in longitudinal rows.

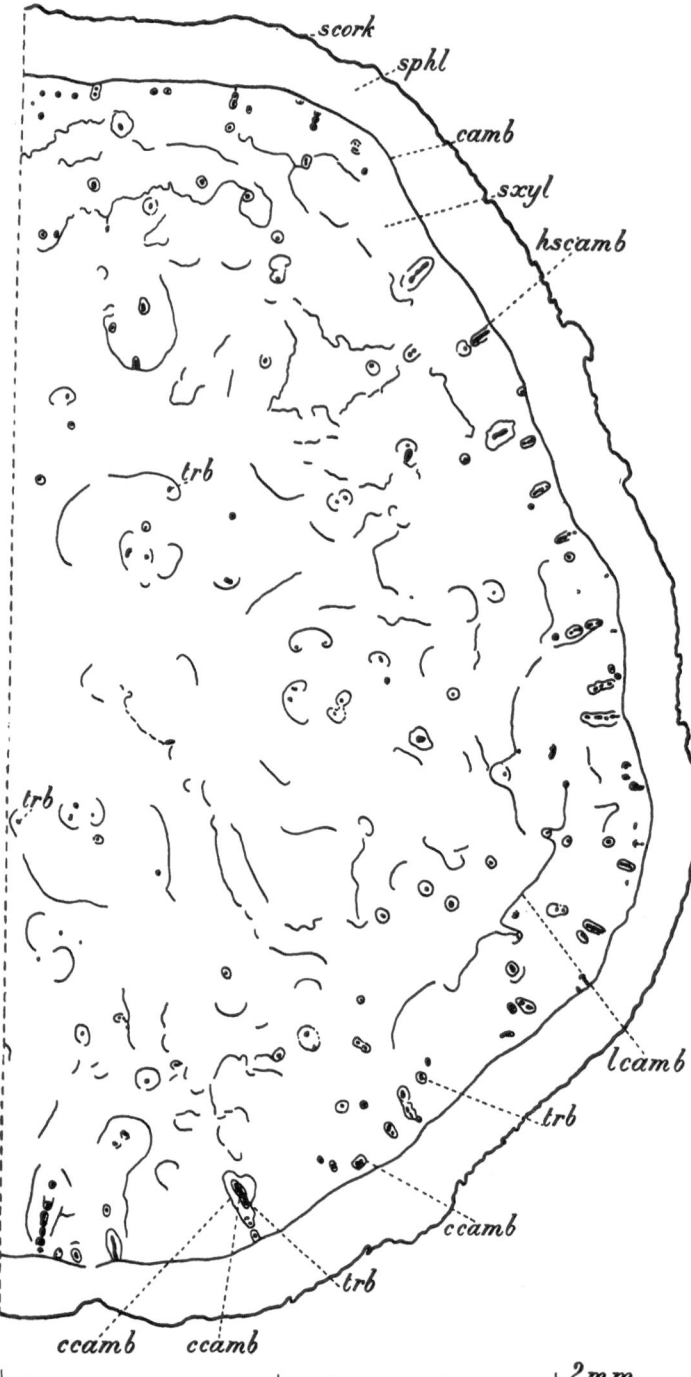

Fig. 81. *Ipomoea Purga*. Tubercule, transverse section. camb Cambium; ccamb Circular cambium layer; hscamb Horse-shoe-shaped cambium layer; lcamb Linear cambium layer; scork Secondary cork tissue; sphl Secondary phloem; sxyl Secondary xylem; trb Tracheal bundles.

Cells of the above. In the drug R. 45 μ, T. 100—300 μ, L. 50 μ; strongly tangentially elongated ellipsoids, often containing 1—6 radial partition-walls. Walls sometimes yellow, then remaining in concentrated sulphuric acid; pitted; showing intercellular spaces. Cell contents: starch grains mostly simple and globular or ellipsoidal, measuring 4—40 μ, mostly with a central, sometimes with an eccentric hilum, sometimes marked by a 3-cornered cleft, stratification clearly discernible; sometimes compound grains, generally 2-adelphous, a few times 3- to 4-adelphous.

Endodermis. In the shape of a protective-sheath, afterwards in roots thicker than 1 m.M. more or less in the shape of a sclerenchyma-sheath; consisting in roots, thick 1 m.M., of a single continuous layer of parenchyma cells, exhibiting suberized radial walls and often a tangential partition wall; in roots somewhat thicker than 1 m.M. walls, except the outer tangential ones, thickened and showing pit canals; in thicker roots, e. g. of 8 m.M. thick, to be found only here and there in the shape of some often flattened cells; in the drug not to be recognized.

S t e l e. In the thinnest material clearly tetrarch; in the drug the primary xylem not to be distinguished.

Pericycle. In the thinnest material consisting of a single layer of common parenchyma cells; in the drug not to be distinguished.

Phloem.

P r i m a r y p h l o e m. In the drug not to be distinguished.

S e c o n d a r y p h l o e m. Consisting of about 35 layers of elements, arranged fairly well in radial rows; all elements more or less flattened in a radial direction. **S i e v e - t u b e s.** In bundles, consisting of 4—10 elements in a transverse section; only near the cambium not flattened. System of **b a s t f i b r e s** wanting. **P a r e n c h y m a t i c s y s t e m** represented by bast parenchyma and medullary rays. Bast parenchyma consisting of common parenchyma cells and idioblasts in 2 kinds [1]: 1. cluster crystal cells like those in the cortex, but less numerous; 2. resin cells very numerous and often arranged in longitudinal rows, composed of about 20 cells. Medullary rays 1- to 2-seriate, high 5—8 cells, most clearly discernible in radial sections.

S i e v e - t u b e s. 5 μ in diameter or 4 by 10 μ, L. of articulations 80 μ; polygonal prisms, often with rounded sides. **B a s t p a r e n c h y m a c e l l s.** R. 45 μ, T. 40 μ, L. 80 μ; rectangular prisms with a longitudinally directed axis and rounded edges, often with a cross partition-wall. Walls pitted; showing intercellular spaces. Cell contents: see the cortical parenchyma. **R e s i n i d i o b l a s t s.** R. 60 μ, T. 50 μ, L. 60 μ; polygonal prisms with a longitudinally directed axis. Walls having a suberized outermost lamella. Contents of nearly all cells: a globular resin mass, often enclosing small more strongly refringent globules; resin gray if not coming into contact with water, soluble in alcohol, insoluble in water, colouring yellow and soon disappearing in potash, becoming lemon-coloured in potassium iodide iodine. **C e l l s o f m e d u l l a r y r a y s.** R. 70 μ, T. 30 μ, L. 40 μ; rectangular prisms with a radially directed axis and rounded edges. See for the rest the bast parenchyma cells.

C a m b i u m. Consisting of 3—4 layers of elements; flattened in the drug.
E l e m e n t s o f t h e a b o v e. R. 10 μ, T. 25 μ, L. 80 μ; rectangular prisms with a longitudinally directed axis.

Xylem.

S e c o n d a r y x y l e m. Already developed in roots thick 1 m.M. and soon forming the bulk of the root. **T r a c h e a l s y s t e m** represented by pitted vessels and tracheid fibres united together in longitudinal small or somewhat larger bundles, only the latter containing both kinds of elements; these bundles varying greatly in number in different roots, often especially in the peripheral

[1] According to M e y e r. l. c. p. 296, P l a n c h o n et C o l l i n. l. c. p. 609, K o c h. Mikr. Anal. d. Drogenpulver. Bd. 2. p. 106 there are also a few sclereids and according to M e y e r sclereids also occur in the endodermis. I did not observe them, but there were found by Moll in the powder, as mentioned beneeth.

parts of the thicker roots some of them near to one another particularly in a
radial direction. S y s t e m o f l i b r i f o r m f i b r e s wanting. P a r e n -
c h y m a t i c s y s t e m consisting of wood parenchyma and medullary rays,
the latter like those of the secondary phloem. Wood parenchyma consisting of
common parenchyma; the cells in the vicinity of the tracheal bundles, giving
rise afterwards to cambia, being of smaller dimensions; cluster crystal idioblasts
and resin cells wanting.

V e s s e l s. R. and T. 20 by 25 μ, up to 40 by 50 μ, L. of articulations 70—170 μ; polygonal prisms. Transverse walls with a single circular perforation. Walls thickened; coloured
somewhat yellow; lignified; showing bordered pits. T r a c h e i d f i b r e s. R. and
T. 15—20 μ, L. 100—170 μ. Walls thickened; coloured somewhat yellow; slightly lignified;

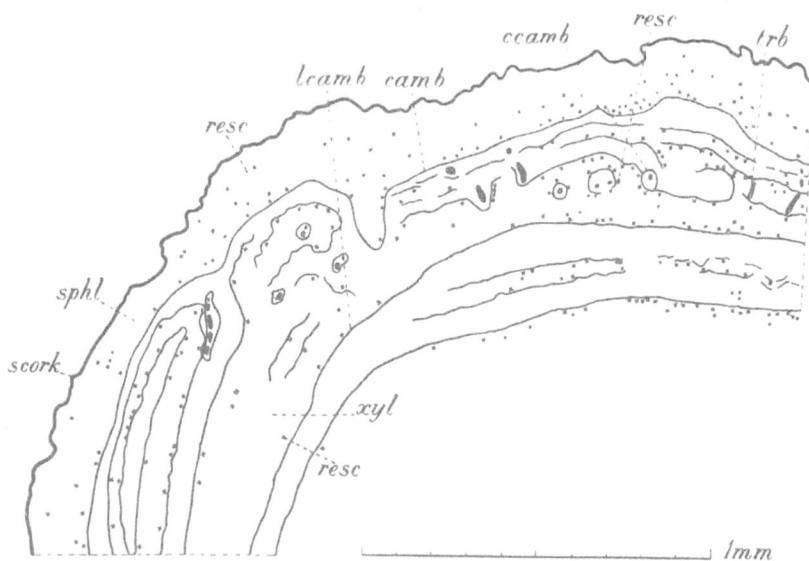

Fig. 82. *Ipomoea Purga.* Tubercule, transverse section showing many linear cambium
layers. camb Cambium; ccamb Circular cambium layer; lcamb Linear cambium layer;
resc Resin cells; scork Secondary cork tissue; sphl Secondary phloem; sxyl Secondary
xylem; trb Tracheal bundles.

with bordered pits. W o o d p a r e n c h y m a c e l l s. R. 70 μ, T. 45 μ, L. 80 μ; rectangular prisms with a longitudinally directed axis and rounded edges. Walls pitted.
Showing intercellular spaces. Cell contents: see bast parenchyma cells.

Especially in the younger parts of this secondary xylem in the vicinity of the
original cambium new cambium layers being produced during the whole secondary growth of the tubercule, hence these cambia and their products scattered
throughout the whole secondary xylem of the drug. These new cambium layers
generally formed in the vicinity of the tracheal bundles and more or less enclosing them. In these cases tertiary phloem being formed on the convex side of
the cambium and tertiary xylem on the concave side. Tertiary phloem mostly
one third in thickness of the tertiary xylem. Tertiary phloem in all principal

points like the secondary phloem. Tertiary xylem without tracheal bundles, thus only consisting of the parenchymatic system.

The formation of tertiary tissues, described above, appearing in the following manners:

1. Formation of cambium layers about circular in a transverse section developing mostly around a single tracheal bundle and at first in closest vicinity to it. The development of these circular cambia often beginning at the inside of tracheal bundles just formed by the original normal cambium of the tubercule, thus forming at first horse-shoe-shaped cambia not touching the original cambium and afterwards forming circular cambia independant from it. In a few cases these circular cambia showing no tracheal bundle in their centre. Now and then these circular cambia developing close to one another often in tangential rows and soon parts of the tertiary phloem produced by them, coming into contact with one another, becoming flattened and the cambium in these parts loosing its meristematic character; the remaining parts of these several cambia fusing together and forming elliptical cambia surrounding a certain not very large number of tracheal bundles. In the same cases the radial portions of such elliptical cambia loosing their meristematic character two linear tangentially directed portions of cambium remaining; the outer one forming tertiary phloem towards the periphery of the tubercule but the inner one towards its centre.

2. Formation of cambia linear in a transverse section and mostly parallel to the original cambium of the tubercule and most abundant in tubercules containing not many tracheal bundles. These linear cambia in 10 per cent. of the examined specimens of the drug much more numerous and extensive than the circular cambia mentioned above. Linear cambia occurring: a. in connection with a circular cambium as a tangent line, touching either the side of a circular (or horse-shoe-shaped) cambium turned towards the centre of the tubercule or the side of a circular cambium, turned towards the periphery of the tubercule; the linear cambia touching the outside of a circular cambium forming phloem towards the periphery of the tubercule and xylem towards the centre, but those linear cambia touching the inside of the circular cambium forming phloem inwards and xylem outwards; b. without any connection with circular cambia. All new cambium layers, whether circular or linear, appearing already in very young tubercules. In both cases also now and then the tertiary xylem producing again a new cambium layer, parallel to the existing one; by means of this new cambium layer formation of quaternary phloem and xylem taking place. The new cambia layers producing quaternary tissues, when appearing in the tertiary xylem of linear cambia, forming phloem and xylem in directions inverted with respect to those of the tertiary tissues, belonging to them.

The various cambia described above, when becoming older, in some of their parts suspending their activity and becoming adult, but in other parts continuing their activity; thence in later stages of development many dislocations of the various tissues leading to the appearance of irregular and curious structures (see the middle part of fig. 81) in the inner parts of large tubercules.

Micrography of the powder. Very much starch, mostly simple about spherical grains, the numerous larger grains about 30 μ in diameter; hilum central or eccentric (½), often conspicuous by a cleft filled with air; stratification often very conspicuous; also compound grains, the main part 2-adelphous, many 3-adelphous, a few 4-adelphous; the starch grains for the rest appearing in all possible stages of swelling. Many resin-spheres of 100 μ or less in diameter, colouring yellow in diluted iodine solution, consisting of an emulsion. Common parenchyma cells sometimes coloured brown; often arranged in rows; not seldom filled with starch, either swollen up or not. Parts of the cork layer consisting of somewhat elongated penta- to hexagonal cells with red-brown walls. Pieces of pitted vessels sometimes scalariform or spiral vessels. Cluster crystals. A few sclereids with not very strongly thickened yellowish walls, exhibiting a conspicuous stratification and pit canals.

February 1901.   J; M; v. E. d. W.

RADIX LIQUIRITIAE.
Liquorice Root.
The rhizome and the roots of Glycyrrhiza glabra, Linn. Sp. Pl. 742, or of Glycyrrhiza glabra, var. glandulifera, Waldst. a. Kit. Pl. Rar. Hung. I. 20. t. 21.

Macroscopic characters.
Pieces 5—10 d.M. in length, 5—25 m.M. thick; firm, though, heavy; terete usually unbranched. Surface wrinkled longitudinally, reddish brown, in many places showing fine greyish brown fragments of cast off tissue. Internally dark yellow; transverse fracture coarsely fibrous; transverse section radiate. The rhizome showing a slightly dark, dirty grey, often angular medulla.
Odourless; taste purely sweet.

Anatomical characters.
LITERATURE. Benecke. Micr. Drogenprakt. 1912. 15. Flückiger. Pharmakogn. 1891. 378. Flückiger & Hanbury. Pharmacographia. 1879. 181. Gilg. Pharmakogn. 1910. 171. Greenish & Collin. Anat. Atl. o. veget. powders. 1904. 244. Hager. Pharm. Praxis. Bd. I. 1900. 1226. Hérail. Mat. Méd. 1912. 18. Karsten u. Oltmanns. Pharmakogn. 1909. 62. Koch. Mikr. Anal. d. Drogenpulver. Bd. II. 1903. 187. Pl. XVII. Koch. Einf. i. d. mikr. Anal. d. Drogenpulver. 1906. 53. Koch. Pharmakogn. Atl. Bd. II. 1914. 48. Pl. VII. Koch u. Gilg. Pharmakogn. Praktik. 1907. 121. Kraemer. Botany a. Pharmacogn. 1910. 735. Marmé. Pharmacogn. 1886. 35. Meyer. Wiss. Drogenk. Bd. I. 1891. 230. Moeller. Pharmakogn. 1889. 331. Moeller. Mikr. phar. Üb. 1901. 290. Oudemans. Aant. o. d. Pharm. neerl. 1854—56. 584. Oudemans. Pharmacogn. 1880. 80. Planchon et Collin. Drogues simples. T. II. 1896. 498. Schneider. Powd. veget. Drugs. 1902. 203. Senft. Über d. s. g. Inklusen in Glycyrrhiza glabra L. u. üb. ihre Funkt. Ber. d. d. bot. Ges. Bd. XXXIV. Heft 9. 1916. 710. Stcherbatscheff. Beitr. z. Entw. gesch. einiger off. Pflanzen. Arch. d. Pharm. Bd. 245. 1907; root p. 59. Tschirch. Pharmacogn. Bd. II. Abt. 1. 1912. 77. Tschirch u. Oesterle. Anat. Atl. 1900. 29. Pl. VIII. Tschirch u. Holfert. Über d. Süssholz. Arch. d. Pharm. Bd. XXVI. Heft 2. 1888. Vogl. Anat. Atl. 1887. 52. Wiesner. Rohstoffe.

1900. 528. Wigand. Pharmakogn. 1879. 38. MATERIAL. The drug. Alcoholic material, thick up to 12—15 m.M., gathered in the Botanic Garden at Groningen, partly June 3rd 1901. REAGENTS. Glycerine, chloral hydrate, potassium iodide iodine, iodine and sulphuric acid 66 per cent., phloroglucin and hydrochloric acid, concentrated sulphuric acid, Schulze's reagent.

MICROGRAPHY.

Root.

Epidermis. Already thrown off in the first year by the development of cork in the pericycle.

Cortex. Thrown off in the same way; in the drug here and there still extant as some wholly dried up cells. Walls colouring red in phloroglucin and hydrochloric acid.

S t e l e. 2- to 4-, mostly 3-arch.

P e r i c y c l e.

Secondary cork tissue. Developed in the young roots already in the first year. *Secondary cork tissue proper.* Initial-celled; dipleuric.

P h e l l e m. Developed as periderm; consisting of 10—15 cell layers, radially arranged.

C e l l s o f t h e a b o v e. R. 9 μ, T. 20—40 μ, L. 12 μ; tetra-, sometimes penta- or hexagonal tables with a radially directed axis. Walls slightly thickened; yellow; suberized; walls of the outermost periderm cells of very young roots colouring red in phloroglucin and hydrochloric acid. Cell contents: brown lumps quite filling the cells. In sulphuric acid the periderm becoming very dark to black.

P h e l l o g e n. Consisting of 1 layer of cells.

P h e l l o d e r m. Consisting of common parenchyma, 3—4 layers of cells, radially arranged; often showing a smaller number of radial walls than the periderm cells. Intercellular spaces wanting.

C e l l s o f t h e a b o v e. Walls somewhat thickened, sometimes slightly collenchymatous; somewhat yellow. Cell contents: small simple starch grains.

Lenticells. Rather numerous; T. 10 m.M., L. 1 m.M.

P h e l l e m. Consisting of 20 periderm layers, composed of closing bands alternating with complementary tissue; at equal distances 3 closing bands, each 3 cells thick. Closing bands showing trigonal intercellular spaces.

C e l l s o f c l o s i n g b a n d s. Similar to the periderm cells. C e l l s o f c o m p l e m e n t a r y t i s s u e. R. 30 μ, T. and L. 20 μ; cylinders or ellipsoids with a radially directed axis. Walls suberized.

P h e l l o g e n. Not to be distinghuished from the outermost phelloderm layers.

P h e l l o d e r m. Consisting of common parenchyma, 8—10 layers of cells, radially arranged.

C e l l s o f t h e a b o v e. R. 9 μ, T. 10—20 μ. Walls slightly thickened, towards the interior more strongly so; colouring red in sulphuric acid; in the outermost layers yellow; suberized. Cell contents: in the innermost layers starch.

P r i m a r y p h l o e m. No more to be distinguished in the drug.

S e c o n d a r y p h l o e m.

Rhytidoma. Scaly; containing in the middle about 10—12 layers of secondary

phloem. Scales about ¾ m.M. in diameter, about circular in surface view.
C e l l s   o f   s e c o n d a r y   p h l o e m. Walls of the innermost cells somewhat thickened;
yellow; colouring red in phloroglucin and hydrochloric acid. Cell contents: in the outermost layers starch, in the innermost layers a yellow granular mass or brown lumps.
C e l l s   o f   s e c o n d a r y   c o r k   t i s s u e. See secondary cork tissue proper and
lenticels.

Secondary phloem proper. Tangential expansion, necessary for keeping up with
the increase in thickness of the root, partly owing to the medullary rays widening outward, partly to the growth and division of the bast parenchyma cells. Consisting of elements of the cribral system, bast fibres and elements of the parenchymatic system. Showing radial structure, due to radial plates of keratenchyma and bands of bast parenchyma. E l e m e n t s   o f   t h e   c r i b r a l s y s t e m. Sieve-tubes and perhaps companion cells, lying in bundles soon reduced to the above mentioned plates of keratenchyma. Elements of the system of bast fibres. Bast fibres arranged in cylindrical bundles, each containing 10—50 fibres in a transverse section and imbedded

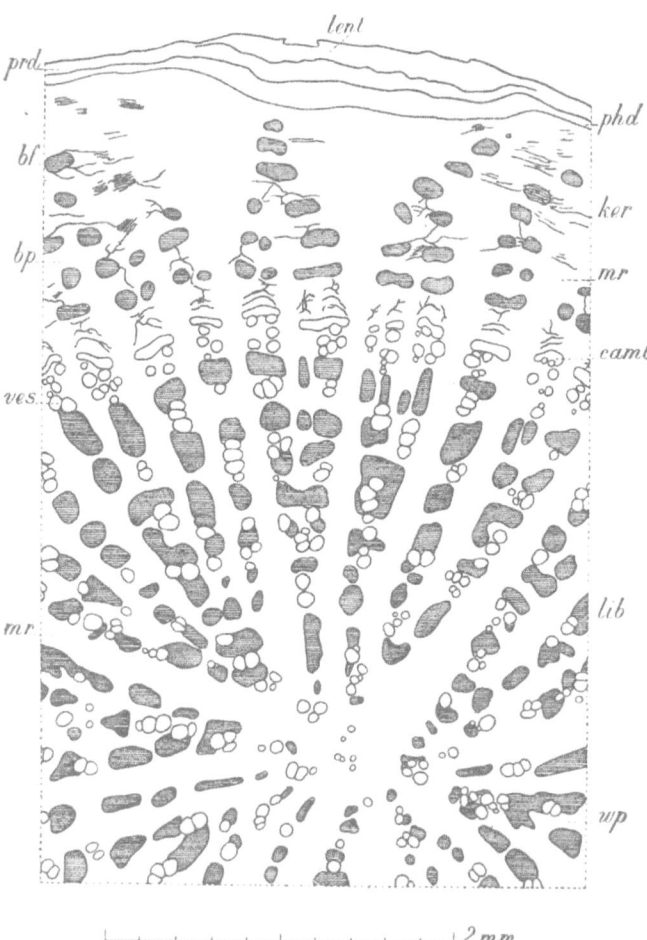

Fig. 83. *Glycyrrhiza glabra.* Root, transverse section. bf Bast fibres;
bp Bast parenchyma; camb Cambium; ker Keratenchyma; lent Lenticels; lib Libriform fibres; mr Medullary rays; phd Phelloderm; prd
Periderm; ves Vessels; wp Wood parenchyma.

in the bast parenchyma bands, in each band mostly 2, sometimes 1 bundle in the
tangential and several ones in the radial direction; bundles anastomosing in a
tangential plane by means of some fibres. Fibres sometimes divided by transverse
walls. Intercellular spaces wanting. E l e m e n t s   o f   t h e   p a r e n c h y m a t i c

s y s t e m. Bast parenchyma, consisting of common bast parenchyma cells and idioblasts in 3 kinds; in bands of 5—15 cells wide and round the bundles of sieve-tubes; only in the inner part cells arranged in radial rows. Intercellular spaces diverging in all directions. Idioblasts: 1°. crystal cells, nearly exclusively in the outer part of the bast parenchyma; 2°. partitioned crystal fibres, forming a sheath of 1 fibre thick round the bundles of bast fibres; fibres divided in up to 8 cells; 3°. sphaero-crystal cells, not numerous. Medullary rays not always quite to be traced as far as the outside of the secondary phloem. In the vicinity of the cambium 2- to 5-seriate, this number increasing towards the outside; very high, up to 100 cells; containing some sphaero-crystal cells.

S i e v e-t u b e s. R. and T. 9 μ, L. of articulations 55—70 μ; tetra- to hexagonal prisms. Walls somewhat thickened; callus conspicuous. B a s t f i b r e s. R. and T. 10 μ, L. 700—1200 μ; tetra- to hexagonal. Walls very thick; consisting of an outer lignified yellow part and a thicker colourless hardly lignified inner part showing beautiful cellulose reaction, often colouring pink in potassium iodide iodine; both parts showing stratification; hardly any cell cavity left, middle lamella discernible as a dark line. B a s t p a r e n-c h y m a c e l l s. In the outer part R. 28 μ, T. 50—100 μ, L. 28 μ; ellipsoids or ellipsoidal polyhedra, often divided in 2 parts by a radial wall. In the inner part R. and T. 25 μ, L. 55 μ; tetra- to hexagonal prisms or cylinders with a longitudinally directed axis. Walls slightly yellow; pitted. Cell contents: many simple starch grains, spherical, ellipsoidal or angular; in alcoholic material only a few small grains always grouped round the nucleus; in many cells also pale yellow transparent granular contents, leaving behind a network in chloral hydrate. I d i o b l a s t s. Crystal cells, shape and dimensions like those of the surrounding parenchyma cells. Walls usually thickened, the simple crystal entirely filling the cavity; slightly lignified, but showing a distinct cellulose reaction. Crystal dissolving exceedingly slowly in concentrated sulphuric acid. Partitioned crystal fibres. R. 12 μ, T. 16 μ, L. 200 μ; tetra- to hexagonal. Walls and crystals like those of the crystal cells. Each constituent cell containing 1 simple crystal. Cells with sphaero-crystals. Quite filled with sphaero-crystals with a beautiful radiate structure; colour slightly brown; dissolving after some days in glycerine gelatine; see for the rest the surrounding cells.

C a m b i u m. Only here and there to be distinguished in the bands of parenchyma; 3 layers of elements thick.

E l e m e n t s o f t h e a b o v e. R. 10 μ, T. 12 μ, L. 100 μ.

S e c o n d a r y x y l e m. Consisting of elements of the tracheal system, libriform fibres and elements of the parenchymatic system.

E l e m e n t s o f t h e t r a c h e a l s y s t e m. Vessels isolated, or a small number of them united in bundles; pitted. Tracheid fibres always immediately adjoining the vessels, pitted, reticulate or showing rhombic thickenings. E l e m e n t s o f t h e l i b r i f o r m s y s t e m. Libriform fibres, see bast fibres. E l e m e n t s o f t h e p a r e n c h y m a t i c s y s t e m. Wood parenchyma consisting of 2 kinds of common parenchyma and idioblasts; the parenchyma cells of the first kind lying between the vessels and showing no intercellular spaces. Medullary rays very high, up to 100 cells.

V e s s e l s. R. and T. 25—75 μ, L. of articulations 70—130 μ; tetra- to hexagonal; transverse walls nearly always horizontal, with a single large perforation. Walls thickened;

yellowish brown; lignified, yet showing some cellulose reaction; strongly swelling up in sulphuric acid, the cavity sometimes disappearing; showing slit-like bordered pits. T r a c h e i d f i b r e s. R. and T. 10—17 μ, L. 100—140 μ; ends obtuse, longitudinal walls somewhat sinuous. Walls slightly thickened; yellow; outermost part lignified, inner thickest part showing cellulose reaction, strongly swelling up in sulphuric acid. L i b r i f o r m f i b r e s. See the bast fibres. W o o d p a r e n c h y m a c e l l s b e t w e e n t h e v e s s e l s. R. 10 μ, T. 15 μ, L. 45 μ; tetra- to hexagonal prisms. Walls somewhat thickened; yellow; chemical constitution like that of the walls of the tracheid fibres; showing very distinct pits. O t h e r w o o d p a r e n c h y m a c e l l s. R. 25 μ, T. 20 μ, L. 60 μ; see for the rest the bast parenchyma cells. C e l l s o f t h e m e d u l l a r y r a y s. Like those of the medullary rays of the phloem in the vicinity of the cambium, but the radial dimensions somewhat larger. I d i o b l a s t s. See those of the secondary phloem.

R h i z o m e.

Differing from the root chiefly by the presence of a smaller or larger medulla. The larger medulla sometimes taking up half of the diameter of the rhizome and indistinctly pentagonal.

M e d u l l a. Consisting of common parenchyma and idioblasts like the secondary phloem of the root; parenchyma cells arranged in longitudinal rows; showing intercellular spaces.

C o m m o n p a r e n c h y m a c e l l s. R. and T. 50 μ, L. 30—40 μ; cylinders or tetra- to hexagonal tables with a longitudinally directed axis. Walls pitted. Cell contents: see those of the bast parenchyma cells.

Micrography of the powder. Very much starch grains; simple, spherical or oblong, the largest ones 20 μ, a great many smaller ones; hilum centric; not much stratification to be seen. Thick-walled very long fibres, mostly united in bundles. Crystal cells often forming a sheath round the bundles; besides many detached crystals; crystals penta- or hexagonal, often oblong. Vessels penta- or hexagonal, showing slit-like bordered pits. Thin-walled parenchyma cells filled with starch. Cork cells occurring occasionally, but wanting in the peeled Russian Liquiritia. In concentrated sulphuric acid colouring orange to yellow. Powder of Radix Liquiritiae should have the sweet taste of the root, without a strange aftertaste.

May 1901.   J; M; L.

RADIX PYRETHRI.
Pyrethrum. Pyrethrum Root.
The dried root of Anacyclus Pyrethrum, D.C. Fl. Fr. Suppl. 480.

LITERATURE. Berg. Anat. Atl. 1865. 15. Pl. IX. Le Blois. Can. Sécrét. e. Poches Sécrétr. Ann. d. Sc. Sér. 7. Bot. T. VI. 1887. 274. Flückiger. Pharmakogn. 1891. 473. Flückiger & Hanbury. Pharmacographia. 1879. 383. Greenish & Collin. Anat. Atl. o. veget. powders. 1904. 252. Hager. Pharm. Praxis. Bd. II. 1902. 702. Hérail. Mat. Méd. 1912. 422. Kraemer. Botany a. Pharmacogn. 1910. 185, 778. Moeller. Mikr. pharm. Üb. 1901. 307. Oudemans. Pharmacogn. 1880. 49. Planchon et Collin. Drogues simples. T. II. 1896. 40. Schneider. Powdered Veget. Drugs. 1902. 267. Solereder. Syst. Anat. d. Dicot. 1889. 515. v. Tieghem. Can. Sécrét. d. Plantes. Ann. d. Sc. Sér. 5. Bot. T. XVI. 1872. 97.

v. Tieghem. Sec. Mém. sur les Can. sécrét. d. Plantes. Ann. d. Sc. Sér. 7. Bot. T. I. 1885. 6. Triebel. Oelbeh. i. Wurz. v. Comp. Nova Acta. Leop. Carol. Ak. Naturf. Bd. L. no. 7. 1885; report in Bot. Jahresber. 1885. 794. Tschirch. Harze u. Harzbeh. 1900. e. g. 343, 360, 361, etc. Wigand. Pharmakogn. 1879. 70. MATERIAL. The drug, 0.8—4.4 c.M. in diameter,soaked in water for one night. REAGENTS. Water, glycerine, potash, iodine in chloral hydrate, phloroglucin and hydrochloric acid, iodine and sulphuric acid 66 per cent., concentrated sulphuric acid, Schulze's macerating mixture.

MICROGRAPHY.

E p i d e r m i s. Thrown off in the drug.

C o r t e x. Mostly thrown off. On 1 of the examined specimens still extant at 4 projecting parts, probably corresponding with the number of radii of the tetrarch root.

Primary cortex proper. Consisting of 2—5 layers of common parenchyma cells, showing intercellular spaces. Cells of the outermost layers having a greater number of radial partition-walls, e. g. 6, and often a transverse partition-wall; cells of the innermost layer or layers lying in radial rows corresponding to the cells of the endodermis. Round the resin canals in the vicinity of the endodermis mostly some layers of radiatingly arranged cells, also corresponding to the cells of the endodermis, having radial partition-walls and showing intercellular spaces. Containing in the outer layers a few schizogenous glands provided with an epithelium and similar to those of the secondary phloem; in the vicinity of the endodermis schizogenous resin canals without an epithelium, the cavity sometimes bounded by the endodermis. Cavity R. 30—50 μ, T. 40—150 μ; contents: a yellowish to brown mass [1]).

C o m m o n p a r e n c h y m a c e l l s. In the outermost part the original, subsequently partitioned cells R. 25—40 μ, T. up to 260 μ, L. 50—80 μ; polygonal prisms with rounded edges and a longitudinally directed axis. Cells of the innermost layer R. 10—20 μ, T. 25 μ, L. 40—80 μ.

Endodermis. Developed as a protective-sheath; consisting of 1 layer of cells, often showing transverse partition-walls; often more or less flattened.

C e l l s o f t h e a b o v e. R. 25 μ, T. 20—30 μ (referring to the constituent cells produced by the subsequent development of radial walls), L. 40—80 μ. Radial walls with a somewhat thickened suberized part (Casparian spots) in the inner half.

S t e l e. Probably tetrarch.

Pericycle. Mostly thrown off, still extant on roots having retained the cortex; 1, perhaps 2—3 layers of cells with 1, sometimes more radial partition-walls.

C e l l s o f t h e a b o v e. R. 30 μ, T. 70 μ, L. 50—80 μ; polygonal prisms with a longitudinally directed axis and rounded edges.

Phloem.

P r i m a r y p h l o e m. Not to be recognized.

S e c o n d a r y p h l o e m.

Rhytidoma. Scaly.

---

[1]) See moreover v. T i e g h e m, T r i e b e l, L e B l o i s and S o l e r e d e r, l.l. c.c.

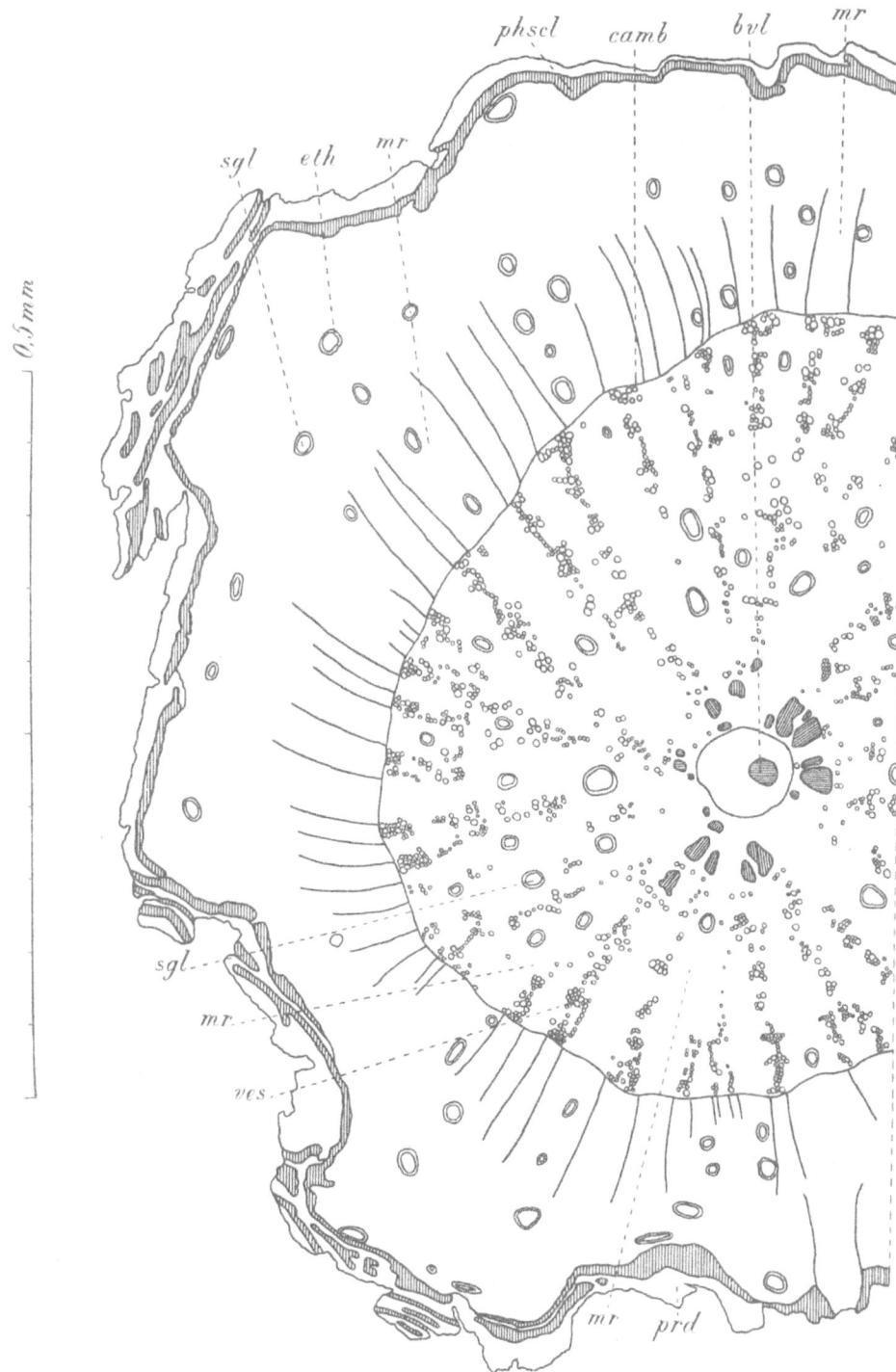

Fig. 84. *Anacyclus Pyrethrum*. Root, transverse section. bvl Vessels and libriform fibres; camb Cambium; eth Epithelium; mr Medullary rays; phscl Sclereids of phellem; prd Periderm; sgl Schizogenous glands; ves Vessels.

Secondary cork layers. Initial-celled; monopleuric.

P h e l l e m. Consisting of 6—15 layers of periderm cells and sclereids; the outermost 2 layers consisting of sclereids and followed usually by 4 layers of periderm cells, the innermost layers consisting generally again of sclereids. Cells lying in radial rows corresponding with those of the secondary phloem. Intercellular spaces wanting throughout the phellem.

P e r i d e r m c e l l s. R. 10 μ, T. 20—40 μ, L. 20—50 μ; polygonal tables with a radially directed axis, often showing a radial and a transverse partition-wall. Walls brown; suberized. S c l e r e i d s. R. 20 μ, T. 20—40 μ, L. 20—50 μ; polygonal tables with a radially directed axis, often somewhat more irregular, showing a radial and a transverse partition-wall. Walls strongly thickened; lignified; showing pit canals.

P h e l l o g e n. Sometimes wanting.

Secondary phloem proper.

T a n g e n t i a l e x p a n s i o n, necessary for keeping up with the increase in thickness of the root, principally due to the medullary rays; thus these rays widening outwards and in consequence cuneiform; the intermediate parts also contributing to the tangential expansion, but nevertheless cuneiform, their edges turned outwards and there 2—3 parenchyma cells wide; sometimes these wedges very narrow. Consisting of elements of the cribral and chiefly of the parenchymatic system; the outermost layers more or less flattened, for the rest showing distinct radial arrangement. E l e m e n t s o f t h e c r i b r a l s y s t e m. Sieve-tubes not very distinct, perhaps with companion cells united in numerous small bundles (potash), one bundle often corresponding to 1 or 2 parenchyma cells in a radial direction and to a single parenchyma cell in tangential direction. E l e m e n t s o f t h e p a r e n c h y m a t i c s y s t e m. Bast parenchyma consisting of substitute fibres and bast parenchyma fibres, sometimes having a radial partition-wall, the latter composed of 2 cells and these also now and then showing a radial partition-wall. Intercellular spaces. Medullary rays consisting of the same elements as the bast parenchyma and quite similar to this; moreover schizogenous glands [1]); only here and there in cross section a radial row of elements showing greater radial dimensions. Glands, often some of them arranged in longitudinal rows; R. 110 μ, T. 80 μ, L. 100—200 μ, smaller near the cambium; ellipsoidal; epithelium consisting of 1 layer of cells; cavity filled with a dark yellow to brown oil.

S i e v e - t u b e s. R. and T. 4—6 μ; polygonal prisms with a longitudinally directed axis. Walls somewhat thickened; sieve-plates not to be distinguished; middle lamella showing no cellulose reaction. Contents: often inuline. S u b s t i t u t e b a s t p a r e n c h y m a f i b r e s. Near the cambium R. 15 μ, T. 18 μ, L. 10—150 μ; at the middle part R. 25 μ, T. 40 μ; in the outermost part R. 25 μ, T. 45 μ, L. 60—80 μ; tetragonal fibres, rectangular in a radial section. Walls, middle lamella showing no cellulose reaction. Contents: inulin. B a s t p a r e n c h y m a f i b r e s. Near the cambium L. 50—75 μ, in the outermost part L. 30—45 μ. See for the rest the substitute fibres. E l e m e n t s o f t h e m e d u l l a r y r a y s. R. 20 μ, T. 18 μ; see for the rest the elements of the bast paren-

---

[1]) See also V a n T i e g h e m, L e B l o i s and T s c h i r c h, l.l. c.c.

chyma. C e l l s o f e p i t h e l i u m o f g l a n d s. Thick 10 μ, wide 20 μ, L. 25 μ; polygonal tables, axis pointing towards the centre of the cavity.

C a m b i u m. Consisting of 3—4 layers of fibres; radially arranged, the radial rows corresponding to those of the parenchymatic system of the secondary phloem and xylem. Sometimes in the medullary rays a cell with 2—3 tangential partition-walls, intercellular spaces occurring between these cells and the adjoining ones.

Xylem.

S e c o n d a r y x y l e m. Consisting of vessels, libriform fibres and chiefly of elements of the parenchymatic system; showing distinct radial arrangement. V e s s e l s in small bundles; most numerous in the outermost part and generally pitted; moreover a central bundle surrounded by a ring of about 7 bundles of mostly reticulate vessels. L i b r i f o r m f i b r e s in the central bundle, in the bundles of the surrounding ring mentioned above, moreover intermixed with the bundles lying farther outwards but within the inner half of the root. Intercellular spaces wanting. E l e m e n t s o f t h e p a r e n c h y m a t i c s y s t e m. Wood parenchyma consisting of substitute fibres and containing a very few glands. Intercellular spaces wanting. Medullary rays consisting of the same elements as the wood parenchyma and quite similar to this; moreover schizogenous glands; only here and there in a cross section a radial row of elements showing greater radial dimensions and mostly corresponding with the analogous rows in the secondary phloem. Intercellular spaces wanting. Glands. See those of the secondary phloem, perhaps somewhat smaller.

V e s s e l s. R. and T. 15—45 μ, in and near the centre R. and T. 10—12 μ, L. of articulations 60—150 μ; polygonal prisms, often with rounded edges and with circular perforations. Walls thickened; somewhat yellow; lignified. Contents: sometimes inulin. L i b r i f o r m f i b r e s. R. and T. 10—20 μ, L. 200—400 μ; polygonal, sometimes with branched ends. Walls strongly thickened; middle lamella conspicuous; lignified, especially the middle lamella; showing pit canals. Contents: sometimes a brown mass. S u b s t i t u t e w o o d p a r e n c h y m a f i b r e s. Near the cambium R. 15 μ, T. 20 μ, L. 80—150 μ, farther towards the interior R. 20 μ, T. 25 μ; rectangular. Walls a little thickened; middle lamella showing no cellulose reaction; often pitted. Contents: inulin. E l e m e n t s o f t h e m e d u l l a r y r a y s. R. and T. 18 μ, L. 70—110 μ; radially elongated elements R. 35—80, T. 20—75 μ; in the innermost part R. and T. 35 μ. Walls especially in the innermost part somewhat thickened; middle lamella showing no cellulose reaction; pitted. P a r e n c h y m a t i c e l e m e n t s a d j o i n i n g t h e c e n t r a l b u n d l e o f v e s s e l s. E. g. R. 18 μ, T. 40 μ; in a cross section rectangular. Walls very thin. Contents: inulin.

P r i m a r y x y l e m. Not clearly to be distinguished from the rest.

November 1903. J; M; L.

RADIX SARSAPARILLAE.

Sarsae Radix. Sarsaparilla.

The dried root of one or more species of the genus Smilax from Central America; known commercially as Honduras Sarsaparilla.

Macroscopic characters.

Length up to 1 M., thick up to 7.5 m.M.; cylindrical, unbranched, more or less sinuous; here and there with a few short fibrils; split up lengthwise into pieces of about 4 c.M. in length; raising dust when breaking; the more superficial layers brittle, the inner layers tough. Surface usually wrinkled longitudinally; varying from light to dark brown or greyish brown. Transverse fracture mealy and white, sometimes more or less horny and light brown. In a transverse section a very thin brown outer layer; within this layer a cortex, thick about half the radius of the root; endodermis visible as a sharp brown line; the outermost woody zone of the stele in its inner parts with many large vessels; large medulla. No odour; taste mealy, somewhat mucilaginous, afterwards pungent.

Anatomical characters.

LITERATURE. Berg. Anat. Atl. 1865. 5. Tafel III, IV. Benecke. Mikr. Drogenprakt. 1912. 21. Biechele. Mikr. Prüf. d. off. Drogen. 1904. 59. Flückiger. Pharmakogn. 1891. 323. Flückiger & Hanbury. Pharmacographia. 1879. 708. Gilg. Pharmakogn. 1910. 50. Greenish & Collin. Anat. Atl. of veget. powders. 1904. 260. Hager. Pharm. Praxis. Bd. 2. 1902. 847. Hérail. Mat. Méd. 1912. 554. Karsten und Oltmanns. Pharmakogn. 1909. 44. Koch und Gilg. Pharmakogn. 1909. 130. Koch. Mikr. Anal. d. Drogenpulver. Bd. 2. 1903. 223. Koch. Pharmakogn. Atlas. Bd. 2. 1914. 71. Kraemer. Botany and Pharmacogn. 1910. 446. Luerssen. Syst. Bot. Bd. 2. 1892. 399. Marmé. Pharmacogn. 1886. 16. Meyer. Wissensch. Drogenkunde. Bd. 1. 1891. 201. Moeller. Pharmakogn. 1889. 285. Moeller. Mikr. pharm. Üb. 1901. 270. Oudemans. Aanteek. o. d. Pharm. neerl. 1854—1856. 60. Oudemans. Pharmacogn. 1880. 21. Planchon et Collin. Drogues simples. T. 1. 1895. 162. Schneider. Powdered Veget. Drugs. 1902. 285. Tschirch. Pharmakogn. Bd. 2. Abt. 2. 1917. 1506. Wigand. Pharmakogn. 1879. 82. MATERIAL. Vera Cruz and Honduras Sarsaparilla, the drug thick about 4 m.M. REAGENTS. Water, glycerine, potash, chloral hydrate, iodine in chloral hydrate, phloroglucin and hydrochloric acid, iodine and sulphuric acid 66 per cent., concentrated sulphuric acid, Schulze's macerating mixture.

MICROGRAPHY of Vera-Cruz sarsaparilla.

Epidermis. Found only here and there and then often flattened. Outer walls nearly always wanting. Root hairs sometimes still found.

E p i d e r m a l c e l l s p r o p e r. R. 20 μ, T. 25 μ, L. 30 μ; tables with a radially directed axis. Walls somewhat thickened except the inner ones.

Cortex.

S c l e r e n c h y m a. 3—4 layers of cells, the innermost cells somewhat larger.

C e l l s o f t h e a b o v e. R. and T. 25 μ, L. 120 μ; tetra- to hexagonal prisms with a longitudinally directed axis. Walls of the two outermost layers strongly thickened, especially the outer ones, the radial ones cuneiform; middle lamella conspicuous; showing stratification; varying from yellow to brown; lignified; with pit canals.

C o m m o n p a r e n c h y m a. About 30 layers of cells; in the drug flattened; cells of the innermost layer with smaller radial and tangential dimensions and

25

having their innermost walls often thickened and pitted; intercellular spaces nearly always filled with a fungus. Rhaphide cells in longitudinally rows; raphides longitudinally directed, L. 70—100 μ.

Common parenchyma cells. Cell contents: in many cells a brown mass, soluble in alcohol; in some roots starch grains, sometimes in a cluster; the grains simple and compound with 2—6 component grains; simple grains spheric or semi-spheric, 10 μ in diameter, in some roots smaller.

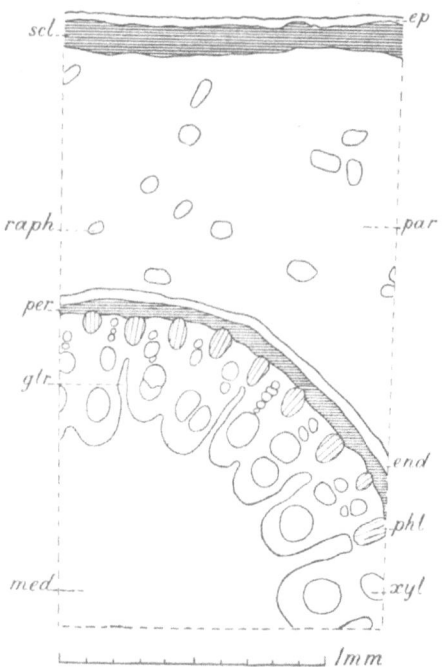

Fig. 85. *Vera-Cruz Sarsaparilla.* Root, transverse section. end Endodermis; ep Epidermis; gtr Medullary commissures; med Medulla; par Common parenchyma; per Pericycle; phl Phloem; raph Raphide cells; scl Sclerenchyma; xyl Xylem.

Endodermis. One layer of cells, developed as sclerenchyma.

Cells of the above. R 30 μ, T 25 μ, L. 100—250 μ; tetra- to hexagonal prisms with a longitudinally directed axis. Walls thickened except the outer ones; middle lamella conspicuous; lignified; showing stratification varying from yellow to brown; having pit canals.

Stele.

Pericycle. Developed as sclerenchyma; 2—4 layers of cells with intercellular spaces.

Cells of the above. R. 20 μ, T. 25 μ, L. 80—120 μ; poly-, mostly hexagonal prisms with a longitudinally directed axis and rounded edges. Walls thickened; sometimes yellow; pitted. Cell contents: sometimes a brown mass.

Phloem. 30—40 bundles consisting of sieve-tubes and cambiform cells, both small in the outer and somewhat larger in the inner parts; sieve-tubes to be distinguished only in the outer part.

Sieve-tubes. R. and T. 8 by 12 μ; polygonal prisms; transverse walls placed obliquely in a radial plane; showing distinct sieve-plates. Cambiform cells. R. and T. 6—30 μ, L. 50—100 μ; polygonal prisms with a longitudinally directed axis.

Xylem. 30—40 bundles; exarch; sometimes more or less blended together; here and there a xylem bundle in the medulla of thicker pieces of the drug. Consisting of vessels and parenchyma cells. Vessels arranged in bundles, usually in a radial row, increasing in size towards the centre; the largest innermost vessel or vessels not lying close to the other ones; reticulate and scalariform; xylem parenchyma cells adjoining the vessels elongated in a cross section in the direction of the periphery of the vessels.

Outermost vessels. R. 15 μ, T. 20 μ, L. of articulations 100—1000 μ, increasing according to the width of the vessels. Transverse walls placed very obliquely in a radial plane. Walls showing stratification; sometimes yellow; lignified, especially the middle lamella. Innermost vessel. R. and T. 100 by 110 μ. Xylem parenchyma cells.

R. and T. 20 by 25 μ, L. 250—450 μ; poly-, mostly hexagonal prisms with a longitudinally directed axis and rounded edges. Walls strongly thickened; yellow; a little lignified, especially the middle lamella; having slit-like pit canals; intercellular spaces. Cell contents: mostly a brown mass. C e l l s s u r r o u n d i n g t h e v e s s e l s. Depth 15 μ, breadth 25 μ, L. 200 μ. Walls with unilateral bordered pits when adjoining vessels and with split-like simple pits when adjoining other parenchyma cells. See for the rest the other parenchyma cells.

M e d u l l a r y c o m m i s s u r e s. Consisting of sclerenchyma cells; 1—3, usually 2 sclerenchyma cells wide; generally extending from the phloem bundles between the xylem bundles to the medulla.

C e l l s o f t h e a b o v e. R. 35 μ, T. 20 μ, L. 150—300 μ; elliptical cylinders with a longitudinally directed axis. Walls: see xylem parenchyma. Cell contents: starch grains; see those of the parenchyma cells of cortex.

M e d u l l a. Cells of the outermost layer similar to those of the medullary commissures; for the rest common parenchyma cells. Intercellular spaces.

C e l l s o f t h e a b o v e. R. and T. 45 μ or 40 by 50 μ; L. 200—400 μ; circular or elliptical cylinders with a longitudinally directed axis. Walls somewhat thickened; varying from slightly yellow to brown when treated with iodine and sulphuric acid; pitted. Cell contents: starch grains, see those of cortex parenchyma.

MICROGRAPHY of Honduras sarsaparilla. In all respects the same as that of Vera Cruz with the following exceptions.

Cortex. Sclerenchyma cells. Outer walls relatively less thickened. Parenchyma cells. Cell contents: always starch grains. Raphide cells less numerous.

Endodermis. Radial and tangential dimensions alike; outer walls somewhat thickened.

Stele. Vessels less regularly arranged.

Novembre 1901.   J; M; L.

RADIX TARAXACI.

Taraxacum. Taraxacum Root.

The fresh and the dried roots of Taraxacum officinale, (Weber in) Wigg. Prim. Fl. Holsat. 56, collected in the autumn.

Macroscopic characters of the fresh root.

Conical root; 2 or more d.M. in length, up to 2 c.M. thick; unbranched or having a few branches near the base; surface brown, rather smooth, base exhibiting fine transverse wrinkles; internally fleshy, whitish; in a transverse section the latex appearing from the secondary phloem in concentric rings, medulla wanting.

Taste bitter, but also sweetish and mucilaginous.

Anatomical characters.

LITERATURE. Berg. Anat. Atl. 1865. 11. Pl. VII. Benecke. Mikr. Drogenprakt. 1912. 20. Flückiger. Pharmakogn. 1891. 438. Flückiger & Hanbury. Pharmacographia. 1879. 393. Gilg. Pharmacogn. 1910. 361. Hager. Pharm. Praxis. Bd. II. 1902. 1015. Karsten und Oltmanns. Pharmakogn. 1909. 83. Koch. Mikr. Anal. d. Drogenpulver. Bd. II. 1903. 243. Koch. Pharmakogn. Atl. Bd. II. 1914. 85. Pl. XII. Kraemer. Botany a. Pharmacogn. 1910. 457, 779. Marmé. Pharmacogn. 1886. 49. Meyer. Wiss. Drogenk.

Bd. I. 258. Moeller. Mikr. d. Nahr. u. Genussm. 1886. 286. Moeller. Mikr. pharm. Üb. 1901. 300. Oudemans. Aant. o. d. Pharm. neerl. 1854—56. 192. Oudemans. Pharmacogn. 1880. 53. Planchon et Collin. Drogues simples. T. II. 1896. 18. Schneider. Powdered Veget. Drugs. 1902. 306. Tschirch. Pharmacogn. Bd. II. Abt. 1. 1912. 211. Tschirch u. Oesterle. Anat. Atl. 1900. 139. Pl. XXXIII. Vogl. Ueber die Intercellularsubstanz und die Milchsaftgefäsze in der Wurzel des gemeinen Löwenzahns. Ber. Wiener Akad. Bd. 48. 1863. 668. Wigand. Pharmakogn. 1879. 75. MATERIAL. The drug. Fresh roots, kept in alcohol, collected in the Botanic Garden at Groningen, thick 2—12 m.M. REAGENTS. Water, glycerine, potash, phloroglucin and hydrochloric acid, iodine and sulphuric acid 66 per cent., concentrated sulphuric acid, Schulze's macerating mixture.

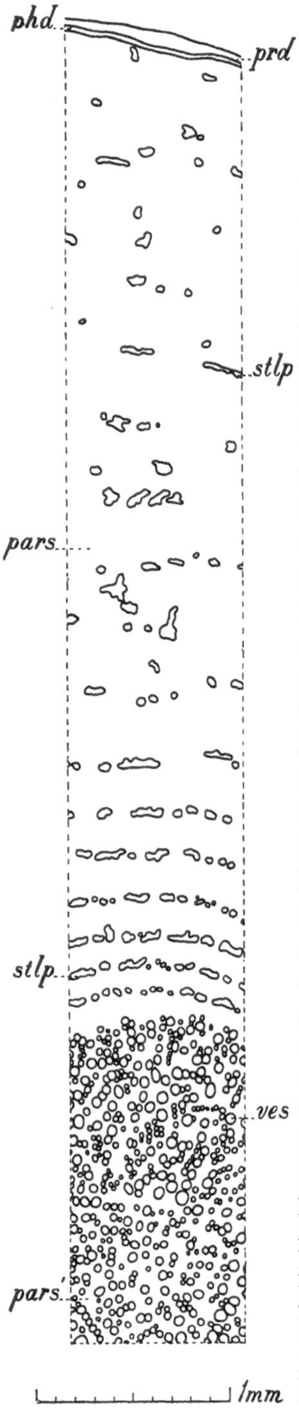

MICROGRAPHY.

Epidermis. Thrown off, even from the thinnest specimen.

Cortex.

Secondary cork tissue. Formation of cork beginning in the second or third layer of cortical cells; cortical cork often thrown off by the formation of cork in the pericycle. Consisting of 3—5 radially arranged layers of periderm cells.

Cells of the above. Walls yellow; suberized.

Primary cortex proper. Consisting of 2—3 layers of common parenchyma cells, lying between the cork tissue and the endodermis. Usually thrown off by the formation of cork in the pericycle.

Endodermis. Developed as a protective-sheath consisting of 1 layer of cells; cells often showing 1 or 2 subsequently formed radial walls and a transverse wall. Persisting a long time after the formation of cork in the pericycle.

Cells of the above. R. 35 μ, T. 40—50 μ, L. 70 μ; tetra- or pentagonal prisms with a longitudinally directed axis. Walls yellow; suberized, especially the radial walls.

Stele of the diarch root.

Pericycle.

Primary pericycle in the young root. Consisting of 1 layer of common parenchyma cells, often showing subsequently formed radial and transverse walls.

Fig. 86. *Taraxacum officinale*. Root, transverse section. pars Elements of the parenchymatic system of the phloem; pars' Elements of the parenchymatic system of the xylem; phd Phelloderm; prd Periderm; stlp Bundles of sieve-tubes, latex vessels and parenchyma cells; ves Vessels.

C e l l s o f t h e a b o v e. R. 25 μ, T. 35 μ, L. 70 μ; tetragonal prisms with a longitudinally directed axis.

S e c o n d a r y p e r i c y c l e in the drug.

Secondary cork tissue. Initial-celled cork; the outer part crushed. Much secondary growth, especially in the secondary phloem, anterior to the formation of cork tissue in the pericycle.

P h e l l e m. Developed as periderm; consisting of 2—4 layers of cells, showing strict radial arrangement. Many cells divided by subsequently formed radial partition-walls.

C e l l s o f t h e a b o v e. R. 10—15 μ, T. 25 or 45 μ, L. 30 μ. Walls varying from yellow to brown; suberized. Cell contents: granules, colourless to yellow.

P h e l l o g e n. No more to be distinguished as such.

P h e l l o d e r m. Here and there distinct as 2—3 layers of cells.

C e l l s o f t h e a b o v e. R. 15 μ, T. 45 μ, L. 60 μ; polygonal tables with a radially directed axis.

Phloem.

P r i m a r y p h l o e m. No more to be recognized in the drug.

S e c o n d a r y p h l o e m. Consisting of elements of the cribral and the parenchymatic system and composed of about 200 layers of elements. E l e m e n t s o f t h e c r i b r a l s y s t e m. Consisting only of sieve-tubes, always united in bundles with latex vessels, sometimes intermixed with cells of the surrounding parenchyma. Bundles containing 10—20 elements in a cross section, the number increasing towards the centre of the root. Bundles arranged in concentric rings. In the outer part bundles less numerous and the arrangement in rings less distinct than in the inner part; towards the centre the concentric rings nearer to each other, number and dimensions of the intermediate cells decreasing. In the outer part the bundles tangentially elongated and sometimes a little flattened. Bundles of the same ring strongly anastomosing; those of different rings never anastomosing. In the outermost 2—4 rings the bundles each containing only 1 latex vessel. E l e m e n t s o f t h e p a r e n c h y m a t i c s y s t e m. Bast parenchyma developed as substitute bast fibres and bast parenchyma fibres, the latter especially in the outer part; with intercellular spaces; especially in the outer part elements showing subsequently formed radial walls; elements surrounding the bundles of sieve-tubes and latex vessels often divided in 2 parts by longitudinal walls. Medullary rays consisting of the same elements as the bast parenchyma; sometimes the 2 original medullary rays still to be seen in the drug and then often only near the cambium; these rays near the cambium 3- to 5-seriate, near the cork tissue 8- to 10-seriate. Cells generally radially elongated in a cross section. Showing intercellular spaces.

S i e v e-t u b e s. R. and T. 10—12 μ and smaller, especially in the inner part of the phloem, L. of articulations 70—120 μ; polygonal prisms; sieve-plates distinct, often with callus plates. Walls slightly thickened. Contents: often inulin, but not in sphaero-crystals. L a t e x v e s s e l s. The outermost vessels R. and T. 12—18 μ, for the rest 10—12 μ and smaller. Walls slightly thickened; cellulose reaction weak. Contents: a yellow granular mass. S u b s t i t u t e   b a s t   p a r e n c h y m a   f i b r e s. In the outer part R. 30 μ, T. 60 μ, in the middle part R. 25 μ, T. 40 μ, in the inner part R. 15 μ, T. 20 μ, L. 70—120 μ; in the outer part polygonal, for the rest tetra- to hexagonal with more or less strongly rounded edges. Walls pitted or showing 2 intercrossing sets of striae, only visible in iodine and sulphuric acid 66 per cent. Cell contents: inulin, either in distinct or in indistinct sphaero-crystals. The intercellular spaces often also containing inulin.

C a m b i u m. Not always distinct; consisting of 2—3 layers of elements. E l e m e n t s   o f   t h e   a b o v e. R. 6 μ, T. 12 μ, L. 70—120 μ.

Xylem.

S e c o n d a r y   x y l e m. Consisting of elements of the tracheal and the parenchymatic system. E l e m e n t s   o f   t h e   t r a c h e a l   s y s t e m. Vessels smaller and larger, reticulate, isolated, sometimes forming bundles of some vessels in a cross section. In the middle part of the xylem the smaller vessels more numerous. Intercellular spaces wanting. E l e m e n t s   o f   t h e   p a r e n - c h y m a t i c   s y s t e m. Wood parenchyma [1]) consisting of substitute fibres and wood parenchyma fibres, the latter composed of 2 cells; without intercellular spaces; elements somewhat larger in the middle part. Medullary rays consisting of the same elements as the wood parenchyma; beginning halfway the xylem; some elements wide; see for the rest the secondary phloem.

V e s s e l s. The largest R. 60 μ, T. 50 μ, the smallest R. 20 μ, T. 15 μ; polygonal prisms with rounded edges or elliptical or circular cylinders; cross walls often oblique with 1 large perforation. Walls thickened; lignified. Contents: often a yellow to brown mass; sometimes sphaero-crystals of inulin, paving the walls. S u b s t i t u t e   w o o d   p a r e n c h y m a   f i b r e s. R. and T. 10—20 μ, L. 75—120 μ, in the middle part R. and T. 20—25 μ, L. 150 μ; polygonal. Walls pitted or showing 2 sets of intercrossing striae only visible in iodine and sulphuric acid 66 per cent. Cell contents: wanting.

P r i m a r y   x y l e m. Not clearly to be distinguished; containing only a few spiral vessels.

December 1901.   J; M; L.

RADIX VALERIANAE.
Valerian Rhizome. Valerian Root. Valeriana. Valerian.
The dried adventitious roots and caudex of Valeriana officinalis, Linn. Sp. Pl. 31, collected in the autumn from plants 2 or 3 years old.

Macroscopic characters.
Numerous fibrous roots, growing on all sides of the caudex, mostly 1 d.M. in length, up to 2 m.M. thick, brittle, wrinkled longitudinally with a relatively very thick cortex and a very thin stele. Caudex up to 5 c.M. in length, up to 2 c.M. thick; sometimes with a few stolones; ob-conical or cylindrical, truncate on both ends; bearing at the top a bud surrounded by remnants of sheaths; internally pale brown with a thin cortex and inside the darker coloured cambium a still thinner, continuous on not continuous ring of lighter coloured xylem bundles; medulla large, often hollow in the internodes and then divided

---

[1]) See O u d e m a n s and T s c h i r c h.

by diaphragms, corresponding with the nodes. Surface of all the parts earthily greyish brown.

The fresh roots nearly odourless; but in drying and keeping by and by a strong characteristic odour developed. Taste aromatic, a little pungent and besides more or less bitter.

Anatomical characters.

LITERATURE. Berg. Anat. Atl. 1865. 31. Pl. XVI. Benecke. Mikr. Drogenpraktik. 1912. 18. Coex. Valeriana Officinalis. Pharm. Weekbl. Jrg. 56. 1919. 747. Dye. Unterird. Org. von Valeriana, Rheum u. Inula. Diss. Bern. 1901. Flückiger. Pharmakogn. 1891. 465. Flückiger & Hanbury. Pharmacographia. 1879. 378. Gilg. Pharmakogn. 1910. 337. Greenish & Collin. Anat. Atlas of veget. powders. 1904. 268. Hager. Pharm. Praxis. Bd. 2. 1902. 1100. Hérail. Pharmacol. 1912. 254. Karsten u. Oltmans. Pharmakogn. 1909. 59. Koch u. Gilg. Pharmkogn. Praktikum. 1907. 133. Koch. Mikr. Anal. der Drogenpulver. Bd. 2. 1903. 251. Koch. Einf. i.d. mikr. Anal. d. Drogenpulver. 1906. 60. Koch. Pharmakogn. Atlas. Bd. 2. 1914. 91. Luerssen. Syst. Bot. Bd. 2. 1882. 1118. Marmé. Pharmacogn. 1886. 96. Meyer. Wiss. Drogenkunde. Bd. 1. 1891. 72, 215. Moeller. Pharmakogn. 1889. 333. Oudemans. Pharmacogn. 1880. Planchon et Collin. Drogues simples. T. 2. 1896. 92. Schneider. Powdered Veget. Drugs. 1902. 314. Tschirch u. Oesterle. Anat. Atl. 1900. Taf. 59. Wigand. Pharmakogn. 1879. 95. Zacharias. Ueber Secret Behälter mit verkorkten Membranen. Bot. Ztg. 1879. 635. MATERIAL. The drug; rhizomes with roots, collected January 1902 in the Botanic Garden at Groningen, fresh, in alcohol and treated with chromic acid; the roots thick 0.8 m.M.— 3.4 m.M., the rhizomes thick 3.5 m.M.—17 m.M. REAGENTS. Water, glycerine, potash, iodine in chloral hydrate, phloroglucin and hydrochloric acid, iodine and sulphuric acid 66 per cent., concentrated sulphuric acid, Schulze's macerating mixture.

MICROGRAPHY of the root.

E p i d e r m i s. Cells sometimes having on their middle part a shorter or longer root hair.

E p i d e r m a l c e l l s p r o p e r. R. 10 μ, T. 18 μ, L. 100 μ; tetra- to hexagonal prisms with a longitudinally directed axis. Outer walls dome-shaped, sometimes also the radial walls; outer walls thickened, thickest in the centre of the dome; thickened walls with stratification, yellow, suberized and lignified. Cell contents: often some yellow grains along the walls. R o o t h a i r s. Walls yellow; suberized and lignified.

C o r t e x.

Primary cork tissue. Generally consisting of 1 layer of periderm cells, in the thickest part sometimes 2—3 layers; cells in longitudinal rows, in a transverse section now and than in radial rows.

C e l l s o f t h e a b o v e. R. 38 μ, T. 42 μ, L. 60—150 μ; polygonal prisms with a longitudinally directed axis. Outer and inner walls curved, radial walls transversally wrinkled; yellow; suberized and a little lignified, moreover at the inner side of the cork lamella a cellulose layer. Cell contents: in each cell one large or several small yellow oildrops; chromic acid and alcoholic material showing in some cells a network (perhaps the protoplasm containing the drops); other cells wholly or partially filled with a granular mass; other cells again showing along the outer or inner wall a homogeneous layer and close by an irregularly limited granular layer [1]).

Common parenchyma. Consisting of 15 layers of cells in longitudinal rows;

---

[1]) C.f. T s c h i r c h. Harze u. Harzbehälter. 390.

cells of the outermost 3—5 layers somewhat smaller than the rest. Intercellular spaces in the outermost layers quite or nearly wanting, in the others very large. C e l l s o f t h e a b o v e. R. and T. 25—40 μ, L. 100—250 μ; in the outermost 3—5 layers polygonal prisms with a longitudinally directed axis, somewhat tangentially elongated, sometimes with a radial partition-wall; in the other layers cylinders or polygonal prisms with a longitudinally directed axis and strongly rounded edges, often with a transverse wall. Walls slightly thickened, those of the outermost layers somewhat collenchymatous; slightly yellow; showing very small pits. Cell contents: sometimes dark yellow sphaerocrystals of unknown composition; in the outermost layers small oil-drops; in the chromic acid material especially in the outermost layers a yellowish green often granular mass, now and then divided in some parts; starch grains smaller and less numerous in the outermost layers than in the rest; grains 5—6 μ in diameter, simple or compound, with up to 5, generally 2—3 component grains with central hilum.

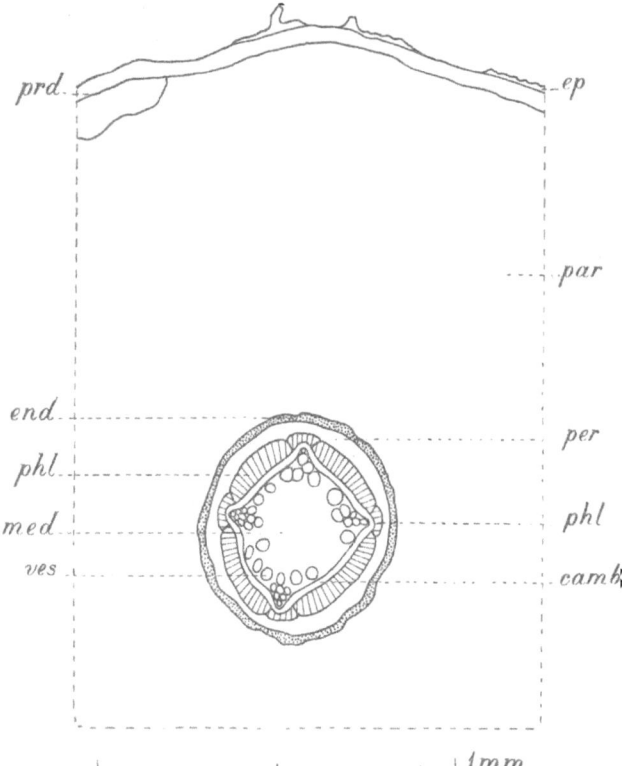

Fig. 87. *Valeriana officinalis*. Root, transverse section. camb Cambium; end Endodermis; ep Epidermis; med Medulla; par Common parenchyma; per Pericycle; phl Phloem; prd Periderm; ves Vessels.

Endodermis. Consisting of 1 layer of cells, developed as a protective-sheath with passage cells. C e l l s o f t h e a b o v e. R. and T. 15 μ, L. 75—175 μ; rectangular prisms with a longitudinally directed axis and rounded edges; very often afterwards forming a radial partition-wall; radial walls, except the radial partition-walls, transversely wrinkled. Walls of nearly all the cells showing an outermost suberized lamella, except the unwrinkled radial partition-walls. A lamella of the old radial walls lignified.

S t e l e.

Pericycle. Generally consisting of 1, sometimes of 2—3 layers of common parenchyma cells.

C e l l s o f t h e a b o v e. R. 17 μ, T. 20 μ, L. 50—70 μ; polygonal prisms with a longitudinally directed axis. Cell contents: see the cortical parenchyma.

Phloem. The vertical roots mostly tetrarch or pentarch (see figure); the horizontal ones usually up to 11-arch (see figure).

P r i m a r y and **s e c o n d a r y p h l o e m** consisting almost exclusively of sieve-tubes; secondary phloem only partially developed even in the thickest roots, the xylem arches of the cambium having formed only parenchyma (see figure 87 of the tetrarch root).

S i e v e-t u b e s of the primary phloem R and T. 6 μ; polygonal prisms with a longitudinally directed axis; of the secondary phloem R. and T. 6 by 12 μ, L. of articulations 100—200 μ; tetra- to hexagonal prisms with a longitudinally directed axis. Transverse walls often oblique, sieve-plates conspicuous. P a r e n c h y m a c e l l s not distinct in longitudinal section.

C a m b i u m. Elements similar to those of the adjoining phloem elements.

Xylem. P r i m a r y x y l e m. Consisting of 4—5 bundles of spiral vessels, adjoining each other in the thinner parts of the roots, afterwards diverging and surrounding a medulla (see figure 87 tetrarch root).

S e c o n d a r y x y l e m. In the thicker parts of the vertical roots consisting of a continuous circle of pitted vessels, one vessel thick in a radial direction (see figure 87). The horizontal roots up to 11-arch, showing secondary xylem only in places corresponding to the primary xylem bundles.

S p i r a l v e s s e l s. R. 5 μ, T. 4 μ. P i t t e d v e s s e l s. R. and T. 30 μ, or 25 by 35 μ, L. of articulation 200—250 μ; polygonal prisms with horizontal transverse walls, each showing a circular perforation. Walls a little thickened; lignified.

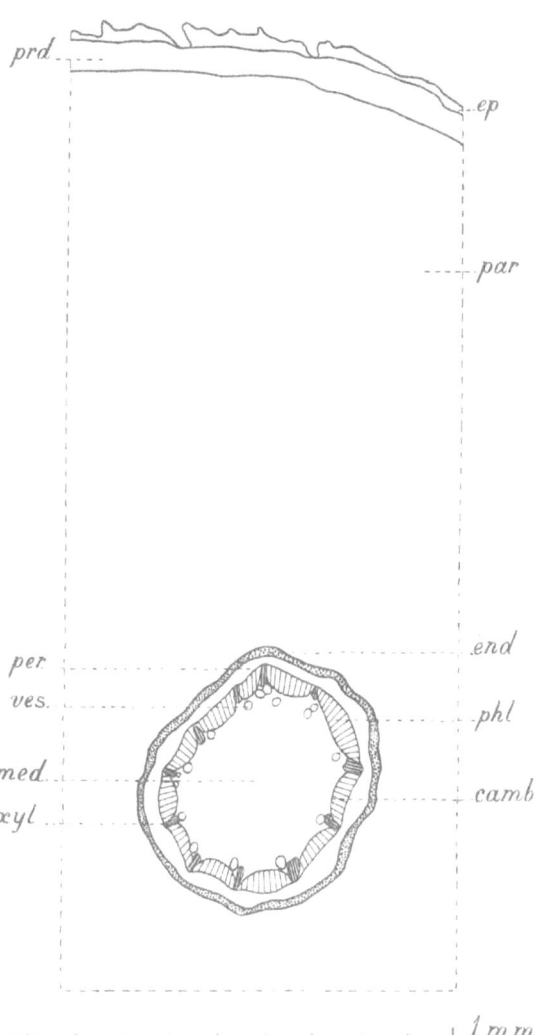

Fig. 88. *Valeriana officinalis*. Root, transverse section. camb Cambium; end Endodermis; ep Epidermis; med Medulla; par Common parenchyma; per Pericycle; phl Phloem; prd Periderm; ves Vessels; xyl Xylem.

Medullary commissures. Consisting of layers of common parenchyma cells, measuring some cells in radial and tangential direction; cells lying in radial rows.

C e l l s   o f   t h e   a b o v e. R. and T. 10 μ, L. 80—120 μ; tetragonal prisms with a longitudinally directed **axis**.

Medulla. Consisting of common parenchyma cells, lying in longitudinal rows.
Intercellular spaces wanting.

C e l l s   o f   t h e   a b o v e. R. and T. 20 by 15 μ, L. 60—100 μ. Cell contents: similar
to those of the cortical parenchyma, only less abundant.

MICROGRAPHY of the rhizome.

E p i d e r m i s. Generally occurring even on the thickest material.

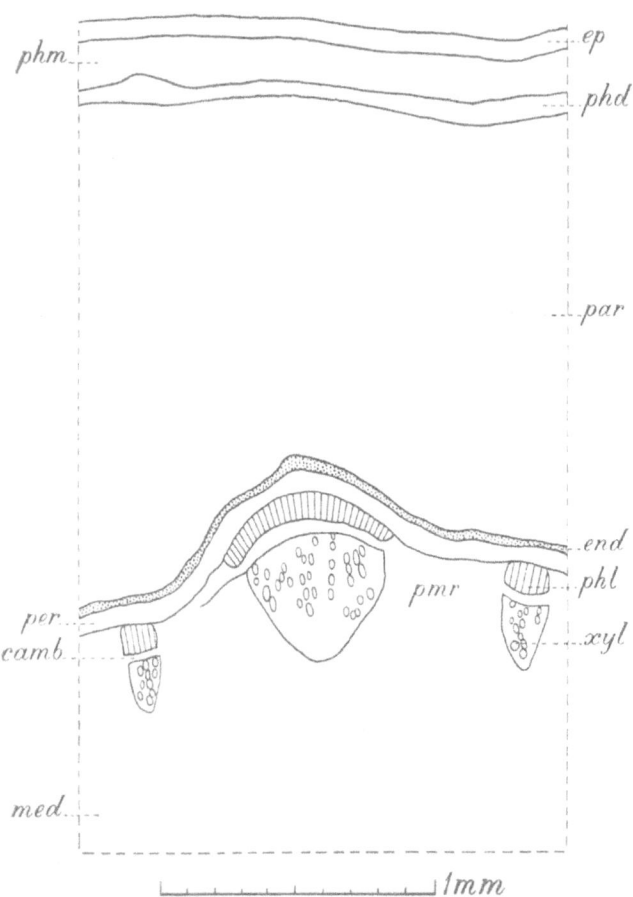

C e l l s   o f   t h e   a b o v e.
R. 45 μ, T. 40 μ, L. 80—
100 μ; polygonal tables
with a radially directed
axis. Outer walls some-
what thickened; showing
coarse, straight, parallel
cuticular striation; all the
walls yellowish brown;
suberized, the middle la-
mella lignified.

C o r t e x.

**Secondary cork tis-
sue.** At the outset storied cork, developed out
of the outermost layer
of the cortex; consisting
of a few stories at the
utmost [1]). Afterwards
formation of dipleuric
initial-celled cork.

P h e l l e m. Scaling off;
consisting of 3—7 lay-
ers of periderm cells.
P e r i d e r m   c e l l s. R.
and T. 40 μ, L. 50—150
μ. Walls yellowish brown;
suberized and the inner-
most lamella lignified;
often pitted. Intercellular
spaces between the origi-
nal cells of the storied
cork still discernible. Cell

Fig. 89. *Valeriana officinalis.* Rhizome, transverse section. camb
Cambium; end Endodermis; ep Epidermis; med Medulla; par Com-
mon parenchyma; per Pericycle; phd Phelloderm; phl Phloem; phm
Phellem; pmr Medullary commissures; xyl Xylem.

contents: one large or some smaller oil-drops.

P h e l l o g e n. Consisting of 1 layer of cells. R. 10 μ.

---

[1]) In the bases of the leaf-sheaths secondary cork tissue is formed in the same way. In places where
t he leaf base is not very thick all the cells become suberized.

P h e l l o d e r m. Mostly consisting of 2—3 layers of slightly collenchymatous-ly thickened parenchyma cells.

C e l l s   o f   t h e   a b o v e. R. 10—20 μ. Walls often a little thickened. Cell contents: small oil drops and starch grains.

Primary cortex proper. Consisting of 40—60 layers of common parenchyma cells; cells often with radial and transverse partition-walls; cells of the layer adjoining the phelloderm often divided by single tangential walls; cells of the inner part smaller. Intercellular spaces.

C e l l s   o f   t h e   a b o v e. R. 40 μ, T. 70 μ, L. 80—190 μ; elliptical cylinders with a longitudinally directed axis or ellipsoids. Walls a little thickened; showing cross-wise striation; pitted. Cell contents: starch grains; a yellowish green granular mass; dark yellow sphaero-crystals like those of the root; small oil drops.

Endodermis. Developed as a protective-sheath [1]). Nearly always consisting of 1, sometimes of 2 layers of cells; the endodermis cells lying either at the outer or at the inner end or again in the middle of radial rows of cells. Passage cells occurring.

C e l l s   o f   t h e   a b o v e. R. 35 μ, T. 40 μ, L. 80 μ; poly-, mostly tetragonal prisms with a longitudinally directed axis. Walls showing a suberized lamella; sometimes, in radial walls always, also a lignified lamella; the innermost lamella consisting usually of cellulose.

S t e l e.

Pericycle. Consisting of some layers of common parenchyma cells strongly resembling those of the innermost layers of the cortical parenchyma; with intercellular spaces. Moreover bundles of collenchyma cells in places correspond-ing to the adjoining phloem.

P a r e n c h y m a   c e l l s. R. 20 μ, T. 30 μ; polygonal prisms with strongly rounded edges or cylinders, both with a longitudinally directed axis. Cell contents: see those of the cortical parenchyma. C o l l e n c h y m a   c e l l s. R. 20 μ, T. 30 μ, L. 70—100 μ; poly-gonal prisms with a longitudinally directed axis, often with rounded edges. Sometimes intercellular spaces.

Vascular bundles. Numerous; arranged in a circle; collateral; showing a cambium.

P h l o e m. Sparingly developed; the primary phloem showing bundles of sieve-tubes; the secondary phloem consisting of some radial rows of elements.

S i e v e - t u b e s   o f   p r i m a r y   p h l o e m. R. and T. 10 μ; polygonal. C o m-p a n i o n   c e l l s. R. and T. 4 by 3 μ; tetragonal prisms. E l e m e n t s   o f   s e c o n d-a r y   p h l o e m. R. 10 μ, T. 18 μ; tetra-, sometimes hexagonal prisms with a longi-tudinally directed axis.

C a m b i u m. Elements similar to the elements of the secondary phloem.

X y l e m. Consisting of smaller spiral vessels, larger reticulate or pitted vessels and common parenchyma. Elements arranged in radial rows.

V e s s e l s. R. and T. 10 by 25 μ, L. of articulations 50—350 μ; the longest articulations in the vessels with the smallest diameter; transverse walls very oblique with one small perforation. Walls a little thickened; somewhat yellow; lignified. Contents: sometimes a

---

[1]) S c h o u t e. Diss. Groningen 1902. 126, mentions in the stem of V. sambucifolia a starch-sheath.

yellow mass. P a r e n c h y m a c e l l s. R. and T. 10 by 15 μ, or smaller. Cell contents: see the cortical parenchyma.

Medullary commissures. Consisting of cells increasing in size towards the inner part.

C e l l s o f t h e a b o v e. R. 25 μ, T. 20 μ; mostly polyhedra; see for the rest the cortical parenchyma.

Medulla. In the diaphragms corresponding with the nodes, and mentioned among the macroscopic characters, cells sometimes flattened and moreover here and there in the centre groups of sclereids of the same shape as the surrounding cells.

S c l e r e i d s. R. and T. 60 by 70 μ; polyhedra. Walls thickened; showing stratification; yellow; lignified, especially the middle lamella; with branched pit canals.

Micrography of the powder of roots, caudex and stolones.

Oblong parenchyma cells, some with thinner, others with thicker walls, sometimes quite filled with starch, some coloured brown. Bundles of pitted, spiral and annular vessels, also of libriform fibres. Square prismatical sclereids with pit canals and sometimes very strongly thickened walls. Rather much starch, simple and compound, up to 6-adelphous grains; single grains 10—15 μ, but also much smaller. To a small amount : epidermal cells with root hairs, wide periderm cells with thin brown walls and cells of primary cork tissue and protective-sheath, with thin strongly and narrowly undulated cell walls.

January 1902. J; M; L.

RHIZOMA ARNICAE.
The dried rhizome and secondary roots of Arnica montana, Linn. Sp. Pl. 884.

Anatomical characters.
LITERATURE. Berg. Anat. Atl. 1865. Tafel XV. 29. Le Blois. Les Canaux Sécréteurs et les poches sécrétrices. Ann. des Sciences. Sér. 7. Botanique. T. VI. 1887. 274. Flückiger. Pharmakogn. 1891. 471. Flückiger & Hanbury. Pharmacographia. 1879. 391. Hager. Pharm. Praxis. Bd. 1. 1900 385. Luerssen. Syst. Bot. Bd. II. 1882. 1145. Marmé. Pharmacogn. 1886. 99. Moeller. Pharmacogn. 1889. 333. Oudemans. Pharmacogn. 1880. 105. Oudemans. Aanteek. o. d. Pharmac. neerl. 1854—56. 175. Planchon et Collin. Drogues simples. T. II. 1896. 73. Schneider. Powdered Veget. Drugs. 1902. 125. Solereder. Syst. Anat. der Dicotyledonen 1899. 515. van Tieghem. Canaux sécrét. des Plantes. Annales des Sciences. Sér. 5. Botanique. T. XVI. 1872. 97. van Tieghem. Sec. Mémoire s. l. Canaux sécrét. des Plantes. Ann. des Sciences. Sér. 7. Botanique. T. I. 1885. 6. Tschirch. Angew. Pfl. Anat. 1889. 296. Tschirch. Harze u. Harzbeh. 1900. e. g. p. 343, 360, 361. etc. Wigand. Pharmakogn. 1879. 96. MATERIAL. Rhizomes, gathered in October 1903 in the vicinity of Groningen, Peizermaden, fresh and in alcohol. REAGENTS. Water, glycerine, potash, iodine in chloral hydrate, phloroglucin and hydrochloric acid, iodine and sulphuric acid 66 per cent., concentrated sulphuric acid, osmic acid, Schulze's macerating mixture, potassium dichromate, copper acetate and iron acetate.

MICROGRAPHY.
E p i d e r m i s. Mostly thrown off or more or less flattened; cells fairly well arranged in longitudinal rows. Hairs only occurring in the leaf-axils and on the

leaf-bases, therefore confined to the leaf-scars, very numerous in the vicinity of the punctum vegetationis; uniseriate; consisting of about 8 cells.

E p i d e r m a l c e l l s p r o p e r. R. 20 μ, T. 22 μ, L. 20—25 μ; nearly always rectangular tables with a radially directed axis. Walls, the outer ones thickened and covered with a cuticle; often brown, especially the outer and lateral ones; the brown walls persisting much longer in concentrated sulphuric acid than e. g. the walls of the cortical parenchyma cells and not becoming blue in iodine and sulphuric acid. Cell contents: mostly a brown mass, consisting of globes; tannin. C e l l s o f h a i r s. 18 μ in thickness and 50—200 μ in length; very long and cylindrical; often flattened. Walls thin.

C o r t e x.

Secondary cork tissue.
Here and there storied cork developed in the outermost collenchymatous layers of the cortical parenchyma. Mostly 3—6 layers of periderm cells, outermost layers more or less flattened. Cells here and there in longitudinal rows; often with a radial and transverse partition-wall, especially in the outermost layers. Intercellular spaces wanting except between the cork tissue and the outermost cortical parenchyma cells.

P e r i d e r m c e l l s. R. 12 μ, T. 20—40 μ, L. 15—40 μ; polygonal tables with a radially directed axis. Walls thin, only inner tangential walls of innermost cell layer collenchymatous.

Primary cortex proper.

C o l l e n c h y m a. Composed of about 5 layers of cells. Close to the cork tissue here and there a cell containing a brown mass. Intercellular spaces increasing in size towards the centre; filled with a mass varying from dark brown to black and insoluble in alcohol, glycerine, potash and concentrated sulphuric acid; those masses more numerous in elder rhizomes; walls surrounding the intercellular spaces covered with a very thin lamella becoming brown in iodine and sulphuric acid. In some rhizomes this collenchyma making quite the impression of phelloderm.

C o l l e n c h y m a c e l l s. R. 22 μ, T. 30—60 μ, L. 35—70 μ; polygonal prisms with a longitudinally directed axis and rounded edges, here and there with a radial partition-wall. Walls collenchymatously thickened, especially those of the 2 outermost layers of cells; sometimes exhibiting stratification; numerous large slit-like pits e. g. extending over the whole width of the cell. Cell contents: in water showing some globules not increasing in number by the addition of e. g. hydrochloric acid; becoming black in osmic acid; a brown mass in potassium dichromate; showing some reddish brown globes in copper acetate, not becoming black in iron acetate.

C o m m o n p a r e n c h y m a. Consisting of an outer part of 10 layers of larger cells and an inner part of 10 layers of smaller cells; showing greater intercellular spaces than the collenchyma. The inner part containing numerous schizogenous glands [1]), developed at an early stage, lying close to the endodermis. Glands surrounded by more or less tangentially elongated parenchyma cells; probably showing many strictures or perhaps wholly divided by transverse cell layers; epithelium consisting of 1 layer of cells, each often divided

---

[1]) According to V a n T i e g h e m and L e B l o i s these glands are produced by a cell adjacent to the endodermis and formed by the tangential division of one of its cells.

by a tangential wall, sometimes strongly bulging out into the cavity. Cavity R. 80—100 μ, T. 80 μ, L. very great; filled with a colourless oil, soluble in alcohol and becoming slowly dark brown in osmic acid.

C o m m o n   p a r e n c h y m a   c e l l s. In the outer part R. 25 μ, T. 35 μ, L. 25—60 μ; polygonal prisms with a longitudinally directed axis, rounded edges and slightly curved sides. Walls pitted like those of the collenchyma cells; those surrounding the intercellular spaces often somewhat thicker than the rest; see for the rest cells of the collenchyma. Cell contents: here and there inulin; becoming black in osmic acid; producing a brown mass in potassium dichromate. In the inner part on the outside of the glands R. 20 μ, T. 35 μ, L. 25—40 μ; polygonal prisms with a longitudinally directed axis, rounded edges and slightly curved sides. Walls, see those of the outer part. Cell contents: when treated with copper acetate containing some brown globes, especially in the cells near the endodermis; see for the rest those of the

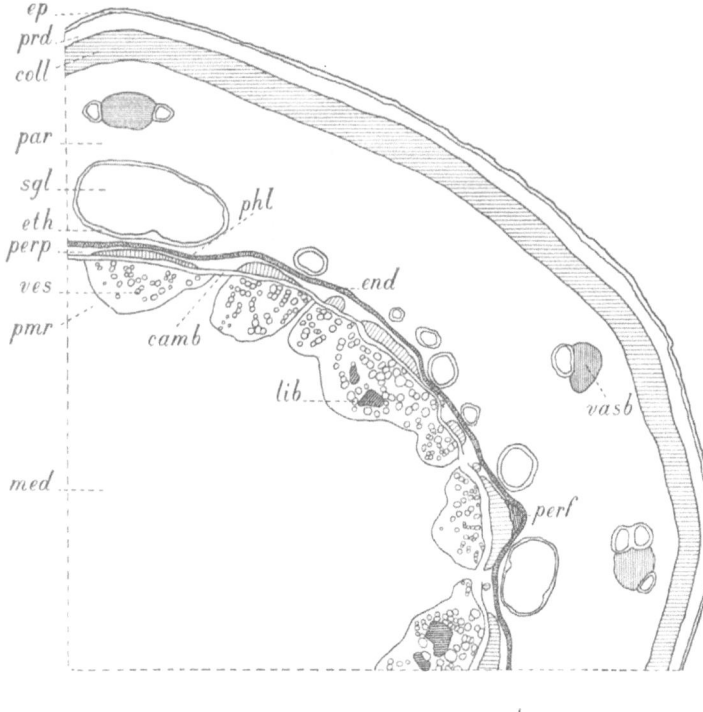

Fig. 90. *Arnica montana*. Rhizome, transverse section. camb Cambium; coll Collenchyma; end Endodermis; ep Epidermis; eth Epithelium; lib Libriform fibres; med Medulla; par Common parenchyma; perf Pericyclic sclerenchyma fibres; perp Pericyclic parenchyma; phl Phloem; pmr Medullary commissures; prd Periderm; sgl Schizogenous glands; vasb Vascular bundles; ves Vessels.

outer part. C e l l s   o f   e p i t h e l i u m   o f   g l a n d s. Depth 20 μ, breadth and L. 20—35 μ; poly-, mostly hexagonal tables with a radially directed axis; often corresponding in length to the surrounding cells. Walls very thin; no intercellular spaces. Cell contents: producing a stronger reaction in osmic acid than those of the other cells.

Endodermis. Consisting of 1 layer of cells; here and there a cell adjoining a gland with a tangential partition-wall.

C e l l s   o f   t h e   a b o v e. R. 15 μ, T. 20—30 μ, L. 30—45 μ; polygonal prisms with a longitudinally directed axis. Walls a little thickened; with suberized thickenings on the innermost parts of the radial and transverse walls; in cells with a partition wall only the walls of the innermost cell suberized; inner walls pitted. Cell contents: see those of the common parenchyma.

S t e l e.

Pericycle. Consisting of 1 or 2 layers of cells; in some parts, corresponding in place to vascular bundles, interrupted by bundles of collenchymatously thickened sclerenchymatic elements, R. and T. 10 µ.

Cells of the above. R. 10 µ, T. 20 µ, L. 30—80 µ; polygonal prisms with a longitudinally directed axis. Walls somewhat thickened; pitted; intercellular spaces wanting. Cell contents: see those of the cortical parenchyma cells.

Vascular bundles. Sometimes developed into fibrovasal bundles; arranged in a ring, in some places of this ring wanting in consequence of vascular bundles bending outwards and being sustituted by parenchyma cells like those of the medullary comissures; sometimes showing the development of cambium; collateral; open; somewhat varying in size. Sometimes a semi-developed sclerenchyma-sheath round the xylem of the largest bundles, consisting of 1 layer of elements.

P h l o e m. Primary phloem bundles showing no radial arrangement. Sieve-tubes probably sometimes lying in small bundles. Secondary phloem more or less developed in some rhizomes by a continuous cambium zone.

Sieve-tubes. R. and T. 5—7 µ, L. of articulations 60—70 µ; polygonal. Transverse walls horizontal with conspicuous sieve-plates. Contents: large nuclei; callus becoming brown in iodine and sulphuric acid. Parenchyma cells. R. and T. 4—10 µ; L. 50—80 µ. Cell contents: a large nucleus, generally conspicuous; producing a reddish brown mass when treated with potassium dichromate.

C a m b i u m. Fascicular cambium occurring in all vascular bundles, generally thick 4—5 layers of elements. Interfascicular cambium wanting in a few rhizomes; in other rhizomes not quite developed, some radial rows of cells of the medullary commissures showing no formation of cambium; most rhizomes having developed a continuous cambium, thick about 3 layers of elements.

X y l e m. Small bundles, containing 1—3 vessels, lying practically in radial rows corresponding with those of the cambium; not numerous; surrounded by parenchyma cells. Vessels in the inner part spiral or reticulate, for the rest pitted. In those rhizomes showing a complete cambium secondary xylem always more or less developed. In the largest vascular bundles generally one or more bundles of libriform fibres.

Vessels. The innermost ones R. and T. 10 µ, the other ones R. and T. 18 µ; those at the sides of the xylem somewhat smaller; L. of articulations 60—130 µ; the innermost vessels showing the longest articulations; polygonal prisms with more or less rounded edges; transverse walls horizontal with circular perforation, sometimes the perforation on one side. Walls lignified. Cell contents: inulin (alcoholic material), perhaps an oily liquid (fresh material). Libriform fibres. R. and T. 10 µ, L. 150—250 µ; polygonal with very irregular ends, sometimes bifurcated. Walls thickened; lignified; with slit-like pit canals; those pit canals somewhat widening in the direction of the middle lamella, hence inclining to become bordered pits; the adjoining slit-like pits of 2 fibres lying crosswise; middle lamella conspicuous; intercellular spaces wanting. Parenchyma cells. R. and T. 10 by 15 µ or 15 by 15 µ, L. 30—50 µ; mostly tetragonal prisms with a longitudinally directed axis, in the innermost part sometimes polygonal; intercellular spaces wanting. Cell contents: see those of the outermost 10 layers of the cortical parenchyma.

Medullary commissures. 2 or more cells wide, intercellular spaces.
C e l l s   o f   t h e   a b o v e. Those corresponding to the phloem R. 10 μ, T. 12 μ; those corresponding to the xylem R. 35 μ, T. 30 μ; polygonal prisms with a longitudinally directed axis and rounded edges. Cell contents: sometimes inulin; for the rest see those of the cortical collenchyma.

Medulla. Consisting of an outer and an inner part different as to the cell contents. Large intercellular spaces similar to those of the cortex.
C e l l s   o f   t h e   a b o v e. R. and T. 40 by 50 μ, L. 20—35 μ; polygonal prisms with a longitudinally directed axis, strongly rounded edges and curved sides. Cell contents: for the outer part see collenchyma of the cortex; for the inner part see parenchyma of the cortex.

October 1903.   J; M; L.

RHIZOMA CALAMI.
Calamus. Sweet Flag Root.
The dried rhizome of Acorus Calamus, Linn. Sp. Pl. 324, collected in the autumn, all appendages removed but not peeled.

Macroscopic characters.
Up to 5 d.M. in length, up to 3 c.M. thick; tough; somewhat curved, cylindrical, flattened at the top, thinner where the annual shoots adjoin each other; these shoots consisting of about 18 internodes up to 1.5 c.M. in length. On the nodes elongately triangular distichous lateral dark brown leaf-scars, on which often dark brown fibrous remnants of the leaf-sheaths covering the internodes; in the axils of the scars occasionally thick-set branches, for the rest rarely ramified. At the lower surface prominent annular scars of the cut off secondary roots placed in irregular, sometimes double zig-zag lines. Surface wrinkled longitudinally, brown. Transverse fracture porously corky or somewhat granular whitish or flesh coloured; cortex 2—4 times thinner than the stele.
Odour peculiarly aromatic; taste aromatic, bitter, somewhat pungent.

Anatomical characters.
LITERATURE. Berg. Anat. Atlas. 1865. 39. Pl. XX. Benecke. Mikr. Drogenpraktikum. 1912. 31. Biechele. Mikr. Prüf. d. off. Drogen. 1904. 62. Biermann. Ueber Bau und Entwicklungsgesch der Oelzellen u. die Oelbildung in ihnen. Diss. Bern. 1898. Flückiger. Pharmakogn. 1891. 348. Flückiger & Hanbury. Pharmacographia. 1879. 676. Gilg. Pharmakogn. 1910. 36. Greenish & Collin. Anat. Atl. of veget. powders. 1904. 226. Hager. Pharm. Praxis. Bd. I. 1900. 536. Karsten u. Oltmanns. Pharmakogn. 1909. 20. Koch. Mikr. Anal. d. Drogenpulver. Bd. II. 1903. 20. Koch. Einf. i. d. mikr. Anal. d. Drogenpulver. 1906. 37. Koch. Pharmakogn. Atlas. Bd. I. 85. Pl. XV. Koch u. Gilg. Pharmakogn. Praktik. 1907. 69. Kraemer. Botany a. Pharmacogn. 1910. 496, 760. Luerssen. Syst. Bot. 1882. 320. Marmé. Pharmacogn. 1886. 71. Meyer. Wiss. Drogenk. Bd. II. 1891. 75. Meyer. Mikr. Unters. v. Pflanzenpulv. 1901. 71. Moeller. Pharmakogn. 1889. 297. Moeller. Mikr. pharm. Üb. 1901. 276. Oudemans. Pharmacogn. 1880. 83. Oudemans. Aant. o. d. Pharm. neerl. 1854—56. 83. Planchon et Collin. Drogues simples. T. II. 1890. 122. Schneider. Powdered Veget. Drugs. 1902. 158. Tschirch. Harze u. Harzbeh. 1900. 390. Tschirch. Pharmakogn. Bd. II. 1911. 969. Tschirch u. Oesterle. Anat. Atl. 1900. 79. Pl. 20. Vogl. Anat. Atl. 1887. Pl. XLIII. Wiesner. Rohstoffe. 1900. 500. Wigand. Pharmakogn. 1879. 105. Zacharias. Über Secretbehälter mit verkorkten Membranen. Bot. Zeit. 1879. 618. MATERIAL. The drug and rhizomes, thick 20 by 23 m.M., collected near Groningen

April 1902, fresh and in alcohol. REAGENTS. Water, glycerine, potash, saffranine, phloroglucin and hydrochloric acid, iodine and sulphuric acid 66 per cent., concentrated sulphuric acid, Schulze's macerating mixture, potassium dichromate, copper acetate and iron acetate, vanillin and hydrochloric acid.

MICROGRAPHY.

E p i d e r m i s. Often thrown off near the roots. Close to the cork tissue the cells often with a tangential partition-wall. Exceptionally in a tangential section a cell with a much smaller cavity and much thicker lateral walls than the other cells.

C e l l s o f t h e a b o v e. R. 20 μ, T. 10 μ, L. 40—60 μ, at the lower surface sometimes wider and shorter; tetra- to hexa-, mostly tetragonal prisms with a radially directed axis; outer walls somewhat dome-shaped, so as to form cuticular folds over the lateral walls. Walls thickened, 6 μ thick; consisting of a thick yellow stratified cuticle and an equally thick cellulose lamella; at the lower surface sometimes all the walls of the epidermal cells yellow and persisting in concentrated sulphuric acid. Cell contents: some chloroplasts and starch grains.

C o r t e x.

Secondary cork tissue. Storied cork, only occurring on the leaf-scars and in the vicinity of the roots. Developed as periderm, 3—6 layers. Outside the cork tissue on the leaf-scars occasionally some layers of common parenchyma cells with brown walls consisting of cellulose. The development of the cork tissue beginning in the layer of cells next to the epidermis by the formation of one or more tangential walls, this process being repeated in the second and sometimes in the third layer; at the same time the radial dimensions becoming much larger and the collenchymatous thickenings disappearing. The intercellular spaces of the original cells yet clearly to be seen.

P e r i d e r m c e l l s. Form and dimensions see the collenchyma cells. Walls yellow; suberized; not pitted.

Primary Cortex.

C o l l e n c h y m a. Consisting of 5—7 layers of collenchyma cells intermixed with idioblasts; showing intercellular spaces. Layer next to the periderm less collenchymatous.

C o l l e n c h y m a c e l l s. R. and T. 25—30 μ, L. 50 μ; in the outermost layer tetragonal prisms with rounded edges, in the other layers circular or elliptical cylinders and both with a longitudinally directed axis. Walls pitted. Cell contents: chloroplasts and small, nearly always simple starch grains.

A e r e n c h y m a. Composed of about 150 layers of common parenchyma cells, intermixed with idioblasts. Large intercellular passages, becoming still larger towards the interior; R. and T. 100—170, L. 400—500 μ; separated from each other by diaphragms of 1 cell wide, except in the outer part. Moreover here and there small triangular intercellular spaces surrounded by thickened walls.

C o m m o n p a r e n c h y m a c e l l s. R. and T. 30—50 μ, L. 40—60 μ; on the lateral sides of the intercellular passages tetragonal, on the corners penta- to hexagonal prisms. The walls bounding the intercellular spaces somewhat curved outwards. Cell contents: small and nearly always simple starch grains; in the outer layers chloroplasts.

26

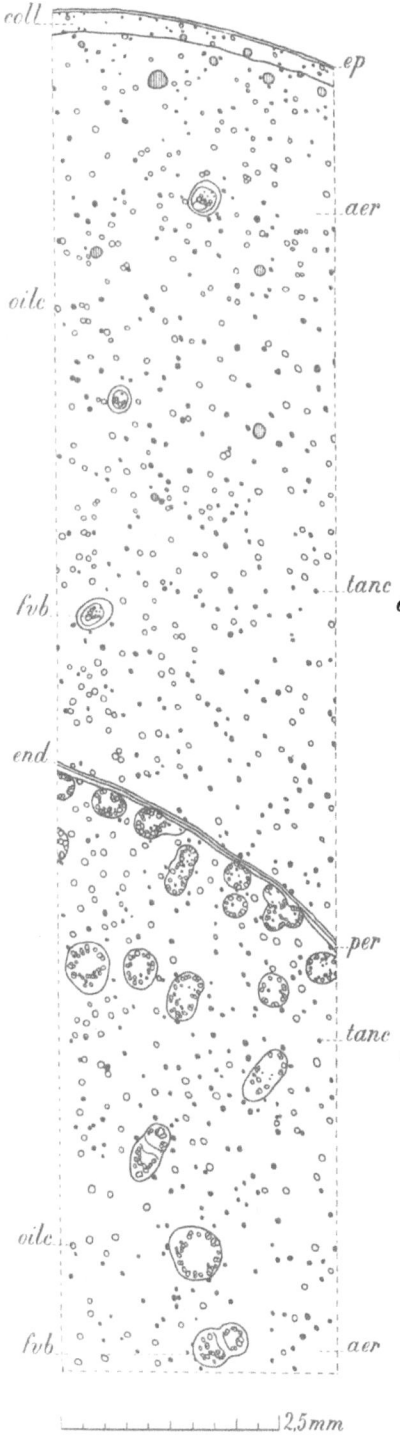

Idioblasts in 3 kinds. 1. Oil cells; mostly lying at the upper or under side of the intercellular passages and always at the intersection of 2 sheets of cells; the surrounding cells somewhat protruding into them. 2. Tannin cells; usually situated like the oil cells. 3. Anthocyanin cells in some rhizomes. O i l c e l l s. R. and T. and L. 70—90 µ; spherical or somewhat irregularly shaped, in the collenchyma nearly having the dimensions and the form of the surrounding cells. Outermost lamella of the walls suberized, the rest not consisting of pure cellulose [1]). Cell contents: in the fresh material mostly 1 colourless to light yellow oil drop quite filling the cell, sometimes some very small drops lying round the large one; in the drug contents wanting [2]). T a n n i n c e l l s. Form, dimensions and walls like those of the surrounding aerenchyma cells. Cell contents: tannin, in the drug a brown mass; contents of both the fresh material and the drug becoming red in vanillin and hydrochloric acid.

Cortical meristeles. Each represented only by a fibrovasal bundle; arranged in 2 irregular rings, both consisting of larger bundles interchanging with very small ones. Sometimes even no xylem or phloem elements inside a bundle of sclerenchyma fibres.

Sclerenchyma sheath. Consisting of 2—4 layers of sclerenchyma fibres; moreover only on the side turned towards the epidermis a very incomplete sheath of crystal fibres divided in cells, each containing a simple crystal. S c l e r e n c h y m a f i b r e s. R. and T. 4—6 µ, L. 300—400 µ; polygonal. Walls thickened, towards the interior of the sheath much less so; middle lamella clearly to be seen; showing stratification; a little lignified; with cross-wise slit-like pits obliquely placed in a spiral.

Vascular bundles. Simple; collateral; closed.
Phloem. Consisting of sieve-tubes and cambiform cells; often situated more or less on one of the lateral sides of the xylem instead of on the outside. Intercellular spaces wanting.

Fig. 91. *Acorus Calamus.* Rhizome, transverse section. aer Aerenchyma; coll Collenchyma; end Endodermis; ep Epidermis; fvb Fibrovasal bundles; oilc Oil cells; per Pericycle; tanc Tannin cells.

---

[1]) See Z a c h a r i a s. Über Secretbehälter mit verkorkten Membranen. Bot. Zeit. 1879. 618.
[2]) See for the formation of the oil T s c h i r c h und O e s t e r l e. Anat. Atl. 81, or T s c h i r c h. Die Harze und Harzbehälter. 1900 387.

S i e v e-t u b e s. R. and T. 6—10 μ, L. of articulations 90—150 μ; polygonal prisms; sieve-plates conspicuous with a very thick callus (the term callus here used in the sense attached to it by L e c o m t e, differing from the usual meaning). C a m b i u m f o r m c e l l s. R. and T. 4—10 μ, L. 40—70 μ; polygonal prisms. Cell contents: in some cells tannin and in other small starch grains.

Xylem. Consisting of vessels, parenchyma cells and spiral cells. The vessels lying in bundles or isolated; annular, spiral, reticulate and scalariform. The spiral cells only occurring in the large vascular bundles and lying in a bundle between the vessels and the sheath.

V e s s e l s. R. and T. 10—30 μ; polygonal prisms often with rounded edges, or cylinders. Walls a little thickened; lignified. P a r e n c h y m a c e l l s. R. and T. 10—12 μ, L. 40—60 μ. Cell contents: here and there very small starch grains. S p i r a l c e l l s. Dimensions strongly varying from R. 25 μ, T. 25 μ, L. 30 μ, to R. 75 μ, T. 75 μ, L. 80 μ. Walls with partial annular lignified thickenings. Cell contents: here and there starch grains.

E n d o d e r m i s. Consisting of 1 layer of cells, developed as a protective-sheath. The endodermis and the pericycle having openings with borders turned inwards for the communication of the vascular bundles of the cortex and the stele; moreover 2 lateral openings for the emission of endodermis, pericycle and vascular bundles into the lateral sprouts.

C e l l s o f t h e a b o v e. R. 15 μ, T. 30—40 μ, L. 40 μ; rectangular prisms with a longitudinally directed axis, both tangential walls somewhat rounded. Radial walls lignified, becoming yellow and showing up much more clearly in potash. Casparian spots in a transverse section. Cell contents: minute starch grains filling the cells.

S t e l e.

Pericycle. Consisting of 1 or 2 layers of common parenchyma cells. See moreover the endodermis.

C e l l s o f t h e a b o v e. R. 20 μ, T. 30 μ; tetra- to hexagonal prisms with a longitudinally directed axis. Walls a little collenchymatous. Cell contents: starch grains.

Fibrovasal bundles. Close to the pericycle a dense ring of bundles, longitudinal or transverse and anastomosing with each other; in the middle part of the stele the bundles more scattered and longitudinal.

S h e a t h. Consisting of 1 or 2 layers of fibres without thickened walls and with a transverse partition-wall; crystal fibres wanting.

F i b r e s o f t h e a b o v e. R. and T. 18—20 μ, L. 200—300 μ; polygonal. Walls pitted. Cell contents: here and there starch grains.

V a s c u l a r b u n d l e s. Almost always amphivasal. Often 2 collateral or 1 collateral and 1 amphivasal or 2 amphivasal bundles blended together. Closed.

Phloem. In the amphivasal bundles often the cribral elements lying in bundles, separated by cambiform; in the internal part often the elements smaller and more or less flattened.

Xylem. Consisting of vessels and parenchyma; vessels lying in a sometimes incomplete ring of 1 vessel thick. Spiral cells wanting in the amphivasal bundles.

V e s s e l s. Diameter 20—30 μ.

Medullary commissures and **medulla.** Consisting of aerenchyma.

Micrography of the powder. Thin-walled parenchyma cells with very conspicuous connecting frames, at the inside of which oval pits; distinct air cavities between these cells, sometimes also small triangular intercellular spaces; contents a great number of starch grains of about 5μ in diameter, most of them simple, about spherical or more oblong, some compound and consisting of up to 4 grains. Many tannin cells of the same shape as the above mentioned, without starch; with yellowish, in the older drug more brownish lumps; these becoming black in iron acetate, bright red in vanillin and hydrochloric acid. Oil cells hardly to be distinguished in the powder; no resin-lumps. Bundles of slender, thick-walled, pitted fibres. Annular, spiral, reticulate and scalariform vessels. Wide, thin-walled, light brown cork cells and narrow oblong epidermal cells with somewhat thickened outer walls, to a small amount.

April 1902.   J; M; L.

RHIZOMA FILICIS.
Filix Mas. Male Fern. Rhizoma Aspidii. Rhizoma Filicis Maris.

The rhizome of Aspidium Filix mas, Swartz, collected late in the autumn; when many young leaves rolled inwards (circinnate) are to be found at the top and the still living fleshy lower portions of the petioles of dead leaves towards the base.

Macroscopic characters.

The commercial material commonly consisting chiefly of detached bases of petioles, mixed with some small irregularly oblong pieces of the rhizome and a very few spirally wound young leaves; all these already nearly entirely freed from roots, dead parts and ramenta as necessary for use. Bases of petioles up to 5 c.M. in length, up to 1.25 c.M. thick; hard, brittle; somewhat curved, tapering towards the top more slowly than towards the base; the concave inner surface flat, the convex outer surface arched in a transverse direction; both faces separated from each other by 2 longitudinal, often more brightly coloured, acute protruding ridges. Sometimes a bud or a scar of a bud at the outer side. Surface showing coarse longitudinal wrinkles, somewhat polished dark brown or of a lighther colour, partly more or less covered with ramenta; these rusty brown, lanceolate to linear, with a slowly tapering apex, having at the margin minute scattered acute teeth, without glands, or having at the utmost 2 glands at the base. Transverse fracture yellow-green, porous, mealy. On the transverse section the meristeles arranged in a curve, opening towards the concave inner side; number in 31 per cent of the petioles 7, in 30 per cent 8, in 22 per cent 9, in 7 percent 5,6 or 10. Pieces of the rhizome often entirely covered by ramenta at the top; fracture the same as that of the petioles.

Odour slight; taste sweetish, somewhat astringent, acrid.

Anatomical characters.

LITERATURE. de Bary. Vergl. Anat. 1877. 125 &c. Berg. Anat. Atl. 1865. 34. Taf. 17.

Deutsch. Arzneib. 5. Ausg. 1910. 433. Flückiger & Hanbury. Pharmacographia. 1879. 734. Flückiger. Pharmakogn. 1891. 313. Hérail. Mat. Méd. 1912. 223. Höhlke. Üb. d. Harzbehälter u. d. Harzbildung b. d. Polypodiaceen u. einigen Phanerogamen. Beitr. z. Bot. Centralbl. Bd. 11. 1901—1902. 8. Karsten u. Oltmanns. Pharmakogn. 1909. 16. Koch u. Gilg. Pharmakognost. Praktikum. 1907. 67. Koch. Die mikrosk. Anal. d. Drogenpulver. Bd. 2. 1903. 27. Koch. Pharmakogn. Atlas. Bd. 1. 1911. 91. Kraemer. Botany a. Pharmacognosy. 1910. 687, 688, 749. Luerssen. Med. Pharmac. Botanik. I. 1879. 563. Marmé. Pharmacogn. 1886. 62. Oudemans. Aanteekeningen op de Pharmacopoea Neerlandica. 1854—56. 14. Oudemans. Pharmacogn. 1880. 11. Schacht. Üb. e. neues Secretions-organ im Wurzelstock von Nephrodium Filix mas. Pringsheim's Jahrb. Bd. 3. 1863. 352. Tschirch. Angew. Planzenanatomie. 1889. 158, 470. Tschirch u. Oesterle. Anat. Atl. 1900. 341. Taf. 79. Vogl. Anat. Atl. 1887. Taf. 40. MATERIAL. The drug; bases of petioles, collected Nov. 1902 in the Botanic Garden at Groningen, fresh and in alcohol. REAGENTS. Water, glycerine, chloral hydrate, potash, absolute alcohol, hydrochloric acid, iodine in chloral hydrate, phloroglucin and hydrochloric acid, iodine and sulphuric acid 66 per cent., concentrated sulphuric acid, osmic acid, Schulze's macerating mixture, cupric acetate and ferric acetate, potassium dichromate.

MICROGRAPHY of base of petiole.

Epidermis. Cells fairly well in longitudinal rows; some cells divided by a tangential partition-wall. Stomata wanting. Trichomes in 2 kinds: 1. ramenta — see fig. 92 — numerous, thick 1 layer of cells, cells fairly well in longitudinal rows; terminating in an apex, some cells in length and 1 cell broad; especially at the margin numerous oblong appendages — 15—30 μ broad at the base, 15—100 μ in length — standing out perpendicularly or somewhat bent towards the base of the ramentum, consist-ing of 2 slender papillae belong-

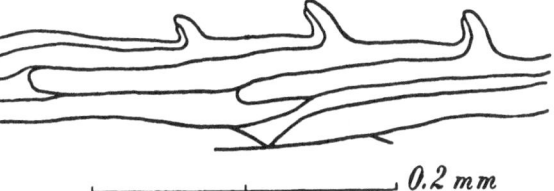

Fig. 92. *Aspidium Filix mas*. Rhizome, margin of ramentum, showing the oblong appendages.

ing to 2 adjacent cells and not separated from these by a transverse wall, one of the papillae always somewhat longer than the other; 2. glandular hairs varying in number, mostly inserted near the bases of the ramenta; unicellular; see for the rest those in the intercellular spaces of the ground tissue.

Epidermal cells proper. R. 20 μ, T. 20—40 μ, L. 30—70 μ; poly-, generally tetragonal tables with a radially directed axis, often with curved lateral walls. Outer walls somewhat thickened, showing stratification, always brown; lateral and inner walls often brown; all walls persisting in concentrated sulphuric acid. Cell contents: generally 1, sometimes 2 or 3 ellipsoidal brown bodies; tannin through the whole cell, excepting a vacuole of about 10 μ in diameter; in some cells a yellow-brown granular mass. Cells of ramenta. H. 20 μ, Lev. 22 and 300—400 μ; tetragonal tables, often somewhat fibre-shaped. Walls persisting in concentrated sulphuric acid. Cell contents: generally a granular mass, close to the transverse walls; nucleus generally close to one of the longitudinal lateral walls. Cells of glandular hairs, head 25 μ in diameter, 30 μ in length; stalk 8 μ in diameter, 10—15 μ in length.

Ground tissue.

Sclerenchyma. In about 10—12 layers of cells everywhere under the epidermis, except in the brightly coloured longitudinal ridges.

C e l l s o f t h e a b o v e. R. 12 μ, T. 20—40 μ, L. 300—700 μ, those of the innermost 2 layers R. 20—30 μ, T. 40—60 μ, L. 300—700 μ; generally fibre-shaped, sometimes having only one pointed end. Walls somewhat sinuously thickened; walls of the outer- and innermost 2 layers of cells colourless; showing stratification; yellow to red-brown; persisting in concentrated sulphuric acid, middle lamellae and gussets lignified; pitted; intercellular spaces wanting. Cell contents: especially in the outer- and innermost layers chloroplasts filled with a starch grain; tannin in 1 larger or some smaller vacuoles; in potash strongly refracting globules, filled with smaller more transparent globules; in concentrated sulphuric acid or osmic acid no oil to be distinguished.

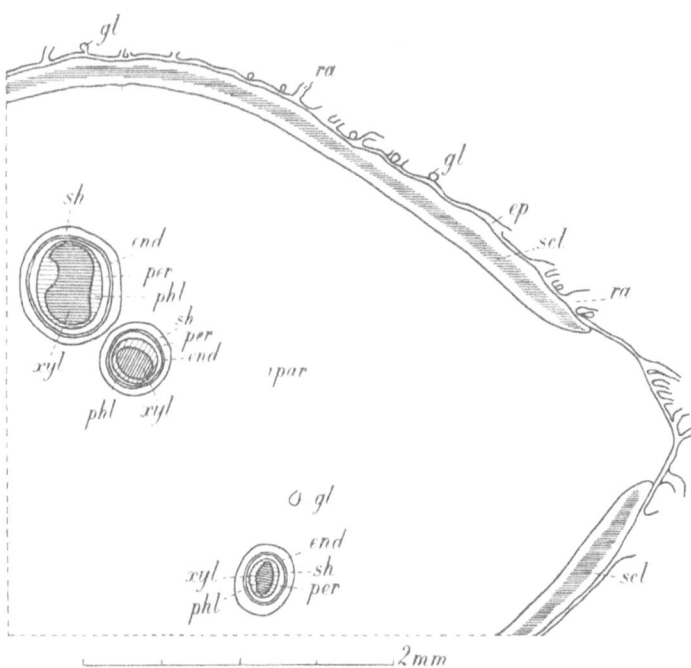

Fig. 93. *Aspidium Filix mas.* Rhizome, transverse section. end Endodermis; ep Epidermis; gl Glandular hairs; ipar Inner parenchyma; per Pericycle; phl Phloem; ra Ramenta; scl Sclerenchyma; sh Sheath of sclerenchymatic parenchyma; xyl Xylem.

Parenchyma filling the brightly coloured longitudinal ridges. About 10—15 layers of cells.

C e l l s o f t h e a b o v e. R. 18 μ, T. 25 μ, L. 50—100 μ; polygonal prisms with a longitudinally directed axis and much rounded edges. Walls sometimes a little collenchymatous; pitted; intercellular spaces. Cell contents: the same as those of the inner and outermost layers of the sclerenchyma.

Inner parenchyma. Constituting the largest part of the base of the petiole. Many smaller and larger longitudinally elongated intercellular spaces containing generally 1 glandular hair, sometimes 2 or 3. Glandular hairs most numerous in the vicinage of the meristeles; unicellular; consisting of a short stalk, 10 μ in diameter, 14 μ in length, and a globular or ellipsoidal head, 40 μ in diameter and 50 μ in length; the head entirely covered with a light green layer, thick up to 15 μ, of a secretion contained in a cuticular bladder, and frequently filling with their secretion the whole, often somewhat irregularly shaped intercellular space; in older petioles the secretion less abundant.

P a r e n c h y m a c e l l s. The outermost layers R. 40 μ, T. 50 μ, L. 120 μ; the inner layers R. 80 μ, T. 100 μ, L. 120—200 μ; polygonal prisms with a longitudinally directed axis and rounded edges. Walls somewhat thickened; those limiting the larger intercellular spaces often covered with a thin cuticle; pitted. Cell contents: in young petioles chloroplasts, almost entirely filled with starch grains; these grains simple, very seldom compound, 3—6 μ in diameter, up to 10 μ in length, globular, ellipsoidal or kidney-shaped; tannin in almost every cell, filling one vacuole, 50—60 μ in diameter, placed somewhat towards one end of the cell and in all cells towards the same side, in some cases this vacuole divided in several smaller ones perhaps by the action of potassium dichromate, in material fixed in alcohol this vacuole sometimes very distinct, the remaining parts of the cell seeming to be filled with chloroplasts; in potash strongly refracting globules, filled with smaller more transparent globules; in concentrated sulphuric acid or osmic acid no oil to be distinguished. C e l l o f g l a n d u l a r h a i r. Wall showing a very thin dark coloured outermost layer and a colourless somewhat thicker inner layer, consisting of cellulose. Secretion covering the wall showing stratification in several media e. g. water, glycerine, potassium dichromate, also in concentrated sulphuric acid and here becoming colourless; after the action of alcohol 96 per cent. or absolute, dissolving a large part of the substance of the secretion, colourless lamellae consisting of cellulose to be dinstinguished; after the action of potash these lamellae also distinct but coloured brown by the presence of tannin; showing in osmic acid no change of colour. Cuticular bladder thin, somewhat thicker in older petioles. Cell contents: in fresh material quite transparent.

S h e a t h o f s c l e r e n c h y m a t i c p a r e n c h y m a adjoining the endodermis. One layer of cells, intercellular spaces wanting between them and the endodermal cells.

C e l l s o f t h e a b o v e. R. 25 μ, T. 50 μ, L. 80—150 μ; poly-, generally tetragonal prisms with a longitudinally directed axis. Walls thickened, especially the inner; yellow to brown; inner walls persisting in concentrated sulphuric acid; pitted. Cell contents: starch grains, somewhat more elongated than in the inner parenchyma; tannin in some small vacuoles, generally close to the inner walls; see for the rest the inner parenchyma.

E n d o d e r m i s. One layer of parenchyma cells.

C e l l s o f t h e a b o v e. R. 5 μ, T. 15 μ, L. 100—180 μ; generally rectangular prisms with a longitudinally directed axis. Walls, radial and transverse walls lignified. Cell contents: a large longitudinally elongated nucleus.

M e r i s t e l e s. Generally cylindrical; at a distance of nearly 1 m.M. from the subepidermal sclerenchyma; those corresponding in place to the protruding ridges being the largest; each containing 1 vascular bundle.

P e r i c y c l e. From 2 to 3 layers of cells; in transverse sections especially the cells of the outer layer in rows with the endodermal cells.

C e l l s o f t h e a b o v e. R. 10 μ, T. 12 μ, L. 70—100 μ; polygonal prisms with a longitudinally directed axis. Cell contents: small chloroplasts filled with starch grains; tannin in numerous small vacuoles; a large longitudinally elongated nucleus.

V a s c u l a r b u n d l e amphicribral.

Phloem. Most strongly developed on 2 diametrically opposed sides of each bundle; consisting of sieve-tubes and cambiform cells; the cambiform cells, adjoining the xylem, being somewhat longer and having very oblique transverse walls.

S i e v e-t u b e s. R. and T. 15—20 μ; polygonal prisms; sieve-plates sometimes distinct in the transverse and lateral walls. C a m b i f o r m c e l l s. R. and T. 6—10 μ, L.

60—90 μ, those of the inner layer L. 125 μ; polygonal prisms with a longitudinally directed axis. Cell contents: some chloroplasts containing some starch grains; tannin; in iodine in chloral hydrate often a red-brown mass filling the cell.

Xylem. Consisting of spiral, reticulate and scalariform vessel tracheids and parenchyma cells; the spiral tracheids sometimes in the interior of the bundle, sometimes peripherically.

V e s s e l   t r a c h e i d s. Diameter 10—50 μ, L. 150—400 μ; polygonal prisms with rounded edges and very oblique transverse walls. Walls somewhat thickened; lignified, also becoming red in sulphuric or hydrochloric acid without addition of phloroglucin [1]), often only after the lapse of some hours. P a r e n c h y m a   c e l l s. Diameter 10 by 15 μ, L. 150 μ; polygonal prisms with a longitudinally directed axis. Cell contents: chloroplasts, containing starch grains; tannin; sometimes red-brown by the action of iodine in chloral hydrate.

November 1902, Juni 1911.   J; M.

RHIZOMA HYDRASTIS.
Hydrastis. Hydrastis Rhizome.
The rhizome with the secondary roots of Hydrastis canadensis, Linn. Syst. ed. x. 1088.

Macroscopic characters.

Rhizomes up to 5 c.M. in length, up to 1 c.M. thick; brittle; sometimes somewhat branched, more or less edgy sinuous; with distinct short internodes; at intervals having scars of died off stems and then being in those places irregularly swollen; on all sides beset with secondary roots or yellow scars corresponding to them; at the summit sometimes remnants of a stem and some leaf-scales. Surface sometimes wrinkled longitudinally, dull yellowish grey or rather brown. Fracture bright yellow, horny or waxy. In a tranverse section up to 20 narrow, radially directed vascular bundles; medulla about $^1/_4$ part of the diameter. Secondary roots 4—5 c.M. in length, up to 1 c.M. thick; brittle; wrinkled longitudinally; brown; internally yellow, with a very thin angular stele. Odour slightly narcotic; taste bitter.

Anatomical characters.

LITERATURE. Benecke. Microsk. Drogenprakt. 1912. 26. Biechele. Mikr. Prüf. d. off. Drogen. 1904. 64. Flückiger. Pharmakogn. 1891. 415. Gilg. Pharmakogn. 1910. 100. Greenish & Collin. Anat. Atl. o. veget. Powders. 1904. 236. Hager. Pharm. Praxis. Bd. II. 1902. 78. Karsten u. Oltmanns. Pharmakogn. 1909. 50. Koch. Mikrosc. Anal. d. Drogenpulver. Bd. II. 1903. 45. Pl. IV. Koch. Pharm. Atl. Bd. I. 1911. 103. Pl. XVIII. Koch u. Gilg. Pharmakogn. Prakt. 1907. 74. Kraemer. Botany a. Pharmacogn. 1910. 500, 739. Planchon et Collin. Drogues Simples. T. II. 1896. 929. Schneider. Powdered Veget. Drugs. 1902. 214. Tschirch. u. Oesterle. Anat. Atl. 1900. 281. Pl. 74. MATERIAL. The drug. Rhizomata, collected August 1909 in the Botanic Garden at Groningen, in alcohol; thick 3 and 9 m.M. REAGENTS. Water, glycerine, potash, iodine in chloral hydrate, phloroglucin and hydrochloric acid, iodine and sulphuric acid 66 per cent., concentrated sulphuric acid, Schulze's macerating mixture.

---

[1]) R. B ö h m. Ueber homologe Phloroglucine aus Filixsäure und Aspidin. L i e b i g. Ann. d. Chem. Bd. 302 and R. B ö h m. Über Filicinsäure. Ibid. Bd. 307.

MICROGRAPHY.

Epidermis. Only extant on the thinnest specimen of the material.

Cells of the above. Poly- mostly hexagonal tables, somewhat longitudinally elongated. Walls yellow; lignified.

Cortex.

Secondary cork tissue. Initial-celled cork, appearing in the outermost cortical layer of the thinnest specimen of the material. In the bases of the leaf-scales sometimes cork tissue developed in the same way as described for Valerian Rhizome.

Phellem. Developed as periderm and consisting of 1—3 layers of cells.

Cells of the above. R. 12 μ, T. 15 μ, L. 30 μ; polygonal tables with a radially directed axis. Walls yellow; suberized; lignified.

Phellogen. Occasionally occurring as 1 layer of cells.

Cells of the above. Radial dimensions somewhat smaller than those of the periderm cells.

Phelloderm. Here and there present as 2 layers of cells.

Cells of the above. Similar to the periderm cells, radial dimensions sometimes a little larger. Walls slightly collenchymatous. Cell contents: starch grains.

Primary cortex proper.

Common parenchyma. Consisting of about 20 layers of cells, increasing in size towards the interior, inclusive of the endodermis and pericycle. Showing intercellular spaces.

Cells of the above. R. 30—40 μ, T. 45—60 μ, L. 50—60 μ. Walls of the outermost layers a little thickened and collenchymatous; pitted; sometimes 2 sets of intercrossing striae to be distinguished without reaction. Cell contents: starch grains, simple and nearly always spherical.

Endodermis. Not to be distinguished.

Stele.

Pericycle. Not to be distinguished.

Vascular bundles. 10—15 in number; varying in size; running irregularly and strongly anastomosing; collateral; open.

Phloem. Cuneiform; sometimes a minute portion of primary phloem at the top of the wedge. Secondary phloem constituting by far the greater part; elements lying in radial rows, not all extending to the top of the wedge; composed of sieve-tubes, having somewhat larger radial dimensions towards the interior, and bast parenchyma cells.

Sieve-tubes. R. 8 μ, T. 20 μ, L. of art. 70—80 μ; tetra- to hexa-, mostly tetragonal prisms, in the outer phloem the edges sometimes slightly rounded; sieve-plates distinct. Bast parenchyma cells. See sieve-tubes.

Cambium. Consisting of some layers of elements; the interfascicular cambium not always developed; if present only forming secondary parenchyma.

Xylem. Cuneiform; the primary xylem endarch and represented by some spiral vessels at the top of the wedge. Secondary xylem: the elements nearly always lying in radial rows; consisting of vessels, libriform fibres and parenchyma. Vessels very numerous in the outer part, rather scarce in the inner part; pitted, sometimes scalariform. Libriform fibres usually lying in a bundle in the centre of the wedge; a few times moreover a smaller bundle near the top. Wood parenchyma showing here and there radially elongated cells; without intercellular spaces.

Vessels. R. and T. 18—22 μ, L. of articulations 60—80 μ, spiral vessels with smaller diameter; tetra- to hexagonal prisms; transverse walls with circular perforations. Walls a little thickened; middle lamella distinct; lignified. Libriform fibres. 10 μ in diameter; tetra- to hexagonal. Walls, middle lamella distinct; lignified, especially the middle lamella; with cross-wise slit-like pits. Wood parenchyma cells. R. 10 μ, T. 20 μ, L. 60—80 μ; in the outer part polygonal, for the rest tetra- to hexa-, mostly tetragonal prisms with a longitudinally directed axis. Cell contents: see the cortical parenchyma.

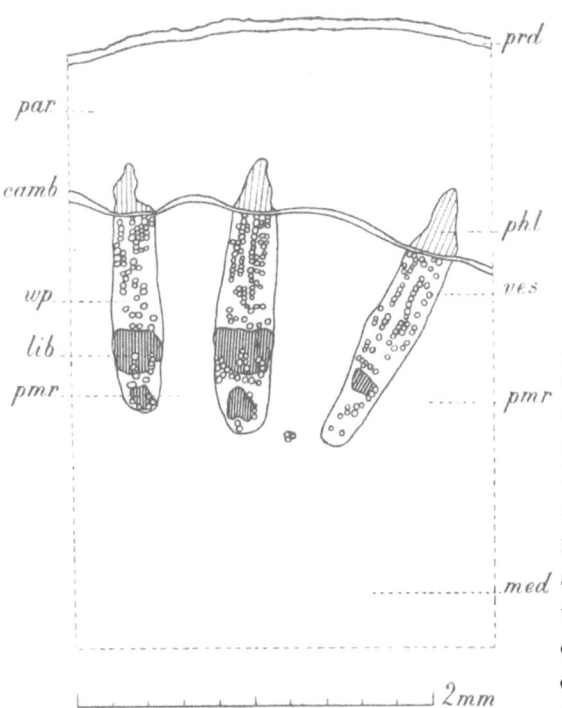

Fig. 94. *Hydrastis canadensis*. Rhizome, transverse section. camb Cambium; lib Libriform fibres; med Medulla; par Common parenchyma; phl Phloem; pmr Medullary commissures; prd Periderm; ves Vessels; wp Wood parenchyma.

Medullary commissures. Cells lying in fairly radial rows; outside the cambium the arrangement less regular and the radial dimensions of the cells smaller. Sometimes the cells hardly to be distinguished from the cortical parenchyma cells.

Cells of the above. R. 50—60 μ, T. 30—40 μ, L. 35 μ; tetragonal prisms with a radially directed axis. Cell contents: see the cortical parenchyma.

Medulla. Consisting of common parenchyma cells; showing intercellular spaces.

Cells of the above. R. and T. 40—50 μ; polyhedra with rounded edges. Cell contents: see the cortical parenchyma.

Micrography of the powder. Most fragments coloured more or less yellow. Thin-walled parenchyma, most cells oblong, quite filled with starch; besides many detached starch grains, usually 4—8 μ in diameter, for the greater part simple and about spherical, also compound ones with 2—4 component grains. Vessels with slit-like bordered pits, not rarely joined in bundles; also short articulations of vessels resembling sclerenchyma cells; a very few spiral vessels. Periderm cells tetragonal, somewhat oblong, with thin very light brown walls. Some root hairs. Occasionnally thick-walled libriform fibres. Chloral hydrate extracting the yellow colouring matter. In iodine in chloral hydrate after some minutes many dark brown crystals, square, oblong, hexagonal or having an other shape, also several ones united together. In strong nitric acid the powder colouring reddish brown. In diluted nitric acid innumerable fine yellow needle-shaped crystals, often somewhat cohering (berberin).

June 1901.   J; M; L.

RHIZOMA PODOPHYLLI.
Podophyllum. Podophyllum Rhizome. Phodopyllum Root.
The rhizome with the secondary roots of Podophyllum peltatum, Linn. Sp. Pl. 505.

Macroscopic characters.
Pieces up to 20 c.M. in length, with enlarged parts at intervals of 5—10 c.M.; from those points sometimes 2—3 branches. The longer thinner parts thick up to 6.5 m.M.; cylindrical, often the upper and the under surface more or less flattened; without nodes or only having at the ends some irregular annular leaf-scars, often obliquely placed. The shorter thickened parts long 1—2 c.M., thick up to 1.6 c.M.; with short internodes and a certain number of circular not very prominent leaf-scars lying close to each other; at the upper surface showing a nearly circular depressed scar of a stalk; at the under surface about 10 scattered fibrous roots or mostly only the dirtily white scars corresponding to them. Also at the ends of the thinner parts sometimes a few roots. Detached secondary roots, about 1.5 m.M. thick, rather numerous. Surface of the rhizome smooth or slightly longitudinally wrinkled, darker or lighter reddish or earthy brown; the roots of a paler colour. Transverse fraction horny and light brown, or mealy and white, not rarely both cases next to each other on the same fraction; outermost layer darker brown, half-way the radius a ring of 40 vascular bundles at the utmost. Odour, especially after being moistened with hot water narcotic; taste slightly bitter and pungent.

Anatomical characters.
LITERATURE. Flückiger & Hanbury. Pharmacographia. 1879. 37. Hager. Pharm. Praxis. Bd. II. 1902. 687. Hérail. Mat. Méd. 1912. 407. Karsten u. Oltmanns. Pharma-kogn. 1909. 48. Marmé. Pharmacogn. 1886. 87. Moeller. Pharmakogn. 1889. 325. Plan-chon et Collin. Drogues Simples. T. II. 1896. 865. Schneider. Powdered Veget. Drugs. 1902. 261. MATERIAL. The drug. Rhizomata collected July 1901 in the Botanic Garden at Groningen; fresh, in alcohol and after treatment with 1 per cent chromic acid in alcohol; thick 3 and 7 m.M. REAGENTS. Water, glycerine, potash, ammonia, iodine in chloral hydrate, phloroglucin and hydrochloric acid, iodine and sulphuric acid 66 per cent., con-centrated sulphuric acid, Schulze's macerating mixture, iron acetate.

MICROGRAPHY.
Epidermis. Only occurring on the thinnest specimen of the material used; consisting of longitudinal rows of cells.
Cells of the above. R. 25 μ, T. 25 μ, L. 85 μ; tetra- to hexagonal tables with a

radially directed axis. Outer and inner walls somewhat thickened; all the walls colourless or varying from yellow to brown; suberized.

C o r t e x.

S e c o n d a r y   c o r k   t i s s u e. Inital-celled cork; monopleuric; developed in the first or second, sometimes in the third cortical layer of the thinnest speci- mens of the material used; in the latter cases the layers outside the cork tissue consisting of common parenchyma cells, sometimes divided by a tangent- ial partition-wall and showing inter- cellular spaces.

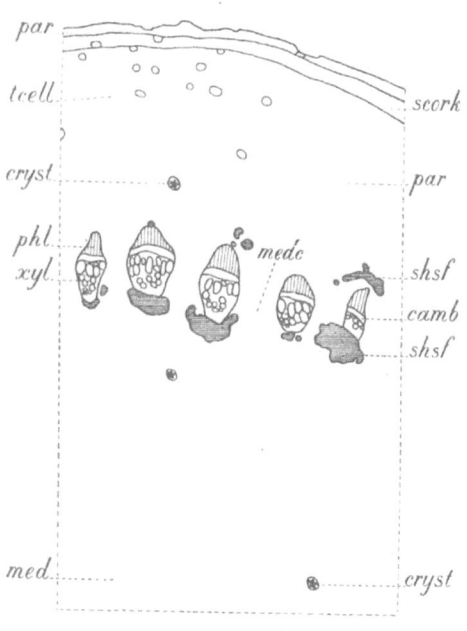

P h e l l e m. Consisting of 2—3 layers of periderm cells, lying in longitudinal rows; in a few places some more layers. P e r i d e r m   c e l l s. R. 30 μ, T. 45 μ, L. 60—80 μ; tetra- to hexagonal tables with a radially directed axis. Walls yellow; suberized, those of the innermost layer also lignified, those of the other layers showing cellulose lamella when treated with iodine and sulphuric acid 66 per cent. Cell contents: sometimes starch grains and in some cells tannin.

P h e l l o g e n. Consisting of 1 layer of cells.

C e l l s   o f   t h e   a b o v e. R. and T. 30 μ; inner walls somewhat bulging out. Walls a little thickened, especially the inner and lateral ones. Cell contents: starch grains, sometimes also tannin.

Fig. 95. *Podophyllum peltatum.* Rhizome, transverse section. camb Cambium; cryst Cells with cluster crystals; med Medulla; medc Medullary commis- sures; par Cortical common parenchyma; par' Com- mon parenchyma outside the cork tissue; phl Phloem; scork Secondary cork tissue; shsf Sheath of sclerenchyma fibres; tcells Cells with tannin; xyl Xylem.

Primary   cortex   proper. Com- mon parenchyma. Consisting of about 20 layers of cells inclusive of the endodermis and perhaps of a part of the peri- cycle; lying in longitudinal rows. Cells often with a transverse partition-wall; close to the phellogen sometimes a cell with a tangential partition-wall, show- ing intercellular spaces. Cells of the middlemost layers larger than the rest. Crystal idioblasts. Here and there a cluster crystal; sometimes some of those cells placed in longitudinal rows; L. 40—70 μ.

C o m m o n   p a r e n c h y m a   c e l l s. Of the layers occasionally occurring outside the secondary cork tissue R. 25 μ, T. 35 μ, L. 90—110 μ; tetra- to hexagonal tables with a radially directed axis. Walls somewhat thickened; varying from yellow to brown; suberized and showing a cellulose layer when treated with iodine and sulphuric acid 66 per cent. Other cortical parenchyma cells R. 60 μ, T. 80 μ, L. 100—150 μ; elliptical cylinders or polygonal prisms with strongly rounded edges and a longitudinally directed axis. Walls of the outermost layers a little thickened and collenchymatous; pitted. Cell contents: starch grains, simple and compound with up to 15 component grains, compound grains

20—25 μ in diameter; in some cells of the outermost layers besides starch grains tannin; in the chromic acid material in many cells of the inner cortical parenchyma yellow globes those globes often containing a lighter coloured smaller globe, dissolving in ammonia; in the fresh material — not in the simplex — becoming yellow im ammonia.

Endodermis. Not to be distinguished.

Stele.

Pericycle. Not to be distinguished; perhaps partially represented by the bundles of sclerenchyma fibres ontside the vascular bundles.

Fibrovasal bundles. About 20, arranged in a ring; collateral; open.

Sclerenchyma sheath. Consisting of fibres and only developed at the outer and the inner side of the vascular bundles; at the outer side the fibres detached or forming 1 or more small bundles, at the inner side about 30 in number in a cross section and always forming 1 bundle, the number varying greatly according to the size of the bundles.

Fibres. On the outer side R. 25 μ, T. 30 μ, L. up to 2 m.M., on the inner side in a transverse section more isodiametrical; polygonal, on the outer side often with strongly rounded edges; sometimes with transverse partition-walls. Walls thickened; lignified, especially the middle lamella; showing stratification; slit-like pit canals lying cross-wise.

Phloem. Consisting of primary and some layers of secondary tissue. The primary phloem exarch; composed of cambiform cells, decreasing in size towards the centre of the rhizome; in the middle part of the primary phloem mostly more or less flattened elements; no intercellular spaces. The secondary phloem nearly exclusively composed of sieve-tubes, here and there with a companion cell; no intercellular spaces.

Sieve-tubes. R. 8—12 μ, T. 15 μ, L. of articulations 125—180 μ; tetra- to hexagonal prisms. Transverse walls a few times oblique; sieve-plates distinct. Companion cells. R. 5 μ, T. 10 μ, L. corresponding with L. of articulations of sieve-tubes; tri- or tetragonal prisms with a longitudinally directed axis. Cambiform cells. R. 20 μ, T. 20 μ, L. 130—160 μ; polygonal prisms with a longitudinally directed axis. Cell contents: some small starch grains; in a very few cases tannin.

Cambium. Consisting of 2 layers of elements.

Elements of the above. R. 5 μ, T. 20 μ, L. corresponding with L. of articulations of sieve-tubes.

Xylem. Primary and secondary xylem not to be distinguished from each other by their arrangement; consisting of spiral and pitted vessels and parenchyma. In the innermost part the vessels with the smallest diameter and the longest articulations. The parenchyma intermixed with the vessels and lying in 2 or 3 layers between the vessels and the inner part of the sclerenchyma sheath; the innermost cells the shortest. Intercellular spaces wanting.

Vessels. Outermost vessels R. 35 μ, T. 25 μ, L. of articultions 170—220 μ; the other R. and T. 15—20 μ or smaller, L. of articulations 290—450 μ; polygonal prisms; transverse walls each with 1 circular perforation. Walls a little thickened; lignified; with slit-like bordered pits, in the spiral vessels with single and double spirals. Parenchyma cells. R. and T. 15—20 μ, L. 90—180 μ; tetra- to hexagonal prisms with a longitudinally directed axis.

Medullary commissures. 6—8 cells wide; consisting of common parenchyma.

Medulla. Consisting of common parenchyma.

Cells of the above. R. and T. 70—80 μ, L. 120—150 μ; polygonal prisms with strongly rounded edges or cylinders, both with longitudinally directed axis; often with transverse partition-walls. See for the rest cells of the cortical parenchyma, except for the globes near the fibrovasal bundles in the chromic acid material.

January 1902.   J; M; L.

RHIZOMA ZINGIBERIS.
Zingiberis. Ginger.

The unpeeled rhizome of Zingiber officinale, Rosc. in Trans. Linn. Soc. VIII. (1807) 348.

Macroscopic characters.

Looking articulate on account of strictures; 2—10 c.M. in length, 2 c.M. high and 1 c.M. wide; sideways flattened, the flat sides slightly arched. Surface coarsely wrinkled, annulated by scars of leaf-scales, ashy grey or pale buff. Transverse fraction short fibrous, with yellowish red dots; cortex about 1 m.M. thick. Odour aromatic; taste hot.

Anatomical characters.

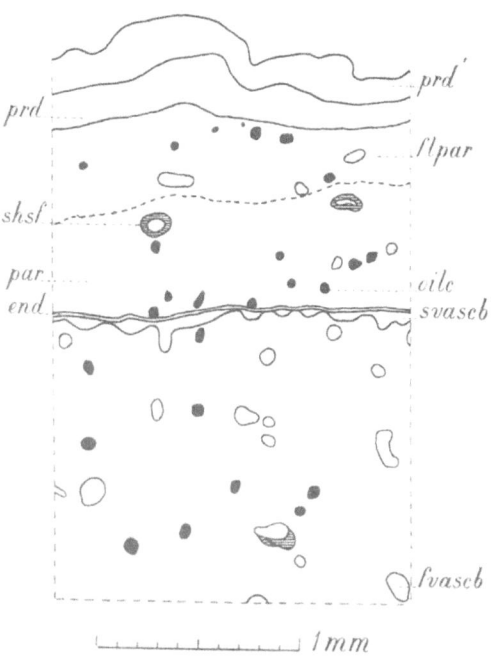

LITERATURE. Berg. Anat. Atl. 1865. 39. Plate XX. Benecke. Mikr. Drogenprakt. 1912. 29. Biechele. Mikr. Prüf. d. off. Drogen. 1904. 68. Erdmann-König. Allg. Warenkunde. 1895. 298. Flückiger. Pharmakogn. 1891. 355. Flückiger & Hanbury. Pharmacographia. 1879. 637. Gilg. Pharmakogn. 1910. 66. Greenish & Collin. Anat. Atl. of veget. powders. 1904. 232. Plate CI. Hager. Pharm. Praxis. Bd. II. 1902. 1176. Hérail. Mat. Méd. 1912. 455. Ingerman. Mikrosk. d. voorn. Handelsw. 1910. 174. Karsten u. Oltmanns. Pharmakogn. 1909. 31. Koch u. Gilg. Pharmakogn. Prakt. 1907. 85. Koch. Mikr. Anal. d. Drogenpulver. Bd. II. 1903. 75. Plate VIII. Koch. Einf. i. d. Mikr. Anal. d. Drogenpulver. 1906. 42. Koch. Pharmakogn. Atl. Bd. I. 1911. 141. Plate XXIV. Luerssen. Syst. Bot. Bd. II. 1882. 457. Marmé. Pharmacogn. 1886. 79. Meyer. Wiss. Drogenk. Bd. II. 63.

Fig. 96. *Zingiber officinale.* Rhizome, transverse section. end Endodermis; flpar Flattened common parenchyma; fvascb Fibrovasal bundles; oilc Oil cells; par Common parenchyma; prd Inner radially arranged layers of periderm cells; prd′ Outer not radially arranged layers of periderm cells; shsf Sclerenchyma sheath of fibrovasal bundles; svascb Small vascular bundles adjacent to the pericycle.

Moeller. Mikrosk. d. Nahr. u. Genuszm. 1886. 361. Moeller. Pharmakogn. 1889. 292. Moeller. Mikr. pharm. Üb. 1901. 278. Oudemans. Aant. o. d. Pharmac. neerl. 1854—56. 75. Oudemans. Pharmacogn. 1880. 97. Planchon et Collin. Drogues simples. T. I. 1895.

223. Schimper. Micr. Unters. d. veget. Nahr. u. Genussm. 1900. 144. Schneider. Powdered Veget. Drugs. 1902. 323. Tschirch. Pharmakogn. Bd. II. Abt. 2. 1917. 1051. Tschirch. Harze u. Harzbeh. 1900. 390. Tschirch u. Oesterle. Anat. Atl. 1900. 109. Pl. 26. Vogl. Veget. Nahr. u. Genussm. 1899. 519. Wiesner. Rohstoffe. 1900. 513. Wigand. Pharmakogn. 1879. 102. MATERIAL. The drug; rhizomes, grown in the Botanic Garden at Groningen, fresh and in alcohol. REAGENTS. Glycerine, chloral hydrate, iodine in chloral hydrate, potash, iodine and sulphuric acid 66 per cent., concentrated sulphuric acid, phloroglucin and hydrochloric acid, aniline sulphate, Schulze's macerating mixture.

MICROGRAPHY.

E p i d e r m i s. Mostly wanting in the drug.

Cells of the above. R. 30 μ, T. 35 μ, L. 45—50 μ; tetra- to hexagonal prisms with a radially directed axis. Walls suberized, with a thin inner cellulose layer [1]. Cell contents: in some cells simple crystals.

C o r t e x.

Secondary cork tissue. Storied cork, developed as periderm; consisting of 10—15 layers of cells; the outermost 5 layers not radially arranged and pro-duced by the suberification of cortical cells without cell-division; intercellular spaces want-ing. The innermost layers pro-duced by tangential division of cortical cells, each cell forming a radial row of 5 cells. The tangential division here and there repeated towards the in-terior in a single layer of cells. Sometimes the whole process repeated.

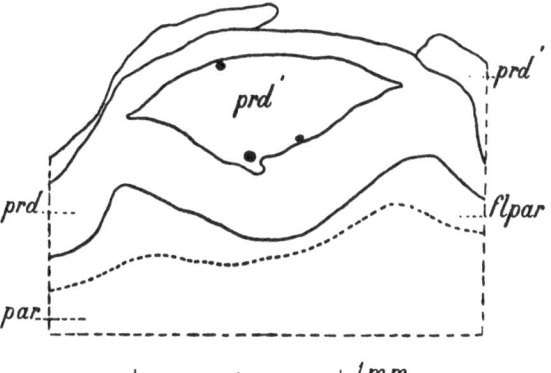

Fig. 97. *Zingiber officinale.* Rhizome, transverse section outer part. flpar Flattened common parenchyma; par Common parenchyma; prd Inner radially arranged layers of periderm cells; prd' Outer not radially arranged layers of periderm cells.

P e r i d e r m   c e l l s. Outer not radially arranged layers R. 50 μ, T. 70 μ, L. 85 μ; polyhedra. Walls light brown; suberized and with a thin inner cellulose layer. Inner radially arranged layers R. 30 μ, T. 70 μ, L. 85 μ; penta- or hexagonal tables with a radially directed axis; see for the rest cells of the outer layers.

Primary cortex proper.

Parenchyma. Common parenchyma in about 40 layers of cells; the outermost 4—5 layers without intercellular spaces, but for the rest similar to the other. The outermost half of this parenchyma in the drug quite flattened. Idioblasts in 2 kinds: numerous oil cells and crystal cells, the latter each containing some small simple crystals.

C o m m o n   p a r e n c h y m a   c e l l s. R. 60 μ, T. 80 μ, L. 90 μ; penta- or hexagonal prisms with a longitudinally directed axis. Cell contents: especially in the outer half of the tissue a great amount of starch; grains simple, flat, irregularly circular or oblong, often

---

[1] The drug never shows cellulose reaction; the alcoholic material on the contrary shows cellulose reaction in all not suberized walls or parts of them.

somewhat angular or with a papilliform protuberance, containing the hilum; hilum eccentric, very close to the periphery; layers hardly to be distinguished, more conspicuous in iodine and sulphuric acid 66 per cent.; diameter of grains on an average 25 μ, thickness .7 μ, in the alcoholic material much smaller. O i l c e l l s. Quite similar to the common parenchyma cells. Walls thin, suberized. Cell contents: a yellow or brown lump of resin.

Cortical meristeles. Each represented only by a fibro vasal bundle. Lying in about 3 rings in the innermost part of the cortex.

S c l e r e n c h y m a - s h e a t h. Strongly developed especially on the outer and inner sides, but wanting round the smaller vascular bundles. Intercellular spaces wanting.

V a s c u l a r b u n d l e s. Collateral, closed, xylem not always turned inwards. S c l e r e n c h y m a f i b r e s. R. 30 μ, T. 20 μ, L. 1000—1400 μ; penta- or hexagonal divided in about 4 partitions by thin transverse walls. Walls somewhat thickened; pale yellow; not lignified; showing slit-like pit canals, obliquely placed.

Phloem. With conspicuous sieve-tubes.

S i e v e - t u b e s. R. 10 μ, T. 10 μ; penta- or hexagonal; sieve-plates distinct.

Xylem. 1—14 vessels, annular, spiral or reticulate. Some parenchyma cells adjoining the sheath. Idioblasts with a brown granular mass, 1—6 in each transverse section.

V e s s e l s. Walls showing no reaction with phloroglucin and hydrochloric acid, neither with aniline sulphate; a greenish blue cellulose reaction of the innermost layer; middle lamella persisting in concentrated sulphuric acid. I d i o b l a s t s. R. 30 μ, T. 30 μ, L. 100—105 μ; cylindrical. Walls without cork layer. Cell contents: a granular brown mass.

Endodermis. Developed as a protective-sheath. In the drug quite flattened.

C e l l s o f t h e a b o v e. In alcoholic material R. 40 μ, T. 50 μ, L. 60 μ; in the drug in potash smaller R. 10 μ, T. 20 μ; penta- or hexagonal prisms. Radial and transverse walls suberized, inner and outer walls consisting of cellulose.

S t e l e.

Pericycle. Consisting of 1 layer of cells, flattened in the drug.

C e l l s o f t h e a b o v e. Penta- or hexagonal prisms, somewhat smaller than the cells of the endodermis. Little or no starch.

Fibrovasal bundles. Arranged according to the type of the monocotyledons; a great number of small vascular bundles without a sclerenchyma-sheath lying in the flattened part of the parenchyma adjacent to the pericycle.

Medullary commissures and **medulla.** Consisting of common parenchyma; some of the outermost layers in the drug flattened. Numerous oil cells, quite similar to those of the cortex.

C o m m o n p a r e n c h y m a c e l l s. R. 90 μ, T. 80 μ, L. 90 μ; similar to those of the inner part of the cortex, but somewhat smaller; flattened.

Micrography of the powder. Much starch grains flat, simple, irregularly circular or oblong, often somewhat angular or with a papilliform protuberance containing the hilum; hilum eccentric, very close to the periphery; layers generally hard to distinguish; on an average 25 μ in diameter, 7 μ thick. The starch contained in parenchyma cells. Oil cells about spherical, quite filled with a reddish brown resin mass. Cylindrical cells, also with brown contents, in the xylem of the vascular bundles. Wide thin-walled cork cells, partly radially arranged. Long, somewhat thick-walled fibres of sheath of vascular bundles. Reticulate vessels.

May 1901.   J; M; L.

SECALE CORNUTUM.

Ergota. Ergot. Sclerotium Clavicipitis purpureae. Ergot of Rye.

The sclerotium of Claviceps purpurea, Tul. in Ann. Sc. nat. Sér. III. T. XX. 1853. 5, collected in the fields from the ears of Secale cereale, shortly before the total ripening of the caryopsides and dried by mild heat.

Macroscopic characters.

Length: M. = 14.6 m.M., thick up to 6.5. m.M.; in the dry state brittle; cylindrical, at the same time somewhat tri- or quadrilateral, bow-shaped, tapering gradually towards apex and base. At the top sometimes a small greyish white cap (Sphacelia). Surface longitudinally furrowed, generally 2 or 3 furrows much deeper than the rest; mostly showing also longitudinal and transverse fissures; dull greyish sometimes purplish black, somewhat covered with bloom. Transverse fracture smooth, dull, pinkish white.

Odour in somewhat large quantities disagreeable; taste feeble, sweetish, afterwards somewhat pungent.

Anatomical characters.

LITERATURE. Berg. Anat. Atl. 1865. 1. Taf. 1. Deutsch. Arzneib. 5. Ausgb. 1910. 453. Flückiger & Hanbury. Pharmacographia. 1879. 744. Flückiger. Pharmakogn. 1891. 292. Gilg. Pharmakogn. 1910. 7. Greenish & Collin. Anat. Atl. of veget. powders. 1904. 276. Hérail. Mat. Méd. 1912. 742. Ingerman. Mikrosk. d. voorn. handelsw. 1910. 58. Koch u. Gilg. Pharmakogn. Praktik. 1907. 21. Mitlacher. Toxikol. od. forens. wicht. Pflanzen u. vegetab. Drogen. 1904. 1. Möller. Mikroskopie d. Nahr. u. Genussm. 1886. 164; in substance the same in Pharmakogn. 1889. 22 and Leitfaden z. mikrosk. pharmakogn. Übungen. 1901. 73. Oudemans. Aanteek. o. d. Pharmacop. Neerland. 1854—56. 30. Oudemans. Pharmacogn. 1880. 5. Tschirch u. Oesterle. Anat. Atl. 1900. 201. Taf. 46. Vogl. Anat. Atl. 1887. Taf. 1. v. Wisselingh. Mikrochem. Unters. üb. d. Zellwände der Fungi. Pringsheim's Jahrb. Bd. 31 .1898. 661. MATERIAL. The drug. REAGENTS. Water, glycerine, iodine in chloral hydrate, phloroglucin and hydrochloric acid, iodine and sulphuric acid 66 per cent., concentrated sulphuric acid, osmic acid.

MICROGRAPHY.

S c l e r o t i u m. Pseudoparenchyma; in the outer part the cells generally in longitudinal rows, showing their origin from hyphae; the inner part consisting of two tissues clearly to be distinguished: the central one being composed of smaller cells and having in transverse sections the shape of an irregular star sometimes showing large and irregular intercellular spaces in the middle part; the peripheral tissue composed of larger cells.

C e l l s   o f   t h e   o u t e r   p a r t. R. and T. 6—10 μ, L. 7—20 μ; polygonal prisms with a longitudinally directed axis. Outer walls of the outmost layer of cells somewhat thickened; walls of the outmost layer or layers of cells deep brown to black, those of the rest pinkish white; walls containing no cellulose but consisting of chitin [1]). Cell contents:

---

[1]) v a n   W i s s e l i n g h   l. c.

bodies with rounded edges, consisting of oil and protein. P e r i p h e r a l c e l l s o f t h e
i n n e r p a r t. Somewhat more elongated and having rounded edges. Walls pinkish
white. See for the rest the cells of the outer part.

S p'h a c e l i a. Consisting of colourless hyphae; these united at the top of the
sclerotium in a greyish white cap and sometimes surrounding the lower
parts of the sclerotium in some flattened layers. Conidia usually to be found.
Some remains of the ovary, especially some epidermal cells, often to be distinguished between the hyphae of the cap, sometimes also in the lower parts.

October 1901, June 1911.   J; M.

SEMEN AMYGDALAE DULCIS.
Amygdalae Dulces. Sweet Almond.
The ripe seeds of Prunus Amygdalus, Stokes, Bot. Mat. Med. III. 111, var. dulcis, Baillon.

Macroscopic characters.
Average length 2.5 c.M., width 1.5 c.M., thickness 0.75 c.M. Somewhat obliquely
oblong-ovate, flat, with an acute apex and an obtuse margin. After swelling in
water the leathery seed-coat easily coming off. Testa coarsely and longitudinally wrinkled, with a cinnamom-brown mealy surface. Hilum extending along
the margin from the vicinity of the top to about half-way the length of the seed,
protruding somewhat in the shape of a crest. Raphe forming a continuation of
the hilum, ending in the chalaza just above the obtuse end of the seed. Chalaza
almost circular, about 3 m.M. in diameter, darker in colour, especially on the
inner side of the seed-coat; 16 darker coloured veins, in connection with the
raphe, arising from the chalaza and extending in an oblique direction from the
chalaza towards the apex of the seed, slightly branched and showing here and
there anastomoses. Tegmen membranaceous, white, even at the chalaza.
Albumen wanting. Embryo having the same shape as the seed; with a short
radicle, lying at the acute end; 2 purely white, not yellowish, convexo-plane,
obovate cotyledons often somewhat differing in size from one another and a
short plumule, provided with about 8 young leaflets. Odour soft; taste sweet
and oily.

Anatomical characters.
LITERATURE. Berg. Anat. Atl. 1865. 89. Tafel XXXXV. Benecke. Mikr. Drogenprakt.
1912. 75. Flückiger. Pharmakogn. 1891. 985. Flückiger & Hanbury. Pharmacographia.
1879. 244. Gilg. Pharmakogn. 1910. 145. Guignard. Sur la localisation des principes de
l'acide cyanhydrique. Journal de Bot. 1890. 3. Hager. Pharm. Praxis. Bd. 1. 1900. 278.
Heut. Beitr. zur Kenntnis des Emulsins. Archiv der Pharmacie. 1901. 581. Johannsen.
Sur la localisation de l'émulsine dans les amandes. Ann. d. Sc. nat. Bot. Sér. VI. 1887.
118. Karsten u. Oltmanns. Pharmakogn. 1909. 238. Kraemer. Botany a. Pharmacogn.
1910. 433, 794. Luerssen. Syst. Bot. Bd. 2. 1882. 851. Marmé. Pharmacogn. 1886. 406.
Meyer. Wiss. Drogenk. Bd. I. 1891. 129. Meyer. Mikr. Unters. v. Pflanzenpulv. 1901. 41.
Moeller. Mikr. d. Nahr. u. Genussm. 1886. 236. Moeller. Pharmakogn. 1889. 198. Molisch.
Mikrochem. d. Pfl. 1913. 172, 290. Oudemans. Aanteek. o.d. Pharmac. neerl. 1854—56.
431. Oudemans. Pharmacogn. 1880. 431. Planchon et Collin. Drogues simples . T. II.
1896. 425. Schimper. Mikr. Unters. d. veget. Nahr. u. Genussm. 1900. 90. Schneider.

Powdered Veget. Drugs. 1902. 113. Tschirch. Angew. Pfl.-anat. 1889. 45. Tschirch. Pharmacogn. Bd. 2. Abt. 1. 1912. 598. Tunmann. Pfl.-mikrochemie. 1913. 360, 431. Wiesner. Rohstoffe. Bd. 2. 1900. 730. Wigand. Pharmakogn. 1879. 299. Wittmack und Buchwald. Die Unterscheidung der Mandeln von ähnlichen Samen. Ber. d. d. bot. Ges. 1901. 584. MATERIAL. The drug. Flowers collected in the Botanic Garden at Groningen, May 1903, in alcohol. Unripe fruits, 0.6—2.3 c.M. in diameter, and nearly ripe fruits, collected in the Botanic Garden at Groningen, in alcohol. REAGENTS. Water, glycerine, potash 50 per cent., chloral hydrate, iodine in chloral hydrate, phloroglucin and hydrochloric acid, iodine and sulphuric acid 66 per cent., concentrated sulphuric acid, Schulze's macerating mixture, osmic acid, iron acetate, oil of cloves, origanum oil, Millon's reagent, sulphuric ether, chloroform.

MICROGRAPHY.

O v u l e. Anatropous, often somewhat semi-anatropous; inserted in the top of the ovary; showing 2 integuments in the uppermost part of the ovule, the lower part showing only a single integument. [1]) Endostomium forming a rather long tube; exostomium still much longer. Epidermis of both integuments with a cuticle. Outer integument consisting of 7—9 layers of cells. Inner integument consisting of 5 layers of cells. Meristele of raphe running through the ground tissue of the lower combined part of the integuments, bending outward rectangularly into the funicle; this also discernible in the unripe seed.

S e e d.

Seed-coat.

S e e d - c o a t   p r o p e r.

Layers produced by the integument.

E p i d e r m i s   o u t e r   s i d e. [2]) Consisting of papillate cells, only the basal parts of the lateral walls adjoining one another; partly thin-walled; other cells either isolated or in groups with thickened walls; the thin-walled cells showing less variation in size than the thick-walled ones, moreover often flattened or thrown off. Epidermal cells in the seed of a fruit of 7 m.M. in diameter still small; in the elder fruit becoming much larger; in the eldest unripe fruit very large and papillate; towards the apex of the seed increasing, towards the base decreasing in size.

C e l l s   o f   t h e   a b o v e. In the dry seed R. 50—200 μ, T. 40—150 μ, L. 50—150 μ; in the chalaza part smaller; polygonal prisms with a radially directed axis. Walls of thick-walled cells yellow to brown; lignified; pit canals in the inner walls and in the basal parts of the lateral walls adjoining one another. In osmic acid or iron acetate these pit canals colouring black. Walls of the thin-walled cells colouring black in osmic acid or iron acetate.

G r o u n d   t i s s u e. After soaking of the seed in water thick about 50 μ; consisting of 12—15 layers of common parenchyma cells; divided in 2 parts:

---

[1]) M e y e r. Wiss. Drogenk. l. c., according to his description and fig. 100 and 102, considers the outer integument as inserted upon the inner one. In the unfertilized ovule I saw the two integuments, but their separating line had disappeared already in the youngest fruits examined by me; this agrees with the fact that M e y e r examined an ovule shortly after fertilization.

[2]) According to W i t t m a c k and B u c h w a l d several authors have considered this epidermis as a part of the pericarp or as hairs of the seed-coat. M e y e r, Wiss. Drogenk. recognized its true character. In the unirpe fruit I found that the inner epidermis of the pericarp is clearly discernible.

1. An outermost one, often thick one third of the ground tissue, consisting of strongly flattened cells with yellow to brown walls, colouring black in osmic acid and in iron acetate. The outermost cell layer sometimes not flattened and in a cross section conspicuous by the regular rectangular shape of its cells, the same shape being observed in unripe seeds; near the chalaza the outermost 2—4 cell layers mostly not flattened. In ripe alcoholic material starch grains in the outermost cell layer wanting; the cells of the next 2—4 cell layers, especially round the meristeles, containing many starch grains. Cluster crystal idioblasts especially on the outside of the meristeles; smaller in size than the surrounding cells.

2. An innermost part, thick two thirds of the ground tissue. Consisting of a very small amount of spongy parenchyma; all cell layers somewhat flattened.
P a r e n c h y m a c e l l s. Those of the outermost cell layer R. 15 μ, T. 25 μ, L. 30 μ; polyhedra with somewhat rounded edges. Those of the following 2—4 cell layers poly-

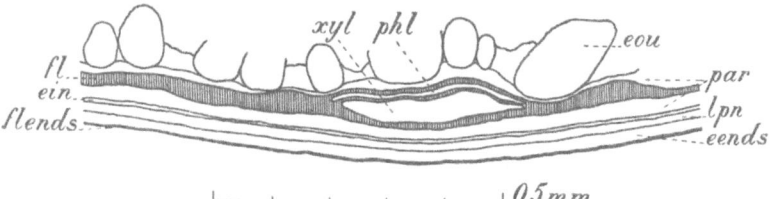

Fig. 98. *Prunus Amygdalus*. Seed, transverse section of the seed-coat proper. eends Epidermis of the layers produced by the endosperm; ein Epidermis inner side of the layers of the seed-coat, produced by the integument; eou Epidermis outer side of the layers of the seed-coat, produced by the integument; fl Flattened cell layers; flends Band of flattened cells of the layers, produced by the endosperm; lpn Flattened cell layers, produced by the nucellus; par Layers of parenchyma cells of the seed-coat, produced by the integument; phl Phloem; xyl Xylem.

hedra. Walls yellow to brown; mostly colouring black in osmic acid or iron acetate. C l u s-
t e r c r y s t a l i d i o b l a s t s. R. 10 μ, T. and L. 12 μ; containing each a single cluster crystal, sometimes a single simple crystal. C e l l s o f s p o n g y p a r e n c h y m a.
R. 15 μ, T. 35 μ, L. 40 μ; polyhedra with rounded edges.
M e r i s t e l e s. Tangentially widened; R. 130 μ, T. 900 μ and containing collateral vascular bundles; phloem strongly flattened, showing elements with somewhat thickened walls; xylem torn asunder, only the inner part showing spiral vessels with lignification.
E p i d e r m i s i n n e r s i d e. At the top round the very small micropyle seeming to consist of some layers of cells with yellow walls.
E p i d e r m a l c e l l s. R. 10 μ, T. 20 μ, L. 25 μ; polygonal, often square tables. Walls: lateral walls somewhat pitted; inner walls thickened; cuticularized. Cell contents: tannin.
C e l l s s u r r o u n d i n g t h e m i c r o p y l e in a cross section rectangular, R. 15 μ, T. 10 μ. Walls yellow. C e l l s p a v i n g t h e m i c r o p y l e in a tangential section 20 by 30 μ; polygonal.
H i l u m p a r t. Containing a very strongly developed meristele.
C h a l a z a p a r t. See Fig. 99.
E p i d e r m i s o u t e r s i d e. Like that of seed-coat proper.

Ground tissue. Consisting of 2 parts. 1. An outer part consisting of cells with colourless walls, the outer 3—4 cell layers mostly not flattened; containing numerous meristeles. 2. An inner part, more or less developed in the shape of a disc, consisting of about 10 layers of parenchyma cells.
Cells of disc. R. 10 μ, T. and L. 18 μ. Walls somewhat thickened; yellow; intercellular spaces wanting. Cell contents: red-brown.
Layers produced by the nucellus. Consisting of a band of flattened tissue, thick when soaked in water about 10 μ; in the unripe seed the development of endosperm and embryo being delayed, the tissue of the nucellus more or less

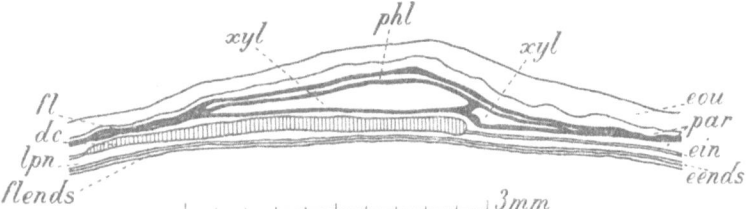

Fig. 99. *Prunus Amygdalus.* Seed, transverse section of chalaza part of seed-coat. dc Disc; eends Epidermis of the cell layers, produced by the endosperm; ein Epidermis inner side of the layers of the seed-coat, produced by the integument; eou Epidermis outer side of the layers of the seed-coat, produced by the integument; fl Flattened cell layers; flends Band of flattened cells of the layers, produced by the endosperm; lpn Flattened cell layers, produced by the nucellus; par Layers of parenchyma cells of the seed-coat, produced by the integument; phl Phloem; xyl Xylem.

persisting in the form of a large celled common parenchyma without contents; the outermost 1—2 cell layers somewhat smaller.
Layers produced by the endosperm. The main part thick 25 μ in a soaked state.
Epidermis. Mostly consisting of a single cell layer; in several places the cells showing a tangential partition-wall; in iodine and sulphuric acid 66 per cent. showing a cuticle.
Ground tissue. Chiefly consisting of a band of flattened cells, only here and there mixed with some layers of not flattened cells; showing a cuticle on its inner surface.
The part in the vicinity of the apex of the seed, showing 7—8 layers of not flattened cells.
Not flattened cells. R. 20 μ, T. 18 μ, L. 20 μ; R. of partitioned epidermal cells 30 μ; poly- often tetragonal prism with a radially directed axis. Walls somewhat thickened; intercellular spaces wanting. Cell contents: numerous small aleurone grains and oil-plasm; in concentrated sulphuric acid the oil drops colouring red to brown.
Nucleus of seed.
Embryo. The majority of its elements in a more or less meristematic stage.
Radicle. Epidermis and cortex consisting of meristematic parenchyma cells filled with oil-plasm and aleurone grains like those of the cotyledons but all without cluster crystals. Stele showing emulsine in pericycle and phloem [1].

---

[1] See also G u i g n a r d. l. c. 25, 26, 27.

Plumule. Leaflets containing much oil and emulsine in pericycle and phloem of the meristeles. [1])

Cotyledons.

E p i d e r m i s   o u t e r   s i d e. Cells fairly well arranged in longitudinal rows. C e l l s   o f   t h e   a b o v e. R. 8 μ, T. 7 μ, L. 20—45 μ; rectangular prisms. Outer walls a little thickened; showing a cuticle. Cell contents: see cells of mesophyll.

E p i d e r m i s   i n n e r   s i d e.

C e l l s. R. 15 μ, T. 25 μ, L. 40 μ; polygonal tables. Outer walls a little thickened; showing a cuticle. Cell contents: see cells of mesophyll.

M e s o p h y l l. Consisting of parenchyma showing intercellular spaces; the cells of the outermost cell layers smaller than the rest. An endodermis, containing emulsine, surrounding each meristele. [2])

C e l l s   o f   t h e   a b o v e. 30 μ in diameter; polyhedra with rounded edges. Cell contents: aleurone grains imbedded in a network of oil-plasm, sometimes 2 of those grains larger than the rest, the largest ones 10-12 μ in diameter, the smaller ones 3—7 μ in diameter. The ground mass of the grains soluble in water and very readily in glycerine [3]). Each aleurone grain containing some globoids, these globoids sometimes seeming to be hollow; the largest grains mostly containing a cluster crystal, 4—6 μ in diameter, or sometimes a simple crystal, the crystals clearly discernible in iodine and chloral hydrate. In osmic acid the network of the plasm colouring black, aleurone grains not discernible. In concentrated sulphuric acid oil-drops colouring red-brown.

M e r i s t e l e s. In a more or less procambial stage; numerous; running through the upper (inner) part of the cotyledon; in a cross section arranged in a sinuous line; towards the top of the cotyledon also some meristeles at the under (outer) side of the mesophyll. Pericycle containing much emulsine [4]) (Millon's reagent). Phloem containing emulsine. Xylem containing already some spiral vessels.

May 1903.   J; M; v. E. d. W.

SEMEN AMYGDALAE AMARAE.
Amygdalae Amarae. Bitter Almond.
The ripe seeds of Prunus Amygdalus, Stokes, Bot. Mat. Med. III. 111, var. amara, Baillon.

Resembling Amygdalae dulces in the large majority of macroscopic and anatomical characters. The following difference may be noticed: Amygdaline occurring in the mesophyll of the cotyledons, but wanting in seed-coat and embryo. [5])

May 1903.   J; M; v. E. d. W.

---

[1]) See also G u i g n a r d. l. c. 25, 26, 27.
[2]) See also G u i g n a r d. l. c. p. 24 and 27.
[3]) According to M e y e r. Mikr. Unters. v. Pflanzenpulv. l. c. the groundmass is also soluble in sodium chloride 10 per cent., sodium carbonate 1 per cent. and potassium ferricyanide 4 per cent.
[4]) See also G u i g n a r d. l. c. p. 24 and 27.
[5]) G u i g n a r d. l. c. p. 25 mentions that Amygdalae amarae show a more distinct emulsine reaction; on p. 22 he says: „M. J o h a n n s e n (Sur la localisation de l'émulsine dans les amandes. Ann. des sc. nat. Bot. Serie 7. t. VI. 1887) a constaté qu'une amande entière renferme une quantité d'emulsine capable de décomposer plus de 40 fois son propre continu d'amygdaline."

SEMEN CARDAMOMI.

Cardamom Seeds. Cardamomum. Cardamom.

The ripe seeds mostly still included in the fruit of Elettaria Cardamomum, Maton, in Trans. Linn. Soc. X. (1811) 254.

Macroscopic characters.

Fruit. Length: n $= 300$, Med $= 11.01$ m.M., $Q_1 = 1.39$ m.M., $Q_3 = 1.18$ m.M., min. $= 6.88$ m.M., max. $= 17.13$ m.M.; breadth: n $= 300$, Med $= 6.49$ m.M., $Q_1 = 0.40$ m.M., $Q_3 = 0.44$ m.M., min. $= 5.13$ c.M., max. $= 9.63$ m.M.; with a short, hollow pedicel. Capsule proper, trigonal prismatic, with rounded edges, apex and base; 3-locular, with membranaceous septa and axile placentae; dehiscent by valves loculicidal; valves leathery. External surface smooth, showing a fine longitudinal striation in consequence of elevated veins, straw-coloured. In each cavity 5—8 seeds, arranged in 2 longitudinal rows and sticking together.

Seeds. Long up to 4 m.M., thick up to 2.5 m.M.; very hard; irregularly poly-hedral, often in the shape of a very steep trigonal truncated pyramid, its base having the shape of an equicrural triangle; bearing at the apex not seldom a membranaceous remainder of the funicle; surrounded by a membranaceous aril-arillode, here and there scaling off and easily to be removed from the soaked seed. Hilum at the top of the largest lateral side; micropyle next to it on the apex in the middle of a little, circular, light coloured operculum; chalaza de-pressed and at the base of the same lateral side; raphe in a sharp groove, con-necting hilum and chalaza. Surface with about 6 coarse transverse wrinkles, dark-brown or more red-brown. Nucleus with a large perisperm; enclosing a much smaller endosperm. Embryo small, with a cylindrical radicle and a some-what flattened conical cotyledon.

Pericarp without either odour or taste; seeds with a penetrating, camphoric, aromatic odour and taste.

Anatomical characters of the seed.

LITERATURE. Attema. De zaadhuid der Angiosp. en Gymnosp. Diss. Groningen. 1901. 178. Berg. Anat. Atl. 1865. 87. Taf. XXXXIV Flückiger. Pharmakogn. 1891. 898. Flückiger & Hanbury. Pharmacographia. 1879. 643. Gilg. Pharmakogn. 1910. 67. Greenish & Collin. Anat. Atl. o. veget. powders. 1904. 148. Hager. Pharm. Praxis. Bd. I. 1900. 636. Karsten u. Oltmanns. Pharmakogn. 1909. 234. Koch u. Gilg. Pharmakogn. Praktik. 1907. 232. Koch. Mikr. Anal. d. Drogenpulver. Bd. 4. 1908. 83. Taf. VIII. Kraemer. Botany a. Pharmacogn. 1910. 581. Luerssen. Syst. Bot. Bd. 2. 1882. 458. Marmé. Pharmacogn. 1886. 275. Meyer. Wiss. Drogenk. Bd. 2. 1892. 388. Meyer. Mikr. Unters. v. Pflanzenpulv. 1901. 36. Moeller. Mikr. d. Nahr. u. Genussm. 1886. 218. Moeller. Pharmakogn. 1889. 154. Moeller. Mikr. pharm. Ueb. 1901. 189. Oudemans. Aanteek. o. d. Pharmac. neerl. 1854—56. 77. Oudemans. Pharmacogn. 1880. 349. Planchon et Collin. Drogues simples. T. I. 1895. 231. Schad. Entwicklungsgeschichtliche Unters. ü. d. Malabar Cardamomum u. vergl. anat. Studien ü. d. Samen einiger anderen Amo-mum- u. Elettaria-Arten. Diss. Bern. 1897. Schimper. Mikr. Unters. d. veget. Nahr. u. Genussm. 1900. 135. Schneider. Powdered Veget. Drugs. 1902. 146. Tschirch. Angew. Pfl. Anat. 1889. 158. Tschirch. Pharmakogn. Bd. 2. Abt. 2. 1917. 1071. Tschirch u.

Oesterle. Anat. Atl. 1900. Taf. 34. Tschirch u. Schad. Schweizer Wochenschr. für Chemie
u. Pharmac. 1897. 476. Vogl. Veget. Nahr. u. Genussm. 1899. 447. Wigand. Pharmakogn.
1879. 287. MATERIAL. The drug. REAGENTS. Water, glycerine, chloral hydrate,
potash 50 per cent., iodine in chloral hydrate, potassium iodide iodine, iodine and sulphuric acid 66 per cent., Millon's reagent.

MICROGRAPHY.

S e e d. [1])

Aril-arillode. Proceeding from the operculum and the funicle and enveloping

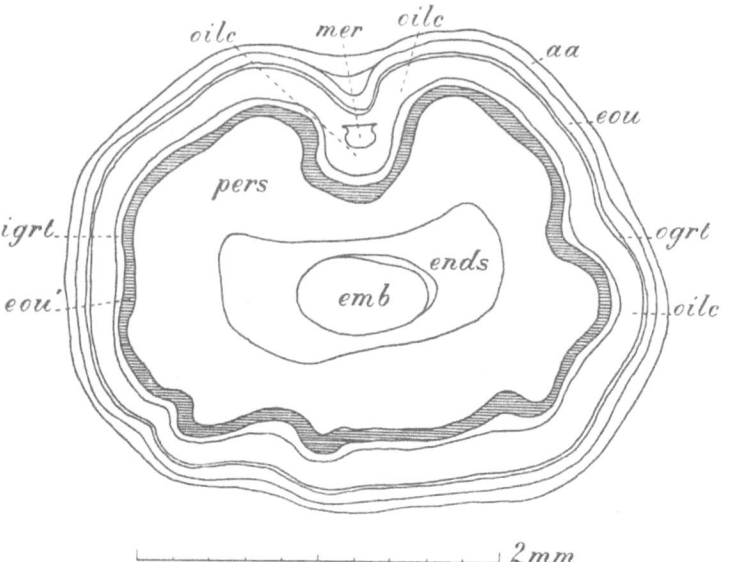

the whole seed;
consisting of
some cell layers,
the outer ones
mostly more or
less flattened,
the cells of the
innermost cell
layer larger in
size than the
rest.

C e l l s o f t h e
a b o v e. Cells of
the main part,
much elongated;
cells of the inner-
most layer R. 10
μ, T. 16 μ; cells of
the part covering
the margin of the
operculum R. and
T. 18 μ, L. 25 μ;
polyhedra; cells of
the part close to
the operculum
with yellow walls,
thickened like

Fig. 100. *Elettaria Cardamomum.* Unripe seed, transverse section. aa Aril-
arillode; emb Embryo; ends Endosperm; eou Epidermis outer side of the
layers of the seed-coat, produced by the outer integument; eou' Epidermis
outer side of the layers of the seed-coat, produced by the inner integument; igrt
Inner layers of parenchyma cells of the ground tissue of the layers of the seed-
coat, produced by the outer integument; mer Meristele of raphe part; ogrt 1
or 2 outer layers of common parenchyma cells of the ground tissue of the
layers of the seed-coat, produced by the outer integument; oilc Oil cells of the
ground tissue of the layers of the seed-coat, produced by the outer integument;
pers Perisperm.

those of the sclerenchyma cells belonging to the epidermis outer side of the inner integument and coloured yellow. Cell contents of all cells: sometimes oil.

Seed-coat.

S e e d - c o a t p r o p e r.

Layers produced by the outer integument.

E p i d e r m i s o u t e r s i d e. Consisting of longitudinally directed fibres
with somewhat thickened walls.

[1]) The ovule and its development was investigated by T s c h i r c h u. O e s t e r l e l. c. and this
study was afterwards completed by S c h a d l. c. and T s c h i r c h u. S c h a d l. c.; it was found
that the ovule is pendulous and anatropous, with 2 integuments; the outer one consisting of 6—8 cell
layers, the inner one consisting of 2, in the vicinity of the micropyle of 3 cell layers. Aril-arillode developing from the funicle and the border of the exostomium; in an ovule of 360 μ in diameter already
conspicuous; in an ovule, measuring 1 m.M. in diameter, surrounding one half of the ovule.

Elements of the above. R. 25 μ, T. 17 μ, L. 250—350 μ. Walls, outer and inner walls thickened; outer walls with a cuticle; innermost lamella, especially that of the lateral walls, colouring yellow in iodine and sulphuric acid 66 per cent., the rest colouring blue.

Ground tissue. Consisting of: 1, one to two outer layers of nearly always flattened common parenchyma cells [1]); 2, one layer of tangentially elongated oil cells; in the concave parts of the coarse wrinkles of the surface of the seed, mentioned among the macroscopic characters, the oil cells somewhat more radially elongated than elsewhere; 3, some inner layers of mostly also flattened parenchyma cells; in the concave parts of the wrinkles, mentioned above, the cells fairly well discernible.

Parenchyma cells. Those of the outer layers R. 5 μ, T. 10 μ. Cell contents: sometimes a brown mass, colouring blue in iron chloride. Those of the inner layers R. 8 μ, T. 16 μ, L. 20 μ; polyhedra. Oil cells. R. 40 μ, T. 70—120 μ, L. 40 μ; mostly rectangular prisms with a radially directed axis. Walls persisting in concentrated sulphuric acid. Cell contents: aetherial oil in drops and in masses of more irregular shape; some remnants of the resinogenous layer here and there still discernible.

Epidermis inner side. Mostly flattened.

Layers produced by the inner integument. Only consisting of an outer and inner epidermis; these layers also covering the truncated top of the pyramid-shaped seed with the exception of a small circular middle part, containing the operculum; at the margin of this aperture bending perpendicularly downwards, the outer epidermis over a distance of 500 μ, the inner epidermis over a distance of 700 μ, then both bending abruptly upwards again, the outer epidermis over a distance of 350 μ, the inner epidermis over a distance of 550 μ and there connected with the upper part of the operculum; hence the fold thus formed consisting for a distance of 200 μ, nearest to the centre of the seed, only of the flattened inner epidermis; the space within the fold and bounded by the outer epidermis filled up with some flattened tissue. This fold separating the upper part of the perisperm on its outside from the upper part of endosperm and embryo on its inside.

Epidermis outer side. Consisting of palisade sclerenchyma.

Epidermis inner side. Quite flattened into a hyaline band; most clearly discernible close to the operculum and there sometimes more or less wrinkled. In iodine and sulphuric acid 66 per cent. colouring brown.

Palisade sclerenchyma cells. R. 20 μ, T. and L. 12 μ; polygonal prisms with a radially directed axis. Inner walls thick 12 μ, other walls not thickened; all walls coloured yellow to red-brown, persisting in concentrated sulphuric acid, not colouring blue in iodine and sulphuric acid 66 per cent.; middle lamella discernible. Cell cavity bowl-shaped; containing a globule, exhibiting small warts on its surface and consisting of silicic

---

[1]) According to Tschirch u. Oesterle l. c. fig. 9 and Tschirch u. Schad l. c. (see also Attema l. c.) these outer cell layers are wanting in the ovule and the layer of oil cells joins upon the outer epidermis. But it is highly improbable that afterwards these parenchyma layers are formed by divisions of the epidermal cells, because the latter very early grow out in a longitudinal direction whilst the parenchyma cells are elongated in an other direction. Schad l. c. e. g. in fig. 9 has drawn the parenchyma layers in accordance with my result as mentioned in the text.

acid [1]), nearly not discernible in glycerine and chloral hydrate. In the part of the integument, bending perpendicularly upwards as mentioned above, in the neighbourhood of the operculum the cells walls not thickened, coloured somewhat brown.

R a p h e p a r t. In many respects like the seed-coat proper, with the following exceptions: in the ground tissue 2 layers of more irregularly shaped oil cells and between these 2 layers a meristele. This meristele extending from the funicle till somewhat beyond the chalaza; exhibiting on the outer side a pericycle, consisting of 1—2 layers of flattened cells; containing more than 1 vascular bundle showing spiral vessels.

C h a l a z a p a r t. Wide 25 µ; containing a part of the meristele as mentioned above, but here not covered with the inner layer of oil cells.

O p e r c u l u m. Corresponding with the layers produced by the inner integument, as mentioned above. Having the shape of a flat stopper; on its outside with a blunt conical part in its centre, this part formed by only a few cells. Consisting of 2—3 layers of palisade sclerenchyma, in all respects like that produced by the outer epidermis of the inner integument. Micropyle discernible as a narrow canal, lined with papillate cells.

Nucleus of seed.

P e r i s p e r m.

E p i d e r m i s. Consisting of starch cells, arranged in longitudinal rows.
C e l l s o f t h e a b o v e. R. 10 µ, T. 9 µ, L. 85 µ; rectangular prisms with somewhat sinuous lateral walls. Cell contents: see those of ground tissue of perisperm.
G r o u n d t i s s u e. In the vicinity of the endosperm somewhat flattened; consisting of starch cells.
C e l l s o f t h e a b o v e. In the outermost part R. 65 µ, T. and L. 30 µ; near the endosperm R. 35 µ, T. 30 µ, L. 35 µ; polyhedra with somewhat sinuous walls. Walls thin; intercellular spaces wanting. Cell contents: starch grains almost quite filling the cavity of the cell, 4 µ in diameter, globular, not swelling up in chloral hydrate or potash 50 per cent, central hilum often clearly discernible; moreover in each cell 1—7 small simple crystals, in the innermost cells of the perisperm these crystals surrounded by some granules, the latter not colouring blue in potassium iodide iodine, not disappearing in mineral acids and not much lighting up in polarized light, crystals and granules more or less clustering together.

E n d o s p e r m. Forming a layer round the embryo, thick towards the operculum only a single cell, becoming thicker downwards at the end of the fold of the seed-coat produced by the inner integument as mentioned above. In a cross section kidney-shaped, with the hollow side turned towards the raphe.
C e l l s o f t h e a b o v e. R. 40 µ, T. and L. 25 µ; in the middle part R. 30 µ, T. and L. 20 µ. Epidermal cells with somewhat thickened outer walls. Cell contents: a hyaline sometimes granular mass, colouring brown in potassium iodide iodine, neither swelling up nor solving in water, potash 50 per cent., chloral hydrate or acetic acid; colouring red in Millon's reagent; probably consisting of albuminous matter.

E m b r y o. Long 1600 µ, conical; radicle joining upon the operculum and sepa-

---

[1]) See S c h a d l. c. p. 23.

rated from the cotyledon by a thinner part. Consisting of small meristematic cells and showing procambium.

C e l l s o f t h e a b o v e. Contents: aleurone grains enclosing protein crystals and globoids [1]).

Micrograyhy of the powder.

Oblong, somewhat polyhedral clusters of starch grains, being the still cohering contents of the parenchyma cells of the perisperm; starch grains up to 4 μ in diameter, about globular; within these clusters often 2 to 7 small simple crystals. Palisade sclerenchyma, prismatic cells with very strongly thickened inner walls, leaving at the outside of the cell only a slight cavity; these cells mostly seen from above or from beneeth; walls yellow-brown to dark red-brown. Epidermal cells elongated, having the shape of fibres, light yellow. Flattened common parenchyma, often without conspicuous cell structure. A few spiral vessels belonging to the raphe.

February 1903.   J; M; v. E. d. W.

SEMEN COLCHICI.
Colchicum Seeds.
The ripe seeds of Colchicum autumnale, Linn. Sp. Pl. 341.

Macroscopic characters.

Long up to 3 m.M.; horny, to be powdered only with some difficulty; about spherical, sometimes more or less angular or oblong; on one side showing a strophiole, pointed or irregularly ridge-shaped, more or less shriveled, brown or whitish. Seed-coat thin, hard, slightly wrinkled, finely pitted, dull, very dark red-brown. Endosperm large, horny, dirty white. Embryo long 0.5 m.M., thick 0.25 m.M., lying in the endosperm about opposite the hilum.

Odour wanting; taste very bitter.

Anatomical characters.

LITERATURE. Barth. Studien ü. d. Nachweis von Alkaloiden in pharm. verwendeten Drogen. Bot. Centrbl. Bd. 75. 1898. 330. Berg. Anat. Atl. 1865. 93. Taf. XXXXVII. Benecke. Mikr. Drogenprakt. 1912. 76. Flückiger. Pharmakogn. 1891. 1000. Flückiger & Hanbury. Pharmacographia. 1879. 702. Gilg. Pharmakogn. 1910. 42. Greenish & Collin. Anat. Atl. o. veget. powders. 1904. 120. Hager. Pharm. Praxis. Bd. I. 1900. 924. Hérail. Mat. Méd. 1912. 710. Karsten u. Oltmanns. Pharmakogn. 1909. 232. Koch u. Gilg. Pharmakogn. Praktik. 1907. 191. Luerssen. Syst. Bot. Bd. II. 1882. 410. Marmé. Pharmacogn. 1886. 348. Meyer. Wiss. Drogenk. Bd. I. 1891. 161. Mitlacher. Toxik. od. Forens. wicht. Pfl. u. Drogen. 1904. 33. Moeller. Pharmakogn. 1889. 171. Moeller. Mikr. pharm. Ueb. 1901. 131. Molisch. Mikroch. d. Pfl. 1913. 273. Oudemans. Aanteek. o. d. Pharmac. neerl. 1854--56. 41. Oudemans. Pharmacogn. 1880. 403. Planchon et Collin. Drogues simples. T. I. 1895. 191. Schneider. Powdered Veget. Drugs. 1902. 167. Tschirch. Angew. Pfl.-anat. 1889. 71. Tunmann. Pfl. microchemie. 1913. 275. Vogl. Anat. Atl. 1887. Taf. 33. Wigand. Pharmakogn. 1879. 315. MATERIAL. The drug. Alcoholic material of flowers, gathered in September 1912 in the Botanic Garden at Groningen. Alcoholic material of unripe seeds, gathered in May 1889 at Utrecht. Unripe seeds, gathered in

---
[1]) See S c h a d l. c. p. 28 and fig. 37.

May and June 1889 at Utrecht, fixed with Flemming's mixture and preserved in alcohol. REAGENTS. Water, glycerine, potash 50 per cent., chloral hydrate, potassium iodide iodine, iodine in chloral hydrate, phloroglucin and hydrochloric acid, iodine and sulphuric acid 66 per cent., concentrated sulphuric acid, Schulze's macerating mixture, osmic acid, iron acetate, oil of cloves, origanum oil, alcoholic solution of tartaric acid 5 per cent.

MICROGRAPHY.

O v u l e. Semi-anatropous; horizontal; the axis of the funicle directed perpendicularly to the axis of the nucellus; with 2 integuments[1]). The inner integument sometimes protruding above the outer one; endostomium in the shape of a rather long tube. Epidermis of both integuments with a cuticle. Outer integument consisting of 3—4 layers of cells, containing starch grains; near the funicle the cell layers more numerous and sometimes containing spiral fibres, the latter often connected with those of the placentae; from the funicle towards the chalaza a band of cells, colouring

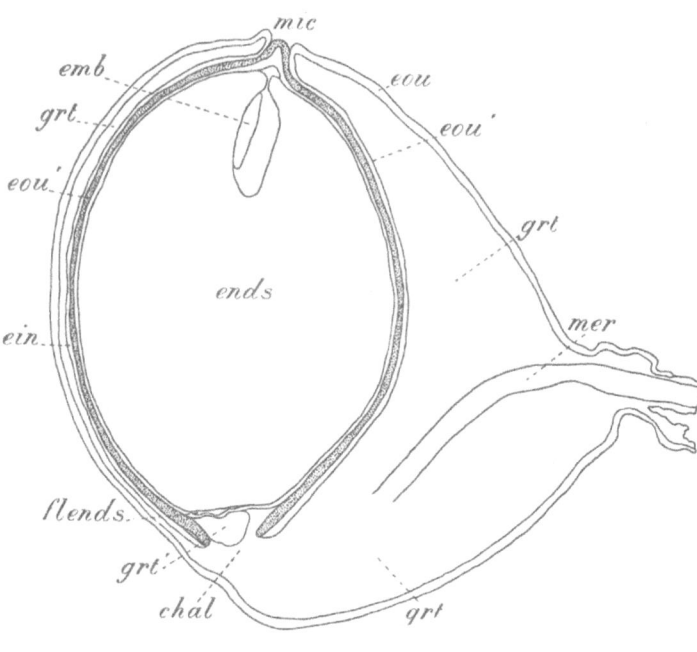

Fig. 101. *Colchicum autumnale.* Unripe seed, fixed with Flemming's mixture, longitudinal section. chal Chalaza; ein Epidermis inner side of the layers of the seed-coat, produced by the outer integument; ein' Epidermis inner side of the layers of the seed-coat, produced by the inner integument; emb Embryo; ends Endosperm; eou Epidermis outer side of the layers of the seed-coat, produced by the outer integument; eou' Epidermis outer side of the layers of the seed-coat, produced by the inner integument; flends Flattened cell layers of the endosperm near the chalaza; grt Ground tissue of the layers of the seed-coat, produced by the outer integument; grt' Ground tissue of the layers of the seed-coat, produced by the inner integument; mer Meristele of raphe part; mic Micropyle.

more yellow in iodine and chloral hydrate than the rest in iodine and chloral hydrate. Inner integument consisting of 2 cell layers; at the top round the endostomium cell layers more numerous. Nucellus consisting of one layer of longitudinally elongated cells, containing polyhedral granules colouring brown in iodine and chloral hydrate. Embryo-sack not discernible. In the youngest material, gathered at Utrecht and not treated with Flemming's mixture, all

[1]) According to B a r t h l. c. p. 334 with only 1 integument.

layers of the seed-coat already clearly differentiated, their correspondence with the layers of the seed-coat of the drug therefore clear; in the same material between the second and third cell layer from the inside of the seed-coat a cuticle. In the elder material, gathered at Utrecht and treated with Flemming's mixture the walls of the cells of the third cell layer from the inside of the seed-coat coloured black, those of the cells of the second cell layer also somewhat darker in colour, in iodine and sulphuric acid 66 per cent. a cuticle not discernible; in this material the connection between the outer

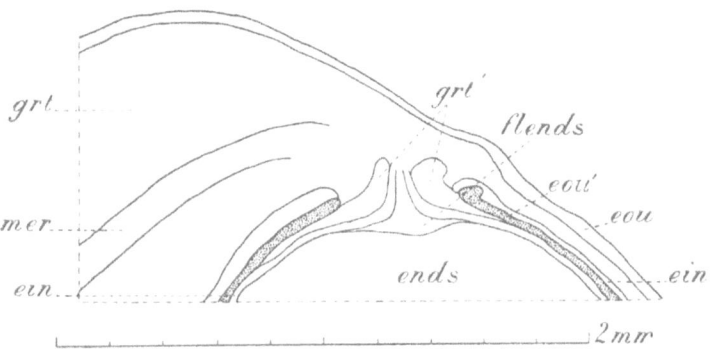

Fig. 102. *Colchicum autumnale.* Unripe seed, fixed with Flemming's mixture; longitudinal section through the chalaza. ein Epidermis inner side of the layers of the seed-coat, produced by the outer integument; ein' Epidermis inner side of the layers of the seed-coat, produced by the inner integument; ends Endosperm; eou Epidermis outer side of the layers of the seed-coat, produced by the outer integument; eou' Epidermis outer side of the layers of the seed-coat, produced by the inner integument; flends Flattened cell-layers of endosperm near the chalaza; grt Ground tissue of the layers of the seed-coat, produced by the outer integument; grt' Ground tissue of the layers of the seed-coat, produced by the inner integument; mer Meristele of raphe part.

epidermis of the seed-coat and the inner epidermis of the outer integument clearly discernible near the micropyle.

S e e d.

Seed-coat.

S e e d - c o a t   p r o p e r.

Layers produced by the outer integument.

E p i d e r m i s   o u t e r   s i d e. Often thrown off; when present consisting of cells quite flattened in a radial direction; near the funicle not flattened and with dome-shaped outer walls; in the material fixed with Flemming's mixture all cells having the latter shape, but towards the funicle smaller and less dome-shaped.

C e l l s   o f   t h e   a b o v e. Near the funicle R. 60 μ, T. 75 μ, L. 80 μ; polygonal tables. Walls often showing striation and thense appearing to have spiral thickenings; walls of all epidermal cells somewhat thickened and coloured yellow to brown. In the material fixed with Flemming's mixture R. 50 μ, T. 100 μ, L. 130 μ; polygonal tables with slightly rounded sides; outer walls showing a very black outermost lamella; small intercellular spaces between these cells and the cells of the outermost layer of ground tissue. Cell contents: a somewhat granular black mass, lying close to the outer walls; nucleus clearly discernible; starch grains often in the cells near the funicle.

G r o u n d   t i s s u e. Consisting of 1—3 layers of common parenchyma cells; these layers sometimes flattened, sometimes thrown off; near the chalaza and funicle consisting of much more numerous and mostly strongly flattened cell layers.

C e l l s o f t h e a b o v e. R. 40 μ, T. and L. 100 μ; polyhedra with rounded edges and with often rounded sides. Walls coloured yellow, but showing a brown colour in the cells of the outermost part and in those adjoining the inner epidermis of this integument; in a few cells showing a cellulose reaction; sometimes showing a structure reminding of spiral thickenings. Cell contents: sometimes some starch grains; in the material fixed June 7th with Flemming's mixture all cells filled with starch grains, simple and compound, 2- to 4-adelphous, with a central hilum; simple grains 4 μ in diameter, globular or polyhedral; compound grains up to 8 μ in diameter; sometimes 3 component grains forming a longitudinal row, measuring 12 by 4 μ; cell nucleus discernible.

E p i d e r m i s i n n e r s i d e. Cells often showing 1 tangential wall, near the funicle sometimes even 2; the majority of the cells also showing a radial partition wall.

P a r t i t i o n s o f e p i d e r m a l c e l l s. R. 15 μ, T. 18 μ, L. 20 μ; poly-, mostly rectangular tables with rounded edges, even those bordering upon the partition-walls; sometimes the partitions of the epidermal cells turned towards the outside of the seed-coat of a more irregular shape and more elongated. Walls coloured yellow to brown; persisting in concentrated sulphuric acid, not colouring blue in iodine and sulphuric acid 66 per cent.; cuticle not discernible. Cell contents: a red-brown granular mass; in sections not treated beforehand with an alcoholic solution of tartaric acid 5 per cent. colouring brown in potassium iodide iodine (colchicine), in sections treated beforehand with the alcoholic solution of tartaric acid 5 per cent. brown precipitate in potassium iodide iodine wanting [1]). In the material fixed June 7th with Flemming's mixture walls coloured black, innermost lamella persisting in concentrated sulphuric acid, also a lamella colouring red in phloroglucin and hydrochloric acid. Cell contents: in the outer 1 or 2 partitions starch grains; nucleus clearly discernible.

Layers produced by the inner integument.

E p i d e r m i s o u t e r s i d e. Flattened in a radial direction and forming a band, thick about 20 μ, together with the inner epidermis; in the main part of this integument being without ground tissue.

C e l l s. R. 15 μ, T. 30 μ, L. 40 μ; poly- often rectangular tables with slightly rounded edges. Walls brown; lateral walls pitted; showing intercellular spaces. Cell contents: tannin along the walls; in the centre of the cell brown masses, at first remaining brown in iron acetate and afterwards colouring black. In the material fixed June 7th with Flemming's mixture walls coloured somewhat black. Cell contents: a dark mass, quite filling up the cell with the exception of 1 colourless vacuole.

G r o u n d t i s s u e. Only occurring near the chalaza.

C e l l s o f t h e a b o v e. Measuring 20 by 20 by 30 μ; polyhedra. In the material, fixed June 7th with Flemming's mixture, walls coloured yellow to black.

E p i d e r m i s i n n e r s i d e. Flattened in a radial direction.

C e l l s o f t h e a b o v e. R. 12 μ, T. 15 μ, L. 40 μ; near the funicle only up to 20 μ; quadrangular tables. Walls often coloured red-brown; tangential walls, turned towards the inside of the seed-coat, strongly thickened; radial and transverse walls cuneiform; the thicker tangential walls, just mentioned, showing towards the inside of the seed-coat a colourless lamella or cuticle, not turning black in iron acetate like all the rest of the cell

---

[1]) According to B a r t h l. c. this cell layer is the principal seat of the colchinine.

walls. In the material, fixed June 7th with Flemming's mixture, walls exhibiting the same structure. In the alcoholic material, gathered in May, thicker tangential walls and cuneiform parts of radial and transverse walls colouring red-brown in iodine and sulphuric acid 66 per cent.

Raphe part. In many respects like the seed-coat proper. Ground tissue of the part produced by the outer integument, in the material fixed June 7th with Flemming's mixture, consisting of 4 layers of cells on the outside as well as on the inside of the meristele. Meristele exhibiting a pericycle, consisting of prismatic cells of various length and with a longitudinally directed axis; vascular bundle containing numerous spiral vessels.

Cells of ground tissue. Those outside of the meristele R. 130 μ, T. and L. 45 μ; polygonal prisms with a radially directed axis; those inside of the meristele and in the vicinity of the funicle measuring e. g. 50 by 70 by 80 μ, often larger; polyhedra. All cells showing intercellular spaces. Cell contents: see seed-coat proper, but starch grains larger, up to 15 μ in diameter and in the drug starch grains here rather numerous.

Nucleus of seed.

Endosperm. Consisting of thick-walled parenchyma cells, exhibiting radial arrangement with many interruptions; near the chalaza cells often flattened. Cells of the above. Epidermal cells R. 60 μ, T. 40 μ, L. 45 μ; cells of the following cell layers R. 70—120 μ, T. 40 μ, L. 45 μ; polygonal prisms with a radially directed axis and somewhat rounded edges; cells near the chalaza measuring 12 by 20 by 20 μ, polyhedra; cells in the centre of the endosperm measuring 25 by 30 by 30 μ, polyhedra. Walls of all cells thick 8—10 μ; not swelling up in water or in chloral hydrate; colouring blue in iodine and sulphuric acid 66 per cent.; middle lamella thin, strongly refringent, colouring somewhat darker blue in iodine and sulphuric acid 66 per cent. and persisting in concentrated sulphuric acid for a longer time than the rest of the cell walls; showing stratification, large circular or elliptical pit canals, 7 μ in diameter, and intercellular spaces; connecting threads of protoplasm, passing through the walls, perhaps discernible in potash. Cell contents: 1, numerous aleurone grains, often polyhedral in consequence of their sides being flattened by reciprocal pressure, surrounded by a very conspicuous network of protoplasm; these grains up to 5 μ in diameter, each containing 1 or more very small globoids; near the chalaza and in the centre of the endosperm aleurone grains nearly wanting, but here the cells often quite filled with a granular mass colouring brown in potassium iodide iodine as well as the aleurone grains; 2, oil plasm; the network of protoplasm colouring black in osmic acid; in iodine and sulphuric acid 66 per cent. oil often collecting into yellow to brown drops, lying at one end of a cell. In potassium iodide iodine contents colouring darker brown in sections not beforehand treated with an alcoholic solution of tartaric acid 5 per cent. than in sections treated beforehand with this reagent (colchicine) [1]. In the material fixed June 7th with Flemming's mixture contents of all cells adjoining the flattened tissue of the chalaza and of a few cells here and there along the periphery of the endosperm, coloured black.

Embryo. Lying in the median plane of the seed, sometimes under the micropyle but often somewhat further from the funicle than the micropyle; directed radially; ellipsoidal; consisting of meristematic cells, containing aleurone grains and oil.

April 1903.   J; M; v. E. d. W.

---

[1] According to B a r t h l. c. the ripe seed in the endosperm also containing colchicine.

SEMEN LINI.
Linum. Linseed. Flaxseed.
The seeds of Linum usitatissimum, Linn. Sp. Pl. 277.

Macroscopic characters.

Long 4—6 m.M., wide 2—3 m.M., thick 0.75—1.5 m.M.; weight 5—10 m.G.; oblong-ovate, pointed, flattened, on both sides somewhat arched, with rather sharp edges. In soaked seeds seed-coat easily splitting into 2 layers, the innermost part (internal seed-coat) also containing the endosperm. External seed-coat yellowish to brown, thin, brittle, glossy, with numerous minute round grooves. Hilum at the edge, in a small curve somewhat below the top, linear, lighter coloured. Micropyle, a small dark spot just over the hilum. Raphe a fine ligther coloured line, running along the edge of the seed from the hilum to the obtuse end. Chalaza, a lighter coloured, narrow oblong spot at the obtuse end. Nucleus showing a thin white endosperm attached to the seed-coat and surrounding the embryo on all sides. Embryo, nearly quite filling up the seed, yellowish white; radicle acute-conical, long 1 m.M.; cotyledons oval, showing an incision at the base, plano-convex; plumule only slightly developed.

When soaked in water very slippery, owing to the epidermis swelling up.

Inodorous; taste mucilaginous, slightly oily, not rancid.

Anatomical characters.

LITERATURE. Attema. Zaadhuid d. Angiosp. en Gymnosp. Diss. Groningen. 1901. 69. Berg. Anat. Atl. 1865. 91. Benecke. Mikr. Drogenprakt. 1912. 73. Biechele. Mikr. Prüfung d. off. Drogen. 1904. 70. Erdmann-König. Allg. Warenk. 1895. 206. Flückiger. Pharmakogn. 1891. 976. Flückiger & Hanbury. Pharmacogr. 1879. 98. Frank. Anat. Bedeut. u. Entstehung d. vegetab. Schleime. Pringsheim's Jahrb. V. 1865. 161. Gilg. Pharmakogn. 1910. 176. Guignard. Recherches s.l. dével d. l. graine etc. Journal de bot. 7. 1893. 100. Hager. Pharm. Praxis. Bd. II. 1902. 295. Hérail. Pharmacogr. 1912. 109. Karsten u. Oltmanns. Pharmakogn. 1909. 265. Koch. Mikrosk. Anal. d. Drogenp. Bd. IV. 1908. 29. Pl. III. Koch. Einf. i. d. mikr. Anal. d. Drogenp. 1906. 130. Koch u. Gilg. Pharmakogn. Praktik. 1907. 208. Kraemer. Botany a. Pharmacogn. 1909. 176, 426, 741. Luerssen. Syst. Botanik. Bd. II. 1882. 674. Marmé. Pharmakogn. 1886. 394. Meyer. Wiss. Drogenk. Bd. I. 1891. 144. Meyer. Mikr. Unters. v. Pfl. pulv. 1901. 34, 38, etc. Moeller. Mikr. d. Nahr. u. Genussm. 1886. 173. Moeller. Mikr. pharm. Üb. 1901. 154. Moeller. Pharmokogn. 1889. 186. Oudemans. Aant. o. d. Pharm. neerl. 1854—56. 551. Oudemans. Pharmacogn. 1880. 426. Planchon et Collin. Drogues simples. T. I. 1895. 684. Schneider. Powdered veg. Drugs. 1902. 233. Tschirch. Pharmakogn. Bd. II. 1911. 314. Tschirch u. Oesterle. Anat. Atl. 1900. 255. Walliczek. Stud. ü. d. Membranschleime veget. Org. Pringsheim's Jahrb. Bd. 25. 1893. 214. Also Diss. Bern. 1893. Vogl. Veget. Nahr. u. Genussm. 1889. 539. Wigand. Pharmakogn. 1879. 305. v. Wisselingh. Bijdr. t. d. Kenn. v. d. Zaadhuid. V. Linaceae. Pharmac. Weekbl. Jrg. 56. 1919. 1437. MATERIAL. Ripe seeds. REAGENTS. Water, alcohol, glycerine, glycerine jelly, chloral hydrate, potash, iodine, iodine and sulphuric acid 66 per cent., phloroglucin and hydrochloric acid, concentrated sulphuric acid, Schulze's reagent, osmic acid.

MICROGRAPHY.

O v u l e. Anatropous; with 2 integuments; the inner one according to

Tschirch und Oesterle consisting of 4 cell layers, according to Guignard of 10 layers.

Seed.

Seed-Coat.

Layers produced by the outer integument.

E p i d e r m i s   o u t e r   s i d e. Consisting of thick-walled cells without a cuticle. C e l l s   o f   t h e   a b o v e. R. 25 μ, when swollen in water R. 60 μ, T. 30 μ, L. 30—35 μ; penta- to septagonal prisms with a radially directed axis. Outer walls strongly thickened; in a somewhat swollen condition showing 2 different parts separated by a cleft [1]): 1, a more solid stratified part; 2, the so-called mucilage membrane with sinuous stratification, this last part producing mucilage when swelling in water. Transverse and radial walls somewhat thickened and folded, also producing some mucilage. Mucilage layer showing no cellulose reaction, but nor colouring brown in iodine and sulphuric acid 66 per cent. Cavity generally not large.

G r o u n d   t i s s u e. Present only in the chalaza part and along the raphe; consisting of some layers of parenchyma cells.

E p i d e r m i s   i n n e r   s i d e. Generally consisting of 2 layers of cells, sometimes of 1 layer; on the side of the seed opposite to the raphe nearly always 1 layer. Between this epidermis and the internal seed-coat distinct triangular spaces. C e l l s   o f   t h e   a b o v e. R. 10 μ, T. 20 μ, L. 20—30 μ; elliptical or circular tables, outer walls of the outermost layer bulging out into the adjoining epidermis of the internal seed-coat; axis radially directed. Walls somewhat thickened, in places of contact of several cells collenchymatously thickened ridges; somewhat yellow. Cell contents: probably some large starch grains.

Layers produced by the inner integument.

E p i d e r m i s   o u t e r   s i d e. Consisting of 1 layer of longitudinally directed and elongated sclereids, arranged in transverse rows. Once noticed 2 layers in the chalaza part of the seed-coat. C e l l s   o f   t h e   a b o v e. R. 15 μ, L. 120—190 μ; along the edges and near the end of the seed R. 17 μ, T. also somewhat larger; tetragonal prisms with a longitudinally directed axis; outer walls re-entering in correspondence to the walls of the preceding layer bulging out. Walls rather thick; brownish yellow; those of the larger cells lignified, those of the other cells not or feebly lignified; pitted.

G r o u n d   t i s s u e. Consisting of flattened parenchyma, the long prismatic cells of the outermost layer sometimes discernible; axis of these cells at right angles to that of the sclereids, the cells being arranged in longitudinal rows. C e l l s   o f   t h e   a b o v e. T. 100—130 μ.

E p i d e r m i s   i n n e r   s i d e. Consisting of 1 layer of pigment cells. Wanting in the Indian linseed. C e l l s   o f   t h e   a b o v e. R. 9 μ, T. 15 μ, L. 18 μ; tetra- or pentagonal tables with a radially directed axis. Walls somewhat thickened, showing reticulate thickenings, those thickenings not discernible on the tangential walls; somewhat yellow. Cell contents: a reddish brown mass showing distinct lumps; tannin.

---

[1]) See T s c h i r c h l. c. 256.

Nucleus of the seed.

E n d o s p e r m. All but enveloping the embryo. At the top of the seed 10—15 cell layers thick, along the edges 2—3 layers, on the sides 5—6 and at the base 3 layers. On the inside of those cell layers some flattened tissue. Epidermis closely resembling the ground tissue. Intercellular spaces wanting.

C e l l s o f t h e a b o v e. R., T. and L. 12 by 15 by 21 μ; polyhedra. Walls somewhat thickened; sometimes colouring blue in iodine whithout sulphuric acid. Cell contents: aleurone grains containing globoids [1]) and protein crystals, the latter sometimes with rounded corners; generally 4 aleurone grains in 1 cell. Fatty oil wanting.

E m b r y o.

C o t y l e d o n s. Already showing the incipient anatomical structure of leaves.

C e l l s o f t h e a b o v e. Walls sometimes colouring blue in iodine without sulphuric acid. Cell contents: aleurone grains, generally only 1 or 2 in each cell, for the rest see those of the endosperm. Fatty oil.

April 1901.   J; M; L.

SEMEN MYRISTICAE.
Myristica. Nutmeg.

The seed nuclei, still surrounded by the tegmen, of Myristica fragrans, Houtt. Handleid. III. 333

Macroscopic characters.

Long up to 3.5 c.M., thick up to 2.8 c.M.; ellipsoidal or somewhat ovate. Hilum indicated by a slight circular elevation, 5 m.M. in diameter, lying somewhat eccentric at the thicker end of the seed and showing about its centre a small perforation corresponding with the micropyle. Chalaza indicated by a smaller circular depression, containing a minute elevation and lying also somewhat eccentric at the opposite end of the seed, but at the other side. Raphe indicated by a wide and shallow groove, connecting hilum and chalaza. Surface for the rest on all sides of the seed showing a network of longitudinal narrow shallow grooves, branched like veins; coloured light brown, hilum still much lighter, chalaza darker; the whole surface covered with powder of lime. Tegmen completely grown together with the seed nucleus, thin, in sections coloured very dark brown, causing a marbled aspect of the endosperm by penetrating into it as far as the centre in the shape of thicker sinuous irregularly branched bluntly ending lamellae; those lamellae somewhat granular, coloured very dark brown and in a fresh section oily. Endosperm bright light brown, with a greasy luster; in the areas confined by the lamellae the inner space circumscribed by a nearly white line. The shriveled remainders of the embryo lying in a cavity below the hilum.

Odour aromatic; taste aromatic, hot and somewhat bitter.

Anatomical characters.

LITERATURE. Attema. De Zaadhuid der Angiosp. en Gymnosp. Diss. Groningen. 1901.

---

[1]) According to T s c h i r c h no or small globoids.

170. Baillon. Sur l'origine du macis de la Muscade et des arilles en général. Compt. Rend. T. 58. Berg. Anat. Atl. 1865. 95. Taf. XXXXVIII. Benecke. Mikr. Drogenprakt. 1912. 70. Biermann. Ueber Bau u. Entwicklungsgeschichte der Oelzellen u. d. Oelbildung in ihnen. Diss. Bern. 1897. 47. Erdmann-König. Allg. Warenk. 1895. 326. Flückiger. Pharmakogn. 1891. 1031. Flückiger & Hanbury. Pharmacographia. 1897. 502. Gilg. Pharmakogn. 1910. 111. Greenish & Collin. Anat. Atl. veget. powders. 1904. 136. Hager. Pharm. Praxis. Bd. 2. 1902. 412. Hallström. Vergleichend-anatom. Studien ü. d. Samen d. Myristicaceen und ihre Arillen. Archiv. d. Pharmacie. 1895. 443. Hérail. Mat. Méd. 1912. 450. Karsten u. Oltmanns. Pharmakogn. 1909. 245. Koch u. Gilg. Pharmakogn. Praktik. 1907. 206. Koch. Mikr. Anal. d. Drogenpulver. Bd. 4. 1908. 37. Tafel IV. Kraemer. Botany a. Pharmacogn. 1910. 439, 771. Luerssen. Syst. Bot. Bd. 2. 1882. 580. Marmé. Pharmacogn. 1886. 358. Mitlacher. Toxik. od. Forens. wicht. Pfl. u. Drogen. 1904. 61. Meyer. Wiss. Drogenk. Bd. 2. 1892. 164. Meyer. Mikr. Unters. v. Pflanzenpulv. 1901. 67. Moeller. Mikr. d. Nahr. u. Genussm. 1886. 268. Moeller. Pharmakogn. 1889. 179. Moeller. Mikr. pharm. Ueb. 1901. 160. Molisch. Mikroch. d. Pfl. 1913. 110 (M. argentea), 112 (M. Ocuba). Oudemans. Pharmacogn. 1880. 415. Planchon et Collin. Drogues simples. T. I. 1895. 400. Schimper. Mikr. Unters. d. veget. Nahr. u. Genussm. 1900. 137. Schneider. Powdered Veget. Drugs. 1902. 246. Tschirch. Angew. Pfl.-anat. 1889. 44. Tschirch. Harze u. Harzbeh. 1900. 390. Tschirch. Pharmakogn. Bd. 2. Abt. 1. 1912. 676. Tschirch u. Oesterle. Anat. Atl. 1900. 245. Tafel 56 u. 57. Vogl. Veget. Nahr. u. Genussm. 1899. 485. Voigt. Ueber den Bau u. d. Entwicklung d. Samens u. d. Samenmantels von Myristica fragrans. Diss. Göttingen. 1885. Wigand. Pharmakogn. 1879. 306. MATERIAL. The drug. REAGENTS. Water, glycerine, potash, chloral hydrate, potassium iodide iodine, phloroglucin and hydrochloric acid, concentrated sulphuric acid, iron acetate, eosine in absolute alcohol and afterwards mounted in oil of cloves (Protein crystals).

MICROGRAPHY.

O v u l e [1]).

S e e d.

Funicle.

Aril. Removed from the ripe seed and brought into trade under the denomination of Macis.

Seed-coat.

Layers produced by the integument. (Testa). Removed in the drug.

Layers produced by the nucellus. (Tegmen. Perisperm).

E p i d e r m i s. Wanting in the drug.

G r o u n d t i s s u e. Composed of an outer primary part and an inner part arising from secondary growth.

1. Outer primary part. Thick about 80 μ, consisting of cells strongly flattened in a radial direction; surface not smooth in consequence of the disappearence of the epidermis and at the chalaza also of the outermost cell layer of ground tissue.

C e l l s o f t h e a b o v e. T. and L. 40 μ; polyhedra with strongly rounded edges.

---

[1]) V o i g t l. c., Hallström l. c., T s c h i r c h u. O e s t e r l e l. c. According to these authors the ovule is anatropous; in the ripe seed with 2 integuments in the uppermost half of the ovule, the lower part showing only a single integument. Outer integument thick about 7 layers of cells. Inner integument much delayed in growth behind the rest of the seed.

Walls often colouring red in phloroglucin and hydrochloric acid (wound gum); showing intercellular spaces. Cell contents: of many cells a brown mass, insoluble in potash 50 per cent., chloral hydrate or concentrated sulphuric acid; of many other cells prismatic or disc-shaped crystals. [1])

2. Inner secondary part. Thick about 40 μ; showing numerous extensive lamel-lae, corresponding with the furrows on the surface of the seed, penetrating into the endosperm and often reaching the centre; in the peripheral part consisting of common parenchyma cells, only here and there intermixed with a single oil cell; in the lamellae mainly consisting of oil cells, at their base and in the middle part containing numerous common pa-

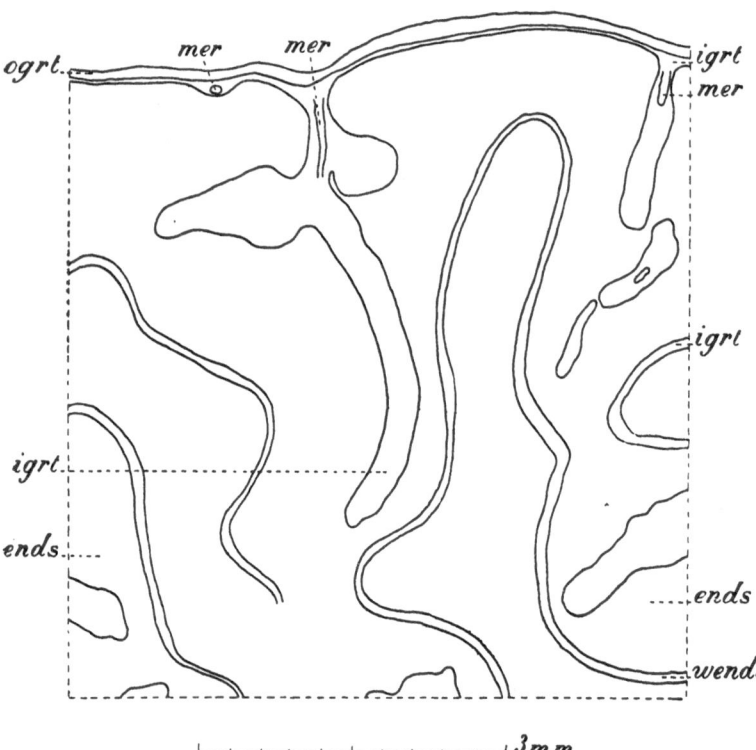

Fig. 103. *Myristica fragrans*. Seed, transverse section. ends Endosperm; igrt Inner secondary part of ground tissue; mer Meristeles; ogrt Outer primary part of ground tissue; wends White lines of endosperm.

renchyma cells; along the surface of the lamellae parenchyma of the peripheral part more or less extending inward in the form of a brown band, consisting of cells without contents; the rest of the lamellae with only a few parenchyma cells.

P a r e n c h y m a c e l l s. Those of the peripheral part R. 6 μ, T. and L. 25 μ; polygonal prisms with a radially directed axis; radially flattened. Walls brown. Cell contents: a dark brown mass, insoluble or nearly so in potash 50 per cent., chloral hydrate and concentrated sulphuric acid. Those of the lamellae at the base of the lamella R. 15 μ, T. and L. 25 μ;

[1]) According to T s c h i r c h u. O e s t e r l e. l. c. p. 250 those crystals soluble in concentrated hydrochloric acid and diluted sulphuric acid, also in water of 100° C., slowly in alkalies; insoluble in cold water, alcohol, sulphuric ether, acetic acid; in concentrated sulphuric acid at first forming needles, these dissolving after addition of water. According to M e y e r. Wiss. Drogenk. l. c. soluble in hydrochloric acid and cold alcohol, altered by potash.

for the rest R. 30 μ, T. 12 μ. Walls sometimes brown; intercellular spaces wanting. Cell contents wanting. O i l c e l l s. Those of the peripheral part R. 12 μ, T. and L. 40 μ; polyhedra with strongly rounded edges. Walls brown; middle lamella persisting in concentrated sulphuric acid. Cell contents: only here and there a small oil drop. Those of the lamella R., T. and L. 50 μ; prisms or polyhedra, often with rounded edges. Walls somewhat thickened; mostly brown; middle lamella persisting in concentrated sulphuric acid. Cell contents: oil drops; along the walls often the resinogenous layer still discernible.

M e r i s t e l e s. 1 or 2 at the insertion of each lamella, extending in a longitudinal direction; a few times penetrating in a radial direction into the lamella. All meristeles continuously connected together and with the bundle of meristeles in the raphe, the latter entering the secondary perisperm through the chalaza [1]). Phloem not very conspicuous. Xylem containing numerous spiral vessels showing lignification.

Nucleus of seed.

E n d o s p e r m. Consisting of common parenchyma cells, intermixed here and there with a tannin cell. The cells irregularly arranged in the outermost part of the endosperm and within the white lines circumscribing the middle part of the areas mentioned under macroscopic characters. The cells on the outside of these white lines (sometimes showing clefts under the microscope), with the exception of the outermost part of the endosperm, arranged in layers perpendicular to the surface of the perisperm lamellae.

P a r e n c h y m a c e l l s. Those irregularly arranged R. 45 μ, T. 50 μ, L. 60 μ; polyhedra; those of the perpendicular layers R. 25 μ, T. and L. 40 μ; polygonal prisms with their axis perpendicular to the surface of the lamellae. Intercellular spaces wanting. Cell contents: 1, starch grains, mostly somewhat yellow, simple and compound, 2- to 10-adelphous, the simple grains mostly being the largest ones e. g. 20 μ in diameter, globular, sometimes showing a slit-like hilum; 2, in each cell one large and mostly also some small aleurone grains, the large aleurone grains containing nearly always a rhombic protein crystal, measuring 12 by 20 μ in cross sections of the seed [2]) and nearly quite filling the grain, the small aleurone grains containing protein crystals of a more irregular shape; 3, crystals of fat, radially arranged, showing double refraction, soluble in absolute alcohol and quite filling up the remaining space in the cells. Those in the middle part of the white lines eventually adjoining the clefts formed in these parts conspicuous, especially in potassium iodide iodine, by their contents only consisting of starch grains. Those of the cell layer adjoining the perisperm containing a small quantity of starch and no aleurone grains. T a n n i n i d i o b l a s t s. Walls often brown, persisting in concentrated sulphuric acid. Cell contents: in the middle part among the starch grains tannin; see for the rest the parenchyma cells.

February 1903.   J; M; v. E. d W.

---

[1]) See T s c h i r c h u. O e s t e r l e. l. c. p. 248.
[2]) According to T s c h i r c h the protein crystals in tangential sections of the seed more or less corroded and having lost their smooth surface.

SEMEN PHYSOSTIGMATIS.
Physostigma. Calabar beans.
The ripe seeds of Physostigma venenosum, Balf. in Trans. Roy. Soc. Edinb. XXII (1861). 310.

LITERATURE. Attema. De zaadhuid der Angiosp. en Gymnosp. Diss. Groningen. 1901. 83. Barth. Stud. üb. d. Nachweis v. Alkaloiden in pharm. verw. Drogen. Bot. Centrbl. 1898. Bd. 75. 371. Also Diss. Zürich. 1898—99. Chalon. La graine d. Légumineuses. 1875. Compton. Invest. of the seedling struct. in the Leguminosae. Journ. of Linn. Soc. T. 411. 1912. 1. Flückiger. Pharmakogn. 1891. 994. Flückiger & Hanbury. Pharmacogr. 1879. 191. Guignard. Embryogénie des Légum. Ann. d. Sc. nat. Bot. 6me Sér. T. XII. 1882. 68; Rech. s. l. Développement de la Graine. Journ. de Bot. 1893. 39. Hérail. Mat. Méd. 1912. 709. Hager. Pharm. Praxis. Bd. II. 1902. 606. Jacquemin. Localis. d. Alcaloides chez les Légumineuses. Rec. de l'Inst. Erréra, Univ. Bruxelles. T. VI. 1906. 257. Karsten u. Oltmanns. Pharmakogn. 1909. 262. Marmé. Pharmacogn. 1885. 419. Maisel. Rech. anat. et taxin. s. l. tégument d. l. graine d. Légumineuses. Autun Bull. soc. hist. nat. T. 22. 1909. 51. Mattirolo e Buscalioni. Ric. anat. fis. sui tegumenti sem. d. Papilionacee. 1892. Nadelmann. Schleimendosp. d. Leguminosen. Pringsheim's Jahrb. T. XXI. 1890. 609. Oudemans. Pharmacogn. 1880. 434. Planchon et Collin. Drogues simples. T. II. 1896. 538. Pfaefflin. Unters. üb. Entwick.-geschichte, Bau etc. Papilionaceen-Samen. Diss. Bern. 1897. Strassburger. Zellbild. u. Zellteil. 1880. 107. Tschirch. Angew. Pfl.-anat. 1889. 200, 204, 305. Tschirch u. Oesterle. Anat. Atl. 1900. Pl. 47 and 48. Walliczek. Stud. üb. veget. Schleime. Diss. Bern. 1893. 11. Also in Pringsheim's Jahrbücher. MATERIAL. The drug. REAGENTS. Water, glycerine, potash 50 per cent., chloral hydrate, potassium iodide iodine, iodine in chloral hydrate, phloroglucin and hydrochloric acid, iodine and sulphuric acid 66 per cent., concentrated sulphuric acid, Schulze's macerating mixture, osmic acid, iron acetate, Millon's reagent.

MICROGRAPHY.

O v u l e. Probably campylotropous; probably with 2 integuments, the inner one being resorbed. [1]

S e e d.

Funicle. Having disappeared for the greater part. Represented only along the bottom of the longitudinal furrow above the hilum by 2 longitudinal bands of palisade sclerenchyma, a very narrow fissure being left between. Radial dimensions of the palisade cells uniform in the greater part of the band; becoming much smaller in the vicinity of the outer margin and much larger again in the outermost part of it; there the cells showing thicker outer ends. There also some further remains of the funicle.

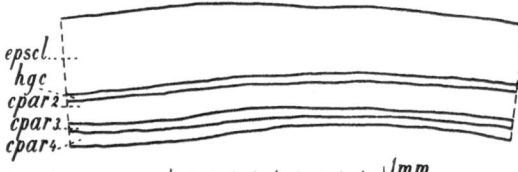

Fig. 104. *Physostigma venenosum.* Seed, transverse section of the seed-coat proper. cpar2 Common parenchyma mentioned under 2; cpar3 Common parenchyma mentioned under 3; cpar4 Common parenchyma mentioned under 4; epscl Epidermal palisade sclerenchyma cells; hgc Hourglass-shaped cells.

C e l l s   o f   t h e   a b o v e. R. 140 μ, T. and L. 10 μ, in the vicinity of the outer margin

---

[1] As far as we know the ovule has not been investigated; for researches on the ovule of the *Leguminosae* and its farther development, also that of the endosperm, see A t t e m a, l. c. 83; G u i g n a r d, l. c. 309; C h a l o n, l. c.; S t r a s s b u r g e r, l. c. 107; T s c h i r c h   u n d   O e s t e r l e, l. c. Pl. 48, etc.

R. 10 μ, T. 15 μ. Lateral walls thickened; the cavity all but disappearing; walls yellowish; inner tangential walls brown, showing a retarded cellulose reaction as compared to the epidermal cells of the seed-coat. „Lichtlinie" not conspicuous; pit canals in the lateral walls. Walls of the cells along the margin less thickened.

Seed-coat.

Seed-coat proper.

Layers produced by the integument.

E p i d e r m i s   o u t e r   s i d e. Consisting of 1 layer of palisade sclerenchyma, in which here and there a small group of thicker cells.

C e l l s   o f   t h e   a b o v e. R. 250 μ, T. and L. 10 μ, thicker cells T. and L. 25 μ; polygonal prisms with a radially directed axis. Lateral walls strongly thickened in the parts turned towards the periphery of the seed, their inner parts about 10 μ in length and the inner walls not thickened; middle lamella conspicuous; thickened parts often showing branched pit canals; thickened parts showing stratification; outer walls with a thin, not always distinct cuticle; throughout the epidermis at a small distance from the cuticle or joining it a more strongly refringent band („Lichtlinie") showing a retarded cellulose reaction as compared to the other parts of the walls and when coloured blue still clearly to be distinguished from them; sometimes a thin innermost part of this band showing a different refraction. Cell contents: a granular mass, somewhat yellow to brown, colouring bluish black in iron acetate, reddish brown in potash.

G r o u n d   t i s s u e. Consisting of 4 different layers of tissue.

1°. Hour-glass-shaped radially directed cells with intercellular spaces between them, mostly in 1, a few times in 2 layers.

C e l l s   o f   t h e   a b o v e. R. 35 μ, T. 25 μ, L. 30 μ; narrow part thick 15 μ; outermost thicker part polygonal in a tangential section, innermost thicker part smaller and with rounded edges. Lateral walls thickened so as to leave the lumen very small in the narrow part; walls yellowish. Cell contents: some small granules, colouring yellow in iodine.

2°. Common parenchyma in about 10 layers of cells, especially the innermost part somewhat flattened; this whole tissue thick about 120 μ in a seed soaked in water. Intercellular spaces wanting.

C e l l s   o f   t h e   a b o v e. In the outermost cell layer R. 32 μ, T. and L. 30 μ; prisms with a radially directed axis; in the other layers R. 20 μ, T. and L. 80 μ; polygonal. Walls yellowish; pitted.

3°. Common parenchyma, strongly flattened; the whole tissue thick about 30 μ in a seed soaked in water.

C e l l s   o f   t h e   a b o v e. Walls brown. Cell contents: here and there a brown granular mass.

4°. Common parenchyma, much shrivelled and flattened; colour yellow; the whole tissue in a seed soaked in water thick 50 μ or less. In the parts of the seed-coat corresponding to the line of contact of the cotyledons this layers less flattened and shrivelled so as to form a ridge on the inner side of the seed-coat; meristeles wanting.

E p i d e r m i s   i n n e r   s i d e. Flattened; only the cuticle fairly distinct in iodine and sulphuric acid 66 per cent.

H i l u m   p a r t. Forming the bottom of the longitudinal furrow running along the edges of the seed.

E p i d e r m i s   o u t e r   s i d e. Interrupted in the middle of the furrow by a very narrow longitudinal cleft. Consisting of palisade sclerenchyma forming a thinner layer than in the seed-coat proper.

C e l l s   o f   t h e   a b o v e. R. 130 μ. Outer walls brown; a brown cuneiform lamella in the outermost parts of the lateral walls; those parts showing no cellulose reaction; „Licht-linie" conspicuous. See for the rest the seed-coat proper.

G r o u n d   t i s s u e. The 4 layers of tissue of the seed-coat proper represented as follows.

1. Hour-glass-shaped cells wanting.

2. A layer of parenchymatic tissue, much thicker than in the seed-coat proper and containing a longitudinal band lying just below the narrow cleft of the epidermis, lenticular in a transverse section and consisting of radially directed elements closely related to tracheid fi-

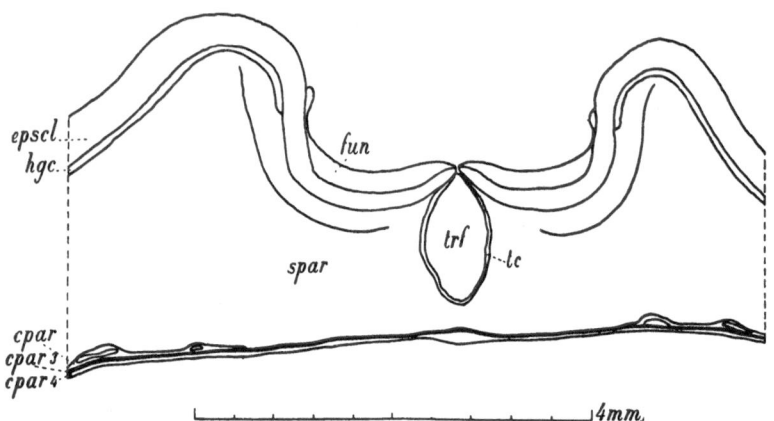

Fig. 105. *Physostigma venenosum*. Transverse section of the seed-coat, hilum part and longitudinal ridges bordering it. cpar Colourless parenchyma; cpar3 Common parenchyma mentioned under 3; cpar4 Common parenchyma mentioned under 4; epscl Epidermal palisade sclerenchyma cells; fun Funicle; hgc Hour-glass-shaped cells; spar Spongy parenchyma; tc Tabular cells; trf Tracheid fibres.

bres. The parenchymatic tissue developed in its outer and principal part as spongy parenchyma with smaller cells towards the tracheid band, the 2 or 3 cell layers surrounding this band consisting of tabular cells in a transverse section of the seed strongly elongated in the direction of the outline of the tracheid band; the inner part consisting of more or less flattened colourless common parenchyma about 50 μ thick in a seed soaked in water.

E l e m e n t s   o f   t r a c h e i d   b a n d. R. 280 μ, T. 45 μ, L. 30 μ or smaller; polyhedra with more or less pointed ends. Walls thickened; somewhat yellow; lignified; showing numerous bordered pits. T a b u l a r   c e l l s surrounding the tracheid band thick 5 μ, wide 40 μ. S p o n g y   p a r e n c h y m a   c e l l s. In the vicinity of the tracheid band R. 20 μ, T. and L. 25 μ; the others e. g. 60 by 180 by 180 μ. For the rest and for the colourless parenchyma cells see the ridges bordering the hilum.

3 and 4. See seed-coat proper.

E p i d e r m i s   i n n e r   s i d e. See seed-coat proper.

L o n g i t u d i n a l   r i d g e s bordering the furrow on both sides.

Epidermis outer side. See seed-coat proper.

Ground tissue. The 4 layers of tissue represented as follows.

1. Hour-glass-shaped cells. See seed-coat proper. Wanting towards the hilum part of the seed-coat.

2. Spongy parenchyma. Forming a very thick layer of tissue; the outermost cell layers consisting of polygonal cells with large intercellular spaces, the inner part consisting of colourless common parenchyma as mentioned for the hilum part of the seed-coat, here containing 2 meristeles on each side of the hilum.

Cells of the above. R. 60 μ, T. and L. 180 μ; in the outermost layer R. and T. 30 μ. Walls thickened; those of the polygonal cells with pit canals; in the outermost layer a very thin brown lamella on the outermost side. Cell contents especially of the large spongy cells: a brown mass becoming black in iron acetate.

3 and 4. See seed-coat proper.

Epidermis inner side. See seed-coat proper.

Chalaza part.

Epidermis outer side. See seed-coat proper.

Ground tissue. The 4 layers represented as follows.

1. Hour-glass-shaped cells wanting.

2. Showing various kinds of tissue. Outermost cell layer consisting of prismatic cells with a radially directed axis and thickened walls; R. 40 μ, T. 20 μ, L. 40 μ. Middle lamella distinct. Some following cell layers of polygonal cells with thickened walls; intercellular paces wanting. R. and T. 20 μ, L. 40—70 μ. Middle lamella distinct. Some following layers of strongly radially elongated polygonal

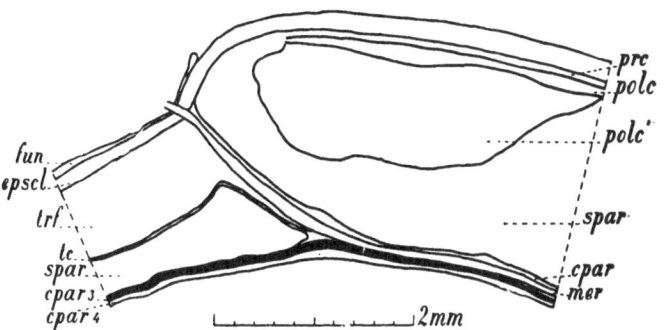

Fig. 106. *Physostigma venenosum*. Longitudinal section of the seed-coat, chalaza part. cpar Colourless parenchyma; cpar3 Common parenchyma mentioned under 3; cpar4 Common parenchyma mentioned under 4; epscl Epidermal palisade sclerenchyma cells; fun Funicle; mer Meristeles; polc Polygonal cells; polc' Polygonal cells forming a trapeziform mass; prc Prismatic cells; spar Spongy parenchyma; tc Tabular cells; trf Tracheid fibres.

cells with thickened walls, forming a mass trapeziform in a longitudinal section of the seed. Cells. R. 300 μ, T. and L. 25 μ. Middle lamella distinct. Spongy parenchyma consisting of radially elongated cells with thickened walls and brown contents becoming black in iron acetate. Colourless parenchyma as described for the hilum part; containing meristeles.

Meristeles. Entering the seed at the end of the band of tracheid fibres, immediately running along it and accompanied in its further course inward along the end of the hilum on its hilum side by a few tracheid fibres, in the chalaza part giving out branches into several directions, 4 of these entering the ridges along the hilum described above.

Phloem. The larger bundles of phloem containing bundles of smaller elements with yellow contents.

Connecting parenchyma. A single layer of elongated cells.

C e l l s   o f   t h e   a b o v e. R. 10—20 μ, T. 15—30 μ, L. 200—600 μ; polygonal prisms with a longitudinally directed axis. Cell contents: a reddish brown mass.

Xylem of the larger branches. In the shape of a semi-elliptical cylinder, the flat side turned outwards; wide 75 μ, thick 200 μ. Consisting of numerous irregularly arranged spiral vessels and a few parenchyma cells.

V e s s e l s. R. and T. 12 μ; polygonal with a single lignified spiral. Middle la mella often dark brown.

3 and 4. See seed-coat proper.

E p i d e r m i s   i n n e r   s i d e. See seed-coat proper.

M i c r o p y l e   p a r t.

Epidermis outer side. See seed-coat proper.

G r o u n d   t i s s u e. The 4 layers represented as follows.

1. Hour-glass-shaped cells wanting.

2. A large group of polygonal cells, becoming larger towards the interior, thick 400—500 μ, surrounded on its lateral and inner sides by spongy parenchyma like that of the hilum part. Intercellular spaces wanting. At the inside of the spongy parenchyma colourless parenchyma like that of the hilum part.

P o l y g o n a l   c e l l s. Outermost ones R., T. and L. 20 by 30 by 40 μ; inner ones 50 by 60 by 60 μ. Walls thickened; yellow to brown; with pit canals. Cell contents: sometimes brownish yellow masses becoming bluish black in iron acetate.

3 and 4. See seed-coat proper.

Epidermis inner side. See seed-coat proper.

Nucleus of the seed.

Albumen wanting.

E m b r y o.

Radicle. The pouch occurring in Phaseolus multiflorus and especially in Pisum sativum [1] here wanting or represented at most by an insignificant excavation in the seed-coat. Starch wanting.

Cotyledons.

E p i d e r m i s   u p p e r   s i d e. Cells nearly everywhere arranged in groups.

C e l l s   o f   t h e   a b o v e. R. 14 μ, T. 35 μ, L. 80 μ; polygonal tables with a radially directed axis. Outer walls a little thickened; showing a cuticle; lateral walls pitted and gussets distinct and resembling interculallar spaces. Cell contents: a granular mass quite filling the cells; in iodine colouring yellow, a dry section giving the same reaction [2]); in Millon's reagent colouring red; oil in the shape of oil-plasm.

E p i d e r m i s   u n d e r   s i d e. Cells nearly everywhere arranged in groups.

---

[1] T s c h i r c h   u.   O e s t e r l e. l. c. 207, 211.
[2] See   B a r t h. l. c. 374.

Cells of the above. R. 10 μ, T. 12 μ, L. 15 μ; polygonal tables. In chloral hydrate the edges of the lateral walls rounded, thus the cells seemingly showing intercellular spaces. For the rest see those of the inner side.

Mesophyll. Consisting of common parenchyma; cells of the outer layers, especially of the outermost one, smaller than the rest.

Cells of the above. Diameter 100 μ, in the outer layer R. 20 μ, T. 30 μ, L. 35 μ; polygonal with rounded edges. Middle lamella somewhat more resistant in sulphuric acid than the other parts of the walls. Cell contents: of the outer layers, especially of the outermost one, like those of the epidermal cells. In the following layers starch grains increasing in number towards the centre; starch grains ellipsoidal, the largest ones long 60—65 μ, thick 30—60 μ, very small grains nearly wanting; hilum conspicuous, often branched, not rarely in the shape of an elongated cleft branched on both ends, stratification not always distinct. See for the rest the epidermal cells.

May 1903.   J; M; L.

SEMEN SINAPIS ALBAE.
White Mustard Seed. White Mustard.
The ripe seeds of Brassica alba, Boiss. Voy. Espagne, II. 39.

LITERATURE. Abraham. Bau u. Entwicklung d. Wandverdickungen i. d. Samen-oberhautzellen einiger Cruciferen. Jarhb. f. wiss. Bot. Bd. XVI. 1885. 599. d'Arbaumont. Nouvelles observations sur les cellules à mucilage d. graines d. Crucifères. Ann. d. Sc. nat. Bot. Sér. 7. T. XI. 1890. 166. Attema. De zaadhuid der Angiosp. en Cymnosp. Diss. Groningen. 1901. 37. Burchard. Ueber den Bau der Samenschalen einiger Brassica- u. Sinapis-Arten. Journ. f. Landw. Jhrg. XLII. 1894. Flückiger. Pharmakogn. 1891. 1029. Flückiger & Hanbury. Pharmacographia. 1879. 68. Gilg. Pharmakogn. 1910. 135. Greenish & Collin. Anat. Atl. o. veget. powders. 1904. 135. Guignard. Recherches s. l. localisation d. Principes actifs d. Crucifères. Journ. de Bot. T. IV. 1890. 385. Guignard. Recherches s. l. développement d. l. graine. Journ. d. Bot. T. VII. 1893. 30. Hager. Pharm. Praxis. Bd. 2. 1902. 907. Hérail. Mat. Méd. 1912. 615. Ingerman. Mikroskopie d. voorn. handelswaren. 1910. 164. Karsten u. Oltmanns. Pharmakogn. 1909. 251. Koch. Mikr. Anal. d. Drogenpulver. Bd. 4. 1908. 51. Koch u. Gilg. Pharmakogn. Praktik. 1907. 221. Kraemer. Botany a. Pharmacogn. 1910. 741. Luerssen. Syst. Bot. Bd. 2. 1882. 625. Marmé. Pharmacogn. 1886. 376. Meyer. Wiss. Drogenk. Bd. I. 1891. 176. Moeller. Mikr. d. Nahr. u. Genussm. 1886. 259. Moeller. Pharmacogn. 1889. 194. Moeller. Mikr. pharm. Ueb. 1900. 153. Molisch. Mikrochem. d. Pfl. 1913. 277. Oudemans. Pharmacogn. 1880. 423. Planchon et Collin. Drogues simples. T. 2. 808. Pecke. Mikrochemischer Nachweis d. Myrosins. Ber. d.d. bot. Ges. Bd. XXXI. 1913. 458—462. Schimper. Mikr. Unters. d. veget. Nahr. u. Genussm. 1900. 111. Spatzier. Ueber das Auftreten u. d. physiologische Bedeutung d. Myrosins in der Pflanze. Pringheim's Jahrb. Bd. 25. 1893. 39. Tschirch. Angew. Pfl-anat. 1889. 199, 204. Tschirch. Pharmakogn. Bd. 2. Abt. 2. 1917. 1495. Tschirch u. Ooestele. Anat. Atl. 1900. 19. Taf. 5. Tunmann. Pflanzenmikrochemie. 433. Vogl. Veget. Nahr. u. Genussm. 1899. 490. Wiesner. Rohstoffe. 1900. 716. Wigand. Pharmakogn. 1879. 303. v. Wisselingh. Bijdr. t.d. Kenn. v. d. Zaadhuid. IV. Over de Zaadhuid der Cruciferen. Pharm. Weekbl. Jrg. 56. 1919. 1245. MATERIAL. The drug. REAGENTS. Water, glycerine, potash, chloral hydrate, potassium iodide iodine, phloroglucin and hydrochloric acid, iodine and sulphuric acid 66 per cent., hyrochloric acid, Millon's reagent.

MICROGRAPHY.

O v u l e. [1]) Campylotropous; with 2 integuments, containing much starch. The outer integument consisting of 3, soon afterwards of 4 layers of common parenchyma cells in consequence of cell division in the middle layer; the inner integument consisting of 6—8 cell layers, becoming flattened in the seed. Before fertilization resorption of the upper $^2/_3$ part of the nucellus taking place. [2])

S e e d.

Seed-coat.

Layers produced by the outer integument.

E p i d e r m i s o u t e r s i d e. In the dry seed compressed into a homogeneous, strongly refringent band of about 20 µ thick; slightly undulating, independent from the underlying cell layers. C e l l s o f s a m e a f t e r s o a k-i n g i n w a t e r. R., T. and L. 50 µ; polygonal prisms. A few cells breaking in consequence of swelling. Walls: [3]) outer walls showing a cuticle, in potash swelling up into conical protuberances; lateral walls thickened, often only the part nearest to the surface of the seed, for the rest thin and pitted. In the mucilaginous mass formed by the epidermal cells during soaking in water and also in the dry seed a column or columella [4]) discernible, extending nearly from the inner nearly to the outer wall, about 10 µ thick, seeming to surround a slight slit-like cavity and consisting of concentric strongly refringent layers; in the soaked seed this column in a tangential section surrounded by a less refringent ring, showing radial striation and thick about 10 µ; in a cross section nothing corresponding with this ring discernible, after deeper focussing many of these radiating lines running towards the lateral walls and looking like pit canals; the rest of the mucilaginous mass showing tangential stratification; towards the lateral walls these layers bending outward. The whole mucilaginous mass colouring blue in iodine and sulphuric acid 66 per cent. Cell contents: some starch grains. In dry seeds the lateral walls in a tangential section not discernible, in a cross section looking like clefts.

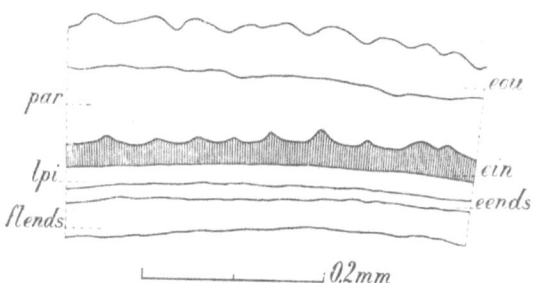

Fig. 107. *Brassica alba*. Seed, transverse section of the seed-coat treated with potash. eends Epidermis of endosperm; ein Epidermis inner side of the layers of the seed-coat, produced by the outer integument; eou Epidermis outer side of the layers of the seed-coat, produced by the outer integument; flends Flattened cell layers of endosperm; lpi Flattened cell layers produced by the inner integument; par Two layers of parenchyma cells of the seed-coat, produced by the outer integument.

G r o u n d t i s s u e. Consisting of 2 layers of parenchyma cells; in the dry seed in a cross section strongly flattened and intercellular spaces, surrounded by thickened walls, only discernible.

C e l l s o f s a m e i n p o t a s h. R. 40 µ, T. 50 µ, L. 55 µ; polygonal tables. Walls collenchymatously thickened. Cell contents: here and there a starch grain.

---

[1]) G u i g n a r d. Recherches s. l. développement d. l. graine etc. 30. T s c h i r c h u. O e s t e r l e. l. c.

[2]) G u i g n a r d. Recherches s. l. développement d. l. graine etc. 28.

[3]) According to d'A r b a u m o n t l. c. only the outer walls strongly thickened, nearly filling the whole cell; strongly swelling up in water.

[4]) See S t r a s b u r g e r. Prakticum. 1897. 564 (Capsella Bursa Pastoris).

E p i d e r m i s i n n e r s i d e. Showing the same structure as the epidermis inner side of Semen Sinapis, but here less distinct.

C e l l s o f t h e a b o v e. R. 20—35 μ, T. and L. 7 μ; polygonal prisms with a radially directed axis. Innermost walls quite thickened. Lateral walls partly thickened, only the portions turned towards the centre of the seed measuring about 10 μ in a radial direction; this thickened part showing in a cross section an undulated outline. The thin-walled outermost parts of the cells often more or less flattened. All walls colourless; the thickened parts lignified, the rest colouring yellow in potash 50 per cent., brown in iodine and sulphuric acid 66 per cent.

Layers produced by the inner integument. All cells flattened into a very thin colourless band.

Layers produced by the endosperm. See Semen Sinapis, but here the band of flattened cells somewhat thicker and here and there a starch grain. In potash 50 per cent. the flattened band colouring light red.

E p i d e r m a l c e l l s. R. 20 μ, T. and L. 25 μ; see for the rest Semen Sinapis.

Nucleus of seed.

E m b r y o. See the embryo of Semen Sinapis nigrae. Cell contents: in potash 50 per cent. first colouring yellow, afterwards orange to red.

Februari 1903.   J; M; v. E. d. W.

SEMEN SINAPIS.

Sinapis nigrae semina. Black Mustard Seed. Black Mustard.

The ripe seeds of Brassica nigra, Koch, in Roehl. Deutschl. Fl. ed. 3. IV. 713.

Macroscopic characters.

Long 1—1.5 m.M.; oblong or about spherical; hard. Hilum represented by a circular white spot at one end; the place of the radicle often indicated by the folds of the seed-coat. Surface with a fine network of ridges enclosing penta- or hexagonal meshes, thence dotted, dark or paler red-brown, here and there with white scales. Albumen wanting. Embryo greenish yellow, curved; the 2 obcordate cotyledons folded both towards the same side against the radicle, moreover folded along the midrib both in the same direction, thus enclosing the radicle in the furrow formed by them.

Inodorous in a dry state, having the odour of mustard oil when powdered and moistened; taste when chewed at first mellowly oily, soon afterwards very pungent.

Anatomical characters.

LITERATURE. Abraham. Bau u. Entwicklung d. Wandverdickungen. i. d. Samen-oberhautzellen einiger Cruciferen. Jahrb. wiss. Bot. Bd. XVI. 1885. 599. d'Arbaumont. Nouvelles observations s. l. cellules à mucilage d. graines de Crucifères. Ann. d. Sc. nat. Bot. Sér. 7. t. XI. 1890. 125. Attema. De zaadhuid der Angiosp. en Gymnosp. Diss. Groningen. 1901. 26. Berg. Anat. Atl. 1865. 91. Tafel 46. Benecke. Mikr. Drogenprakt. 1912. 71. Burchard. Ueber den Bau der Samenschalen einiger Brassica- u. Sinapis-Arten. Journ. f. Landw. Jhrg. XLII. 1894. Flückiger. Pharmkogn. 1891. 1024. Flückiger & Hanbury. Pharmacographia. 1879. 64. Gilg. Pharmakogn. 1910. 132. Greenish & Collin. Anat. Atl. o. veget. powders. 1904. 132. Guignard. Recherches s. l. localisation

d. Principes actifs d. Crucifères. Journ. d. Bot. T. IV. 1890. 385. Guignard. Recherches s. l. développement d. l. graine. Journ. d. Bot. T. VII. 1893. 27. Hager. Pharm. Praxis. Bd. 2. 1902. 903. Hérail. Mat. Méd. 1912. 612. Ingerman. Mikroskopie d. voorn. handelsw. 1910. 164. Karsten u. Oltmanns. Pharmakogn. 1909. 249. Koch. Mikr. Anal. d. Drogenpulver. Bd. 4. 1908. 45. Tafel V. Koch u. Gilg. Pharmakogn. Praktik. 1907. 215. Kraemer. Botany a. Pharmacogn. 1910. 743. Luerssen. Syst. Bot. Bd. 2. 1882. 623. Marmé. Pharmacogn. 1886. 371. Meyer. Wiss. Drogenk. Bd. I. 1891. 146. Meyer. Mikr. Unters. v. Pflanzenpulv. 1901. 71. Tafel 1 u. 2. Moeller. Mikr. d. Nahr. u. Genussm. 1886. 259. Moeller. Pharmakogn. 1889. 194. Moeller. Mikr. pharm. Ueb. 1900. 150. Molisch. Mikrochem. d. Pfl. 1913. 288. Oudemans. Aanteek. o. d. Pharmac. neerl. 1854—56. 429. Oudemans. Pharmacogn. 1880. 421. Pecke. Mikrochemischer Nachweis d. Myrosins. Ber. d.d. bot. Ges. Bd. XXXI. 1913. 458—462. Planchon et Collin. Drogues simples. T. 2. 803. Schimper. Mikr. Unters. d. veget. Nahr. u. Genussm. 1900. 111. Spatzier. Ueber das Auftreten u. d. physiologische Bedeutung des Myrosins in der Pflanze. Pringsheim's Jahrb. Bd. 25. 1893. 39. Tschirch. Angew. Pfl.-anat. 1889. 71, 204, 327, 449. Tschirch. Pharmacogn. Bd. 2. Abt. 2. 1917. 1483. Tschirch u. Oesterle. Anat. Atl. 1900. 17. Tunmann. Pflanzenmikrochemie. 433. Vogl. Veget. Nahr. u. Genussm. 1899. 490. Wiesner. Rohstoffe. 1900. 716. Wigand. Pharmakogn. 1879. 302. v. Wisselingh. Bijdr. t. d. Kenn. v. d. Zaadhuid. IV. Over de zaadhuid der Cruciferen. Pharm. Week-

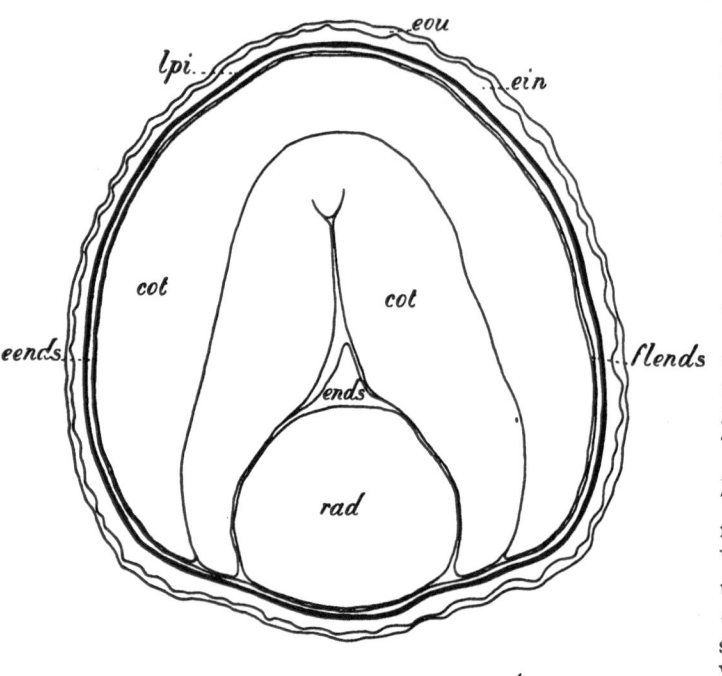

Fig. 108. *Brassica nigra*. Seed, transverse section. cot Cotyledonen; eends Epidermis of endosperm; ein Epidermis inner side of the layers of the seedcoat, produced by the outer integument; ends Endosperm; eou Epidermis outer side of the layers of the seed-coat, produced by the outer integument; flends Flattened cell layers of endosperm; lpi Cell layers produced by the inner integument; rad Radicle.

bl. Jrg. 56. 1919. 245. MATERIAL. The drug and unripe seeds. REAGENTS. Water, glycerine, potash, chloral hydrate, iodine in chloral hydrate, phloroglucin and hydrochloric acid, iodine and sulphuric acid 66 per cent., concentrated sulphuric acid, iron chloride, osmic acid, hydrochloric acid, Millon's reagent, oil of cloves, origanum oil.

MICROGRAPHY.

O v u l e.[1]) Campylotropous; with 2 integuments, containing much starch.

---

[1]) G u i g n a r d. Recherches s. l. développement d. l. graine etc. 27. T s c h i r c h u. O e s t e r l e, l. c.

The outer integument consisting of 3 layers of common parenchyma cells; the inner integument consisting of 6—8 layers of cells, showing somewhat larger cells in the outermost layer. During the development of the ovule into a seed the cell layers of the inner integument with the exception of the innermost one becoming flattened. [1]). Before fertilization resorption of the upper $^2/_3$ part of the nucellus taking place. [2])

Seed.

Seed-coat.

Layers produced by the outer integument.

E p i d e r m i s   o u t e r   s i d e. Often off in the form of scales; in the dry seed forming a continuous layer of about 8 μ thick, showing no discernible cell structure; quite following the surface formed by the inner epidermis of this integument.

C e l l s   o f   s a m e   a f t e r   s o a k i n g   i n   w a t e r. R. 20 μ, T. 40 μ, L. 50 μ; polygonal tables. Outer walls [3]) with a cuticle; slightly swelling up in water and then filling the whole cell; in alcoholic material also nearly quite filling the cell with the exception of a slit-like cavity adjoining the inner wall; in potassium iodide iodine the thickened walls not or nearly not colouring blue, in iodine and sulphuric acid 66 per cent. colouring blue; stratification not easily discernible. Cell contents in the unripe seed: starch grains, mostly 2- to 3-adelphous.

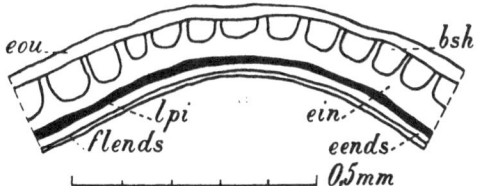

Fig. 109. *Brassica nigra*. Seed, transverse section of the seed-coat treated with potash. bsh Bowl-shaped cells of the layers of the seed-coat, produced by the outer integument; eends Epidermis of endosperm; ein Epidermis inner side of the layers of the seed-coat, produced by the outer integument; eou Epidermis outer side of the layers of the seed-coat, produced by the outer integument; flends Flattened cell layers of endosperm; lpi Cell layers produced by the inner integument.

G r o u n d   t i s s u e. Consisting of a single layer of common parenchyma cells, quite following the surface formed by the inner epidermis of the outer integument; in the dry seed the outer walls of the cells closely adjacent to the inner walls; after soaking in water cells bowl-shaped with their concave side turned outwards, often not adjoining one another and in these places cells of the third layer of the integument adjoining the epidermal cells. C e l l s   o f   t h e   a b o v e. R. 35 μ, T. and L. 60—100 μ; polygonal. Cell contents in the unripe seed: starch grains like those in the epidermal cells.

E p i d e r m i s   i n n e r   s i d e. Consisting of a single layer of partly sclerotic cells; much varying in radial dimension and thus forming on the outside of the layer bowl-shaped cavities, causing the network of ridges at the surface of the seed mentioned among the macroscopic characters; each bowl-shaped cavity containing a single cell of the ground tissue.

[1]) According to G u i g n a r d, the innermost cell layer also later on becoming flattened.
[2]) G u i g n a r d. Recherches s.l. développement d.l. graine etc. 28.
[3]) According to T s c h i r c h   u n d   O e s t e r l e l. c. outer and lateral walls strongly thickened, outer walls pitted; according to F l ü c k i g e r and M o e l l e r ll. cc. inner and lateral walls strongly thickened. I saw no pits in the outer walls though I payed sufficient attention to it.

C e l l s o f t h e a b o v e. R. 18—50 μ, T. 6 μ, L. 8 μ; polygonal prisms with a radially directed axis. Inner walls quite thickened. Lateral walls partly thickened, only the portions turned towards the centre of the seed and measuring about 10 μ in a radial direction, independent from the various radial dimensions of the whole cells; the thickness of this portion increasing towards the periphery of the seed. The thin-walled outermost parts of the cells often more or less flattened. All walls brown; not colouring red with phloroglucin and hydrochloric acid, though somewhat changing in colour; persisting in concentrated sulphuric acid; middle lamella not discernible. Cell contents: wanting in the ripe seed; very small starch grains filling the cells still without thickenings of the walls in the unripe seed.

Layers produced by the inner integument. [1]) About 6 in number; common parenchyma cells; quite flattened; forming together a continuous brown band, thick about 6 μ; not changing its colour in potash 50 per cent. or iron chloride. Cells in the vicinity of the hilum less flattened; rectangular; showing colourless walls and brown contents, not soluble in potash 50 per cent.

Layers produced by the endosperm. Forming a part of the seed-coat in the shape of a continuous band, adjoining the layers produced by the inner integument; consisting of the epidermis and some layers of quite flattened cells. Epidermis not clearly discernible in the vicinity of hilum, micropyle and chalaza. Layers of flattened cells, some of the innermost of these layers enclosing the radicle and filling up the cuneiform cavity between the radicle and the innermost cotyledon.

E p i d e r m a l c e l l s. H. 15 μ, T. 20 μ, L. 25 μ; polygonal tables. Walls showing cellulose reaction. Outer walls thickened. Cell contents: abundant oil, like the endosperm of Semen Strychni; also protein. C e l l s o f f l a t t e n e d c e l l l a y e r s. In potash 50 per cent. ultimately swelling up, then showing rectangular shape in a cross section. Walls showing cellulose reaction.

Nucleus of seed.

E m b r y o. Quite consisting of meristematic cells.

Radicle.

E p i d e r m a l c e l l s. Hexagonal prisms. Outer walls somewhat thickened; showing a cuticle. Cell contents: see those of the cortex.

C o r t e x. Consisting of about 5 layers of parenchyma cells; those of the innermost layers radially arranged. Here and there myrosine idioblasts showing in glycerine homogeneous yellow contents.

P a r e n c h y m a c e l l s. Polygonal tables with a longitudinally directed axis and rounded edges; showing intercellular spaces. Cell contents: greenish yellow oil like that of the endosperm cells; numerous aleurone grains, long up to 25 μ, much varying in form and size, with numerous small globoids; in potash 50 per cent. colouring yellow. M y r o s i n e c e l l s. Sooner and more distinctly colouring red in Millon's reagent than the other cells, but in hydrochloric acid not showing the purple myrosine reaction.

S t e l e. Conspicuously diarch; phloem and xylem bundles clearly discernible.

---

[1]) By most authors this layer is called a pigment layer and thought to consist of cells with somewhat thickened walls and brown contents, soluble in potash and coloured blue in iron chloride. P l a n c h o n et C o l l i n and G u i g n a r d however obtained the same result as mentioned in the text.

Cotyledons.

E p i d e r m a l c e l l s. H. 10 μ, Lev. 15 μ; polygonal tables. Outer walls a little thickened; showing a cuticle. Cell contents: see those of cortex.

M e s o p h y l l. At the upper side consisting of 3 layers of palisade cells; the rest, about 4 layers, also of palisade-shaped cells; cells arranged in longitudinal rows. Cell contents: see those of cortex.

M e r i s t e l e s. Present in the form of procambium bundles; oval in a cross section.

Micrography of the powder.

Many drops of oil (in chloral hydrate); for the rest the powder quite composed of 2 ingredients: colourless parts of the embryo and mostly red-brown, larger and smaller pieces of the seed-coat. Parts of the embryo quite consisting of angular common parenchyma cells, often various tissues especially palisade tissue and epidermis to be distinguished; all cells filled with a large number of aleurone grains; these long up to 25 μ, irregularly oblong, finely granular by the presence of many small globoids (in oil of cloves). In potash the tissue of the embryo colouring bright light yellow. The pieces of the seed-coat mostly showing fairly completely all component layers, viewed from above, from beneath or in cross section. Beginning at the outside the following parts to be distinguished: 1, epidermis; cells flat, angular, with strongly thickened walls, easily becoming mucilaginous, colourless; 2, one layer of very large cells, their outlines corresponding with the network on the seed-coat, thin-walled, colourless; 3, one layer of palisade cells, strongly elongated in a radial direction, forming towards the outside, in consequence of their gradual variation in length, bowl-shaped cavities or furrows, each containing a cell of the second layer; inner walls and the lower part of the lateral walls, especially upwards, strongly thickened, red-brown; upper part of the lateral walls and the outer walls thin; on this cell layer the boundaries of the bowl-shaped cavities indicated by a dark brown network; 4, some layers of flattened dark brown cells, mostly not conspicuous in the powder; 5, one layer of cells, containing much oil, polygonal, rather deep, with somewhat thickened walls; this layer often present in a detached state in the powder; 6, layers of flattened endosperm cells, polygonal, thin-walled, colourless, also present between the parts of the embryo and often attached to the latter. Many starch grains 1—2 μ in diameter, often together in not very large numbers in the parenchyma cells of the embryo.

Januari 1902. J; M.

SEMEN STRAMONII.
Stramonium Seeds.
The ripe seeds of Datura Stramonium, Linn. Sp. Pl. 179.

LITERATURE. Attema. De zaadhuid d. Angiosp. en Gymnosp. Diss. Groningen. 1901. Barth. Pflanzenmicrochem. Diss. Zürich. 1898. 25; also Bot. Centrlbl. Bd. 75. 1898. No. 225. Berg. Anat. Atl. 1865. Pl. 47. Clautriau. Localis. e. signif. d. Alcaloides d. quelques graines. Ann. d. l. Soc. belge d. Micr. T. XVIII. 1894. 35. Chatin. Dével. d. l'Ovule et d. l. Graine d. l. Scrophularinées, Solanées, Boraginées e. Labiées. Ann. d. Sc. nat. Sér. V. T. XIX. 1894. Feldhaus. Quant. Unters. d. Vert. d. Alkaloides i. d. Org. v. D. Stramonium. Diss. Marburg. 1903. Flückiger. Pharmakogn. 1891. 1013. Flückiger & Hanbury. Pharmacogr. 1879. 461. Hager. Pharm. Praxis. Bd. I. 1900. 1015. Hartwich. Samenschale d. Solaneen. Viertelj. schr. Naturf. Gesellsch. Zürich. XLI. 1896. Marmé. Pharmakogn. 1886. 425. Meyer. Mikr. Unters. v. Pflanzenpulv. 1901. 38, 69. Mitlacher. Toxik. od. Forens. wicht. Pfl. u. Drogen. 1904. 156. Molisch. Mikrochemie d. Pfl. 1913. 257. Molle. Rech. d. Microchémie comp. s.l. localis. d. Alcaloides d. l. Solanées. Mém. d. l'Acad. d. Sc. d. Belgique. T. LIII. 1896. Oudemans. Aant. o. d. Pharm. neerl. 1854—56. 325.

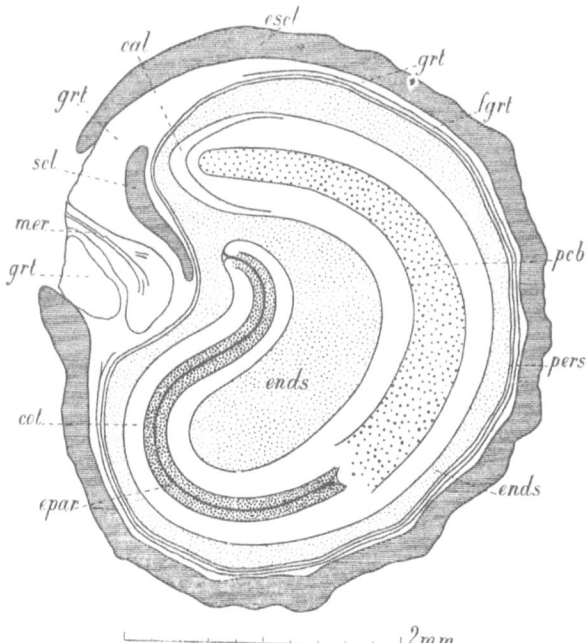

Fig. 110. *Datura Stramonium*. Seed, longitudinal section. cal Calyptra; cot Cotyledons; ends Endosperm; epar Epidermis and layer of palisade parenchyma of the cotyledons; escl Epidermal sclerenchyma cells; fgrt Flattened innermost layer of ground tissue; grt Ground tissue; mer Meristele of the chalaza; pcb Procambium bundle; pers Perisperm; scl Sclerenchyma.

Oudemans. Pharmacogn. 1880. 407. Planchon et Collin. Drogues simples. T. I. 1896. 590. Schneider. Powd. veget. Drugs. 1902. 302. Schlotterbeck. Beitr. z. Entw.-gesch. pharmakogn. wicht. Samen. Diss. Bern. 1896. Tschirch. Angew. Pfl.-anat. 1889. 44. Tschirch u. Oesterle. Anat. Atl. 1900. 285. Pl. 65. Tunmann. Pflanzenmikrochemie. 1913. 323. Wigand. Pharmakogn. 1879. 311. MATERIAL. The drug. REAGENTS. Water, glycerine, potash, chloral hydrate, potassium iodide iodine, phloroglucin and hydrochloric acid, iodine and sulphuric acid 66 per cent., concentrated sulphuric acid, oil of cloves, origanum oil, osmic acid.

MICROGRAPHY.

Ovule.

Campylotropous. 1 integument consisting of 9 layers of cells; this number afterwards strongly increasing, especially by cell division in the innermost cell layers.

Seed.

Seed-coat.

Seed-coat proper.

Layers produced by the integument.

Epidermis outer side. Consisting of 1 layer of sclerenchyma cells with strongly sinuous lateral walls and a cavity, sand-glass-shaped in a radial section, often showing a re-entering part of its innermost half; the outer cell walls showing on their outside numerous warts, in a section cut dry and afterwards treated with water imbedded in a mucilage layer thick 25 μ, covering these parts so as to form a smooth surface; hence the lines of demarcation of the cells generally undiscernible in a tangential section. Radial dimensions of the cells varying; hence the surface of the seed presenting a network of projecting ridges, with the exception of some places in the vicinity of the hilum(Schwiele").

C e l l s o f t h e a b o v e. R. 80—180 μ, T. 75 μ, L. 110 μ, in the smooth places without a projecting network R. 250 μ. Walls strongly thickened, lateral and inner walls most thickened in their middle part; yellow to brown; lignified; outer walls with a cuticle; the mucilage layer in a dry transverse section hardly discernible; this layer in iodine not colouring yellow, in iodine and sulphuric acid 66 per cent colouring slightly blue; all the walls showing stratification; in sections lateral walls showing a sinuous line along the middle lamella. Cell contents: mostly a dark brown mass.

G r o u n d t i s s u e. Common parenchyma consisting of 3—4 layers of cells strictly arranged in tangential planes; with intercellular spaces; on the inside a layer of strongly flattened yellow cells; the innermost layer of the integument being often resorbed, according to Tschirch und Oesterle. In the vicinity of hilum, chalaza and micropyle the ground tissue thicker and inner cell layers not flattened. In the upper part of the seed, extending towards the micropyle, a crest-shaped mass of sclereids, without intercellular spaces; at its base high and wide 4—6 cells and quite imbedded in the common parenchyma of the ground tissue; this parenchyma forming some layers on the inner and the outer side of the crest; parenchyma cells surrounding the crest somewhat larger than the others, especially in a radial dimension.

E p i d e r m i s i n n e r s i d e. Flattened or resorbed.

P a r e n c h y m a c e l l s. R. 16 μ, T. 30 μ, L. 40 μ; polygonal tables with a radially directed axis and strongly rounded edges. Walls yellow to brown; lignified; walls of the flattened cells also lignified. Cell contents: in potassium iodide iodine a dark brown granular precipitate, not crystaline (alcaloid [1])). Yellow colour produced by alcali not clearly discernible, perhaps owing to the yellow colour of the cell walls. S c l e r e i d s. R. 35 μ, T. 40 μ, L. 45 μ; polyhedra. Walls strongly thickened; yellow to brown; lignified; showing wide pit canals. Contents of some cells a dark granular mass.

C h a l a z a p a r t of the seed-coat.

In the ground tissue a small meristele with spiral vessels.

C e l l s o f t h e c o m m o n p a r e n c h y m a o f t h e g r o u n d t i s s u e e. g. R. 25 μ, T. and L. 30 μ. Some of the outermost cells filled with crystal sand.

Tissue produced by the nucellus (perisperm). Consisting of 1 layer of cells.

C e l l s o f t h e a b o v e. R. 9 μ, T. 20 μ, L. 25 μ; polygonal tables with a radially directed axis. Radial walls a little thickened; all the walls yellow; lignified. Cell contents: see the parenchyma cells of the ground tissue mentioned above.

Nucleus of the seed. Much more distinctly reniform than the seed itself, owing to the greater thickness of the seed-coat in the vicinity of hilum, chalaza and micropyle.

A l b u m e n. Only consisting of endosperm.

Endosperm.

---

[1]) See C l a u t r i a u, M o l l e, B a r t h, M o l i s c h and T u n m a n n, ll. cc.

E p i d e r m i s. Consisting of 1 cell layer.

C e l l s o f t h e a b o v e. R. 35 μ, T. 30 μ, L. 35 μ; polygonal prisms. Outer walls strongly thickened; outermost half of these walls and the middle lamella of the lateral walls cuticularized, colouring yellow to reddish brown in iodine and sulphuric acid 66 per cent., persisting in concentrated sulphuric acid; in phloroglucin and hydrochloric acid a red layer separating the cuticularized part of the outer walls from their inner layers, moreover a red layer bordering the cavity. Cell contents: see ground tissue of endosperm; aleurone grains somewhat smaller.

G r o u n d t i s s u e. Consisting of common parenchyma; round the embryo some layers of tangentially elongated cells. Intercellular spaces wanting.

C e l l s o f t h e a b o v e. R. 35 μ, T. and L. 40 μ; polyhedra. Walls a little thickened; with feeble cellulose reaction. Cell contents: aleurone grains, diameter 5 μ, globular or ellipsoidal, containing 1 or 2 protein crystals and some globoids not always very distinct; protoplasmic stroma to be distinguished in water, oil diffused throughout the protoplasm, perhaps also in the aleurone grains; all the cell contents colouring homogeneously black in osmic acid, this reaction somewhat feebler in the vicinity of the embryo. In some cells near the micropyle a dark granular mass.

E m b r y o. Strongly curved and following the outer surface of the endosperm at a distance of about 150 μ; the radicle, thick 550 μ, nearly meeting the micropyle; surface of contact of the cotyledons radially directed, their tops bending inwards and afterwards outwards again, ending very near the top of the radicle.

R a d i c l e. Consisting of meristematic tissue. Calyptra, epidermis and central procambium bundle already clearly differentiated. Cell contents: aleurone grains and oil like those of the endosperm.

C o t y l e d o n s. Consisting of meristematic tissue. Showing an epidermis, a layer of palisade parenchyma and some procambium bundles already clearly differentiated. Cell contents: aleurone grains and oil like those of the endosperm.

February 1903.   J; M; L.

SEMEN STRYCHNI.

Nux Vomica.

The ripe seeds of Strychnos Nux-vomica, Linn. Sp. Pl. 189.

Macroscopic characters.

Disc-shaped, about circular, often shallow bowl-shaped or otherwise curved, up to 2.8 c.M. in diameter, thick up to 6.5 m.M.; a slightly elevated, keel-shaped ridge along the rounded margin. Seed-coat thin. Hilum in the middle of one of the flat sides, mostly in the shape of a slight circular elevation; in the middle of the other flat side a slight circular depression or just as on the other side a slight elevation. Micropyle situated on a slight nipple-shaped elevation at the margin; in many seeds a somewhat protruding narrow ridge, connecting hilum and micropyle, not clearly discernible in soaked seeds. Surface grey-brown, sometimes slightly greenish, quite covered with fine thickset adpressed hairs, turning their tops towards the periphery; these hairs causing a somewhat silky luster and a certain fattiness to the touch. Endosperm having the shape of the seed; internally with a circular flat cavity, reaching till 1 or 2 m.M. from the margin; hard, horny, white, somewhat transparent. Embryo long up to 7.5 m.M. Radicle reaching, through the endosperm, as far as the micropyle, cylindrical. Cotyledons lying in the cavity of the endosperm, cordate, palminerved, with 5 to 7 veins.

Odour wanting; taste very bitter.

Anatomical characters.

LITERATURE. Attema. De zaadhuid der Angiosp. en Gymnosp. Diss. Groningen. 1901. 126. Barth. Studien ü. d. microchem. Nachweis v. Alkaloiden in pharm. verwendeten Drogen. Bot. Centrblt. Bd. 75. 1898. 225, 374. Berg. Anat. Atl. 1865. 93. Tafel XXXXVII. Benecke. Mikr. Drogenprakt. 1902. 75. Clautriau. Localisation et signification des alcaloides dans quelques graines. Ann. Soc. Belge de Microsc. T. XVIII. 1894. 35. Errera. Sur la distinction microchimique des alcaloides et des matières protéiques. Ann. d. l. Soc. Belge de Microsc. (Mémoires). T. XIII. 1889. 73. Flückiger. Pharmakogn. 1891. 1015. Flückiger & Hanbury. Pharmacographia. 1879. 428. Gerock u. Skippari. Ueber den Sitz der Alkaloide im Strychnossamen. Arch. d. Pharm. Bd. CCXXX. 1892. 555. Gilg. Pharmakogn. 1910. 270. Greenish & Collin. Anat. Atl. o. veget. powders. 1904. 138. Hager. Pharm. Praxis. Bd. 2. 1902. 982. Hartwich. Strychnos-Drogen . Festschr. 50 jähr. Stift. des Schweiz. Apoth.-Ver. in Zürich. 1893. Hérail. Mat. Méd. 1912. 687. Karsten u. Oltmanns. Pharmakogn. 1909. 289. Klein. Ueber den mikrochem. Nachweis von Strychnin u. Brucin im Samen von Strychnos Nux-vomica. Anz. ksl. Akad. Wiss. Wien. math.-nat. Kl. III. 1914. 39—40. Koch u. Gilg. Pharmakogn. Praktik. 1907. 224. Koch. Mikr. Anal. d. Drogenpulver. Bd. 4. 1908. 53. Tafel VI. Kraemer. Botany a. Pharmacogn. 1910. 416. Lindt. Ueber den mikrochem. Nachweis von Brucin u. Strychnin. Zeitschr. f. Wissensch. Mikrosk. I. 1884. 237. Lotsy. Ueber die Auffindung eines neuen Alkaloids in Strychnos-Arten auf mikrochem. Wege. Rec. d. Travaux Bot. Néerlandais. Vol. II. 1905. 1. Leurssen. Syst. Bot. Bd. 2. 1882. 1058. Marmé. Pharmacogn. 1886. 429. Meyer. Wiss. Drogenk. Bd. I. 1891. 152. Meyer. Mikr. Unters. v. Pflanzenpulv. 1901. Mitlacher. Toxik. od. Forens. wicht. Pfl. u. Drogen. 1904. 131. Moeller. Pharmakogn. 1889. 174. Moeller. Mikr. pharm. Ueb. 1901. 140. Molisch. Mikroch. d. Pfl. 1913. 267. Oudemans. Aanteek. o. d. Pharmac. neerl. 1854—56. 267. Oudemans. Pharmacogn. 1880. 408. Planchon et Collin. Drogues simples. T. 1. 1895. 656. Rosoll. Beitr. z. Histochemie d. Pflanzen. Ber. d. Wiener Akad. d. Wissenschaften. 1883. I. 137. Sauvan. Unters. ü. d. Lokalisation von Brucin u. Strychnin i. d. Samen von Strychnos Nux-vomica. u. a. Journ. de Bot. X. 1896. 133. Schneider. Powdered Veget. Drugs. 1902. 249. Tschirch. Angew. Anat. 1889. 130, 259. Tschirch. Indische Fragmente. Archiv. d. Pharmacie. 1890. 1. Tschirch u. Oesterle. Anat. Atl. 1900. 149. Tafel 35. Tunmann. Planzenmikrochemie. 1913. 320. Tunmann. Ueber die Alkaloide in Strychnos Nux-vomica L. während der Keimung. Archiv. d. Pharm. CCXLVIII. 644—657. Verschaffelt. De imbibitie van Strychnoszaad. Pharm. Weekblad. 1913. No. 24. Vogl. Anat. Atl. 1887. Tafel 32. Wasicky. Der mikrosk. Nachweis von Strychnin u. Brucin im Samen von Strychnos Nux-vomica. L. Zeitsch. allg. oesterr. Apoth. Ver. LII. 1914. 35, 41—42, 53—55, 67—69. Wigand. Pharmakogn. 1879. 309. MATERIAL. The drug. REAGENTS. Water, glycerine, potash, iodine in chloral hydrate, potassium iodide iodine, phoroglucin and hydrochloric acid, iodine and sulphuric acid 66 per cent., concentrated sulphuric acid, Schulze's macerating mixture, iron acetate, copper sulphate and potash 50 per cent., osmic acid, nitric acid, Millon's reagent, tartaric acid 50 per cent in absolute alcohol, alcohol 96 per cent. with 0.2 per cent hydrochloric acid of 25 per cent [1]).

[1]) See Errera l. c.

MICROGRAPHY.
O v u l e.[1]
S e e d.
Seed-coat.
S e e d - c o a t p r o p e r.
Layers produced by the integument.

E p i d e r m i s o u t e r s i d e. Composed of thick-walled hairs showing a short prismatic base and a long slightly conical upper part, directed at a right angle with the base and radiating towards the margin of the seed. The protruding narrow ridge, connecting hilum and micropyle as mentioned under macroscopic characters, caused exclusively by the quite irregular direction of the conical upper parts of the hairs in this place. On the keel-shaped ridge along the margin of the seed these hairs not knee-shaped but straight and radiating from the centre of the seed.

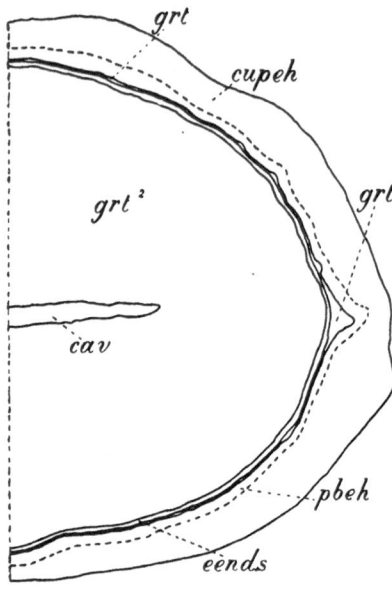

Fig. 111. *Strychnos Nux-vomica.* Seed, transverse section. cav Cavity; cupeh Conical upper part of epidermal hairs; eends Epidermis of endosperm; grt Ground tissue of seed-coat; grt′ Ground tissue of keel-shaped ridges along the margin of the seed; grt2 Ground tissue of endosperm; pheb Prismatic base of epidermal hairs.

E p i d e r m a l h a i r s. Prismatic basal parts R. 55 μ, T. 30 μ, L. 35 μ; polygonal prisms with strongly sinuous lateral walls. Walls strongly thickened; often coloured brown; lignified; showing stratification, a clearly discernible middle lamella and a few pit canals, slit-like and sometimes branched. Cell cavity slit-like, often branched. Conical upper parts thick at their base 25 μ, long up to 800 μ; near the micropyle twisted and long up to 1000 μ. Walls showing about 10 considerable longitudinal ridge-like thickenings; these ridges branching and anastomosing even up to the tops of the hairs, lignified, with very thin parts of the walls between them and thence easily separable from one another.

G r o u n d t i s s u e. Only the outermost cell layers represented by a brown band of flattened common parenchyma cells; the rest resorbed. The keel-shaped ridge along the margin of the seed formed by mostly about 8 less flattened layers of common parenchyma cells.

C e l l s o f t h e a b o v e. R. 25 μ, T. 30—70 μ, L. 40 μ; polyhedra. Walls somewhat brown.

E p i d e r m i s i n n e r s i d e. Resorbed.

H i l u m p a r t. Conspicuous by many ruptured spiral vessels, extending on

---

[1] According to T s c h i r c h u. O e s t e r l e l. c. showing a single integument, thick some layers of cells; the inner cell layers of this integument are resorbed and have disappeared in the ripe seed. See also M e y e r. Wiss. Drogenk. l. c.

all sides a short way into the ground tissue but not reaching as far as the chalaza.

Nucleus of seed.

E n d o s p e r m.

E p i d e r m i s.

C e l l s. R. 25—35 μ, T. 10 μ, L. 12 μ; polygonal prisms with a radially directed axis. Outer walls and outer parts of the lateral walls thickened, cuticularized, colouring yellow in iodine and sulphuric acid, persisting in concentrated sulphuric acid. Cell contents: see ground tissue.

G r o u n d   t i s s u e. Consisting of very thick-walled parenchyma cells.

C e l l s o f t h e a b o v e. Of the outermost cell layers R. 30 μ, T. 20 μ, L. 25 μ; of the rest R. 40—60 μ, T. 30 μ, L. 35 μ; polyhedra or polygonal prisms with a radially directed axis. Walls, especially those of the more central cells, strongly thickened, swelling up in water and then showing stratification; middle lamella and tertiary layer always clearly discernible, with iodine and sulphuric acid 66 per cent. both showing a deeper blue cellulose reaction than the rest of the walls [1]); with numerous fine perforations, containing connecting threads to be mentioned below. Cell cavity sometimes becoming slit-like after swelling up of the endosperm in water. Cell contents: protoplasts united together with those of adjacent cells by means of connecting threads running through the fine perforations of the walls, mentioned above; these connecting threads especially visible in sections soaked during a long time in alcohol, either with or without a subsequent treatment with potassium iodide iodine. For the rest: 1, oil-plasm; 2, numerous aleurone grains, 15—20 μ in diameter, the larger ones in the central cells less numerous; 3, globoids nearly wanting, only a very few to be found [2]); 4, granular masses, especially in the central cells; 5, alkaloids: in potassium iodide iodine a red-brown praecipitate, especially in the protoplasm, in the connecting threads in the form of rows of red-brown granules, in concentrated nitric acid coloured red to orange. [3]).

E m b r y o. The majority of its elements in a more or less meristematic stage. Cell contents: like those of the parenchyma cells of the endosperm; protein reaction nearly wanting [4])

Micrography of the powder.

Parenchyma cells of the endosperm, angular, sometimes oblong, with strongly thickened colourless walls, showing sometimes a clearly discernible stratification and also connecting threads; cell cavity filled with a granular mass. Thickened ridges belonging to the cell walls of the unicellular hairs separated from one another in powdering the drug; these ridges in the shape of much elongated rods, often curved and knee-shaped, pale yellow, forming the bulk of powders proceeding from unpeeled seeds, often present in relatively very small quantities in powders proceeding from seeds after removal of the seed-coat.

---

[1]) According to T s c h i r c h u. O e s t e r l e l. c. the cell walls of some parts of the endosperm colouring yellow in iodine and sulphuric acid 66 per cent.; this reaction indicating the occurrence of true mucilage.

[2]) T s c h i r c h u. O e s t e r l e l. c. p. 152, M e y e r. Wiss. Drogenk. p. 156 and nearly all other authors mention globoids.

[3]) According to B a r t h l. c., whoms direction I followed, all cells of the endosperm contain strychnine as well as brucine.

[4]) According to B a r t h l. c. brucine wanting in embryo.

A few basal parts of the hairs, thick-walled and not split up. Small oil drops (in chloral hydrate). Starch wanting. Powder colouring orange after addition of nitric acid.

January 1903.   J; M; v. E. d. W.

STIGMATA CROCI.

Crocus. Saffron.

The dried stigmata of Crocus sativus, Linn. Sp. Pl. 36, in a few cases still united together 3 in number by the tops of the styles.

Macroscopic characters.

The separated stigmata long 2 to 4 c.M., wide at the apex up to 4 m.M.; much elongatedly wedge-shaped, rolled up and somewhat gutter-shaped, thus at the apex even funnel-shaped, fibrous, flattened, somewhat curved, at the upper margin irregularly fissured and provided with stigmatic papillae, dark orange-brown, somewhat polished and somewhat fatty to the touch. Stigmata soaked for a short time in water and then superficially dried by filter paper, colouring at once dark blue in concentrated sulphuric acid; this colour communicating itself to the fluid and passing after about half a minute into red-brown.

Odour strongly aromatic; taste bitter and aromatic.

Anatomical characters.

LITERATURE. Berg. Anat. Atl. 1865. 81. Benecke. Mikr. Drogenprakt. 1912. 69. Bie-chele. Mikr. Prüf. d. off. Drogen. 1904. 32. Erdmann-König. Allg. Warenk. 1895. 304. Flückiger. Pharmakogn. 1891. 773. Flückiger & Hanbury. Pharmacographia. 1879. 663. Gilg. Pharmakogn. 1910. 52. Greenish & Collin. Anat. Atl. veget. powders. 1904. 106. Hager. Pharm. Praxis. Bd. 1. 1900. 965. Hérail. Mat. Méd. 1912. 561. Ingerman. Mikr. d. voorn. handelsw. 1910. 171. Karsten u. Oltmanns. Pharmakogn. 1909. 197. Koch. Mikr. Anal. d. Drogenpulver. Bd. 3. 1906. 245. Koch u. Gilg. Pharmakogn. Praktik. 1907. 178. Luerssen. Syst. Bot. Bd. 2. 1882. 441. Molisch. Mikrochem. d. Pfl. 1913. 244. Marmé. Pharmacogn. 1886. 243. Meyer. Wiss. Drogenk. Bd. 2. 1892. 344. Meyer. Mikr. Unters. v. Pflanzenpulv. 1901. 222. Mitlacher. Toxik. od. Forens. wicht. Pfl. u. Drogen. 1904. 46. Moeller. Mikr. d. Nahr. u. Genussm. 1886. 58. Moeller. Pharmakogn. 1889. 110. Moeller. Mikr. pharm. Ueb. 1900. 121. Oudemans. Aanteek. o. d. Pharmac. neerl. 1854—56. 55. Oudemans. Pharmacogn. 1880. 316. Planchon et Collin. Drogues simples. T. 1. 1895. 206. Schimper. Mikr. Unters. d. veget. Nahr. u. Genussm. 1900. 115. Tschirch. Pharmacogn. Bd. 2. Abt. 2. 1917. 1453. Tschirch u. Oesterle. Anat. Atlas. 1900. Tafel. 23. Tunmann. Pfl.-microchemie. 1913. 363. Vogl. Veget. Nahr. u Genuszm. 1899. 353. Wiesner. Rohstoffe. 1900. 639. Wigand. Pharmakogn. 1879. 253. MATERIAL. The drug. REAGENTS. Water, glycerine, chloral hydrate, potash, phloroglucin and hydrochloric acid, iodine and sulphuric acid 66 per cent., concentrated sulphuric acid.

MICROGRAPHY.

Epidermis.

Inner side. Cells arranged in longitudinal rows.

Cells of the above. H. 10 μ, Lev. B. 10—20 μ, Lev. L. 80—150 μ; tetragonal tables, sometimes forming very small papillae. Walls yellow to brown; strongly swelling up in water, especially the outer walls; outer walls a little thickened, with a thin cuticle. Cell contents: a yellow to brown mass.

O u t e r   s i d e.

C e l l s   o f   t h e   a b o v e. H. 14 μ, Lev. B. 10—20 μ, Lev. L. surpassing that of the cells at the inner side; nearly all cells forming a smaller or larger papilla. Outer walls with a cuticle, forming often a parallel longitudinal striation. See for the rest the cells of the inner side.

U p p e r   m a r g i n. Showing some, mostly 3 rows of much longer papillae. Between these here and there a pollen grain coloured yellow to brown; exine dotted.

P a p i l l a e. At their base 25 μ in diameter, length 90—120 μ; conical with a blunt apex. Outer walls strongly swelling up in water; with a cuticle showing dots or short striae. Cell contents: almost without colouring substances.

G r o u n d   t i s s u e. Consisting of 8—10 layers of common parenchyma cells; showing intercellular spaces.

C e l l s   o f   t h e   a b o v e. H. and Lev. B. 20 μ, Lev. L. 150—250 μ; polygonal prisms with a longitudinally directed axis and strongly rounded edges. Walls somewhat swelling up in water; yellow to brown; showing a distinct cellulose reaction, especially the inner layers. Cell contents: a yellow to brown mass, in concentrated su·phuric acid colouring blue to black but soon passing into reddish brown, in potash colouring brown, soluble in water, less so in alcohol; some oil, forming small drops in concentrated sulphuric acid; sometimes a crystaline body.

M e r i s t e l e. At the base of the stigma a single one ⸱ ranching towards the top; anastomoses of the branches nearly wanting.

Vascular bundle. Collateral; containing spiral vessels with lignification.

J; M; v. E. d. W.

GLOSSARY

of anatomical terms. To the explanation of each term there is appended an abbreviated indication of the subdivision of anatomy to which the term belongs. Thus: Cyt. = cytology; Hist. = histology; Micr. Anat. = general microscopical anatomy; App. of surf. = anatomy of the appendages of the surface; Micr. Anat. root, stem, stamen or pistil = special microscopical anatomy of these parts.

Adult cells : showing the highest grade of differentiation belonging to them. Hist.

Adventitious meristem : meristem mostly arising from permanent cells. Micr. Anat.

Adventitious vacuoles : small vacuoles in cells containing also one or more sap cavities. Cyt.

Aerenchyma : parenchyma containing large intercellular passages and chambers, the enlargement of these chiefly ensues by reiterated divisions of the cells originally in contact with one another Hist.

Air cavity : see respiratory cavity.

Alburnum : younger, living, only slightly coloured outer part of secondary xylem. Micr. Anat.

Aleurone grains : the albuminous contents of small and mostly numerous vacuoles in a dry state.

Amphicribral vascular bundle : concentric bundle with phloem on the outside. Micr. Anat.

Amphivasal vascular bundle : concentric bundle with xylem on the outside. Micr. Anat.

Amyloid substances : cell wall substances colouring blue in a dilute solution of iodine without addition of an auxiliary reagent. Cyt.

Amylum body : protoplasmic body occurring in plastids, and often containing one or more pyrenoids and amylum grains. Cyt.

Annular cells and fibres : with annular thickenings of the walls. Hist.

Annular rhytidoma : see ringed rhytidoma.

Annular vessels : xylem vessels with annular wall thickenings. Hist.

Anthocyanin : collective name for pigments mostly blue or greenish in an alkaline and red in an acid cell sap; generally dissolved, sometimes in a crystalline or amorphous form.

Apical cell : single initial cell at the top of a meristem. Micr. Anat.

Apical meristem : full meristem at the top of the part it produces. Micr. Anat.

Articulations : parts of vessels corresponding to the original cells from which they were derived. Hist.

Back chamber : part of porus in stoma above the lower projecting ridges. Micr. Anat.

Balance hairs : consisting of a stalk bearing a crossbeam. App. of surf.

Bands of protoplasm : traversing the vacuoles. Cyt.

Bark : see rhytidoma.

Bar-like thickening of the cell wall : ingrowths sometimes branched, protruding into the cell cavity. Cyt.

Basal meristem : full meristem at the base of the part it produces. Micr. Anat.

Bast fibre class or system of elements : occurring only in secondary phloem; often with longitudinal growth, sometimes septate; walls mostly thick, with simple pits; contents sometimes air. Hist.

Bast parenchyma cells : elements of the parenchymatic class, sometimes sclerenchymatous; those arising from one and the same cambial fibre forming a bast parenchyma fibre. Hist.

Bast parenchyma fibres : see bast parenchyma cells.

Bicollateral vascular bundle : simple bundle containing one xylem bundle and a phloem bundle on either side. Micr. Anat.

Bi-, tri-, etc. seriate hairs : consisting of more than one row of cells. App. of surf.

Bi-, tri-, up to multiseriate medullary rays : more than one cell in breath. Micr. Anat.

Bordered pit : pit canal widening suddenly towards the non-thickened part of the cell wall. Cyt.

Bordered-pit-like margins of perforations in xylem vessels : bordered perforation pits. Hist.

Border of bordered pit : see halo.

Bristle trichomes : rigid hairs. App. of surf.

Callus plate : artificial product, often covering one side of the partition walls in sieve-tubes. Hist.

Cambiform : parenchymatic tissue of primary phloem with the exception of companion cells. Micr. Anat.

Cambium : a kind of meristem forming no complete parts of plants but only special tissues; having the shape of a layer parallel to the surface; by repeated parallel cell divisions throwing off tissues either on one or on both sides. Micr. Anat.

Canal of bordered pit: not widened part of pit canal, being more or less extended. Cyt.

Capitate hairs: provided with a thicker head. App. of surf.

Carotinoids: collective name for red or yellow pigments in plastids. Cyt.

Cell: unit, constituting all organisms either by itself or in conjunction with others. Cyt.

Cell plasm, general: see cytoplasm.

Cells of protective sheath: cell walls characterized by a narrow strip of lignified matter giving rise to Casparian dots. Hist.

Cellulose: in the ordinary course of micrography this term is used for all cell wall substances coloured blue after treatment with iodine followed by sulphuric acid 66 per cent or another auxiliary reagent. Cyt.

Central cylinder: see stele.

Chitin: in micrography cell wall substances are called chitin if they are coloured violet by an iodine solution acidulated with sulphuric acid after previous heating up to 160° C. in a solution of potash. Cyt.

Chromatophores: see plastids.

Chromoplasts: plastids containing pigments only of the carotinoid group, coloured yellow to brick red. Cyt.

Chlorenchyma cells: containing abundant chlorophyll. Hist.

Chlorophyll: collective name for green pigments in plastids. Cyt.

Chloroplast: plastid containing besides carotinoid pigment chlorophyll and generally coloured green; in some Algae moreover containing other pigments and coloured otherwise. Cyt.

Chloroplasts improper: see spurious chloroplasts.

Chloroplasts proper: see ordinary chloroplasts.

Closed vascular bundle: simple bundle without cambium. Micr. Anat.

Closing bands: layers of coherent cells with intercellular spaces in lenticels. Micr. Anat.

Closing membrane of pits: the thin part of the cell wall separating from one another 2 pit canals, which belong to adjacent cells and correspond in place. Cyt.

Cluster crystals: aggregates of crystals more or less in the form of a morning-star. Cyt.

Collateral vascular bundle: simple bundle containing one xylem and phloem bundle. Micr. Anat.

Collecting cells: cells of the upper layer of spongy chlorenchyma each having contact with a group of somewhat converging palisade chlorenchyma cells. Micr. Anat. leaf.

Collenchyma: tissue consisting of collenchymatous elements. Hist.

Collenchymatous elements: provided with collenchymatous thickenings of the walls; with living contents. Hist.

Collenchymatous thickenings of the cell wall: walls principally thickened in the corners of the cells where more than 2 cells come together; these thickenings mostly show a very conspicuous cellulose reaction. Cyt.

Combined bordered pits: much elongated pits fusing together by means of furrows on the inside of the cell wall, which are formed by the sharp edged apertures or inner apertures of the canals belonging to adjacent pits. Cyt.

Common parenchyma: tissue consisting of thin-walled parenchyma cells. Hist.

Companion cells: thin sister cells of the articulations of sieve-tubes with living contents and cell-nuclei. Hist.

Complementary tissue: layers of cells, isolated from one another, between closing bands in lenticels. Micr. Anat.

Component grain: the simple grains of a compound starch grain. Cyt.

Compound medullary rays: composed of vertically alternating uniseriate and pluriseriate stories. Micr. Anat.

Compound starch grain: formed by an aggregate of simple grains; called bi-, tri-, to poly-adelphous according to the number of component grains. Cyt.

Compound tissue: tissue consisting of more then one kind of cells, but never forming by combination whole parts of plants. Micr. Anat.

Compound vascular bundle: consisting of simple vascular bundles separated by medullary commissures. Micr. Anat.

Concentric starch grain: with a central hilum. Cyt.

Concentric vascular bundle: simple bundle with xylem and phloem surrounding one another. Micr. Anat.

Conductive tissue: strands of thick-walled cells in pistils. Micr. Anat. pistil.

Conjugated elements: with tubular lateral processes, those of adjoining cells meeting one another. Hist.

Conjugation tubes: the lateral processes of conjugated elements. Hist.

Connecting frames (cadres d'union of Mangin): ridges of pectin substances filling up the corners of the intercellular spaces and surrounding those parts of the cell walls common to adjacent cells. Cyt.

Connecting threads: fine cytoplasmic filaments uniting protoplasts, which are separated by a cell wall; either confined to the pit membrane or penetrating the whole thickness of the cell wall. Cyt.

Copulated elements: see conjugated elements.

Cork cambium: see phellogen.

Corky substances: mixture of fats and some other substances, phellonic acid playing often a prominent part among their constituents; in the ordinary course of micrography the term is used for all cell wall substances which, without being lignified, withstand the action of concentrated sulphuric acid. Cyt.

Cortex: ground tissue in not disciform monostelous parts. Micr. Anat.

Cribral class or system of elements: occurring only

in secondary phloem; with slight longitudinal growth and no septa; walls thin, with sieve-plates or simple pits; living contents. Hist.

Cryptoporous stomata: guard-cells more or less overarched by subsidiary cells. Micr. Anat.

Crystal sand: a mass of very small more or less irregular crystals. Cyt.

Crystal-sheath: consisting of crystal cells. Hist.

Crystal skins: pouches enveloping crystals; consisting of organic cell wall substances and connected with the cell wall. Cyt.

Cuticle: continuous layer covering the surface of the epidermis in many organs; it consists of cutin and contains no cellulose. Cyt.

Cuticularized cells: provided with a cuticle, sometimes also with cuticular layers. Hist.

Cuticular layers: cellulose layers, lying beneath the cuticle and containing cutin. Cyt.

Cuticular striation: fine ridges on the outside of the cuticle. Cyt.

Cutin: corky cell wall substance, containing no phellonic acid and occurring at the surface of many organs. Cyt.

Cystoliths: peculiar ingrowths of the cell wall, mostly having a thick clustered appearance and impregnated with calcium carbonate. Cyt.

Cytoplasm: the main body of the protoplast. Cyt.

Dipleuric cambium: cambium producing tissue on both sides. Micr. Anat.

Di-, tri-, up to poly-arched root: with two, three, up to numerous xylem, respectively phloem strands. Micr. Anat., root.

Duramen: older, dead, more strongly coloured inner part of secondary xylem. Micr. Anat.

Ectoplasm: the thin layer of hyaloplasm covering the protoplast on the outside. Cyt.

Emergences: appendages of the surface formed by the superficial and internal cell layers. App. of surf.

Emissaries: places where fluid water is excreted by the plant. Micr. Anat.

Endarch primary xylem: protoxylem turned inwards in the stele, metaxylem outwards. Micr. Anat.

Endodermis: inner layer of ground tissue. Micr. Anat.

Epidermis: tissue system forming the surface of the plants. Micr. Anat.

Epithelium: layers of glandular cells either at the surface of the plant or bordering on the cavity of an internal gland. Hist.

Epithema: water tissue with intercellular spaces and occurring in some emissaries. Hist.

Erect cells: medullary ray cells with their largest diameter longitudinally directed. Hist.

Exarch primary phloem: protophloem turned outwards in the stele, metaphloem inwards. Micr. Anat.

Exarch primary xylem: protoxylem turned outwards in the stele, metaxylem inwards. Micr. Anat.

Excentric starch grain: with an excentric hilum. For indicating the degree of excentricity see page 23. Cyt.

Exine: outermost cell wall layer of pollen grains, consisting of cutin and containing no cellulose. Cyt.

Exodermis: outer layer of cortex in roots, either primary periderm or sclerotic. Micr. Anat. root.

Exosporium: outermost cell wall layer of spores consisting of cutin and containing no cellulose Cyt.

Fascicular cambium: cambium in a simple vascular bundle producing secondary xylem and phloem. Micr. Anat. stem and leaf.

Fascicular secondary phloem: formed by the cambium of a simple vascular bundle. Micr. Anat. stem.

Fascicular secondary xylem: formed by the cambium of a simple vascular bundle. Micr. Anat., stem.

Fibres: prosenchymatic elements. Hist.

Fibre tracheides: see tracheid fibres.

Fibrous layer: consisting of parenchyma cells provided with more or less reticulate thickenings of the walls; occurring in anthers below the epidermis. Micr. Anat. stamen.

Fibrovasal bundle: vascular bundle surrounded by a sheath of sclerenchyma fibres. Micr. Anat.

File trichomes: covered with small knobs formed by cuticle or small crystals. App. of surf.

Front chamber: part of porus in stoma below the upper projecting ridges. Micr. Anat.

Full meristem: meristem forming a complete organographical part. Micr. Anat.

Gelatinous layers of cell walls: mostly not lignified secondary layers, often folded and more or less detached from the other layers; occurring only in fibres of secondary xylem, phloem and pericycle. Cyt.

Glandular cells: excreting special substances often in a fluid state, but not pure water. Hist.

Glandular hairs: capitate hairs; the heads secreting certain substances, often in a cuticular bladder. App. of surf.

Glandular hairs, type of compositae: consisting of about 4 stories, each of 2 cells. App. of surf.

Glandular hairs, type of labiatae: head consisting of a single cell layer. App. of surf.

Gluten cells: containing numerous minute aleurone grains. Hist.

Globoids: small globular inclusions of aleurone grains, consisting chiefly of phosphate of calcium and magnesium.

Granular protoplasm: hyaloplasm containing microsomes, thereby less transparent. Cyt.

Ground tissue: as a general term standing for the tissue system between epidermis and stele; moreover used as a special term when not disciform parts are schizostelous and when the stelar tissue system is wanting. Micr. Anat.

Growth rings: occasioned by periodical differences

in the elements formed in different seasons; in secondary xylem, secondary phloem and phellem Micr. Anat.

Guard-cells: cells of stoma exhibiting the porus between them. Micr. Anat.

Gum: in micrography practically characterized by its tendency to swell up in water. Cyt.

Gusset: dilated part of the middle lamella where more than 2 cells meet together; sometimes containing an intercellular space. Cyt.

Hairs: cylindrical, thin and soft trichomes. App. of surf.

Half compound starch grain: consisting of 2 or more simple grains, held together by common enveloping layers. Cyt.

Halo of bordered pit: boundary of the widest part of the pit chamber. Cyt.

Heart wood: see duramen.

Hilum of starch grain: organic centre of the grain. Cyt.

Horizontal partition-walls of vessels: at right angles to the longitudinal axis of the vessel. Hist.

Hyaloplasm: protoplasm proper, mostly homogeneous and transparent. Cyt.

Hyphae: much elongated parenchymatic elements, often loosely interwoven. Hist.

Hyphal tissue: consisting of hyphae. Hist.

Hypoderma: layers of thick-walled elements adjoining the epidermis. Hist.

Idioblasts: isolated elements showing characters differing from those of the surrounding tissue. Hist.

Initial celled cambium: showing an initial celled growth. Micr. Anat.

Initial celled growth: one or more cells, destined to form a tissue, continue to divide as long as the formation of tissue lasts. Hist.

Initial celled secondary cork: formed by initial celled growth. Micr. Anat.

Initial cells: meristematic cells continuing to divide as long as the action of the meristem lasts. Micr. Anat.

Inner aperture of canal of bordered pit: leading into the cell cavity. Cyt.

Intercalary meristem: meristem situated between permanent tissues. Micr. Anat.

Intercellular spaces: cavities, mostly filled with air, between the cells; they are formed in the middle lamella and in the gussets. Cyt.

Interfascicular cambium: cambium in a medullary commissure producing secondary xylem and phloem. Micr. Anat. stem and leaf.

Interfascicular secondary phloem: formed by the cambium of a medullary commissure. Micr. Anat. stem and leaf.

Interfascicular secondary xylem: formed by the cambium of a medullary commissure. Micr. Anat. stem and leaf.

Intermediate bands: see closing bands.

Intermediate parietal cell layer: below the fibrous layer in antherae. Micr. Anat. stamen.

Internal glands: groups of glandular cells deposi-

ting their secretion in an cavity surrounded by those cells. Micr. Anat.

Intervenium: area lying between one or more veins or veinlets in a leaf-blade.

Intine: inner part of the cell wall of a pollen grain, containing cellulose. Cyt.

Keratenchyma: tissue consisting of entirely compressed elements of the cribral class. Hist.

Large secondary medullary rays: reaching from pith to cambium or from pericycle to cambium; interfascicular. Micr. Anat.

Latex: an emulsion white or otherwise coloured. Cyt.

Latex vessels: containing latex in their sap-cavity. Hist.

Layers of elements: plates thick one or more elements. Hist.

Leaf-trace: part of one or more vascular bundles, passing from a leaf into the stem. Micr. Anat.

Lens-shaped pit cavity: the combination of the 2 pit chambers of a bilateral bordered pit. Cyt.

Lenticels: portions of secondary cork tissue provided with intercellular spaces and having a peculiar structure. Micr. Anat.

Leucoplasts: plastids containing no pigment but generally a large amount of starch. Cyt.

Libriform class or system of elements: occurring only in secondary xylem; often with abundant longitudinal growth and often septate; cavity mostly small; walls mostly thick with not abundant either simple or bordered pits and without spiral bands; contents: living protoplasm accompanying simple pits, water and air accompanying bordered pits. Hist.

Lignification: state of cell walls, containing substances which make them hard and brittle and resist the action of concentrated sulphuric acid; in the ordinary course of micrography cell walls are called lignified if woody and coloured red after treatment with phloroglucin and hydrochloric acid. Cyt.

Limiting lamella: see closing membrane.

Mature cells: see adult cells.

Medullary commissure: medullary rays between the vasal bundles of the stele. Micr. Anat.

Medullary rays: radial layers of cells in actinomorphous parts. Hist.

Medullary ray cells: elements of the parenchymatic class showing no fibrous shape. Hist.

Meristeles: branches of stele in schizostelous parts. Micr. Anat.

Meristem: tissue consisting of meristematic cells. Hist.

Meristematic cells: showing a low grade of differentiation, but large cell-nuclei. Hist.

Mesophyll: ground tissue in disciform, mostly schizostelous parts, especially leave-blades. Micr. Anat.

Metacribral bands of bast parenchyma: tangential layers. Micr. Anat.

Metaphloem: part of primary phloem formed by

cell divisions during longitudinal growth but differentiated after the end of it. Micr. Anat.

Metatracheal bands of wood parenchyma : tangential layers. Micr. Anat.

Metaxylem : part of primary xylem formed by cell divisions during longitudinal growth but differentiated after the end of it. Micr. Anat.

Microsomes : small particles of various shape and character occurring in granular protoplasm. Cyt.

Middle lamella : a homogeneous, mostly thin lamella seemingly belonging in common to 2 adjoining cells and consisting of the primitive membrane forming the original cell walls, fused together with the primary thickening layers of both adjoining cells. Cyt.

Middle layer of cytoplasm : the bulk of the cytoplasm, between the ectoplasm and the tonoplast, generally consisting of granular protoplasm. Cyt.

Monopleuric cambium : cambium producing tissue only on one side. Micr. Anat.

Monostelous part : containing an unbranched solid stele occupying a central position. Micr. Anat.

Monostely : the state of a monostelous part. Micr. Anat.

Mucilage : see gum.

Mucilage cells : characterized by the occurrence of mucilage, either arising from the cell walls or contained in the cell cavity. Hist.

Nucleus of the cell : well defined protoplasmic body in the cytoplasm, showing a complex structure. Mostly one in number. Cyt.

Oblique partition-walls of vessels : slanting with respect to the longitudinal axis of the vessel. Hist.

Oblito-schizogenous glands : schizogenous glands with obliterated secernating cells. Micr. Anat.

Oil cells : containing fatty or aetherial oils. Hist.

Open vascular bundle : simple bundle with cambium. Micr. Anat.

Ordinary chloroplast : plastids containing mostly small though sometimes numerous starch grains soon disappearing in darkness. Cyt.

Outer aperture of canal of bordered pit : leading into the pit chamber. Cyt.

Papillae : short trichomes mostly unicellular and not separated by a partition wall from the cells producing them. App. of surf.

Palisade chlorenchyma : palisade tissue consisting of chlorenchyma cells. Hist.

Palisade epithelium : palisade tissue consisting of glandular cells. Hist.

Palisade sclerenchyma : palisade tissue consisting of sclerenchyma cells. Hist.

Palisade tissues : layers parallel to the surface of the parts containing them and consisting of cylindrical cells of the same length. Hist.

Paraphyses : hairs standing between organs of reproduction. App. of surf.

Paratracheal wood parenchyma : arranged in groups round the vessels. Micr. Anat.

Parenchyma : tissue consisting of parenchyma cells. Hist.

Parenchyma cells : parenchymatic elements in contrast with hyphae. Hist.

Parenchymatic class or system of elements : occurring only in secondary phloem and xylem; without longitudinal growth, mostly septate; cavity mostly large; walls relatively thin with abundant simple pits; contents living protoplasm. Hist.

Parenchymatic elements : showing many shapes from isodiametrical to much elongated, but never having pointed ends. Hist.

Parietal cytoplasm : the layer of cytoplasm lining the cell wall in cells containing 1 or more sap cavities. Cyt.

Partition-walls of vessels : the perforated walls between the articulations. Hist.

Passage cells : sheath-cells differentiated so as not to impede communication between the tissues on either side of the sheath. Hist.

Pectin substances : distinct chemical substances, occurring in all layers of the cell wall and especially conspicuous in the middle lamella; they promote the formation of intercellular spaces by passing into a state of solution. Cyt.

Perforation pits in xylem vessels : giving rise to perforations by the disappearance of their closing membrane. Hist.

Pericambium : see pericycle.

Pericycle : outer non vasal part of the stele lying between the endodermis and the bundles of vasal elements. Micr. Anat.

Periderm : tissue consisting of periderm cells. Hist.

Periderm cells : walls provided with a suberin layer, intercellular spaces generally wanting. Hist.

Phaeophyll : a brown pigment in Phaeophyceae. Cyt.

Phaneroporous stomata : lying in the level of the epidermis. Micr. Anat.

Phellem : tissue formed in initial celled secondary cork on the outside of the phellogen. Micr. Anat.

Phelloderm : tissue formed in initial celled secondary cork on the inside of the phellogen. Micr. Anat

Phellogen : cambium producing secondary cork tissue. Micr. Anat.

Phloem arch of cambium : part of cambium on the inside of a phloem bundle in roots. Micr. Anat. root.

Phloem bundles : composed of cribral elements (sieve-tubes, companion cells) and parenchyma called cambiform. Micr. Anat.

Phloem vessels : see sieve-tubes.

Phycocyanin : a blue pigment in Schizophyceae. Cyt.

Phycoerithrin : a red pigment in Rhodophyceae. Cyt.

Pit canal : cavity traversing the cell wall transversely where a pit is formed. Cyt.

Pit chamber of bordered pit : much widened outer part of the pit canal. Cyt.

Pith : central non vasal part of the stele. Micr. Anat.

Pits : gaps in the internal thickening of the cell wall Cyt.

Pitted cell walls and cells : walls and cells exhibiting pits. Cyt.

Pitted vessels : xylem vessels with pitted walls; mostly bordered pits. Hist.

Plasmodesms : see connecting threads.

Plastids : well defined protoplasmic bodies in the cytoplasm, often coloured and mostly occurring in large numbers. Cyt.

Porus : air-passage of stoma divided by projecting ridges and the bulging out parts of the guard-cells. Micr. Anat.

Primary tissue : tissue formed during longitudinal growth. Micr. Anat.

Primary meristem : persisting part of primitive meristem. Micr. Anat.

Primitive elements : first developed elements of protoxylem and -phloem. Micr. Anat.

Primitive meristem : meristem still constituting the whole plant. Micr. Anat.

Primordial elements : see primitive elements.

Procambium : strands of elongated cells in a meristem. Micr. Anat.

Procumbent cells : medullary ray cells with their largest diameter radially directed. Hist.

Prosenchyma : tissue consisting of fibres. Hist.

Prosenchymatic elements : mostly spindle-shaped and always with two or more pointed ends. Hist.

Protective sheath : see cells of protective sheath. Hist.

Protophloem : part of phloem bundle, earliest developed and differentiated during longitudinal growth. Micr. Anat.

Protoplasm : the living substance constituting the protoplast, considered from a chemical and physical point of view. Cyt.

Protoplast : the organism constituting the cell. Cyt.

Protoxylem : part of xylem bundle, earliest developed and differentiated during longitudinal growth. Micr. Anat.

Pseudoparenchyma : tissue consisting of pseudoparenchymatic elements. Hist.

Pseudoparenchymatic elements : hyphae in close contact to one another and divided into short cells. Hist.

Pyrenoid : albumen crystal in amylum body. Cyt.

Ramenta : scaly trichomes of ferns. App. of surf.

Raphides : needle shaped crystals pointed at both ends, often in bundles and surrounded by a mucilaginous substance. Cyt.

Ray-tracheids : medullary ray cells showing bordered pits and other characters of the tracheal class. Hist.

Respiratory cavity : intercellular space below a stoma and corresponding with its porus. Micr. Anat.

Reticulate cells : with reticulate thickenings of the walls. Hist.

Reticulate vessels : xylem vessels with reticulate wall thickenings. Hist.

Rhytidoma : outer part of secondary phloem consisting of dead phloem portions enclosed by secondary cork layers. Micr. Anat.

Ringed rhytidoma : with annular secondary cork layers. Micr. Anat.

Root hairs : unicellular and not separated by a partition-wall from the cell producing them. App. of surf.

Sap cavities : large vacuoles; a single one or a few forming the bulk of the cell body. Cyt.

Sap wood : see alburnum.

Scalariform arrangement of sieve-plates : elongated sieve-plates placed in a row in the partition walls of sieve-tubes. Hist.

Scalariform bordered pits : strongly slit-like, transversely directed and arranged in a longitudinal row. Cyt.

Scalariform perforations of xylem vessels : slit-like, numerous, placed in rows. Hist.

Scalariform vessels : xylem vessels with scalariform pits. Hist.

Scaly rhytidoma : with scale-like secondary cork layers. Micr. Anat.

Schizogenous glands : internal glands with a cavity being an intercellular space. Micr. Anat.

Schizo-lysigenous glands : internal glands with a cavity arising as an intercellular space but soon enlarged by disorganisation of surrounding cells. Micr. Anat.

Schizostelous part : containing a branched, sometimes hollow stele; the branches often showing an incomplete structure. Micr. Anat.

Schizostely : the state of a schizostelous part. Micr. Anat.

Sclereids : parenchymatous sclerenchyma cells. Hist.

Sclerenchyma : tissue consisting of sclerenchymatous elements. Hist.

Sclerenchyma sheath : consisting of sclereids or sclerenchyma fibres. Hist.

Sclerenchymatous cell walls : strongly thickened over a considerable area, moreover hard and mostly lignified. Cyt.

Sclerenchymatous elements : provided with sclerenchymatous thickenings of the walls. Hist.

Sclerotic cell walls : see sclerenchymatous cell walls Cyt.

Secondary thickening layers of cell walls : layers adjoining the middle lamella on both sides, mostly forming the bulk of the cell walls. Cyt.

Secondary tissue : tissue formed after the end of longitudinal growth. Micr. Anat.

Semi-bordered pit : see unilateral bordered pit.

Septate bast fibres : divided into cells by very thin cross walls. Hist.

Septate libriform fibres : divided into cells by very thin unpitted cross walls; side walls with simple pits; contents starch or crystals. Hist.

Sharp-edged aperture of bordered pit : not widened part of pit canal, being very short. Cyt.

Sheaths : layers of elements enclosing other tissues. Hist.

Sieve-plates : areas in the walls of sieve-tubes, exhibiting sieve-pores. Cyt.

Sieve-pores : fine perforations in the transverse or lateral walls of sieve-tubes. Cyt.

Sieve structure in closing membrane of bilateral and unilateral bordered pits : the membrane finely dotted like sieve-plates. Cyt.

Sieve-tubes : vessels with thin, not lignified walls, sometimes exhibiting sieve-plates; partition walls either having disappeared or provided with sieve-plates; contents living protoplasm, often starch and albumen. Hist.

Simple medullary rays : not composed of vertically alternating different parts. Micr. Anat.

Simple pits : pit canal equally wide throughout or widening towards the cell cavity. Cyt.

Simple starch grain : with a single hilum. Cyt.

Simple tissue : tissue consisting of only one kind of cells. Hist.

Simple vascular bundle : containing no medullary rays. Micr. Anat.

Slit : the narrowest portion of porus in stoma. Micr. Anat.

Slit-like bordered pits : showing in a surface view a more or less elongated shape of 1 or more of the following parts : halo, sharp-edged aperture, inner and outer aperture. Cyt.

Slit-like simple pits : pit canal, or a part of it, towards the cell cavity more or less elongated in a surface view. Cyt.

Smaller secondary medullary rays : not reaching to pith or pericycle. Micr. Anat.

Smooth margins of perforations in xylem vessels : simple perforation pits. Hist.

Sphaerites : see sphaero-crystals.

Sphaero-crystals : spheroidal bodies composed of radiating crystal needles often arranged in concentric layers. Cyt.

Spiral cells and fibres : with spiral thickenings of the walls. Hist.

Spiral vessels : xylem vessels with spiral wall thickenings. Hist.

Spongy chlorenchyma : spongy tissue consisting of chlorenchyma cells. Hist.

Spongy tissue : parenchyma with enlarged intercellular spaces; their enlargement ensuing by subsequent growth, without cell divisions of the cells, between which they were originally formed. Hist.

Spurious chloroplasts : plastids containing large mostly not very numerous starch grains, not disappearing after a short time in darkness. Cyt.

Starch grains : in the ordinary course of micrography all grains colouring blue in a solution of iodine containing a certain amount of water. Cyt.

Starch-sheath : consisting of cells containing more abundant and mostly larger starch grains then the adjoining cells on either side. Hist.

Stele : central tissue system consisting of a vasal and a parenchymatic part, sometimes branched and hollow. Micr. Anat.

Stellate hairs : either uni- or pluricellular. App. of surf.

Steps : portions between scalariform perforations in xylem vessels. Hist.

Stigmatic papillae : covering the stigmata. App. of surf.

Stoma : intercellular passage perforating the epidermis. Micr. Anat.

Stone cells : see sclereids.

Storied arrangement of secondary xylem and phloem : in consequence of the cambial fibres being arranged in transverse rows. Micr. Anat.

Storied cambium : cambium showing a storied growth. Micr. Anat.

Storied growth : cells, destined to form a tissue, do not divide simultaneously but one after another shows a limited number of divisions. Hist.

Storied secondary cork : formed by storied growth. Micr. Anat.

Stratification of cell walls : concentric layers; thicker dense ones with higher refractive power alternating with thinner less dense ones. Cyt.

Stratification of starch grain : concentric layers surrounding the hilum. Cyt.

Striation of cell walls : delicate striae to be observed in surface view, running obliquely in different directions and often causing a lattice work appearance. Cyt.

Strings of protoplasm : see bands of protoplasm.

Suberin : corky cell wall substance containing phellonic acid. Cyt.

Suberized : containing suberin; in the ordinary course of micrography this term is used for not lignified cell walls persisting in concentrated sulphuric acid. Cyt.

Subsidiary cells : epidermal cells taking part in the formation of a stoma. Micr. Anat.

Substitute parenchyma fibres : non-septate wood or bast parenchyma fibres. Hist.

Tapetal cell layer : below the intermediate layer in antherae. Micr. Anat. stamen.

Tapetum : see tapetal cell layer.

Tertiary thickening layers of cell walls : the innermost layers bordering upon the cell cavity and adjoining the secondary layers; mostly thin and more highly refractive; not always to be distinguished. Cyt.

Tertiary tissue : tissue due to the renewed action of secondary tissue. Micr. Anat.

Thickness of cell walls adjoining those of other cells : to be determined for each cell separately. Cyt.

Thin-walled bast fibres : sometimes becoming thick-walled at a late stage. Hist.

Thyloses : parenchyma cells occurring in the cavities of vessels in secondary xylem and arising as outgrowths of adjacent parenchyma cells through the pits of the vessels. Hist.

Tier-like arrangement : see storied arrangement.

Tier-like cambium : see storied cambium.

Tier-like growth : see storied growth.

Tissue : group of cellular elements considered as belonging together. Hist.

Tissue systems : tissues constituting by their combination whole parts of plants and generally consisting of simple and compound tissues. Micr. Anat.

Tonoplast : thin layer of hyaloplasm, surrounding a vacuole. Cyt.

Torus of bordered pit : local, plano-convex thickening of the thin cell wall portion, that forms the bottom of the pit chamber; in a bilateral bordered pit it is bi-convex. Cyt.

Tracheae : see xylem vessels.

Tracheal class or system of elements : occurring only in secondary xylem; with mostly slight longitudinal growth and no septa; cavity rather large; walls chiefly with bordered pits, sometimes with spiral bands; contents water and air. Hist.

Tracheidal medullary ray cells : see ray-tracheids.

Tracheide fibres : always fibre-shaped, longer, thicker-walled and less abundantly pitted then vascular tracheides, mostly shorter then libriform fibres and never septate. Hist.

Tracheids : elements of the tracheal class not being vessels, but either vascular tracheides or tracheide fibres. Hist.

Transfusion tissue : consisting of parenchymatic cells with bordered pits. Hist.

Trichomes : appendages of the surface formed only by the superficial cell layer. App. of surf.

Unilateral bordered pit : a combination of a simple pit and a bordered pit belonging to 2 adjacent elements and corresponding in place to one another. Cyt.

Uniseriate hairs : consisting of a single row of cells. App. of surf.

Uniseriate medullary rays : only a single cell in breadth. Micr. Anat.

Upright cells : see erect cells.

Vacuoles : cavities in the cytoplasm filled with some not living substance, mostly in a fluid state. One or more in number. Cyt.

Vascular bundle : combination of one or more xylem and phloem bundles, sometimes separated by parenchyma or a cambium and sometimes containing medullary rays. Micr. Anat.

Vascular tracheides : see vessel tracheides.

Vegetative cone : see apical meristem.

Vessels : elements arising from the fusion of single cells by disappearance of partition walls. Hist.

Vessel tracheides : articulations of xylem vessels without perforations or only perforated at one end. Hist.

Vittae : schizogenous glands in pericarps of Umbelliferae, lined with vittin. Micr. Anat.

Vittin : corky cell wall substance, containing no phellonic acid and occurring in the vittae of Umbelliferae, lining the cavity, forming transverse partition walls in it and impregnating the epithelium cell walls which adjoin the cavity. Cyt.

Water-pores : large stomata on emissaries. Micr. Anat.

Water-reservoir cells : colourless cells able to contain much water and yield it to adjacent cells. Hist.

Water tissue : consisting of water-reservoir cells. Hist.

Waxy substances : cell wall substances, in micrography chiefly characterized by a melting point beneeth 100° C., and by being soluble in several of the same reagents as bees wax. Cyt.

Wood parenchyma cells : elements of the parenchymatic class, those arising from a single cambial fibre forming a wood parenchyma fibre. Hist.

Wood parenchyma fibres : see wood parenchyma cells.

Xylem arch of cambium : part of cambium on the outside of a xylem bundle in roots. Micr. Anat. root.

Xylem bundles : composed of tracheal elements (vessels or vessel tracheides) and parenchyma. Micr. Anat.

Xylem parenchyma : parenchymatic tissue of primary xylem. Hist.

Xylem vessels : very long; walls thickened, lignified and always provided with relief figures; partition-walls perforated or having entirely disappeared; contents water and air. Hist.

INDEX

of names and of interesting anatomical structures described in this work.